Fifth Edition

Computer Organization, Design, and Architecture

Fifth Edition

Computer Organization, Design, and Architecture

Sajjan G. Shiva

CRC Press
Taylor & Francis Group
Boca Raton   London   New York

CRC Press is an imprint of the
Taylor & Francis Group, an **informa** business

CRC Press
Taylor & Francis Group
6000 Broken Sound Parkway NW, Suite 300
Boca Raton, FL 33487-2742

Version Date: 20131104

International Standard Book Number-13: 978-1-4665-8554-6 (Hardback)

Visit the Taylor & Francis Web site at
http://www.taylorandfrancis.com

and the CRC Press Web site at
http://www.crcpress.com

To
My students
1971–2013

Contents

Preface

This book covers the organization, design, and architecture of computers. Architecture is the "art or science of building; a method or style of building," according to *Webster's*. A computer architect develops the functional and performance specifications for the various blocks of a computer system and defines the interfaces between those blocks in consultation with hardware and software designers. A computer designer, on the other hand, refines those building-block specifications and implements those blocks with an appropriate mix of hardware, software, and firmware. It is my belief that the capabilities of an architect would be greatly enhanced if he or she were to be exposed to the design aspects of a computer system. Computer organization deals with providing just enough details on the operation of the computer system for sophisticated users and programmers. The backbone of this book is a description of the complete design of a simple but complete hypothetical computer. The book describes the architectural features of contemporary computer systems as enhancements to the structure of the simple computer.

Books on digital systems' architecture fall into four categories: (1) logic design books that cover the hardware logic design in detail but fail to provide the details of computer hardware design, (2) books on computer organization that deal with the computer hardware from a programmer's viewpoint, (3) books on computer hardware design that are suitable for an electrical engineering curriculum, and (4) books on computer system architecture with no detailed treatment of hardware design aspects. I have tried to capture the important attributes of the four book categories to create a comprehensive text that includes pertinent hardware, software, and system aspects.

The first edition of the book, published in 1985, was a result of my teaching a sequence of computer architecture courses at the senior undergraduate and beginning graduate levels for several years to both computer science and electrical engineering students. The second edition, published in 1991, included several additional topics in response to the comments from the users of the first edition. The third edition, published in 2000, expanded the topical coverage of the second edition and contained additional contemporary architectures as examples. The fourth edition, published in 2008, expanded the book to include organization aspects, embedded systems, and performance evaluation. This edition updates the architectural features of contemporary systems, includes a chapter on mobile processors, and introduces cloud computing. The book does not assume prior knowledge of computer programming, although exposure to programming with high-level and assembly languages makes the reading easier. Exposure to electronics is not required as a prerequisite for understanding this book.

Chapter 1 briefly outlines digital computer system terminology, traces the evolution of computer structures, and introduces performance evaluation.

Chapter 2 describes number systems, computer codes, data representation, storage, and transfer concepts.

Chapter 3 covers the analysis and design of combinational logic circuits, along with logic minimization procedures. Popular off-the-shelf integrated circuits (ICs) and designing with ICs are introduced in this chapter. Appendix A at the end of the book provides brief descriptions of selected ICs.

Chapter 4 covers the analysis and design of synchronous sequential circuits. Concepts of registers and register-transfer logic are introduced along with a primitive register-transfer language (RTL). This RTL is used in the rest of the book, although it can be substituted with any other available RTL. Programmable logic devices are introduced in this chapter.

Chapter 5 provides a programmer's view of A Simple (hypothetical) Computer (ASC). ASC organization, instruction set, assembly-language programming, and details of an assembler are provided along with an introduction to program linking and loading.

Chapter 6 details the hardware design of ASC, including both hardwired and microprogrammed control units. Although the development of such detailed design is tedious, it is my belief that system designers, architects, and system programmers should go through the steps of such detailed design to better appreciate the inner workings of a digital computer system.

Chapter 7 enhances the input/output subsystem of ASC from the programmed input/output structure to the concept of input/output processors through interrupt processing, direct memory access, and input/output channels. System structures of several popular commercial systems are detailed.

Chapter 8 covers popular data representations and instruction set architectures, along with example systems.

Chapter 9 expands the memory model introduced in Chapter 4 to include commonly used memories and describes various memory devices and organizations with an emphasis on semiconductor memory design. Enhancements to the memory system in terms of cache and virtual memory organizations are also described in this chapter.

Chapter 10 details enhancements to the arithmetic/logic unit (ALU) to cover the concepts of stack-based ALUs, pipelined ALUs, and parallel processing with multiple functional units.

Chapter 11 is devoted to enhancements to the control unit and covers the topics of pipelining and parallel instruction execution, along with performance issues.

Chapter 12 introduces a popular architecture classification. A brief introduction to dataflow and systolic architectures is provided in this chapter.

Chapter 13 covers embedded system architectures. A microcontroller and ARM (Advanced Risc Machines Ltd) architectures are used as examples.

Chapter 14 introduces mobile processors and the system-on-chip (SoC) concept. ARM architectures and a SoC are provided as examples.

Chapter 15 introduces computer networks, distributed processing, grid architectures, and cloud computing.

Chapter 16 introduces performance evaluation.

I have utilized the architectural features found in practical computer systems as examples in Chapters 7 through 16. Only pertinent details of commercial systems are used rather than complete descriptions of commercial architectures. Problems are given at the end of every chapter. The list of references provided in the Bibliography section at the end of each chapter may be consulted by readers for further details on topics covered in that chapter. A solutions manual and all the figures and tables in electronic form are available from the publisher for use of instructors.

The following method of study may be followed for a single-semester course:

Computer organization: Chapter 1, Chapter 2, Section 3.1, Section 3.2, Section 3.4, Section 4.1, Section 4.9, Chapter 5, Sections 6.1 through 6.5, Sections 7.1 through 7.3, and selected topics from the remaining chapters

Computer architecture (for computer science students with no logic design background): Chapter 1, Chapter 2 (review), Chapters 3 through 7, and selected topics from the remaining chapters

Computer architecture (for computer engineering students): Chapter 1, Chapters 2 through 4 (review), Chapters 5 through 10, and selected topics from the remaining chapters

A two-semester course in computer architecture may follow this method of study:

Semester 1: Chapters 1 through 7 and selected topics from the remaining chapters
Semester 2: Chapters 8 through 16 and case studies of selected contemporary systems

The book covers all the following topics suggested by ACM/IEEE Draft Curriculum 2013 on Architecture and Organization (AR) for computer science and computer engineering:

AR 1: Digital logic and digital systems
AR 2: Machine level representation of data
AR 3: Assembly level machine organization

AR 4: Memory system organization and architecture
AR 5: Interfacing and communication
AR 6: Functional organization
AR 7: Multiprocessing and alternative architectures
AR 8: Performance enhancements

Several of my colleagues and students in my architecture classes over the years have contributed immensely toward the contents of this book; I would like to thank them all. I would also like to thank the users of previous editions (more than 100 universities) both in the United States and abroad for their comments and suggestions for improvement, which have resulted in this new edition. It is a pleasure to acknowledge the critical reviews and support of Abdullah Abuhussein, Harkeerat Bedi, Ramya Dharam, Vivek Shandilya, and Chris Simmons during the preparation of this edition. I am indebted to several individuals at the Taylor & Francis Group: Nora Konopka for encouraging me to prepare this edition, Amber Donley, Laurie Schlags, and Kyle Meyer for their production support, and John Gandour for a great cover design. Thanks are also due to Christine Selvan and staff at SPi Global for their superb support in the production of this book. I thank my family, Kalpana, Sruti and Uday, Sweta and Ashish, and the two welcome distractions (my grandsons Karthik and Navik) for their love and forbearance when the manuscript preparation clashed with family commitments.

Sajjan G. Shiva

CHAPTER 1

Introduction

Recent advances in microelectronic technology have made computers an integral part of our society. Each step in our everyday lives is influenced by computer technology: we awake to a digital alarm clock's beaming of preselected music at the right time, drive to work in a digital processor–controlled automobile, work in an extensively automated office, shop for computer-coded grocery items, and return to rest in the computer-regulated heating and cooling environment of our homes. It may not be necessary to understand the detailed operating principles of a jet plane or an automobile to use and enjoy the benefits of these technical marvels. Computer systems technology has also reached the level of sophistication wherein an average user need not be familiar with all the intricate technical details of an operation to use them efficiently. Computer scientists, engineers, and application developers, however, require a fair understanding of the operating principles, capabilities, and limitations of digital computers to enable the development of complex yet efficient and user-friendly systems. This book is designed to give such an understanding of the operating principles of digital computers. This chapter begins by describing the organization of a general-purpose digital computer system and briefly traces the evolution of computers.

1.1 COMPUTER SYSTEM ORGANIZATION

The primary function of a digital computer is to process data input to it to produce results that can be better used in a specific application environment. For example, consider a digital computer used to control the traffic light at an intersection. The *input* data are the number of cars passing through the intersection during a specified time period, the processing consists of the computation of red–yellow–green time periods as a function of the number of cars, and the *output* is the variation of the red–yellow–green time intervals based on the results of processing. In this system, the data input device is a sensor that can detect the passing of a car at the intersection. Traffic lights are the output devices. The electronic device that keeps track of the number of cars and computes the red–yellow–green time periods is the processor. These physical devices constitute the hardware components of the system. The processing hardware is programmed to compute the red–yellow–green time periods according to some rule. This rule is the *algorithm* used to solve the particular problem. The algorithm (a logical sequence of steps to solve a problem) is translated into a program (a set of instructions) for the processor to follow in solving the problem. Programs are written in a language "understandable" by the processing hardware. The collection of such programs constitutes the software component of the computer system.

1.1.1 Hardware

The traffic light controller is a very simple special-purpose computer system requiring only a few of the physical hardware components that constitute a general-purpose computer system (see Figure 1.1). The four major hardware blocks of a general-purpose computer system are its memory unit (MU), arithmetic and logic unit (ALU), input/output unit (IOU), and control unit (CU). Input/output (I/O) devices input and output data into and out of the MU. In some systems, I/O devices send and receive data into and from the ALU rather than the MU. Programs reside in the MU. The ALU processes the data taken from the MU (or the ALU) and stores the processed data back in the MU (or the ALU). The CU coordinates the activities of the other three units. It retrieves instructions from programs resident in the MU, decodes these instructions, and directs the ALU to perform corresponding processing steps. It also oversees I/O operations.

Some representative I/O devices are shown in Figure 1.1. A keyboard and a mouse are the most common input devices nowadays. Touch screens are taking over as common interaction devices. A video display and a printer are the most common output devices. Scanners are used to input data from hard copy sources. Magnetic tapes and disks are used as I/O devices. These devices are also used as memory devices to increase the capacity of the MU. The console is a special-purpose I/O device that permits the system operator to interact with the computer system. In modern-day computer systems, the console is typically a dedicated terminal.

1.1.2 Software

The hardware components of a computer system are electronic devices in which the basic unit of information is either a 0 or a 1, corresponding to two states of an electronic signal. For instance, in one of the popular hardware technologies, a 0 is represented by 0 V, while a 1 is represented by 5 V. Programs and data must therefore be expressed using this binary alphabet consisting of 0 and 1. Programs written using only these binary digits are *machine language* programs. At this level of programming, operations such as ADD and SUBTRACT are each represented by

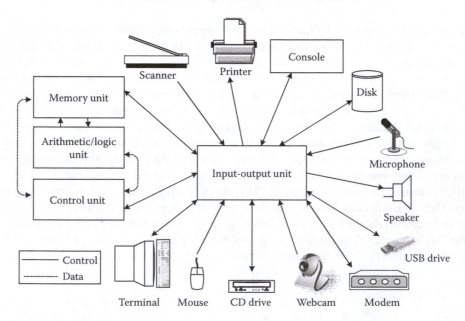

Figure 1.1 Typical computer system.

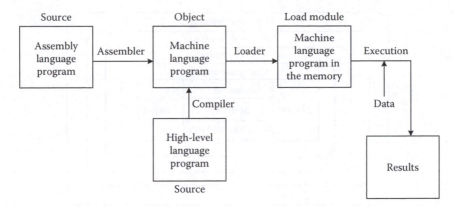

Figure 1.2 Program translation and execution.

a unique pattern of 0s and 1s, and the computer hardware is designed to interpret these sequences. Programming at this level is tedious since the programmer has to work with sequences of 0s and 1s and needs to have very detailed knowledge of the computer structure.

The tedium of machine language programming is partially alleviated by using symbols such as ADD and SUB rather than patterns of 0s and 1s for these operations. Programming at the symbolic level is called *assembly language* programming. An assembly language programmer is also required to have a detailed knowledge of the machine structure, because the operations permitted in the assembly language are primitive and the instruction format and capabilities depend on the hardware organization of the machine. An assembler program is used to translate assembly language programs into machine language.

Use of high-level programming languages such as FORTRAN, COBOL, C, and JAVA further reduces the requirement of an intimate knowledge of the machine organization. A compiler program is needed to translate a high-level language program into the machine language. A separate compiler is needed for each high-level language used in programming the computer system. Note that the assembler and the compiler are also programs written in one of those languages and can translate an assembly or high-level language program, respectively, into the machine language.

Figure 1.2 shows the sequence of operations that occurs once a program is developed. A program written in either the assembly language or a high-level language is called a source program. An assembly language source program is translated by the assembler into the machine language program. This machine language program is the object code. A compiler converts a high-level language source into an object code. The object code ordinarily resides on an intermediate device such as a magnetic disk or tape. A loader program loads the object code from the intermediate device into the MU. The data required by the program will be either available in the memory or supplied by an input device during the execution of the program. The effect of program execution is the production of processed data or results.

1.1.3 System

Operations such as selecting the appropriate compiler for translating the source into object code; loading the object code into the MU; and starting, stopping, and accounting for the computer system usage are automatically done by the system. A set of supervisory programs that permit such automatic operation is usually provided by the computer system manufacturer. This set, called the *operating system*, receives the information it needs through a set of command language statements from the user and manages the overall operation of the computer system.

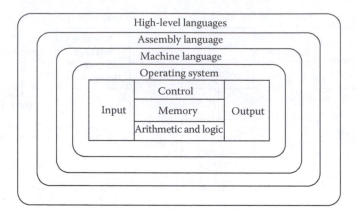

Figure 1.3 Hardware and software components.

Operating system and other utility programs used in the system may reside in a memory block that is typically read only. Special devices are needed to write these programs into read-only memory (ROM). Such programs and commonly used data are termed firmware. Figure 1.3 is a simple rendering of the complete hardware–software environment of a general-purpose computer system.

1.2 COMPUTER EVOLUTION

Man has always been in search of mechanical aids for computation. The development of the abacus around 3000 BC introduced the positional notation of number systems. In seventeenth-century France, Pascal and Leibnitz developed mechanical calculators that were later developed into desk calculators. In 1801, Jacquard used punched cards to instruct his looms in weaving various patterns on cloth.

In 1822, Charles Babbage, an Englishman, developed the difference engine, a mechanical device that carried out a sequence of computations specified by the settings of levers, gears, and cams. Data were entered manually as the computations progressed. Around 1820, Babbage proposed the analytical engine, which would use a set of punched cards for program input, another set of cards for data input, and a third set of cards for output of results. The mechanical technology was not sufficiently advanced and the analytical engine was never built; nevertheless, the analytical engine as designed probably was the first computer in the modern sense of the word.

Several unit-record machines to process data on punched cards were developed in the United States in 1880 by Herman Hollerith for census applications. In 1944, Mark I, the first automated computer, was announced. It was an electromechanical device that used punched cards for input and output of data and paper tape for program storage. The desire for faster computations than those Mark I could provide resulted in the development of the electronic numerical integrator and computer (ENIAC), the first electronic computer built out of vacuum tubes and relays by a team led by Americans Eckert and Mauchly. ENIAC employed the *stored-program concept* in which a sequence of instructions (i.e., the program) is stored in the memory for use by the machine in processing data. ENIAC had a control board on which the programs were wired. A rewiring of the control board was necessary for each computation sequence.

John von Neumann, a member of the Eckert–Mauchly team, developed the electronic discrete variable automatic computer (EDVAC), the first stored-program computer. At the same time, Wilkes developed the electronic delay storage automatic calculator (EDSAC), the first operational stored-program machine, which also introduced the concept of primary and secondary memory hierarchy.

von Neumann is credited for developing the stored-program concept, beginning with his 1945 first draft of EDVAC. The structure of EDVAC established the organization of the stored-program computer (von Neumann machine), which contains

1. An input device through which data and instructions can be entered
2. A storage unit into which results can be entered and from which instructions and data can be fetched
3. An arithmetic unit to process data
4. A CU to fetch, interpret, and execute the instructions from the storage
5. An output device to deliver the results to the user

All contemporary computers are von Neumann machines, although various alternative architectures have evolved.

1.2.1 von Neumann Model

Figure 1.4 shows the von Neumann model, a typical uniprocessor computer system consisting of the MU, the ALU, the CU, and the IOU. The MU is a single-port device consisting of a memory address register (MAR) and a memory buffer register (MBR)—also called a memory data register (MDR). The memory cells are arranged in the form of several memory words, where each word is the unit of data that can be read or written. All the read and write operations on the memory utilize the memory port. The ALU performs the arithmetic and logic operations on the data items in the accumulator (ACC) and/or MBR and typically the ACC retains the results of such operations. The CU consists of a program counter (PC) that contains the address of the instruction to be fetched and an instruction register (IR) into which the instructions are fetched from the memory for execution. Two registers are included in the structure. These can be used to hold the data and address values during computation. For simplicity, the I/O subsystem is shown to input to and output from the ALU subsystem. In practice, the I/O may also occur directly between the memory and I/O devices without utilizing any processor registers. The components of the system are interconnected by a multiple-bus structure on which the data and addresses flow. The CU manages this flow through the use of appropriate control signals.

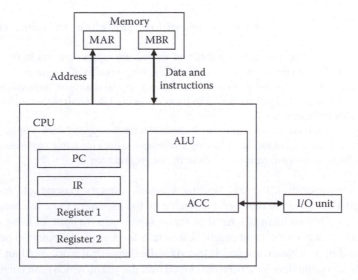

Figure 1.4 von Neumann architecture.

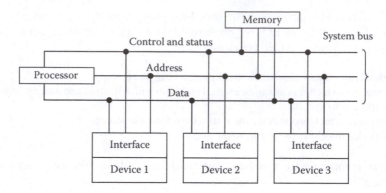

Figure 1.5 General computer system.

Figure 1.5 shows a more generalized computer system structure representative of modern-day architectures. The processor subsystem (i.e., the central processing unit—CPU) now consists of the ALU and CU and various processor registers. The processor, memory, and I/O subsystems are interconnected by the system bus, which consists of data, address, control, and status lines.

Practical systems may differ from the single-bus architecture of Figure 1.5, in the sense that they may be configured around multiple buses. For instance, there may be a memory bus that connects the processor to the memory subsystem and an I/O bus to interface I/O devices to the processor, forming a two-bus structure. Further, it is possible to configure a system with several I/O buses wherein each bus may interface one type of I/O device to the processor. Since multiple-bus structures allow simultaneous operations on the buses, a higher throughput is possible, compared to single-bus architectures. However, because of the multiple buses, the system complexity increases. Thus, a speed/cost trade-off is required to decide on the system structure. The processor subsystem may also consist of several processors. Each processor may be dedicated to a specific computational task (arithmetic/logic, graphics and display, I/O handling, etc.), or processors as a group may share the overall computational load as the load changes. The memory subsystem may also contain several modules and types of memory.

It is important to note the following characteristics of the von Neumann model that make it inefficient:

1. Programs and data are stored in a single sequential memory, which can create a memory access "bottleneck."
2. There is no explicit distinction between data and instruction representations in the memory. This distinction has to be brought about by the CPU during the execution of programs.
3. High-level language programming environments utilize several data structures (such as single and multidimensional arrays and linked lists). The memory, being 1D, requires that such data structures be linearized for representation.
4. The data representation does not retain any information on the type of data. For instance, there is nothing to distinguish a set of bits representing floating-point data from that representing a character string. Such distinction has to be brought about by the program logic.

Because of these characteristics, the von Neumann model is overly general and requires excessive mapping by compilers to generate the code executable by the hardware from the programs written in high-level languages. This problem is termed as the *semantic* gap. In spite of these deficiencies, the von Neumann model has been the most practical structure for digital computers. The central concept in the model devised by von Neumann was inspired by the theoretical work of Allan Turing and Kurt Godel. Several efficient compilers have been developed over the years, which have narrowed the semantic gap to the extent that it is almost invisible for a high-level language programming environment.

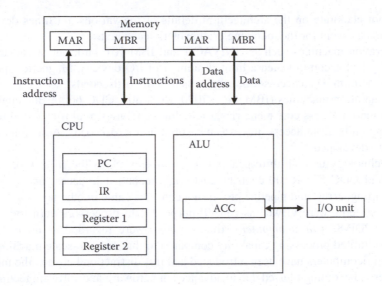

Figure 1.6 Harvard architecture.

Note that the von Neumann model of Figure 1.4 provides one path for addresses and a second path for data and instructions, between the CPU and the memory. An early variation of this model is the Harvard architecture shown in Figure 1.6. This architecture provides independent paths for data addresses, data, instruction addresses, and instructions. This allows the CPU to access instruction and data simultaneously. The name *Harvard architecture* is due to Howard Aiken's work on Mark I through Mark IV computers at Harvard University. These machines had separate storage for data and instructions. Current Harvard architectures do not use separate storage for data and instructions but have separate paths and buffers to access data and instructions simultaneously.

Early enhancements to the von Neumann model were mainly concentrated on increasing the speed of the basic hardware structure. As the hardware technology progressed, efforts were made to incorporate as many high-level language features as possible into hardware and firmware in an effort to reduce the semantic gap.

Note that the hardware enhancements alone may not be sufficient to attain the desired performance. The architecture of the overall computing environment starting from the algorithm development to execution of programs needs to be analyzed to arrive at the appropriate hardware, software, and firmware structures. If possible, these structures should exploit the parallelism in the algorithms themselves. Thus, performance enhancement is one reason for *parallel processing*. There are also other reasons such as reliability, fault tolerance, expandability, and modular development that dictate parallel processing structures. We will introduce parallel processing concepts further later in the book.

1.2.2 Generations of Computer Technology

Commercial computer system development has followed development of hardware technology and is usually divided into five generations:

1. First generation (1945–1955)—vacuum tube technology
2. Second generation (1955–1965)—transistor technology
3. Third generation (1965–1980)—integrated circuit (IC) technology
4. Fourth generation (1980s–present)—very large-scale integrated (VLSI) circuit technology
5. Fifth generation (present and beyond)—artificial intelligence-based computing devices, extensive parallel processing structures, using superconductors, nanotechnology, and quantum computing

We will not elaborate on the architectural details of the various machines developed during these generations, except for the following brief evolution account.

First-generation machines such as UNIVAC 1 and IBM 701, built out of vacuum tubes, were slow and bulky and accommodated a limited number of I/O devices. Magnetic tape was the predominant I/O medium. Data access time was measured in milliseconds.

Second-generation machines (IBM 1401, 7090; RCA 501; CDC 6600; Burroughs 5500; DEC PDP-1) used random-access core memories, transistor technology, multifunctional units, and multiple processing units. Data access time was measured in microseconds. Assembler and high-level languages were developed.

The IC technology used in third-generation machines such as the IBM 360, UNIVAC 1108, ILLIAC-IV, and CDC STAR-100 contributed to nanosecond data access and processing times. Multiprogramming, array, and pipeline processing concepts came into being.

Computer systems were viewed as general-purpose data processors until the introduction in 1965 of DEC PDP-8, a *minicomputer*. Minicomputers were regarded as dedicated application machines with limited processing capability compared to that of large-scale machines. Since then, several new minicomputers have been introduced and this distinction between the mini- and large-scale machines is becoming blurred due to advances in hardware and software technology.

The development of *microprocessors* in the early 1970s allowed a significant contribution to the third class of computer systems: microcomputers. Microprocessors are essentially computers on an IC chip that can be used as components to build a dedicated controller or processing system. Advances in IC technology leading to the current VLSI era have made microprocessors as powerful as minicomputers of the 1970s. VLSI-based systems are called fourth-generation systems since their performance is so much higher than that of third-generation systems.

Modern computer system architecture exploits the advances in hardware and software technologies to the fullest extent. Due to advances in IC technology that make the hardware much less expensive, the architectural trend is to interconnect several processors to form a high-throughput system.

We are now witnessing the development of fifth-generation systems. There is no accepted definition of what a fifth-generation computer is. Fifth-generation development efforts in the United States involve building supercomputers with very high computational capability, large memory capacity, and flexible multiple-processor architectures, employing extensive parallelism. Japanese fifth-generation activities aimed toward building artificial intelligence-based machines with very high numeric and symbolic processing capabilities, large memories, and user-friendly natural interfaces. Some attribute fifth generation to biology-inspired (neural networks, DNA) computers and optical computer systems.

The current generation of computer systems exploits parallelism in algorithms and computations to provide high performance. The simplest example of parallel architecture is the Harvard architecture, which utilizes two buses operating simultaneously. Parallel processing architectures utilize a multiplicity of processors and memories operating concurrently. We will describe various parallel and pipelined architecture structures later in this book.

1.2.3 Moore's Law

The progress in hardware technology resulting in the current VLSI era has given us the capability to fabricate millions of transistors on an IC (chip). This has allowed us to build chips with powerful processing structures with large memory systems. The early supercomputers such as CDC 6600 had about 128 kB of memory and could perform 10 million instructions per second. Today's supercomputers contain thousands of processors and terabytes of memory and perform trillions of operations per second. All this is possible because of the miniaturization of transistors. How far can this miniaturization continue? In 1965, Gordon Moore (Founder of Intel Corporation) stated that "the density of transistors in an IC will double every year." This so-called Moore's law is now modified to state that "the density of chips doubles every 18 months." This law has held true for more than 40 years and many believe that it will continue to hold for a few more decades. Arthur Rock proposed a corollary to Moore's law that states that "the cost of capital equipment to manufacture ICs will double every four years." Indeed, the cost of building new IC fabrication facilities has escalated from about $10,000 in the 1960s to $10 million in the 1990s to about $3 billion today. Thus, the cost might get prohibitive even though the technology allows building denser chips. Newer technologies and processing paradigms are continually being invented to achieve this cost/technology compromise.

1.3 ORGANIZATION VERSUS DESIGN VERSUS ARCHITECTURE

Computer organization addresses issues such as types and capacity of memory, control signals, register structure, and instruction set. It answers the question "How does a computer work?" basically from a programmer's point of view. Architecture is the art or science of building, a method or style of building. Thus, a computer architect develops the performance specifications for various components of the computer system and defines the interconnections between them. A computer designer, on the other hand, refines these component specifications and implements them using hardware, software, and firmware elements. An architect's capabilities are greatly enhanced if he or she is also exposed to the design aspects of the computer system.

A computer system can be described at the following levels of detail:

1. Processor-memory-switch (PMS) level, at which an architect views the system. It is simply a description of system components and their interconnections. The components are specified to the block diagram level.
2. Instruction set level, at which level the function of each instruction is described. The emphasis of this description level is on the behavior of the system rather than the hardware structure of the system.
3. Register transfer level, at which level the hardware structure is more visible than previous levels. The hardware elements at this level are registers that retain the data that are processed until the current phase of processing is complete.
4. Logic gate level, at which level the hardware elements are logic gates and flip-flops. The behavior is now less visible, while the hardware structure predominates.
5. Circuit level, at which level the hardware elements are resistors, transistors, capacitors, and diodes.
6. Mask level, at which level the silicon structures and their layout that implements the system as an IC are shown.

As one moves from the first level of description toward the last, it is evident that the behavior of the machine is transformed into a hardware–software structure. A computer architect concentrates on the first two levels described earlier, whereas the computer designer takes the system design to the remaining levels.

1.4 PERFORMANCE EVALUATION

Several measures of performance have been used in the evaluation of computer systems. The most common ones are million instructions per second (MIPS), million operations per second (MOPS), million floating-point operations per second (MFLOPS or megaflops), billion floating-point operations per second (GFLOPS or gigaflops), and million logical inferences per second (MLIPS). Machines capable of trillion floating-point operations per second (teraflops) are now available. Table 1.1 lists the common prefixes used for these measures. Power-of-10 prefixes shown

Table 1.1   Common Prefixes Used in Computer Systems Measurements

Power of 10	Power of 2	Prefix	Symbol
Thousand (10^3)	2^{10}	Kilo	k
Million (10^6)	2^{20}	Mega	M
Billion (10^9)	2^{30}	Giga	G
Trillion (10^{12})	2^{40}	Tera	T
Quadrillion (10^{15})	2^{50}	Peta	P
Quintillion (10^{18})	2^{60}	Exa	E
Sextillion (10^{21})	2^{70}	Zetta	Z
Septillion (10^{24})	2^{80}	Yotta	Y
Thousandth (10^{-3})	2^{-10}	Milli	m
Millionth (10^{-6})	2^{-20}	Micro	μ
Billionth (10^{-9})	2^{-30}	Nano	n
Trillionth (10^{-12})	2^{-40}	Pico	p
Quadrillionth (10^{-15})	2^{-50}	Femto	f
Quintillionth (10^{-18})	2^{-60}	Atto	a
Sextillionth (10^{-21})	2^{-70}	Zepto	z
Septillionth (10^{-24})	2^{-80}	Yocto	y

in the first column are typically used for power, frequency, voltage, and computer performance measurements. Power-of-2 prefixes shown in the second column are typically used for memory, file, and register sizes. Columns three and four of the table show the prefixes and symbols used in denoting the power-of-2 prefixes.

The measure used depends on the type of operations one is interested in, for the particular application for which the machine is being evaluated. As such, these measures have to be based on the mix of operations representative of their occurrence in the application.

The performance rating could be either the peak rate (i.e., the MIPS rating the CPU cannot exceed) or the more realistic average or sustained rate. In addition, a comparative rating that compares the average rate of the machine to that of other well-known machines is also used. In addition to the performance, other factors considered in evaluating architectures are generality (how wide is the range of applications suited for this architecture), ease of use, and expandability or scalability. One feature that is receiving considerable attention now is the openness of the architecture. The architecture is said to be open if the designers publish the architecture details such that others can easily integrate standard hardware and software systems to it. The other guiding factor in the selection of architecture is the cost.

Several analytical techniques are used in estimating the performance. All these techniques are approximations, and as the complexity of the system increases, most of these techniques become unwieldy. A practical method for estimating the performance in such cases is using benchmarks.

1.4.1 Benchmarks

Benchmarks are standardized batteries of programs run on a machine to estimate its performance. The results of running a benchmark on a given machine can then be compared with those on a known or standard machine, using criteria such as CPU and memory utilization and throughput and device utilization.

Benchmarks are useful in evaluating hardware as well as software and single processor as well as multiprocessor systems. They are also useful in comparing the performance of a system before and after certain changes are made.

As a high-level language host, a computer architecture should execute efficiently those features of a programming language that are most frequently used in actual programs. This ability is often measured by benchmarks. Benchmarks are considered to be representative of classes of applications envisioned for the architecture.

There are several benchmark suites in use today and more are being developed. It is important to note that the benchmarks provide only a broad performance guideline. It is the responsibility of the user to select the benchmark that comes close to his application and further evaluate the machine based on scenarios expected in the application for which the machine is being evaluated. One simple example for benchmark of general-purpose computers is drawing polygons, typically triangles, with single/multiple colors at a specified frame rate. This benchmark exercises the computer's processing capacity related to generating diagrams and also colors. Another very popular example is calculating the fast Fourier transform (FFT) of standard functions. This exercises all the common arithmetic operations like addition and subtraction. Chapter 16 provides further details on benchmarks and performance evaluation.

1.5 SUMMARY

This chapter has introduced the basic terminology associated with modern computer systems. The five common levels of architecture abstractions were introduced along with a trace of the evolution of computer generations. Performance issues and measures were briefly introduced. Subsequent chapters of the book expand on the concepts and issues highlighted in this chapter.

PROBLEMS

1.1   What is the difference between hardware, software, and firmware?
1.2   Look up the definition of the following terms: parallel computing, computer networks, and distributed computing.
1.3   Does Moore's law hold forever? What are the limitations?
1.4   Look up the characteristics that distinguish between the following:
         Supercomputers
         Mainframe (large-scale) computers
         Minicomputers
         Microcomputers
         Desktops
         Laptops
         Tablet computers
1.5   What are the important factors that influence buying a laptop and tablet computers in today's technology? Scan the advertisements from vendors and vendor catalogs to list the characteristics and their importance.
1.6   Compilers translate the program into machine language so that it can be executed by hardware. Some systems use an "interpreter" rather than a compiler. How is this environment different?
1.7   Assume that a transistor in current technology is 1 μm in diameter. How large would this transistor be 2 years from today, according to Moore's law? (Hint: Consider the transistors are fabricated side by side.)
1.8   How many kilobytes are in a megabyte of memory? How many megabytes are in a gigabyte memory? Express your answers as powers of 2.
1.9   Look up the details on the following popular benchmarks: Whetstone, Dhrystone, PERFECT, SLALOM, and GeekBench.
1.10 What is a "virtual machine" architecture?

1.11 Does the main memory size in the system affect the speed of operation, that is, execution of the programs, in a considerable way? How and why?

1.12 What evolutionary compulsion drove the invention of high-level languages? With the high-level languages available, what is the relevance of low-level languages today?

1.13 What are the strengths and weaknesses of the high-level languages and low-level languages? In the foreseeable future, will either make the other extinct?

1.14 Compare the capabilities and limitations of a typical tablet PC to those of a typical laptop.

1.15 There are several "apps" available now for use on tablet computers and smartphones. What is an "app"? Why does an app designed for one type of smartphone not work on another type of smartphone?

BIBLIOGRAPHY

Burks, A.W., Goldstine, H.H., and von Neumann, J., Preliminary discussion of the logical design of electrical computing instrument, U.S. Army Ordnance Department Report, 1946.

Godfrey, M.D. and Hendry, D.F., The computer as von Neumann planned it, *IEEE Annals of the History of Computing*, 15(1), 11–21, 1993.

Goldstine, H.H., *The Computer from Pascal to von Neumann*, Princeton, NJ: Princeton University Press, 1972.

Grace, R., *The Benchmark Book*, Upper Saddle River, NJ: Prentice Hall, 1996.

Price, W.J., A benchmarking tutorial, *IEEE Microcomputer*, 9(5), 28–43, 1989.

Schaller, R., Moore's law: Past, present and future, *IEEE Spectrum*, 34(6), 52–59, 1997.

Shiva, S.G., *Advanced Computer Architectures*, Boca Raton, FL: CRC Press/Taylor & Francis, 2006.

CHAPTER 2

Number Systems and Codes

As mentioned earlier, the elements in the discrete data representation correspond to discrete voltage levels or current magnitudes in the digital system hardware. If the digital system is required to manipulate only numeric data, for instance, it will be best to use 10 voltage levels, with each level corresponding to a decimal digit. But the noise introduced by the circuitry dealing with multiple levels of voltage for representation makes such representation impractical. Therefore, digital systems typically use a two-level representation, with one voltage level representing a 0 and the other representing a 1. To represent all 10 decimal digits using this *binary* (two-valued) alphabet of 0 and 1, a unique pattern of 0s and 1s is assigned to each digit. For example, in an electronic calculator, each keystroke should produce a pattern of 0s and 1s corresponding to the digit or the operation represented by that key.

Because the data elements and operations are all represented in binary form in all practical digital systems, a good understanding of the binary number system and data representation is basic to the analysis and design of digital system hardware. In this chapter, we will discuss the binary number system in detail. In addition, we will discuss two other widely used systems: *octal* and *hexadecimal*. These two number systems are useful in representing binary information in a compact form. When the human user of the digital system works with data manipulated by the system, either to verify it or to communicate it to another user, the compactness provided by these systems is helpful. As we will see in this chapter, data conversion from one number system to the other can be performed in a straightforward manner.

The data to be processed by the digital system are made up of decimal digits, alphabetic characters, and special characters, such as +, −, and *. The digital system uses a unique pattern of 0s and 1s to represent each of these digits and characters in the binary form. The collection of these binary patterns is called the *binary code*. Various binary codes have been devised by digital system designers over the years. Some popular codes will be discussed in this chapter.

2.1 NUMBER SYSTEMS

Let us review the decimal number system, the system with which we are most familiar. There are 10 symbols (0 through 9), called digits, in the system along with a set of symbols defining the operations of addition (+), subtraction (−), multiplication (×), and division (/). The total number of digits in a number system is called the *radix* or *base* of the system. The digits in the system range in value from 0 through $r - 1$, where r is the radix. For the decimal system, $r = 10$ and the digits range in value from 0 through $(10 - 1) = 9$.

In the so-called positional notation of a number, the radix point separates the "integer" portion of the number from the "fraction" portion. If there is no fraction portion, the radix point is not explicitly shown in the positional notation. Furthermore, each position in the representation has a weight associated with it. The weight of each position is equivalent to the radix raised to a power. The power starts with a 0 at the position immediately to the left of the radix point and increases by 1 as we move each position toward the left and decreases by 1 as we move each position toward the right. A typical number in the decimal system is shown in Example 2.1.

Example 2.1

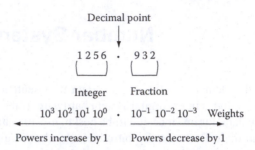

This number can also be represented as a polynomial:

$$1\times10^3 + 2\times10^2 + 5\times10^1 + 6\times10^0 + 9\times10^{-1} + 3\times10^{-2} + 2\times10^{-3}.$$

We can thus generalize these two representations to any number system. The general positional notation of a number N is

$$N = (a_n \ldots a_3 a_2 a_1 a_0 \cdot a_{-1} a_{-2} a_{-3} \ldots a_{-m})_r, \qquad (2.1)$$

where
  r is the radix of the number system
  a_{-1}, a_0, a_1, a_2, etc., are digits such that $0 \le a_i \le (r-1)$ for all i
  a_n is the most significant digit (MSD)
  a_{-m} is the least significant digit (LSD)

The polynomial representation of the earlier number is

$$N = \sum_{i=-m}^{n} a_i r^i. \qquad (2.2)$$

There are $n + 1$ integer digits and m fraction digits in the number shown earlier.

Consider an integer with n digits. A finite range of values can be represented by this integer. The smallest value in this range is 0 and corresponds to each digit of the n-digit integer being equal to 0. When each digit corresponds in value to $r - 1$, the highest digit in the number system, the n-digit number attains the highest value in the range. This value is equal to $r^n - 1$. Table 2.1 lists the first few numbers in various systems. We will discuss binary, octal, and hexadecimal systems next.

2.1.1 Binary System

In this system, the radix is 2 and the two allowed digits are 0 and 1. *BInary digiT* is abbreviated as BIT. A typical binary number is shown in the positional notation in Example 2.2.

Table 2.1   Number Systems

Decimal $r = 10$	Binary $r = 2$	Ternary $r = 3$	Quaternary $r = 4$	Octal $r = 8$	Hexadecimal $r = 16$
0	0	0	0	0	0
1	1	1	1	1	1
2	10	2	2	2	2
3	11	10	3	3	3
4	100	11	10	4	4
5	101	12	11	5	5
6	110	20	12	6	6
7	111	21	13	7	7
8	1000	22	20	10	8
9	1001	100	21	11	9
10	1010	101	22	12	A
11	1011	102	23	13	B
12	1100	110	30	14	C
13	1101	111	31	15	D
14	1110	112	32	16	E
15	1111	120	33	17	F
16	10000	121	100	20	10
17	10001	122	101	21	11
18	10010	200	102	22	12
19	10011	201	103	23	13
20	10100	202	110	24	14

Example 2.2

$$N = (11010 \quad \cdot \quad 1101)_2$$

$$2^4\,2^3\,2^2\,2^1\,2^0 \quad \cdot \quad 2^{-1}2^{-2}2^{-3}2^{-4} \quad \text{Weights}$$

$$16\ 8\ \ 4\ 2\ 1 \quad \cdot \quad \tfrac{1}{2}\ \tfrac{1}{4}\ \tfrac{1}{8}\ \tfrac{1}{16} \quad \text{Weights in decimal}$$

Weights double for each move   Weights are halved for each move to
to left from the binary point   right from the binary point

In polynomial form, this number is

$$N = 1 \times 2^4 + 1 \times 2^3 + 0 \times 2^2 + 1 \times 2^1 + 0 \times 2^0 + 1 \times 2^{-1} + 1 \times 2^{-2}$$

$$+ 0 \times 2^{-3} + 1 \times 2^{-4}$$

$$= 16 + 8 + 0 + 2 + 0 + \frac{1}{2} + \frac{1}{4} + 0 + \frac{1}{16}\text{(decimal)}$$

$$= 26 + \frac{1}{2} + \frac{1}{4} + \frac{1}{16}\text{(decimal)}$$

$$= \left(26\frac{13}{16}\right)_{10}.$$

As we can see from the polynomial expansion and summation shown here, the positions containing a 0 do not contribute to the sum. To convert a binary number into decimal, we can simply accumulate the weights corresponding to each nonzero bit of the number.

Table 2.2   Binary Numbers

$n = 2$	$n = 3$	$n = 4$
1 | 0	2 | 1 | 0	3 | 2 | 1 | 0 ←bit position
00	000	0000
01	001	0001
10	010	0010
11	011	0011
	100	0100
	101	0101
	110	0110
	111	0111
		1000
		1001
		1010
		1011
		1100
		1101
		1110
		1111

Each bit can take either of the two values: 0 or 1. With 2 bits, we can derive 2^2, or 4, combinations: 00, 01, 10, and 11. The decimal values of these combinations (binary numbers) are 0, 1, 2, and 3, respectively. Similarly, with 3 bits we can derive 2^3, or 8, combinations ranging in value from 000 (0 in decimal) to 111 (7 in decimal). In general, with n bits it is possible to generate 2^n combinations of 0s and 1s, and these combinations when viewed as binary numbers range in value from 0 to $(2^n - 1)$. Table 2.2 shows some binary numbers for various values of n. The 2^n combinations possible for any n are obtained by starting with n 0s and counting in binary until the number with n 1s is reached. A more mechanical method of generating these combinations is described herein.

The first combination has n 0s and the last has n 1s. As we can see from Table 2.2, the value of the least significant bit (LSB), that is, bit position 0, alternates in value between 0 and 1 every row, as we move from row to row. Similarly, the value of the bit in position 1 alternates every two rows (i.e., two 0s followed by two 1s). In general, the value of the bit in position i alternates every 2^i rows starting from 0s. This observation can be utilized in generating all the 2^n combinations.

2.1.2 Octal System

In this system, $r = 8$, and the allowed digits are 0, 1, 2, 3, 4, 5, 6, and 7. A typical number is shown in positional notation in Example 2.3.

Example 2.3

$$N = (4\,5\,2\,6 \cdot 2\,3)_8$$

$$8^3 8^2 8^1 8^0 \cdot 8^{-1} 8^{-2} \qquad \text{(Weights)}$$

$$= 4 \times 8^3 + 5 \times 8^2 + 2 \times 8^1 + 6 \times 8^0 + 2 \times 8^{-1} + 3 \times 8^{-2}$$

$$\text{(Polynomial form)}$$

$$= 2048 + 320 + 16 + 6 + \frac{2}{8} + \frac{3}{64} \qquad \text{(Decimal)}$$

$$= \left(2390\,\frac{19}{64}\right)_{10}$$

2.1.3 Hexadecimal System

In this system, $r = 16$, and the allowed digits are 0, 1, 2, 3, 4, 5, 6, 7, 8, 9, A, B, C, D, E, and F. Digits A through F correspond to decimal values 10 through 15, respectively. A typical number is shown in Example 2.4.

Example 2.4

$$N = (A\ 1\ F \cdot 1\ C)_{16}$$

$$16^2 16^1 16^0 \cdot 16^{-1} 16^{-2} \qquad \text{(Weights)}$$

$$= A \times 16^2 + 1 \times 16^1 + F \times 16^0 + 1 \times 16^{-1} + C \times 16^{-2}$$

$$\text{(Polynomial form)}$$

$$= 10 \times 16^2 + 1 \times 16^1 + 15 \times 16^0 + 1 \times 16^{-1} + 12 \times 16^{-2}$$

$$= \left(2591\, \frac{28}{256}\right)_{10} \qquad \text{(Decimal)}$$

2.2 CONVERSION

To convert numbers from a nondecimal system to decimal, we simply expand the given number as a polynomial and evaluate the polynomial using decimal arithmetic, as shown in Examples 2.1 through 2.4. When a decimal number is converted to any other system, the integer and fraction portions of the number are handled separately. The radix divide technique is used to convert the integer portion, and the radix multiply technique is used for the fraction portion.

2.2.1 Radix Divide Technique

1. Divide the given integer successively by the required radix, noting the remainder at each step. The quotient at each step becomes the new dividend for subsequent division. Stop the division process when the quotient becomes zero.
2. Collect the remainders from each step (last to first) and place them left to right to form the required number.

Examples 2.5 through 2.7 illustrate the procedure.

Example 2.5

$$(245)_{10} = (?)_2 \text{ (i.e., convert } (245)_{10} \text{ to binary)}$$

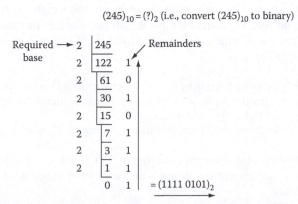

Here, 245 is first divided by 2, generating a quotient of 122 and a remainder of 1. Next 122 is divided, generating 61 as the quotient and 0 as the remainder. The division process is continued until the quotient is 0, with the remainders noted at each step. The remainder bits from each step (last to first) are then placed left to right to form the number in base 2.

To verify the validity of the radix divide technique, consider the polynomial representation of a 4-bit integer $A = (a_3 a_2 a_1 a_0)$:

$$A = \sum_{i=0}^{3} a_1 \cdot 2^i.$$

This can be rewritten as

$$A = (2(2(2(a_3) + a_2) + a_1) + a_0).$$

From this form, it can be seen that the bits of the binary number correspond to the remainder at each divide-by-two operation. Examples 2.6 and 2.7 show the application of the techniques to other number systems.

Example 2.6

$$(245)_{10} = (?)_8$$

$$
\begin{array}{r|rl}
8 & 245 & \\
8 & 30 & 5 \\
8 & 3 & 6 \\
& 0 & 3 \quad = (365)_8
\end{array}
$$

Example 2.7

$$(245)_{10} = (?)_{16}$$

$$
\begin{array}{r|rl}
16 & 245 & \\
16 & 15 & 5 = 5 \\
& 0 & 15 = F \quad = (F5)_{16}
\end{array}
$$

2.2.2 Radix Multiply Technique

As we move each position to the right of the radix point, the weight corresponding to each bit in the binary fraction is halved. The radix multiply technique uses this fact and multiplies the given decimal number by 2 (i.e., divides the given number by half) to obtain each fraction bit. The technique consists of the following steps:

1. Successively multiply the given fraction by the required base, noting the integer portion of the product at each step. Use the fractional part of the product as the multiplicand for subsequent steps. Stop when the fraction either reaches 0 or recurs.
2. Collect the integer digits at each step from first to last and arrange them left to right.

If the radix multiplication process does not converge to 0, it is not possible to represent a decimal fraction in binary exactly. Accuracy, then, depends on the number of bits used to represent the fraction. Some examples follow.

Example 2.8

$$(0.250)_{10} = (?)_2$$

$$
\begin{array}{l}
0.25 \\
\underline{\times 2} \\
0.50 \\
\underline{\times 2} \\
1.00
\end{array} \quad = (0.01)_2
$$

Example 2.9

$$(0.345)_{10} = (?)_2$$

$$
\begin{array}{l}
0.345 \\
\underline{\times 2} \\
0.690 \\
\underline{\times 2} \\
1.380 \\
\underline{\times 2} \\
0.760 \qquad \text{Multiply fractions only}\\
\underline{\times 2} \\
1.520 \\
\underline{\times 2} \\
1.040 \\
\underline{\times 2} \\
0.080
\end{array}
$$

$$= (0.010110)_2$$

The fraction may never reach 0; stop when the required number of fraction digits is obtained; the fraction will not be accurate.

Example 2.10

$$(0.345)_{10} = (?)_8$$

$$
\begin{array}{l}
0.345 \\
\underline{\times 8} \\
2.760 \\
\underline{\times 8} \\
6.080 \\
\underline{\times 8} \\
0.640 \\
\underline{\times 8} \\
5.120
\end{array} \quad = (0.2605)_8
$$

Example 2.11

$$(242.45)_{10} = (?)_2$$

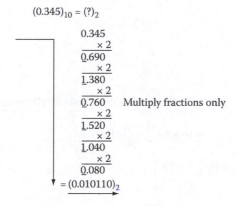

2	242	
2	121	0
2	60	1
2	30	0
2	15	0
2	7	1
2	3	1
2	1	1
	0	1

$$
\begin{array}{l}
0.45 \\
\underline{\times 2} \\
0.90 \\
\underline{\times 2} \\
1.80 \\
\underline{\times 2} \\
1.60 \\
\underline{\times 2} \\
1.20 \\
\underline{\times 2} \\
0.40 \\
\underline{\times 2} \\
0.80 \\
\underline{\times 2} \\
1.60
\end{array} \quad {}^{*}\text{Repeats}
$$

$$= (1111\ 0010.01\ \overline{1100})_2$$

The radix divide and multiply algorithms are applicable to the conversion of numbers from any base to any other base. When a number is converted from base p to base q, the number in base p is divided (or multiplied) by q in base p arithmetic. Because of our familiarity with decimal arithmetic, these methods are convenient when $p = 10$. In general, it is easier to convert a base p number to base q ($p \neq 10$, $q \neq 10$) by first converting the number to decimal from base p and then converting that decimal number to base q (i.e., $(N)_p \rightarrow (?)_{10} \rightarrow (?)_q$), as shown by the following example.

Example 2.12

$$(25.34)_8 = (?)_5.$$

Convert to base 10:

$$(25.34)_8 = 2 \times 8^1 + 5 \times 8^0 + 3 \times 8^{-1} + 4 \times 8^{-2} \text{(Decimal)}$$

$$= 16 + 5 + \frac{3}{8} + \frac{4}{64} \text{(Decimal)}$$

$$= \left(21\frac{28}{64}\right)_{10}$$

$$= (21.4375)_{10}.$$

Convert to base 5:

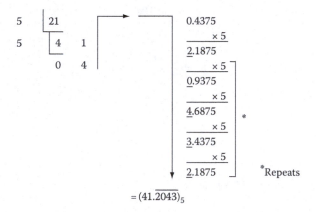

$$= (41.\overline{2043})_5$$

2.2.3 Base 2^k Conversion

Each of the eight octal digits can be represented by a 3-bit binary number. Similarly, each of the 16 hexadecimal digits can be represented by a 4-bit binary number. In general, each digit of the base p number system, where p is an integral power k of 2, can be represented by a k-bit binary number.

In converting a base p number to base q, if p and q are both integral powers of 2, the base p number can first be converted to binary, and this can in turn be converted to base q by inspection. This conversion procedure is called the *base 2^k conversion*.

Example 2.13

$$(4\,2\,A\,5\,6 \cdot F\,1)_{16} = (?)_8$$

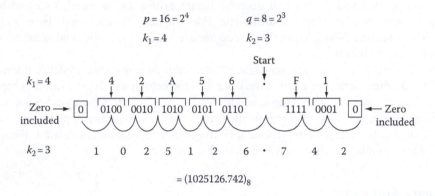

$p = 16 = 2^4$ $q = 8 = 2^3$

$k_1 = 4$ $k_2 = 3$

$= (1025126.742)_8$

Example 2.14

$$(AF5.2C)_{16} = (?)_4$$

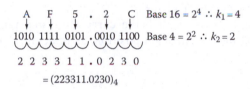

$= (223311.0230)_4$

Example 2.15

$$(567.23)_8 = (?)_{16}$$

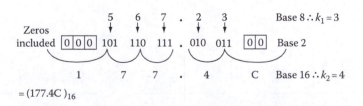

$= (177.4C)_{16}$

It is thus possible to represent binary numbers in a very compact form by using octal and hexadecimal systems. The conversion between these systems is also straightforward. Because it is easier to work with fewer digits than with a large number of bits, digital system users prefer to work with octal or hexadecimal systems when understanding or verifying the results produced by the system or communicating data between users or among users and the machine.

2.3 ARITHMETIC

Arithmetic in all other number systems follows the same general rules as in decimal. Binary arithmetic is simpler than decimal arithmetic since only two digits (0 and 1) are involved. Arithmetic in octal and hexadecimal systems requires some practice because of the general unfamiliarity with those systems. In this section, we will describe binary arithmetic in detail, followed by a brief discussion of octal and hexadecimal arithmetic. For simplicity, integers will be used in all the examples in this section. Nonetheless, the procedures are valid for fractions and numbers with both integer and fraction portions.

In the so-called fixed-point representation of binary numbers in digital systems, the radix point is assumed to be either at the right end or the left end of the field in which the number is represented. In the first case, the number is an integer, and in the second it is a fraction. Fixed-point representation is the most common type of representation. In scientific computing applications, in which a large range of numbers must be represented, floating-point representation is used. Floating-point representation of numbers is discussed in Section 2.6.

2.3.1 Binary Arithmetic

Figure 2.1 illustrates the rules for binary addition, subtraction, and multiplication.

2.3.1.1 Addition

In Figure 2.1a, note that $0 + 0 = 0$, $0 + 1 = 1$, $1 + 0 = 1$, and $1 + 1 = 10$. Thus, the addition of two 1s results in a SUM of 0 and a CARRY of 1.

When two binary numbers are added, the carry from any position is included in the addition of bits in the next most significant position, as in decimal arithmetic. Example 2.16 illustrates this.

Example 2.16

```
          4 3 2 1 0   Bit position

          0 1 1 0     CARRY    CARRY in the LSB position is 0
          1 0 1 1 0   Augend
        + 0 0 1 1 1   Addend
          1 1 1 0 1   SUM
```

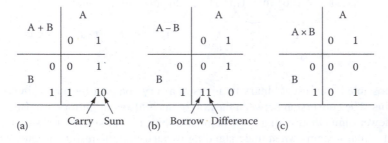

A + B		A	
		0	1
	0	0	1
B	1	1	10

(a) Carry Sum

A − B		A	
		0	1
	0	0	1
B	1	11	0

(b) Borrow Difference

A × B		A	
		0	1
	0	0	0
B	1	0	1

(c)

Figure 2.1 Binary arithmetic. (a) Addition. (b) Subtraction. (c) Multiplication.

Here, bits in the LSB position (i.e., position 0) are first added, resulting in a sum bit of 1 and a carry of 0. The carry is included in the addition of bits at position 1. The 3 bits in position 1 are added using two steps (0 + 1 = 1, 1 + 1 = 10), resulting in a sum bit of 0 and a carry bit of 1 to the next most significant position (position 2). This process is continued through the most significant bit (MSB).

In general, the addition of two n-bit numbers results in a number that is $n + 1$ bits long. If the number representation is to be confined to n bits, the operands of the addition should be kept small enough so that their sum does not exceed n bits.

2.3.1.2 Subtraction

From Figure 2.1b, we can see that $0 - 0 = 0$, $1 - 0 = 1$, $1 - 1 = 0$, and $0 - 1 = 1$ with a BORROW of 1. That is, subtracting a 1 from a 0 results in a 1 with a borrow from the next most significant position, as in decimal arithmetic. Subtraction of two binary numbers is performed stage by stage as in decimal arithmetic, starting from the LSB to the MSB. Some examples follow.

Example 2.17

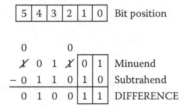

Bit position 1 requires a borrow from bit position 2. Because of this borrow, minuend bit 2 becomes a 0. The subtraction continues through the MSB.

Example 2.18

	7	6	5	4	3	2	1	0	Bit position

```
          0 0      0
        X̸ X̸ 0̸ 0̸ X̸ 0 1 1   Minuend
            1 1
      – 0 1 1 0 1 1 1 0   Subtrahend
        0 1 0 1 1 1 0 1   DIFFERENCE
```

Bit 2 requires a borrow from bit 3; after this borrow, minuend bit 3 is 0. Then, bit 3 requires a borrow. Because bits 4 and 5 of the minuend are zeros, borrowing is from bit 6. In this process, the intermediate minuend bits 4 and 5 each attain a value of 1 (compare this with the decimal subtraction). The subtraction continues through the MSB.

2.3.1.3 Multiplication

Binary multiplication is similar to decimal multiplication. From Figure 2.1c, we can see that $0 \times 0 = 0$, $0 \times 1 = 0$, $1 \times 0 = 0$, and $1 \times 1 = 1$.

An example follows.

Example 2.19

$$
\begin{array}{rl}
1011 & \text{Multiplicand} \\
\times\ 1100 & \text{Multiplier} \\
\hline
\end{array}
$$

Multiplier bits

$$
\begin{array}{rl}
0000 & (1011) \times 0 \\
0000 & (1011) \times 0 \\
1011 & (1011) \times 1 \\
1011 & (1011) \times 1 \\
\hline
10000100 & \text{PRODUCT}
\end{array}
$$

In general, the product of two n-bit numbers is $2n$ bits long. In Example 2.19, there are two nonzero bits in the multiplier, one in position 2 corresponding to 2^2 and the other in position 3 corresponding to 2^3. These 2 bits yield partial products whose values are simply that of the multiplicand shifted left 2 and 3 bits, respectively. The 0 bits in the multiplier contribute partial products with 0 values. Thus, the following shift-and-add algorithm can be adopted to multiply two n-bit numbers A and B, where $B = (b_{n-1}\ b_{n-2}...b_1 b_0)$.

1. Start with a $2n$-bit product with a value of 0.
2. For each $b_i\ (0 \leq i \leq n-1) \neq 0$, shift A i positions to the left and add to the product.

This procedure reduces the multiplication to repeated shift and addition of the multiplicand.

2.3.1.4 Division

The longhand (trial-and-error) procedure of decimal division can also be used in binary, as shown in Example 2.20.

Example 2.20

$$110101 \div 111 = ?$$

$$
\begin{array}{r}
0111 \quad \text{Quotient} \\
111\,\overline{\big)\ 110,101} \\
-000 \\
\hline
1101 \\
-111 \\
\hline
1100 \\
-111 \\
\hline
1011 \\
-111 \\
\hline
100 \quad \text{Remainder}
\end{array}
$$

$110 < 111 \therefore q_1 = 0$

$1101 > 111 \therefore q_2 = 1$

$1100 > 111 \therefore q_3 = 1$

$1011 > 111 \therefore q_4 = 1$

In this procedure, the divisor is compared with the dividend at each step. If the divisor is greater than the dividend, the corresponding quotient bit is 0; otherwise, the quotient bit is 1, and the divisor is subtracted from the dividend. The compare-and-subtract process is continued until the LSB of the dividend. The procedure is formalized in the following steps:

1. Align the divisor (Y) with the most significant end of the dividend. Let the portion of the dividend from its MSB to its bit aligned with the LSB of the divisor be denoted X. We will assume that there are n bits in the divisor and $2n$ bits in the dividend. Let $i = 0$.
2. Compare X and Y. If $X \geq Y$, the quotient bit is 1: perform $X - Y$. If $X < Y$, the quotient bit is 0.
3. Set $i = i + 1$. If $i \geq n$, stop. Otherwise, shift Y 1 bit to the right and go to step 2.

For the purposes of illustration, this procedure assumed the division of integers. If the divisor is greater than the dividend, the quotient is 0, and if the divisor is 0, the procedure should be stopped since dividing by 0 results in an error.

As we can see from these examples, multiplication and division operations can be reduced to repeated shift and addition (or subtraction). If the hardware can perform shift, add, and subtract operations, it can be programmed to perform multiplication and division as well. Older digital systems used such measures to reduce hardware costs. With the advances in digital hardware technology, it is now possible to implement these and more complex operations in an economical manner.

2.3.1.5 *Shifting*

Generally, shifting a base r number left by one position (and inserting a 0 into the vacant LSD position) is equivalent to multiplying the number by r. Shifting the number right by one position (inserting a 0 into the vacant MSD position) generally is equivalent to dividing the number by r.

In the binary system, each left shift multiplies the number by 2, and each right shift divides the number by 2, as shown in Example 2.21.

Example 2.21

	Binary	Decimal	
N	01011.11	11¾	
$2 \cdot N$	10111.1 $\boxed{0}$ INSERT	23½	
$N \div 2$ INSERT $\boxed{0}$ 0101.11		5¾	(inaccurate, since only 2-bit accuracy is retained)

If the MSB of an n-bit number is not 0, shifting it left would result in a number larger than the magnitude that can be accommodated in n bits, and the 1 shifted out of the MSB position cannot be discarded. If nonzero bits shifted out of the LSB position during a right shift are discarded, the accuracy is lost. Later in this chapter, we will discuss shifting in further detail.

2.3.2 Octal Arithmetic

Table 2.3 shows the octal addition and multiplication tables. The examples that follow illustrate the four arithmetic operations in octal and their similarity to decimal arithmetic. (Table 2.3 can be used to look up the result at each stage in the arithmetic.) An alternate method is used in the following examples. The operation is first performed in decimal and then the result is converted into octal, before proceeding to the next stage, as shown in the scratch pad.

Example 2.22: Addition

```
   1 1 1        ←─ Carries        Scratch pad
   1 4 7 6                              6
 + 3 5 5 4                             +4
 ─────────                       ────────────────
   5 2 5 2        SUM             (10)₁₀ = (12)₈
```

Scratch pad:

$$6$$
$$+4$$
$$(10)_{10} = (12)_8$$
$$1$$
$$+7$$
$$+5$$
$$(13)_{10} = (15)_8$$
$$1$$
$$+4$$
$$+5$$
$$(10)_{10} = (12)_8$$
$$1$$
$$+1$$
$$+3$$
$$(5)_{10} = (5)_8$$

Table 2.3   Octal Arithmetic

| | Addition | | | | | | | |
| | A | | | | | | | |
A + B	0	1	2	3	4	5	6	7
0	0	1	2	3	4	5	6	7
1	1	2	3	4	5	6	7	10
2	2	3	4	5	6	7	10	11
B 3	3	4	5	6	7	10	11	12
4	4	5	6	7	10	11	12	13
5	5	6	7	10	11	12	13	14
6	6	7	10	11	12	13	14	15
7	7	10	11	12	13	14	15	16

| | Multiplication | | | | | | | |
| | A | | | | | | | |
A × B	0	1	2	3	4	5	6	7
0	0	0	0	0	0	0	0	0
1	0	1	2	3	4	5	6	7
2	0	2	4	6	10	12	14	16
B 3	0	3	6	11	14	17	22	25
4	0	4	10	14	20	24	30	34
5	0	5	12	17	24	31	36	43
6	0	6	14	22	30	36	44	52
7	0	7	16	25	34	43	52	61

Example 2.23: Subtraction

$$
\begin{array}{r}
4\ 14 \\
\cancel{5}\ \cancel{4}\ 7\ 5 \\
-\ 3\ 7\ 6\ 4 \\
\hline
1\ 5\ 1\ 1
\end{array}
$$

Digit position 2 required a borrow from position 3.

∴

Octal	Decimal
14_8	12
-7_8	-7
5_8	$5_{10} = 5_8$

Example 2.24: Subtraction

$$
\begin{array}{r}
3\ 7\ 7 \\
5\ \cancel{4}\ \cancel{0}\ 0\ 4\ 5 \\
-\ 3\ 2\ 5\ 6\ 5\ 4 \\
\hline
2\ 1\ 2\ 1\ 7\ 1
\end{array}
$$

The intermediate 0s become $r - 1$ or 7 when borrowed.

Example 2.25: Multiplication

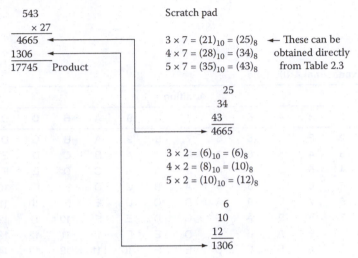

$$
\begin{array}{r}
543 \\
\times\ 27 \\
\hline
4665 \\
1306 \\
\hline
17745 \quad \text{Product}
\end{array}
$$

Scratch pad

$3 \times 7 = (21)_{10} = (25)_8$ ← These can be
$4 \times 7 = (28)_{10} = (34)_8$ obtained directly
$5 \times 7 = (35)_{10} = (43)_8$ from Table 2.3

$$
\begin{array}{r}
25 \\
34 \\
43 \\
\hline
4665
\end{array}
$$

$3 \times 2 = (6)_{10} = (6)_8$
$4 \times 2 = (8)_{10} = (10)_8$
$5 \times 2 = (10)_{10} = (12)_8$

$$
\begin{array}{r}
6 \\
10 \\
12 \\
\hline
1306
\end{array}
$$

Example 2.26: Division

$$543 \div 7$$

$$
\begin{array}{r}
062 \\
7\ \overline{)\ 543} \\
0 \\
\hline
54 \\
52 \\
\hline
23 \\
16 \\
\hline
5
\end{array}
$$

Use the multiplication table in Table 2.3 to derive the quotient digit (by trial and error).

2.3.3 Hexadecimal Arithmetic

Table 2.4 shows the addition and multiplication tables. The following examples illustrate hexadecimal arithmetic.

Example 2.27: Addition

Scratch pad

Decimal

$$
\begin{array}{r}
1\ \ 1 \\
1\ 5\ F\ C \\
+\ 2\ 4\ 5\ D \\
\hline
3\ A\ 5\ 9
\end{array}
$$

$C = 12$
$D = 13$

$$
\begin{array}{r}
16\ \overline{)\ 25} = (19)_{16} \quad \leftarrow \text{These can be obtained} \\
16\ \overline{)\ 1} \qquad 9 \qquad \text{directly from Table 2.4} \\
0 \qquad 1
\end{array}
$$

Decimal

$$
\begin{array}{r}
1\ =\ 1 \\
F\ =\ 15 \\
5\ =\ 5 \\
\hline
21 = (15)_{16}
\end{array}
$$

Table 2.4 Hexadecimal Arithmetic

Addition															
+	0	1	2	3	4	5	6	7	8	9	A	B	C	D	E	F
0	0	1	2	3	4	5	6	7	8	9	A	B	C	D	E	F
1	1	2	3	4	5	6	7	8	9	A	B	C	D	E	F	10
2	2	3	4	5	6	7	8	9	A	B	C	D	E	F	10	11
3	3	4	5	6	7	8	9	A	B	C	D	E	F	10	11	12
4	4	5	6	7	8	9	A	B	C	D	E	F	10	11	12	13
5	5	6	7	8	9	A	B	C	D	E	F	10	11	12	13	14
6	6	7	8	9	A	B	C	D	E	F	10	11	12	13	14	15
7	7	8	9	A	B	C	D	E	F	10	11	12	13	14	15	16
8	8	9	A	B	C	D	E	F	10	11	12	13	14	15	16	17
9	9	A	B	C	D	E	F	10	11	12	13	14	15	16	17	18
A	A	B	C	D	E	F	10	11	12	13	14	15	16	17	18	19
B	B	C	D	E	F	10	11	12	13	14	15	16	17	18	19	1A
C	C	D	E	F	10	11	12	13	14	15	16	17	18	19	1A	1B
D	D	E	F	10	11	12	13	14	15	16	17	18	19	1A	1B	1C
E	E	F	10	11	12	13	14	15	16	17	18	19	1A	1B	1C	1D
F	F	10	11	12	13	14	15	16	17	18	19	1A	1B	1C	1D	1E

Multiplication															
×	0	1	2	3	4	5	6	7	8	9	A	B	C	D	E	F
0	0	0	0	0	0	0	0	0	0	0	0	0	0	0	0	0
1	0	1	2	3	4	5	6	7	8	9	A	B	C	D	E	F
2	0	2	4	6	8	A	C	E	10	12	14	16	18	1A	1C	1E
3	0	3	6	9	C	F	12	15	18	1B	1E	21	24	27	2A	2D
4	0	4	8	C	10	14	18	1C	20	24	28	2C	30	34	38	3C
5	0	5	A	F	14	19	1E	23	28	2D	32	37	3C	41	46	4B
6	0	6	C	12	18	1E	24	2A	30	36	3C	42	48	4E	54	5A
7	0	7	E	15	1C	23	2A	31	38	3F	46	4D	54	5B	62	69
8	0	8	10	18	20	28	30	38	40	48	50	58	60	68	70	78
9	0	9	12	1B	24	2D	36	3F	48	51	5A	63	6C	75	7E	87
A	0	A	14	1E	28	32	3C	46	50	5A	64	6E	78	82	8C	96
B	0	B	16	21	2C	37	42	4D	58	63	6E	79	84	8F	9A	A5
C	0	C	18	24	30	3C	48	54	60	6C	78	84	90	9C	A8	B4
D	0	D	1A	27	34	41	4E	5B	68	75	82	8F	9C	A9	B6	C3
E	0	E	1C	2A	38	46	54	62	70	7E	8C	9A	A8	B6	C4	D2
F	0	F	1E	2D	3C	4B	5A	69	78	87	96	A5	B4	C3	D2	E1

Example 2.28: Subtraction

Scratch pad

$$
\begin{array}{c}
1\ 13\ 15 \\
3
\end{array}
$$

						Decimal	
2̷	4̷	3̷	D	Minuend	$(15)_{16}$ =	21	
−1	5	F	C	Subtrahend	$-(F)_{16}$ =	−15	
0	E	6	1	Difference		6 = $(6)_{16}$	
					$(13)_{16}$ =	19	
					$-(5)_{16}$ =	−5	
						14 = $(E)_{16}$	

Example 2.29: Multiplication

Scratch pad

$$
\begin{array}{r}
1\,E\,4\,A \\
\times\,FA2 \\
\hline
1\,2 \\
03\,C\,94 \\
+\quad 12\,E\,E\,4 \\
+\ 1\,C\,6\,5\,6 \\
\hline
1\,D\,9\,8\,0\,D\,4
\end{array}
$$

With P_1, P_2, P_3:

Decimal				Hexadecimal
A × 2	=	20	=	1 4
4 × 2	=	8	=	0 8
E × 2	=	28	=	1 C
1 × 2	=	2	=	0 2
				0 3 C 9 4 = P_1
A × A	=	100	=	6 4
4 × A	=	40	=	2 8
E × A	=	140	=	8 C
1 × A	=	10	=	A
				1 2 E E 4 = P_2
A × F	=	150	=	9 6
4 × F	=	60	=	3 C
E × F	=	210	=	D 2
1 × F	=	15	=	D
				1 C 6 5 6 = P_2

Example 2.30: Division

$$
\begin{array}{r}
\quad\ 0\,1\,E\,C \quad \text{Quotient} \\
E\ \overline{\big)\,1\,A\,F\,3} \\
0 \\
\overline{1\,A} \\
E \\
\overline{\ C\,F} \\
C\,4 \\
\overline{\ \ B\,3} \\
A\,8 \\
\hline
B \quad \text{Remainder}
\end{array}
$$

The examples shown so far have used only positive numbers. In practice, a digital system must represent both positive and negative numbers. To accommodate the sign of the number, an additional digit, called the sign digit, is included in the representation, along with the magnitude digits. Thus, to represent an n-digit number, we would need $n + 1$ digits. Typically, the sign digit is the MSD. Two popular representation schemes have been used: the sign–magnitude system and the complement system.

2.4 SIGN–MAGNITUDE SYSTEM

In this representation, $n + 1$ digits are used to represent a number, where the MSD is the sign digit and the remaining n digits are magnitude digits. The value of the sign digit is 0 for positive numbers and $r - 1$ for negative numbers, where r is the radix of the number system. Some sample representations follow.

Example 2.31

Here, we assume that five digits are available to represent each number. The sign and magnitude portions of the number are separated by ";" for illustration purposes only. The ";" is not used in the actual representation.

Number	Representation	
$(-2)_2$	1,0010	All numbers are shown as five-digit numbers.
$(+56)_8$	0,0056	
$(-56)_8$	7,0056	
$(+1F)_{16}$	0,001F	
$(-1F)_{16}$	F,001F	

Sign   Magnitude

The sign and magnitude portions are handled separately in arithmetic using sign–magnitude numbers. The magnitude of the result is computed and then the appropriate sign is attached to the result, just as in decimal arithmetic. The sign–magnitude system has been used in such small digital systems as digital meters and typically when the decimal mode of arithmetic is used in digital computers. The decimal (or binary-coded decimal [BCD]) arithmetic mode will be described later in this chapter. Complement number representation is the most prevalent representation mode in modern-day computer systems.

2.5 COMPLEMENT NUMBER SYSTEM

Consider the subtraction of a number A from a number B. This is equivalent to adding $(-A)$ to B. The complement number system provides a convenient way of representing negative numbers (i.e., complements of positive numbers), thus reducing the subtraction to an addition. Because multiplication and division correspond, respectively, to repeated addition and subtraction, it is possible to perform the four basic arithmetic operations using only the hardware for addition when the negative numbers are represented in complement form. The two popular complement number systems are radix complement and diminished radix complement.

The *radix complement* of a number $(N)_r$ is defined as

$$[N]_r = r^n - (N)_r \quad \text{if} (N)_r \neq 0$$
$$= 0 \quad \text{if} (N)_r = 0 \tag{2.3}$$

where
$[N]_r$ is the radix complement
n is the number of digits in the integer portion of the number $(N)_r$

This system is commonly called either 2s complement or 10s complement, depending on which number system is used. This section will describe the 2s complement system. Because the 10s complement system displays the same characteristics as the 2s complement system, it will not be discussed here.

Example 2.32

a. The 2s complement of $(01010)_2$ is

$$2^5 - (01010) = 100000 - 01010 = 10110.$$

Here, $n = 5$ and $r = 2$.

b. The 2s complement of $(0.0010)_2$ is

$$2^1 - (0.0010) = 10.0000 - 0.0010 = 1.1110.$$

Here, $n = 1$ and $r = 2$.

c. The 10s complement of $(4887)_{10}$ is

$$10^2 - 4887 = 5113.$$

Here, $n = 4$ and $r = 10$.

d. The 10s complement of $(48.87)_{10}$ is

$$10^2 - 48.87 = 51.13.$$

Here, $n = 2$ and $r = 10$.

As can be verified by Example 2.32, there are two other methods for radix complement of a number.

Method 1: Complement and Add 1

$[01010]_2 = ?$	
10101	a. Complement each bit (i.e., change each 0 to 1 and 1 to 0)
+1	b. Add 1 to the LSB to get the 2s complement.
10110	

Method 2: Copy and Complement

$[010\vdots 10]_2 = ?$	
$\vdots 10$	a. Copy the bits from the LSB until and including the first nonzero bit.
$101\vdots$	b. Complement the remaining bits through the MSB to get the 2s complement.
$101\vdots 10$	

The *diminished radix complement* $[N]_{r-1}$ of a number $(N)_r$ is defined as

$$[N]_{r-1} = r^n - r^{-m} - (N)_r, \tag{2.4}$$

where n and m are, respectively, the number of digits in integer and fraction portion of the number. Note that

$$[N]_r = [N]_{r-1} + r^{-m}, \tag{2.5}$$

that is, the radix complement of a number is obtained by adding $(r - 1)$ to the LSB of the diminished radix complement form of the number.

The diminished radix complement is commonly called the 1s complement or 9s complement, depending on whether the binary or decimal number system is used, respectively.

Example 2.33

	$(N)_r$	r	n	M	$[N]_{r-1}$
(a)	1001	2	4	0	$= 2^4 - 2^0 - 1001$
					$= 1000 - 1 - 1001$
					$= 1111 - 1001 = 0110$
(b)	100.1	2	3	1	$= 2^3 - 2^{-1} - 100.1$
					$= 1000 - 0.1 = 100.1$
					$= 111.1 - 100.1 = 011.0$
(c)	486.7	10	3	1	$= 10^3 - 10^{-1} - 486.7$
					$= 1000 - 0.1 - 486.7$
					$= 999.9 - 486.7 = 513.2$

From Example 2.33, it can be seen that the 1s complement of a number is obtained by subtracting each digit from the largest digit in the number system. In the binary system, this is equivalent to complementing (i.e., changing 1 to 0 and 0 to 1) each bit of the given number.

Example 2.34

$N = 10110.110$
1s complement of $N = 11111.111$
-10110.110
01001.001

which can also be obtained by complementing each bit of N.

As in sign–magnitude representation, a sign bit is included in the representation of numbers in complement systems as well. Because the complement of a number corresponds to its negative, positive numbers that are represented in complement systems remain in the same form as in the sign–magnitude system. Only negative numbers are represented in the complement form as shown by the following example.

Here, we assume that 5 bits are available for representation and that the MSB is the sign bit.

Example 2.35

Decimal	Sign–Magnitude	2s Complement	1s Complement
+5	0,0101	0,0101	0,0101
−5	1,0101	1,1011	1,1010
+4	0,0100	0,0100	0,0100
−4	1,0100	1,1100	1,1011

To obtain the complement of a number, we can start with the sign–magnitude form of the corresponding positive number and adopt the complementing procedures discussed here. In Example 2.35, the sign bit is separated from the magnitude bits by a "," for illustration purposes only. This separation is not necessary in complement systems since the sign bit also participates in the arithmetic as though it were a magnitude bit (as we will see later in this section).

Table 2.5 shows the range of numbers that can be represented in 5 bits, in the sign–magnitude, 2s complement, and 1s complement systems. Note that the sign–magnitude and 1s complement systems have two representations for 0 (+0 and −0), whereas the 2s complement system has a unique representation for 0. Note also the use of the combination 10,000 to represent the largest negative

Table 2.5   Three Representation Schemes

Decimal	Sign–Magnitude	2s Complement	1s Complement
+15	01111	01111	01111
+14	01110	01110	01110
+13	01101	01101	01101
+12	01100	01100	01100
+11	01011	01011	01011
+10	01010	01010	01010
+9	01001	01001	01001
+8	01000	01000	01000
+7	00111	00111	00111
+6	00110	00110	00110
+5	00101	00101	00101
+4	00100	00100	00100
+3	00011	00011	00011
+2	00010	00010	00010
+1	00001	00001	00001
+0	00000	00000	00000
−0	10000	00000	11111
−1	10001	11111	11110
−2	10010	11110	11101
−3	10011	11101	11100
−4	10100	11100	11011
−5	10101	11011	11010
−6	10110	11010	11001
−7	10111	11001	11000
−8	11000	11000	10111
−9	11001	10111	10110
−10	11010	10110	10101
−11	11011	10101	10100
−12	11100	10100	10011
−13	11101	10011	10010
−14	11110	10010	10001
−15	11111	10001	10000
−16		10000[a]	

[a] 2s complement uses 1000 to expand the range to (−16).

number in the 2s complement system. In general, the ranges of integers that can be represented in an n-bit field (using 1 sign bit and $n - 1$ magnitude bits) in the three systems are

1. Sign–magnitude: $-(2^{n-1} -1)$ to $+(2^{n-1} -1)$
2. 1s complement: $-(2^{n-1} -1)$ to $+(2^{n-1} -1)$
3. 2s complement: $-(2^{n-1})$ to $+(2^{n-1} -1)$

We will now illustrate the arithmetic in these systems of number representation.

2.5.1 2s Complement Addition

Example 2.36 illustrates the addition of numbers represented in 2s complement form.

Example 2.36

(a) Decimal	Sign–Magnitude	2s Complement
5	0,0101	0,0101
4	0,0100	<u>0,0100</u>
		0,1001 = 9

Here the sign–magnitude and 2s complement representations are the same, since both numbers are positive. In 2s complement addition, the sign bit is also treated as a magnitude bit and participates in the addition process. In this example, the sign and magnitude portions are separated for illustration purposes only.

(b) Decimal	Sign–Magnitude	2s Complement
5	0,0101	0,0101
-4	1,0100	<u>1,1100</u>
		10,0001 = $+(0001)_2$
	Carry from the sign. Position is ignored.	

Here the negative number is represented in the complement form and the two numbers are added. The sign bits are also included in the addition process. There is a carry from the sign bit position, which is ignored. The sign bit is 0, indicating that the result is positive.

(c) Decimal	Sign–Magnitude	2s Complement
4	0,0100	0,0100
-5	1,0101	<u>1,1011</u>
		1,1111 = $-(0001)_2$
	The result is negative; no carry.	

Here, no carry is generated from the MSB during the addition. The result is negative since the sign is 1; the result is in the complement form and must be complemented to obtain the sign–magnitude representation.

(d) Decimal	Sign–Magnitude	2s Complement
−5	1,0101	1,1011
−4	1,0100	1,1100
		11,0111 = −(1001)$_2$

Ignore the carry; the result is negative.

When subtraction is performed in decimal arithmetic (and in the sign–magnitude system), the number with the smaller magnitude is subtracted from the one with the larger magnitude, and the sign of the result is that of the larger number. Such comparison is not needed in the complement system as shown in Example 2.36.

In summary, in 2s complement addition, the carry generated from the MSB is ignored. The sign bit of the result must be the same as that of the operands when both operands have the same sign. If it is not, the result is too large to fit into the magnitude field and hence an overflow occurs. When the signs of the operands are different, a carry from the sign bit indicates a positive result; if no carry is generated, the result is negative and must be complemented to obtain the sign–magnitude form. The sign bit participates in the arithmetic.

2.5.2 1s Complement Addition

Example 2.37 illustrates 1s complement addition. It is similar to that in 2s complement representation, except that the carry generated from the sign bit is added to the LSB of the result to complete the addition.

Example 2.37

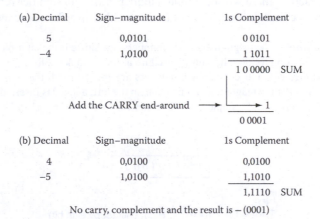

2.5.3 Shifting Revisited

As we have seen earlier, shifting a binary number left 1 bit is equivalent to multiplying it by 2. Shifting a binary number right 1 bit is equivalent to dividing it by 2. Example 2.38 illustrates the effect of shifting an unsigned number.

Example 2.38

Consider the number N with six integer bits and four fraction bits:

$$N \quad \boxed{0\,0\,1\,0\,1\,1\,.\,1\,0\,1\,0}$$

Shifting N 1 bit left, with a 0 inserted into the LSB:

$$2N \quad \boxed{0\,1\,0\,1\,1\,1\,.\,0\,1\,0\,0} \longleftarrow \text{Inserted}$$

Shifting left again:

$$4N \quad \boxed{1\,0\,1\,1\,1\,0\,.\,1\,0\,0\,0} \longleftarrow \text{Inserted}$$

If we shift this number left again, the 1 in the MSB would be lost, thereby resulting in an overflow.

Shifting N right 1 bit:

$$N/2 \quad \boxed{0\,0\,0\,1\,0\,1\,.\,1\,1\,0\,1}$$

Shifting right again:

Inserted

$$N/4 \quad \boxed{0\,0\,0\,0\,1\,0\,.\,1\,1\,1\,0\,1}$$

The 1 in the LSB is lost because of this shift, thereby resulting in a less accurate fraction. If there are enough bits to retain all the nonzero bits in the fraction through the shift operation, no loss of accuracy will result. In practice, there will be a finite number of bits for number representation. Hence, care must be taken to see that shifting does not result in either overflow or inaccuracy.

Sign–magnitude shifting: When sign–magnitude numbers are shifted, the sign bit is not included in shift operations. Shifting follows the same procedure as in Example 2.38.

2s complement shifting: When 2s complement numbers are shifted right, the sign bit value is copied into the vacant (MSB of the magnitude) bit position on the left, and a 0 is inserted in the vacant LSB position during the left shift. Example 2.39 illustrates this.

Example 2.39

$$\boxed{1 \; \vert \; 0\,0\,1\,0\,1\,0\,0\,0\,0}$$

1 1 0 0 1 0 1 0 0 0   Right shift (copy sign bit)

1 1 1 0 0 1 0 1 0 0   Right shift (copy sign bit)

1 1 0 0 1 0 1 0 0 $\boxed{0}$ Left shift (insert 0)

1 0 0 1 0 1 0 0 0 $\boxed{0}$ Left shift (insert 0)

A change in the value of the sign bit during a left shift indicates that there is an overflow (i.e., the result is too large).

1s Complement Shifting: When 1s complement numbers are shifted, a copy of the sign bit is inserted in the vacant LSB position during the left shift or in the MSB position of the magnitude bits during the right shift. The sign bit receives the MSB of the magnitude during the left shift.

Example 2.40

	(a)		(b)	
N	1 1001		0 0001	
$2N$	1 0011		0 0010	Insert sign bit
$4N$	0 0111		0 0100	
		Overflow		
$N/2$	1 1100		0 0000	Copy sign bit
$N/4$	1 1110			

2.5.4 Comparison of Complement Systems

Figure 2.2 summarizes the operations in both the complement systems. The 2s complement system is now used in almost all digital computer systems. The 1s complement system has been used in older computer systems. The advantage of the 1s complement system is that the 1s complement can be obtained by inverting each bit of the original number from 1 to 0 and 0 to 1, which can be done very easily by the logic components of the digital system. The conversion of a number to 2s complement system requires an addition operation after the 1s complement of the number is obtained or a scheme to implement the copy/complement algorithm described earlier in this chapter. 2s complement is the most widely used system of representation.

One other popular representation-biased or excess-radix representation is used to represent floating-point numbers. This representation is described in Section 2.6.

2.6 FLOATING-POINT NUMBERS

Fixed-point representation is convenient for representing numbers with bounded orders of magnitude. For instance, in a digital computer that uses 32 bits to represent numbers, the range of integers that can be used is limited to $\pm (2^{31} - 1)$, which is approximately $\pm 10^{12}$. In scientific computing

Operation	If the carry from the MSB is	1s Complement — Then perform	2s Complement — Then perform	Sign bit of the result
Add	0	(Result is in complement form) complement to convert to sign–magnitude form		1
	1	(Result is in sign–magnitude form) add 1 to the LSB of the result	(Result is in sign–magnitude form) neglect the carry	0
Left shift		Copy sign bit into the LSB	Insert 0 into the LSB	Sign bit = MSB of magnitude
Right shift		Copy sign bit into the MSB of the magnitude		Sign bit unchanged

Figure 2.2   Comparison of complement number systems.

environments, a wider range of numbers may be needed, and floating-point representation may be used. The general form of a floating-point representation number N is

$$N = F \times r^E, \tag{2.6}$$

where
  F is the fraction (or mantissa)
  r is the radix
  E is the exponent

  Consider the number

$$N = 3560000$$

$$= (0.356) \times 10^7$$

$$= (0.0356) \times 10^8$$

$$= (3.56) \times 10^6.$$

The last three forms are valid floating-point representations. Among them, the first two forms are preferred, however, since with these forms there is no need to represent the integer portion of the mantissa, which is 0. The first form requires the fewest digits to represent the mantissa, since all the significant zeros have been eliminated from the mantissa. This form is called the normalized form of floating-point representation. Note from the example earlier that the radix point floats within the mantissa, incrementing the exponent by 1 for each move to the left and decrementing the exponent by 1 for each move to the right. This shifting of the mantissa and scaling of the exponent is frequently done in the manipulation of floating-point numbers.

  Let us now concentrate on the representation of floating-point numbers. The radix is implied by the number system used and hence is not shown explicitly in the representation. The mantissa and the exponent can be either positive or negative. Hence, the floating-point representation consists of the four components E, F, SE, and SF, where SE and SF are signs of the exponent and mantissa, respectively.

  The F is represented in the normalized form. The true binary form is used in the representation of F (rather than any of the complement forms), and 1 bit is used to represent SF (0 for positive, 1 for negative). Since the MSB of the normalized mantissa is always 1, the range of mantissa is

$$0.5 \leq F < 1. \tag{2.7}$$

The floating-point representation of 0 is an exception and contains all 0s.

  When two floating-point numbers are added, the exponents must be compared and equalized before the addition. To simplify the comparison operation without involving the sign of the exponent, the exponents are usually converted to positive numbers by adding a bias constant to the true exponent. The bias constant is usually the magnitude of the largest negative number that can be represented in the exponent field. Thus, in a floating-point representation with q bits for the exponent field, if a 2s complement representation is used, the unbiased exponent E_n will be in the range

$$-2^{q-1} \leq E_n \leq 2^{q-1} - 1. \tag{2.8}$$

If we add the bias constant of 2^{q-1}, the biased exponent E_b will be in the range

$$0 \le E_b \le 2^{q-1}. \tag{2.9}$$

The true (unbiased) exponent is obtained by subtracting the bias constant from the biased exponent. That is,

$$E_n = E_b - 2^{q-1}. \tag{2.10}$$

For example, if $q = 9$, then the bias constant is 256. Three values of the exponent are shown in the following:

	−256	0	+256
E_n:	100000000	000000000	011111111
E_b:	000000000	1000000000	111111111

As long as the mantissa is 0, theoretically the exponent can be anything, thereby making it possible to have several representations for floating-point representation numbers. In fixed-point representation, we represented a 0 by a sequence of 0s. To retain the uniqueness of 0 representation for both fixed- and floating-point representations, the mantissa is set to all 0s and the exponent is set to the most negative exponent in biased form (i.e., all 0s).

2.6.1 IEEE Standard

In 1985, the Institute of Electrical and Electronics Engineers (IEEE) published a standard for floating-point numbers. This standard, officially known as IEEE-754 (1985), specifies how single-precision (32 bits) and double-precision (64 bits) floating-point numbers are to be represented, as well as how arithmetic should be carried out on them. Binary floating-point numbers are stored in a sign–magnitude form shown in the following:

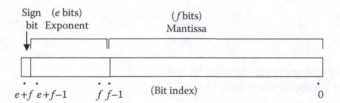

where the MSB is the sign bit, exponent is the biased exponent, and mantissa is the significand minus the MSB.

Exponents are signed small and huge values. The 2s complement representation, the usual representation for signed values, would make comparison of exponent values harder. To solve this problem, the exponent is biased by $2^{e-1} - 1$ before being stored. This makes its value to be in an unsigned range suitable for comparison.

The MSB of the mantissa is determined by the value of exponent. If $0 < \text{exponent} < 2^e - 1$, the MSB of the mantissa is 1, and the number is said to be normalized. If the exponent is 0, the MSB of the mantissa is 0 and the number is said to be denormalized. Three special cases arise:

1. If the exponent is 0 and mantissa is 0, the number is ±0 (depending on the sign bit).
2. If the exponent = $2^e - 1$ and mantissa is 0, the number is ±infinity (again depending on the sign bit).
3. If the exponent = $2^e - 1$ and mantissa is not 0, the number represented is not a number (NaN).

This can be summarized as

Type	Exponent	Mantissa
Zeroes	0	0
Denormalized numbers	0	Nonzero
Normalized numbers	1 to $2^e - 2$	Any
Infinities	$2^e - 1$	0
NaNs	$2^e - 1$	Nonzero

2.6.1.1 Single Precision

A single-precision binary floating-point number is stored in a 32-bit word:

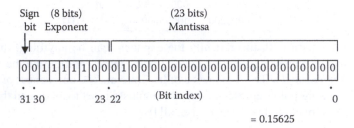

$$= 0.15625$$

The exponent is biased by $2^{8-1} - 1 = 127$ in this case. In an 8-bit exponent field, we can represent values in the range -126 to $+127$. An exponent of -127 would be biased to the value 0, but this is reserved to encode that the value is a denormalized number or zero. An exponent of 128 would be biased to the value 255, but this is reserved to encode an infinity or NaN. For normalized numbers, the most common, exponent field contains the biased exponent value and the mantissa field is the fractional part of the significand. The number has the value

$$v = s \times 2^e \times m,$$

where
$s = +1$ (positive numbers) when the sign bit is 0
$s = -1$ (negative numbers) when the sign bit is 1
$e = \text{Exp} - 127$
$m = 1$

Fraction in binary (i.e., the significand is the binary number 1 followed by the radix point followed by the binary bits of fraction). Therefore, $1 \le m < 2$.

Note that

1. Demoralized numbers are the same except that $e = -126$ and m is 0. Fraction. (e is not -127: The significand has to be shifted to the right by one more bit, in order to include the leading bit, which is not always 1 in this case. This is balanced by incrementing the exponent to -126 for the calculation.)
2. -126 is the smallest exponent for a normalized number.
3. There are two zeroes: $+0$ (S is 0) and -0 (S is 1).
4. There are two infinities: $+\infty$ (S is 0) and $-\infty$ (S is 1).
5. NaNs may have a sign and a significand, but these have no meaning other than for diagnostics; the first bit of the significand is often used to distinguish signaling NaNs from quiet NaNs.
6. NaNs and infinities have all 1s in the Exp field.

7. The smallest nonzero positive and largest nonzero negative numbers (represented by the denormalized value with all 0s in the Exp field and the binary value 1 in the fraction field) are

$$\pm 2^{-149} \approx \pm 1.4012985 \times 10^{-45}.$$

8. The smallest nonzero positive and largest nonzero negative normalized numbers (represented with the binary value 1 in the Exp field and 0 in the fraction field) are

$$\pm 2^{-126} \approx \pm 1.175494351 \times 10^{-38}.$$

9. The largest finite positive and smallest finite negative numbers (represented by the value with 254 in the Exp field and all 1s in the fraction field) are

$$\pm (2^{128} - 2^{104}) \approx \pm 3.4028235 \times 10^{38}.$$

Example 2.41

To represent the decimal number −118.625 using the IEEE 754 system:

1. Since this is a negative number, the sign bit is "1."
2. The number (without the sign) in binary is 1110110.101.
3. Moving the radix point left, leaving only a 1 at its left: $1110110.101 = 1.110110101 \times 2^6$. This is a normalized floating-point number. The mantissa is the part at the right of the radix point, filled with 0s on the right to make it 23 bits. That is 11011010100000000000000.
4. The exponent is 6, the bias is 127, and hence the exponent field will be 6 + 127 = 133. In binary, this is 10000101.

The representation is thus

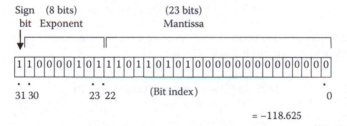

$$= -118.625$$

Some additional representations are shown in the following.

Example 2.42

0 00000000 00000000000000000000000	= 0
1 00000000 00000000000000000000000	= −0
0 11111111 00000000000000000000000	= Infinity
1 11111111 00000000000000000000000	= −Infinity
0 11111111 00000100000000000000000	= NaN
1 11111111 00100010001001010101010	= NaN
0 10000000 00000000000000000000000	= +1 * $2^{(128-127)}$ * 1.0 = 2
0 10000001 10100000000000000000000	= +1 * $2^{(129-127)}$ * 1.101 = 6.5
1 10000001 10100000000000000000000	= −1 * $2^{(129-127)}$ * 1.101 = −6.5
0 00000001 00000000000000000000000	= +1 * $2^{(1-127)}$ * 1.0 = 2**(−126)
0 00000000 10000000000000000000000	= +1 * $2^{(-126)}$ * 0.1 = 2**(−127)
0 00000000 00000000000000000000001	= +1 * $2^{(-126)*}$
0.00000000000000000000001	= $2^{(-149)}$ (smallest positive value)

2.6.1.2 Double Precision

Double-precision format is essentially the same except that the fields are wider:

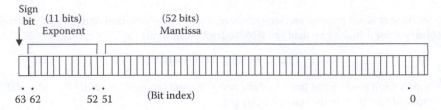

NaNs and infinities are represented with exponent being all 1s (2047). For normalized numbers, the exponent bias is +1023 (so e is Exp −1023). For denormalized numbers, the exponent is −1022 (the minimum exponent for a normalized number is not −1023 because normalized numbers have a leading 1 digit before the binary point and denormalized numbers do not). As earlier, both infinity and zero are signed.

Note that

1. The smallest nonzero positive and largest nonzero negative numbers (represented by the denormalized value with all 0s in the Exp field and the binary value 1 in the fraction field) are

$$\pm 2^{-1074} \approx \pm 5 \times 10^{-324}.$$

2. The smallest nonzero positive and largest nonzero negative normalized numbers (represented by the value with the binary value 1 in the Exp and 0 in the fraction field) are

$$\pm 2^{-1022} \approx \pm 2.2250738585072020 \times 10^{-308}.$$

3. The largest finite positive and smallest finite negative numbers (represented by the value with 1022 in the Exp field and all 1s in the fraction field) are

$$\pm (2^{1024} - 2^{971}) \approx \pm 1.7976931348623157 \times 10^{308}.$$

2.6.1.3 Comparison

This representation makes comparisons of some subsets of numbers possible on a byte-by-byte basis, if they share the same byte order and the same sign, and NaNs are excluded. For example, for two positive floating-point numbers a and b, a comparison between a and b gives identical results as the comparison of two signed (or unsigned) binary integers with the same bit patterns and same byte order as a and b. In other words, two positive floating-point numbers (known not to be NaNs) can be compared with a signed (or unsigned) binary integer comparison using the same bits, provided the floating-point numbers use the same byte order.

2.6.1.4 Rounding

The IEEE standard has four different rounding modes:

1. Unbiased, which rounds to the nearest value, if the number falls midway it is rounded to the nearest value with an even (zero) LSB. This mode is required to be default.
2. Toward zero.
3. Toward positive infinity.
4. Toward negative infinity.

The standard behavior of computer hardware is to round the ideal (infinitely precise) result of an arithmetic operation to the nearest representable value and give that representation as the result. In practice, there are other options. IEEE-754-compliant hardware allows one to set the rounding mode to any of the following:

1. Round to nearest (the default; by far the most common mode).
2. Round up (toward $+\infty$ negative results round toward zero).
3. Round down (toward $-\infty$ negative results round away from zero).
4. Round toward zero (sometimes called "chop" mode; it is similar to the common behavior of float-to-integer conversions, which convert −3.9 to −3).

In the default rounding mode, the IEEE 754 standard mandates the round-to-nearest behavior described earlier for all fundamental algebraic operations, including square root. ("Library" functions such as cosine and log are not mandated.) This means that IEEE-compliant hardware's behavior is completely determined in all 32 or 64 bits.

The mandated behavior for dealing with overflow and underflow is that the appropriate result is computed, taking the rounding mode into consideration, as though the exponent range were infinitely large. If that resulting exponent cannot be packed into its field correctly, the overflow or underflow action described earlier is taken.

The arithmetic distance between two consecutive representable floating-point numbers is called an "ULP," for Unit in the Last Place. For example, the numbers represented by 45670123 and 45670124 hexadecimal are one ULP. An ULP is about 10^{-7} in single precision and 10^{-16} in double precision. The mandated behavior of IEEE-compliant hardware is that the result be within one-half of an ULP.

2.6.1.5 Accuracy

Because floating-point numbers cannot faithfully mimic the real numbers, and floating-point operations cannot faithfully mimic true arithmetic operations, there are many problems that arise in writing mathematical software that uses floating point. First, although addition and multiplication are both commutative ($a + b = b + a$ and $a \times b = b \times a$), they are not associative ($(a + b) + c = a + (b + c)$). Using 7-digit decimal arithmetic,

$$1234.567 + 45.67844 = 1280.245,$$
$$1280.245 + 0.0004 = 1280.245.$$

But

$$45.67844 + 0.0004 = 45.67884,$$
$$45.67884 + 1234.567 = 1280.246.$$

They are also not distributive ($(a + b) \times c = a \times c + b \times c$):

$$1234.567 \times 3.333333 = 4115.223,$$
$$1.234567 \times 3.333333 = 4.115223,$$
$$4115.223 + 4.115223 = 4119.338.$$

But

$$1234.567 + 1.234567 = 1235.802,$$
$$1235.802 \times 3.333333 = 4119.340.$$

Aside from that, the rounding actions that are performed after each arithmetic operation lead to inaccuracies that can accumulate in unexpected ways. Consider the 24-bit (single-precision) representation of (decimal) 0.1 that was given previously:

$$e = -4; \quad s = 110011001100110011001101,$$

which is

$$0.100000001490116119384765625$$

exactly; squaring this number gives

$$0.01000000029802322609739917425031308084726333618164 0625$$

exactly, for which $0.010000000070780515670776367 1875$ or

$$e = -7; \quad s = 101000111101011100001011$$

is the nearest representable number but $0.009999999977648258209228515 6250$ or

$$e = -7; \quad s = 101000111101011100001010$$

is the representable number closest to 0.01.

Thus, in binary floating point, the expectation that 0.1 squared equals 0.01 is not met. Similarly, division by 10 will not always give the same results as multiplication by 0.1 even though division by 4 and multiplication by 0.25 does, just as with decimal arithmetic, division by 3 does not give the same results as multiplication by 0.33333 so long as only a finite number of digits are considered.

In addition to loss of significance, inability to represent numbers such as π and 0.1 exactly, and other slight inaccuracies, the following phenomena may occur:

1. *Cancellation*: Subtraction of nearly equal operands may cause extreme loss of accuracy. This is perhaps the most common and serious accuracy problem.
2. *Conversions to integer are unforgiving*: Converting (63.0/9.0) to integer yields 7, but converting (0.63/0.09) may yield 6. This is because conversions generally truncate rather than rounding.
3. *Limited exponent range*: Results might overflow, yielding infinity.
4. *Testing for safe division is problematical*: Checking that the divisor is not zero does not guarantee that a division will not overflow and yield infinity.
5. *Equality is problematical*: Two computational sequences that are mathematically equal may well produce different floating-point values. Programmers often perform comparisons within some tolerance (often a decimal constant, itself not accurately represented), but that does not necessarily make the problem go away.

2.6.1.6 Exceptions

In addition to the "infinity" value that is produced when an overflow occurs, there is a special value NaN that is produced by such operations as taking the square root of a negative number. NaN is encoded with the reserved exponent of 128 (or 1024) and a significand field that distinguishes it from infinity.

The intention of the infinity and NaN values is that, under the most common circumstances, they can just propagate from one operation to the next (any operation with NaN as an operand produces NaN as a result), and they only need to be attended to at a point that the programmer chooses.

In addition to the creation of exceptional values, there are "events" that may occur, though some of them are quite benign:

1. Overflow occurs as described previously, producing an infinity.
2. Underflow occurs as described previously, producing a denorm.

3. Zerodivide occurs whenever a divisor is zero, producing an infinity of the appropriate sign. (The sign of zero is meaningful here.) Note that a very small but nonzero divisor can still cause an overflow and produce an infinity.

4. "Operand error" occurs whenever a NaN has to be created. This occurs whenever any operand to an operation is a NaN, or some other obvious thing happens, such as sqrt(−2.0) or log(−1.0).

5. "Inexact" event occurs whenever the rounding of a result changed that result from the true mathematical value. This occurs almost all the time and is usually ignored. It is looked at only in the most exacting applications.

Computer hardware is typically able to raise exceptions (traps) when these events occur. How these are presented to the software is very language and system dependent. Usually all exceptions are masked (disabled). Sometimes overflow, zerodivide, and operand error are enabled.

2.7 BINARY CODES

So far we have seen various ways of representing numeric data in binary form. A digital system requires that all information be in binary form. The external world, however, uses various other symbols such as alphabetic characters and special characters (e.g., +, =, −) to represent information. In order to represent these various symbols in binary form, a unique pattern of 0s and 1s is assigned to represent each symbol. This pattern is the code word corresponding to that symbol.

As we have seen earlier, it is possible to represent 2^n elements with a binary string containing n bits. That is, if q symbols are to be represented in binary form, the minimum number of bits n required in the code word is given by

$$2^{n-1} < q \leq 2^n. \tag{2.11}$$

The code word might possibly contain more than n bits to accommodate error detection and correction. Once the number of bits in the code word is set, the assignment of the code words to the symbols of information to be represented could be arbitrary (in which case a table associating each element with its code word is needed) or might follow some general rule.

For example, if the code word is required to represent four symbols—say, dog, cat, tiger, and camel—we can use the four combinations of 2 bits (00, 01, 10, and 11). Assignments of these combinations to the four symbols can be arbitrary.

To represent the 26 letters of the alphabet, we would need a 5-bit code. With 5 bits, it is possible to generate 32 combinations of 0s and 1s. Any of the 26 out of these 32 combinations can be used to represent the alphabet. Similarly, a 4-bit code is needed to represent the 10 decimal digits. Any 10 of the 16 combinations possible can be used to represent the decimal digits (we will examine some possibilities later in this section).

The codes designed to represent only numeric data (i.e., decimal digits 0 through 9) can be classified into two categories: weighted and nonweighted. The alphanumeric codes can represent both alphabetic and numeric data. A third class of codes is designed for error-detection and error-correction purposes.

2.7.1 Weighted Codes

As stated, we will need at least 4 bits to represent the 10 decimal digits. But with 4 bits it is possible to represent 16 elements. Since only 10 of the 16 possible combinations are used, numerous distinct codes are possible. Table 2.6 shows some of these possibilities. Note that all of these codes are weighted codes, since each bit position in the code has a weight associated with it. The sum of weights corresponding to each nonzero bit in the code is the decimal digit represented by it.

Note that each word in the (8 4 2 1) code in the table is a binary number whose decimal equivalent is the decimal digit it represents. This is a very commonly used code and is known as the

Table 2.6   Some Weighted Codes

Weights Digit	8 4 2 1 Code	2 4 2 1 Code	6 4 2 –3 Code
0	0000	0000	0000
1	0001	0001	0101
2	0010	0010	0010
3	0011	0011	1001
4	0100	0100	0100
5	0101	1011	1011
6	0110	1100	0110
7	0111	1101	1101
8	1000	1110	1010
9	1001	1111	1111

Note: The (8 4 2 1) code is the popular
BCD code.

BCD code. When a BCD-encoded number is used in arithmetic operations, each decimal digit is represented by 4 bits. For example, $(567)_{10}$ is represented in BCD as

$$5 \quad 6 \quad 7$$
$$(0101\ 0110\ 0111)_{BCD}$$

During the arithmetic, each 4-bit unit is treated as a digit and the arithmetic is performed on a digit-by-digit basis, as shown in Example 2.43.

Example 2.43

Decimal	BCD	
532	0101 0011 0010	
+126	0001 0010 0110	Binary arithmetic on
658	0110 0101 1000	each 4-bit unit

When the sum of two digits is greater than 9, the resulting 4-bit pattern is not a valid code word in BCD. In such cases, a correction factor of 6 is added to that digit, to derive the valid code word.

Example 2.44

Decimal	BCD		
532	0101	0011	0010
+268	0010	0110	1000
	0111	1001	1010
			+ 0110 Correction
	0111	0101	0000
		+ 0110	Correction
	1000	0000	0000

Here the LSD of the sum exceeds 9. The correction of that digit results in the next significant digit exceeding 9. The correction of that digit yields the final correct sum.

It is important to understand the difference between binary and BCD representations of numbers. In BCD, each decimal digit is represented by the corresponding 4-bit code word. In binary, the complete number is converted into binary pattern with the appropriate number of bits. For example, the binary representation of $(567)_{10}$ is (1000110111), whereas its BCD representation is (0101 0110 0111).

In the 2 4 2 1 code shown in Table 2.6, decimal 6 is represented by 1100. Another valid representation of 6 in this code is 0110. We have chosen the combinations in Table 2.6 to make this code self-complementing. A code is said to be self-complementing if the code word of the 9s complement of N (i.e., $9-N$) can be obtained by complementing each bit of the code word for N. This property of the code makes taking the complement easy to implement in digital hardware. A necessary condition for a code to be self-complementing is that the sum of weight of the code is 9. Thus, both the (2 4 2 1) and (6 4 2 −3) codes are self-complementing, while BCD is not.

2.7.2 Nonweighted Codes

Table 2.7 shows two popular codes. They do not have any weight associated with each bit of the code word.

Excess-3 is a 4-bit self-complementing code. The code for each decimal digit is obtained by adding 3 to the corresponding BCD code word. Excess-3 code has been used in some older computer systems. In addition to being self-complementing, thereby making subtraction easier, this code enables simpler arithmetic hardware. This code is included here for completeness and is no longer commonly used.

The Gray code (named after Frank Gray) is a 4-bit code in which the 16 code words are selected such that there is a change in only 1-bit position as we move from one code word to the subsequent code word. Such codes are called cyclic codes. Because only 1-bit changes from code word to code word, it is easier to detect an error if there is a change in more than 1 bit. For example, consider the case of a shaft position indicator. Assume that the shaft position is divided into 16 sectors indicated by the Gray code. As the shaft rotates, the code words change. If at any time there is a change in 2 bits of the code word compared with the previous one, there is an error. Several cyclic codes have been devised and are commonly used.

Table 2.7   Nonweighted Codes

Decimal	Excess-3 Code	Gray Code
0	0011	0000
1	0100	0001
2	0101	0011
3	0110	0010
4	0111	0110
5	1000	0111
6	1001	0101
7	1010	0100
8	1011	1100
9	1100	1101
10	Nd	1111
11	Nd	1110
12	Nd	1010
13	Nd	1011
14	Nd	1001
15	Nd	1000

Note: Nd, not defined.

2.7.3 Error-Detection Codes

Errors occur during digital data transmission as a result of the external noise introduced by the medium of transmission. For example, if a digital system uses BCD code for data representation and if an error occurs in the LSB position of the data 0010, the resulting data will be 0011. Because 0011 is a valid code word, the receiving device assumes that the data are not in error. To guard against such erroneous interpretations of data, several error-detection and error-correction schemes have been devised. As the names imply, an error-detection scheme simply detects that an error has occurred, whereas an error-correction scheme corrects the errors. We will describe a simple error-detection scheme using parity checking here. For information on more elaborate schemes, such as cyclic redundancy check (CRC), check sums, and XModem protocols, refer to the books by Kohavi (2009) and Ercegovac and Lang (2003) listed in the Bibliography section at the end of the chapter.

In the simple parity check error-detection scheme, an extra bit (known as a parity bit) is included in the code word. The parity bit is set to 1 or 0, depending on the number of 1s in the original code word, to make the total number of 1s even (in an even parity scheme) or odd (in an odd parity scheme). The sending device sets the parity bit. The receiving device checks the incoming data for parity. If the system is using an even parity scheme, an error is detected if the receiver detects an odd number of 1s in the incoming data (and vice versa). A parity bit can be included in the code words of each of the codes described earlier.

Table 2.8 shows two error-detection codes. The first code is the even parity BCD. The fifth bit is the parity bit, and it is set for even parity. The second code is known as the 2-out-of-5 code. In this code, 5 bits are used to represent each decimal digit. Two and only 2 bits out of 5 are 1s. Out of the 32 combinations possible using 5 bits, only 10 are utilized to form this code. This is also an even parity code. If an error occurs in transmission, the even parity is lost and detected by the receiver.

In this simple parity scheme, if 2 bits are in error, the even parity is maintained and we will not be able to detect that an error has occurred. If the occurrence of more than one error is anticipated, more elaborate parity-checking schemes using more than one parity bit are used. In fact, it is possible to devise codes that not only detect errors but also correct them, by including enough parity bits. For example, if a block of words is being transmitted, each word might include a parity bit and the last word in the block might be a parity word, each bit of which checks for an error in the corresponding bit position of each word in the block (see Problem 2.20). This scheme is usually referred to as cross-parity checking or vertical and horizontal redundancy check and is a coding scheme that detects and corrects single errors. Hamming (1950) invented a single error-detecting/ error-correcting scheme using a distance-3 code. That is, any code word in this scheme differs from other code words in at least 3 bit positions.

Table 2.8   Error-Detection Codes

Decimal	Even Parity BCD	2 Out of 5
0	0000 0 ← parity bit	11000
1	0001 1	00011 no specific parity bit
2	0010 1	00101
3	0011 0	00110
4	0100 1	01001
5	0101 0	01010
6	0110 0	01100
7	0111 1	10001
8	1000 1	10010
9	1001 0	10100

2.7.4 Alphanumeric Codes

If alphabetic characters, numeric digits, and special characters are used to represent information processed by the digital system, an alphanumeric code is needed. Two popular alphanumeric codes are extended BCD interchange code (EBCDIC) and American standard code for information interchange (ASCII). Table 2.9 shows these codes. ASCII is more commonly used, and EBCDIC is used primarily in large IBM computer systems. Both EBCDIC and ASCII are 8-bit codes and hence can represent up to 256 elements.

In a general computer system, each and every component of the system need not use the same code for data representation. For example, in the simple calculator system shown in Figure 2.3, the keyboard produces ASCII-coded characters corresponding to each keystroke. These ASCII-coded numeric data are then converted by the processor into BCD for processing. The processed data are then reconverted into ASCII for the printer. Such code conversion is common, particularly when devices from various vendors are integrated to form a system.

2.8 DATA STORAGE AND REGISTER TRANSFER

Let us now examine the operation of a digital computer in more detail, to better understand the data representation and manipulation schemes. The binary information is stored in digital systems, in devices such as flip-flops (discussed in Chapter 4). We call such storage devices as storage cells. Each cell can store 1 bit of data. The content (or state) of the cell can be changed from 1 to 0 or 0 to 1 by the signals on its inputs, whereas the content of the cell is determined by sensing its outputs. A collection of storage cells is called a register. An n-bit register can thus store n-bit data. The number of bits in the most often manipulated data unit in the system determines the word size of the system. That is, a 16-bit computer system manipulates 16-bit numbers most often and its word size is 16 bits. Computer systems with 8-, 16-, 32-, and 64-bit words are common. An 8-bit unit of data is commonly called a byte, and a 4-bit unit is a nibble. Once the word size of the machine is defined, half-word and double-word designations are also used to designate data with half or twice the number of bits in the word.

A digital system manipulates the data through a set of register transfer operations. A register transfer is the operation of moving the contents of one register (i.e., the source) to another register (i.e., the destination). The source register contents remain unchanged after the register transfer, whereas the contents of the destination register are replaced by those of the source register.

Let us now examine the set of register transfer operations needed to bring about the addition of two numbers in a digital computer. The memory unit of the digital computer is composed of several words, each of which can be viewed as a register. Some of the memory words contain data, and others contain instructions for manipulating the data. Each memory word has an address associated with it.

In Figure 2.4, we have assumed that the word size is 16 bits and the memory has 30 words. The program is stored in memory locations 0 through 10, and only two data words at addresses 20 and 30 are shown. Word 0 contains the instruction ADD A TO B, where A and B are the operands stored in locations 20 and 30, respectively. The instruction is coded in binary, with the 6 MSBs representing the add operation and the remaining 10 bits used to address the two operands, 5 bits for each operand address.

The control unit fetches the instruction ADD A TO B from the memory word 0. This fetch operation is a register transfer. The instruction is analyzed by the control unit to decode the operation called for and the operand address required. The control unit then sends a series of control signals to the processing unit to bring about the set of register transfers needed.

We will assume that the processing unit has two operand registers, R_1 and R_2, and a results register, R_3. The adder adds the contents of R_1 and R_2 and stores the result in R_3.

Table 2.9 Alphanumeric Codes

Character	EBCDIC Code	ASCII Code
Blank	0100 0000	0010 0000
.	0100 1011	0010 1110
(	0100 1101	0010 1000
+	0100 1110	0010 1011
$	0101 1011	0010 0100
*	0101 1100	0010 1010
)	0101 1101	0010 1001
–	0110 0000	0010 1101
/	0110 0001	0010 1111
;	0110 1011	0010 0111
'	0111 1101	0010 1100
=	0111 1110	0011 1101
A	1100 0001	0100 0001
B	1100 0010	0100 0010
C	1100 0011	0100 0011
D	1100 0100	0100 0100
E	1100 0101	0100 0101
F	1100 0110	0100 0110
G	1100 0111	0100 0111
H	1100 1000	0100 1000
I	1100 1001	0100 1001
J	1101 0001	0100 1010
K	1101 0010	0100 1011
L	1101 0011	0100 1100
M	1101 0100	0100 1101
N	1101 0101	0100 1110
O	1101 0110	0100 1111
P	1101 0111	0101 0000
Q	1101 1000	0101 0001
R	1101 1001	0101 0010
S	1110 0010	0101 0011
T	1110 0011	0101 0100
U	1110 0100	0101 0101
V	1110 0101	0101 0110
W	1110 0110	0101 0111
X	1110 0111	0101 1000
Y	1110 1000	0101 1001
Z	1110 1001	0101 1010
0	1111 0000	0011 0000
1	1111 0001	0011 0001
2	1111 0010	0011 0010
3	1111 0011	0011 0011
4	1111 0100	0011 0100
5	1111 0101	0011 0101
6	1111 0110	0011 0110
7	1111 0111	0011 0111
8	1111 1000	0011 1000
9	1111 1001	0011 1001

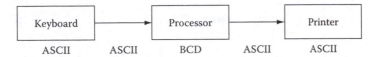

Figure 2.3 Coding in a simple calculator system.

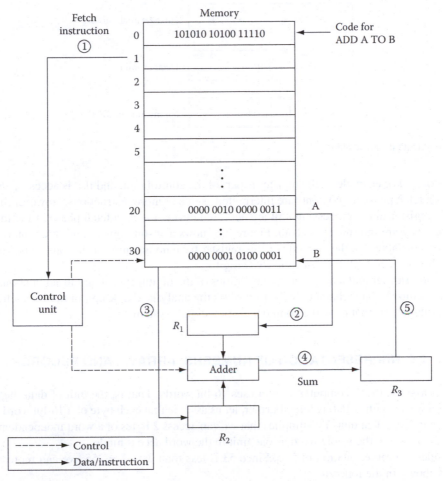

Figure 2.4 Register transfers for an ADD operation. *Note:* Numbers in circles indicate the sequence of operations.

To carry out the ADD A TO B instruction, the control unit

1. Transfers the contents of memory word 20 (operand A) to R_1
2. Transfers the contents of memory word 30 (operand B) to R_2
3. Commands the adder to add R_1 and R_2 and sends the result to R_3
4. Transfers the contents of R_3 to memory word 30 (operand B)

This description is obviously very much simplified compared with the actual operations that take place in a digital computer. Nevertheless, it illustrates the data-flow capabilities needed in a digital system.

As seen from this example, the binary data stored in a register can be interpreted in various ways. The context in which the register content is examined determines the meaning of the bit

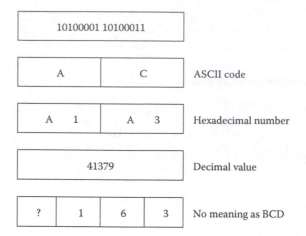

Figure 2.5   Data interpretations.

pattern stored. For example, if the register is part of the control unit, and if it is accessed during the instruction fetch phase, its contents are interpreted as one or more instructions. However, if the contents of a register in the processing unit are accessed during the data fetch phase of the instruction execution, they are interpreted as data. Figure 2.5 shows a 16-bit register and three interpretations of its contents. Note that the contents of this register have no meaning if they are to be considered as BCD digits, since 1010 is not a valid BCD digit.

The data transfer and manipulative capabilities of the digital system are brought about by digital logic circuits. In later chapters, we will discuss the analysis and design of the logic circuits and memory subsystems that form the components of a digital system.

2.9 REPRESENTATION OF NUMBERS, ARRAYS, AND RECORDS

Let us assume that a computer system uses 16-bit words. That is, the unit of data the machine most often works with is 16 bits long. Further, let us assume that each byte of a 16-bit word is accessible as an independent unit. This implies that we can access 2 bytes of a word independently in the byte access mode or the whole word as one unit in the word access mode.

Consider the representation of 53_{10}. Since 53 is less than $2^8 - 1 = 255$, we can represent 53 in 8 bits, as shown in the following:

16-bit word

0 0 0 0 0 0 0 0	0 0 1 1 0 1 0 1
Byte 1	Byte 0

The left half of the word (Byte 1) contains all zeros. Now consider the representation for 300_{10} that requires the 16-bit word as shown in the following:

16-bit word

0 0 0 0 0 0 0 1	0 0 1 0 1 1 0 0
Byte 1	Byte 0

This representation of the number where the least significant part of the number occupies Byte 0 and the most significant part occupies Byte 1 is called *little Endien*. If we swap the contents of the two bytes previously, we get the *big Endien* representation. Both the representations are used in practice. When accessing the number, it is important to remember which Endien representation is used so that the interpretation results in the appropriate value for the number represented.

2.9.1 BCD Numbers

Most financial applications represent data in BCD rather than binary, to retain the accuracy of calculations. For instance, 53 is represented as two BCD digits (each 4-bit long):

$$5-0101$$
$$3-0011$$

In the 16-bit word, the representation is

16-bit word

	5	3
0 0 0 0 0 0 0 0	0 1 0 1	0 0 1 1

We can pack up to four BCD digits in the 16-bit word.

2.9.2 Floating-Point Numbers

Using the IEEE standard, 53 is represented as

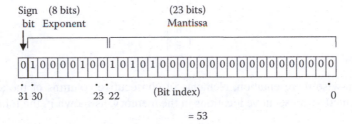

= 53

2.9.3 Representation of Strings

Strings are arrays of characters. Each character is represented as a byte in either ASCII or EBCDIC format. Representation of the string COMPUTER is shown in the following using the 32-bit machine architecture. The end of the string is usually denoted by a special character called null, which is represented as "\0."

C	O	M	P
U	T	E	R
\0			

Alternatively, an integer value indicating the number of characters in the string can be made part of the string representation, instead of a null terminating character.

2.9.4 Arrays

Consider the array of integers:

$$A = [10, 25, 4, 76, 8].$$

Here $N = 5$, number of elements in the array. In an 8-bit machine, it is represented as

0 0 0 0 0 1 0 1	N
0 0 0 0 1 0 1 0	$A[0]$
0 0 0 1 1 0 0 1	$A[1]$
0 0 0 0 0 1 0 0	$A[2]$
0 1 0 0 1 1 0 0	$A[3]$
0 0 0 0 1 0 0 0	$A[4]$

Consider the 2D array:

$$A = \begin{bmatrix} 4 & 6 \\ 8 & 9 \end{bmatrix}.$$

There are two representations for this array. In the row-wise representation, elements of consecutive rows of the array (each a 1D array) are represented in consecutive locations in the memory, as shown in the following:

4
6
8
9

In the column-wise representation, elements of consecutive columns of the array (each a 1D array) are represented in consecutive locations in the memory, as shown in the following:

4
8
6
9

2.9.5 Records

A record typically contains one or more fields. The data in these fields may be of different types. The number of bits needed to represent each field may vary. Consider, for instance, the records of students in a university containing the fields: ID, Name, Phone Number, and GPA. The ID field could be treated as either numeric or alphanumeric depending on the processing performed on it. The Name field is a string and might be partitioned into subfields for first, last, and middle names. The Phone Number could be treated as numeric or alphanumeric, and the GPA is a floating-point number. The detailed characteristic of the record indicating the field lengths and the type of data in each field needs to be maintained to access data appropriately.

2.10 SUMMARY

The topics covered in this chapter form the basis for all the data representation schemes that are used in digital systems. We have presented the most common number systems and conversion procedures. The most basic arithmetic schemes in these number systems and popular representation schemes have been examined. Various binary codes encountered in digital systems have been discussed. We have also included a brief introduction to the operation of a digital computer based on the register transfer concept. The floating-point representation and associated accuracy problems have been introduced.

PROBLEMS

2.1 Convert the following decimal numbers to base 3, base 5, base 8, and base 16: 245, 461, 76.5, 46.45, 232.78, 1023.25.

2.2 List the first 20 decimal numbers in base 7 and base 9.

2.3 Assume that your car's odometer shows the mileage in octal. If the current reading of the odometer is 24,516, how many miles (in decimals) has the car been driven? What will be the new reading if the car is driven 23 (decimal) miles?

2.4 What will be the current odometer reading in the problem given earlier, if the odometer uses a 5-digit hexadecimal representation? What will be the new reading after driving 23 (decimal) miles?

2.5 Find the 1s and 2s complements of the following binary numbers:

    (a) 10010   (b) 110010   (c) 0010101

    (d) 10110.0101   (e) 1102.1100   (f) 111010.0011

    (g) 1002.0001   (h) 110100.0100   (i) 1010110.111

2.6 Find the 9s and 10s complement of the following decimal numbers:

    (a) 465   (b) 09867   (c) 42678

    (d) 8976   (e) 423.76   (f) 562.876

    (g) 463.90   (h) 1786.967   (i) 12356.078

2.7 Determine $X + Y$, $X - Y$, $X \times Y$, and X/Y in each of the following sets of binary numbers:

    (a) $X = 1101010$ (b) $X = 101101$ (c) $X = 1001$

        $Y = 10111$ $Y = 1111$ $Y = 1111$

    (d) $X = 110.11$ (e) $X = 1110.101$ (f) $X = 1011.00$

        $Y = 10.11$ $Y = 1011.10$ $Y = 1100$

2.8 Determine $X + Y$, $X - Y$, $X \times Y$, and X/Y in each of the following sets of octal numbers:

    (a) $X = 533$ (b) $X = 46537$ (c) $X = 26$

        $Y = 234$ $Y = 234$ $Y = 533$

    (d) $X = 123.2$ (e) $X = 234.6$

        $Y = 234$ $Y = 156.7$

2.9 Determine $X + Y$, $X - Y$, $X \times Y$, and X/Y in each of the following sets of hexadecimal numbers:

    (a) $X = 1CF$ (b) $X = 1B59A$ (c) $X = B6$

        $Y = B6$ $Y = C23$ $Y = 1CF$

    (d) $X = 2ECD$ (e) $X = 234F.16$

        $Y = 4321$ $Y = 456E$

2.10 Perform the following conversions:

    (a) $(234)_{10} = (?)_2$ (b) $(3345)_6 = (?)_2$ (c) $(875)_9 = (?)_{11}$

    (d) $(0.3212)_4 = (?)_{10}$ (e) $(87.35)_9 = (?)_{11}$

2.11 Perform the following conversions using base 2^k conversion technique:

    (a) $(10110100.00101)_2 = (?)_4$

    (b) $(AB143)_{16} = (?)_4$

    (c) $(2347.45)_8 = (?)_{16}$

    (d) $(110111110.010000011)_2 = (?)_{16}$

2.12 Following the conversion technique of the preceding problem, convert $(2574)_9$ to base 3.

2.13 If $(130)_X = (28)_{10}$, find the value of X (X is a positive integer).

2.14 Perform the following operations:
(a) $11101 + 1111 + 1011$
(b) $111000 - 10101$
(c) $11001101/101$
(d) 11010×11001

2.15 Use (a) 2s complement and (b) 1s complement arithmetic to perform the following operations:
(a) $1011010 - 10101$
(b) $10101 - 1011010$

2.16 Use (a) 9s complement and (b) 10s complement arithmetic to perform the following operations:
(a) $1875 - 924$
(b) $924 - 1875$

2.17 Use an 8-bit 2s complement representation (with 1 sign bit and 7 magnitude bits) to perform the following operations:
(a) $113 - 87$ (b) $87 - 113$ (c) $43 + 26$
(d) $96 - 22$ (e) $46 - 77$

2.18 Perform the following operations:
(a) $(7256)_8 \times (23)_8 = (?)_8$
(b) $(56)_8 \times_8 (AF)_{16} = (?)_4$ (base 8 multiplication)

2.19 Represent the following numbers in the IEEE standard floating-point format:
(a) $(1 \quad 1010.010)_2$ (b) $(432.26)_{10}$
(c) $-(10100112.1001)_2$ (d) $-(236.77)_{10}$

2.20 The following four code words were transmitted. The LSB of each code word (row) is a parity bit, and odd parity is used. The last word is a parity word across all the earlier words so that even parity is adopted in each bit position (column). Determine whether an error has occurred. If there is an error, correct the bit in error:

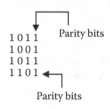

2.21 Determine the code words for each of the 10 decimal digits in the weighted code $(6\ 3 - 1\ 1)$. The code should be self-complementing.

2.22 Design a 4-bit code to represent each of the six digits of the base six number system. Make the code self-complementing by selecting appropriate weights.

2.23 A 16-bit register contains the following:
0100100101010111
Interpret the contents as
(a) a BCD number   (b) a binary number
(c) an excess-3 number   (d) two ASCII characters

2.24 Represent the following in a 16-bit register:
(a) $(356)_{10}$ (b) $(356)_{BCD}$ (c) $(A1)_{ASCII}$

2.25 Express $(746)_{10}$ in (a) BCD, (b) binary, and (c) ASCII, using the minimum number of bits in each case.

2.26 Convert the following IEEE standard floating-point representation into binary:
(a) 0 10000010 10000010...000
(b) 1 01111000 01000110...000

2.27 Just as in BCD arithmetic, when two excess-3 digits are added, a correction is needed. Determine the correction factor.

2.28 Assume that you have a 64-bit computer. That means, all data items are represented in a 64-bit field.
   a. What is the decimal range of values that can be represented using fixed-point and IEEE standard floating-point representation?
   b. Which one has a larger range?
   c. Why do we need to use the one with a smaller range?

2.29 Consider the number $A = 3.7$ in base 10. Assume that you have a 16-bit computer.
   a. What is the binary value that will be stored to represent A?
   b. What is the value in decimal base if it is converted back from the binary form stored?
   c. Is the answer to (b) rational?
   d. Compare the values of the answers to (a) and (b). Explain your observation.

BIBLIOGRAPHY

ANSI/IEEE Standard 754-1985, *Standard for Binary Floating Point Arithmetic*, http://en.wikipedia.org/wiki/IEEE_754.

Chu, Y., *Computer Organization and Programming*, Englewood Cliffs, NJ: Prentice Hall, 1972.

Ercegovac, M.D. and Lang, T., *Digital Arithmetic*, San Francisco, CA: Morgan Kaufmann, 2003.

Hamming, R.W., Error detecting and correcting codes, *The Bell System Technical Journal*, 29, 147–160, April 1950.

Hwang, K., *Computer Arithmetic*, New York: John Wiley, 1979.

IEEE, IEEE standard for binary floating-point arithmetic, *SIGPLAN Notices*, 22(2), 9–25, 1985.

Kohavi, Z., *Switching and Hardware Designs*, Oxford, U.K.: Oxford University Press, 2009.

Oppenheim, A.V. and Schafer, R.W., *Discrete-Time Signal Processing*, 3rd edn., Upper Saddle River, NJ: Prentice Hall, 2009.

Shannon, C.E., A mathematical theory of communication, *The Bell System Technical Journal*, 27(3), 379–423, 623–656, 1948.

Shiva, S.G., *Introduction to Logic Design*, New York: Marcel Dekker, 1998.

CHAPTER 3

Combinational Logic

Each hardware component of a computer system is built of several logic circuits. A logic circuit is an interconnection of several primitive logic devices to perform a desired function. It has one or more inputs and one or more outputs. This chapter introduces some logic devices that are used in building one type of logic circuit called a *combinational circuit*. Each output of a combinational circuit is a function of all the inputs to the circuit. Further, the outputs at any time are each a function of inputs at that particular time, and so the circuit does not have a memory. A circuit with a memory is called a *sequential circuit*. The output of a sequential circuit at any time is a function of not only the inputs at the time but also the state of the circuit at that time. The state of the circuit is dependent on what has happened to the circuit prior to that time, and hence the state is also a function of previous inputs and states. This chapter is an introduction to the analysis and design of combinational circuits. Details of sequential circuit analysis and design are given in Chapter 4.

3.1 BASIC OPERATIONS AND TERMINOLOGY

Example 3.1

Consider the addition of two bits:

0 plus 0 = 0
0 plus 1 = 1
1 plus 0 = 1
1 plus 1 = 10 (i.e., a SUM of 0 and a CARRY of 1)

The addition of two single-bit numbers produces a SUM bit and a CARRY bit. The previously mentioned operations can be arranged into the following table to separate the two resulting bits SUM and CARRY:

		A plus B	
A	*B*	SUM	CARRY
0	0	0	0
0	1	1	0
1	0	1	0
1	1	0	1

A and B are the two operands. Each can take a value of either 0 or 1. The first two columns show the four combinations of values possible for the two operands A and B, and the last two columns represent the sum of the two operands represented as a SUM and a CARRY bit.

Note that the CARRY is 1 only when A is 1 and B is 1, whereas the SUM bit is 1 when one of the following two conditions is satisfied: A is 0 and B is 1; A is 1 and B is 0. That is, SUM is 1 if

A	B	A · B
0	0	0
0	1	0
1	0	0
1	1	1

AND

"0" if at least A or B is "0."

"1" only if both A and B are "1".

A	B	A + B
0	0	0
0	1	1
1	0	1
1	1	1

OR

"0" only if both A and B are "0."

"1" if at least A or B is "1."

A	A'
0	1
1	0

NOT

A' is the COMPLEMENT of A.

Figure 3.1 Basic operations: AND, OR, and NOT.

(A is 0 and B is 1) or (A is 1 and B is 0). Let us say A' (pronounced "not A") represents the opposite condition of A; that is, A is 0 if A' is 1, and vice versa; similarly, B' represents the opposite condition of B. Then we can say SUM is 1 if (A' is 1 and B is 1) or (A is 1 AND B' is 1). A shorthand notation for this is as follows:

$$SUM = (A' \cdot B) + (A \cdot B')$$

$$CARRY = A \cdot B. \tag{3.1}$$

where
+ represents the OR operation (not arithmetic addition)
· represents the AND operation
′ represents the NOT operation (complement operation)
– is also used to represent the NOT operation, as in $\overline{A}, \overline{B}$, etc.

The definitions of AND, OR, and NOT operations are shown in Figure 3.1. The right sides of Equation 3.1 are Boolean expressions. A and B are Boolean variables. Possible values for the Boolean variables in the earlier examples are 0 and 1 (could also be true or false). An expression is formed by combining variables with operations. The value of SUM depends on the values of A and B. That is, SUM is a function of A and B and so is CARRY.

Example 3.2

Consider the following statement: Subtract if and only if an add instruction is given and the signs are different, or a subtract instruction is given and the signs are alike.
Let

S represent the "subtract" action
A represent "add instruction given" condition
B represent "signs are different" condition
C represent "subtract instruction given" condition

Then, the earlier statement can be expressed as

$$S = (A \cdot B) + (C \cdot B'). \tag{3.2}$$

Usually, the "·" and "()" are removed from expressions when there is no ambiguity. Thus, the earlier function can be written as

$$S = AB + CB'. \tag{3.3}$$

3.1.1 Evaluation of Expressions

Knowing the value of each of the component variables of an expression, we can find the value of the expression itself. The hierarchy of operations is important in the evaluation of expressions. We always perform NOT operations first, followed by AND, and lastly OR, in the absence of parentheses. If there are parentheses, the expressions within the parentheses are evaluated first, observing the earlier hierarchy of operations, and then the remaining expression is evaluated. That is,

---

Perform

    Parenthesis grouping (if any), Then

    NOT operation first,

    AND operation next,

    OR operation last

while evaluating an expression.

The following examples illustrate expression evaluation. The sequential order in which the operations are performed is shown by the numbers below each operation.

Example 3.3

$$A \cdot B' + B \cdot C' \cdot D$$

                                          Insert "·"

    1   2_____ Scan 1 for NOT operations

 3   4 5_____Scan 2 for AND operations

    6_____Scan 3 for OR operations

Example 3.4

$$A \cdot (B + C' \cdot D) + A \cdot B' + C' \cdot D' \qquad \text{Insert "·"}$$

1	NOT ⎫	
2	AND ⎬ Within parentheses	
3	OR ⎭	
4 5 6	NOT	
7 8 9	AND	
10 11	OR	

Example 3.5

Evaluate the function $Z = AB'\,C + (A'\,B)\,(B + C')$, given $A = 0, B = 1, C = 1$:

$$Z = (A \cdot B' \cdot C) + (A' \cdot B) \cdot (B + C') \qquad \text{Insert "·"}$$
$$= (0 \cdot 1' \cdot 1) + (0' \cdot 1) \cdot (1 + 1') \qquad \text{Substitute values}$$
$$= (0 \cdot 0 \cdot 1) + (1 \cdot 1) \cdot (1 + 0) \qquad \text{Evaluate NOT}$$
$$= (0) + (1) \cdot (1) \qquad \text{Evaluate parenthetical expressions}$$
$$= 0 + 1 \qquad \text{AND operation}$$
$$= 1 \qquad \text{OR operation (value of } Z \text{ is 1)}$$

3.1.2 Truth Tables

Figure 3.1 shows truth tables for the three primitive operations AND, OR, and NOT. A truth table indicates the value of a function for all possible combinations of the values of the variables of which it is a function. There will be one column in the truth table corresponding to each variable and one column for the value of the function. Since each variable can take either of the two values (0 or 1), the number of combinations of values increases exponentially with the number of component variables having the base as 2 (the number of values each variable can take). For instance, if there are two variables, there will be $2 \times 2 = 4$ combinations of values and hence four rows in a truth table. In general, there will be 2^N rows in a truth table for a function with N component binary variables. If the expression on the right-hand side of a function is complex, the truth table can be developed in several steps. The following example illustrates the development of a truth table.

Example 3.6

Draw a truth table for $Z = AB' + A'C + A'B'C$.

There are three component variables, A, B, and C. Hence, there will be 2^3 or eight combinations of values of A, B, and C. The eight combinations are shown on the left-hand side of the truth table in Figure 3.2. These combinations are generated by changing the value of C from 0 to 1 and from 1 to 0 as we move down from row to row while changing the value of B once every two (i.e., 2^1) rows and changing the value for A once every four (i.e., 2^2) rows. These combinations are thus in a numerically increasing order in the binary number system, starting with $(000)_2$ or $(0)_{10}$ to $(111)_2$ or $(7)_{10}$, where the subscripts denote the base of the number system.

In general, if there are N component variables, there will be 2^N combinations of values ranging in their numeric value from 0 to $2^N - 1$.

To evaluate Z in the example function, knowing the values for A, B, and C at each row of the truth table in Figure 3.2, the values of A' and B' are first generated; the values for (AB'), $(A'C)$, and $(A'B'C)$ are then generated by ANDing the values in appropriate columns at each row; and finally the value of Z is found by ORing the values in the last three columns at each row. Note that evaluating $A'B'C$ corresponds to ANDing A' and B' values, followed by ANDing the value of C. Similarly, if more than two values are to be ORed, they are ORed two at a time. The columns corresponding to A', B', (AB'), $(A'C)$, and $(A'B'C')$ are not usually shown in the final truth table.

3.1.3 Functions and Their Representation

There are two constants in the logic alphabet: 0 and 1 (true or false). A variable such as A, B, X, or Y can take the value of either 1 or 0 at any time. There are three basic operations: AND, OR, and NOT. When several variables are ANDed together, we get a *product term* (conjunction).

A	B	C	A'	B'	AB'	$A'C$	$A'B'C$	Z
0	0	0	1	1	0	0	0	0
0	0	1	1	1	0	1	1	1
0	1	0	1	0	0	0	0	0
0	1	1	1	0	0	1	0	1
1	0	0	0	1	1	0	0	1
1	0	1	0	1	1	0	0	1
1	1	0	0	0	0	0	0	0
1	1	1	0	0	0	0	0	0

Figure 3.2 Truth table for $Z = AB' + A'C + A'B'C$.

Example 3.7

$$AB, AB'C, AB'\,XY'Z'$$

When several variables are ORed together, we get a *sum term* (disjunction).

Example 3.8

$$(A + B + C'), (X + Y'), (P + Q + R')$$

Each occurrence of a variable either in true form or in complemented (inversed, NOT) form is a *literal*. For example, the product term $XY'Z$ has three literals; the sum term $(A' + B' + C + D)$ has four literals.

A product term (sum term) X is included in another product term (sum term) Y if Y has each literal that is in X.

Example 3.9

XY' is included in XY'. $XY'Z$ is included in $XY'ZW$. $(X' + Y)$ is included in $(X' + Y + W)$. $X'Y$ is not included in XY. (Why?)

If the value of the variable Q is dependent on the value of several variables (say, A, B, C)—that is, Q is a function of $A, B,$ and C—then Q can be expressed as a sum of several product terms in $A, B,$ and C.

Example 3.10

$Q = AB' + A'C + B'C$ is the *sum of products* (SOP) form.

If none of the product terms is included in the other product terms, we get a *normal SOP* form.

Example 3.11

$Q = AB + AC$, $Q = X + Y$, and $P = AB'C + A'CD + AC'D'$ are in normal SOP form.

Similarly, we can define a *normal product of sums* (POS) form.

Example 3.12

$P = (X + Y') \cdot (X' + Y' + Z')$ and $Q = (A + B') \cdot (A' + B + C') \cdot (A + B + C)$ are in normal POS form.

A truth table can be used to derive the function SOP or POS forms, as detailed later.

Example 3.13

Consider the following truth table for Q, a function of $A, B,$ and C:

A	B	C	Q
0	0	0	0
0	0	1	1
0	1	0	0
0	1	1	1
1	0	0	1
1	0	1	1
1	1	0	0
1	1	1	0

Note: This is same as Figure 3.2.

From the truth table, it can be seen that Q is 1 when $A = 0$, $B = 0$, and $C = 1$. That is, Q is 1 when $A' = 1$, $B' = 1$, and $C = 1$, which means Q is 1 when $(A' \cdot B' \cdot C)$ is 1. Similarly, corresponding to the other three 1s in the Q column of the table, Q is 1 when $(A'BC)$ is 1 or $(AB'C')$ is 1 or $(AB'C)$ is 1. This argument leads to the following representation for Q:

$$Q = A'B'C + A'BC + AB'C' + AB'C,$$

which is the normal SOP form.

In general, to derive an SOP form from the truth table, we can use the following procedure:

1. Generate a product term corresponding to each row where the value of the function is 1.
2. In each product term, consider the individual variables uncomplemented if the value of the variable in that row is 1 and complemented if the value of the variable in that row is 0.

The POS form for the function can be derived from the truth table by a similar procedure:

1. Generate a sum term corresponding to each row where the value of the function is 0.
2. In each sum term, consider the individual variables complemented if the value of the variable in that row is 1 and uncomplemented if the value of the variable in that row is 0.

$Q = (A + B + C) \cdot (A + B' + C) \cdot (A' + B' + C) \cdot (A' + B' + C')$ is the POS form for Q, in Example 3.13.

The SOP form is easy and natural to work with compared to the POS form. The POS form tends to be confusing to the beginner since it does not correspond to the algebraic notation that we are used to.

Example 3.14

The derivation of SOP and POS forms of representation for another three-variable function P is shown here:

	A	B	C	P	
0	0	0	0	1	← $A'B'C'$
1	0	0	1	0	← $(A + B + C')$
2	0	1	0	0	← $(A + B' + C)$
3	0	1	1	0	← $(A + B' + C')$
4	1	0	0	1	← $AB'C'$
5	1	0	1	1	← $AB'C$
6	1	1	0	0	← $(A' + B' + C)$
7	1	1	1	0	← $(A' + B' + C')$

SOP form: $P = A'B'C' + AB'C' + AB'C$.
POS form: $P = (A + B + C') \cdot (A + B' + C) \cdot (A + B' + C') \cdot (A' + B' + C) \cdot (A' + B' + C')$.

3.1.4 Canonical Forms

The SOP and POS forms of the functions derived from a truth table by the earlier procedures are canonical forms. In a canonical SOP form, each component variable appears in either complemented or uncomplemented form in each product term.

Example 3.15

If Q is a function of A, B, and C, then $Q = A'B'C + AB'C' + A'B'C'$ is a canonical SOP form, while $Q = A'B + AB'C + A'C'$ is not, because in the first and last product terms, all three variables are not present.

A canonical POS form is similarly defined. A canonical product term is also called a *minterm*, while a canonical sum term is called a *maxterm*. Hence, functions can be represented either in sum of minterm or in product of maxterm formats.

Example 3.16

From the truth table of Example 3.14:

$$P(A, B, C) = A'B'C' + AB'C' + AB'C$$

$$\begin{array}{ccc} 0\,0\,0 & 1\,0\,0 & 1\,0\,1 \end{array} \leftarrow \text{Input combinations (0 for a complemented variable; 1 for an uncomplemented variable)}$$

$$\begin{array}{ccc} 0 & 4 & 5 \end{array} \leftarrow \text{Decimal values}$$

$$= \Sigma m\,(0, 4, 5) \qquad \leftarrow \text{Minterm List form}$$

The minterm list form is a compact representation for the canonical SOP form.

$$P\,(A, B, C) = (A + B + C')\cdot(A + B' + C)\cdot(A + B' + C')\cdot(A' + B' + C)\cdot(A' + B' + C')$$

$$\begin{array}{ccccc} 0\,0\,1 & 0\,1\,0 & 0\,1\,1 & 1\,1\,0 & 1\,1\,1 \end{array}$$

Input combinations (1 for a complemented variable and 0 for an uncomplemented variable)

$$\begin{array}{ccccc} 1 & 2 & 3 & 6 & 7 \end{array} \longleftarrow \text{Decimal values}$$

$$= \Pi M\,(1, 2, 3, 6, 7) \qquad \longleftarrow \text{Maxterm list form}$$

The maxterm list form is a compact representation for the canonical POS form. Knowing one form, the other can be derived as shown by the following example.

Example 3.17

Given $Q\,(A, B, C, D) = \Sigma m\,(0, 1, 7, 8, 10, 11, 12, 15)$.

Q is a four-variable function. Hence, there will be 2^4 or 16 combinations of input values whose decimal values range from 0 to 15. There are eight minterms. Hence, there should be $16 - 8 = 8$ maxterms: that is,

$$Q(A,B,C,D) = \prod M(2,3,4,5,6,9,13,14).$$

Also, note that the complement of Q is represented as

$$Q'(A,B,C,D) = \sum m(2,3,4,5,6,9,13,14)$$

$$= \prod M(0,1,7,8,10,11,12,15).$$

Note that for an n-variable function,

$$\text{(Number of minterms)} + \text{(number of maxterms)} = 2^n. \tag{3.4}$$

3.2 BOOLEAN ALGEBRA (SWITCHING ALGEBRA)

In 1854, George Boole introduced a symbolic notation to deal with symbolic statements that take a binary value of either true or false. The symbolic notation was adopted by Claude Shannon to analyze logic functions and has since come to be known as Boolean algebra or switching algebra. The definitions, theorems, and postulates of this algebra are described here.

Definition: Boolean algebra is a closed algebraic system containing a set K of two or more elements and two binary operators "+" (OR) and "·" (AND); that is, for every X and Y in set K, $X \cdot Y$ belongs to K, and $X + Y$ belongs to K. In addition, the following postulates must be satisfied:

Postulates	
P1 Existence of 1 and 0	(a) $X + 0 = X$ (b) $X \cdot 1 = X$
P2 Commutativity	(a) $X + Y = Y + X$ (b) $X \cdot Y = Y \cdot X$
P3 Associativity	(a) $X + (Y + Z) = (X + Y) + Z$ (b) $X \cdot (Y \cdot Z) = (X \cdot Y) \cdot Z$
P4 Distributivity	(a) $X + (Y \cdot Z) = (X + Y) \cdot (X + Z)$ (b) $X \cdot (Y + Z) = X \cdot Y + X \cdot Z$
P5 Complement	(a) $X + X' = 1$ (X' is the complement of X) (b) $X \cdot X' = 0$

Definition: Two expressions are said to be equivalent if one can be replaced by the other.

Definition: The "dual" of an expression is obtained by replacing each "+" in the expression by "·", each "·" by "+", each 1 by 0, and each 0 by 1.

The principle of duality states that if an equation is valid in a Boolean algebra, its dual is also valid. Note that part (b) of each of the postulates is the dual of the corresponding part (a), and vice versa.

Example 3.18

Given $X + YZ = (X + Y) \cdot (X + Z)$, its dual is $X \cdot (Y + Z) = (X \cdot Y) + (X \cdot Z)$.

Theorems: The following theorems are useful in manipulating Boolean functions. They are traditionally used for converting Boolean functions from one form to another, deriving canonical forms, and minimizing (reducing the complexity of) Boolean functions. These theorems can be proven by drawing truth tables for both sides to see if the left-hand side has the same values as the right-hand side, for each possible combination of component variable values.

Theorems	
T1 Idempotency	(a) $X + X = X$ (b) $X \cdot X = X$
T2 Properties of 1 and 0	(a) $X + 1 = 1$ (b) $X \cdot 0 = 0$
T3 Absorption	(a) $X + (XY) = X$ (b) $X \cdot (X + Y) = X$
T4 Absorption	(a) $X + X'Y = X + Y$ (b) $X \cdot (X' + Y) = X' \cdot Y$
T5 De Morgan's law	(a) $(X + Y)' = X' \cdot Y'$ (b) $(X \cdot Y)' = X' + Y'$
T6 Consensus	(a) $XY + X'Z + YZ = XY + X'Z$ (b) $(X + Y) \cdot (X' + Z) \cdot (Y + Z) = (X + Y) \cdot (X' + Z)$

We can summarize some important properties thus:

$$X + 0 = X, \quad X + 1 = 1$$

$$X \cdot 0 = 0, \quad X \cdot 1 = X$$

$$0' = 1, \quad 1' = 0, \quad X'' = (X')' = X.$$

Algebraic proofs for the previously mentioned theorems can be found in any of the references listed in the Bibliography section at the end of this chapter.

3.3 MINIMIZATION OF BOOLEAN FUNCTIONS

Theorems and postulates of Boolean algebra can be used to simplify (minimize) Boolean functions. A minimized function yields a less complex circuit than a nonminimized function. In general, the complexity of a gate increases as the number of inputs increases. Hence, a reduction in the number of literals in a Boolean function reduces the complexity of the complete circuit. In designing ICs, there are other considerations, such as the area taken up by the circuit on the silicon wafer used to fabricate the IC and the regularity of the structure of the circuit from a fabrication point of view. For example, a programmable logic array (PLA) implementation (discussed in Chapter 4) of the circuit yields a more regular structure than the random logic (i.e., using gates) implementation. Minimizing the number of literals in the function may not yield a less complex PLA implementation. However, if some product terms in the SOP form can be completely eliminated from the function, the PLA size can be reduced.

Minimization using theorems and postulates is tedious. Two other popular minimization methods are (1) using Karnaugh maps (K-maps) and (2) the Quine–McCluskey procedure. These two methods are described in this section.

3.3.1 Venn Diagrams

Truth tables and canonical forms were used earlier in this chapter to represent Boolean functions. Another method of representing a function is by using Venn diagrams. The variables are represented as circles in a universe that is a rectangle. The universe corresponds to 1 (everything), and 0 corresponds to null (nothing). Figures 3.3 and 3.4 show typical logic operations using Venn diagrams. In these diagrams, the NOT operation is identified by the area NOT belonging to the particular variable, the OR operation is the union of two areas (i.e., the area that belongs to either or both) corresponding to the two operands, and the AND operation is the intersection of the areas (i.e., the area that is common to both) corresponding to the two operands. The unshaded area is the one in which the expression is 0. Note that

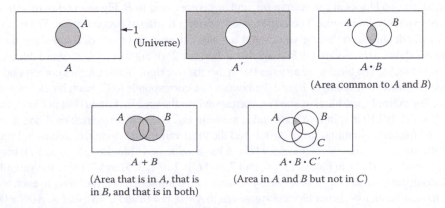

Figure 3.3   Logic operations using Venn diagrams.

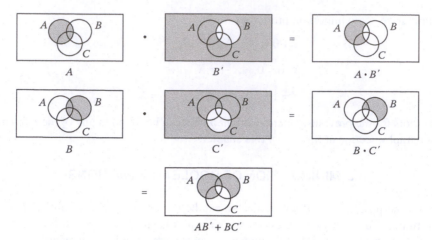

Figure 3.4 Representation of $AB' + BC'$ using Venn diagrams.

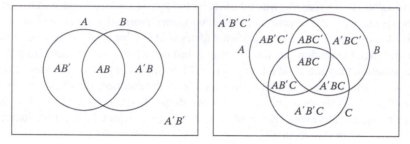

Figure 3.5 Venn diagram designating all possible combinations of two and three variables.

all the combinations shown in the truth tables can also be shown in the Venn diagrams. Figure 3.5 shows all the combinations corresponding to two- and three-variable functions.

3.3.2 Karnaugh Maps

Karnaugh maps (K-maps) are modified Venn diagrams. Consider the two-variable Venn diagram shown in Figure 3.6a. All four combinations of the two variables are shown. The four areas are identified by the four minterms in Figure 3.6b, and Figure 3.6c shows the Venn diagram rearranged such that the four areas are equal. Also note that the two right-hand blocks of the diagram correspond to A (m_2 and m_3), and the two blocks at the bottom (m_1 and m_3) correspond to B. Figure 3.6d marks the areas A, A', B, and B' explicitly, and Figure 3.6e is the usual form for a K-map of two variables. The two variables A and B are distributed such that the values of A are along the top and those of B are along the side.

Figure 3.7 shows a three-variable K-map. Since there are 2^3 or eight combinations of three variables, we need eight blocks. The blocks are arranged such that the two right-hand columns correspond to A, the two middle columns correspond to B, and the bottom row corresponds to C. Each block corresponds to a minterm. For example, the block named m_6 corresponds to the area in A and B but not in C, that is, the area ABC', which is 110 in minterm code and is minterm m_6. The first two variables A and B are represented by the four combinations along the top and the third variable C along the side as in Figure 3.7b. Note that the area A consists of the blocks where A has a value of 1 (blocks 4, 5, 6, and 7), irrespective of B and C; similarly, B is 1 in blocks 2, 3, 6, and 7, and C is 1 in 1, 3, 5, and 7. Once the variable values are listed along the top and side, it is very easy to identify the minterm corresponding to each block. For example, the left-hand, top-corner block corresponds to A = 0, B = 0, and C = 0; that is, ABC = 000 = m_0.

A four-variable Karnaugh map is shown in Figure 3.8. The areas and minterms are also identified.

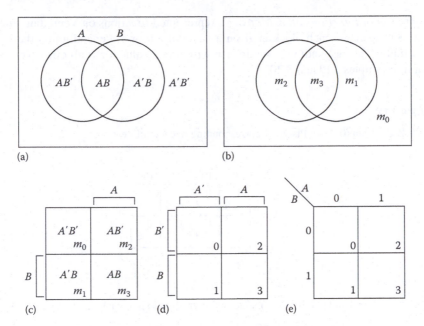

Figure 3.6 Two-variable Karnaugh map. *Note: m_0, m_1, m_2, and m_3 are minterms.*

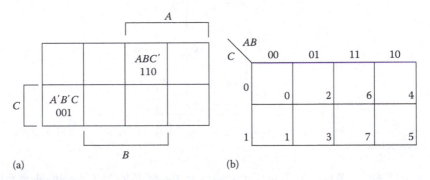

Figure 3.7 Three-variable Karnaugh map.

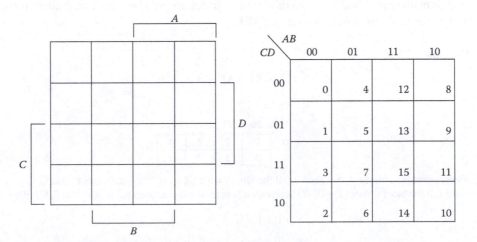

Figure 3.8 Four-variable Karnaugh map.

Representation of Functions on K-Maps: We represented functions on Venn diagrams by shading the areas. On K-maps, each block is given a value of a 0 or 1, depending on the value of the function. Each block corresponding to a minterm will have a value for 1; all other blocks will have 0s as shown in Examples 3.19 and 3.20.

Example 3.19

$f(X, Y, Z) = \Sigma m\ (0, 1, 4)$. Place a 1 corresponding to each minterm:

Z \ XY	00	01	11	10
0	1 0	0 2	0 6	1 4
1	1 1	0 3	0 7	0 5

Example 3.20

$f(A, B, C, D) = \Pi M\ (1, 4, 9, 10, 14)$. Place a 0 corresponding to each maxterm:

CD \ AB	00	01	11	10
00	1 0	0 4	1 12	1 8
01	0 1	1 5	1 13	0 9
11	1 3	1 7	1 15	1 11
10	1 2	1 6	0 14	0 10

Usually, 0s are not shown explicitly on the K-map. Only 1s are shown and a blank block corresponds to a 0.

Plotting SOP Form: When the function is given in the SOP form, the equivalent minterm list can be derived by the method described earlier in this chapter and the minterms can be plotted on the K-map. An alternative and faster method is to intersect the areas on the K-map corresponding to each product term, as illustrated in Example 3.21.

Example 3.21

$$F(X,Y,Z) = XY' + Y'Z':$$

Z \ XY	00	01	11	10
0	0	2	6	4
1	1	3	7	5

X corresponds to blocks 4, 5, 6, and 7 (all the blocks where X is 1); Y' corresponds to blocks 0, 1, 4, and 5 (all the blocks where Y is 0): XY' corresponds to their intersection; that is, $XY' = 4, 5$. Similarly,

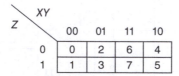

$$Y' = 0,1,4,5$$

$$Z' = 0,2,4,6 \quad \therefore Y'Z' = \underline{\underline{0,4}}.$$

Therefore, the K-map will have 1 in the union of (4, 5) and (0, 4), which is (0, 4, 5):

$$X'Y' + Y'Z'$$

Alternatively, XY' corresponds to the area where $X = 1$ and $Y = 0$, which is the last column: $Y'Z'$ corresponds to the area where both $Y = 0$ and $Z = 0$, which is blocks 0 and 4. Hence, the union of the two corresponds to blocks 0, 4, and 5.

Note also that in this three-variable K-map, if a product term has two variables missing (as in Y), we use four 1s corresponding to the four minterms that can be generated out of a single-variable product term in the representation.

In general, a product term with n missing variables will be represented by 2^n 1s on the K-map. An example follows.

Example 3.22

$$P(A,B,C,D) = AB' + A'BC + C'D'.$$

$$P(A, B, C, D) = AB' + A'BC + C'D'$$

Plotting POS Form: The procedure for plotting a POS expression is similar to that for the SOP form, except that 0s are used instead of 1s.

Example 3.23

$$F(X,Y,Z) = (X + Y')(Y' + Z').$$

Z \ XY	00	01	11	10
00		0		
01		0		

$(X + Y') = 0$ only if $X = 0$ and $Y' = 0$; that is, $X = 0$ and $Y = 1$ or the area $(X'Y)$.

Z \ XY	00	01	11	10
00				
01		0	0	

$(Y' + Z') = 0$ only if $Y' = 0$ and $Z' = 0$; that is, $Y = 1$ and $Z = 1$ or the area (YZ).

Z \ XY	00	01	11	10
00		0		
01		0	0	

$F(X, Y, Z)$ is 0 when either $(X + Y')$ is 0 or $(Y' + Z') = 0$ or the area $(X'Y) + (YZ)$.

Z \ XY	00	01	11	10
0	1		1	1
1	1			1

$F(X,Y,Z)$

Minimization: Note that the combination of variable values represented by any block on a K-map differs from that of its adjacent block only in one variable, that variable being complemented in one block and true (or uncomplemented) in the other. For example, consider blocks 2 and 3 (corresponding to minterms m_2 and m_3) of a three-variable K-map: m_2 corresponds to 010 or $A'BC'$, and m_3 corresponds to 011 or $A'BC'$. The values for A and B remain the same, while C is different in these adjacent blocks. The property where the two terms differ by only one variable is called logical adjacency. In a K-map then, physically adjacent blocks are also logically adjacent. In the three-variable K-map, block 2 is physically adjacent to blocks 0, 3, and 6. Note that m_2 is also logically adjacent to m_0, m_3, and m_6. This adjacency property can be used in the simplification of Boolean functions.

Example 3.24

Consider the following K-map for a four-variable function:

CD \ AB	00	01	11	10
00	0	4	1 12	1 8
01	1	1 5	1 13	1 9
11	3	7	15	11
10	2	6	14	10

Blocks 8 and 12 are adjacent:

$$m_8 = 1000 = AB'C'D'$$

$$m_{12} = 1100 = ABC'D'.$$

Also,

$$AB'C'D' + ABC'D' = AC'D'(B' + B) \quad \text{P4b}$$

$$= AC'D'.(1) \qquad \text{P5a}$$

$$= AC'D' \qquad \text{P1b.}$$

That is, we can combine m_8 and m_{12}. This combination is shown later by the grouping of 1s on the K-map. Note that by this grouping, we eliminated the variable B because it changes in value between these two blocks.

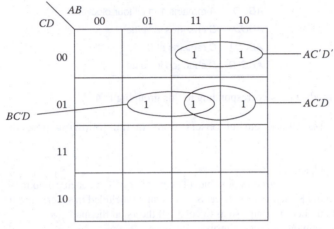

Similarly, the grouping of m_9 and m_{13} yields $AC'D$, and the grouping m_5 and m_{13} yields $BC'D$.
    If we combine $AC'D$ with $AC'D'$,

$$AC'D + AC'D' = AC'(D + D') \quad \text{P4b}$$

$$= AC' \cdot (1) \qquad \text{P5a}$$

$$= AC' \qquad \text{P1b.}$$

This in effect is equivalent to grouping all four 1s in the top-right corner of the K-map, as shown here:

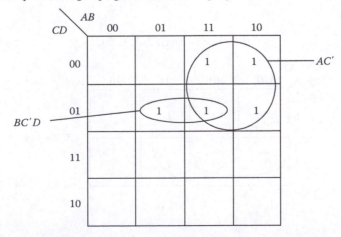

By forming a group of two adjacent 1s, we eliminated 1 literal from the product term; by grouping four adjacent 1s, we eliminated 2 literals. In general, if we group 2^n adjacent 1s, we can eliminate n literals. Hence, in simplifying functions, it is advantageous to form as large a group of 1s as possible. The number of 1s in any group must be a power of 2, that is, 1, 2, 4, and 8. Once the groups are formed, the product term corresponding to each group can be derived by the following general rules:

1. Eliminate the variable that changes in value within the group (move from block to block within the group to observe this change) from a product term containing all the variables of the function.
2. A variable that has a value of 0 in all blocks of the group should appear complemented in the product term.
3. A variable that has a value of 1 in all blocks of the group should appear uncomplemented in the product term.

For the group of four 1s in the aforementioned K-map,

$ABCD$	Start with all the variables.
$ABCD$	A remains 1 in all four blocks.
$A\!\!\!/BCD$	B changes in value
$A\!\!\!/BC'D$	C remains at 0
$A\!\!\!/BC'\!\!\!/D$	D changes in value

Therefore, the product term corresponding to this grouping is AC'.

We can summarize all the earlier observations in the following procedure for simplifying functions:

1. Form groups of adjacent 1s.
2. Form each group to be as large as possible. (The number of 1s in each group must be a power of 2.)
3. Cover each 1 on the K-map at least once. Same 1 can be included in several groups if necessary.
4. Select the least number of groups so as to cover all the 1s on the map.
5. Translate each group into a product term.
6. OR the product terms, to obtain the minimized function.

To recognize the adjacencies on a three-variable map, the right-hand edge is considered to be the same as the left-hand edge, thus making block 0 adjacent to block 4 and 1 adjacent to 5. Similarly, on a four-variable map, the top and bottom edges can be brought together to form a cylinder. The two ends of the cylinder are brought together to form a toroid (like a donut). The following examples illustrate the grouping on the K-maps and corresponding simplified functions.

Example 3.25

$$F(X, Y, Z) = \Sigma m(1, 2, 3, 6, 7).$$

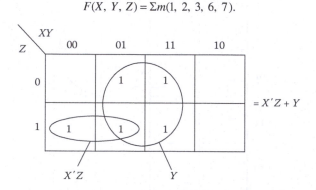

Example 3.26

$$F(A,B,C,D) = \Sigma m(2, 4, 8, 9, 10, 11, 13, 15).$$

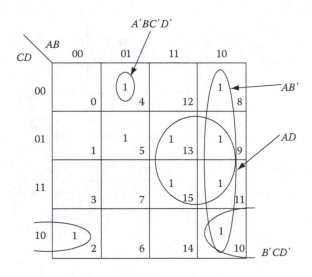

$$F(A,B,C,D) = AB' + AD + B'CD' + A'BC'D'.$$

Example 3.27

$$F(X,Y,Z,W) = \Sigma m(0, 4, 5, 8, 12, 13).$$

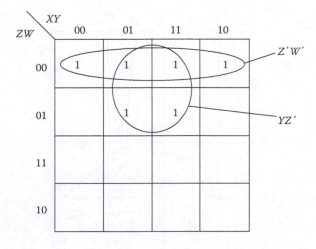

$$F(X,Y,Z,W) = YZ' + Z'W'.$$

Example 3.28

$$F(A,B,C,D) = \Sigma m(0,\ 1,\ 2,\ 7,\ 8,\ 9,\ 10).$$

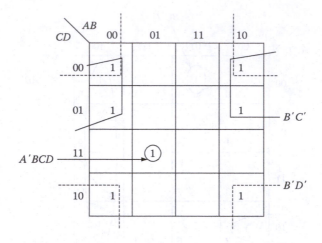

$$F(A,B,C,D) = A'BCD + B'D' + B'C'.$$

Example 3.29

$$F(A,B,C,D) = ABC' + ABC + BCD' + BCD + AB'D' + A'B'D' + A'BC'D.$$

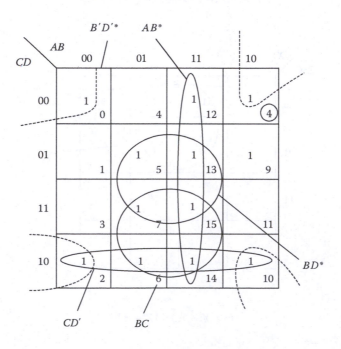

1. Groupings marked by an "*" are "essential." m_{12} is covered only by (AB); m_0 and m_8 are covered only by $(B'D')$; m_5 is covered only by (BD).
2. Once the aforementioned three groups are chosen, the only minterms left uncovered are m_6 and m_{14}. To cover these, we can choose either (BC) or (CD'). Hence, there are two simplified forms:

$$F(A,B,C,D) = AB + B'D' + BD + BC$$

$$= AB + B'D' + BD + CD'.$$

Either of the aforementioned is a satisfactory form, since each contains the same number of literals.

Simplified Functions in POS Form: To obtain the simplified function in POS form, perform the following:

1. Plot the function F on the K-map.
2. Derive the K-map for F' (by changing 1 to 0 and 0 to 1).
3. Simplify the K-map for F' to obtain F' in SOP form.
4. Use De Morgan's laws to obtain F.

Example 3.30

$$F(P,Q,R,S) = P'Q'R' + P'Q'RS + P'RS' + PQ\,RS' + PQ'R'.$$

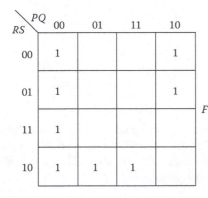
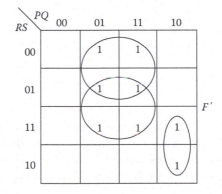

$$F' = QR' + QS + PQ'R$$

$$F = F'' = (QR' + QS + PQ'R)'$$

$$= (QR')' \cdot (QS)' \cdot (PQ'R)' \qquad \text{T5a}$$

$$= (Q' + R)(Q' + S')(P' + Q + R') \qquad \text{T5b.}$$

Minimization Using Don't Cares: In designing logic circuits, if we know that certain input combinations do not occur, corresponding outputs from the circuit can be ignored. It is also possible that we do not care what the circuit output would be even when certain input combinations occur. Such conditions are termed *don't cares*.

Example 3.31

Consider a circuit that converts the BCD code input into corresponding excess-3 code. This BCD-to-excess-3 decoder expects as inputs only the combinations corresponding to decimals 0 through 9. The other six inputs (10–15) will never occur. Hence, the output corresponding to each of these six inputs is a don't care.

Don't cares are indicated by a "*d*" on the K-map. Each can be treated as either a 1 or a 0. It is not necessary to cover all the don't cares while grouping: that is, don't cares not covered are treated as 0s.

The following maps illustrate the use of don't cares in simplifying the output functions of the decoder.

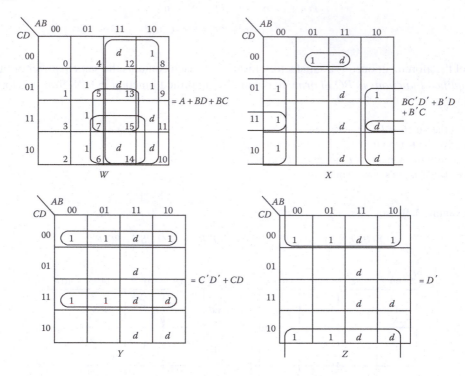

K-maps are useful for functions with up to four or five variables. Figure 3.9 shows a five-variable K-map. Since the number of blocks doubles for each additional variable, minimization using K-maps becomes complex for functions with more than five variables. The Quine–McCluskey procedure described in the next section is useful in such cases.

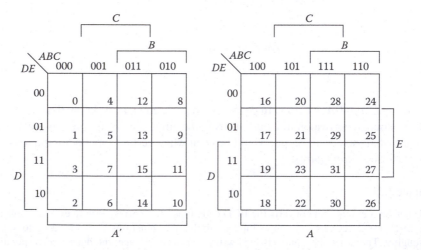

Figure 3.9 Five-variable Karnaugh map. *Note:* The two parts of the map are treated as two planes, one super-imposed over the other. The blocks at the same position on each plane are also logically adjacent.

3.3.3 Quine–McCluskey Procedure

The Quine–McCluskey procedure also uses the logical adjacency property to reduce the Boolean function. Two minterms are logically adjacent if they differ in the open position only, that is, if one of the variables is in uncomplemented form in one minterm and complemented form in the other. Such a variable is eliminated by combining the two minterms. The Quine–McCluskey procedure compares each minterm with all the others and combines them if possible. The procedure uses the following steps:

1. Classify the minterms (and don't cares) of the function into groups such that each term in a group contains the same number of 1s in the binary representation of the term.
2. Arrange the groups formed in step 1 in the increasing order of number of 1s. Let the number of groups be *n*.
3. Compare each minterm in group *i* (*i* = 1 to *n* − 1) with those in group (*i* + 1); if the two terms are adjacent, form a combined term. The variable thus eliminated is represented as "−" in the combined term.
4. Repeat the matching operation of step 3 on the combined terms until no more combinations can be done. Each combined term in the final list is called a *prime implicant* (PI). A PI is a product term that cannot be combined with others to yield a term with fewer literals.
5. Construct a PI chart in which there is one column for each minterm (only minterms; don't cares are not listed) and one row for each PI. An "*X*" in a row–column intersection indicates that the PI corresponding to the row covers the minterm corresponding to the column.
6. Find all the essential PIs (i.e., the PIs that each cover at least one minterm that is not covered by any other PI).
7. Select a minimum number of PIs from the remaining, to cover those minterms not covered by the essential PIs.
8. The set of PIs thus selected forms the minimum function.

This procedure is illustrated by Example 3.32.

Example 3.32

$$F(A,B,C,D) = \sum m\underbrace{(0,2,4,5,6,9,10)}_{\text{Minterms}} + \sum d\underbrace{(7,11,12,13,14,15)}_{\text{Don't cares}}.$$

Steps 1 and 2:

√	0	0000	Group 0: terms with no 1s
√	2	0010	
√	4	0100	Group 1: terms with one 1
√	5	0101	
√	6	0110	
√	9	1001	Group 2: terms with two 1s
√	10	1010	
√	12	1100	
√	7	0111	
√	11	1011	
√	13	1101	Group 3: terms with three 1s
√	14	1110	
√	15	1111	Group 4: terms with four 1s

The "√" indicates that the term is used in forming a combined term at least once.

Step 3:

√	(0, 2)	00–0	Obtained by matching groups 0 and 1
√	(0, 4)	0–00	
√	(2, 6)	0–10	
√	(2, 10)	–010	
√	(4, 5)	010–	Obtained by matching groups 1 and 2
√	(4, 6)	01–0	
√	(4, 12)	–100	
√	(5, 7)	01–1	
√	(5, 13)	–101	
√	(6, 7)	011–	Obtained by matching groups 2 and 3
√	(6, 14)	–110	
√	(9, 11)	10–1	
√	(9, 13)	1–01	
√	(10, 11)	101–	
√	(10, 14)	1–10	
√	(12, 13)	110–	
√	(12, 14)	11–0	
√	(7, 15)	–111	
√	(11, 15)	1–11	Obtained by matching groups 3 and 4
√	(13, 15)	11–1	
√	(14, 15)	111–	

Step 4:

	(0, 2, 4, 6)	0– – 0	Same as (0, 4, 2, 6)	
	(2, 6, 10, 14)	– – 10	Same as (2, 10, 6, 14)	
√	(4, 5, 6, 7)	01 – –	Same as (4, 6, 5, 7)	
√	(4, 5, 12,13)	– 10 –	Same as (4, 12, 5, 13)	
√	(4, 6, 12, 14)	– 1–0	Same as (4, 12, 6, 14)	
√	(5, 7, 13, 15)	– 1–1	Same as (5, 13, 7, 15)	
√	(6, 7, 14, 15)	– 11 –	Same as (6, 14, 7, 15)	
	(9, 11, 13, 15)	1– –1	Same as (9, 13, 11, 15)	
	(10, 11, 14, 15)	1–1 –	Same as (10, 14, 11, 15)	
√	(12, 13, 14, 15)	11– –	Same as (12, 14, 11, 15)	
	(0, 2, 4, 6)	0 – – 0		PI_1
	(2, 6, 10, 14)	– – 10		PI_2
	(4, 5, 6, 7, 12, 13, 14, 15)	– 1– –		PI_3
	(9, 11, 13, 15)	1 – – 1		PI_4
	(10, 11, 14, 15)	1 – 1 –		PI_5

No further reduction possible.

Step 5: PI chart

Minterms

	√ 0	√ 2	√ 4	√ 5	√ 6	√ 9	√ 10
PI_1	⊗	×	×		×		
PI_2		×			×		×
PI_3			×	⊗	×		
PI_4						⊗	
PI_5							×

Step 6: PI_1, PI_3, and PI_4 are "essential" since minterms 0, 5, and 9, respectively, are covered by only these PIs. These PIs together also cover minterms 2, 4, and 6.

Step 7: To cover the remaining minterm 10, we can select either PI_2 or PI_5.

Step 8: The reduced function is

$$F(A,B,C,D) = PI_1 + PI_3 + PI_4 + PI_2 \text{ or } PI_5$$

$$= 0--0+-1--+1--1+--10 \text{ or } 1-1-$$

$$= (A'D' + B + AD + CD') \text{ or } (A'D' + B + AD + AC).$$

The Quine–McCluskey procedure can be programmed on a computer and is efficient for functions of any number of variables. Several other techniques to simplify Boolean function have been devised. The interested reader is referred to the books listed under references. The automation of Boolean function minimization was an active research area in the 1970s. Advances in IC technology have contributed to a decline of interest in the minimization of Boolean functions. The minimization of the number of ICs and the efficient interconnections between them is of more significance than the saving of a few gates in the present-day design environment.

3.4 PRIMITIVE HARDWARE BLOCKS

A logic circuit is the physical implementation of a Boolean function. The primitive Boolean operations AND, OR, and NOT are implemented by electronic components known as *gates*. The Boolean constants 0 and 1 are implemented as two unique voltage levels or current levels. A gate receives these logic values on its inputs and produces a logic value that is a function of its inputs on its output. Each gate has one or more inputs and an output. The operation of a gate can be described by a truth table. The truth tables and standard symbols used to represent the three primitive gates are shown in Figure 3.10.

The NOT gate will always have one input and one output. Only two inputs are shown for the other gates in Figure 3.10 for convenience. The maximum number of inputs allowed on the gate is limited by the electronic technology used to build the gate. The number of inputs on the gate is termed its *fan-in*. We will assume that there is no restriction on the fan-in. A four-input AND gate is shown in the succeeding text.

Figure 3.10 shows three other popular gates. The utility of these gates will be discussed later in this chapter.

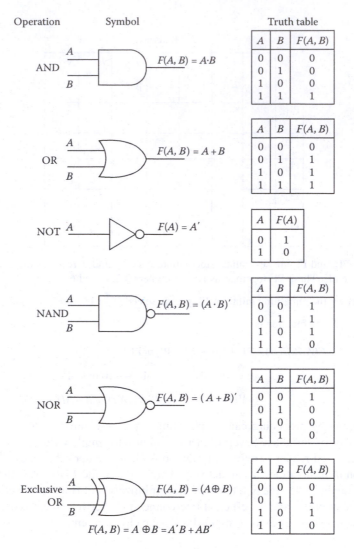

Figure 3.10 Gates and truth tables.

3.5 FUNCTIONAL ANALYSIS OF COMBINATIONAL CIRCUITS

The *functional analysis* of a combinational circuit is the process of determining the relations of its outputs to its inputs. These relations are expressed either as a set of Boolean functions, one for each output, or as a truth table for the circuit. The functional analysis is usually performed to verify the stated function of a combinational circuit. If the function of the circuit is not known, the analysis determines it. Two other types of analysis are commonly performed on logic circuits. They are *loading* and *timing* analyses. We will discuss functional analysis in this section and describe the other types of analyses in Section 3.9.

Consider the combinational circuit with n input variables and m outputs shown in Figure 3.11a. Since there are n inputs, there are 2^n combinations of input values. For each of these input combinations, there are unique combinations of output values. A truth table for this circuit will have 2^n rows and $(n + m)$ columns, as shown in Figure 3.11b. We will need m Boolean functions to describe this circuit. We will demonstrate the analysis procedure in the following example.

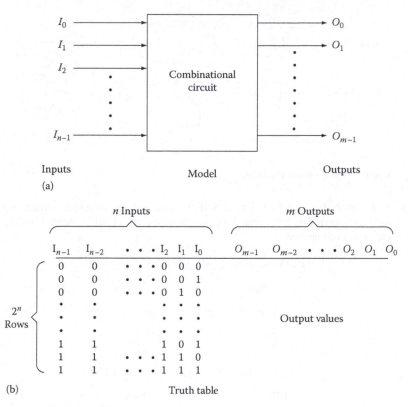

Figure 3.11 Combinational circuit.

Example 3.33

Consider the circuit shown in Figure 3.12. The three-input variables to the circuit are X, Y, and Z; the two outputs are P and Q.

In order to derive the outputs P and Q as functions of X, Y, and Z, we can trace through the signals from inputs to the outputs. To facilitate such a tracing, we have labeled the outputs of all the gates in the circuit with the arbitrary symbols R_1, R_2, R_3, and R_4. Note that R_1 and R_2 are functions of only input variables; R_3 and R_4 are functions of R_1, R_2, and the input variables. P is a function of R_2 and R_3, and Q is a function of P and R_4. Tracing through the circuit, we note that

$$R_1 = Y' \text{ and } R_2 = Z'$$

$$R_3 = X \cdot R_1$$

$$= X \cdot Y' = XY'$$

$$R_4 = R_2 \cdot X$$

$$= Z' \cdot X = XZ'$$

$$P = R_3 + R_2$$

$$= XY' + Z'$$

$$Q = P \oplus R_4$$

$$= (XY' + Z') \oplus XZ'.$$

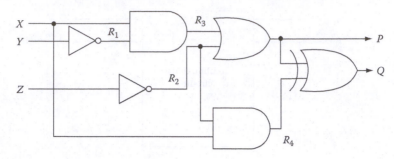

Figure 3.12   Circuit with three-input variables.

We can transform the function for Q into an SOP form using theorems and postulates of Boolean algebra, as shown later. The theorems and postulates used are identified with T and P, respectively, along with their numbers:

$$Q = (XY' + Z')' \cdot (XZ') + (XY' + Z') \cdot (XZ')'$$

$$\text{T5a} \qquad\qquad\qquad\qquad \text{T5b}$$

$$= (XY')' \cdot (Z')' \cdot (XZ') + (XY' + Z') \cdot (X' + Z'').$$

$$\qquad \text{T5b} \qquad\qquad\qquad\qquad \text{T5b}$$

$$= (X' + Y'') \cdot Z \cdot (XZ') + (XY' + Z') \cdot (X' + Z)\,\text{T5b}$$

$$\qquad \text{T5b} \qquad\qquad\qquad \text{P4b}$$

$$= (X' + Y) \cdot Z \cdot Z'X + (XY' + Z')X' + (XY' + Z'))Z$$

$$\qquad\qquad\qquad\qquad \text{P4b} \qquad \text{P4b}$$

$$= 0 \qquad + XY'X' + Z'X' + XY'Z + ZZ'$$

$$\qquad\qquad 0 \qquad\qquad\qquad\quad 0$$

$$= X'Z' + XY'Z.$$

Thus,

$$P = XY' + Z'$$

$$Q = X'Z' + XY'Z.$$

Figure 3.13 shows the derivation of the truth table for the aforementioned circuit. The truth table can be drawn from the earlier functions for P and Q or by tracing through the circuit. Since there are three inputs, the truth table will have eight rows. The eight combinations of input values are shown in Figure 3.13b. We can now impose each combination of values on the inputs of the circuit and not the output values, tracing through the circuit. Figure 3.13a shows the condition of the circuit corresponding to the input conditions $X = 0$, $Y = 0$, and $Z = 0$. Tracing through the circuit, we note that $R_1 = 1$, $R_2 = 1$, $R_3 = 0$, $R_4 = 0$, $P = 1$, and $Q = 1$. We repeat this process with the other seven input combinations to derive the complete truth table shown in Figure 3.13b.

We have shown the column corresponding to intermediate outputs R_1, R_2, R_3, and R_4 in the truth table for convenience only. These columns are not usually retained in the final truth table.

We summarize the functional analysis procedures as follows.

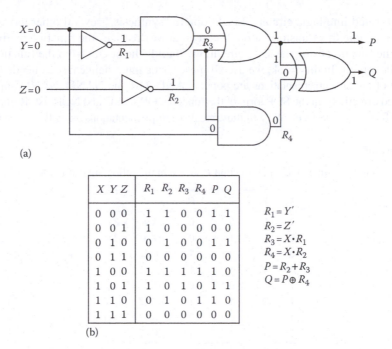

(a)

X Y Z	R_1 R_2 R_3 R_4 P Q
0 0 0	1 1 0 0 1 1
0 0 1	1 0 0 0 0 0
0 1 0	0 1 0 0 1 1
0 1 1	0 0 0 0 0 0
1 0 0	1 1 1 1 1 0
1 0 1	1 0 1 0 1 1
1 1 0	0 1 0 1 1 0
1 1 1	0 0 0 0 0 0

$R_1 = Y'$
$R_2 = Z'$
$R_3 = X \cdot R_1$
$R_4 = X \cdot R_2$
$P = R_2 + R_3$
$Q = P \oplus R_4$

(b)

Figure 3.13 Derivation of truth table. (a) Circuit with 000 input condition. (b) Truth table.

To obtain the output functions from a logic diagram, perform the following:

1. Label outputs of all gates in the circuit with arbitrary variable names.
2. Express the outputs of the first level of gates (i.e., the gates whose inputs are circuit input variables) as functions of their inputs.
3. Express the outputs of the next level of gates as functions of their inputs (which are circuit inputs and outputs from the previous level of gates).
4. Continue the process of step 3 until the circuit outputs are obtained.
5. Substitute the functions corresponding to each intermediate variable into the output functions, to eliminate the intermediate variables from the functions.
6. Simplify the output functions (if possible) using the methods described in Section 3.3.

To obtain the truth table from the circuit diagram, perform the following:

1. Determine the number of inputs n in the circuit. List the binary numbers from 0 through $(2^n - 1)$ forming the input portion of the truth table with 2^n rows.
2. Label all the outputs of all the gates in the circuit with arbitrary symbols.
3. Determine the outputs of the first level of gates for each input combination. Each first-level output forms a column in the truth table.
4. Using the combination of input values and the values of intermediate outputs already determined, derive the values for the outputs of the next level of gates.
5. Continue the process in step 4 until the circuit outputs are reached.

3.6 SYNTHESIS OF COMBINATIONAL CIRCUITS

Synthesis is the process of transforming the word statement of the function to be performed into a logic circuit. The word statement is first converted into a truth table. This step requires the identification of the input variables and the output values corresponding to each combination of input values. Each input can then be expressed as a Boolean function of input variables. The functions are then transformed into logic circuits. In addition to "synthesis," several other terms are used in the literature to denote this transformation. We say that we "realize" the function by the circuit, we "implement" the function using the logic circuit, we "build" the circuit from the function, or simply we "design" the circuit. In this book, we use all these terms interchangeably as needed.

Four types of circuit implementations are popular. AND–OR and NAND–NAND implementations can be generated directly from the SOP form of the function: OR–AND and NOR–NOR implementations evolve directly from the POS form. We will illustrate these implementations using the following example.

Example 3.34

Build a circuit to implement the function P shown in the following truth table:

X	Y	Z	P	
0	0	0	0	$(X + Y + Z)$
0	0	1	0	$(X + Y + Z')$
0	1	0	1	$X'YZ'$
0	1	1	1	$X'YZ$
1	0	0	0	$(X' + Y + Z)$
1	0	1	1	$XY'Z$
1	1	0	0	$(X' + Y' + Z)$
1	1	1	1	XYZ

$$\underbrace{\qquad\qquad\qquad}_{\text{Inputs}} \quad \overset{\uparrow}{\underset{\text{Output}}{}}$$

3.6.1 AND–OR Circuits

From the earlier truth table, P in SOP form is

$$P = X' \cdot Y \cdot Z' + X' \cdot Y \cdot Z + X \cdot Y' \cdot Z + X \cdot Y \cdot Z.$$

$P(X, Y, Z)$ is the sum of four product terms, so we use an OR gate with four inputs to generate P (see Figure 3.14a). Each of the inputs to this OR gate is a product of three variables. Hence, we use four AND gates, each realizing a product term. The outputs of these AND gates are connected to the four inputs of the OR gate, as shown in Figure 3.14b.

A complemented variable can be generated using a NOT gate. Figure 3.14c shows the circuit needed to generate X', Y', and Z'. The final task in building the circuit is to connect these complemented and uncomplemented signals to appropriate inputs of AND gates. Often, the NOT gates are not specifically shown in the circuit. It is then assumed that the true and complemented values of variables are available. The logic circuit is usually shown as in Figure 3.14b. This type of circuit, designed using the SOP form of Boolean function as the starting point, is called a two-level AND–OR circuit because the first level consists of AND gates and the second level consists of OR gates.

3.6.2 OR–AND Circuits

An OR–AND circuit can be designed starting with the POS form of the function:

$$P = (X + Y + Z) \cdot (X + Y + Z') \cdot (X' + Y + Z) \cdot (X' + Y' + Z).$$

The design is carried out in three steps. The first two are shown in Figure 3.15a. The third step of including NOT gates is identical to that required in AND–OR circuit design.

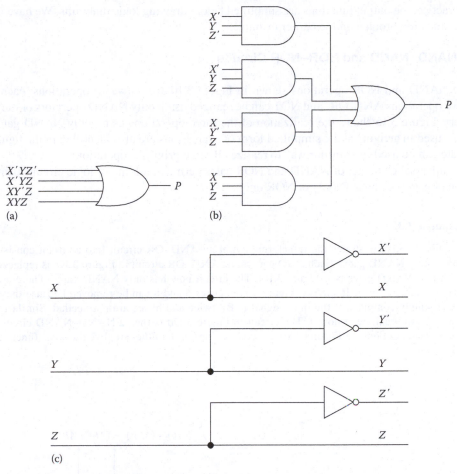

Figure 3.14 AND–OR circuit. (a) OR. (b) AND-OR. (c) NOT.

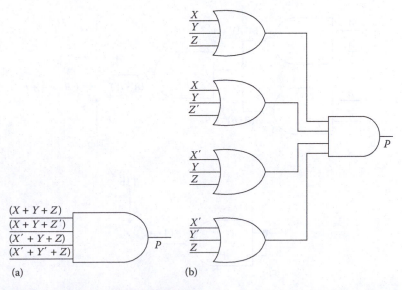

Figure 3.15 OR–AND circuit. (a) AND. (b) OR–AND.

In practice, the output functions are simplified before drawing logic diagrams. We have ignored the simplification problem in the earlier example.

3.6.3 NAND–NAND and NOR–NOR Circuits

The NAND and NOR operations shown in Figure 3.10 are *universal* operations: each of the primitive operations AND, OR, and NOT can be realized using only NAND operators or only NOR operators. Figure 3.16 shows the realization of the three operations using only NAND gates. The theorems used in arriving at the simplified form of expressions are also identified in the figure. The NOR gate can be used in a similar way to realize all three primitive operations.

The universal character of NAND and NOR gates permits building of logic circuits using only one type of gate (i.e., NAND only or NOR only).

Example 3.35

Figure 3.17 illustrates the transformation of an AND–OR circuit into a circuit consisting of only NAND gates. Each AND gate in the AND–OR circuit in Figure 3.17a is replaced with two NAND gates (see Figure 3.16). The circuit now has only NAND gates. There are some redundant gates in circuit (Figure 3.17c). Gates 5 and 8 can be removed because these gates simply complement the input signal $(AB)'$ twice and hence are not needed. Similarly, gates 7 and 9 can be removed. The circuit in Figure 3.17d is then a NAND–NAND circuit. The circuits in Figure 3.17a and d are equivalent since both of them realize the same function $(AB + A'C)$.

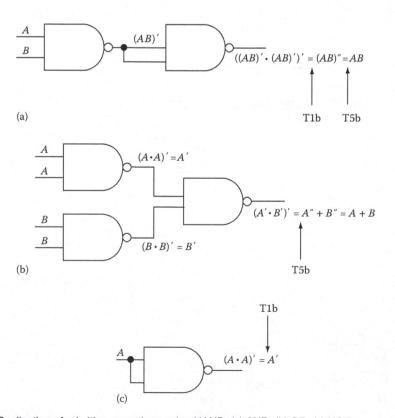

Figure 3.16 Realization of primitive operations using NAND. (a) AND. (b) OR. (c) NOT.

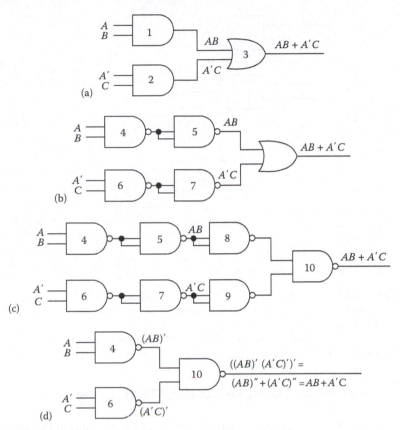

Figure 3.17 NAND–NAND transformation. (a) An AND–OR circuit. (b) Replace AND gates. (c) Replace OR gate. (d) Remove redundant gates (NAND–NAND circuit).

A NAND–NAND implementation can thus be derived from an AND–OR circuit by simply replacing each gate in the AND–OR circuit with a NAND gate having the same number of inputs as that of the gate it replaces. A NOR–NOR implementation likewise can be obtained by starting with an OR–AND implementation and replacing each gate with a NOR gate.

Any input literal feeding the second level directly must be inverted. Consider the circuit in Figure 3.18a. Here, C is fed directly to the second level. Hence, it must be inverted to derive the correct NAND–NAND implementation shown in Figure 3.18b.

These implementations are feasible because the gates are available commercially in the form of ICs in packages containing several gates of the same type. Using the same types of gates eliminates the need for different types of ICs, and using the same type of ICs usually allows more efficient use of the ICs and reduces the IC package count. Further, the NAND and NOR circuits are primitive circuit configurations in major IC technologies, and the AND and OR gates are realized by complementing the outputs of NAND and NOR, respectively. Thus, NAND and NOR gates are less complex to fabricate and more cost efficient than the corresponding AND and OR gates.

We now summarize the combinational circuit design procedure:

1. From the specification of the circuit function, derive the number of inputs and outputs required.
2. Derive the truth table.
3. If either an AND–OR or NAND–NAND form of circuit is required,
   a. Derive the SOP form of function for each output from the truth table.
   b. Simplify the output functions.
   c. Draw the logic diagram with the first level of AND (or NAND) gates and the second level of an OR (or NAND) gate, with appropriate number of inputs.

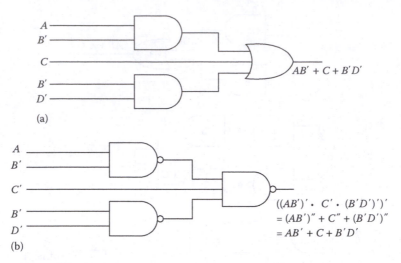

Figure 3.18 AND–OR to NAND–NAND transformation.

4. If either an OR–AND or NOR–NOR form of circuit is required,
   a. Derive the POS form of function for each output, from the truth table.
   b. Simplify the output functions if possible.
   c. Draw the logic diagram with the first level of OR (or NOR) gates and the second level of an AND (or NOR) gate, with appropriate number of inputs.

3.7 SOME POPULAR COMBINATIONAL CIRCUITS

We describe several most commonly used combinational logic circuits in this section. These are available as IC components from various manufacturers and are used as components in designing logic systems.

3.7.1 Adders

Addition is the most common arithmetic operation performed by processors. If a processor has hardware capable of addition of two numbers, the other three primitive arithmetic operations can also be performed using the addition hardware. Subtraction is performed by adding the subtrahend expressed in either 2s or 1s complement form to the minuend, multiplication is repeated addition of multiplicand to itself by multiplier number of times, and division is the repeated subtraction of divisor from dividend.

Consider the addition of two 4-bit numbers A and B:

		c_2	c_1	c_0	
A		a_3	a_2	a_1	a_0
B		b_3	b_2	b_1	b_0
SUM	c_3	s_3	s_1	s_1	s_0

Bits a_0 and b_0 are LSBs; a_3 and b_3 are MSBs. The addition is performed starting with the LSB position. Adding a_0 and b_0 will produce a SUM bit s_0 and a CARRY c_0. This CARRY c_0 is now used in the addition of next significant bits a_1 and b_1, producing s_1 and c_1; this addition process is carried out through the MSB position.

A *half-adder* is a device that can add 2 bits producing a SUM bit and a CARRY bit as its outputs. A *full adder* adds 3 bits, producing a SUM bit and a CARRY bit as its outputs. To add two

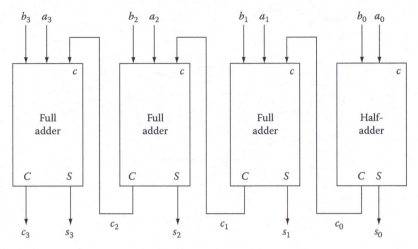

Figure 3.19 Ripple-carry adder.

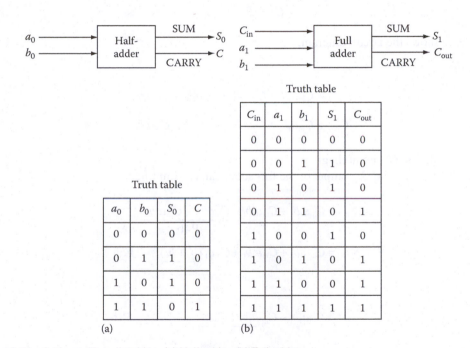

Figure 3.20 Adders with truth tables. (a) Half-adder. (b) Full adder.

n-bit numbers, we thus need one half-adder and $n - 1$ full adders. Figure 3.19 shows the half-adder and full-adder arrangement to perform 4-bit addition. This is called a *ripple-carry adder* since the carry ripples through the stages of the adder starting at LSB to MSB. The time needed for this carry propagation is in proportion to the number of bits. Since the sum is of correct value only after the carry appears at MSB, the longer the carry propagation time, the slower the adder will be. There are several schemes to increase the speed of this adder. Some of them are discussed in Chapter 10.

Figure 3.20 shows the block diagram representation and truth tables for full adders and half-adders. From truth tables, we can derive the SOP form functions for the outputs of the adders. They are

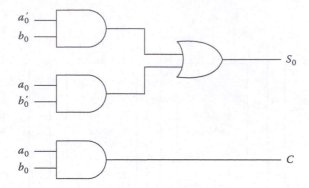

Figure 3.21 Half-adder circuits.

Half-adder:

$$S_0 = a_0'b_0 + a_0b_0'$$

$$C = a_0b_0.$$

(3.5)

Figure 3.21 shows the circuit diagram.

Full adder:

$$S_1 = C_{in}'a_1'b_1 + C_{in}'a_1b_1' + C_{in}a_1'b_1' + C_{in}a_1b_1$$

$$C_{out} = C_{in}'a_1b_1 + C_{in}a_1'b_1 + C_{in}a_1b_1' + C_{in}a_1b_1.$$

(3.6)

Figure 3.22 shows the circuit diagram.

The equation for the C_{out} output of the full adder can be simplified:

$$C_{out} = C_{in}'a_1b_1 + C_{in}a_1'b_1 + \underbrace{C_{in}a_1b_1' + C_{in}a_1b_1}_{P4b}$$

$$= C_{in}'a_1b_1 + C_{in}a_1'b_1 + C_{in}a_1\underbrace{(b_1' + b_1)}_{1} \qquad P5a$$

$$= C_{in}'a_1b_1 + C_{in}\underbrace{a_1'b_1 + C_{in}a_1}_{} \qquad P1b$$

$$= C_{in}'a_1b_1 + C_{in}\underbrace{(a_1'b_1 + a_1)}_{T4a} \qquad P4b$$

$$= C_{in}'a_1b_1 + C_{in}(b_1 + a_1) \qquad P4b$$

$$= C_{in}'a_1b_1 + C_{in} \cdot b_1 + C_{in} \cdot a_1$$

$$= \underbrace{(C_{in}'a_1 + C_{in})}_{T4a}b_1 + C_{in}a_1 \qquad P4b$$

$$= (a_1 + C_{in})b_1 + C_{in}a_1 \qquad P4b.$$

$$= a_1b_1 + C_{in}b_1 + C_{in}a_1.$$

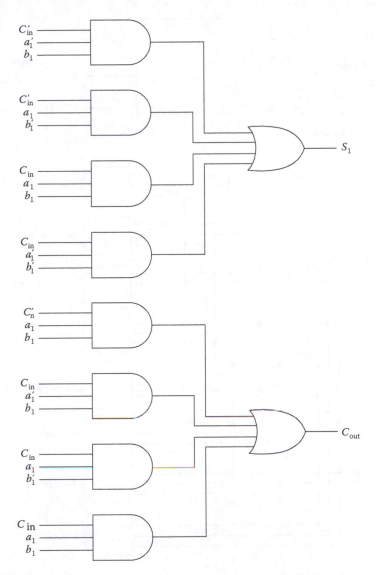

Figure 3.22 Full-adder circuits.

This equation has only 6 literals compared to the 12 literals of the original equation. This simplified equation can be realized with three two-input AND gates and one three-input OR gate.

3.7.2 Decoders

A *code word* is a string of a certain number of bits, as described in Chapter 2. An n-bit binary string can take 2^n combinations of values. An n-to-2^n decoder shown in Figure 3.23 is a circuit that converts the n-bit input data into 2^n outputs (at the maximum). At any time, only one output line corresponding to the combination on the input lines will be 1; all the other outputs will be 0. The outputs are usually numbered from 0 to $(2^n - 1)$. If, for example, the combination on the input of a 4-to-2^4 decoder is 1001, only output 9 will be 1; all other outputs will be 0.

It is not usually necessary to draw a truth table for a decoder. There would be a single 1 in each output column of the truth table, and the product (or SUM) term corresponding to that 1 could be

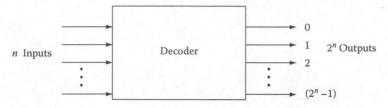

Figure 3.23 n-to-2^n decoder.

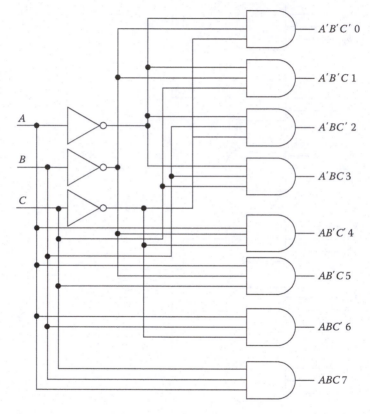

Figure 3.24   3-to-8 Decoder circuits.

easily derived. Figure 3.24 shows the circuit diagram of a 3-to-8 decoder. The three inputs are designated A, B, and C, with C as the LSB. The outputs are numbered 0 through 7.

3.7.3 Code Converters

A code converter translates an input code word into an output bit pattern corresponding to a new code word. A decoder is a code converter that changes an n-bit code word into a 2^n-bit code word. The design of a code converter for BCD into excess-3 code conversion was provided earlier in this chapter.

3.7.4 Encoders

An encoder generates an n-bit code word as a function of the combination of values on its input. At the maximum, there can be 2^n inputs (Figure 3.25).

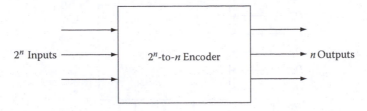

Figure 3.25 2^n-to-n encoder.

Inputs				Outputs	
W	X	Y	Z	D_0	D_1
1	0	0	0	0	0
0	1	0	0	0	1
0	0	1	0	1	0
0	0	0	1	1	1

(a)   (b)

Figure 3.26 4-to-2 encoder. (a) Truth table. (b) Circuit.

The design of an encoder is executed by first drawing a truth table that shows the n-bit output needed for each of the 2^n combinations of inputs. The circuit diagrams are then derived for each output bit.

Example 3.36

A partial truth table for a 4-to-2 line encoder is shown in Figure 3.26a. Although there are 16 combinations of 4 inputs, only 4 are used because the 2-bit output supports only 4 combinations. The output combinations identify which of the four input lines is at 1 at a particular time.

These functions may be simplified by observing that it is sufficient to have $W = 0$ and $X = 0$ for D_0 to be 1, no matter what the values of Y and Z are. Similarly, $W = 0$ and $Y = 0$ are sufficient for D_1 to be 1. Such observations, although not always straightforward, help in simplifying functions, thus reducing the amount of hardware needed. Alternatively, the truth table in Figure 3.26a can be completed by including the remaining 12 input combinations and entering don't cares for the outputs corresponding to those inputs. D_0 and D_1 can then be derived from the truth table and simplified.

The output functions as can be seen from this truth table are

$$D_0 = W'X', \quad \text{and}$$

$$D_1 = W'Y'.$$

(3.7)

Figure 3.26b shows the circuit diagram for the 4-to-2 encoder.

3.7.5 Multiplexers

A multiplexer is a switch that connects one of its several inputs to the output. A set of n control inputs is needed to select one of the 2^n inputs that is to be connected to the output (Figure 3.27).

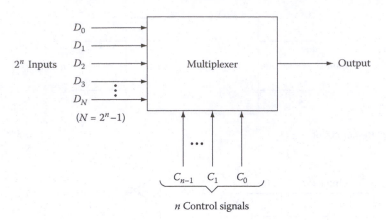

Figure 3.27   Multiplexer.

Example 3.37

The operation of a multiplexer with four inputs ($D_0 - D_3$) and hence two control signals (C_0, C_1) can be described by the following table:

C_1	C_0	Output
0	0	D_0
0	1	D_1
1	0	D_2
1	1	D_3

Although there are six inputs, a complete truth table with 2^6 rows is not required for designing the circuit, since the output simply assumes the value of one of the four inputs depending on the control signals C_1 and C_0. That is,

$$\text{Output} = D_0 \cdot C_1'C_0' + D_1 \cdot C_1'C_0 + D_2 \cdot C_1C_0' + D_3 \cdot C_1C_0. \qquad (3.8)$$

The circuit for realizing the earlier multiplexer is shown in Figure 3.28.

Each of the inputs D_0, D_1, D_2, and D_3 and the output in this multiplexer circuit are single lines. If the application requires that the data lines to be multiplexed have more than 1 bit each, the aforementioned circuit has to be duplicated once for each bit of data.

3.7.6 Demultiplexers

A demultiplexer has one input and several outputs. It switches (connects) the input to one of its outputs based on the combination of values on a set of control (select) inputs. If there are n control signals, there can be a maximum of 2^n outputs (Figure 3.29).

Example 3.38

The operation of a demultiplexer with four outputs ($O_0 - O_3$) and hence two control signals (C_1, C_0) can be described by the following table:

C_1	C_0	O_3	O_2	O_1	O_0
0	0	0	0	0	I
0	1	0	0	I	0
1	0	0	I	0	0
1	1	I	0	0	0

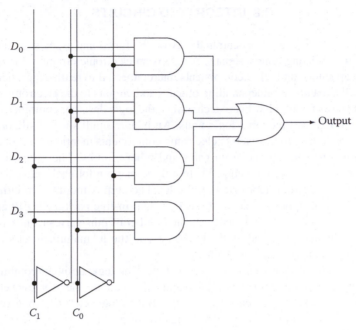

Figure 3.28 4-to-1 multiplexer.

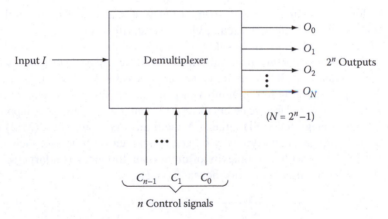

Figure 3.29 Demultiplexer.

It is not necessary to draw a truth table with eight rows for this circuit, although there are three inputs to it, since the input I is directed to only one of the four outputs based on the four combinations of values on the control signals C_1 and C_0. Thus,

$$O_0 = IC_1'C_0' \quad O_1 = IC_1'C_0$$
$$O_2 = IC_1C_0' \quad O_3 = IC_1C_0.$$

A typical application for a multiplexer is to connect one of the several input devices (as selected by a device number) to the input of a computer system. A demultiplexer can be used to switch the output of the computer onto one of the several output devices.

3.8 INTEGRATED CIRCUITS

So far in this chapter, we have concentrated only on the functional aspects of gates and logic circuits in terms of manipulating binary signals. A gate is an electronic circuit made up of transistors, diodes, resistors, capacitors, and other components interconnected to realize a particular function. In this section, we will expand our understanding of gates and circuits to the electronic level of detail.

At the current state of digital hardware technology, the logic designer combines ICs that perform specific functions to realize functional logic units. An IC is a small slice of silicon semiconductor crystal called a *chip*, on which the discrete electronic components mentioned earlier are chemically fabricated and interconnected to and from gates and other circuits. These circuits are accessible only through the pins attached to the chip. There will be one pin for each input signal and one for each output signal of the circuit fabricated on the IC. The chip is mounted in either a metallic or a plastic package. Various types of packages, such as dual-in-line package (DIP) and flat package, are used. DIP is the most widely used package. The number of pins varies from 8 to 64. Each IC is given a numeric designation (printed on the package), and the IC manufacturer's catalog provides the functional and electronic details on the IC.

Each IC contains one or more gates of the same type. The logic designer combines ICs that perform specific functions to realize functional logic units. As such, the electronic-level details of gates usually are not needed to build efficient logic circuits. But as logic circuit complexity increases, the electronic characteristics become important in solving the timing and loading problems in the circuit.

Figure 3.30 shows the details of an IC that comes from the popular transistor–transistor logic (TTL) family of ICs. It has the numeric designation of 7400 and contains four two-input NAND gates. There are 14 pins. Pin 7 (ground) and pin 14 (supply voltage) are used to power the IC. Three pins are used by each gate (two for the input and one for the output). On all ICs, a "notch" on the package is used to reference pin numbers. Pins are numbered counterclockwise starting from the notch.

Digital and *linear* are two common classifications of ICs. Digital ICs operate with binary signals, whereas linear ICs operate with continuous signals. In this book, we will be dealing only with digital ICs.

Because of the advances in IC technology, it is now possible to fabricate a large number of gates on a single chip. According to the number of gates it contains, an IC can be classified as a small-, medium-, large-, or very-large-scale circuit. An IC containing a few gates (approximately 10) is called a small-scale integrated (SSI) circuit. A medium-scale integrated (MSI) circuit has a complexity of around 100 gates and typically implements an entire function, such as an adder or a decoder, on a chip. An IC with a complexity of more than 100 gates is a large-scale integrated (LSI) circuit, while a VLSI circuit contains thousands of gates.

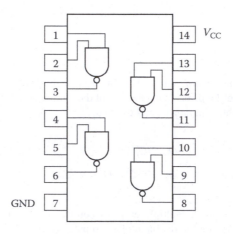

Figure 3.30   Typical IC (TTL 7400).

There are two broad categories of IC technology, one based on *bipolar* transistors (i.e., *p-n-p* or *n-p-n* junctions of semiconductors) and the other based on the *unipolar* metal-oxide-semiconductor field-effect transistor (MOSFET). Within each technology, several logic families of ICs are available. The popular bipolar logic families are TTL and emitter-coupled logic (ECL). P-channel MOS (PMOS), N-channel MOS (NMOS), and complementary MOS (CMOS) are all popular MOS logic families. A new family of ICs based on gallium arsenide has been introduced recently. This technology has the potential of providing ICs that are faster than ICs in silicon technology.

In the following discussion, we will examine functional-level details of ICs. These details are adequate to build circuits using ICs. Various performance characteristics of ICs will be introduced, without electronic-level justification.

In addition to the details of the type provided in Figure 3.30, the IC manufacturer's catalog contains such information as voltage ranges for each logic level, fan-out, propagation delays, and so forth, for each IC. In this section, we will introduce the major symbols and notation used in the IC catalogs and describe the most common characteristics in selecting and using ICs.

3.8.1 Positive and Negative Logic

As mentioned earlier, two distinct voltage levels are used to designate logic values 1 and 0, and in practice, these voltage levels are range of voltages, rather than fixed values. Figure 3.31 shows the voltage levels used in the TTL technology: the high level corresponds to the range of 2.4–5 V, and the low level corresponds to 0–0.4 V. These two levels are designated H and L.

In general, once the voltage levels are selected, the assignment of 1 and 0 to those levels can be arbitrary. In the so-called *positive logic* system, the higher of the two voltages denotes logic-1, and the lower value denotes logic-0. In the *negative logic* system, the designations are the opposite. The following table shows the two possible logic value assignments.

	Positive Logic	Negative Logic
Logic-1	H	L
Logic-0	L	H

Note that H and L can both be positive, as in TTL, or both negative, as in ECL ($H = -0.7$ to -0.95 V, $L = -1.9$ to -1.5 V). It is the assignment of the relative magnitudes of voltages to logic-1 and logic-0 that determines the type of logic, rather than the polarity of the voltages.

Because of these "dual" assignments, a logic gate that implements an operation in the positive logic system implements its dual operation in the negative logic system. IC manufacturers describe the function for gates in terms of H and L, rather than logic-1 and logic-0. As an example, consider the TTL 7408 IC, which contains four two-input AND gates (Figure 3.32a). The function of this IC

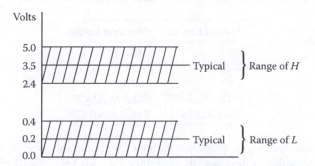

Figure 3.31 TTL voltage levels.

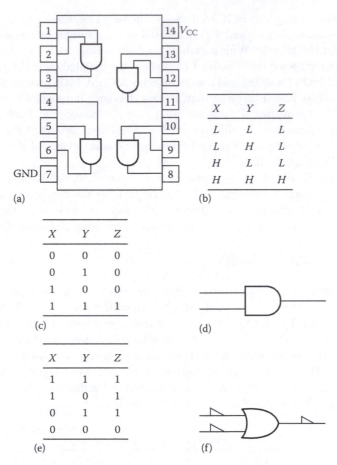

Figure 3.32 Positive and negative logic. (a) TT I 7408. (b) Voltage table. (c) Positive logic truth table. (d) Positive logic AND. (e) Negative logic truth table. (f) Negative logic OR.

as described by the manufacturer in terms of H and L is shown in the voltage table. Using positive logic, the voltage table in Figure 3.32b can be converted into the truth table in Figure 3.32c representing the positive logic AND, as shown by the gate in Figure 3.32d. Assuming negative logic, the table in Figure 3.32b can be converted into truth table in Figure 3.32e, which is the truth table for OR. The negative logic OR gate is shown in Figure 3.32f. The half-arrows on the inputs and the output designate them to be negative logic values. Note that the gates in Figure 3.32d and f both correspond to the same physical gate and function either as positive logic AND or as negative logic OR. Similarly, it can be shown that the following dual operations are valid:

Positive Logic	Negative Logic
OR	AND
NAND	NOR
NOR	NAND
EXCLUSIVE-OR	EQUIVALENCE
EQUIVALENCE	EXCLUSIVE-OR

We will assume the positive logic system throughout this book and as such will not use negative logic symbolism. In practice, a designer may combine the two logic notations in the same circuit (mixed logic), so long as the signal polarities are interpreted consistently.

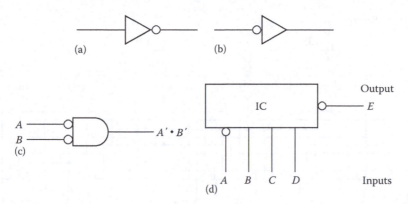

Figure 3.33 Bubble notation. (a) NOT gate. (b) NOT with active-low input. (c) AND with active-low inputs. (d) A typical IC.

3.8.2 Signal Inversion

We have used a "bubble" in NOT, NOR, and NAND gate symbols to denote signal inversion (complementation). This bubble notation can be extended to any logic diagram. Some examples are shown in Figure 3.33. The NOT gate symbol in Figure 3.33a implies that when the input X is asserted (e.g., at H), the output is low (L). The input is said to be *active high*, and the output is said to be *active low*. An alternative symbol for a NOT gate, with an active-low input, is shown in Figure 3.33b. An AND gate with active-low inputs is shown in Figure 3.33c. The output of this gate is high only when both inputs are low. Note that this is the INVERT–AND or a NOR gate. A typical IC with four inputs and one output is shown in Figure 3.33d. Input A and output E are active low, and inputs B, C, and D are active high. An L on input A would appear as an H internal to the IC, and an H corresponding to E internal to the IC would appear as an L external to the IC. That is, an active-low signal is active when it carries a low logic value, whereas an active-high signal is active when it carries a high logic value. A bubble in the logic diagram indicates an active-low input or output. For example, the BCD-to-decimal decoder IC (7442) shown in Figure 3.34 has four active-low outputs. Only the output corresponding to the decimal value of the input BCD number is active at any time. That is, the output corresponding to the decimal value of the BCD input to the IC will be L, and all other outputs will be H.

When two or more ICs are used to the circuit, the active-low and active-high designations of input and output signals must be observed for the proper operation of the circuit, although ICs of the same logic family generally have compatible signal-active designations.

It is important to distinguish between the negative logic designation (half-arrow) and the active-low designation (bubble) in a logic diagram. These designations are similar in effect, as shown by the gate symbols in Figure 3.35, and as such can be replaced by each other. In fact, a half-arrow following the bubble cancels the effect of the bubble on the signal, and hence both the half-arrow and the bubble can be removed. For example, in the last two equivalent representations in Figure 3.35, if the half-arrow and the bubble are removed from the output, the inputs represent negative logic and the outputs represent positive logic polarities.

3.8.3 Other Characteristics

Other important characteristics to be noted while designing with ICs are the active-low and active-high designations of signals, voltage polarities, and low and high voltage values, especially when ICs of different logic families are used in the circuit. ICs of the same family are usually

Logic diagram

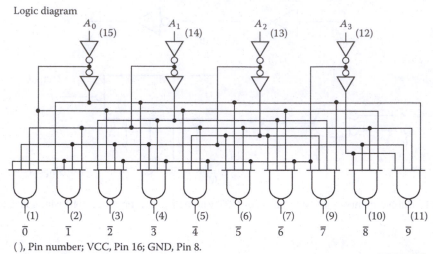

( ), Pin number; VCC, Pin 16; GND, Pin 8.

LD02750S

Function table

A_3	A_2	A_1	A_0	$\bar{0}$	$\bar{1}$	$\bar{2}$	$\bar{3}$	$\bar{4}$	$\bar{5}$	$\bar{6}$	$\bar{7}$	$\bar{8}$	$\bar{9}$
L	L	L	L	L	H	H	H	H	H	H	H	H	H
L	L	L	H	H	L	H	H	H	H	H	H	H	H
L	L	H	L	H	H	L	H	H	H	H	H	H	H
L	L	H	H	H	H	H	L	H	H	H	H	H	H
L	H	L	L	H	H	H	H	L	H	H	H	H	H
L	H	L	H	H	H	H	H	H	L	H	H	H	H
L	H	H	L	H	H	H	H	H	H	L	H	H	H
L	H	H	H	H	H	H	H	H	H	H	L	H	H
H	L	L	L	H	H	H	H	H	H	H	H	L	H
H	L	L	H	H	H	H	H	H	H	H	H	H	L
H	L	H	L	H	H	H	H	H	H	H	H	H	H
H	L	H	H	H	H	H	H	H	H	H	H	H	H
H	H	L	L	H	H	H	H	H	H	H	H	H	H
H	H	L	H	H	H	H	H	H	H	H	H	H	H
H	H	H	L	H	H	H	H	H	H	H	H	H	H
H	H	H	H	H	H	H	H	H	H	H	H	H	H

H, HIGH voltage levels; *L*, LOW voltage levels.

Figure 3.34 TTL 7442 BCD-to-decimal decoder.

compatible with respect to these characteristics. Special ICs to interface circuits built out of different IC technologies are also available.

Designers usually select a logic family on the basis of the following characteristics:

1. Speed
2. Power dissipation
3. Fan-out
4. Availability
5. Noise immunity (noise margin)
6. Temperature range
7. Cost

Power dissipation is proportional to the current that is drawn from the power supply. The current is inversely proportional to the equivalent resistance of the circuit, which depends on the values

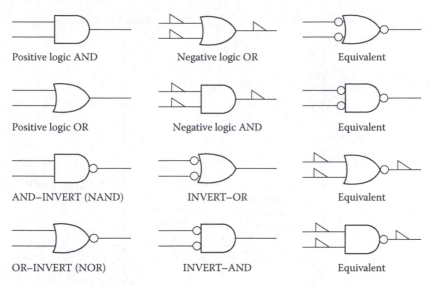

Figure 3.35 Equivalent symbols.

of individual resistors in the circuit, the load resistance, and the operating point of the transistors in the circuit. To reduce the power dissipation, the resistance should be increased. However, increasing the resistance increases the rise times of the output signal. The longer the rise time, the longer it takes for the circuit output to "switch" from one level to the other. That is, the circuit is slower. Thus, a compromise between the speed (i.e., switching time) and power dissipation is necessary. The availability of various versions of TTL, for instance, is the result of such compromises.

A measure used to evaluate the performance of ICs is the *speed–power product*. The smaller this product, the better the performance. The speed–power product of a standard TTL with a power dissipation of 10 mW and a propagation delay of 10 ns is 100 pJ.

The *noise margin* of a logic family is the deviation in the H and L ranges of the signal that is tolerated by the gates in the logic family. The circuit function is not affected so long as the noise margins are obeyed. Figure 3.36 shows the noise margins of TTL. The output voltage level of the gate stays either below $V_{L\,max}$ or above $V_{H\,min}$. When this output is connected to the input of another gate, the input treats any voltage above $V_{IH\,min}$ (2.0 V) as high and any voltage below $V_{IL\,min}$ (0.8 V) as low. Thus, TTL provides a *guaranteed noise margin* of 0.4 V. A supply voltage V_{CC} of 5 V is required. Depending on the IC technology, as the load on the gate is increased (i.e., as the number of inputs connected to the output is increased), the output voltage may enter the *forbidden region*, thereby contributing to the improper operation. Care must be taken to ensure that the output levels are maintained by obeying the fan-out constraint of the gate or by using gates with special outputs or with higher fan-out capabilities.

The fan-out of a gate is the maximum number of inputs of other gates that can be connected to its output without degrading the operation of the gate. The fan-out is a function of the current sourcing and sinking capability of the gate. When a gate provides the driving current to gate inputs connected to its output, the gate is a current *source*, whereas it is a current *sink* when the current flows from the gates connected to it into its output. It is customary to assume that the driving and driven gates are of the same IC technology and can be determined according to the current sourcing and sinking capabilities of the gates in both technologies.

When the output of a gate is high, it acts as a current source for the inputs it is driving (see Figure 3.37). As the current increases, the output voltage decreases and may enter the forbidden

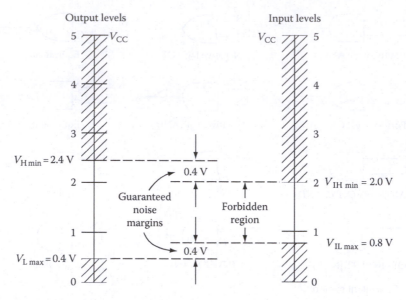

Figure 3.36 TTL noise margins.

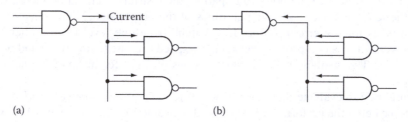

Figure 3.37 Fan-out. (a) Output high, current source. (b) Output low, current sink.

region. The driving capability is thus limited by the voltage drop. When the output is low, the driving gate acts as a current sink for the inputs. The maximum current that the output transistor can sink is limited by the *heat dissipation* limit of the transistor. Thus, the fan-out is the minimum of these driving capabilities.

A standard TTL gate has a fan-out of 10. A standard TTL buffer can drive up to 30 standard TTL gates. Various versions of TTL are available. Depending on the version, the fan-out ranges between 10 and 20. The typical fan-out of ECL gates is 25, and that of CMOS is 50.

The popularity of the IC family helps the availability. The cost of ICs comes down when they are produced in large quantities. Therefore, very popular ICs are generally available. The other availability measure is the number of types of ICs in the family. The availability of a large number of ICs in the family makes for more design flexibility.

Temperature range of operation is an important consideration, especially in such environments as military applications, automobiles, and so forth, where temperature variations are severe. Commercial ICs operate in the temperature range of 0°C–70°C, while ICs for military applications can operate in the temperature range of –55°C to +125°C.

The cost of an IC depends on the quantity produced. Popular off-the-shelf ICs have become very inexpensive. As long as off-the-shelf ICs are used in a circuit, the cost of the circuit's other components (e.g., the circuit board, connectors, and interconnections) currently is higher than that of the ICs themselves.

Table 3.1   Characteristics of Some Popular Logic Families

Characteristics	TTL	ECL	CMOS
Supply voltage (V)	5	−5.2	3 to 18
High-level voltage (V)	2 to 5	−0.95 to −0.7	3 to 18
Low-level voltage (V)	0 to 0.4	−1.9 to −1.5	0 to 0.5
Propagation delay (ns)	5 to 10	1 to 2	25
Fan-out	10 to 20	25	50
Power dissipation (mW) per gate	2 to 20	25	0.1

Circuits that are required in very large quantities can be *custom designed* and fabricated as ICs. Small quantities do not justify the cost of custom design and fabrication. There are several types of *programmable* ICs that allow a *semicustom* design of ICs for special applications. Table 3.1 summarizes the characteristics of the popular IC technologies.

3.8.4 Special Outputs

Several ICs with special characteristics are available and useful in building logic circuits. We will briefly examine these special ICs in this section, at a functional level of detail.

A gate in the logic circuit is said to be "loaded" when it is required to drive more inputs than its fan-out. Either the loaded gate is replaced with a gate of the same functionality but of a higher fan-out (if available in the IC family) or a *buffer* is connected to its output. The ICs designated as buffers (or "drivers") provide a higher fan-out than a regular IC in the logic family. For example, following are some of the TTL 7400 series of ICs that are designated as buffers:

7406 Hex inverter buffer/driver
7407 Hex buffer/driver (noninverting)
7433 Quad two-input NOR buffer
7437 Quad two-input NAND buffer

These buffers can drive approximately 30 standard TTL loads, compared to a fan-out of 10 for nonbuffer ICs.

In general, the outputs of two gates cannot be connected without damaging those gates. Gates with two special types of outputs are available that, under certain conditions, allow their outputs to be connected to realize an AND or an OR function of the output signals. The use of such gates thus results in reduced complexity of the circuit. The need for such gates is illustrated by Example 3.39.

Example 3.39

We need to design a circuit that connects one of the four inputs A, B, C, and D to the input Z. There are four control inputs (C_1, C_2, C_3, and C_4) that determine whether A, B, C, or D is connected to Z, respectively. It is also known that only one of the inputs is connected to the output at any given time. That is, only one of the control inputs will be active at any time. The function of this circuit can thus be represented as

$$Z = P \cdot C_1 + Q \cdot C_2 + R \cdot C_3 + S \cdot C_4.$$

Figure 3.38a shows the AND–OR circuit implementation of the function.

If each of the AND gates in Figure 3.38a is such that their outputs can be connected to form an OR function, the four-input OR gate can be eliminated from the circuit. Furthermore, as the number of inputs increases (along with the corresponding increase in control inputs), the circuit can be

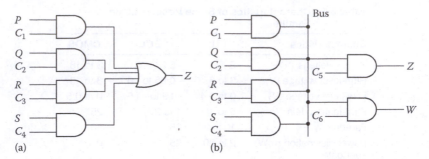

Figure 3.38   Data transfer circuits. (a) Data transfer circuit. (b) Bus.

expanded, simply by connecting the additional AND gate outputs to the common output connection. This way of connecting outputs to realize the OR function is known as the *wired-OR* connection.

In fact, the circuit can be generalized to form a *bus* that transfers the selected source signal to the selected destination. The bus shown in Figure 3.38b is simply a *common path* that is shared by all the source-to-destination data transfers. W is an additional destination. In the circuit, only one source and one destination can be activated at any given time. For example, to transfer the data from R to Z, the control signals C_3 and C_5 are activated simultaneously; similarly, Q is connected to W when C_2 and C_6 are activated and so on. All the sources are wired-OR to the bus, and only one of them will be active at any given time. However, more than one destination can be activated simultaneously if the same source signal must be transferred to several destinations. To transfer P to both Z and W, control signals C_1, C_5, and C_6 are activated simultaneously. Buses are commonly used in digital systems when a large number of source and destinations must be interconnected. Using gates whose outputs can be connected to form the wired-OR reduces the complexity of the bus interconnection.

Two types of gates are available with special outputs that can be used in this mode: (1) gates with *open-collector* (or free collector) outputs and (2) gates with *tristate outputs*.

Figure 3.39 shows two circuits. When the outputs of the TTL open-collector NAND gates are tied together, an AND is realized, as shown in Figure 3.39a. The second level will not have a gate. This fact is illustrated with the dotted gate symbol. Note that this wired-AND capability results in the realization of an AND–OR–INVERT circuit (i.e., a circuit with a first level of AND gates and an OR–INVERT or NOR gate in the second level) with only one level of gates. Similarly, when open-collector ECL NOR gates are used in the first level, we realize an OR when the outputs are tied together, as shown in Figure 3.39b. This *wired-OR* capability results in the realization of an OR–AND–INVERT circuit (i.e., a circuit with a first level of OR gates and a second level consisting of one AND–INVERT or NAND gate) with just one level of gates.

There is a limit to the number of outputs (typically about 10 in TTL) that can be tied together. When this limit is exceeded, ICs with tristate outputs can be used in place of open-collector ICs.

In addition to providing two logic levels (0 and 1), the output of these ICs can be made to stay at a *high-impedance* state. An *enable* input signal is used for this purpose. The output of the IC is either at logic 1 or 0 when it is enabled (i.e., when the enable signal is active). When it is not enabled, the output will be at the high-impedance state and is equivalent in effect to the IC not being in the circuit. It should be noted that the high-impedance state is not one of the logic levels but rather is a state in which the gate is not electrically connected to the rest of the circuit.

The outputs of tristate ICs can also be tied together to form a wired-OR as long as only one IC is enabled at any time. Note that in the case of a wired-OR (or wired-AND) formed using open-collector gates, more than one output can be active simultaneously.

Figure 3.40 shows the bus circuit of Example 3.39 using tristate gates. The control inputs now form the enable inputs of the tristate gates. Tristate outputs allow more outputs to be connected together than the open-collector ICs do.

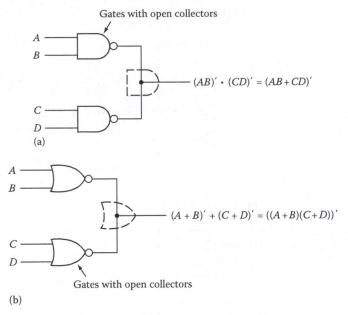

Gates with open collectors

$$(AB)' \cdot (CD)' = (AB + CD)'$$

(a)

$$(A + B)' + (C + D)' = ((A + B)(C + D))'$$

Gates with open collectors

(b)

Figure 3.39 Open-collector circuits. (a) TTL NANDs, WIRED–AND, AND–OR-INVERT. (b) ECL NORs, WIRED–OR, OR–AND–INVERT.

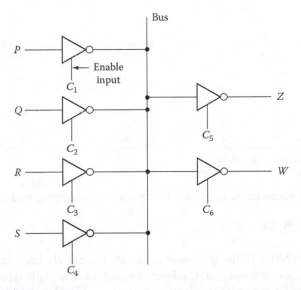

Figure 3.40 Bus using tristate gates.

Figure 3.41 shows a tristate IC (TTL 74241). This IC has eight tristate buffers, four of which are enabled by the signal on pin 1 and the other four by the signal on pin 19. Pin 1 is active-low enable, while pin 19 is active-high enable. Representative operating values are shown in Figure 3.41b.

ICs come with several other special features. Some ICs provide both the truth and complement outputs (e.g., ECL 10107); some have a STROBE input signal that needs to be active in order for the gate output to be active (e.g., TTL 7425). Thus, STROBE is an enable input. Figure 3.42 illustrates these ICs. For further details, consult the IC manufacturer manuals listed at the end of this chapter; they provide a complete listing of the ICs and their characteristics. Refer to Appendix A for details on some popular ICs.

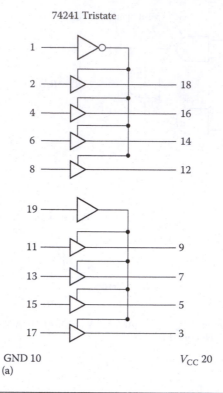

74241 Tristate

GND 10
(a)

V_{CC} 20

Control	Input	Output	Control	Input	Output
Pin 1	2	18	Pin 19	11	9
L	L	L	L	L	Z
L	H	H	L	H	Z
H	L	Z	H	L	L
H	H	Z	H	H	H

Z, High impedance.
(b)

Figure 3.41 TTL 74241 tristate buffer. (a) Circuit. (b) Representative operating values.

3.8.5 Designing with ICs

NAND–NAND or NOR–NOR implementations are extensively used in designing with ICs. NAND and NOR functions are basic to IC fabrication, and using a single type of gate in the implementation is preferable because several identical gates are available on one chip. Logic designers usually choose an available IC (decoder, adder, etc.) to implement functions rather than implement at the gate level as discussed earlier in this chapter.

Several nonconventional design approaches can be taken in designing with ICs. For example, since decoders are available as MSI components, the outputs of a decoder corresponding to the input combination where a circuit provides an output of 1 can be ORed to realize a function.

Example 3.40

An implementation of a full adder using a 3-to-8 decoder (whose outputs are assumed to be high active) is shown in Figure 3.43. (Can you implement the full adder if the outputs of the decoder are low-active as in TTL 74166?)

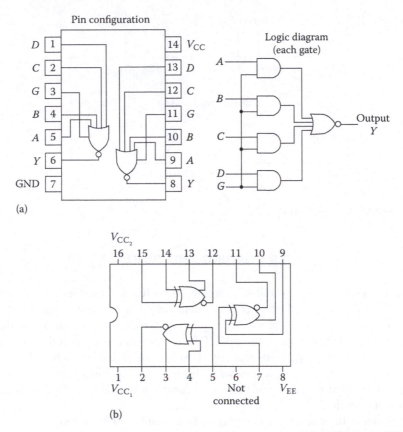

Figure 3.42 Some special ICs. (a) TTL 7425 positive NOR gates with STROBE (*G*). (b) ECL 10107 triple EXCLUSIVE-OR/NOR.

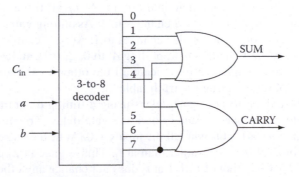

Figure 3.43 Full-adder implementation using a decoder.

Example 3.41

As another example, the BCD-to-excess-3 code converter can be realized using a 4-bit parallel binary adder IC (TTL 7483), as shown in Figure 3.44.

The BCD code word is connected to one of the two 4-bit inputs of the adder, and a constant of 3 (i.e., 0011) is connected to the output input. The output is the BCD input plus 3, which is the excess-3 code.

In practice, logic designers select available MSI and LSI components first to implement as much of the circuit as possible and use SSI components as required to complete the design.

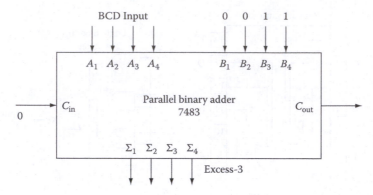

Figure 3.44   BCD-to-excess-3 converter.

3.9 LOADING AND TIMING

Two main problems to be resolved in designing with ICs are *loading* and *timing*. A loading problem occurs in the event that the output of one gate cannot *drive* the subsequent gates connected to it. In practice, there is a limit to the number of gate inputs that can be connected to the output of a gate. This limit is called the fan-out of the gate. If the fan-out limit is exceeded, the signals degrade, and hence the circuit does not perform properly. This loading problem can be solved by providing a buffer at the output of the loaded gate, either by using a separate inverting or noninverting buffer or by replacing the loaded gate with one that has a higher fan-out. The number of inputs to the gate is referred to as its fan-in.

Timing problems in general are not critical in a simple combinational circuit. However, a timing analysis is usually necessary in any complex circuit. Timing diagrams are useful in such analysis. Figure 3.45 shows the timing characteristics of a NOT gate. The x-axis indicates time. Logic values 1 and 0 are shown (as magnitudes of a voltage) on the y-axis. Figure 3.46 shows the timing diagram for a simple combinational circuit. At t_0, all three inputs A, B, and C are at 0. Hence, Z_1, Z_2, and Z are all 0. At t_1, B changes to 1. Assuming gates with no delays (ideal gates), Z_1 changes to 1 at t_1 and hence Z also changes to 1. At t_2, C changes to 1, resulting in no changes in Z_1, Z_2, or Z. At t_3, A changes to 1, pulling A' to 0, Z_1 to 0, and Z_2 to 1; Z remains at 1. This timing diagram can be expanded to indicate all the other combinations of inputs. It will then be a graphic way of representing the truth table.

We can also analyze the effects of gate delays using a timing diagram. Figure 3.47 is such an analysis for the aforementioned circuit, where the gate delays are shown as T_1, T_2, T_3, and T_4. Assume that the circuit starts at t_0 with all the inputs at 0. At t_1, B changes to 1. This change in B results in a change in Z_1 at $(t_1 + T_2)$, rather than at t_1. This change in Z_1 causes Z to change T_4 later (i.e., at $t_1 + T_2 + T_4$). Changing of C to 1 at t_2 does not change any other signal value. When A rises to 1 at t_3, A' falls to 0 at $(t_3 + T_1)$, Z_1 falls to 0 at $(t_3 + T_1 + T_2)$, and Z_2 rises to 1 at $(t_3 + T_3)$. If $T_3 > (T_1 + T_2)$, there is a time period in which both Z_1 and Z_2 are 0, contributing a "glitch" at Z. Z rises back to 1, and T_4 after Z_2 rises to 1. This momentary transition of Z to 0 might cause some problems in a complex circuit. Such hazards are the results of unequal delays in the signal paths of a circuit. They can be prevented by adding additional circuitry. This analysis indicates the utility of a timing diagram.

The hazard in the earlier example is referred to as *static 1-hazard*, since that output momentarily goes to 0 when it should remain at 1. This hazard occurs when the circuit is realized from the SOP form of the function. When the circuit is realized from the POS form, a *static 0-hazard* may occur, wherein the circuit momentarily gives an output of 1 when it should have remained at 0. A *dynamic*

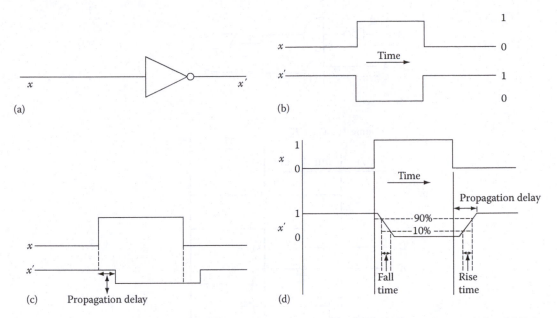

Figure 3.45 Timing characteristics and models of an IC. (a) NOT gate. (b) Ideal gate characteristics. (c) Gate with delay. (d) Timing characteristics.

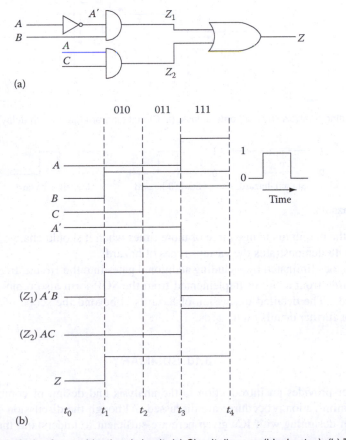

Figure 3.46 Timing analysis of a combinational circuit. (a) Circuit diagram (ideal gates). (b) Timing diagram.

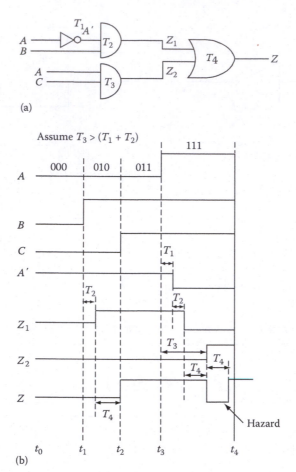

(a)

(b)

Figure 3.47 Timing analysis showing gate delays. (a) Circuit diagram (gates with delay). (b) Timing diagram.

Figure 3.48 Hazards.

hazard causes the output to change three or more times when it should change from 1 to 0 or from 0 to 1. Figure 3.48 demonstrates the various types of hazards.

Hazards can be eliminated by including additional gates into the circuit. In general, the removal of static 1-hazards from a circuit implemented from the SOP form also removes the static 0- and dynamic hazards. The detailed discussion of hazards is beyond the scope of this book. Refer to Shiva (1998) for further details on hazards.

3.10 SUMMARY

This chapter provides an introduction to the analysis and design of combinational logic circuits. Logic minimization procedures are discussed. Although the discussion on IC technology is brief, details on designing with ICs given here are sufficient to understand the information in an

IC vendor's catalog and start building simple circuits. A complete understanding of the timing and loading problem helps, but is not mandatory, to understand the rest of the material in the book. Logic circuits can also be implemented using programmable logic components such as Programmable logic array (PLA), programmable array logic (PAL), gate array (GA), and field programmable gate array (FPGA). Chapter 4 provides the details of programmable logic design.

PROBLEMS

3.1 If $A = 0$, $B = 1$, $C = 1$, and $D = 0$, find the value of F in each of the following:
   a. $F = AC' + B'$
   b. $F = A'D + B'C' + BC' + D$
   c. $F = BD' (A' + B + CD) + C'D$
   d. $F = (AB(C + D)')' (A' + B')(C' + BD')$
   e. $F = (D + ((A + B)C)')AB' + C'D (D + A(A'B + CD'))$

3.2 Draw a truth table for each of the following:
   a. $Q = X'Z + Y'Z' + XYZ$
   b. $Q = (X + Y')(X + Z)(X' + Y' + Z')'$
   c. $Q = AC'D' + D' (A + C) (A' + B)$
   d. $Q = ABC'D + A' + B' + D' + BCD$
   e. $Q = M(N + P')(N' + P)$

3.3 State if the following identities are true or false:
   a. $(X + Y + Z) (X + Y + Z') = X + Y$
   b. $A'BC' + ABC' + BC'D + BC + B' = 1$
   c. $A'C + ABC' + A'B + AB' = B'C + AvBC$
   d. $PR' + Q'R' + PR + Q'R = P + Q + R$
   e. $((P + Q')(P' + Q))' = PQ$

3.4 State if the following statements are true or false:
   a. $P' + Q' + R'$ is a sum term
   b. $AB + AC + AD$ is a canonical SOP form
   c. XYZ is included in $X + Y + Z$
   d. $(XY)' = X' + Y'$ is De Morgan's law
   e. NPQ' is a product term
   f. $X' + Y'$ is a conjunction
   g. AB is not included in $BD + AC$

3.5 State if the following functions are in normal POS form, normal SOP form, canonical POS, or canonical SOP:
   a. $F(A, B, C) = ABC + A'B'C'$
   b. $F(X, Y, Z) = Y(X' + Z')(X + Y')$
   c. $F(P, Q, R) = (PQR')(P'QR)(PQR)$
   d. $F(A, B, C) = ABC + A'B'C + B$
   e. $F(A, B, C) = AB'C + BC$
   f. $F(W, X, Y, Z) = (X + Y)(W + Y + Z')(W + X + Y + Z)$
   g. $F(P, Q, R) = PQ + QR + PR + P'Q' + Q'R' + P'R'$
   h. $F(X, Y) = X'Y + XY'$

3.6 Express each of the following functions in (1) canonical SOP form and (2) canonical POS form (Hint: Draw the truth table for each function.):
   a. $F(P, Q, R) = PQR' + P'R' + Q'R'$
   b. $F(A, B, C, D) = A'B' (C + D) + C' (A + B'C')$
   c. $F(W, X, Y, Z) = (X' + Y + Z)(W + Z') + WX'$
   d. $F(A, B, C) = (AB + ABC')(BD)'$

3.7 Express F in minterm list form in each of the following:
    a. $F(X, Y, Z) = X(Y' + Z') + YZ$
    b. $F(A, B, C, D) = (B' + C)(A + B + D')(B' + C' + D')$
    c. $F(A, B, C, D) = \Pi M(3, 5, 6, 12, 14, 15)$
    d. $F(P, Q, R) = PQ' + R' + P'R$
    e. $F(X, Y, Z) = \Pi M(3, 5, 6)$

3.8 Express F in maxterm list form in each of the following:
    a. $F(P, Q, R) = P'R' + Q(P + R)$
    b. $F(A, B, C, D) = AB(C + D') + A'C'D + B'CD'$
    c. $F(W, X, Y, Z) = \Sigma m(0, 7, 8, 9, 13)$
    d. $F(P, Q, R) = (Q + R')(P + Q' + R)$
    e. $F(A, B, C, D) = \Sigma m(10, 12, 13, 14, 15)$

3.9 Note that if two functions are equal, they will have the same minterms and the same maxterms. Use this fact to solve Problem 3.3.

3.10 Transform the following AND–OR circuit into a circuit consisting of only three NOR gates. Which circuit is more efficient? Why?

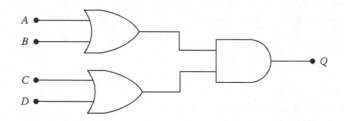

3.11 Determine the output expression for the following circuit and simplify it using algebraic manipulations.

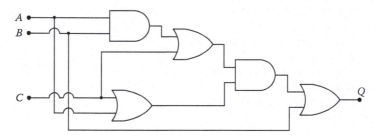

3.12 Given,

$$F_1(X, Y, Z) = \sum m(2, 4, 6, 7) + d(1)$$

$$\uparrow \qquad\qquad\qquad \uparrow$$

Minterms   Don't cares

$$\downarrow \qquad\qquad\qquad \downarrow$$

$$F_2(X, Y, Z) = \sum m(1, 4, 5, 6) + d(3, 7).$$

Find:
    a. F_1' in minterm list form
    b. F_1' in maxterm list form
    c. $F_1' \cdot F_2'$
    d. $F_1' + F_2'$

3.13 What is the output of the following XNOR gate?

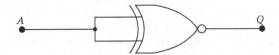

3.14 Design a combinational circuit that checks a 4-digit binary number to determine if it is an integer multiple of 3 and outputs 1 if it is and 0 otherwise.

3.15 Design a combinational circuit that outputs 1 if its 4-bit input contains an even number of 1s and 0 otherwise. Consider 0 an even number.

3.16 One way to detect single-bit errors when transmitting data is to attach a parity bit to the transferred data that indicates the number of 1s contained in the data. If the number of 1s in the received data does not agree with the parity bit, then the data were corrupted and an error output is produced. The circuit you designed in Problem 1.15 could be used as a parity bit generator. Design the parity bit checker circuit that receives the 4-bit data and the parity bit and detects single-bit errors by checking the total number of 1s in the data and the parity bit. Your circuit output (**ERROR**) should be 1 if the number of 1s in its 5-bit input is an even number and 0 if the number of 1s is an odd number.

3.17 Design a 4-to-7 decoder circuit that converts a 4-bit excess-3 into BCD on a seven segment display. There are seven outputs, each of which drives one segment of the seven segment display shown later to form the BCD digit corresponding to the input code. Assume that the display device requires a 1 to turn a segment on and a 0 to turn it off.

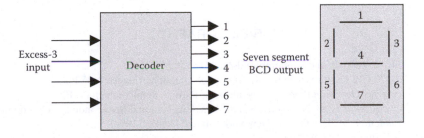

3.18 Design a 3-bit parallel adder using only half-adders and OR gates.

3.19 Show how you can use three full adders to design a 3-bit subtractor using the 2s complement system.

3.20 Two or more IC adders like TTL7483 can be cascaded to add large binary numbers. How many 7483 chips are needed to add two 12-bit numbers? Show how to connect these chips.

3.21 Design a one-digit BCD adder using two 4-bit adders (TTL7483). Remember that the value of a BCD digit cannot exceed 9 (1001). If the sum of the two added digits is greater than 9, a correction factor of 6 (0110) should be added to the sum bits and a carry of 1 is generated. Your circuit should include the logic needed to detect whenever the sum is greater than 9 so that the correction can be added in. Use one of the 4-bit adders to add the two digits and the other one to add the correction factor if needed.

3.22 Modify the 3-to-8 decoder circuit in Figure 3.15 so that all the eight outputs are active-low and add an active-high ENABLE input. When the ENABLE input is high, the decoder should function normally. When the ENABLE input is low, all the outputs of the decoder should be high.

3.23 $Q(A, B, C, D) = ABC' + A'BD + C'D' + AB' + B'C$

a. Implement the previous function using logic gates.

b. Implement the same function using a 16 input multiplexer (74150) only. (Hint: Draw the truth table for Q.)

3.24 What is the advantage of using multiplexers to implement logic functions?

3.25 Implement the following logic function using 3-to-8 decoder (74138) and any necessary gates. (Hint: Express Q in minterms.)

$$Q = AB'C + A'C' + BC'.$$

3.26 Design a circuit that can divide two 2-bit numbers using a 4–16 decoder (74154) and any other neces- sary logic gates. The circuit has two 2-bit inputs A_1A_2 and B_1B_2 and generates two 2-bit outputs, the quotient Q_1Q_2, and the remainder R_1R_2. Ignore the cases where $B_1B_2 = 00$ (division by zero).

3.27 Draw the truth table and design an octal to binary (8-to-3) encoder.

3.28 Determine the output of the encoder in the previous problem when two inputs are simultaneously active. Can you think of a way to improve your encoder?

3.29 Design a 3-to-8 decoder using 2-to-4 decoders.

3.30 Apply the following input waveforms to the circuit below and draw the output waveform (Q).

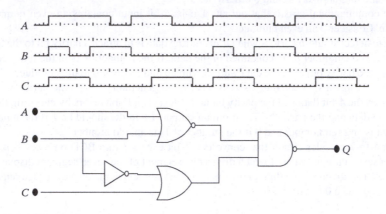

BIBLIOGRAPHY

Altera CPLDs, http://www.altera.com/literature/ds/m7000.pdf.

Cypress FLASH 370 CPLDs, http://hep.physics.lsa.umich.edu/alpha/images/7c372.pdf.

FAST TTL Logic Series Data Handbook, Sunnyvale, CA: Phillips Semiconductors, 1992.

National Advanced Bipolar Logic Databook, Santa Clara, CA: National Semiconductor Corporation, 1995.

Shiva, S.G., *Introduction to Logic Design*, New York: Marcel Dekker, 1998.

TTL Data Manual, Sunnyvale, CA: Signetics, 1987.

Wakerly, J.F., *Digital Design Principles and Practices*, Upper Saddle River, NJ: Prentice Hall, 2000.

Xilinx SRAM-based FPGAs, http://www.xilinx.com/appnotes/FPGA_NSREC98.pdf.

CHAPTER 4

Synchronous Sequential Circuits

The digital circuits we have examined so far do not possess any memory. That is, the output of the combinational circuit at any time is a function of the inputs at that time. In practice, most digital systems contain memory elements in addition to the combinational logic portion, thus making them *sequential* circuits. The output of a sequential circuit at any time is a function of its external inputs and the internal *state* at that time. The state of the circuit is defined by the contents of the memory elements in the circuit and is a function of previous states and inputs to the circuit.

Figure 4.1 shows a block diagram of a sequential circuit with m inputs, n outputs, and p internal memory elements. The output of the p memory elements combined constitutes the state of the circuit at time t (i.e., the present state). The combinational logic determines the output of the circuit at time t and provides the next-state information to the memory elements based on the external inputs and the present state. Based on the next-state information at t, the contents of all the memory elements change to the next state, which is the state at time $(t + \Delta t)$, where Δt is a time increment sufficient for the memory elements to make the transition. We will denote $(t + \Delta t)$ as $(t + 1)$ in this chapter.

There are two types of sequential circuits: *synchronous* and *asynchronous*. The behavior of a synchronous circuit depends on the signal values at discrete points of time. The behavior of an asynchronous circuit depends on the order in which the input signals change, and these changes can occur at any time.

The discrete time instants in a synchronous circuit are determined by a controlling signal, usually called a *clock*. A clock signal makes 0 to 1 and 1 to 0 transitions at regular intervals. Figure 4.2 shows two clock signals (one is the complement of the other), along with the various terms used to describe the clock. A pair of 0-to-1 and 1-to-0 transitions constitutes a *pulse*. That is, a pulse consists of a *rising edge* and a *falling edge*. The time between these transitions (edges) is the *pulse width*. The *period* (T) of the clock is the time between two corresponding edges of the clock, and the clock frequency is the reciprocal of its period. Although the clock in Figure 4.2 is shown with a regular period T, the intervals between two pulses do not need to be equal.

Synchronous sequential circuits use flip-flops as memory elements. A flip-flop is an electronic device that can store either a 0 or a 1. That is, a flip-flop can stay in one of the two logic states, and a change in the inputs to the flip-flop is needed to bring about a change of state. Typically, there will be two outputs from a flip-flop: one corresponds to the normal state (Q) and the other corresponds to the complement state (Q'). We will examine four popular flip-flops in this chapter.

Asynchronous circuits use time-delay elements (*delay lines*) as memory elements. The delay line shown in Figure 4.3a introduces a propagation delay (Δt) into its input signal. As shown in Figure 4.3b, the output signal is the same as the input signal, except that it is delayed by Δt. For instance, the 0-to-1 transition of the input at t_1 occurs on the output at t_2, Δt later. Thus, if delay lines are used as memory elements, the present-state information at time t forms their input and the next state is achieved at $(t + \Delta t)$. In practice, the propagation delays introduced by the combinational circuit's logic gates may be sufficient to produce the needed delay, thereby not necessitating a physical

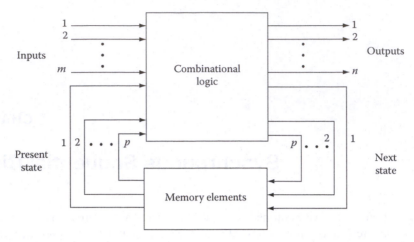

Figure 4.1 Block diagram of a sequential circuit.

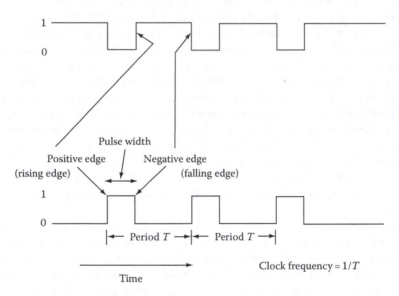

Figure 4.2 Clock.

time-delay element. In such cases, the model of Figure 4.1 reduces to a combinational circuit with feedback (i.e., a circuit whose outputs are fed back as inputs). Thus, an asynchronous circuit may be treated as a combinational circuit with feedback. Because of the feedback, the changes occurring in the output as a result of input changes may in turn contribute to further changes in inputs—and the cycle of changes may continue to make the circuit unstable if the circuit is not properly designed. In general, asynchronous circuits are difficult to analyze and design. If properly designed, however, they tend to be faster than synchronous circuits.

A synchronous sequential circuit generally is controlled by pulses from a *master clock*. The flip-flops in the circuit make a transition to the new state only when a clock pulse is present at their inputs. In the absence of a single master clock, the operation of the circuit becomes unreliable, since two clock pulses arriving from different sources at the inputs of the flip-flops cannot be guaranteed to arrive at the same time (because of unequal path delays). This phenomenon is called *clock skewing*. Clock skewing can be avoided by analyzing the delay in each path from the clock source and inserting additional gates in paths with shorter delays to make the delays of all paths equal.

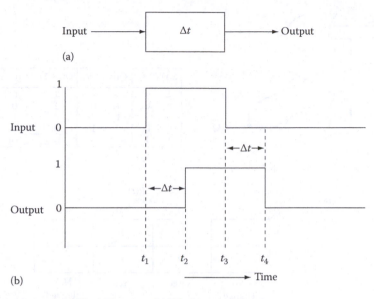

Figure 4.3 Delay element. (a) Block diagram. (b) I/O characteristics.

In this chapter, we describe the analysis and design procedures for synchronous sequential circuits. Refer to the books listed in the Bibliography section of this chapter for details on asynchronous circuits.

4.1 FLIP-FLOPS

As mentioned earlier, flip-flop is a device that can store either a 0 or a 1. When the flip-flop contains a 1, it is said to be set (i.e., $Q = 1$, $Q' = 0$) and when it contains a 0 it is reset (i.e., $Q = 0$, $Q' = 1$). We will introduce the logic properties of four popular types of flip-flops in this section.

4.1.1 Set–Reset Flip-Flop

A set–reset (SR) flip-flop has two inputs: S for setting and R for resetting the flip-flop. An ideal SR flip-flop can be built using a cross-coupled NOR circuit, as shown in Figure 4.4a. The operation of this circuit is illustrated in Figure 4.4b. When inputs $S = 1$ and $R = 0$ are applied at any time t, Q' assumes a value of 0 (one gate delay later). Since Q' and R are both at 0, Q assumes a value of 1 (another gate delay later). Thus, in two gate delay times, the circuit settles at the set state. We will denote the two gate delay times as Δt. Hence, the state at time $(t + \Delta t)$ or $(t + 1)$, designated at $Q(t + 1)$, is 1. If S is changed to 0, as shown in the second row of Figure 4.4b, an analysis of the circuit indicates that the Q and Q' values do not change. If R is then changed to 1, the output values change to $Q = 0$ and $Q' = 1$. Changing R back to 0 does not alter the output values. When $S = 1$ and $R = 1$ are applied, both outputs assume a value of 0, regardless of the previous state of the circuit. This condition is not desirable, since the flip-flop operation requires that one output always be the complement of the other. Further, if now the input condition changes to $S = 0$ and $R = 0$, the state of the circuit depends on the order in which the inputs change from 1 to 0. If S changes faster than R, the circuit attains the reset state; otherwise, it attains the set state.

Thus, the cross-coupled NOR gate circuit forms an SR flip-flop. The input condition $S = 1$ and $R = 0$ *sets* the flip-flop; the condition $S = 0$ and $R = 1$ *resets* the flip-flop. $S = 0$ and $R = 0$ constitute a "no change" condition. (The input condition $S = 1$ and $R = 1$ is not permitted to occur on the inputs.)

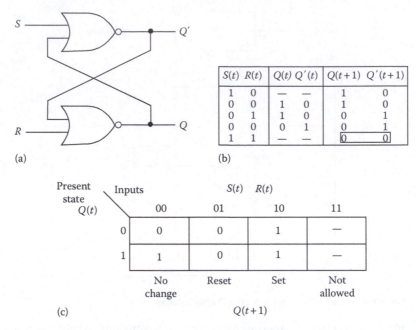

S(t)	R(t)	Q(t)	Q'(t)	Q(t+1)	Q'(t+1)
1	0	—	—	1	0
0	0	1	0	1	0
0	1	1	0	0	1
0	0	0	1	0	1
1	1	—	—	0	0

(a)   (b)

Present state Q(t) \ Inputs S(t) R(t)	00	01	10	11
0	0	0	1	—
1	1	0	1	—
	No change	Reset	Set	Not allowed

(c)   Q(t+1)

Figure 4.4 *SR* flip-flop. (a) Logic diagram. (b) Partial truth table. (c) State table.

The transitions of the flip-flop from the present state $Q(t)$ to the next state $Q(t + 1)$ for various input combinations are summarized in Figure 4.4c. This table is called the state table. It has four columns (one corresponding to each input combination) and two rows (one corresponding to each state the flip-flop can be in).

Recall that the outputs of the cross-coupled NOR gate circuit in Figure 4.4 do not change instantaneously once there is a change in the input condition. The change occurs after a delay of Δt, which is the equivalent of at least two gate delays. This is an asynchronous circuit, since the outputs change as the inputs change. The circuit is also called the *SR* latch. As the name implies, such a device is used to latch (i.e., store) the data for later use. In the circuit of Figure 4.5, the data (1 or a 0) on the INPUT line are latched by the flip-flop.

A clock input can be added to the circuit of Figure 4.4 to construct a clocked *SR* flip-flop, as shown in Figure 4.6a. As long as the clock stays at 0 (i.e., in the absence of the clock pulse), the

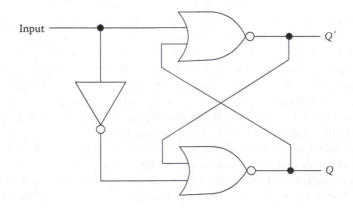

Figure 4.5 Latch.

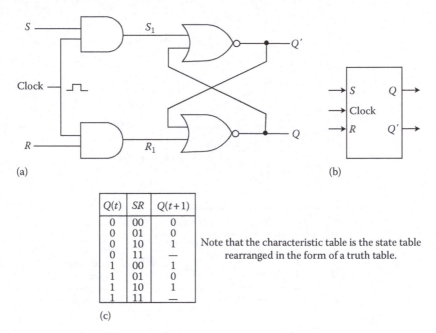

$Q(t)$	SR	$Q(t+1)$
0	00	0
0	01	0
0	10	1
0	11	—
1	00	1
1	01	0
1	10	1
1	11	—

Note that the characteristic table is the state table rearranged in the form of a truth table.

(c)

Figure 4.6 Clocked *SR* flip-flop. (a) Circuit. (b) Graphic symbol. (c) Characteristic table.

outputs of the two AND gates (S_1 and R_1) are 0, and hence, the state of the flip-flop does not change. The S and R values are impressed on the flip-flop inputs (S_1 and R_1) only during the clock pulse. Thus, the clock controls all the transitions of this synchronous circuit. The graphic symbol for the clocked *SR* flip-flop is shown in Figure 4.6b.

Given the present state and the S and R input conditions, the next state of the flip-flop can be determined, as shown in the characteristic table in Figure 4.6c. This table is obtained by rearranging the state table in Figure 4.4c so that the next state can be determined easily once the present state and the input condition are known.

Figure 4.7 shows an *SR* flip-flop formed by cross-coupling two NAND gates along with the truth table. As can be seen by the truth table, this circuit requires a 0 on its input to change the state, unlike the circuit of Figure 4.4, which required a 1.

As mentioned earlier, it takes at least two gate delay times for the state transition to occur after there has been a change in the input condition of the flip-flop. Thus, the pulse width of the clock controlling the flip-flop must be at least equal to this delay, and the inputs should not change until the transition is complete. If the pulse width is longer than the delay, the state transitions resulting from the first input condition change are overridden by any subsequent changes in the inputs during the clock pulse. If it is necessary to recognize all the changes in the input conditions, however, the pulse width must be short enough. The pulse width and the clock frequency thus must be adjusted to accommodate the flip-flop circuit transition time and the rate of input change. (The timing characteristics and requirements of flip-flops are further discussed later in this chapter.)

Figure 4.8 shows the graphic symbol for an *SR* flip-flop with both clocked (S and R) and asynchronous inputs (*preset* and *clear*). A clock is not required to activate the flip-flop through the asynchronous inputs. Asynchronous (or direct) inputs are not used during the regular operation of the flip-flop. They generally are used to initialize the flip-flop to either the set or the reset state. For instance, when the circuit power is turned on, the state of the flip-flops cannot be determined. The direct inputs are used to initialize the state, either manually through a "master clear" switch or through a power-up circuit that pulses the direct input of all the flip-flops in the circuit.

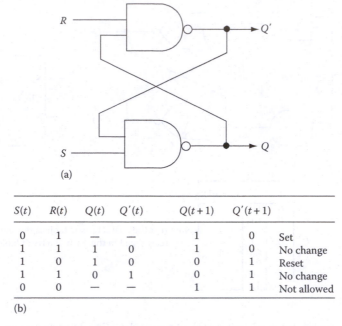

$S(t)$	$R(t)$	$Q(t)$	$Q'(t)$	$Q(t+1)$	$Q'(t+1)$	
0	1	—	—	1	0	Set
1	1	1	0	1	0	No change
1	0	1	0	0	1	Reset
1	1	0	1	0	1	No change
0	0	—	—	1	1	Not allowed

(b)

Figure 4.7   *SR* flip-flop formed by cross-coupled NAND gates. (a) Circuit. (b) Truth table.

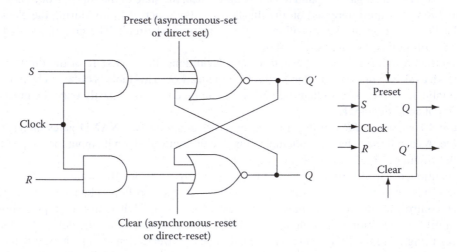

Figure 4.8   Clocked *SR* flip-flop with preset/clear.

We will now examine three other commonly used flip-flops. The preset, clear, and clocked configurations discussed earlier apply to these flip-flops as well. In the remaining sections of this chapter, if a reference to a signal does not show a time associated with it, it is assumed to be the current time *t*.

4.1.2 *D* Flip-Flop

Figure 4.9 shows a *D* (delay or data) flip-flop and its state table. The *D* flip-flop assumes the state of the *D* input; that is, $Q(t+1) = 1$ if $D(t) = 1$ and $Q(t+1) = 0$ if $D(t) = 0$. The function of this

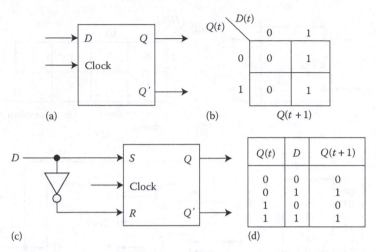

Figure 4.9 *D* flip-flop. (a) Graphic symbol. (b) State table. (c) *D* flip-flop realized from an *SR* flip-flop. (d) Characteristic table.

flip-flop is to introduce a unit delay (Δt) in the signal input at *D*. Hence, this flip-flop is known as a *delay flip-flop*. It is also called a *data flip-flop*, since it stores the data on the *D* input line.

The *D* flip-flop is a modified *SR* flip-flop that is obtained by connecting *D* to an *S* input and *D'* to an *R* input, as shown in Figure 4.9c. A clocked *D* flip-flop is also called a gated *D*-latch, in which the clock signal gates the data into the latch.

The next state of the *D* flip-flop is the same as the data input at any time, regardless of the present state. This is illustrated by the characteristic table shown in Figure 4.9d.

4.1.3 *JK* Flip-Flop

The *JK* flip-flop is a modified *SR* flip-flop in that the $J = 1$ and $K = 1$ input combination is allowed to occur. When this combination occurs, the flip-flop complements its state. The *J* input corresponds to the *S* input, and the *K* input corresponds to the *R* input of an *SR* flip-flop. Figure 4.10 shows the graphic symbol, the state table, the characteristic table, and the realization of a *JK* flip-flop using an *SR* flip-flop. (See Problem 4.1 for a hint on how to convert one type of flip-flop into another.)

4.1.4 *T* Flip-Flop

Figure 4.11 shows the graphic symbol, state table, and the characteristic table for a *T* (toggle) flip-flop. This flip-flop complements its state when $T = 1$ and remains in the same state as it was when $T = 0$. A *T* flip-flop can be realized by connecting the *J* and *K* inputs of a *JK* flip-flop, as shown in Figure 4.11d.

4.1.5 Characteristic and Excitation Tables

The characteristic table of a flip-flop is useful in the analysis of sequential circuits, since it provides the next-state information as a function of the present state and the inputs. The characteristic tables of all the flip-flops are given in Figure 4.12 for ready reference.

The *excitation tables* (or the *input tables*) shown in Figure 4.13 for each flip-flop are useful in designing sequential circuits, since they describe the excitation (or input condition) required to

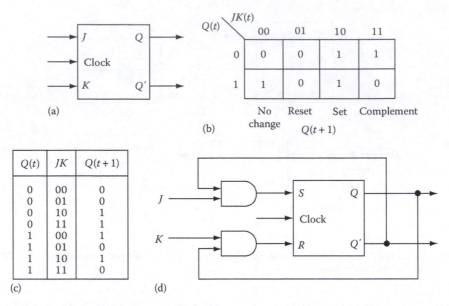

Figure 4.10 *JK* flip-flop. (a) Graphic symbol. (b) State table. (c) Characteristic table. (d) Realization of *JK* flip-flop using an *SR* flip-flop.

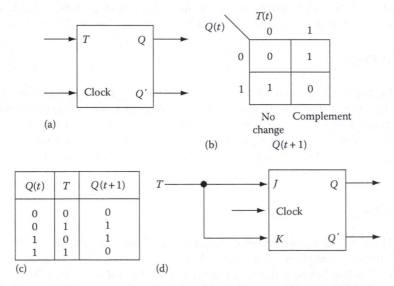

Figure 4.11 *T* flip-flop. (a) Graphic symbol. (b) State table. (c) Characteristic table. (d) *T* flip-flop realized from a *JK* flip-flop.

bring the state transition of the flip-flop from $Q(t)$ to $Q(t+1)$. These tables are derived from the state tables of the corresponding flip-flops. Consider the state table for the *SR* flip-flop shown in Figure 4.4. For a transition of the flip-flop from state 0 to 0 (as shown by the first row of the state table), the input can be either $SR = 00$ or 01. That is, an *SR* flip-flop makes a transition from 0 to 0 as long as S is 0 and R is either 1 or 0. This excitation requirement is shown as $SR = 0d$ in the first row of the

Q(t)	SR	Q(t+1)
0	00	0
0	01	0
0	10	1
0	11	—
1	00	1
1	01	0
1	10	1
1	11	—

(a)

Q(t)	D	Q(t + 1)
0	0	0
0	1	1
1	0	0
1	1	1

(b)

Q(t)	JK	Q(t + 1)
0	00	0
0	01	0
0	10	1
0	11	1
1	00	1
1	01	0
1	10	1
1	11	0

(c)

Q(t)	T	Q(t + 1)
0	0	0
0	1	1
1	0	1
1	1	0

(d)

Figure 4.12 Characteristic tables. (a) SR flip-flop. (b) D flip-flop. (c) JK flip-flop. (d) T flip-flop.

Q(t)	Q(t + 1)	SR	D	JK	T
0	0	0d	0	0d	0
0	1	10	1	1d	1
1	0	01	0	d1	1
1	1	d0	1	d0	0

Figure 4.13 Excitation tables. *Note: d*, don't care (0 or 1).

excitation table. A transition from 0 to 1 requires an input of $SR = 10$, a transition from 1 to 0 requires $SR = 01$, and that from 1 to 1 requires $SR = d0$. Thus, the excitation table accounts for all four possible transitions. The excitation tables for the other three flip-flops are similarly derived.

4.2 TIMING CHARACTERISTICS OF FLIP-FLOPS

Consider the cross-coupled NOR circuit forming an SR flip-flop. Figure 4.14 shows a timing diagram, assuming that the flip-flop is at state 0 to begin with. At t_1, input S changes from 0 to 1. In response to this, Q changes to 1 at t_2, a delay of Δt after t_1. Δt is the time required for the circuit to settle to the new state. At t_3, S goes to 0, with no change in Q. At t_4, R changes to 1, and hence, Q changes to 0, Δt time later at t_5. At t_6, R changes to 0, with no effect on Q.

Note that the S and R inputs should each remain at their new data value at least for time Δt for the flip-flop to recognize the change in the input condition (i.e., to make the state transition). This time is called the hold time. Now consider the timing diagram for the clocked SR flip-flop shown in Figure 4.15. The clock pulse width is $w = t_8 - t_1$. S changes to 1 at t_2, and in response to it,

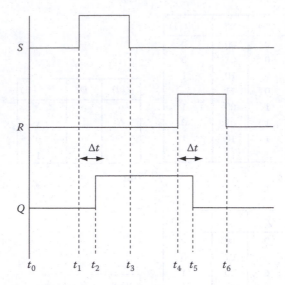

Figure 4.14 Timing diagram for an *SR* flip-flop (unclocked).

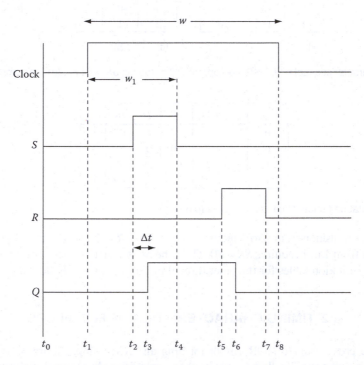

Figure 4.15 Timing diagram for an *SR* flip-flop (clocked).

Q changes to 1 at t_3, Δt later. Since the clock pulse is still at 1 when R changes to 1 at t_5, Q changes to 0 at t_6. If the pulse width were to be $w_1 = t_4 - t_1$, only the change in S would have been recognized. Thus, in the case of a clocked *SR* flip-flop, the clock pulse width should at least equal Δt for the flip-flop to change its state in response to a change in the input. If the pulse width is greater than Δt, S and R values should change no more than once during the clock pulse, since the flip-flop circuit will keep

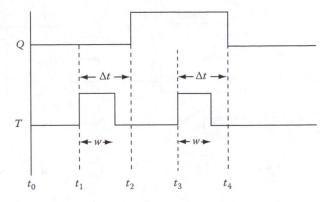

Figure 4.16   Timing diagram for a T flip-flop.

changing states as a result of each input change and registers only the last input change. As such, the clock pulse width is a critical parameter for the proper operation of the flip-flop.

Consider the timing diagram of Figure 4.16 for a T flip-flop. When T changes to 1 at t_1, the flip-flop changes from its original state of 0 at t_2, Δt time later. Since the T flip-flop circuit contains a feedback path from its outputs to its input, if the T input stays at 1 longer (i.e., beyond t_2), the output would be fed back to the input and the flip-flop changes state again. To avoid this oscillation, w must always be less than Δt.

In order to avoid such problems resulting from clock pulse width, flip-flops in practice are designed either as *master–slave flip-flops* or as *edge-triggered flip-flops*, which are described next.

4.2.1 Master–Slave Flip-Flops

The master–slave configuration is shown in Figure 4.17a. Here, two flip-flops are used. The clock controls the separation and connection of the circuit inputs from the inputs of the master, and the inverted clock controls the separation and connection of slave inputs from the master outputs. In practice, the clock signal takes a certain amount of time to make the transition from 0 to 1 and 1 to 0, as shown by t_R and t_F, respectively, in the timing diagram (Figure 4.17b). As the clock changes from 0 to 1, at point A, the slave stage is disconnected from the master stage; at point B, the master is connected to the circuit inputs and changes its state based on the inputs; at point C, as the clock makes its transition from 1 to 0, the master stage is isolated from the inputs; and at D, the slave inputs are connected to the outputs of the master stage. The slave flip-flop changes its state based on its inputs, and the slave stage is isolated from the master stage at A again. Thus, the master–slave configuration results in at most one state change during each clock period, thereby avoiding the race conditions resulting from clock pulse width.

Note that the inputs to the master stage can change after the clock pulse, while the slave stage is changing its state without affecting the operation of the master–slave flip-flop, since these changes are not recognized by the master until the next clock pulse. Master–slave flip-flops are especially useful when the input of a flip-flop is a function of its own output.

Consider the timing diagram of Figure 4.17c for a master–slave flip-flop. Here, S and R initially are both 0. The flip-flop should not change its state during the clock pulse. However, a glitch in the S line while clock is high sets the master stage, which in turn is transferred to the slave stage, resulting in an erroneous state. This is called one's catching problem and can be avoided by ensuring that all the input changes are complete and the inputs stable well before the leading edge of the clock. This timing requirement is known as the setup time (t_{setup}). That is, $t_{setup} > w$, the clock pulse width. This can be achieved either by a narrow clock pulse width (which

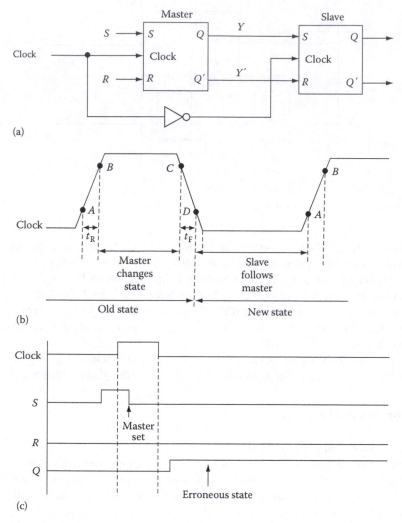

Figure 4.17   Master–slave flip-flop. (a) Circuit. (b) Timing diagram. (c) One's catching problem.

is difficult to guarantee) or by a large setup time (which reduces the flip-flop's operating speed). Edge-triggered flip-flops are preferred over master–slave flip-flops because of the one's catching problem associated with the latter.

4.2.2 Edge-Triggered Flip-Flops

Edge-triggered flip-flops are designed so that they change their state based on input conditions at either the rising or the falling edge of the clock. The rising edge of the clock triggers a positive edge-triggered flip-flop (as shown in Figure 4.15), and the falling edge of the clock triggers a negative edge-triggered flip-flop. Any change in input values after the occurrence of the triggering edge will not bring about a state transition in these flip-flops until the next triggering edge.

Figure 4.18a shows the most common trailing edge-triggered flip-flop circuit, built out of three cross-coupled NOR flip-flops. Flip-flops 1 and 2 serve to set the inputs to the third flip-flop at appropriate values based on the clock and D inputs. Consider the clock and D input transitions shown in Figure 4.18b. Flip-flop 3 is reset initially (i.e., $Q = 0$). When the clock goes to

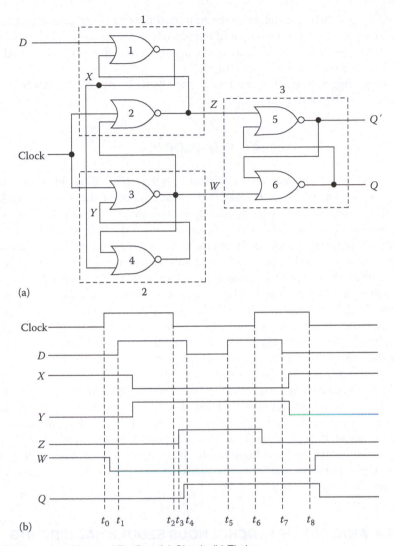

Figure 4.18   Trailing edge-triggered flip-flop. (a) Circuit. (b) Timing.

1 at t_0, point W goes to 0 (after one gate delay). Since Z remains at 0, flip-flop 3 does not change its state. While the clock pulse is at 1, X and Y follow D (i.e., $X = D'$ and $Y = D$), as at t_1. When the clock changes to 0 at t_2, Z changes to 1 (after a delay) at t_3, but W remains 0. Consequently, flip-flop 3 changes its state to 1 (after a delay). Thus, the state change is brought about by the trailing edge of the clock.

While the clock is at 0, change in the D input does not change either Z or W, as shown at t_4 and t_5.

Z is 1 and W is 0 at t_6 when the clock changes to 1 and Z goes to 0. At t_7, the D input changes to 0. Since the clock is at 1, X and Y change accordingly (after a delay). These changes result in changing W to 1 at the trailing edge of the clock at t_8. Since $Z = 0$ and $W = 1$, flip-flop 3 changes to 0.

As can be seen by the timing diagram shown in Figure 4.18b, after the trailing edge of the clock pulse, either W or Z becomes 1. When Z is 1, D is blocked at gate 1. When W is 1, D is blocked at gates 2 and 4. This blocking requires one gate delay after the trailing edge of the clock, and hence, D should not change until this blocking occurs. Thus, the hold time is one gate delay.

Note that the total time required for the flip-flop transition is three gate delays after the trailing edge—one gate delay for W and Z to change and two gate delays after that for Q and Q' to change.

Thus, if we add t_{setup} of two gate delays to the transition time of three gate delays, the minimum clock period is five gate delays if the output of the flip-flop is fed directly back to its input. If additional circuitry is in the feedback path, as is usually the case with most sequential circuits, the minimum clock period increases correspondingly.

A leading edge-triggered flip-flop can be designed using cross-coupled NAND circuits along the lines of the circuit shown in Figure 4.18.

4.3 FLIP-FLOP ICs

Appendix A provides the details of several flip-flop ICs. TTL 7474 is a dual-positive edge-triggered D flip-flop IC. The triangle at the clock input in the graphic symbol indicates positive edge triggering. Negative edge triggering is indicated by a triangle along with a bubble at the input as shown in the case of 74LS73. S_D and R_D are active-low asynchronous set and reset inputs, respectively, and operate independent of the clock. The data on the D input are transferred to the Q output at the positive clock edge. The D input must be stable one setup time (20 ns) prior to the positive edge of the clock. The positive transition time of the clock (i.e., from 0.8 to 2.0 V) should be equal to or less than the clock-to-output delay time for the reliable operation of the flip-flop.

The 7473 and 74LS73 are dual master–slave JK flip-flop ICs. The 7473 is positive pulse triggered (note the absence of the triangle on the clock input in the graphic symbol). JK information is loaded into the master while the clock is high and transferred to the slave during the clock high-to-low transition. For the conventional operation of this flip-flop, the JK inputs must be stable while the clock is high. The flip-flop also has direct set and reset inputs.

The 74LS73 is a negative edge-triggered flip-flop. The JK inputs must be stable one setup time (20 ns) prior to the high-to-low transition of the clock. The flip-flop has an active-low direct reset input.

The 7475 has four bistable latches. Each 2-bit latch is controlled by an active-high enable input (E). When enabled, the data enter the latch and appear at the Q outputs. The Q outputs follow the data inputs as long as the enable is high. The latched outputs remain stable as long as the enable input stays low. The data inputs must be stable one setup time prior to the high-to-low transition of the enable, for the data to be latched.

4.4 ANALYSIS OF SYNCHRONOUS SEQUENTIAL CIRCUITS

The analysis of a synchronous sequential circuit is the process of determining the functional relation that exists between its outputs, its inputs, and its internal states. The contents of all the flip-flops in the circuit combined determine the internal state of the circuit. Thus, if the circuit contains n flip-flops, it can be in one of the 2^n states. Knowing the present state of the circuit and the input values at any time t, we should be able to derive its next state (i.e., the state at time $t + 1$) and the output produced by the circuit at t.

A sequential circuit can be described completely by a state table that is very similar to the ones shown for flip-flops in Figures 4.4 through 4.11. For a circuit with n flip-flops, there will be 2^n rows in the state table. If there are m inputs to the circuit, there will be 2^m columns in the state table. At the intersection of each row and column, the next state and the output information are recorded. A state diagram is a graphical representation of the state table, in which each state is represented by a circle and the state transitions are represented by arrows between the circles. The input combination that brings about the transition and the corresponding output information are shown on the arrow. Analyzing a sequential circuit thus corresponds to generating the state table and the state diagram for the circuit. The state table or state diagram can be used to determine the output sequence generated by the circuit for a given input sequence if the *initial state* is known. It is important to note that for proper operation, a sequential circuit must be in its initial state before the inputs to it can be applied. Usually the power-up circuits are used to initialize the circuit to the appropriate state when the power is turned on. The following examples will illustrate the analysis procedure.

Example 4.1

Consider the sequential circuit shown in Figure 4.19a. There is one circuit input X and one output Z, and the circuit contains one D flip-flop. To analyze the operation of this circuit, we can trace through the circuit for various input values and states of the flip-flop to derive the corresponding output and the next-state values. Since the circuit has one flip-flop, it has two states (corresponding to $Q = 0$ and $Q = 1$). The present state is designated as Y (in this circuit, $Y = Q$). The output Z is a function of the state of the circuit Y and the input X at any time t. The next state of the circuit $Y(t + 1)$ is determined by the value of the D input at time t. Since the memory element is a D flip-flop, $Y(t + 1) = D(t)$.

Assume that $Y(t) = 0$ and $X(t) = 0$. Tracing through the circuit, we can see that $Z = 0$ and $D = 0$. Hence, $Z(t) = 0$ and $Y(t + 1) = 0$. The aforementioned state transition and output are shown in the top left blocks of the next-state and output tables in Figure 4.19b. Similarly, when $X(t) = 0$ and $Q(t) = 1$, $Z(t)$ and $D(t) = 1$, making $Y(t + 1) = 1$, as shown in the bottom left blocks of these tables. The other two entries in these tables are similarly derived by tracing through the circuit. The two tables are merged into one, entry by entry, as shown in Figure 4.19c to form the so-called transition table for the circuit. Each block of this table corresponds to a present state and an input combination. Corresponding next-state and output information is entered in each block, separated by a slash mark. From the table in Figure 4.19c, we can see that if the state of the circuit is 0, it produces an output of 0 and stays in the state 0 as long as the input values are 0s. The first 1 input condition sends the circuit to a 1 state with an output of 1. Once it is in the 1 state, the circuit

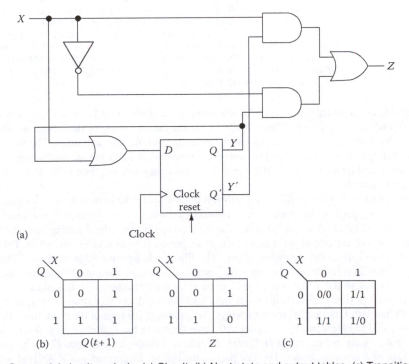

Figure 4.19 Sequential circuit analysis. (a) Circuit. (b) Next-state and output tables. (c) Transition table.

(continued)

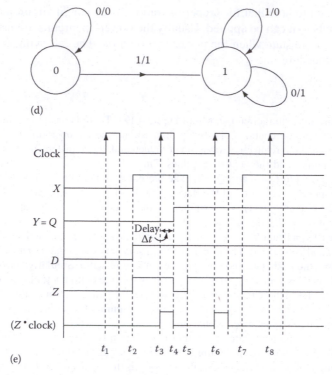

(d)

(e)

Figure 4.19 (continued) Sequential circuit analysis. (d) State diagram. (e) Timing diagram for level input. *Notes*: (1) Arrows are shown on the rising edge of the clock to emphasize that the circuit uses a flip-flop triggered by that edge. (2) The clock at t_3 triggers Q to change from 1 to 0, since X is 1. But the state change is delayed by Δt and occurs before the corresponding falling edge of the clock. For simplicity, we have assumed that Δt = clock width. Hence, the transition of Q is shown at t_4. In practice, the clock width will be slightly larger than Δt. (3) We will also ignore the data setup and hold times in timing diagrams in this chapter for simplicity. We will simply use the value of the flip-flop inputs at the triggering clock edge to determine the transition of the flip-flop.

remains in that state regardless of what the input is, but the output is the complement of the input X. The state diagram in Figure 4.19d illustrates the operation of the circuit graphically. Here, each circle represents a state, and an arrow represents the state transition. The input value corresponding to that transition and the output of the circuit at that time are represented on each arrow, separated by a slash. We will generalize the state diagram and state table representations in the next example.

Since the flip-flop is a positive edge-triggered flip-flop, the state transition takes place only when the rising edge of the clock occurs. However, this fact cannot be explicitly shown in the aforementioned tables. A timing diagram can be used to illustrate these timing characteristics. The operation of the circuit for a four-clock pulse period is shown in Figure 4.19e. The input X is assumed to make the transition shown. The flip-flop is positive edge triggered. The state change occurs as a result of the rising edge but takes a certain amount of time after the edge occurs. It is assumed that the new state is attained by the falling edge of the clock. Assuming an initial state of 0, at t_1, D is 0 and Z is 0, thus not affecting Y. At t_2, X changes to 1, and hence, D and Z change to 1, neglecting the gate delays. At t_3, the positive edge of the clock starts the state transition, making Q reach 1 by t_4. Z goes to 0, since Q goes to 1 at t_4. X changes to 0 at t_5, thereby bringing Z to 1 but not changing D. Hence, Y remains 1 through t_8, as does D. Corresponding transitions of Z are also shown. The last part of the timing diagram shows Z ANDed with the clock to illustrate the fact that output Z is valid only during the clock pulse. If X is assumed to be valid only during clock pulses, the timing diagram represents an input sequence of $X = 0101$ and

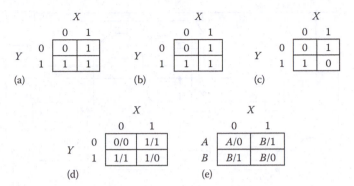

Figure 4.20 Sequential circuit analysis. (a) $D = X + Y$. (b) $Y(t + 1)$. (c) $Z = XY' + X'Y$. (d) Transition table. (e) State table.

the corresponding output sequence of $Z = 0110$. Note that the output sequence is the 2s complement of the input sequence with the LSB occurring first and the MSB last.

The circuit-tracing procedure discussed here can be adopted for the analysis of simple circuits. As the circuit becomes more complex, the tracing becomes cumbersome. We will now illustrate a more systematic procedure for the derivation of the state table (and hence the state diagram).

From the analysis of the combinational circuit of Figure 4.19a, the flip-flop *input equation* (or *excitation equation*) is

$$D = X + Y$$

and the circuit *output equation* is

$$Z = XY' + X'Y.$$

These equations express the inputs of the flip-flops in the circuit and the circuit output as functions of the circuit inputs and the state of the circuit at time t.

Figure 4.20a shows the truth table for D in the form of a table whose rows correspond to the present state Y of the circuit and whose columns correspond to the combination of values for the circuit input X. Knowing the value of D, at each block in this table (i.e., at each present-state–input pair), we can determine the corresponding next state $Y(t + 1)$ of the flip-flop, using the D flip-flop excitation table. In the case of the D flip-flop, the next state is the same as D; hence, the next-state table shown in Figure 4.20b is identical to Figure 4.20a. The truth table for the output Z is represented by the table in Figure 4.20c. Tables in Figure 4.20b and c are merged to form Figure 4.20d, which is usually called a transition table, since it shows the state and output transitions in the binary form. In general, each state in the circuit is designated by an alphabetic character. The transition table is converted into the state table by the assignment of a letter to each state. The state table obtained by assigning A to 0 and B to 1 is shown in Figure 4.20e.

Example 4.2 illustrates the analysis procedure for a more complex circuit.

Example 4.2

Consider the sequential circuit shown in Figure 4.21a. There are two clocked flip-flops, one input line X and one output line Z. The Q outputs of the flip-flops (Y_1, Y_2) constitute the present state of the circuit at any time t. The signal values J and K determine the next state $Y_1(t + 1)$ of the JK flip-flop. The value of D determines the next state $Y_2(t + 1)$ of the D flip-flop. Since both flip-flops are triggered by the same clock, their transitions occur simultaneously.

In practice, only one type of flip-flop is used in a circuit. Since an IC generally contains more than one flip-flop, using a single type of flip-flop in the circuit reduces the component count and

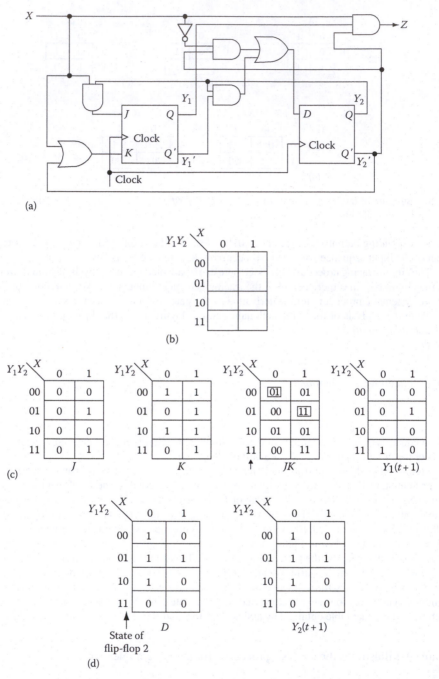

Figure 4.21 Sequential circuit. (a) Circuit diagram. (b) State table format. (c) *JK* flip-flop transitions. (d) *D* flip-flop transitions.

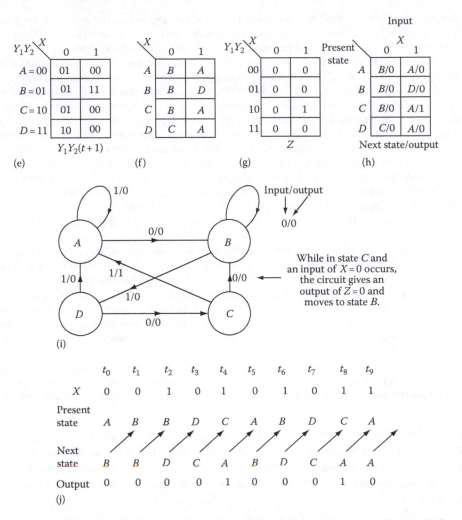

Y_1Y_2\X	0	1
$A = 00$	01	00
$B = 01$	01	11
$C = 10$	01	00
$D = 11$	10	00

$Y_1Y_2(t+1)$

(e)

\X	0	1
A	B	A
B	B	D
C	B	A
D	C	A

(f)

Y_1Y_2\X	0	1
00	0	0
01	0	0
10	0	1
11	0	0

Z

(g)

Input Present state \ X	0	1
A	$B/0$	$A/0$
B	$B/0$	$D/0$
C	$B/0$	$A/1$
D	$C/0$	$A/0$

Next state/output

(h)

While in state C and an input of $X = 0$ occurs, the circuit gives an output of $Z = 0$ and moves to state B.

(i)

	t_0	t_1	t_2	t_3	t_4	t_5	t_6	t_7	t_8	t_9
X	0	0	1	0	1	0	1	0	1	1
Present state	A	B	B	D	C	A	B	D	C	A
Next state	B	B	D	C	A	B	D	C	A	A
Output	0	0	0	0	1	0	0	0	1	0

(j)

Figure 4.21 (continued) Sequential circuit. (e) Transition table. (f) Next state table. (g) Output. (h) State table. (i) State diagram. (j) Input/Output sequence.

hence the cost of the circuit. Different types of flip-flops have been used in the examples in this chapter for illustration purposes only.

By analyzing the combination portion of the circuit, we can derive the flip-flop input (or excitation) equations:

$$J = XY_2 \quad \text{and} \quad K = X + Y_2'$$

and

$$D = Y_1'Y_2 + X'Y_2'.$$

The circuit output equation is

$$Z = XY_1Y_2'.$$

Because there are two flip-flops, there will be four states, and hence, the state table shown in Figure 4.21b will have four rows. The rows are identified with the state vectors $Y_1Y_2 = 00, 01, 10,$ and 11. Input X can be 0 or 1, and hence, the state table will have two columns.

The next-state transitions of the JK flip-flop are derived in Figure 4.21c. Note again that the tables shown in Figure 4.21c for J and K are simply the rearranged truth tables that reflect the combination of input values along the columns and the combination of present-state values along the rows. Tables for J and K are then merged, entry by entry, to derive the composite JK table that makes it easier to derive the state transition of the JK flip-flop. Although both $Y_1(t)$ and $Y_2(t)$ values are shown in this table, only the $Y_1(t)$ value is required to determine $Y_1(t + 1)$ once the J and K values are known. For example, in the boxed entry at the top left of the table, $J = 0$ and $K = 1$; hence, from the characteristic table for the JK flip-flop (Figure 4.13), the flip-flop will reset, and $Y_1(t + 1)$ will equal 0. Similarly, in the boxed entry in the second row, $J = 1$ and $K = 1$. Hence, the flip-flop complements its state. Since the $Y_1(t)$ value corresponding to this entry is 0, $Y_1(t + 1) = 1$. This process is repeated six more times to complete the $Y_1(t + 1)$ table.

The analysis of D flip-flop transitions is shown in Figure 4.21d; $Y_2(t + 1)$ is derived from these transitions.

The transition tables for the individual flip-flops are then merged, column by column, to form the transition table in Figure 4.21e for the entire circuit. Instead of denoting the states by primary state vectors, letter designations can be used for each state, as shown in Figure 4.21e, and the next-state table shown in Figure 4.21f is derived. The output table in Figure 4.21g is derived from the circuit output equation shown earlier. The output and next-state tables are then merged to form the state table Figure 4.21h for the circuit. The state table thoroughly depicts the behavior of the sequential circuit. The state diagram for the circuit derived from the state table is shown in Figure 4.21i.

Assuming a starting (or initial) state of A, the input sequence and the corresponding next-state and output sequences are shown in Figure 4.21j. Note that the output sequence indicates that the output is 1 only when the circuit input sequence is 0101. Thus, this is a 0101 sequence detector.

Note that once a sequence is detected, the circuit goes into the starting state A and another complete 0101 sequence is required for the circuit to produce an output of 1. Can the state diagram in the aforementioned example be rearranged to make the circuit detect *overlapping* sequences? That is, the circuit should produce an output if a 01 occurs directly after the detection of a 0101 sequence. For example,

$$X = 000101010100101$$

$$Z = 000001010100001.$$

The state diagram shown in Figure 4.22 accomplishes this.

If the *starting* or the initial state of the circuit and the input sequence are known, the state table and the state diagram for a sequential circuit permit a functional analysis whereby the circuit's behavior can be determined. Timing analysis is required when a more detailed analysis of the circuit parameters is needed. Figure 4.23 shows the timing diagram for the first five clock pulses in Example 4.2.

This heuristic analysis can be formalized into the following step-by-step procedure for the analysis of synchronous sequential circuits:

1. Analyze the combinational part of the circuit to derive excitation equations for each flip-flop and the circuit output equations.
2. Note the number of flip-flops (p) and determine the number of states (2^p). Express each flip-flop input equation as a function of circuit inputs and the present state and derive the transition table for each flip-flop, using the characteristic table for the flip-flop.
3. Derive the next-state table for each flip-flop and merge them into one, thus forming the transition table for the entire circuit.

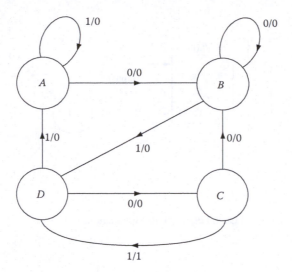

Figure 4.22 0101 overlapped sequence detector.

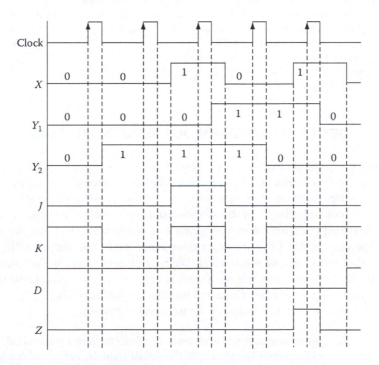

Figure 4.23 Timing diagram for Example 4.2.

4. Assign names to state vectors in the transition table to derive the next-state table.
5. Using the output equations, draw a truth table for each output of the circuit and rearrange these tables into the state table form. If there is more than one output, merge the output tables column by column to form the circuit output table.
6. Merge the next-state and output tables into one to form the state table for the entire circuit.
7. Draw the state diagram.

It is not always necessary to follow this analysis procedure. Some circuits yield a more direct analysis, as shown in Example 4.3.

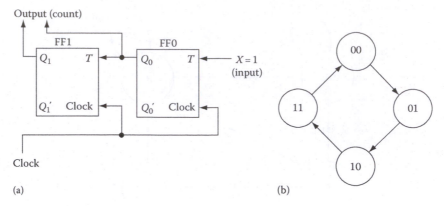

Figure 4.24 Modulo-4 counter. (a) Circuit diagram. (b) State diagram.

Example 4.3

Figure 4.24a shows a sequential circuit made up of two T flip-flops. Recall that the T flip-flop complements its state when the T input is 1. Hence, if input X is held at 1, FF0 complements at each clock pulse, while FF1 complements only when Q_0 is 1 (i.e., every second clock pulse). The state diagram is shown in Figure 4.24b. Note that the output of the circuit is the state itself. As can be seen in Figure 4.24b, this is a modulo-4 counter. Refer to Example 4.5 for another modulo-4 counter design. (What would be the count sequence if FF0 and FF1 in Figure 4.23 were falling-edge triggered?)

4.5 DESIGN OF SYNCHRONOUS SEQUENTIAL CIRCUITS

The design of a sequential circuit is the process of deriving a logic diagram from the specification of the circuit's required behavior. The circuit's behavior is often expressed in words. The first step in the design is then to derive an exact specification of the required behavior in terms of either a state diagram or a stable table. This is probably the most difficult step in the design, since no definite rules can be established to derive the state diagram or a stable table. The designer's intuition and experience are the only guides. Once the description is converted into the state diagram or a state table, the remaining steps become mechanical. We will examine the classical design procedure through the examples in this section. It is not always necessary to follow this classical procedure, as some designs lend themselves to more direct and intuitive design methods.

The classical design procedure consists of the following steps:

1. Deriving the state diagram (and state table) for the circuit from the problem statement.
2. Deriving the number of flip-flops (p) needed for the design from the number of states in the state diagram by the formula $2^{p-1} < n \leq 2^p$, where n is the number of states.
3. Deciding on the types of flip-flops to be used. (This often simply depends on the type of flip-flops available for the particular design.)
4. Assigning a unique p-bit pattern (state vector) to each state.
5. Deriving the state transition table into p tables, one for each flip-flop.
6. Separating the state transition table into p tables, one for each flip-flop.
7. Deriving an input table for each flip-flop input using the excitation tables (Figure 4.13).
8. Deriving input equations for each flip-flop input and the circuit output equations.
9. Drawing the circuit diagram.

This design procedure is illustrated by the following examples.

Example 4.4

Design a sequential circuit that detects an input sequence of 1011. The sequences may overlap. A 1011 sequence detector gives an output of 1 when the input completes a sequence of 1011. Because overlap is allowed, the last 1 in the 1011 sequence could be the first bit of the next 1011 sequence, and hence, a further input of 011 is enough to produce an output of 1. That is, the input sequence 1011011 consists of two overlapping sequences.

Figure 4.25a shows a state diagram. The sequence starts with a 1. Assuming a starting state of A, the circuit stays in A as long as the input is 0, producing an output of 0 waiting for an input of 1 to occur. The first 1 input takes the circuit to a new state B. So long as the inputs continue to be 1, the circuit has to stay in B waiting for a 0 to occur to continue the sequence and hence to move to a new state C. While in C, if a 0 is received, the sequence of inputs is 100, and the current sequence cannot possibly lead to 1011. Hence, the circuit returns to state A. But if a 1 is received while in C, the circuit moves to a new state D, continuing the sequence. While in D, a 1 input completes the 1011 sequence. The circuit gives a 1 output and goes to B in preparation for a 011 for a new sequence. A 0 input while at B creates the possibility of an overlap, and hence, the circuit returns to C so that it can detect the 11 subsequence required completing the sequence.

Drawing a state diagram is purely a process of trial and error. In general, we start with an initial state. At each state, we move either to a new state or to one of the already-reached states, depending on the input values. The state diagram is complete when all the input combinations are tested and accounted for at each state. Note that the number of states in the diagram cannot be predetermined and various diagrams typically are possible for a given problem statement. The amount of hardware needed to synthesize the circuit increases with the number of states. Thus, it is desirable to reduce the number of states if possible.

The state table for the example is shown in Figure 4.25b. Since there are four states, we need two flip-flops. The four 2-bit patterns are arbitrarily assigned to the states, and the transition table in Figure 4.25c and the output table in Figure 4.25d are drawn. From the output table, we can see that

$$Z = XY_1Y_2.$$

We will use an SR flip-flop and a T flip-flop. It is common practice to use one kind of flip-flop in a circuit. Different kinds are used here for illustration purposes only. The transitions of flip-flop 1 (SR) extracted from Figure 4.25c are shown in the first table $Y_1(t + 1)$ of Figure 4.25e. From these transitions and using the excitation tables for the SR flip-flop (Figure 4.14), S and R excitations are derived (e.g., the 0-to-0 transition of the flip-flop requires that $S = 0$ and $R = 0$, and a 1-to-0 transition requires that $S = 0$ and $R = 1$). The S and R excitations (which are functions of X, Y_1, and Y_2) are separated into individual tables, and the excitation equations are derived. These equations are shown in Figure 4.25e. The input equation for the second flip-flop (T) is similarly derived and is shown in Figure 4.25f. The circuit diagram is shown in Figure 4.25g.

The complexity of a sequential circuit can be reduced by simplifying the input and output equations. In addition, a judicious allocation of state vectors to the states also reduces circuit complexity.

Example 4.5

Modulo-4 up–down counter: The modulo-4 counter will have four states: 0, 1, 2, and 3. The input X to the counter controls the direction of the count: up if $X = 0$ and down if $X = 1$. The state of the circuit (i.e., the count) itself is the circuit output. The state diagram is shown in Figure 4.26.

Derivation of a state diagram for this counter is straightforward, since the number of states and the transitions are completely defined by the problem statement. Note that only input values are shown on the arcs—the output of the circuit is the state of the circuit itself. We will need two flip-flops, and the assignment of 2-bit vectors for states is also defined to be 00, 01, 10, and 11 to correspond to 0, 1, 2, and 3, respectively. The state table and transition table are shown in Figure 4.26b and c, respectively. The control signal is X; $X = 0$ indicates "count up"; $X = 1$ indicates "count down."

We will use a JK flip-flop and a D flip-flop. The input equations are derived in Figure 4.26d, and Figure 4.26e shows the circuit.

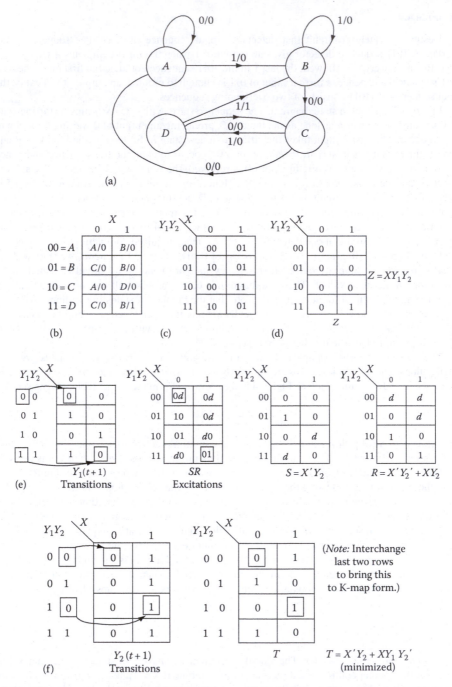

Figure 4.25 1011 sequence detector. (a) State diagram. (b) State table. (c) Transition table. (d) Output table. (e) Flip-flop 1 (*SR*). (f) Flip-flop 2 (*T*).

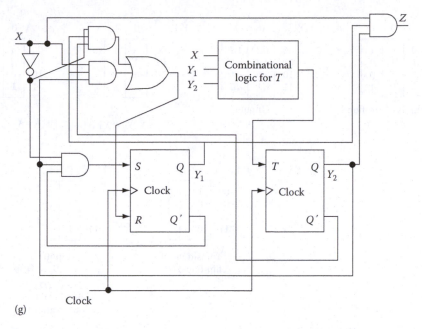

(g)

Figure 4.25 (continued) 1011 sequence detector. (g) Circuit diagram. *Note:* Interchange the last two rows of *S* and *R* tables to bring them in to the form of K-map and then minimize.

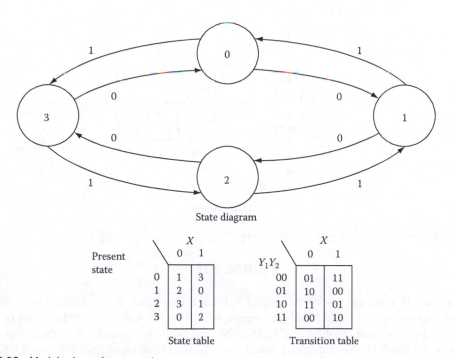

Figure 4.26 Modulo-4 up–down counter.

(*continued*)

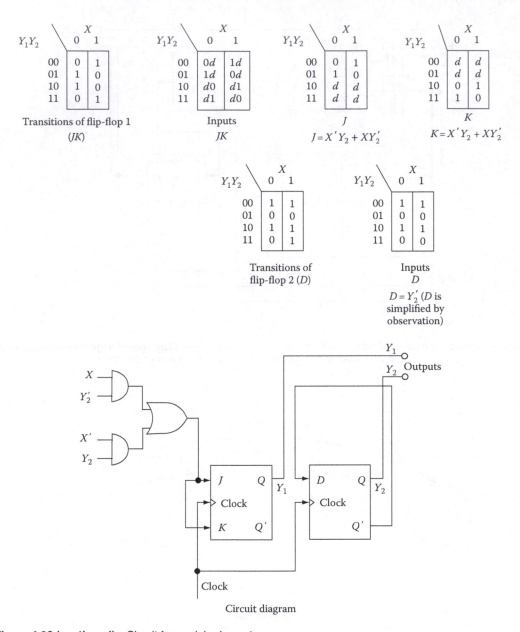

Figure 4.26 (continued)   Circuit for modulo-4 counter.

4.6 REGISTERS

A register is a storage device capable of holding binary data; it is a collection of flip-flops. An n-bit register is built of n flip-flops. Figure 4.27 shows a 4-bit register built out of four D flip-flops. There are four input lines, IN_1, IN_2, IN_3, and IN_4, each connected to the D input of the corresponding flip-flop. When a clock pulse occurs, the data from input lines IN_1 through IN_4 enter the register. The clock thus loads the register. The loading is in parallel, since all four bits enter the register simultaneously. Q outputs of flip-flops are connected to output lines OUT_4, and hence, all four bits of data (i.e., contents of the register) are available simultaneously (i.e., in parallel) on the output lines. Hence, this is a parallel-input (parallel-load), parallel-output register.

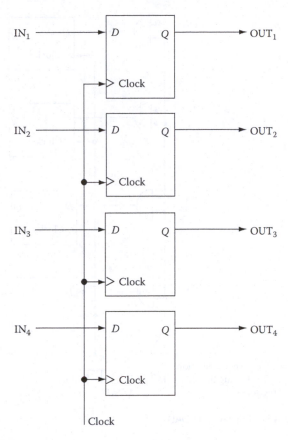

Figure 4.27   4-bit register.

At each clock pulse, a 4-bit data input enters the register from input lines IN_1 through IN_4 and remains in the register until the next clock pulse. The clock controls the loading of the register, as shown in Figure 4.28. LOAD must be 1 for data to enter the register. The CLEAR signal shown in Figure 4.28 leads zeros into the register (i.e., *clears* the register). Clearing a register is a common operation and is normally done through the asynchronous clear input (RESET) provided on flip-flops. Thus, when asynchronous inputs are used, a clearing operation can be done independent of the clock. The CLEAR signal shown in Figure 4.29 clears the register asynchronously. In this scheme, the CLEAR input must be set to 1 for clearing the register and should be brought to 0 to deactivate RESET and allow resumption of normal operation.

In the circuit of Figure 4.29, the data on the input lines enter the register at each rising edge of the clock. Therefore, we need to make sure that the data on input lines are always valid. Figure 4.30 shows a 4-bit register built out of *JK* flip-flops in which the data on input lines enter the register at the rising edge of the clock only when the LOAD is 1. When LOAD is 0, since both *J* and *K* will be 0, the contents of the register remain unchanged.

Note that we have used two AND gates to gate the data in each flip-flop in Figure 4.30. If we try to eliminate one of these gates by gating the IN line rather than *J* and *K* lines, we will be introducing unequal delays on *J* and *K* lines due to the extra NOT gate on the *K* line. As long as the clock edge appears after both the inputs are settled, this unequal delay would not present a problem. If it does not appear, the circuit will not operate properly. (Can you replace the *JK* flip-flops in Figure 4.30 with *D* flip-flops? What changes are needed to retain the contents of the register unaltered when LOAD is 0, in this case?)

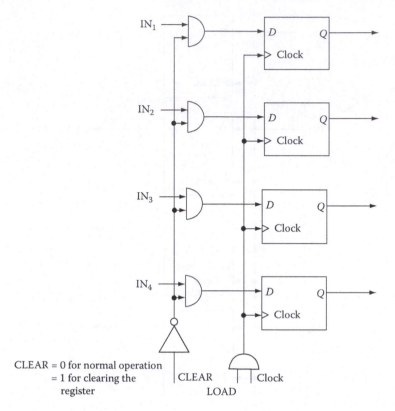

Figure 4.28 4-bit register with CLEAR and LOAD.

Choice of the type of flip-flop used in the circuit depends on the mode of implementation of the synchronous circuit. In designing ICs, it is often more efficient to use JK flip-flops with gated inputs. In implementing the circuit with MSI parts, it is common to use D flip-flops with gated clocks.

A common operation on the data in a register is to shift it either right or left. Figure 4.31 shows a 4-bit shift register built out of D flip-flops. The Q output of each flip-flop is connected to the D input of the flip-flop to its right. At each clock pulse, content of d_1 moves to d_2, content of d_2 moves to d_3, and that of d_3 moves into d_4, simultaneously. Hence, this is a right-shift register. The output of the shift register at any time is the content of d_4. If the input is set to 1, a 1 is entered into d_1 at each shift pulse. Similarly a 0 can be loaded by setting the input to 0.

An n-bit shift register can be loaded serially in n clock pulses, and the contents of the register can be output serially using the output 1 in n clock pulses. Note that in loading an n-bit right-shift register serially, the least significant bit must be entered first, followed by more significant bit values. Also, if the output line of the shift register is connected to its input line, the contents of the register "circulate" at each shift pulse.

Figure 4.32 shows a shift register with a serial-input, serial-output, parallel-output, circulate (left or right), and shift (left or right) capabilities. Each D input receives data from the flip-flop to the right or left of it, depending on whether the DIRECTION signal is 0 or 1, respectively. Since right and left signals are complements, the register can shift only in one direction at any time. The register performs shift or circulate based on the value of the MODE signal. When in shift mode, the data on left input enter the register if DIRECTION is 1, and the data on right input enter the register if DIRECTION is 0. The content of the register can be output in parallel through $0_1 0_2 0_3 0_4$ or in a serial mode through 0_4.

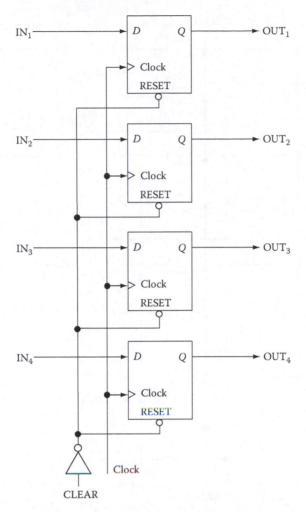

Figure 4.29 4-bit register with asynchronous CLEAR. *Note:* Low-active reset.

A 3-bit shift register using *SR* flip-flops with right-shift and parallel or serial-input capabilities is shown in Figure 4.33. Refer to Appendix A for details of shift register ICs. The following examples illustrate the utility of shift registers in sequential circuit design.

Example 4.6: 1011 Sequence Detector Using a Shift Register

It is possible to design sequential circuits without following the classical design procedure discussed earlier in this chapter. We will illustrate this by designing the 1011 sequence detector of Example 4.4, using a 4-bit right-shift register. The circuit is shown in Figure 4.34a. The input is Z_1 when the shift register contains the sequence 1011 (i.e., 1101 left to right, since the input enters the shift register from the left). Z_1 is gated by the clock to produce Z. The same clock is used as the shift control for the shift register. Note that the shift register is activated by the rising edge of the clock.

The operation of the circuit is illustrated by the timing diagram in Figure 4.34b. At t_1, X is 1. Hence, a 1 enters the shift register. We will assume that the shift register contents are settled by t_2. Similarly, at t_3, a 0 enters the shift register, and at t_5 and t_7, 1 enters, resulting in the sequence 1011 being the content of the shift register. Hence, Z_1 goes to 1 by t_8. Since a 0 enters

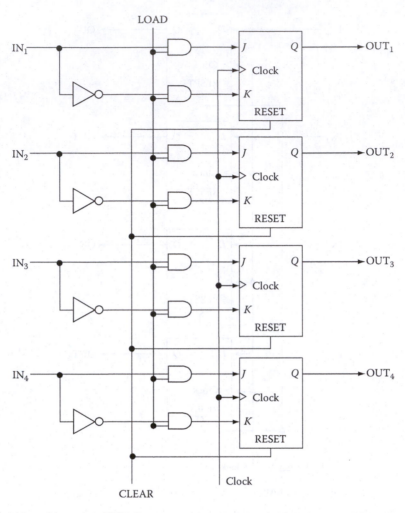

Figure 4.30   4-bit register using *JK* flip-flops.

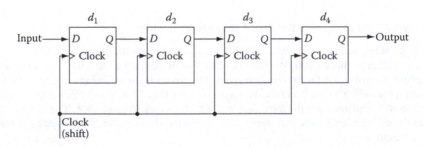

Figure 4.31   4-bit shift register.

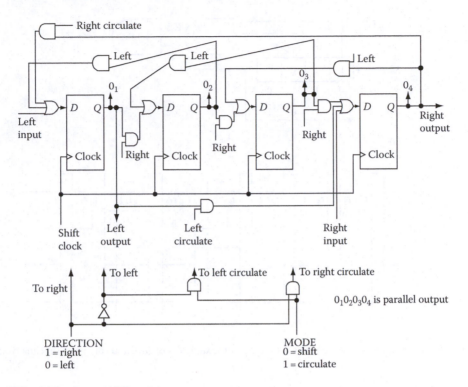

Figure 4.32 4-bit universal shift register.

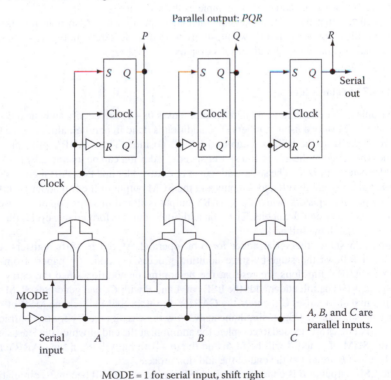

Figure 4.33 3-bit shift register.

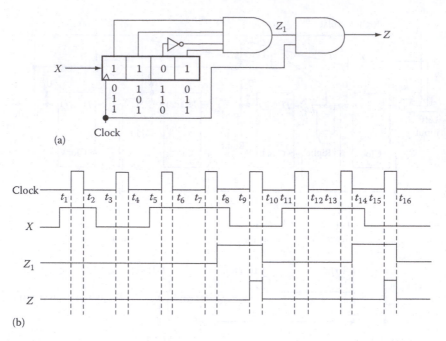

Figure 4.34 1011 Sequence detector using a shift register (Example 4.6). (a) Circuit diagram. (b) Timing diagram.

the shift register at t_9, Z_1 will be 0 by t_{10}. Thus, Z is 1 during t_9 and t_{10}. Note that Z will be 1 again during t_{15} and t_{16}, since the entry of 011 completes the sequence.

It is required that the shift register be cleared to begin with. Also note that this circuit requires four flip-flops compared to two flip-flops needed by the circuit in Figure 4.25, while the combinational portion of this circuit is less complex.

Example 4.7: Serial Adder

The ripple adder circuit described in Chapter 3 uses $(n - 1)$ full adders and one half-adder to generate the SUM of two n-bit numbers. The addition is done in parallel, although the CARRY has to propagate from the LSB position to the MSB position. This CARRY propagation delay determines the speed of the adder. If a slower speed of addition can be tolerated by the system, a serial adder can be utilized. The serial adder uses one full adder and two shift registers. The bits to be added are brought to full adder inputs and the SUM output of the full adder is shifted into one of the operand registers, while the CARRY output is stored in a flip-flop and is used in the addition of next most significant bits. The n-bit addition is thus performed in n cycles (i.e., n clock pulse times) through the full adder.

Figure 4.35 shows the serial adder for 6-bit operands stored in shift registers A and B. The addition follows the stage-by-stage addition process (as done on paper) from LSB to MSB. The CARRY flip-flops are reset at the beginning of addition since the carry into the LSB position is 0. The full adder adds the LSBs of A and B with C_{in} and generates SUM and C_{out}. During the first shift pulse, C_{out} enters the CARRY flip-flop, SUM enters the MSB of A, and A and B registers are shifted right, simultaneously. Now the circuit is ready for the addition of the next state. Six pulses are needed to complete the addition, at the end of which the least significant n-bit of the SUM of A and B will be in A, and the $(n + 1)$th bit will be in the CARRY flip-flop. Operands A and B are lost at the end of the addition process.

If the LSB output of B is connected to its MSB input, then B will become a circulating shift register. The content of B is unaltered due to addition, since the bit pattern in B after the sixth

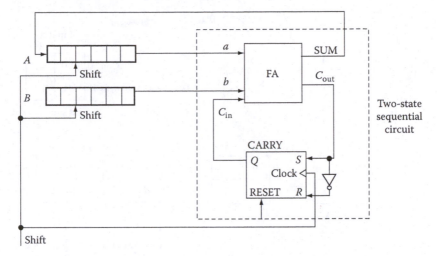

Figure 4.35 Serial adder.

shift pulse will be the same as that before addition began. If the value of A is also required to be preserved, A should be converted into a circulating shift register, and the SUM output of the full adder must be fed into a third shift register.

The circuit enclosed by dotted lines in Figure 4.35 is a sequential circuit with one flip-flop and hence two states, two input lines (a and b), and one output line (SUM). C_{in} is the present-state vector and C_{out} is the next-state vector.

Example 4.8: Serial 2s Complementer

A serial 2s complementer follows the COPY–COMPLEMENT algorithm for 2s complementing the contents of a register (see Chapter 2). Recall that the algorithm examines the bits of the register starting from the LSB. All consecutive zero bits as well as the first nonzero bit are "copied" as they are and the remaining bits until and including MSB are "complemented" to convert a number into its 2s complement. An example is as follows:

1 0 1 1 0 1 0	1 0 0 0 An 11-bit number
COMPLEMENT	COPY
0 1 0 0 1 0 1	1 0 0 0 Its 2s complement

There are two distinct operations in this algorithm: COPY and COMPLEMENT. Further, the transition from a copying mode to complementing mode is brought about by the first nonzero bit. The serial complementer circuit must be a sequential circuit because the mode of operation at any time depends on whether the nonzero bit has occurred or not. There will be two states and hence one flip-flop in the circuit; one input line on which bits of the number to be complemented are entering starting with LSB and one output line that is either the copy or the complement of the input. The circuit starts in the COPY state and changes to COMPLEMENT state when the first nonzero bit enters through the input line. At the beginning of each 2s complement operation, the circuit must be set to the COPY state.

Figure 4.36 shows the design of this circuit. From the state diagram in Figure 4.36a, the state table in Figure 4.36b is derived, followed by the output equation in Figure 4.36c and the input equations for the SR flip-flop in Figure 4.36e. The circuit is shown in Figure 4.36f. The circuit is set to the COPY state by resetting the flip-flop before each complementation.

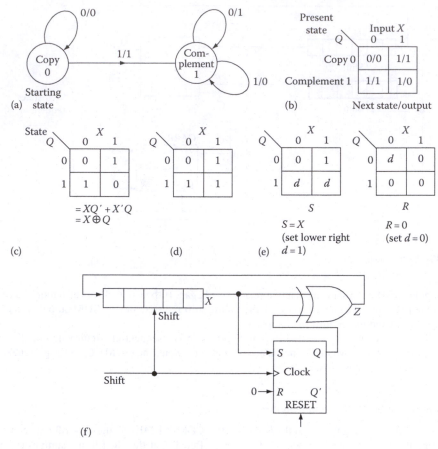

Figure 4.36 Serial 2s complementer. (a) State diagram. (b) State table. (c) Output Z. (d) Next state. (e) Input equations. (f) Circuit diagram.

4.7 REGISTER TRANSFER LOGIC

Manipulation of data in most digital systems involves the movement of data between registers. The data movement can be accomplished either in serial or in parallel. Transfer of an n-bit data from one register to the other (each of n-bit) takes n shift pulses if done in serial mode, while it is done in one pulse time in parallel mode. A data path that can transfer 1-bit between the registers is sufficient for serial mode operation. This path is repeatedly used for transferring all n-bit one at a time. For a parallel, transfer scheme, n such data paths are needed. Thus, a serial transfer scheme is less expensive in terms of hardware and slower than the parallel scheme.

Figure 4.37 shows the parallel transfer scheme from a 4-bit register A to a 4-bit register B. Here, X is a control signal. The data transfer occurs at the rising edge of the clock pulse only when X is 1. When X is 0, the J and K inputs of all the flip-flops in register B are at 0, and hence, the contents of register B remain unchanged even if the rising edge of the clock pulse occurs. In a synchronous digital circuit, control signals such as X are also synchronized with the clock. The timing requirements for the proper operation of the parallel transfer circuit are shown in Figure 4.37b. At t_1, the control signal X goes to 1, and at t_2, the register transfer occurs. X can be brought to 0 at t_2.

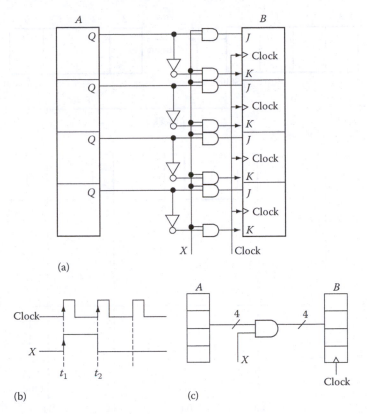

(a)

(b)

(c)

Figure 4.37 Parallel transfer scheme. (a) Circuit. (b) Timing. (c) Symbol.

We will represent the aforementioned transfer scheme by the diagram in Figure 4.37c. Each register is represented by a rectangle along with the clock input. The inputs and the outputs of the registers are shown as required. The number 4 shown next to the / indicates the number of bits transferred and hence the number of parallel lines needed, each line controlled by X. This is a common convention used to represent multiple bits of any signal in a circuit diagram.

Figure 4.38 shows the serial transfer scheme. Here, A and B are two 4-bit shift registers. They shift right in response to the shift clock. As seen by the timing diagram in Figure 4.38b, we need four clock pulses to complete the transfer, and the control signal X must stay at 1 during all the four clock pulses.

All data processing done in the processing unit of a computer is accomplished by one or more register transfer operations. It is often required that data in one register be transferred into several other registers or a register receives its inputs from one or more other registers. Figure 4.39 shows two schemes for transferring the contents of either register A or register B into register C. When the control signal "A to C" is on, contents of A are moved into C. When "B to C" signal is on, contents of B are moved into C. Only one control signal can be active at any time. This can be accomplished by using the true and the complement of the same control signal to select one of the two transfer paths.

Note that it takes at least two gate delay times after the activation of the control signal for the data from either A or B to reach the inputs of C. The control signal must stay active during this time and until the occurrence of the rising edge of the clock pulse that gates the data into C.

Figure 4.39b shows the use of a 4-line 2-to-1 multiplexer to accomplish the register transfer required in Figure 4.39a.

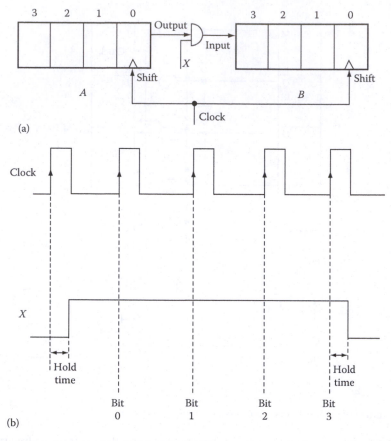

Figure 4.38 Serial transfer scheme. (a) Circuit. (b) Timing.

4.8 REGISTER TRANSFER SCHEMES

When it is required to transfer data between several registers to complete a processing sequence in a digital computer, one of two transfer schemes is generally used: (1) point-to-point and (2) bus. In a point-to-point scheme, there will be one transfer path between each of the two registers involved in the data transfer. In a bus scheme, one common path is time shared for all register transfers.

4.8.1 Point-to-Point Transfer

The hardware required for a point-to-point transfer between three 3-bit registers A, B, and C is shown in Figure 4.40. Only a few of the paths are shown. "A to C" and "B to C" are control signals used to bring the data transfer. This scheme allows more than one transfer to be made at the same time (in parallel) because independent data paths are available. For example, the control signals "A to C" and "C to B" can both be enabled at the same time. The disadvantage of the scheme is that the amount of hardware required for the transfer increases rapidly as additional registers are included, and each new register is connected to other registers through newer data paths. This growth makes the scheme too expensive; hence, a point-to-point scheme is used only when fast, parallel operation is desired.

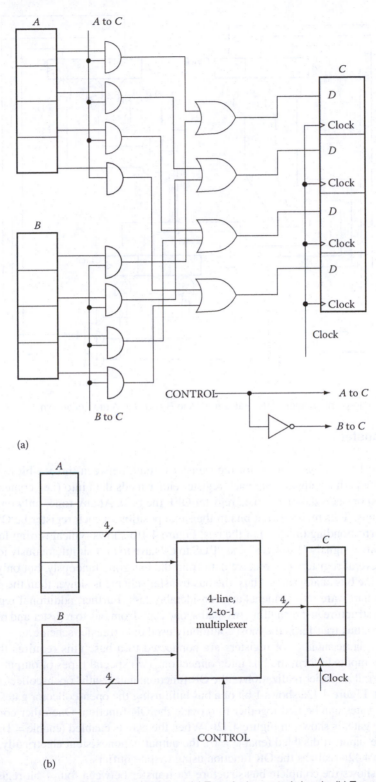

Figure 4.39 Transfer from multiple-source registers. (a) Transfer by control signals. (b) Transfer by multiplexer.

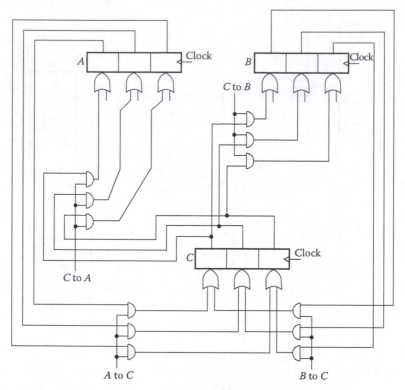

Figure 4.40 Point-to-point transfer. *Note:* Paths from *A* to *B* and *B* to *A* are not shown.

4.8.2 Bus Transfer

Figure 4.41a shows a bus scheme for the transfer of data between three 3-bit registers. A bus is a common data path (highway) that each register either feeds data into (i.e., contents of the register ON the bus) or takes data from (i.e., register OFF the bus). At any time, only one register can put data on the bus. This requires that bits in the same position in each register be ORed and connected to the corresponding bit (line) of the bus. Figure 4.41b shows typical timing for the transfer from *A* to *C*. Control signals "*A* to BUS" and "BUS to *C*" have to be 1 simultaneously for the transfer to take place. Several registers can receive data from the bus simultaneously, but only one register can put data on the bus at any time. Thus, the bus transfer scheme is slower than the point-to-point scheme, but the hardware requirements are considerably less. Further, additional registers can be added to the bus structure just by adding two paths, one each from bus to register and register to bus. For these reasons, bus transfer is the most commonly used data transfer scheme.

In practice, a large number of registers are connected to a bus. This requires the use of OR gates with many inputs to form the bus interconnection. Two special types of outputs available on certain gates permit an easier realization of the OR function: gates with "open-collector" output and "tristate" output. Figure 4.42a shows 1 bit of a bus built using the open-collector gates. The outputs of these special gates can be tied together to provide the OR function. One other commonly used device, a tristate gate, is shown in Figure 4.42b. When the gate is enabled (enable = 1), the output is a function of the input; if disabled (enable = 0), the output is nonexistent electrically. The scheme shown in Figure 4.42a realizes the OR function using tristate buffers.

Figure 4.43 shows the complete bus structure for transfer between four 4-bit registers *A*, *B*, *C*, and *D*. The SOURCE register is connected to the bus by enabling the appropriate tristate, as selected by the outputs of the source control 2-to-4 decoder. The DESTINATION register is selected by the

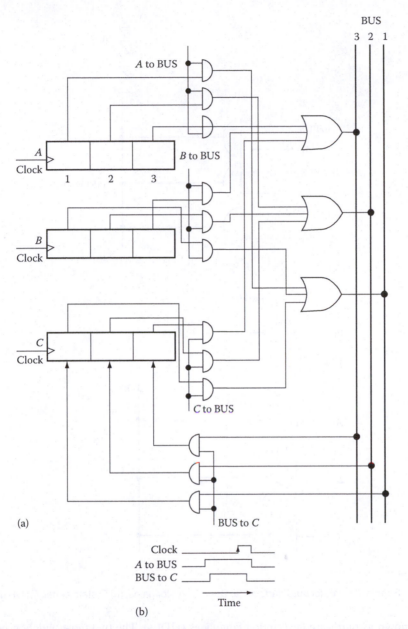

Figure 4.41 Bus transfer. (a) Bus structure. (b) Timing. *Note:* BUS to *A* and BUS to *B* are not shown.

outputs of the destination control 2-to-4 decoder. Note that a 4-line 4-to-1 multiplexer could also be used to form the connections from the registers to the bus.

4.9 REGISTER TRANSFER LANGUAGES

Since register transfer is the basic operation in a digital computer, several register transfer notations have evolved over the past decade. These notations, complete enough to describe any digital computer at the register transfer level, have come to be known as register transfer languages. Since they are used to describe the hardware structure and behavior of digital systems, they are more

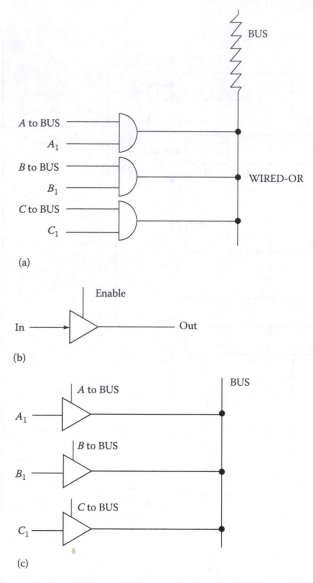

Figure 4.42   Special devices for bus interface. (a) Open-collector gates. (b) Tristate buffer. (c) OR using tristates.

generally known as hardware description languages (HDLs). The two most widely used HDLs are very-high-speed IC (VHSIC)-hardware design language (VHDL) and Verilog. References listed in the Bibliography section at the end of this chapter provide details on these languages. For the purposes of this book, a relatively simple HDL is used and the details are shown later.

Tables 4.1 and 4.2 show the basic operators and constructs of our HDL. The general format of a register transfer is

$$\text{Destination} \leftarrow \text{Source},$$

where
   "Source" is a register or an expression consisting of registers and operators
   "Destination" is a register or a concatenation (linked series) of registers

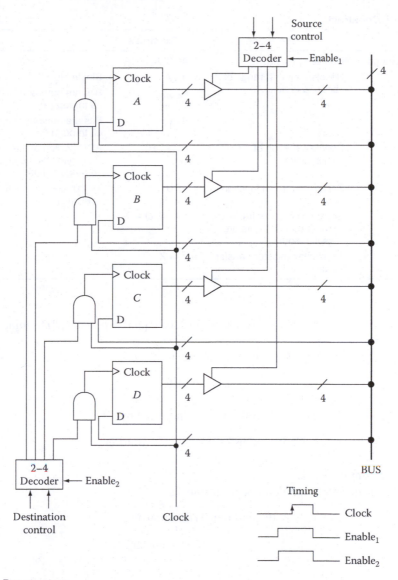

Figure 4.43 Bus structure.

Table 4.1 HDL Operators

Operator	Description	Examples
Left arrow, ←	Transfer operator	$Y \leftarrow X$; Contents of register X are transferred to register Y
Plus, +	Addition	$Z \leftarrow X + Y$
Minus, −	Subtraction	$Z \leftarrow X - Y$
¢	Concatenation	$C \leftarrow A \mathbin{¢} B$
Prime, '	Complement	$D \leftarrow A'$
∧	Logical AND	$C \leftarrow A \wedge B$
∨	Logical OR	$C \leftarrow A \vee B$
SHL	Shift left 1 bit; zero filled on right	$A \leftarrow$ SHL (A)
SHR	Shift right 1 bit; copy most significant bit on left	$A \leftarrow$ SHR (A)

Table 4.2 HDL Constructs

Construct	Description	Examples	
Capital-letter strings	Denote registers.	ACC, A, MBR	
Subscripts	Denote a bit or a range of bits of a register.	A_0, A_{15}	Single bit.
		A_{5-15}	Bits are numbered left to right, bits 5 through 15.
		A_{5-0}	Bits are numbered right to left, bits 0 through 5.
Parentheses ()	Denote a portion of a register (subregister).	IR (ADR)	ADR portion of the register IR; this is a symbolic notation to address a range of bits.
Colon :	Serves as control function delimiter.	ADD:	Terminates the control signal definition.
Comma ,	Separates register transfers; implies that transfers are simultaneous.	$Y \leftarrow X, Q \leftarrow P$	
Period .	Terminates register transfer statement.	$Y \leftarrow X.$	

The number of bits in source and destination must be equal. A period (".") terminates a register transfer statement.

A transfer controlled by a control signal has the format

$$\text{Control:transfer.}$$

Multiple transfers controlled by a control signal are indicated by

$$\text{Control:transfer}_1, \text{transfer}_2, \ldots, \text{transfer}_n.$$

The transfers are simultaneous.

The general format of a conditional register transfer is

$$\text{IF condition THEN transfer}_1$$

$$\text{ELSE transfer}_2.$$

where

"condition" is a Boolean expression
"transfer$_1$" occurs if condition is TRUE (or 1)
"transfer$_2$" occurs if condition is FALSE (or 0)

The ELSE clause is optional. Thus,

$$\text{IF condition THEN transfer.}$$

is valid.

A control signal can be associated with a conditional register transfer:

$$\text{Control:IF condition THEN transfer}_1$$

$$\text{ELSE transfer}_2.$$

Figure 4.44 illustrates the features of the HDL.

$B \leftarrow A.$ A and B must have the same number of bits.

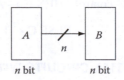

$C \leftarrow A + B' + 1.$ 2s complement of B added to A, transferred to C.

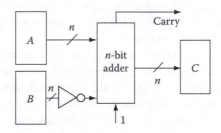

$B \leftarrow A, A \leftarrow B.$ A and B exchange. (A and B must be formed using master–slave flip-flops to accomplish this exchange.)

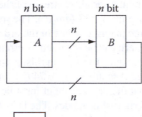

$T1: B \leftarrow A.$

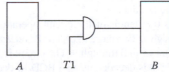

$A \leftarrow C \phi D.$ The total number of bits in C and D must be equal to that in A.

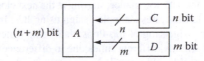

$C \phi D \leftarrow A.$ Reverse operation of the above

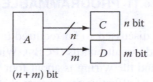

Figure 4.44 HDL features.

The transfers in Figure 4.39 can be described by the statement

$$\text{If control THEN C} \leftarrow \text{A}$$

$$\text{ELSE C} \leftarrow \text{B.}$$

4.10 DESIGNING SEQUENTIAL CIRCUITS WITH INTEGRATED CIRCUITS

Appendix A shows some small- and medium-scale ICs from the transistor–transistor logic (TTL) family. The reader is referred to IC manufacturer catalogs for further details on these ICs. A sequential circuit can be designed by following the classical design procedure described in this chapter. As a final step in the design, the circuit components (flip-flops, registers, etc.) are selected by referring to manufacturer catalogs.

It is often possible to design sequential circuits without following the classical design procedure. The serial adder design (see Example 4.7) is one example. Where the number of states in practical circuits becomes so large that the classical design procedure becomes impractical, the circuit functions are usually partitioned, and each partition is separately designed. Ad hoc methods of design based on familiarity with available ICs may be used in designing a partition or the complete circuit. Example 4.9 illustrates the design process using ICs.

Example 4.9: Parallel-to-Serial Data Converter

The object of our design is a parallel-to-serial data converter that accepts 4-bit data in parallel and produces as its output a serial-bit stream of the data input into it. The input consists of the sign bit (a_0) and three magnitude bits $(a_1 a_2 a_3)$. The serial device expects to receive the sign bit a_0 first, followed by the three magnitude bits in the order $a_3 a_2 a_1$, as shown in Figure 4.45a.

Note that the output bit pattern can be obtained by circulating the input data right three times and then shifting right 1 bit at a time. To perform this, a 4-bit shift register that can be loaded in parallel and can be right shifted if required. TTL 7495 is one such circuit. From the 7495 circuit diagram, it can be deduced that the "mode" input must be 1 for the parallel-load operation and has to be 0 for serial-input and right-shift modes. The D_s output must be connected to the "serial" input line for circulating the data.

Figure 4.45b shows the details of the circuit operation. The complete operation needs eight steps, designated 0 through 7. Two more idle steps 8 and 9 are shown, since a decade counter (7490) is available that can count from 0 through 9. The circuit is shown in Figure 4.45c. The 4-bit output of the decade counter 7490 is decoded using a BCD-to-decimal decoder (7442). Since the outputs of 7442 are low active, output 0 will have a value of 0 during the 0 time step and a value of 1 during other times. Hence, it can be used as mode control signal for 7495. Inputs CP_1 and CP_2 of 7495 must be tied together so that the circuit receives "clock" in both modes. Output 3 of 7442 is used to alert the serial device for data acceptance, starting at the next clock edge; output 8 indicates the idle state.

This example illustrates a simple design using ICs. In practice, the timing problems will be more severe. The triggering of flip-flops, data setup times, clock skews (i.e., arrival of clock on parallel lines at slightly different times due to differences in path delays), and other timing elements must be considered in detail.

4.11 PROGRAMMABLE LOGIC

The logic implementations discussed so far required the interconnection of selected SSI, MSI, and LSI components on printed circuit boards (PCBs). With the current hardware technology, the cost of the PCB, connectors, and the wiring is about four times that of the ICs in the circuit. Yet this implementation mode is cost-effective for circuits that are built in small quantities. With the progress in IC technology leading to the current VLSI era, it is now possible to fabricate a very complex digital system on a chip. But, for such implementation to be cost-effective, large quantities of the circuit or the system would be needed, since the IC fabrication is a costly process. To manage

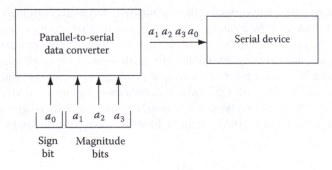

(a)

(b)

Count	Event	Action	Mode
0	Load the register (parallel)	Parallel in	1
1	↑	Shift circular	0
2	Circulate	"	0
3	↓	"	0
4	↑	Shift right	0
5	Serial	"	0
6	output	"	0
7	↓	"	0
8	Idle	Idle	d
9	Idle	Idle	d

(c)

Figure 4.45 Parallel-to-serial data converter. (a) Requirements. (b) Operation details. (c) Circuit.

costs while exploiting the capabilities of the technology, three implementation approaches are currently employed, based on the quantities of the circuits needed: custom, semicustom, and programmable logic. Custom implementations are for circuits needed in large quantities, semicustom for medium quantities, and the programmable logic mode is used for small quantities. In this section, we first provide a brief description of the IC fabrication process and examine the relative merits of each of these approaches. We will then provide details on programmable logic design.

Figure 4.46 shows the steps in a typical IC manufacturing process. The process starts off with a thin (10 mil thick, with 1 mil = 1/1000 of an inch) slice of p-type semiconductor material, about

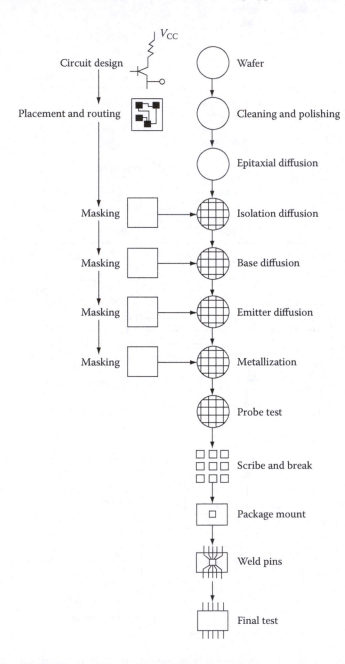

Figure 4.46 IC manufacturing process.

2–5 in. in diameter, called a wafer. Hundreds of identical circuits are fabricated on the wafer using a multistep process. The wafer is then cut into individual dies, each die corresponding to an IC.

The fabrication process consists of the following steps:

1. Wafer surface preparation
2. Epitaxial growth
3. Diffusion
4. Metallization
5. Probe test
6. Scribe and break
7. Packaging
8. Final test

The circuit designer first develops the circuit diagram. Before fabrication, the circuit must be at the transistor, diode, and resistor level of detail. Today, however, the circuit designer does not need to deal with this level of detail, since there are computer-aided design (CAD) tools that translate a gate-level design to the circuit level. Once the circuit is designed, it is usually simulated to verify its functional, timing, and loading characteristics. If these characteristics are acceptable, the circuit is brought to the placement and routing stage.

Placement is the process of placing the circuit components at the appropriate positions so that the interconnections can be routed properly. Automatic placement and routing tools are available. For a very complex circuit, this step is the most time-consuming one, even with CAD tools. At the end of this step, the circuit layout reflects its structure on the silicon.

The wafer is made of a high-resistivity, single-crystal, p-type silicon material. The wafer is first cleaned, and both sides of it are polished, with one side finished to mirror smooth.

The epitaxial diffusion is the process of adding minute amounts of n- or p-type impurities (dopants) to achieve the desired low resistivity. The epitaxial layer forms the collector of the transistors. This layer is then covered with a layer of silicon dioxide formed by exposing the wafer to an oxygen atmosphere around 1000°C.

The most critical part of the fabrication process is the preparation of masks to transfer the circuit layout into silicon. The mask plates are first drawn according to the circuit layout. These plates are then reduced by photographic techniques to the size of the final chip. The masks are then placed on the wafer, one mask for each die, and a photoresist coating on the wafer is exposed to ultraviolet light. The unexposed surface is etched chemically, leaving the desired pattern on the wafer. The surface pattern is then subjected to diffusion.

Although the steps vary depending on the process and the technology used to manufacture the IC, the diffusion can be classified into isolation, base, and emitter diffusion stages. These are the stages in which the corresponding terminals of the transistors are fabricated; each diffusion stage corresponds to one or more masks and etch operation.

By now, the wafer contains several identical dies with all the circuit components formed on each die. The wafer is then subjected to photo etching to open the windows to provide the connections between the components. The interconnections are made (i.e., through metallization) by vacuum deposition of a thin film of aluminum over the entire wafer, followed by another mask and etch operation to remove unnecessary interconnections.

Among the dies now on the wafer, some may be defective as a result of imperfections in the wafer or in the fabrication process. Selected dies are now tested to mark the failing ones. The percentage of good dies obtained is called the "yield" of the fabrication process.

The wafer is now scribed by a diamond-tipped tool along the boundaries of the dice to separate them into individual chips. Each die is then tested and mounted on a leader, and pins are attached and packaged.

Circuit layout and mask preparation are the most time-consuming and error-prone stages in the fabrication process and hence contribute most to the design cost. When the circuit is very complex, the circuit layout requires thousands of labor-hours, even with the use of CAD tools.

VHDL and Verilog are the prominent HDLs today. There are several commercial and open-source CAD tools available to use descriptions in these languages or others to aid the IC design and fabrication. Some of the prominent CAD tools for VLSI design are the following:

1. Espresso for Boolean logic minimization for two-level combinational circuits.
2. Sequential Interactive System (SIS) for optimization and implementation of both 2-level and multi-level combinational and sequential circuits.
3. Magic is an interactive editor for VLSI layouts.
4. IRSIM is an event-driven logic-level simulator for MOS (both N and P) transistor circuits.
5. Simulation Program for Integrated Circuits (SPICE) is a general-purpose electric and semiconductor circuit simulation program.
6. Bipolar simulation models are provided by BSIM3 and BSIM4 for nmos and pmos level = 9 and level = 14 for spice3f5 and spice3e2.
7. ALLIANCE supports both construction tools and validation tools. The design flow is divided into five parts: capture and simulation of the behavioral view, capture and validation of the structural view, physical design, verification, and coverage evaluation.
8. ELECTRIC is a robust CAD tool using only connectivity for circuit designs speeding up network-oriented operations.

For an introductory survey and comparative study of some of the open-source tools, refer to http://ilin.asee.org/Conference2010/Papers/A1_Liu_Anan.pdf.

With so many available tools, choosing the right one for any designer is not an obvious task. The needs and priority of one user like a multinational company with a fabrication business would vary from that of another like a university course designer for pedagogy. Usually the following factors are considered in selecting a suitable tool:

1. The file formats that software supports natively and through plug-ins
2. The functionalities like automatic layout generation and automatic wire arrangement
3. Model building method
4. The output analysis features like the model checking function, design rule check, and electric rule check
5. The availability of help and documentation from the support groups and authors
6. Operational difficulties
7. Installation difficulties
8. System requirements
9. License
10. Stability
11. Version variation and update frequency

With the advancement in the design and fabrication of ICs, the number of transistors per IC chip has been increasing though not at the rate foreseen by Moore's law, especially after 2000. The main limitation for this is caused by the power consumption constraints. They are of two types, static power leakage and dynamic power leakage. The dynamic power leakage happens due to the capacitance and rapid switching of the transistors, while static power leakage happens when a transistor drains some power even if it is switched off. These two power leakages affect the number of transistors on the IC. Many models, estimations, and predictions are done regarding the limiting condition of transistor density due to these constraints. These predictions are generically called power law. Since fabrication is a very active research area, everyday newer techniques are developed to overcome these constraints in creative ways. For more related information, further reading is suggested at the reference.

4.11.1 Circuit Implementation Modes and Devices

In the custom-design mode of circuit implementation, all the steps in the IC fabrication process are unique to the application at hand. Although this mode offers the smallest chip size and highest speed, the design costs do not justify this mode for low-volume applications. Typically, the annual sales volume of the IC should be around 5–10 times the nonrecurring engineering (NRE) costs of the design. This makes the custom-design mode beyond the reach of most applications.

In the semicustom-design mode, the initial steps in the fabrication process remain standard for all applications. Only the last step (metallization) is unique to each application. This is accomplished using fixed arrays of gates predesigned on the chip real estate. These gates are interconnected to form the application-specific IC (ASIC). The NRE cost for these ICs is an order of magnitude smaller than that for the custom-design mode, thus making them cost-effective for low-volume applications. Because of the use of standard gate patterns on the IC and simpler design rules employed, these ICs cannot use the chip area as efficiently and their speeds are also lower compared to custom-designed ICs. The ICs used in semicustom-design mode are mask programmable. That means, to make these ICs application specific, the user supplies the interconnection pattern (i.e., the program) to the IC manufacturer, who in turn prepares masks to program the IC. Obviously, once programmed, the function of these ICs cannot be altered.

In the programmable-design mode, ICs known as programmable logic devices (PLDs) are used. There are several types of PLDs, each with their own pattern of gates and interconnection paths prefabricated on the chip. These ICs are programmed by the user with special programming equipment (called PLD programmers). That is, they are field programmable. Some PLDs allow erasing the program to reprogram them and some do not.

In the early development stages of the digital system, field-programmable devices are typically used to allow the flexibility of design alterations. Once the design is tested and the circuit's performance is deemed acceptable, mask-programmable ICs can be used to implement the system.

Because PLDs are available off-the-shelf, no fabrication expenses are involved. The designers simply program the IC to create the ASIC. PLDs typically replace several components from the typical SSI-/MSI-based design, thereby reducing the cost through reduced number of external interconnections, PCBs, and connectors. There are four popular PLDs in use today:

1. Read-only memory (ROM)
2. Programmable logic array (PLA)
3. Programmable array logic (PAL)
4. Gate arrays (GAs)

The first three types are based on the two-level AND–OR circuit structure, while the last uses a more general gate structure to implement circuits. All these devices are available in both field- and mask-programmable versions. We will use the circuit in Figure 4.47 to illustrate the difference between the ROM, PLA, and PAL structures.

Example 4.10

Consider the two-level implementation of the functions $F1 = AB + A'B'$ and $F2 = AB' + A'B$ shown in Figure 4.47. The first level is implemented by an AND array. Each column corresponds to an input variable or its complement, and each row corresponds to a minterm of the input variables. The second level is implemented by an OR array. Each column corresponds to the OR combination of selected minterms generated by the AND array. For the purposes of this example, assume that each row–column intersection of these arrays can be programmed. That is, the electronic devices at these intersections can be used to either connect or disconnect the row

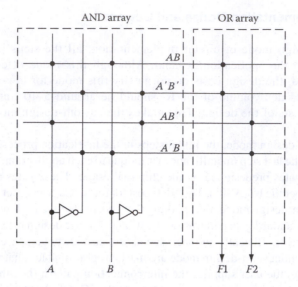

Figure 4.47   AND–OR array implementation of functions.

line to the column line, and when connected, the intersection realizes either an AND or an OR operation (depending on the array in which the intersection is located). A dot at the intersections indicates that the two lines are connected.

In a ROM, the AND array is fabricated so that all the minterms corresponding to the input variables are available, and hence, it is not programmable. The OR array is programmable to realize the circuit. With PAL, the OR array is completely fabricated and is not programmable, while the AND array is programmable. In a PLA, both arrays are programmable. We will defer the description of ROMs to Chapters 5 and 9. We provide brief descriptions of the other three PLDs in the following sections.

4.11.2 Programmable Logic Arrays

PLAs are LSI devices with several inputs, outputs, an AND array, and an OR array. An SOP form of the function to be implemented is used in designing with PLAs. Connecting the appropriate inputs (and their complements) in the AND array yields the product terms. The product terms are then combined by the OR array to yield the outputs. The AND and OR operations are realized using wired logic rather than discrete gates. Since only the required sum and product terms are generated, the PLA implementation is economical. Example 4.11 illustrates the structure and operation of PLAs.

Example 4.11

It is required to implement the following function using a PLA:

$$F1(A,B,C) = \Sigma m(2,3,7),$$

$$F2(A,B,C) = \Sigma m(1,3,7).$$

The K-maps of Figure 4.48a minimize this function. Note that the minimization procedure should reduce the number of product terms rather than just the number of literals in order to simplify the PLA.

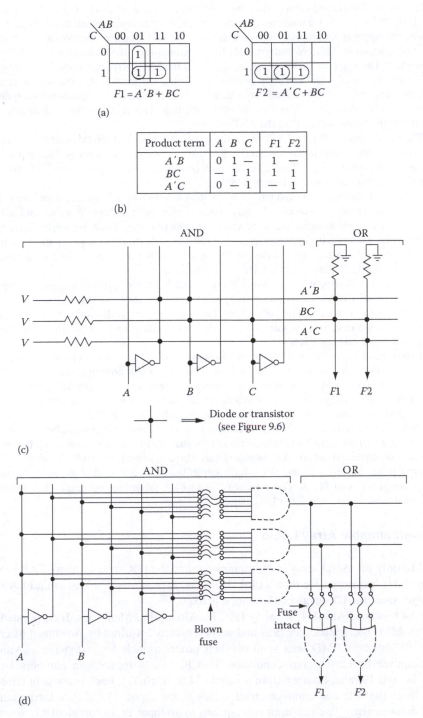

Figure 4.48 PLA. (a) K-maps. (b) PLA table. (c) PLA logical structure. (d) Actual PLA.

Only three product terms are needed to implement this two-output function, since BC is common to both. A modified truth table is shown in Figure 4.48b. The first column lists the product terms. The second column lists the circuit inputs. For each product term, the inputs are coded "0," "1," or "—." A 0 indicates that the variable appears complemented, a 1 indicates that the variable appears uncomplemented, and a "—" indicates that the variable does not appear in the product term. This column provides the programming information for the AND array of the PLA. There are six inputs (A, A', B, B', C, C') to each AND gate in the array, as shown in Figure 4.48c, each input having a fusible link. Thus, a 0 implies that the link is retained corresponding to the complemented variable, a 1 implies that the link is retained corresponding to the truth variable, and "—" implies "blowing" the link. This retaining/blowing operation on the links is the "programming" of the AND array.

The last column in Figure 4.48b shows the circuit outputs. A "1" in this column in any row indicates that the product term corresponding to that row is a product term for the output corresponding to the column; a "—" indicates no connection. Thus, this column is useful in programming the OR array, as shown in Figure 4.48c.

In Figure 4.48c, the AND and OR gates are shown dotted, to indicate the wired logic. Each AND gate has six inputs to receive the three circuit inputs and their complements. But only the required connections, as indicated by the second column of Figure 4.48b, are made. There is one AND gate for each required product term. Similarly, the OR gates are shown with three inputs each, one for each product term in the circuit. Again, only the required product terms, as indicated by the third column of Figure 4.48b, are connected.

The actual PLA structure is shown in Figure 4.48d. Each dot in this figure corresponds to a connection using a switching device (a diode or a transistor), and the absence of a dot at a row–column intersection indicates no connection. The arrow shown in the connection detail at the bottom of the figure can be considered the diode positive direction to verify that this structure implements the AND and OR logic.

Two types of PLAs are available: mask and field programmable. A PLA program is provided by the designer so that the IC manufacturer can fabricate a mask-programmed PLA, which can never be altered once programmed. Special devices are used to program the field-programmable PLA (FPLA). For this type of PLA, switching devices are fabricated at each row–column intersection, and the connections are established by blowing the fusible links.

PLAs with 12–20 inputs, 20–50 product terms, and 6–12 outputs are available off-the-shelf. In addition, programmable logic sequencers (PLSs) that are PLAs with storage capabilities that are useful in sequential circuit implementations are also available. In recent PLAs, the AND–OR structure is augmented with additional logic capabilities. Several CAD tools are available to enable designing with PLAs. These tools generate PLA programs, optimize PLA layout for custom fabrication, and simulate PLA designs.

4.11.3 Programmable Array Logic

In a PAL, only the AND array is programmable and the OR array is fixed. PAL is thus easier to program and is less expensive than a PLA. However, PAL is less flexible than a PLA in terms of circuit design, since the OR-array configuration is fixed.

Figure 4.49 shows PAL14H4, an early PAL IC. Monolithic Memories, Incorporated, invented PAL devices. MMI was founded in 1969 and was eventually acquired by Advanced Micro Devices (AMD) in 1987, though AMD later spun off their programmable logic division as Vantis, which was then acquired by Lattice Semiconductor. This IC has 14 inputs (pin numbers 1 through 9, 11, 12, 13, 18, and 19) and 4 outputs (pin numbers 14 through 17). Each input is buffered, and the buffer produces the true and complemented values of the input. The device layout consists of a 63 row, 32 column array. Each column corresponds to an input or its complement. Since there are only 14 inputs, only 28 of the 32 columns are utilized. Each row corresponds to a product term. Each output in this IC is realized by ORing four product terms. The product terms are realized by

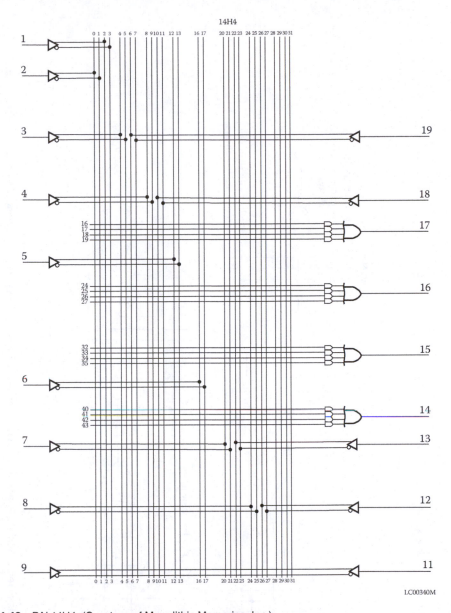

Figure 4.49   PAL14H4. (Courtesy of Monolithic Memories, Inc.)

wired logic. The AND gates shown at the inputs of OR gates thus are symbolic representations only. Only 16 of the 64 rows are used by this device.

Figure 4.50 shows the symbology used in PALs. In Figure 4.50a, the product term (ABC) is realized by retaining the fusible links at the row–column intersections of the AND array, as shown by an X. Absence of an X at any intersection indicates a blown link and hence no connection. Realization of the function $AB' + A'B$ is shown in Figure 4.50b. The unprogrammed PALs will have all the fuses intact. Fuses are blown during programming the PAL to realize the required function. A shorthand notation to indicate that all the fuses along a row are intact is shown in Figure 4.50c. This simply implies that a particular row is not utilized. A sample implementation using this convention is shown in Figure 4.50d.

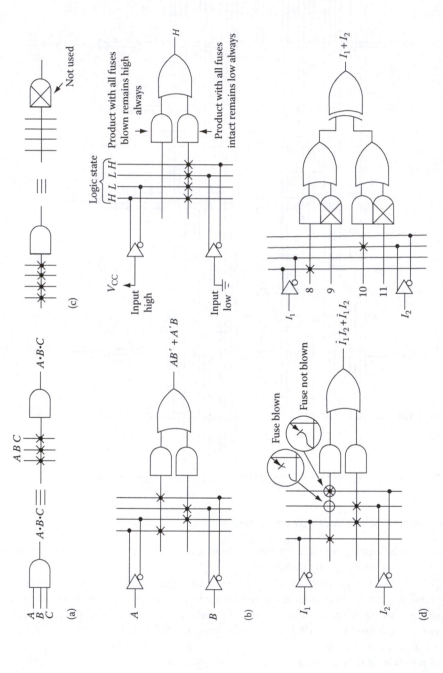

Figure 4.50 PAL symbology. (a) Product term realization. (b) Realization of $A \oplus B$. (c) All fuses intact. (d) Examples. (Courtesy of Monolithic Memories, Inc.)

In realizing multiple-output circuits using a PLA, common product terms can be shared among the various outputs. With PAL, since the OR array cannot be programmed, such sharing is not possible. As such, each output function must be minimized separately before the implementation. Further, since the number of inputs to the OR gates at the output of the PAL is fixed, alternate ways of realizing the circuit may need to be explored when the number of product terms in the minimized function exceeds the number of inputs available. Example 4.12 shows a circuit implementation using PAL.

Example 4.12

Implement the following function using PAL:

$$F1(A,B,C,D) = \Sigma m(0,1,5,7,10,11,13,15),$$

$$F2(A,B,C,D) = \Sigma m(0,2,4,5,10,11,13,15),$$

$$F3(A,B,C,D) = \Sigma m(0,2,3,7,10,11,12,13,14).$$

Figure 4.51a shows the K-maps and the corresponding simplified functions, and Figure 4.51b shows the implementation using the PAL14H4. The implementation of $F1$ and $F2$ does not present any problem, since the number of product terms in these functions is less than or equal to four. Since $F3$ has five product terms, one of the PAL outputs is used to realize the first four product terms (Z). Z is then fed into the PAL as an input and combined with the remaining product term of $F3$, to realize $F3$. Obviously, the implementation of this simple circuit using the PAL14H4 is not economical, since we did not use mostly the inputs and the product terms possible. Nevertheless, the example illustrates the design procedure.

Several PAL ICs are available with 10–20 inputs, 1–10 outputs (in the various configurations), and 2–16 inputs per output-OR gate. Typically, PALs are programmed with PROM programmers that use a PAL personality card. During programming, half of the PAL outputs are selected for programming, while the other inputs and outputs are used for addressing. The outputs are then switched to program the unprogrammed locations. One of the early PAL design aids available was PALASM software (from Monolithic Memories, Inc.). PALASM accepts the PAL design specification and verifies the design against an optional function table and generates the fuse plot required to program the PAL.

4.11.3.1 Altera Complex PLDs (CPLDs)

This section is extracted from Altera Max 7000 Series Data Sheets at http://www.altera.com/literature/ds/m7000.pdf.

Altera has developed three families of chips that fit within the CPLD category: MAX 5000, MAX 7000, and MAX 9000. MAX 7000 Series represents the widely used technology that offers state-of-the-art logic capacity and speed performance. MAX 5000 represents an older technology that offers a cost-effective solution, and MAX 9000 is similar to MAX 7000, except that MAX 9000 offers higher logic capacity.

The general architecture of the Altera MAX 7000 Series is depicted in Figure 4.52. It consists of an array of blocks called logic array blocks (LABs), and interconnect wires called a programmable interconnect array (PIA). The PIA is capable of connecting any LAB input or output to any other LAB. Also, the inputs and outputs of the chip connect directly to the PIA and to LABs. LAB is a complex SPLD-like structure, and so the entire chip can be considered to be an array of SPLDs. MAX 7000 devices are available both based in EPROM and EEPROM technology. Until recently, even with EEPROM, MAX 7000 chips could be programmable only "out-of-circuit" in a special-purpose programming unit; however, in 1996, Altera released the 7000S series, which is reprogrammable "in-circuit."

$$F1 = (A, B, C, D) = \Sigma m\ (0, 1, 5, 7, 10, 11, 13, 15)$$

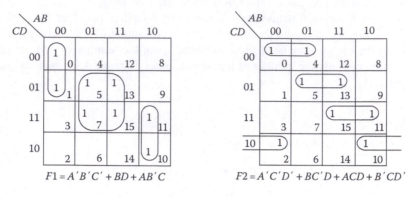

$$F1 = A'B'C' + BD + AB'C$$

$$F2 = A'C'D' + BC'D + ACD + B'CD'$$

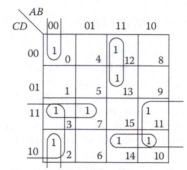

$$F3 = A'B'D' + A'CD + ABC' + ACD' + B'C$$

(a)

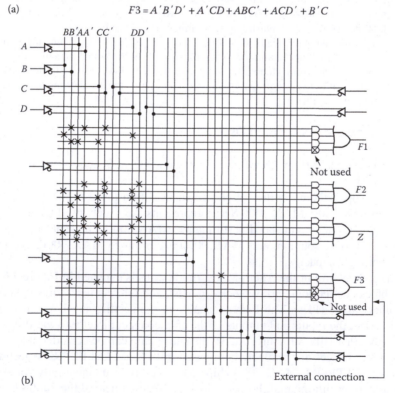

(b)

Figure 4.51   Implementation of functions in PAL. (a) Simplifications. (b) Implementation.

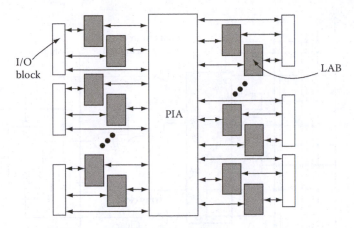

Figure 4.52   Altera MAX 7000 Series. (From http://www.altera.com/literature/ds/m7000.pdf. With permission.)

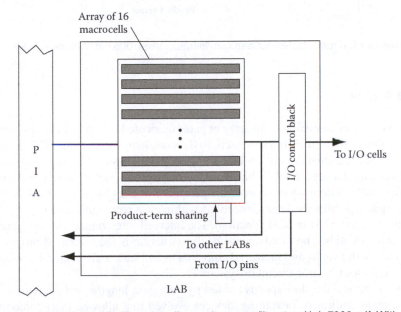

Figure 4.53   Altera MAX 7000 LAB. (From http://www.altera.com/literature/ds/m7000.pdf. With permission.)

The structure of a LAB is shown in Figure 4.53. Each LAB consists of two sets of eight macrocells (shown in Figure 4.54), where a macrocell comprises a set of programmable product terms (part of an AND plane) that feeds an OR gate and a flip-flop. The flip-flops can be configured as *D* type, *JK, T, SR*, or can be transparent. The number of inputs to the OR gate in a macrocell is variable; the OR gate can be fed from any or all of the 5 product terms within the macrocell, and in addition can have up to 15 extra product terms from macrocells in the same LAB. This product term flexibility makes the MAX 7000 Series LAB more efficient in terms of chip area because typical logic functions do not need more than five product terms, and the architecture supports wider functions when they are needed. It is interesting to note that variable-sized OR gates of this sort are not available in basic SPLDs.

Besides Altera, several other companies produce devices that can be categorized as CPLDs. For example, AMD manufactures the Mach family, Lattice has the (i) pLSI series, and Xilinx produces a CPLD series that they call XC7000 and has announced a new family called XC9500.

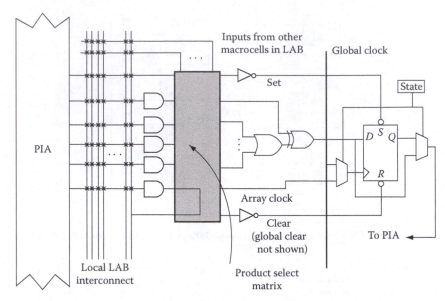

Figure 4.54 Macrocell. (From http://www.altera.com/literature/ds/m7000.pdf. With permission.)

4.11.4 Gate Arrays

GAs are LSI devices consisting of an array of gates fabricated over the chip area along with wire routing channels to facilitate their interconnection. GAs originated in the late 1970s as replacements for SSI- and MSI-based circuits built on PCBs. Figure 4.55 shows the structure of a typical GA. Each shaded rectangular area on the chip is an array of gates. The channels between these are the wire routing paths. The array of gates at the periphery of the chip is the IO pads. In using this device, the designer specifies the interconnections within each rectangular area (i.e., cell) to form a function (equivalent to an SSI or MS1 function). The intercell interconnections are generated using PCB routing software aids. The disadvantage of this structure is the increased propagation delays as a result of long path lengths and increased chip area to fabricate a given circuit, compared with a custom-design chip for the same circuit or system.

In order to overcome the slow speeds caused by long path lengths and decreased density as a result of large areas dedicated for routing, devices evolved that allowed the interconnection over the GA area, rather than through dedicated channels. Figure 4.56 shows one such GA (Signetics 8A1260). This device uses Integrated Schottky Logic (ISL) NAND gates, arranged in 2 arrays of 26 rows and 22 columns. There are 52 Schottky buffers driving multiload enable signals. The 60 LSTTL IO buffers can be programmed as inputs or bidirectional paths or as totem pole, tristate, or open-collector outputs.

By using a combination of appropriately configured NAND gates, any function can be realized. In fact, SSI and MSI functions from the TTL manuals can be copied. Unnecessary functionalities (such as multiple chip enables) provided in TTL ICs can be eliminated during the copying to make those circuits more efficient.

In designing with GAs, the designer (user) interacts with the IC manufacturer (supplier) extensively through the CAD tools (see Figure 4.57). The user generates the logic circuit description and verifies the design through logic and timing simulation. A set of tests to uncover faults is also generated by the user and provided to the supplier, along with the design specifications (e.g., wire lists, schematics). The supplier performs the automatic placement and routing, mask

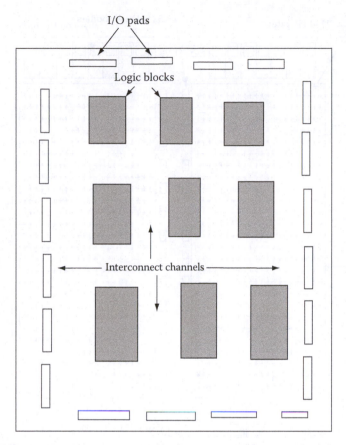

Figure 4.55   GA structure.

generation, and IC prototype fabrication. Once the performance of the prototype is accepted by the user, the production run of the ICs is initiated.

It is no longer necessary to design at gate level of detail, while designing with GAs. The GA manufacturers provide a set of standard cells (functions) as part of the CAD environment. These standard cells are utilized in configuring the GA very much like using the SSI/MSI components in the design of the system. The standard cells in each library correspond to gates of various types and configurations and MSI-type cells such as multiplexers, counters, adders, and arithmetic/logic units.

As can be seen from the discussion earlier, GAs offer much more design flexibility compared to other three PLDs. But their disadvantage is that the turnaround time from design to prototype is several weeks, although GA manufacturers have started offering 1–2 day turnarounds. The disadvantage of the other three PLDs is their design inflexibility, since only two-level AND–OR realizations are possible.

Several devices that combine the features of GA flexibility and programmability are now being offered. These devices are essentially PLDs with enhanced architectures, some departing completely from the AND–OR structure and some enhancing the AND–OR structure with additional programmable logic blocks (macro cells). We will collectively call all such devices field-programmable GAs (FPGA).

There are two basic categories of FPGAs on the market today: SRAM-based and antifuse-based FPGAs. Xilinx, Altera, and AT&T are the leading manufacturers in terms of number of users of SRAM-based FPGAs. Actel, Quicklogic, Cypress, and Xilinx offer antifuse-based FPGAs. We now provide a brief description of one FPGA.

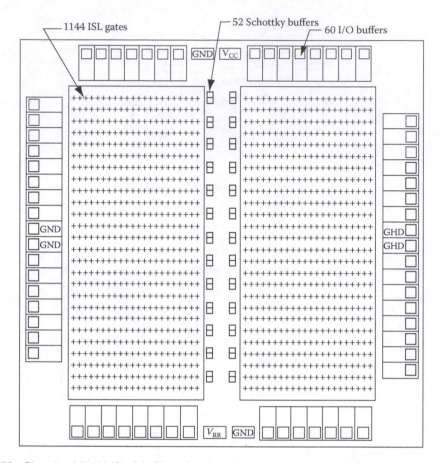

Figure 4.56   Signetics 8A1260 ISL GA. (From *Computer Design*, August 20, 1981.)

4.11.4.1 Xilinx SRAM-Based FPGAs

This section is extracted from Xilinx SRAM-based FPGA Data Sheets at http://www.xilinx.com/appnotes/FPGA_NSREC98.pdf.

The basic structure of Xilinx FPGAs is array based, with each chip comprising of a 2D array of logic blocks that can be interconnected via horizontal and vertical routing channels. Xilinx introduced the first FPGA family (XC2000 series), in about 1985, and now offers three more generations: XC3000, XC4000, and XC5000. Xilinx has recently introduced an FPGA family based on antifuses (XC8100).

The Xilinx 4000 family devices range in capacity from about 2,000 to more than 15,000 equivalent gates. The XC4000 features a configurable logic block (CLB) that is based on lookup tables (LUTs). A LUT is a small 1-bit wide memory array, where the address lines for the memory are inputs of the logic block and the 1-bit output from the memory is the LUT output. A LUT with K inputs would then correspond to a $2^K \times 1$-bit memory and can realize any logic function of its K inputs by programming the logic function's truth table directly into the memory. The XC4000 CLB contains three separate LUTs, in the configuration shown in Figure 4.58. There are two 4-input LUTS that are fed by CLB inputs, and the third LUT can be used in combination with the other two. This arrangement allows the CLB to implement a wide range of logic functions of up to nine inputs, two separate functions of four inputs, or other possibilities.

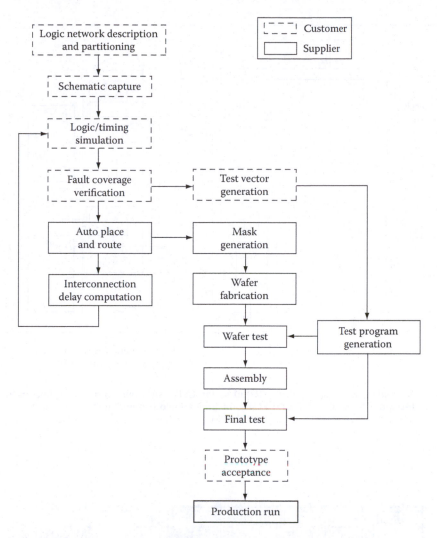

Figure 4.57 Semicustom logic design procedure. (Courtesy of Signetics Corporation.)

Each CLB also contains two flip-flops. Toward the goal of providing high-density devices that support the integration of entire systems, the XC4000 chips have "system-oriented" features. For instance, each CLB contains circuitry that allows it to efficiently perform arithmetic (i.e., a circuit that can implement a fast carry operation for adder-like circuits) and also the LUTs in a CLB can be configured as read/write RAM cells. A new version of this family, the 4000E, has the additional feature that the RAM can be configured as a dual port RAM with a single write and two read ports. In the 4000E, RAM blocks can be synchronous RAM. Also, each XC4000 chip includes very wide AND planes around the periphery of the logic block array to facilitate implementing circuit blocks such as wide decoders.

The XC4000 interconnect is arranged in horizontal and vertical channels. Each channel contains some number of short wire segments that span a single CLB, longer segments that span two CLBs, and very long segments that span the entire length or width of the chip. Programmable switches are available to connect the inputs and outputs of the CLBs to the wire segments or to connect one wire segment to another. A small section of a routing channel representative of an XC4000 device is shown in Figure 4.59. The figure shows only the wire segments in a horizontal channel and does not show the vertical routing channels, the CLB inputs and outputs, or the routing switches.

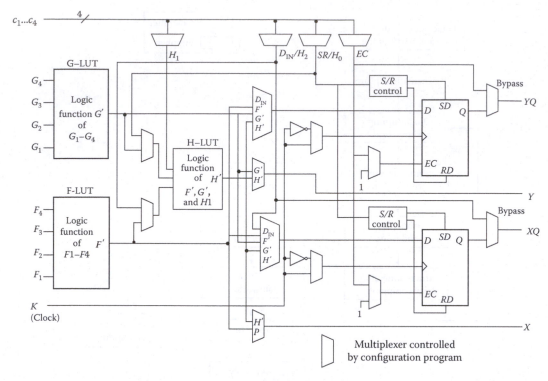

Figure 4.58 Simplified block diagram of XC4000 series CLB (RAM and Carry logic function not shown). (Adapted from XC4000E and XC4000X Series Field Programmable Gate Arrays, http://direct. xilinx.com/bvdocs/publications/4000.pdf.)

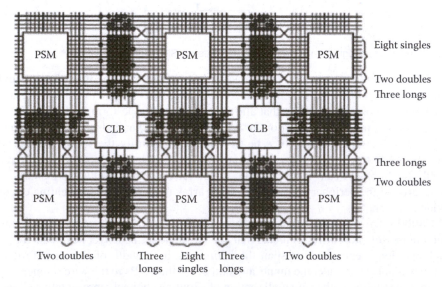

Figure 4.59 Spartan/XL CLB routing channels and interface block diagram. (Adapted from Spartan and Spartan-XL Families Field Programmable Gate Arrays, http://www.xilinx.com/bvdocs/ publications/ds060.pdf.)

The signals must pass through switches to reach one CLB from another, and the total number of switches traversed depends on the particular set of wire segments used. Thus, speed performance of an implemented circuit depends in part on how the wire segments are allocated to individual signals by CAD tools.

It should be noted that the circuits designed using PLDs are in general slower than custom-designed ICs. They also do not use the silicon area on the chip very efficiently and hence tend to be less dense than their custom-designed counterparts. Nevertheless, these designs are more cost-effective for low-volume applications. As with any IC technology, newer PLDs are continually being introduced by IC manufacturers. Their architecture and characteristics also vary. Because of the rapid rate of introduction of new devices, any description of these devices becomes obsolete by the time it is published in a book. As such, it is imperative that the designer consult the manufacturer's manuals for the most up-to-date information. The periodicals listed in the references section should also be consulted for new IC announcements.

4.12 SUMMARY

The analysis and design of synchronous sequential circuits described in this chapter are given as an overview of the subject. The reader is referred to the logic design texts listed at the end of this chapter for further details on these topics and also on asynchronous circuit analysis and design. IC manufacturer catalogs are important sources of information for logic designers, although the detailed electrical characteristics given in these catalogs are not required for the purposes of this book. Register transfer logic concepts described in this chapter will be used extensively in Chapters 5 and 6 in the logical design of a simple computer. Chapter 13 on embedded systems expands on the programmable logic design concepts introduced in this chapter.

PROBLEMS

4.1 You are given a D flip-flop. Design the circuitry around it to convert it into a
a. T flip-flop
b. SR flip-flop
c. JK flip-flop
Hint: A flip-flop is a sequential circuit. Start the design with the state table of the required flip-flop. Use a D flip-flop in the design.

4.2 A set-dominate flip-flop is similar to an SR flip-flops, except that an input $S = R = 1$ will result in setting the flip-flop. Draw the state table and excitation table for the flip-flop.

4.3 Shown later is a 3-bit counter made up of negative edge-triggered JK flip-flops. Draw the timing diagram for outputs Q_A, Q_B, and Q_C in respect to the clock input.

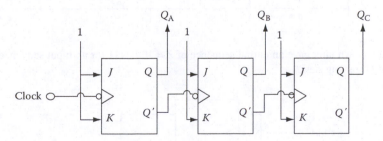

4.4 For the following circuit, derive the state table. Use the assignments

$$Y_1Y_2: \ 00 = A, \ 01 = B, \ 10 = C, \ 11 = D.$$

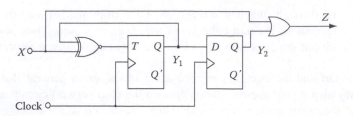

4.5 For the following circuit,
  a. Complete the timing diagram starting with $Y_1Y_2 = 00$ (at time = 0).
  b. Derive excitation tables.
  c. Derive the state table, using Y_1Y_2: $00 = D$, $01 = C$, $10 = B$, and $11 = A$. Assume that the flip-flops are triggered by the raising edge of the clock.

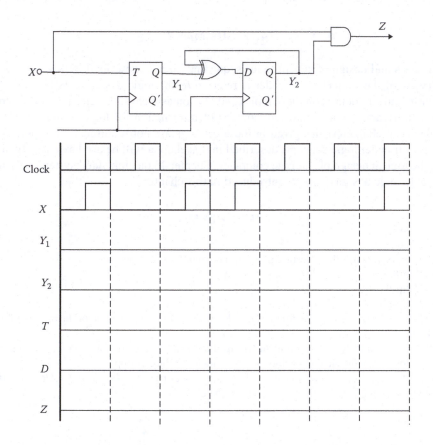

4.6 The circuit shown later gave an output sequence of $Z = 10011111$ for an input sequence $X = 11101010$. What was Q starting state?

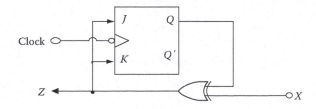

4.7 Construct state diagram for sequence detectors that can detect the following sequences:
   a.   10010 (sequences may overlap)
   b.   10010 (overlap not allowed)
   c.   1011 or 1001 (overlap allowed)
   d.   1011 or 1001 (overlap not allowed)

4.8 Implement the state diagrams from Problem 4.7 using *JK* flip-flops.

4.9 Design the circuit for a newspaper vending machine. Each item costs 40 cents. The machine accepts nickels, dimes, and quarters. Assume a coin sorter that accepts coins and provides three signals, one for each of the three types of coins. Assume that the signals on these three lines are separated far enough so that the other circuits can make a state transition between pulses received on any two of these lines. Correct change must be released. The sequential circuit must generate signals to release correct change (one signal for each type of coin) and the newspaper. Use *JK* flip-flops in your design.

4.10 There are two 4-bit registers *A* and *B*, built out of *JK* flip-flops. There is a control signal *C*. The following operations are needed:
   If $C = 0$, set the content of *B* to 1.
   If $C = 1$, send the content of *A* to *B*.
   Draw the circuit to perform these functions
   a.   In parallel mode
   b.   In serial mode

4.11 Design the circuit in Problem 4.10a to send $A + 1$ to *B* when $C = 1$.

4.12 There are three 2-bit registers *A*, *B*, and *C*. Design the logic to perform

$$\text{XOR} : C \leftarrow A \oplus B.$$

$$\text{NAND} : C \leftarrow (AB)'.$$

XOR and NAND are control signals. Each bit in the register is a *D* flip-flop.

4.13 In your design in Problem 4.12, what happens if both the control signals are at 0 and a clock pulse comes along? Redesign the circuit (if necessary) to prevent the clearing of register *C* under the aforementioned conditions. (Hint: Feed the output of *C* to its input.) What type of flip-flop should *C* be made of?

4.14 Implement the circuit of Problem 4.13 using appropriate multiplexers.

4.15 A 2-bit counter *C* controls the register transfers shown here:

$$C = 0 : B \leftarrow A \wedge B \quad C = 2 : B \leftarrow 1.$$

$$C = 1 : B \leftarrow A \vee B \quad C = 3 : B \leftarrow A - B.$$

A and *B* are 2-bit registers. Draw the circuit. Use 4-to-1 multiplexers in your design. Show the details of register *B*.

4.16 Draw a bus structure to perform the operations in Problem 4.15.

4.17 Connect four 5-bit registers *A*, *B*, *C*, and *D* using a bus structure capable of performing the following:

$$C_0 : B \leftarrow A' \qquad C_4 : B \leftarrow A \wedge C$$

$$C_1 : D \leftarrow A \vee B \quad C_5 : B \leftarrow C' \vee D$$

$$C_2 : A \leftarrow B + D \quad C_6 : D \leftarrow B$$

$$C_3 : C \leftarrow A + C' \quad C_7 : A \leftarrow B' \wedge D.$$

C_0 through C_7 are control signals. Use a 3-bit counter and a decoder to generate those signals. Assume tristate output for each register.

4.18 Assume regular outputs for each register in Problem 4.17. How many OR gates are needed to implement the bus structure?

4.19 Design a 4-bit serial in/parallel out shift register using TTL 7474 D flip-flops. Include a RESET control input so that all bits are set to 0 when RESET is high.

4.20 TTL 74174 is a 6-bit parallel in/parallel out register that has six inputs D_0–D_5 and six outputs Q_0–Q_5. Show how to connect TTL 74174 so that it operates as a serial shift register where serial data enters D_0 and outputs at Q_5.

4.21 Use your design in Problem 4.20 to build an 18-bit shift register with three TTL 74174.

4.22 TTL 74194 is a 4-bit universal shift register. Show how to connect TTL 74194 to left shift the input discarding the leftmost bit and setting the rightmost bit to 0.

4.23 Show how to connect TTL 74194 to right circulate the input.

4.24 Design a 4-bit serial binary adder using three 4-bit shift registers, one full adder, and a D flip-flop. Two unsigned numbers are stored in two of the shift registers, A and B. Bits are added one pair at a time sequentially starting with the least significant bits. The carry out of the full adder is transferred to the D flip-flop in which the output is used as an input carry for the next pair of bits. The sum bit from the S output of the full adder is transferred into the third shift register C.

4.25 Repeat Problem 4.24 using only two 4-bit registers by shifting the sum into A while the bits of A are shifted out.

BIBLIOGRAPHY

Altera Max 7000 Series, http://www.altera.com/literature/ds/m7000.pdf.

Brown, S. and Vranesic, Z., *Fundamentals of Digital Logic with VHDL*, New York: McGraw-Hill, 2008.

Composite Cell Logic Data Sheets, Sunnyvale, CA: Signetics, 1981.

Computer, New York: IEEE Computer Society (monthly).

Davis, G.R., ISL gate arrays operate at low power schottky TTL speeds, *Computer Design*, 20, 183–186, August 1981.

FAST TTL Logic Series Data Handbook, Sunnyvale, CA: Phillips Semiconductors, 1992.

Greenfield, J.D., *Practical Design Using ICs*, New York: John Wiley, 1983.

Jin, L., Liu, C., and Anan, M. Open-source VLSI CAD tools: A comparative study, *American Society for Engineering Education-Indiana Section Conference*, Vol. 36, No. 12, Indianapolis, IN, 2010.

Kim, N.S., Leakage current: Moore's law meets static power, *Computer*, IEEE Computer Society, December 2003.

National Advanced Bipolar Logic Databook, Santa Clara, CA: National Semiconductor Corporation, 1995.

Perry, D.L., *VHDL*, New York: McGraw-Hill, 1991.

PAL Device Data Book, Sunnyvale, CA: AMD and MMI, 1988.

Programmable Logic Data Manual, Sunnyvale, CA: Signetics, 1986.

Programmable Logic Data Book, San Jose, CA: Cypress, 1996.

Programmable Logic Devices Data Handbook, Sunnyvale, CA: Philips Semiconductors, 1992.

Shiva, S.G., *Introduction to Logic Design*, New York: Marcel Dekker, 1998.

Technical Staff of Monolithic Memories, *Designing with Programmable Array Logic*, New York: McGraw-Hill, 1981.

Thomas, D.E. and Moorby, P.R., *The VERILOG Hardware Description Language*, Boston, MA: Kluwer, 1998.

TTL Data Manual, Sunnyvale, CA: Signetics, 1987.

Xilinx SRAM-based FPGAs, http://www.xilinx.com/appnotes/FPGA_NSREC98.pdf.

CHAPTER 5

A Simple Computer: Organization and Programming

The purpose of this chapter is to introduce the terminology and basic functions of a simple but complete computer, mainly from a programmer's (user's) point of view. We call the simple hypothetical computer a simple computer (ASC). Although ASC appears very primitive in comparison with any commercially available machine, its organization reflects the basic structure of the most modern computers. The instruction set is limited but complete enough to write powerful programs. Assembly language programming and understanding of assembly process are a must for a system designer. We will not outline the trade-offs involved in selecting the architectural features of a machine in this chapter. Subsequent chapters of this book, however, deal with such trade-offs. The detailed hardware design of ASC is provided in Chapter 6. Chapters 7 through 16 examine selected architectural attributes of commercially available computer systems.

5.1 A SIMPLE COMPUTER

Figure 5.1 shows the hardware components of ASC. We will assume that ASC is a 16-bit machine; hence, the unit of data manipulated by and transferred between various registers of the machine is 16 bits long. Recall that a register is a storage device capable of holding certain number of bits. Each bit corresponds to a flip-flop. The data bits written (loaded) into the register remain in the register until new data are loaded, as long as the power is on. The data in a register can be read and transferred to other registers. The read operation does not change the content of the register.

Figure 5.2 shows the model of a random-access memory (RAM) used as the main memory of ASC. Other types of memory systems are described in Chapter 9. In a RAM, any addressable location in the memory can be accessed in a random manner. That is, the process of reading from and writing into a location in a RAM is the same and consumes an equal amount of time, no matter where the location is physically in the memory. The two types of RAM available are *read/write memory* (RWM) and *read-only memory* (ROM).

The most common type of main memory is the RWM, whose model is shown in Figure 5.2. In an RWM, each memory register or memory location has an "address" associated with it. Data are input into (written into) and output from (read from) a memory location by accessing the location using its "address." The memory address register (MAR) stores such an address. With n-bit in the MAR, 2^n locations can be addressed, and they are numbered from 0 through $2^n - 1$.

Transfer of data in and out of memory is usually in terms of a set of bits known as a memory word. Each of the 2^n words in the memory has m bits. Thus, this is a $(2^n \times m)$-bit memory. This is a common notation used to describe RAMs. In general, an $(N \times M)$ unit memory contains N words of M units each. A "unit" is either a bit, a byte (8 bits), or a word of a certain number of bits. A memory

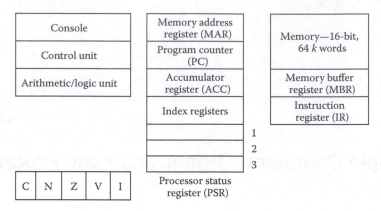

Figure 5.1 ASC hardware components.

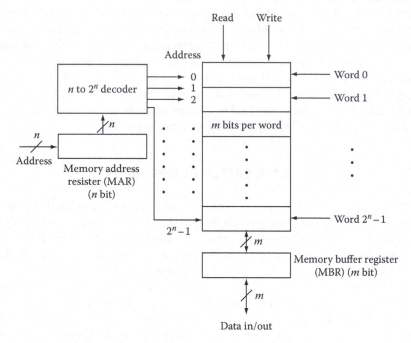

Figure 5.2 Random access (read/write) memory.

buffer register (MBR) is used to store the data to be written into or read from a memory word. To read the memory, the address of the memory word to be read from is provided in MAR and the read signal is set to 1. A copy of the contents of the addressed memory word is then brought by the memory logic into the MBR. The content of the memory word is thus not altered by a read operation. To write a word into the memory, the data to be written are placed in MBR by external logic; the address of the location into which the data are to be written is placed in MAR; and the write signal is set to 1. The memory logic then transfers the MBR content into the addressed memory location. The content of the memory word is thus altered during a write operation. The MBR has also been called a memory data register (MDR). We will stay with the designation as MBR in this book.

A memory word is defined as the most often accessed unit of data. The typical word sizes used in memory organizations of commercially available machines are 6, 16, 32, 36, and 64 bits. In addition to addressing a memory word, it is possible to address a portion of it (e.g., half-word, quarter-word) or a multiple of it (e.g., double word, quad word), depending on the memory organization.

In a "byte-addressable" memory, for example, an address is associated with each byte (8 bits/byte) in the memory, and a memory word consists of one or more bytes.

The literature routinely uses the acronym RAM to mean RWM. We follow this popular practice and use RWM only when the context requires us to be more specific. We have included MAR and MBR as components of the memory system in this model. In practice, these registers may not be located in the memory subsystem, but other registers in the system may serve the functions of these registers.

ROM is also a RAM, except that data can only be read from it. Data are usually written into a ROM either by the memory manufacturer or by the user in an off-line mode; that is, by special devices that can write (burn) the data pattern into the ROM. A ROM is also used as main memory and contains data and programs that are not usually altered in real time during the system operation. Chapter 9 provides further description of ROM and other memory systems.

With a 16-bit address, we can address 2^{16} (i.e., $2^6 \times 2^{10} = 64 \times 1024$) 64 k memory words, where $k = 2^{10}$ or 1024. For ASC we will assume a memory with 64 k, 16-bit words. A 16-bit long MAR is thus required. The MBR is also 16 bits long, as we have assumed ASC to be a 16-bit machine. MAR stores the address of a memory location to be accessed, and MBR receives the data from the memory word during a memory read operation and retains the data to be written into a memory word during a memory write operation. These two registers are not normally accessible by the programmer.

ASC is a stored-program machine. That is, programs are stored in the memory. During the execution of the program, each instruction from the stored program is first *fetched* from the memory into the control unit and then the operations called for by the instruction are performed (i.e., the instruction is *executed*). Two special registers are used for performing fetch–execute operations: a *program counter* (PC) and an *instruction register* (IR). The PC contains the address of the instruction to be fetched from the memory and is usually incremented by the control unit to point to the next instruction address at the end of an instruction fetch. The instruction is fetched into IR. The circuitry connected to IR decodes the instruction and generates appropriate control signals to perform the operations called for by the instruction. PC and IR are both 16 bits long in ASC.

There is a 16-bit *accumulator register* (ACC) used in all arithmetic and logic operations. As the name implies, it accumulates the result of arithmetic and logic operations.

There are three *index registers* (INDEX 1, 2, and 3) in ASC that are used in manipulation of addresses. We will discuss the function of these registers later in this chapter.

There is a 5-bit *processor status register* (PSR) whose bits represent carry (C), negative (N), zero (Z), overflow (V), and interrupt-enable (I) conditions. If an operation in the arithmetic/logic unit results in a carry from the most significant bit of the accumulator, then the carry bit is set. Negative and zero bits indicate the status of the accumulator after each operation that involves the accumulator. If the accumulator contains a negative value, the negative bit (N) is set, else it will be reset. Similarly, if the accumulator contains a zero value, the zero bit (Z) is set, else it will be reset. The interrupt-enable flag, when set, indicates that the processor can accept an interrupt. Interrupts are discussed in Chapter 7. The overflow bit is provided to complete the PSR illustration but is not used further in this chapter.

A *console* is needed to permit operator interaction with the machine. ASC console permits the operator to examine and change the contents of memory locations and initialize the PC. Power ON/OFF and START/STOP controls are also on the console. The console has a set of 16 switches through which a 16-bit data word can be entered into ASC memory. There are 16 lights (monitors) that can display 16-bit data from either a memory location or a specified register. To execute a program, the operator first loads the programs and data into the memory, then sets the PC contents to the address of the first instruction in the program and STARTs the machine. The concept of a console is probably old fashioned, since most of the modern machines are designed to use one of the I/O devices (such as a terminal) as the system console. A console is necessary during the debugging phase of the computer design process. We have included a console to simplify the discussion of program loading and execution concepts.

During the execution of the program, additional data input (or output) is done through an input (or output) device. For simplicity, we assume that there is one input device that can transfer a 16-bit data word into the ACC and one output device that can transfer the 16-bit content of the ACC to an output medium. It could very well be that the keyboard of a terminal is the input device and its display is the output device. Note that the data in the ACC are not altered due to the output, but an input operation replaces the original ACC content with the new data.

5.1.1 Data Format

ASC memory is an array of up to 64 k 16-bit words. Each of these 16-bit words will be either an instruction or a 16-bit unit of data. The exact interpretation depends on the context in which the machine accesses a particular memory word. The programmer should be aware (at least at the assembly and machine language programming levels) of the data and program segments in the memory and should make certain that a data word is not accessed during a phase in which the processor is accessing an instruction and vice versa.

Only fixed-point (integer) arithmetic is allowed on ASC. Figure 5.3a shows the data format: the most significant bit is the sign bit followed by 15 magnitude bits. Since ASC uses 2s complement representation, the sign and magnitude bits are treated alike in all computations. Note also the four-digit hexadecimal notation used to represent the 16-bit data word. We use this notation to denote a 16-bit quantity, irrespective of whether it is data or an instruction.

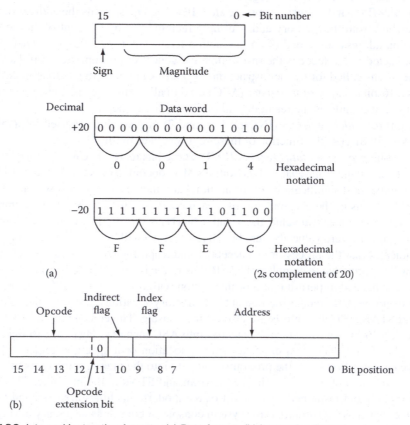

Figure 5.3 ASC data and instruction formats. (a) Data format. (b) Instruction format.

5.1.2 Instruction Format

Each instruction in an ASC program occupies a 16-bit word. An instruction word has four fields, as shown in Figure 5.3b. Bits 15 through 11 of the instruction word are used for the operation code (opcode). The *opcode* is a unique bit pattern that encodes a primitive operation the computer can perform. Thus, ASC can have a total of $2^5 = 32$ instructions. We use an instruction set with only 16 instructions for simplicity in this book. The opcodes for these 16 instructions occupy bits 15 through 12, and bit 11 is set to 0. If the instruction set were to be expanded beyond the current set of 16, the opcodes for the new instructions would have a 1 in bit 11. Bit 10 of the instruction word is the *indirect* flag. This bit will be set to 1 if indirect addressing is used; otherwise it is set to 0. Bits 9 and 8 of the instruction word select one of the three index registers when indexed addressing is called for or if the instruction manipulates an index register:

Bit 9	Bit 8	Index Register Selected
0	0	None
0	1	1
1	0	2
1	1	3

Bits 7 through 0 are used to represent the memory address in those instructions that refer to memory. If the instruction does not refer to memory, the indirect, index, and memory address fields are not used; the opcode field represents the complete instruction.

With only 8 bits in the address representation, ASC can directly address only $2^8 = 256$ memory locations. That means the program and data must always be in the first 256 locations of the memory. Indexed and indirect addressing modes are used to extend the addressing range to 64 k. Thus, ASC has direct, indirect, and indexed addressing modes. When both indirect and indexed addressing mode fields are used, the addressing mode can be interpreted either as indexed-indirect (preindexed-indirect) or as indirect-indexed (postindexed-indirect). We assume that ASC allows only the indexed-indirect mode. We will describe the addressing modes further after the description of the instruction set that follows.

5.1.3 Instruction Set

Table 5.1 lists the complete instruction set of ASC. Column 2 shows the most significant four bits of the opcode in hexadecimal form. The fifth bit being 0 is not shown. Each opcode is also identified by a *symbolic name*, or *mnemonic*, shown in column 1.

We add one more construct to our hardware description language (HDL) described in Chapter 4. The memory is designed as M. A memory read operation is shown as

$$MBR \leftarrow M[MAR].$$

and a memory write operation is shown as

$$M[MAR] \leftarrow MBR.$$

Table 5.1   ASC Instruction Set

Mnemonic	Opcode[a] (Hexadecimal)	Description
HLT	0	Halt
LDA	1	ACC ← M[MEM][b]
STA	2	M[MEM] ← ACC
ADD	3	ACC ← ACC + M[MEM]
TCA	4	ACC ← ACC' + 1 (2s complement)
BRU	5	Branch unconditional
BIP	6	Branch if ACC > 0
BIN	7	Branch if ACC < 0
RWD	8	Read a word into ACC
WWD	9	Write a word from ACC
SHL	A	Shift left ACC once
SHR	B	Shift right ACC once
LDX	C	INDEX ← M[MEM]
STX	D	M[MEM] ← INDEX
TIX	E	Test index increment
		INDEX ← INDEX + 1
		Branch if INDEX = 0
TDX	F	Test index decrement
		INDEX ← INDEX − 1
		Branch if INDEX ≠ 0

[a] Most significant 4 bits of opcode only. Bit 11 is 0.
[b] MEM refers to a memory word; that is, the symbolic address of a
    memory word. M[MEM] refers to the contents of the memory word
    MEM when used as a source and to the memory word when used
    as a destination. INDEX refers to one of the three index registers.

The operand within the [ ] can be

1. A register—the content of the register is a *memory address.*
2. A symbolic address—the *symbolic address* will eventually be associated with an *absolute address.*
3. An absolute address.

Thus,

$$ACC \leftarrow M[27],$$

$$M[28] \leftarrow ACC,$$

$$IR \leftarrow M[Z1], \text{ and}$$

$$M[Z1] \leftarrow ACC \text{ (Z1 is a symbolic address)}$$

are all valid data transfers. Further,

$$ACC \leftarrow Z1$$

implies that the absolute address value corresponding to Z1 is transferred to ACC. Thus,

$$Z1 \leftarrow ACC$$

is not valid.

The ASC instruction set consists of the following three classes of instructions:

1. Zero address (TCA, HLT, SHL, and SHR)
2. One address (LDA, STA, ADD, BRU, BIP, BIN LDX, STX, TIX, and TDX)
3. Input/output (I/O) (RWD, WWD)

A description of instructions and their representation follows. In this description, hexadecimal numbers are distinguished from decimal numbers with a preceding "#H."

Zero-Address Instructions: In this class of instructions, the opcode represents the complete instruction. The operand (if needed) is implied to be in the ACC. The address field, the index flag, and the indirect flag are not used.

Opcode	-	--	-------

A description of each instruction follows:

HLT Stop Halt

The HLT instruction indicates the logical end of a program and hence stops the machine from fetching the next instruction (if any).

TCA ACC $\leftarrow$ ACC' + 1 2s complement accumulator

TCA complements each bit of the ACC to produce the 1s complement and then a 1 is added to produce the 2s complement. The 2s complement of the ACC is stored back into the ACC.

SHL $ACC_{15-1} \leftarrow ACC_{14-0}$ Shift left
$ACC_0 \leftarrow 0$

The SHL instruction shifts the contents of the ACC 1 bit to the left and fills a 0 into the least significant bit of the ACC.

SHR $ACC_{14-0} \leftarrow ACC_{15-1}$ Shift right
$ACC_{15} \leftarrow ACC_{15}$

The SHR instruction shifts the contents of the ACC 1 bit to the right and the most significant bit of the ACC remains unchanged. The contents of the last significant bit position are lost.

One-Address Instructions: These instructions use all 16 bits of an instruction word. In the following, MEM is a *symbolic address* of an arbitrary memory location. An *absolute address* is the physical address of a memory location, expressed as a numeric quantity. A symbolic address is mapped to an absolute address when an assembly language problem is translated into machine language. The description assumes a direct-addressing mode in which MEM is the *effective address* (the address of the operand). The 8-bit address is usually modified by the indirect and index operations to generate the effective address of a memory operand for each of these instructions.

| Opcode | x | xx | xxxxxxxx |

The description of one-address instructions follows:

| LDA MEM   ACC ← M[MEM]. Load accumulator |

LDA loads the ACC with the contents of the memory location (MEM) specified. Contents of MEM are not changed, but the contents of the ACC before the execution of this instruction are replaced by the contents of MEM.

| STA MEM   M[MEM] ← ACC. Store accumulator |

STA stores the contents of the ACC into the specified memory location. ACC contents are not altered.

| ADD MEM   ACC ← ACC + M[MEM]. Add |

ADD adds the contents of the memory location specified to the contents of the ACC. Memory contents are not altered.

| BRU MEM   PC ← MEM. Branch unconditional |

BRU transfers the program control to the address MEM. That is, the next instruction to be executed is at MEM.

| BIP MEM   IF ACC > 0 THEN PC ← MEM. Branch if ACC is positive |

The BIP instruction tests the N and Z bits of the PSR. If both of them are 0, then the program execution resumes at the address (MEM) specified; if not, execution continues with the next instruction in sequence. Since the PC must contain the address of the instruction to be executed next, the branching operation corresponds to transferring the address into PC.

| BIN MEM   IF ACC < 0 THEN PC ← MEM. Branch if accumulator negative |

The BIN instruction tests the N-bit of the PSR; if it is 1, program execution resumes at the address specified; if not, the execution continues with the next instruction in sequence.

| LDX MEM, INDEX   INDEX ← M[MEM]. Load index register |

The LDX loads the index register (specified by INDEX) with the contents of memory location specified. In the assembly language instruction format, INDEX will be 1, 2, or 3.

| STX MEM, INDEX   M[MEM] ← INDEX. Score index register |

The STX stores a copy of the contents of the index register specified by the index flag into the memory location specified by the address. The index register contents remain unchanged.

TIX MEM, INDEX INDEX ← INDEX + 1	Test index
IF INDEX = 0 THEN PC ← MEM.	increment

TIX increments the index register content by 1. Next, it tests the index register content; if it is 0, the program execution resumes at the address specified; otherwise, execution continues with the next sequential instruction.

TDX MEM, INDEX INDEX ← INDEX – 1	Test index
IF INDEX ≠ 0 THEN PC ← MEM.	decrement

TDX decrements the index register content by 1. Next it tests the index register content; if it is not equal to 0, the program execution resumes at the address specified; otherwise, execution continues with the next sequential instruction.

It is important to note that LDX, STX, TDX, and TIX instructions "refer" to an index register as one of the operands. Indexed mode of addressing is thus not possible with these instructions since the index field is used for the index register reference. Only direct and indirect modes of addressing can be used. For example,

LDA Z, 3 adds the contents of index register 3 to Z to compute the effective address EA. Then the contents of memory location EA are loaded into the ACC. Index register is not altered.	LDX Z, 3 loads the index register 3 from the contents of memory location Z.

Input/Output Instructions: Since ASC has one input and one output device, the address, index, and indirect fields in the instruction word are not used. Thus, these are also zero-address instructions.

RWD   ACC ← Input data. Read a word

RWD instruction reads a 16-bit word from the input device into the ACC. The contents of the ACC before RWD are thus lost.

WWD   Output ← ACC. Write a word

WWD instruction writes a 16-bit word from the ACC onto the output device. ACC contents remain unaltered.

5.1.4 Addressing Modes

Addressing modes allowed by a machine are influenced by the programming languages and corresponding data structures that the machine uses. ASC instruction format allows the most common addressing modes. Various other modes are used in machines commercially available. They are described in Chapter 8.

ASC addressing modes are described here with reference to the load accumulator (LDA) instruction. Here, Z is assumed to be the symbolic address of the memory location 10. For each mode, the assembly language format is shown first, followed by the instruction format encoded in binary (i.e., the machine language). The *effective address* calculation and the effect of the instruction are also illustrated. Note that the effective address is the address of the memory word where the operand is located.

Direct addressing
Instruction format: LDA Z

| 00010 | 0 | 00 | 00001010 |

Effective address: Z or #HA

Effect: ACC ← M[Z].

The effect of this instruction is illustrated in Figure 5.4a. We will use hexadecimal notation to represent all data and addresses in the following diagrams.

Note: Contents of register and memory are shown in hexadecimal.

Indexed addressing
Instruction format: LDA Z, 2

| 00010 | 0 | 10 | 00001010 |

Effective address   Z + index register 2
Effect   ACC ← M[Z + index register 2].

The number in the operand field after the comma denotes the index register used. Assuming that index register 2 contains 3, Figure 5.4b illustrates the effect of this instruction. The numbers in circles show the sequence of operations. Contents of index register 2 are added to Z to derive the effective address Z + 3. Contents of location Z + 3 are then loaded into the accumulator, as shown in Figure 5.4b.

Note that the *address field* of the instruction refers to Z and the contents of the index register specify an offset from Z. Contents of an index register can be varied by using LDX, TIX, and TDX instructions, thereby accessing various memory consecutive memory locations dynamically, by changing the contents of the index register. Further, since index registers are 16 bits wide, the effective address can be 16 bits long, thereby extending the memory addressing range to 64k from the range of 256 locations possible with 8 address bits.

The most common use of indexed addressing mode is in referencing the elements of an array. The address field in the instruction points to the first element. Subsequent elements are referenced by incrementing the index register.

Indirect addressing
Instruction format: LDA* Z

| 00010 | 1 | 00 | 00001010 |

Effective address   M[Z]
Effect   MAR ← M[Z].
                              ACC ← M[MAR].
                              i.e., ACC ← M[M[Z]].

The asterisk next to the mnemonic denotes the indirect addressing mode. In this mode, the address field points to a location where the address of the operand can be found (see Figure 5.4c).

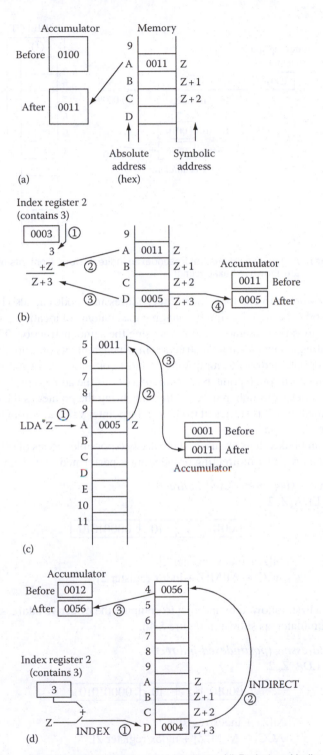

Figure 5.4 Addressing modes. (a) Direct. (b) Indexed. (c) Indirect. (d) Preindexed indirect.

(continued)

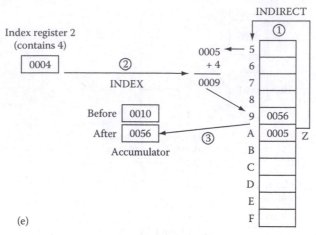

(e)

Figure 5.4 (continued)   Addressing modes. (e) Postindexed indirect. *Note:* Contents of register and memory are shown in hexadecimal.

Since a memory word is 16 bits long, the indirect addressing mode can also be used to extend the addressing range to 64 *k*. Further, by simply changing the contents of location Z in the previous illustration, we can refer to various memory addresses using the same instruction. This feature is useful, for example, in creating a multiple-jump instruction in which contents of Z are dynamically changed to refer to the appropriate address to jump to. The most common use of indirect addressing is in referencing data elements through pointers. A *pointer* contains the address of the data to be accessed. The data access takes place through indirect addressing, using the pointer as the operand. When data are moved to other locations, it is sufficient to change the pointer value accordingly, in order to access the data from the new location.

If both indirect and index flags are used, there are two possible modes of effective address computation, depending on whether indirecting or indexing is performed first. They are illustrated here.

Indexed-indirect addressing (preindexed-indirect)
Instruction format: LDA*Z, 2

| 00010 | 1 | 10 | 00001010 |

Effective address	M[Z + index register 2]
Effect	ACC ← M[M[Z + index register 2]].

Indexing is done first, followed by indirect to compute the effective address whose contents are loaded into the accumulator, as shown in Figure 5.4d.

Indirect-indexed addressing (postindexed-indirect)
Instruction format: LDA*Z, 2

| 00010 | 1 | 10 | 00001010 |

Effective address	M[Z] + index register 2
Effect	ACC ← M[M[Z] + index register 2].

Indirect is performed first, followed by indexing to compute the effective address whose contents are loaded into the accumulator, as shown in Figure 5.4e.

Note that the instruction formats in the two previous modes are identical. ASC cannot distinguish between these two modes. The indirect flag must be expanded to 2 bits if both of these modes are to be allowed. Instead, we assume that ASC always performs preindexed-indirect, and postindexed-indirect is not supported.

The earlier addressing modes are applicable to all single-address instructions. The only exceptions are the index reference instructions (LDX, STX, TIX, and TDX) in which indexing is not permitted.

Consider an array of pointers located in consecutive locations in the memory. The preindexed-indirect addressing mode is useful in accessing the data elements since we can first index to a particular pointer in the array and indirect on that pointer to access the data element. On the other hand, the postindexed-indirect mode is useful in setting up pointers to an array since we can access the first element of the array by indirecting on the pointer and access subsequent elements of the array by indexing over the pointer value.

5.1.5 Other Addressing Modes

Chapter 8 describes several other addressing modes that are employed in practice. For example, it is convenient sometimes in programming to include data as part of the instruction. *Immediate addressing mode* is used in such cases. Immediate addressing implies that the data are part of the instruction itself. This mode is not allowed in ASC, but the instruction set can be extended to include instructions such as load immediate (LDI) and add immediate (ADI). In such instructions, the opcode field will contain a 5-bit opcode, and the remaining 11 bits will contain the data. For instance,

LDI 10 would imply loading 10 into ACC
ADI 20 would imply adding 20 to the ACC

Since ASC does not permit this addressing mode, the ASC assembler is designed to accept the so-called literal addressing mode, which simulates the immediate addressing mode on ASC. Refer to the following section for further details.

5.1.6 Addressing Limitations

As discussed earlier, ASC instruction format restricts the direct-addressing range to the first 256 locations in the memory. Thus, if the program and data can fit into locations 0 through 255, no programming difficulties are encountered. If this is not possible, the following programming alternatives can be used:

1. The program resides in the first 256 locations, and the data reside in higher-addressed locations in the memory. In this case, all instruction addresses can be represented by the 8-bit address field. Since data references require an address field longer than 8 bits, all data references are handled using indexed and indirect addressing modes. For example, the data location 300 can be loaded into the ACC by either of the following instructions:
   a. LDA 0,2 assuming that index register 2 contains 300
   b. LDA* 0 assuming that location 0 in the memory contains 300
2. Data reside in the first 256 locations, and the program resides beyond location 255. In this case, all data reference instructions (such as LDA, STA) can use direct, indirect, and/or indexed modes, but all other memory reference instructions (such as BRU, BIP, BIN) must use indexed and/or indirect modes.
3. If the program and data both reside beyond location 255, all memory reference instructions must be indirect and/or indexed.

Recall that the index reference instructions can use only the direct and indirect modes of addressing.

5.1.7 Machine Language Programming

It is possible to write a program for ASC using absolute addresses (actual physical memory addresses) and opcodes only, since the instruction set and instruction and data formats are now known. Such programs are called *machine language programs*. They need not be further translated for the hardware to interpret them since they are already in binary form. Programming at this level is tedious, however. A program in ASC machine language to add two numbers and store the sum in a third location is shown in the following:

0001	0000	0000	1000
0011	0000	0000	1001
0010	0000	0000	1010

Can you decode these instructions and determine what the program is doing?

Modern-day computers are seldom programmed at this level. All programs must be at this level, however, before execution of the program can begin. Translators (assemblers and compilers) are used in converting programs written in assembly and high-level languages into this machine language. We will discuss a hypothetical assembler for ASC assembly language in the next section.

5.2 ASC ASSEMBLER

An assembler that translates ASC assembly language program into machine language programs is available. We will provide details of the language as accepted by this assembler and outline the assembly process in this section.

An assembly language program consists of a sequence of statements (instructions) coded in mnemonics and symbolic addresses. Each statement consists of four fields: label, operation (mnemonic), operand, and comments, as shown in Figure 5.5.

The *label* is a symbolic name denoting the memory location where the instruction itself is located. It is not necessary to provide a label for each statement. Only those statements that are referenced from elsewhere in the program need labels. When provided, the label is a set of alphabetic and numeric characters, the first of which must be an alphabetic character. The *mnemonic* field contains the instructions mnemonic. An "*" following the instruction mnemonic denotes indirect addressing. The *operand* field consists of symbolic addresses, absolute addresses, and index register designations. Typical operands are shown in Figure 5.6.

The *comments* fields start with a ".". This optional field consists only of comments by the programmer. It does not affect the instruction in any way and is ignored by the assembler. An "*" as the first character in the label field designates that the complete statement is a comment.

Each instruction in an assembly language program can be classified as either an *executable instruction* or an *assembler directive* (or a pseudo-instruction). Each of the 16 instructions in ASC instruction set is an executable instruction. The assembler generates a machine language instruction corresponding to each such instruction in the program. A pseudo-instruction is a directive to the assembler. This instruction is used to control the assembly process, to reserve memory locations,

Label	Mnemonic	Operand	Comments[a]
Consists of alphabetic and numeric characters.	Comprises three-character standard symbolic opcodes.	Consists of absolute and symbolic addresses.	Starts with a ".". Assembler ignores it.
First character must be alphabetic.	An "*" as fourth character signifies indirect addressing.	Index register designations following a ",".	
An "*" as first character denotes that complete statement is a comment.			

[a] Optional field.

Figure 5.5 Assembly language statement format. *Note:* A space is required between label and mnemonic fields and between mnemonic and operand fields.

Operand	Description	Memory
25	Decimal 25 (by default).	
#H25	Hexadecimal 25.	
#O25	Octal 25.	
#B1001	Binary 1001.	
Z	Symbolic address Z.	
Z, 1	Z indexed with index register 1.	
Z + 4	Four locations after Z.	
Z + 4, 1	Address Z + 4 indexed with register 1.	
Z − 4	Address Z − 4 (four locations before Z).	
Z − P	Z and P are symbolic addresses; Z − P yields an absolute address; Z must be at a higher physical address than P for Z − P value to be positive.	

Memory (right portion):

```
7        Z − 4
8        Z − 3
9        Z − 2
A        Z − 1
B        Z
C        Z + 1
D        Z + 2
E        Z + 3

Absolute    Symbolic
address     address
```

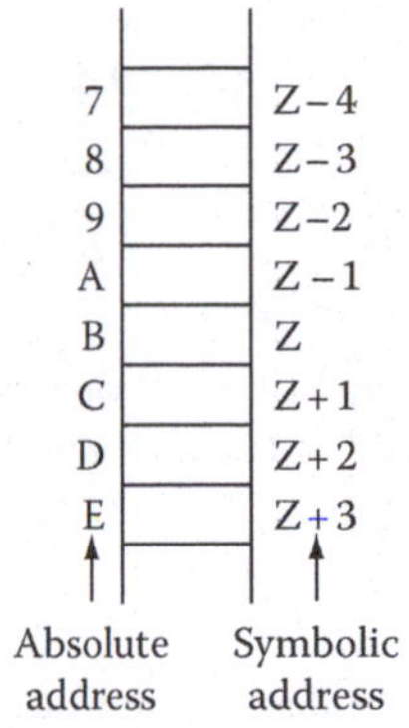

Figure 5.6 ASC operands.

and to establish constants required by the program. The pseudo-instructions when assembled do not generate machine language instructions and as such are not executable. Care must be taken by the assembly language programmer to partition the program such that an assembler directive is not in the execution sequence. A description of ASC pseudo-instructions follows:

ORG   Address   Origin

The function of ORG directive is to provide the assembler the memory address where the next instruction is to be located. ORG is usually the first instruction in the program. Then the operand field of ORG provides the starting address (i.e., the address where the first instruction in the program is located). If the first instruction in the program is not ORG, the assembler defaults to a starting address of 0. There can be more than one ORG directive in a program.

END   Address   Physical end

END indicates the physical end of the program and is the last statement in a program. The operand field of the END normally contains the label of the first executable statement in the program.

EQU   Equate

EQU provides a means of giving multiple symbolic names to memory locations, as shown by the following example:

Example 5.1

A	EQU	B	A is another name for B. (B must already be defined.)
A	EQU	B + 5	A is the name of location B + 5.
A	EQU	10	A is the name of the absolute address 10.

BSS   Block storage starting

BSS is used to reserve blocks of storage locations for intermediate or final results.

Example 5.2

Z	BSS	5	Reserve five locations, the first of which is named Z.	

Z	
Z+1	
Z+2	
Z+3	
Z+4	

The operand field always designates the number of locations to be reserved. Contents of these reserved locations are not defined.

BSC	Block storage of constants

BSC provides a means of storing constants in memory locations in addition to reserving those locations. The operand field consists of one or more operands (separate by a ";"). Each operand requires one memory word.

Example 5.3

Z	BSC 5	Reserves one location named Z containing a 5
P	BSC 5, –6, 7	Reserves three locations: P containing 5, P + 1 containing –6, and P + 2 containing 7

Literal Addressing Mode: It is convenient for the programmer to be able to define constants (data) as a part of the instruction. This feature also makes an assembly language program more readable. Literal addressing mode enables this. A literal is a constant preceded by an " = ". For example,

$$LDA = 2$$

implies loading a constant 2 (decimal) into the accumulator, and

$$ADD = \#H10$$

implies adding a #H10 to the accumulator. ASC assembler recognizes such literals, reserves an available memory location for the constant in the address field, and substitutes the address of the memory location into the instruction.

We will now provide some assembly language programs as examples.

Example 5.4

Figure 5.7 shows a program to add three numbers located at A, B, and C and save the result at D. The program is ORiGined to location 10. Note how the HLT statement separates the program logic from the data block. A, B, and C are defined using BSC statements, and one location is reserved for D using a BSS.

Example 5.5

Figure 5.8 shows a program to accumulate five numbers stored starting at location X in memory and store the result at Z. Here, index register 1 is first set to 4 so as to point to the last number in the block (i.e., at X + 4). The ACC is set to 0. TDX is used to access numbers one at a time from the last to first and to terminate the loop after all the numbers are accumulated.

```
* Program to add three numbers
                ORG        10
    BEGIN       LDA        A
                ADD        B
                ADD        C
                STA        D
                HLT
    A           BSC        5
    B           BSC        7
    C           BSC        -3
    D           BSS        1
                END        BEGIN
```

Figure 5.7 Program to add three numbers. *Note:* BEGIN points to the first executable instruction in the program, and the operand of the END instruction is BEGIN.

```
* A program to accumulate the values of
* five numbers located at x
            ORG    0
    START   LDX    =4,1
            LDA    =0              .Zero accumulator
    LOOP    ADD    X,1             .ADD loop
            TDX    LOOP,1
            ADD    X               .ADD the remaining
                                    value
            STA    Z               .Store the result
                                    in Z
            HLT
    X       BSC    5,35,26,-7,4
    Z       BSS    1
            END    START
```

Figure 5.8 Program to accumulate a block of numbers.

Example 5.6

Division can be treated as the repeated subtraction of the divisor from the dividend until a zero or negative result is obtained. The quotient is equal to the maximum number of times the subtraction can be performed without yielding a negative result. Figure 5.9 shows the division routine.

The generation of the object code from the assembly language programs is described in the following section.

5.2.1 Assembly Process

The major functions of the assembler program are (1) to generate an address for each symbolic name in the program and (2) to generate the binary equivalent of each assembly instruction. The assembly process is usually carried out in two scans over the source program. Each of these scans is called a *pass* and the assembler is called a *two-pass assembler.* The first pass is used for allocating a memory location for each symbolic name used in the program; during the second pass, references to these symbolic names are resolved. If the restriction is made that each symbol must be defined before it can be referenced, one pass will suffice.

Details of the ASC two-pass assembler are given in Figures 5.10 and 5.11. The assembler uses a counter known as *location counter* (LC) to keep track of the memory locations used. If the first instruction in the program is ORG, the operand field of ORG defines the initial value of LC;

```
0001                    *        SOURCE PROGRAM
0002                    *                          . A DIVISION ALGORITHM FOR ASC
                                                     [A/B]
0003                    *                          . A AND B ARE NON-NEGATIVE
                                                     INTEGERS
0004                             ORG    0
0005   0000   8000  BEGIN   RWD                    . READ A
0006   0001   201E          STA    A
0007   0002   6004          BIP    NEXT
0008   0003   501B          BRU    OUT             . IF A=0, QUOTIENT=0
0009   0004   8000  NEXT    RWD                    . READ B
000A   0005   201F          STA    B
000B   0006   6008          BIP    INIT
000C   0007   501B          BRU    OUT             . IF B=0, THEN BY DEFN QUO=0
000D   0008   1021  INIT    LDA    ZERO            . INITIALIZE QUOTIENT TO 0
000E   0009   2020          STA    COUNT
000F   000A   101F          LDA    B
0010   000B   4000          TCA
0011   000C   201F          STA    B               . STORE NEGATIVE B IN B
0012   000D   101F  LOOP    LDA    B
0013   000E   301E          ADD    A               . A- (I*B). I=NUMBER OF LOOPS
0014   000F   201E          STA    A
0015   0010   7015          BIN    FINISH          . RESULT NEGATIVE, FINISHED
0016                    *                          . DO NOT INCREMENT COUNTER
0017   0011   6017          BIP    OTHER
0018   0012   1020          LDA    COUNT           . REMAINDER IS ZERO
0019   0013   3022          ADD    ONE
001A   0014   501C          BRU    OUT1
001B   0015   1020  FINISH  LDA    COUNT           . LOAD FINAL COUNT
001C   0016   501C          BRU    OUT1
001D   0017   1020  OTHER   LDA    COUNT           . INCREMENT QUOTIENT
001E   0018   3022          ADD    ONE             . RESULT IS 0 OR POSITIVE
001F   0019   2020          STA    COUNT
0020   001A   500D          BRU    LOOP
0021   001B   1021  OUT     LDA    ZERO
0022   001C   9000  OUT1    WWD                    . WRITE RESULT
0023   001D   0000          HLT
0024   001E   0000  A       BSS    1
0025   001F   0000  B       BSS    1
0026   0020   0000  COUNT   BSS    1
0027   0021   0000  ZERO    BSC    0 .
0028   0022   0001  ONE     BSC    1               . END
0029                        END    BEGIN
```

Figure 5.9 ASC division program. *Note:* The assembler is assumed to assemble 0s for undefined bits in BSS and zero-address instructions.

otherwise, LC is set to 0. LC is incremented appropriately during the assembly process. Content of LC at any time is the address of the next available memory location.

The assembler performs the following tasks during the first pass:

1. Enters labels into a symbol table along with the LC value as the address of the label
2. Validates mnemonics
3. Interprets pseudo-instructions completely
4. Manages the LC

The major activities during the second pass are

  1. Evaluation of the operand field
  2. Insertion of opcode, address, and address modifiers into the instruction format
  3. Resolution of literal addressing

The assembler uses an opcode table to extract opcode information. The opcode table is a table storing each mnemonic, the corresponding opcode, and any other attribute of the instruction useful for the assembly process. The symbol table created by the assembler consists of two entries for each symbolic name: the symbol itself and the address in which the symbol will be located. We will illustrate the assembly process in Example 5.7.

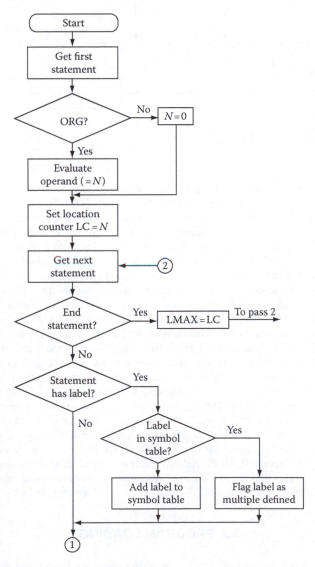

Figure 5.10 Assembler pass 1.

(continued)

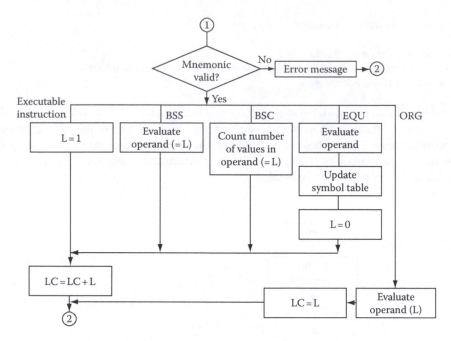

Figure 5.10 (continued) Assembler pass 1.

Example 5.7

Consider the program shown in Figure 5.12a. The symbol table is initially empty. LC starts at the default value of 0. The first instruction is ORG. Its operand field is evaluated and the value (0) is entered into LC. The label field of the next instruction is BEGIN. BEGIN is entered into symbol table and assigned the address of 0. The mnemonic field has LDX, which is a valid mnemonic. Since this instruction takes up one memory word, LC is incremented by 1. The operand field of LDX instruction is not evaluated during the first pass. This process of scanning the label and mnemonic fields, entering labels (if any) into the symbol table, validating mnemonic, and incrementing LC continues until END instruction is reached. Pseudo-instructions are completely evaluated during this pass. LC values are shown in Figure 5.12a along with symbol table entries at the end of pass 1 in Figure 5.12b. By the end of the first pass, the LC will have advanced to E, since BSS 4 takes up four locations (A through D).

During the second pass, machine instructions are generated using the source program and the symbol table. The operand fields are evaluated during this pass and instruction format fields are appropriately filled for each instruction. Starting with LC = 0, the label field of instruction at 0 is ignored and the opcode (11000) is substituted for the mnemonic LDX. Since this is a one-address instruction, the operand field is evaluated. There is no "*" next to the mnemonic, and hence the indirect flag is set to 0. The absolute address of C is obtained from the symbol table and entered into the address field of the instruction, and the index flag is set to 01. This process continues for each instruction until END is reached. The object code is shown in Figure 5.9c in binary and hexadecimal formats. The symbol table shown in Figure 5.9d has one more entry corresponding to the literal = 0. Note that the instruction LDX = 0,2 has been assembled as LDX #HE, 2 with the location #HE containing a 0. Contents of the words reserved in response to BSS are not defined, and unused bits of HLT instruction words are assumed to be 0s.

5.3 PROGRAM LOADING

The object code must be loaded into the machine memory before it can be executed. ASC console can be used to load programs and data into the memory. Loading through the console is tedious and time consuming, however, especially when programs are large. In such cases, a small program that reads the object code statements from the input device and stores them in appropriate memory locations is first written, assembled, and loaded (using the console) into machine memory. This loader program is then used to load the object code or data into the memory.

Figure 5.13 shows a loader program for ASC. Instead of loading this program each time through the console, it can be stored in a ROM that forms part of the 64 k memory space of ASC. Then loading can be initiated by setting the PC to the beginning address of the loader (using the console) and starting the machine. Note that the loader occupies locations 224 through 255. Hence, care must be taken to make sure that other programs do not overwrite this space.

Note also that the assembler itself must be in the binary object code form and loaded into ASC memory before it can be used to assemble other programs. That means the assembler must be translated from the source language to ASC binary code either manually or by implementing the

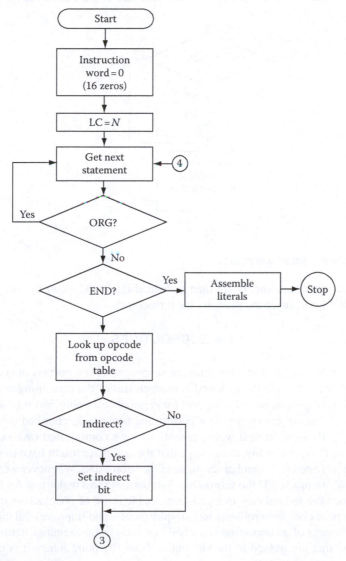

Figure 5.11 Assembler pass 2.

(continued)

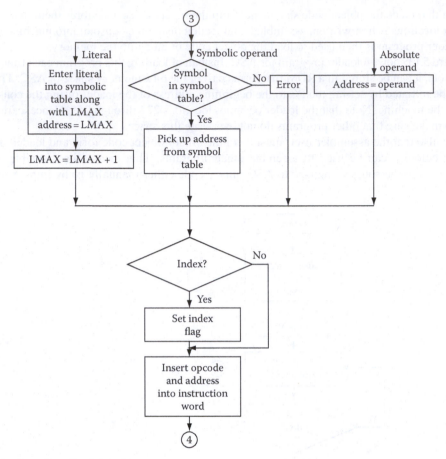

Figure 5.11 (continued)   Assembler pass 2.

assembler on some other machine. It can then be loaded into ASC either by using the loader or by retaining it in a ROM portion of the memory for further use.

5.4 SUBROUTINES

A subroutine (function, method, procedure, or subprogram) is a portion of code within a larger program, which performs a specific task and is independent of the remaining code. The syntax of many programming languages includes support for creating self-contained subroutines.

A subroutine consists of instructions for performing some task, chunked together and given a name. "Chunking" allows us to deal with a potentially very complicated task as a single concept. Instead of worrying about the many, many steps that the computer might have to go through to perform that task, we just need to remember the name of the subroutine. Whenever we want our program to perform the task, we just "call" the subroutine. Subroutines are a major tool for handling the complexity. They reduce the redundancy in a program, enable reuse of code across multiple programs, allow us to decompose complex problems into simpler pieces, and improve readability of a program.

Typical components of a subroutine are a body of code to be executed when the subroutine is called, parameters that are passed to the subroutine from the point where it is called, and one or more values that are returned to the point where the call occurs.

Some programming languages, like Pascal and FORTRAN, distinguish between functions, which return values, and subroutines or procedures, which do not. Other languages, like C and

LC		Instruction	
		Org	0
0	Begin	LDX	C,1
1		LDX	=0,2
2	LOOP	LDA	D,2
3		ADD	SUM
4		STA	SUM
5		TIX	TEMP, 2
6	TEMP	TDX	LOOP, 1
7		HLT	
8	SUM	BSC	0
9	C	BSC	4
A	D	BSC	1, 2, 3, 4
		END	BEGIN

(a)

Symbol	Address (H)
BEGIN	0
LOOP	2
TEMP	6
SUM	8
C	9
D	A

LMAX = #HE

(b)

LC	Opcode	Indirect	Index	Address	Object Code (HEX)
0	1100 0	0	01	0000 1001	C109
1	1100 0	0	10	0000 1110	C203
2	0010 0	0	10	0000 1010	120A
3	0110 0	0	00	0000 1000	3008
4	0010 0	0	00	0000 1000	2008
5	1110 0	0	10	0000 0110	E206
6	1111 0	0	01	0000 0010	C102
7	0000 0	-	--	---- ----	0000
8	0000 0000 0000 0000				0000
9	0000 0000 0000 0100				0004
A					dddd
B	4 locations reserved				dddd
C					dddd
D					dddd
E	0000 0000 0000 0000				0000

(c)

Symbol	Address (H)
BEGIN	0
LOOP	2
TEMP	6
SUM	8
C	9
D	A
=0	E

LMAX = #HF

(d)

Undefined

Figure 5.12 Assembly process. (a) Program. (b) Symbol table after pass 1. (c) Object code. (d) Symbol table after pass 2.

LISP, do not make this distinction and treat those terms as synonymous. The name method is commonly used in connection with object-oriented programming, specifically for subroutines that are part of objects.

Calling a subroutine means jumping to the first instruction in the subroutine, using a JMP instruction. The execution of the subroutine will end with a jump back to the same point in the program from which the subroutine was called so that the program can pick up where it left off before calling the subroutine. This is known as returning from the subroutine. If the subroutine is to be reusable in a meaningful sense, it must be possible to call the subroutine from many different places in a program. If this is the case, how does the computer know what point in the program to return to when the subroutine ends? The answer is that the return point has to be recorded somewhere before the subroutine is called. The address in memory to which the computer is supposed to return after the subroutine ends is called the *return address*. Before jumping to the start of the subroutine, the program must store the return address in a place where the subroutine can find it. When the subroutine has finished performing its assigned task, it ends with a jump back to the return address.

Label	Mnemonic	Operand	Comment
	ORG	244	
START	RWD		.Read starting address
	STA	SAVE	
	LDX	SAVE,2	.Starting address in index register 2
	RWD		.Input number of statements N
	STA	SAVE	.SAVE contains N
	LDX	SAVE,1	.Load (N) into index register 1
LOOP	RWD		.Input an object code statement
	STA	0,2	.Store the object code statement
LL	TIX	LL,2	.Increment index register 2
	TDX	LOOP,1	.Decrement N by 1 and loop
	HLT		
SAVE	BSS	1	
	END	START	

Figure 5.13 Loader for ASC.

Let us expand ASC instruction set to include two instructions to handle subroutines: jump to subroutine (JSR) and return from subroutine (RET). The following program illustrates the subroutine call and return operations.

	ORG	0
BEGIN	LDA	X
	JSR	SUB1
A	WWD	
	HLT	
X	BSC	23
SUB1	STA	Z
	ADD	Z
	RET	
	END	BEGIN

The main program calls the subroutine SUB1 through JSR. The program control transfers to SUB1, and the two instructions in the subroutine are executed followed by RET. The effect of executing RET is to return control to the return address A in the main program. The main program executes WWD followed by HLT.

To enable this operation, the JSR instruction has to store the return address (i.e., the address of the instruction following JSR) somewhere, before jumping to the subroutine. When RET is executed, the return address is retrieved and the program transfers to the instruction following the call.

The return address can be stored in a dedicated register or a dedicated memory location. If the subroutine calls another subroutine (i.e., nested call), the return address corresponding to the first call will be lost and hence the program cannot return properly. We typically push the return address on to a stack (see Appendix B for details on stack implementation) during the call and it is popped from the stack during return. This allows subroutine calls to be nested.

The return address is not the only item of information that the program has to send to the subroutine. If the task of the subroutine is to multiply a number by seven, the main program has to tell the subroutine which number to multiply by seven. This information is said to be a parameter of the subroutine. Similarly, the subroutine has to get its answer, the result of multiplying the parameter value by seven back to the main program. This answer is called the return value of the subroutine. In the previous program, the parameter value is in the accumulator before calling the subroutine. The subroutine knows to look for it there. Before it jumps back to the main program, the subroutine puts its return value in the accumulator. The main program knows to look for it there. Passing parameter values and return values back and forth in a register, such as the accumulator, is a very simple and efficient method of communication between a subroutine and the rest of a program. In ASC, the ACC and the three index registers can be used for parameter passing. A more common way of passing parameters is for the main program to push them onto the stack just prior to calling the subroutine and the subroutine to retrieve them from the stack, operate on them, and return the results back on the stack, for the main program.

5.5 MACROS

A macro is a group of instructions that are codified only once and can be used as many times as necessary in a program. Unlike subroutines, when a macro is called from a program, the macro call is replaced by the group of instructions that constitute the macro body. Thus, each call to the macro results in the substitution of the macro body. We can also pass parameters to a macro. When a macro call is executed, each parameter is substituted by the name or value specified at the time of the call.

An example of macro definition is shown here:

	MACRO	ADD2
	BIP	POS
	ADD	= 1
POS	ADD	= 1
	ENDM	

MACRO is an assembler directive indicating the macro definition. ADD2 is the macro name. ENDM signifies the end of macro definition. The instructions in the macro body add 1 to the accumulator if it is positive; otherwise, they add 2 to the accumulator.

To call the macro, we simply use ADD2 as an instruction in the program. The assembler will replace ADD2 with the three instructions corresponding to the macro body every time ADD2 is used in the program. Usually this substitution (macro expansion) is done prior to the first pass of the assembler.

A macro to accumulate two values A and B and store the result in C is defined after. A, B, C are parameters to the macro.

MACRO	ABC (A, B, C)
LDA	A
ADD	B
STA	C
ENDM	

A call to this macro would be

ABC (X, Y, Z)

This call would result in the replacement of A, B, and C with X, Y, and Z during the macro expansion. Note that once the macro is expanded, A, B, and C are not visible in the program. Also, X, Y, and Z must be defined.

Macros allow the programmer to define repetitive code blocks once and use the macro call to expand them into the program. Thus, macros convert into the code in line with the program they are called. The control overhead required by the subroutines (to save and retrieve return address) is thus not needed. The advantage of subroutines is that the subroutine body is not expanded into the calling program, even with multiple calls. Subroutines save instruction memory at the cost of run time overhead. Macros consume instruction memory but eliminate subroutine call/return overhead.

One of the facilities that the use of macros offers is the creation of libraries, which are groups of macros that can be included in a program from a different file. The creation of these libraries is very simple. We first create a file with all the macro definitions and save it as a text file such as MACROS1. To call these macros, it is only necessary to use an instruction such as Include MACROS1 at the beginning of the program.

5.6 LINKERS AND LOADERS

In practice, an executable program is composed of a main program and several subroutines (*modules*). The modules can either come from a predefined library or developed by the programmer for the particular application. Assemblers (and compilers) allow independent translation of program modules into corresponding machine code. A linker is a program that takes one or more modules generated by assemblers and compilers and assembles them into a single executable program. During the linking process, object files and static libraries are assembled into a new library or an executable program.

The program modules contain machine code and information for the linker. This information comes mainly in the form of two types of symbol definitions:

1. Defined or exported symbols are functions or variables that are present in the module represented by the object and that should be available for use by other modules.
2. Undefined or imported symbols are functions or variables that are called or referenced by this object, but not internally defined.

The linker's job is to resolve references to undefined symbols by finding out which other module defines a symbol in question and replacing placeholders with the symbol's address.

The linker also takes care of arranging the modules in a program's address space. This may involve *relocating* code that assumes a specific base address to another base. Since an assembler (compiler) seldom knows where a module will reside, it often assumes a fixed base location. Relocating machine code may involve retargeting of absolute jumps, loads, and stores. For instance, in case of ASC programs, the ORG directive defines the beginning address of the program module. If all the modules are assembled with their own origins, the linker has to make sure that they do not overlap in memory when put together into a single executable program. Even after the linking is done, there is no guarantee that the executable will reside at its specified origin when loaded into the machine memory, since other programs may be residing in the memory space at the time of loading. Thus, the program may have to be relocated.

Linkers and loaders perform several related but conceptually separate actions:

Program loading: Copying a program from secondary storage into main memory so that it is ready to run. In some cases loading just involves copying the data from disk to memory, in others it involves allocating storage, setting protection bits, or arranging for virtual memory to map virtual addresses to disk pages.

Relocation: As mentioned earlier, relocation is the process of assigning load addresses to the various parts of the program, adjusting the code and data in the program to reflect the assigned addresses.

In many systems, relocation happens more than once. It is quite common for a linker to create a program from multiple subprograms and create one linked output program that starts at zero, with the various subprograms relocated to locations within the big program. Then, when the program is loaded, the system picks the actual load address, and the linked program is relocated as a whole to the load address.

Symbol resolution: When a program is built from multiple subprograms, the references from one subprogram to another are made using symbols; a main program might use a square root routine called `sqrt`, and the math library defines `sqrt`. A linker resolves the symbol by noting the location assigned to `sqrt` in the library and patching the caller's object code such that the call instruction refers to that location.

Although there is considerable overlap between linking and loading, it is reasonable to define a program that does program loading as a loader and one that does symbol resolution as a linker. Either can do relocation, and there have been all-in-one linking loaders that do all three functions.

5.6.1 Dynamic Linking

Modern operating system environments allow *dynamic linking*, that is, the postponing of the resolving of some undefined symbols until a program is run. That means that the executable still contains undefined symbols, plus a list of modules or libraries that will provide definitions for these. Loading the program will load these modules/libraries as well and perform a final linking. This approach offers two advantages:

1. Often-used libraries (e.g., the standard system libraries) need to be stored in only one location, not duplicated in every single binary.
2. If an error in a library function is corrected by replacing the library, all programs using it dynamically will immediately benefit from the correction. Programs that included this function by static linking would have to be relinked first.

Dynamic linking means that the data in a library are not copied into a new executable or library at compile time but remain in a separate file on disk. Only a minimal amount of work is done at compile time by the linker—it only records what libraries the executable needs and the index names or numbers. The majority of the work of linking is done at the time the application is loaded (load time) or during the execution of the program (run time). At the appropriate time, the loader finds the relevant libraries on disk and adds the relevant data from the libraries to the program's memory space.

Some operating systems can only link in a library at load time, before the program starts executing; others may be able to wait until after the program has started to execute and link in the library just when it is actually referenced (i.e., during run time). The latter is often called "delay loading." In either case, such a library is called a *dynamically linked library*.

Dynamic linking was originally developed in the Multics operating system, starting in 1964. It was also a feature of the Michigan Terminal System (MTS), built in the late 1960s. In Microsoft Windows, dynamically linked libraries are called dynamic link libraries or "DLLs."

One wrinkle that the loader must handle is that the location in memory of the actual library data cannot be known until after the executable and all dynamically linked libraries have been loaded into memory. This is because the memory locations used depend on which specific dynamic libraries have been loaded. It is not possible to store the absolute location of the data in the executable, not even in the library, since conflicts between different libraries would result: if two of them specified the same or overlapping addresses, it would be impossible to use both in the same program. This might change with increased adoption of 64-bit architectures, which offer enough virtual memory addresses to give every library ever written its own unique address range.

It would theoretically be possible to examine the program at load time and replace all references to data in the libraries with pointers to the appropriate memory locations once all libraries have been loaded, but this method would consume unacceptable amounts of either time or memory. Instead, most dynamic library systems link a symbol table with blank addresses into the program at compile time. All references to code or data in the library pass through this table, the *import directory*. At load time, the table is modified with the location of the library code/data by the loader/linker. This process is still slow enough to significantly affect the speed of programs that call other programs at a very high rate, such as certain shell scripts.

The library itself contains a table of all the methods within it, known as *entry points*. Calls into the library "jump through" this table, looking up the location of the code in memory and then calling it. This introduces overhead in calling into the library, but the delay is usually so small as to be negligible.

Dynamic linkers/loaders vary widely in functionality. Some depend on explicit paths to the libraries being stored in the executable. Any change to the library naming or layout of the file system will cause these systems to fail. More commonly, only the name of the library (and not the path) is stored in the executable, with the operating system supplying a system to find the library on disk based on some algorithm.

Most UNIX-like systems have a "search path" specifying file system directories in which to look for dynamic libraries. On some systems, the default path is specified in a configuration file; in others, it is hard coded into the dynamic loader. Some executable file formats can specify additional directories in which to search for libraries for a particular program. This can usually be overridden with an environment variable, although it is disabled for setuid and setgid programs, so that a user cannot force such a program to run arbitrary code. Developers of libraries are encouraged to place their dynamic libraries in places in the default search path. On the downside, this can make installation of new libraries problematic, and these "known" locations quickly become home to an increasing number of library files, making management more complex.

Microsoft Windows will check the registry to determine the proper place to find an ActiveX DLL, but for other DLLs it will check the directory that the program was loaded from; the current working directory (only on older versions of Windows); any directories set by calling the SetDllDirectory( ) function; the System 32, System, and Windows directories; and finally the directories specified by the PATH environment variable.

One of the biggest disadvantages of dynamic linking is that the executables depend on the separately stored libraries in order to function properly. If the library is deleted, moved, or renamed, or if an incompatible version of the DLL is copied to a place that is earlier in the search, the executable could malfunction or even fail to load; damaging vital library files used by almost any executable in the system will usually render the system completely unusable.

Dynamic loading is a subset of dynamic linking where a dynamically linked library loads and unloads at run time on request. The request to load such a dynamically linked library may be made implicitly at compile time or explicitly by the application at run time. Implicit requests are made by adding library references, which may include file paths or simply file names, to an object file at compile time by a linker. Explicit requests are made by applications using a run time linker application program interface (API).

Most operating systems that support dynamically linked libraries also support dynamically loading such libraries via a run-time linker API. For instance, Microsoft Windows uses the API functions LoadLibrary, LoadLibraryEx, FreeLibrary, and GetProcAddress with Microsoft Dynamic Link Libraries; POSIX-based systems, including most UNIX and UNIX-like systems, use dlopen, dlclose, and dlsym.

5.7 SUMMARY

This chapter is a programmer's introduction to ASC organization. Details of an assembler for ASC, assembly language programming, and the assembly process have been described. A brief introduction to program loaders and linkers has been given. Various components of ASC have been assumed to exist, and no justification has been given as to why a component is needed. Architectural trade-offs used in selecting the features of a machine are discussed in subsequent chapters of the book. Chapter 6 provides the detailed hardware design of ASC. Further details on these topics can be found in the References.

PROBLEMS

5.1 For each of the following memory systems, determine the number of bits needed in MAR and MBR, assuming a word-addressable memory:
a. $64\,k \times 8$
b. $64\,k \times 32$
c. $32\,k \times 16$
d. $32\,k \times 32$
e. $128\,k \times 64$

5.2 Answer Problem 5.1, assuming that the memory is byte addressable.

5.3 Determine the content of the MAR of a $16\,k$, word-addressable memory for each of the following:
a. Word 48
b. Word 341
c. Lower half of the memory (i.e., words 0 through $8\,k - 1$)
d. Upper half of the memory (i.e., words $8\,k$ through $16\,k - 1$)
e. Even memory words (0, 2, 4, etc.)
f. Any of the 8 words 48 through 55

5.4 Calculate the effective address for each of the following instructions. (Remember: An "*" as the fourth symbol denotes indirect addressing.)
STA Z
STA Z,3
STA* Z
STA* Z,3 (Preindex)
STA* Z,3 (Postindex)
Assume index register 3 contains 5: Z is memory location #H10 and is the first location of a memory block containing 25, 7, 4, 86, 46, and 77.

5.5 Determine the contents of the ACC after each of the following instructions, assuming that the ACC is set to 25 before each instruction is executed:
SHL
ADD = 5
TCA
ADD = #H3
SHR

5.6 Determine the content of the ACC at the end of each instruction in Problem 5.5, assuming that the ACC is set to 25 to begin with and the effect of each instruction is cumulative; that is, the ACC is not reset to 25 at the beginning of second and subsequent instructions.

5.7 What is the difference (if any) between
a. An LC and a PC
b. END and HLT
c. BSS and BSC
d. ORG and EQU
e. Executable and pseudo-instructions

5.8   Show the contents (in hexadecimal) of the memory locations reserved by the following program:

	ORG	200
	BSS	3
Z	BSC	3, –3, 5
	BSS	24
M	EQU	Z + 1
	END	

5.9   What is the effect of inserting ORG 300 instruction after the instruction at Z, in the program of Problem 5.8?

5.10   There are two ways of loading the accumulator with a number:

	LDA	X	LDI 5
X	BSC	5	

a.   Which of the two executes faster? Why?

b.   What is the range of the data value that can be loaded by each method?

5.11   NOP (No operation) is a common instruction found in almost all instruction sets. How does the assembler handle this instruction? Why is it needed?

5.12   Another popular addressing mode is PC-relative addressing. An example of such an instruction is BRU + 5, meaning add 5 to the current PC value to determine the target address. Similarly, the target address of BRU − 5 is determined by subtracting 5 from the current PC value.

a.   How does the assembler handle this instruction?

b.   What is the range of relative jump addresses possible in ASC?

c.   What are the advantages and disadvantages of this mode compared to other jump addressing modes?

5.13   Give the symbol table resulting from the assembly of the following program:

	ORG	200
BEGIN	RWD	
	STA	A
	BIP	TEMP 1
	BIN	TEMP 1
	BRU	OUT
TEMP1	LDX	B, 1
	LDX	= 0, 2
LOOP	LDA	C, 2
	ADD	SUM
	STA	SUM
	TIX	TEMP2, 2
TEMP2	TDX	OUT, 1
	BRU	LOOP
OUT	HLT	
	ORG	500
SUM	BSC	0
A	BSS	1
B	BSC	4
C	BSS	4
	END	BEGIN

5.14 The dump of locations 64–77 of ASC memory is shown here. Decode the program segment represented by this object code.

1000	0000	0000	0000
0010	0000	1000	0001
0110	0000	0100	0101
0111	0000	0100	0101
0101	0000	0100	1101
1100	0001	1000	0010
1100	0010	1000	0111
0001	0010	1000	0011
0011	0000	1000	0000
0010	0000	1000	0000
1110	0010	0100	1011
1111	0001	0100	1101
0101	0000	0100	0111
0000	0000	0000	0000

5.15 Write ASC assembly language programs for the following. Start programs at locations #H0, using ORG 0 Statement.

a. Subtract an integer stored at memory location A from that at B and store the result at C.

b. Read several integers from the input device one at a time and store them in a memory location starting at Z. The input process should stop when the integer read has a value of 0.

c. Change the program in (b) to store only positive integers at a location starting at POS.

d. Modify the program in (b) to store the positive integers starting at POS and the negative integers starting at NEG.

e. Location #H50 contains an address pointing to the first entry in a table of integers. The table is also in the memory and the first entry is the number of entries in the table, excluding itself. Store the maximum and minimum valued integers at memory locations MAX and MIN, respectively.

f. SORT the entries in a table of *n* entries in increasing order of *magnitude*.

g. Multiply integers stored at memory locations A and B and store the result in C. Assume the product is small enough and can be represented in 16 bits. Note that multiplication is the repeated addition of multiplicand to itself multiplier times.

h. Compute the absolute value of each of the 50 integers located at the memory block starting at A and store them at the block starting at B.

i. Read a sequence of numbers and compute their minimum, maximum, and average values. Reading a value of 0 should terminate the reading process. What problems arise in computing the average?

5.16 Assemble each program in Problem 5.15 and list the object code in binary and hexadecimal forms.

5.17 What restrictions are to be imposed on the assembly language if a single-pass assembler is needed?

5.18 Numbers larger than $(2^{15} - 1)$ are required for certain applications. Two ASC words can be used to store each such number. What changes to the ASC instruction set are needed to enable addition of numbers that are each stored in two consecutive memory words?

5.19 SHR instruction must be enhanced to allow multiple shifts. The address field can be used to represent the shift count; for example,

SHR 5

implies a shift right by 5 bits. Discuss the assembler modifications needed to accommodate multiple-shift SHR. The hardware is capable of performing only one shift at a time.

5.20 Write a subroutine to subtract two numbers in ASC. Use this subroutine in a program that inputs 10 pairs of numbers and outputs the difference of each pair.

5.21 Write a MACRO to perform the subtraction of two numbers. Use it in a program that inputs 10 pairs of numbers and outputs the difference of each pair.

5.22 Compare the solutions in Problems 5.21 and 5.22 in terms of speed and memory requirements.

5.23 Consider the two programs that follow, written in pseudocode.

      Pseudocode-1
      Inputs in locations A and B
      Output in C
      C = A, A = B, B = C
      Pseudocode-2
      A = A + B
      B = A – B
      A = A – B

    a. What is the result of executing each of them?
    b. Use ASC instructions to code both. What is the difference between the two implementations in terms of program size and execution time?

5.24 Write a program for ASC to compute the value of pi to four decimal places.

BIBLIOGRAPHY

Aho, A.V., Sethi, R., and Ullman, J.D., *Compilers: Principles, Techniques and Tools*, Reading, MA: Addison-Wesley, 1986.

Calingaert, P., *Assemblers, Compilers and Program Translation*, Rockville, MD: Computer Science Press, 1979.

Stallings, W., *Operating Systems*, New York: Macmillan, 2005.

Tannenbaum, A.S., Woodhull, A.S., and Woodhull, A., *Operating Systems: Design and Implementation*, Englewood Cliffs, NJ: Prentice Hall, 1997.

Watt, D.A., *Programming Language Processors: Compilers and Interpreters*, Englewood Cliffs, NJ: Prentice Hall, 1993.

CHAPTER 6

A Simple Computer: Hardware Design

The organization of a simple computer (ASC) provided in Chapter 5 is the programmer's view of the machine. We illustrate the complete hardware design of ASC in this chapter. Assume that the design follows the sequence of eight steps:

1. Selection of an instruction set
2. Word size selection
3. Selection of instruction and data formats
4. Register set and memory design
5. Data and instruction flow-path design
6. Arithmetic and logic unit (ALU) design
7. Input/output (I/O) mechanism design
8. Generation of control signals and design of control unit (CU)

In practice, design of a digital computer is an iterative process. A decision made early in the design process may have to be altered to suit some parameter at a later step. For example, instruction and data formats may have to be changed to accommodate a better data or instruction flow-path design.

A computer architect selects the architecture for the machine based on cost and performance trade-offs, and a computer designer implements the architecture using the hardware and software components available. The complete process of the development of a computer system thus can also be viewed as consisting of two phases, design and implementation. Once the architect derives an architecture (design phase), each subsystem in the architecture can be implemented in several ways depending on the available technology and requirements. We will not distinguish between these two phases in this chapter. We restrict this chapter to the design of hardware components of ASC and use memory elements, registers, flip-flops, and logic gates as components in the design.

Architectural issues of concern at each stage in the design are described in subsequent chapters with reference to architectures of commercially available machines. We first describe the program execution process in order to depict the utilization of each of the registers in ASC.

6.1 PROGRAM EXECUTION

Once the object code is loaded into the memory, it can be executed by initializing the program counter (PC) to the starting address and activating the START switch on the console. Instructions are then fetched from the memory and executed in sequence until an HLT instruction is reached or an error condition occurs. The execution of an instruction consists of two phases:

1. Instruction fetch
2. Instruction execute

6.1.1 Instruction Fetch

During instruction fetch, the instruction word is transferred from the memory into the *instruction register* (IR). To accomplish this, the contents of the PC are first transferred into the memory address register (MAR), a memory read operation is performed to transfer the instruction into the memory buffer register (MBR), and the instruction is then transferred to the IR. While memory is being read, the CU uses its internal logic to add 1 to the contents of the PC so that the PC points to the memory word following the one from which the current instruction is fetched. This sequence of operations constitutes the fetch phase and is the same for all instructions.

6.1.2 Instruction Execution

Once the instruction is in the IR, the opcode is decoded and a sequence of operations is brought about to retrieve the operands (if needed) from the memory and to perform the processing called for by the opcode. This is the execution phase and is unique for each instruction in the instruction set. For example, for the load accumulator (LDA) instruction the effective address is first calculated, and then the contents of the memory word at the effective address are read and transferred into ACC. At the end of the execute phase, the machine returns to the fetch phase.

ASC uses an additional phase to compute the effective address if the instruction uses the indirect addressing mode, since such computation involves reading an address from the memory. This phase is termed *defer phase.*

The fetch, defer, and execute phases together form the so-called *instruction cycle*. Note that an instruction cycle need not contain all the three phases. The fetch and execute phases are required for all instructions, and the defer phase is required only if the indirect addressing is called for.

> **Example 6.1**
>
> Figure 6.1 gives a sample program and shows the contents of all the ASC registers during the execution of the program.

6.2 DATA, INSTRUCTION, AND ADDRESS FLOW

A detailed analysis of the flow of data, instructions, and addresses during instruction fetch, defer, and execute phases for each instruction in the instruction set is required to determine the signal flow paths needed between registers and memory. Such an analysis for ASC, based on the execution process shown in Figure 6.1, is as follows.

6.2.1 Fetch Phase

Assuming that the PC is loaded with the address of the first instruction in the program, the following set of register transfers and manipulations is needed during the fetch phase of each instruction:

MAR ← PC.	
Read memory.	Instruction is transferred to MBR, i.e., MBR ← M[MAR].
IR ← MBR.	Transfer instruction to IR.
PC ← PC + 1.	Increment PC by 1.

Location (hexadecimal)		Instruction	
		ORG	0
0000	BEGIN	LDX	TWO,1
0001		LDA	ZERO
0002		ADD	X
0003	LOOP	ADD	X,1
0004		TDX	LOOP,1
0005		STA*	Y
0006		HLT	
0007	TWO	BSC	2
0008	ZERO	BSC	0
0009	X	BSC	5
000A		BSC	7
000B		BSC	34
000C	Y	BSC	76
		END	BEGIN

(a)

Phase	Operations	Comments
INITIALIZE	PC ← 0.	
FETCH	MAR ← #H0000.	
	READ MEMORY.	
	MBR ← Instruction LDX TWO, 1.	
	IR ← MBR.	Instruction in IR
	PC ← #H0001.	Increment PC
EXECUTE	MAR ← TWO.	Address of TWO to MAR
	READ MEMORY.	
	MBR ← M[TWO].	
	Index1 ← MBR.	Load 2 into Index Register 1
FETCH	MAR ← #H0001.	
	READ MEMORY.	
	MBR ← Instruction LDA ZERO.	
	IR ← MBR.	Instruction in IR
	PC ← #H0002.	Increment PC
EXECUTE	MAR ← ZERO.	Address of ZERO to MAR
	READ MEMORY.	
	MBR ← M[ZERO].	
	ACC ← MBR.	Load 0 into ACC

(b)

Figure 6.1 Execution process. (a) Source program. (b) Execution trace.

(*continued*)

Phase	Operations	Comments
FETCH	MAR ← #H0002.	
	READ MEMORY.	
	MBR ← Instruction ADD X.	
	IR ← MBR.	Instruction in IR
	PC ← #H0003.	Increment PC
EXECUTE	MAR ← X.	Address of X to MAR
	READ MEMORY.	
	MBR ← M[X].	
	ACC ← MBR + ACC.	Add 5 to ACC
FETCH	MAR ← #H0003.	
	READ MEMORY.	
	MBR ← Instruction ADD X, 1.	
	IR ← MBR.	Instruction in IR
	PC ← #H0004.	Increment PC
EXECUTE	MAR ← X + (INDEX1).	X + 2 to MAR
	READ MEMORY.	
	MBR ← M[X+2].	
	ACC ← MBR + ACC.	Add 34 to ACC
FETCH	MAR ← #H0004.	
	READ MEMORY.	
	MBR ← Instruction TDX LOOP, 1.	
	IR ← MBR.	Instruction in IR
	PC ← #H0005.	Increment PC
EXECUTE	INDEX1 ← INDEX1 − 1.	Decrement Index1
	PC ← #H0003.	Jump to LOOP since Index1 is not 0
FETCH	MAT ← #H0003.	
	READ MEMORY.	
	MBR ← Instruction ADD X, 1.	
	IR ← MBR.	Instruction in IR
	PC ← #H0004.	Increment PC
EXECUTE	MAR ← X + (INDEX1).	Address X + 1 to MAR
	READ MEMORY.	
	MBR ← M[X + 1].	
	ACC ← MBR + ACC.	Add 7 to ACC
FETCH	MAR ← #H0004.	
	READ MEMORY.	
	MBR ← Instruction TDX LOOP, 1.	
	IR ← MBR.	Instruction in IR
	PC ← #H0005.	Increment PC

(b)

Figure 6.1 (continued) Execution process. (b) Execution trace.

Phase	Operations	Comments
EXECUTE	INDEX1 ← INDEX − 1.	Decrement Index 1, Index1 is 0, Do not jump to Loop
FETCH	MAR ← #H0005.	
	READ MEMORY.	
	MBR ← Instruction STA* Y.	
	IR ← MBR.	Instruction in IR
	PC ← #H0006.	Increment PC
DEFER	MAR ← Y.	Indirect addressing
	READ MEMORY.	
	MBR ← 76.	Effective address in MBR
EXECUTE	MAR ← MBR.	Address 76 to MAR
	MBR ← ACC.	
	WRITE MEMORY.	
	M[76] ← MBR.	ACC stored at 76
FETCH	MAR ← #H0006.	
	READ MEMORY.	
	MBR ← Instruction HLT.	
	IR ← MBR.	Instruction in IR
	PC ← #H0007.	Increment PC
EXECUTE	Halt.	Stops execution

(b)

Figure 6.1 (continued) Execution process. (b) Execution trace.

6.2.2 Address Calculations

For one-address instructions, the following address computation capabilities are needed:

Direct	MAR ← IR_{7-0}.	MAR ← IR(ADRS).
Indexed	MAR ← IR_{7-0} + INDEX.	(INDEX refers to the index register selected by the index flag.)
Indirect	MAR ← IR_{7-0}.	Get the new address.
	READ memory.	
	MAR ← MBR.	Address to MAR.
Preindexed indirect	MAR ← IR_{7-0} + INDEX.	Get the new address.
	READ memory.	
	MAR ← MBR.	

In the previous computation, *effective address* is assumed to be in MAR at the end of address calculation, just before the execution of the instruction begins. Using the concatenation operator "¢," the transfer MAR ← IR_{7-0} should be further refined as

$$MAR \leftarrow 00000000 \ \text{¢} \ IR_{7-0},$$

which means that the most significant 8 bits of MAR receive 0s. The concatenation of 0s to the address field of IR as given earlier is assumed whenever the contents of the IR address field are transferred to a 16-bit destination or the IR address field is added to an index register. The memory operations can also be designated as

$$\text{READ memory} \qquad \text{MBR} \leftarrow \text{M[MAR], and}$$

$$\text{WRITE memory} \qquad \text{M[MAR]} \leftarrow \text{MBR.}$$

Note that if the contents of the index register are 0,

$$\text{MAR} \leftarrow \text{IR}_{7-0} + \text{INDEX}$$

can be used for direct address computation also. This fact will be used later in this chapter.

Assuming that all arithmetic is performed by one arithmetic unit, the instruction and address flow paths required for fetch cycle and address computation in ASC are as shown in Figure 6.2 (for a list of mnemonic codes, see Table 5.1).

6.2.3 Execution Phase

The detailed data flow during the execution of each instruction is determined by the analysis shown in Table 6.1.

6.3 BUS STRUCTURE

The data and address transfers shown in Figure 6.2 can be brought about either by a point-to-point interconnection of registers or by a bus structure connecting all the registers in ASC. Bus structures are commonly used because a point-to-point interconnection becomes complex as the number of registers to be connected increases.

The most common bus structures are *single bus* and *multibus*. In a single-bus structure, all data and address flow is through one bus. A multibus structure typically consists of several buses, each dedicated to certain transfers; for example, one bus could be a data bus and the other could be an address bus. Multibus structures provide the advantage of tailoring each bus to the set of transfers it is dedicated to and permits parallel operations. Single-bus structures have the advantage of uniformity of design. Other characteristics to be considered in evaluating bus structures are the amount of hardware required and the data-transfer rates possible.

Figure 6.3 shows a bus structure for ASC. The bus structure is realized by recognizing that two operands are required for arithmetic operations using the arithmetic unit of ASC. Each operand will be on a bus (BUS1 and BUS2) feeding the arithmetic unit. The output of the arithmetic unit will be on another bus (BUS3). For transfers that do not involve any arithmetic operation, two direct paths are assumed through the arithmetic unit. These paths either transfer BUS1 to BUS3 or transfer BUS2 to BUS3. The contents of each bus during other typical operations are listed here:

Operation	BUS1	BUS2	BUS3
Increment PC	PC	1 (constant)	PC + 1
Indexing	00000000 ϕ IR$_{7-0}$	INDEX	Indexed address
ADD	ACC	MBR	ACC + MBR

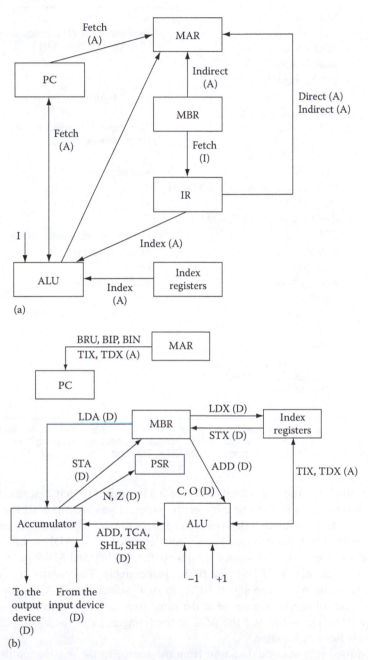

Figure 6.2 Register transfer paths of ASC. (a) Instruction and address flow during fetch phase and address computation phase. (b) Address and data flow during execution. *Note:* A indicates address, I indicates instruction, and D indicates data. All paths are 16 bits wide unless indicated otherwise. Input 1 into the arithmetic unit is a constant input required to increment PC. Constant inputs are required for incrementing and decrementing PC and index registers. Refer to Table 5.1, ASC instruction set.

Table 6.1   Analysis of Data Flow during Execution

Mnemonic	Operations	Comments
HLT	RUN ← 0.	Stop.
LDA	READ memory.	Data into MBR. (The effective address of
	ACC ← MBR.	the operand is in MAR.)
STA	MBR ← ACC.	
	WRITE memory.	
ADD	READ memory.	Data into MBR.
	ACC ← ACC + MBR.	
TCA	ACC ← ACC.$'$	Two-step computation of the 2s complement.
	ACC ← ACC + 1.	
BRU	PC ← MAR.	Address goes to PC.
BIP	IF ACC_{15} = 0 and ACC_{14-0} ≠ 0 THEN	
	PC ← MAR.	
BIN	If ACC_{15} ≠ 0 THEN PC ← MAR.	
RWD	ACC ← Input.	Input device data buffer contents to ACC.
WWD	Output ← ACC.	ACC to output device.
SHL	ACC ← ACC_{14-0} ¢ 0.	Zero fill.
SHR	ACC_{14-0} ← ACC_{15-1}.	ACC_{15} not altered.
LDX	READ memory.	
	INDEX ← MBR.	
STX	MBR ← INDEX.	
	WRITE memory.	
TIX	INDEX ← INDEX + 1.	
	IF INDEX = 0 THEN PC ← MAR.	
TDX	INDEX ← INDEX – 1.	
	IF INDEX ≠ 0 THEN PC ← MAR.	

Note: The data paths required for the operations given in this table are shown in Figure 6.2. All paths are 16-bit wide unless indicated otherwise. Also note that indexing is not allowed in index register reference instructions LDX, STX, TIX, and TDX.

The operation of the bus structure shown in Figure 6.3 is similar to that described in Section 4.8. The data enter a register at the rising edge of the clock. The clock and all the other required control signals to select source and destination registers during a bus transfer are generated by the CU. Figure 6.4a shows the detailed bus transfer circuit for the ADD operation. The timing diagram is shown in Figure 6.4b. Control signals ACC to BUS1 and MBR to BUS2 select ACC and MBR as source registers for BUS1 and BUS2, respectively. The control signal ADD enables the add operation in the ALU. The signal BUS3 to ACC selects the ACC as the destination for BUS3. All these control signals are active at the same time and stay active long enough to complete the bus transfer. The data enter the ACC at the rising edge of the clock pulse, after which the control signals become inactive.

The time required to transfer a data unit from the source to the destination through the ALU is the *register transfer* time. The clock frequency must be such that the slowest register transfer is accomplished in a clock period. (In ASC, the slowest transfer is the one that involves an ADD operation.) Thus, the register transfer time dictates the speed of transfer on the bus and hence the processing speed (*cycle time*) of the processor.

Note also that the contents of the ACC and MBR cannot be altered until their sum enters the ACC, since the ALU is a combinational circuit. To accommodate this feedback operation on the ACC,

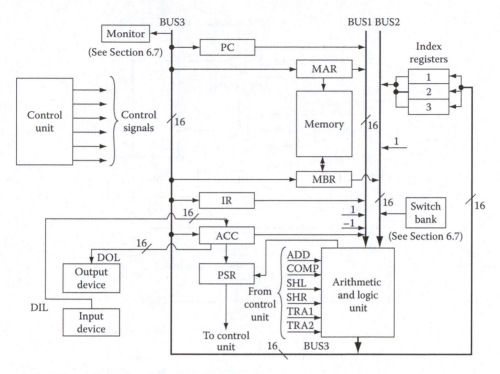

Figure 6.3 ASC bus structure. *Note:* Each bit of PC and ACC is a master–slave flip-flop. 1 and −1 represent constant registers connected to the respective bus; they are 16-bit constants represented in 2s complement form. Only the control signals to ALU are shown; other control signals are not shown.

it must be configured using the master–slave flip-flops. Similarly, to accommodate the increment PC operation, the PC must also be configured using master–slave flip-flops.

Input and output transfers are performed on two separate 16 bit paths: data-input lines (DILs) and data-output lines (DOLs) connected to and from the accumulator. This I/O scheme was selected for simplicity. Alternatively, DIL and DOL could have been connected to one of the three buses.

A single-bus structure is possible for ASC. In this structure, either one of the operands of the two operand operations or the result must be stored in a buffer register before it can be transmitted to the destination register. Thus, there will be some additional transfers in the single-bus structure, and some operations will take longer to complete, thereby making the structure slower than the multibus structure.

Transfer of data, instructions, and addresses on the bus structure is controlled by a set of control signals generated by the CU of the machine. Detailed design of the CU is illustrated later in this chapter.

6.4 ARITHMETIC AND LOGIC UNIT

The ALU of ASC is the hardware that performs all arithmetic and logical operations. The instruction set implies that the ALU of ASC must perform addition of two numbers, compute the 2s complement of a number, and shift the contents of the accumulator either right or left by 1 bit. Additionally, the ASC ALU must directly transfer either of its inputs to its output to support data-transfer operations such as IR ← MBR and MAR ← IR.

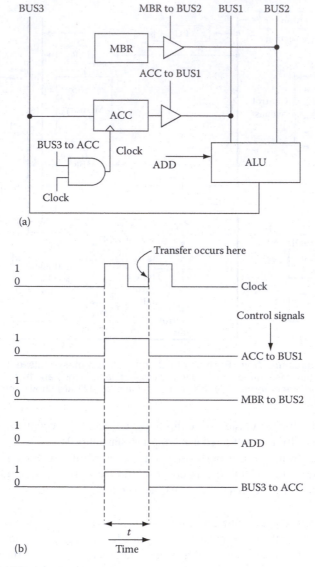

Figure 6.4 Timing of ADD operation on the bus structure. (a) Circuit. (b) Timing.

We will assume that the CU of the machine provides the appropriate control signals to enable the ALU to perform one of these operations. Since BUS1 and BUS2 are the inputs and BUS3 is the output of ALU, the following operations must be performed by the ALU:

$$\text{ADD} : \text{BUS3} \leftarrow \text{BUS1} + \text{BUS2}.$$

$$\text{COMP} : \text{BUS3} \leftarrow \text{BUS1}'.$$

$$\text{SHR} : \text{BUS3} \leftarrow \text{BUS1}_{15} \, \text{¢} \, \text{BUS1}_{15-1}.$$

$$\text{SHL} : \text{BUS3} \leftarrow \text{BUS}_{14-0} \, \text{¢} \, 0.$$

$$\text{TRA1} : \text{BUS3} \leftarrow \text{BUS1}.$$

$$\text{TRA2} : \text{BUS3} \leftarrow \text{BUS2}.$$

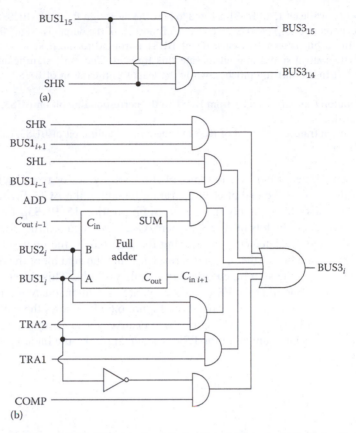

Figure 6.5 Logic diagram of ASC ALU: (a) sign-bit circuit and (b) typical bit. *Note:* During SHL, all lines connected to BUS3$_0$ are 0, and hence the LSB is automatically zero filled; no special circuitry is needed. During SHR, BUS1$_0$ is not connected to any bit of BUS3 and hence is lost.

ADD, COMP, SHR, SHL, TRA1, and TRA2 are the control signals generated by the CU, and bit positions of the buses are numbered 15 through 0, left to right. Each of the control signals activates a particular operation in ALU. Only one of these control signals may be active at any time. Figure 6.5 shows a typical bit of ALU and its connections to the bits of BUS1, BUS2, and BUS3. A function description of the ALU follows corresponding to the previous control signals.

Add: The addition circuitry consists of 15 full adders and one half-adder for the least significant bit (0). The sum output of each adder is gated through an AND gate with the ADD control signal. The carry output of each adder is input to the carry-in of the adder for the next most significant bit. The half-adder for bit 0 has no carry-in and the carry-out from bit 15 is the CARRY flag bit in the PSR. The accumulator contents on BUS1 are added to the contents of MBR on BUS2, and the result is stored in the accumulator.

Comp: The complement circuitry consists of 16 NOT gates, one for each bit on BUS1. Thus, the circuitry produces the 1s complement of a number. The output of each NOT gate is gated through an AND gate with the COMP control signal. The operand for complement is the contents of the accumulator and the result is stored in the accumulator. TCA (2s complement accumulator) command is accomplished by taking the 1s complement of an operand first and then adding 1 to the result.

SHR: For shifting a bit pattern right, each bit of BUS1 is routed to the next least significant bit of BUS3. This transfer is gated by SHR control signal. The least significant bit of BUS1 (BUS1$_0$) is lost in the shifting process, while bit 15 of BUS1 is gated to both the most significant bit (BUS3$_{15}$) and the next least significant bit (BUS3$_{14}$) of BUS3. Thus, the leftmost bit of the output is "sign" filled.

SHL: For shifting a bit pattern left, each bit of BUS1 is routed to the next most significant bit of BUS3. SHL control signal is used to gate this transfer. The most significant bit on BUS1 (BUS1$_{15}$) is lost in the shifting process, while the least significant bit of BUS3 (BUS3$_0$) is zero filled.

TRA1: This operation transfers each bit from BUS1 to the corresponding bit on BUS3, each bit gated by TRA1 signal.

TRA2: This operation transfers each bit of BUS2 to the corresponding bit of BUS3, each bit gated by TRA2 control signal.

The carry, negative, zero, and overflow bits of the PSR are set or reset by the ALU, based on the content of the ACC at the end of each operation involving the ACC. Figure 6.6 shows the circuits needed. The carry bit is simply the C_{out} from bit position 15. BUS3$_{15}$ forms the negative bit. Zero bit is set only if all the bits of BUS3 are zero. Overflow occurs if the sum exceeds ($2^{15}-1$) during addition. This can be detected by observing the sign bits of the operands and the result. Overflow occurs if the sign bits of the operands are each 1 and the sign bit of the result is 0 or vice versa. Also, during SHL if the sign bit changes, an overflow results. This is detected by comparing BUS1$_{14}$ with BUS1$_{15}$. Note that the PSR bits are updated simultaneously with the updating of the ACC contents with the results of the operation. Figure 6.6 also shows the circuit to derive the accumulator-positive condition (i.e., ACC is neither negative nor zero) required during the BIP instruction execution. The interrupt bit of the PSR is set and reset by the interrupt logic. Interrupts are discussed in Chapter 7.

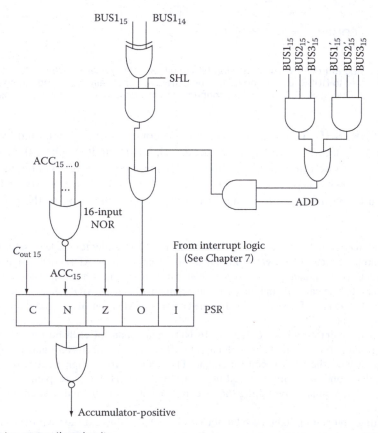

Figure 6.6 Status generation circuits.

6.5 INPUT/OUTPUT

ASC is assumed to have one input device and one output device. Both these functions could very well be performed by a terminal with a keyboard for input and a display or printer for the output. We assume that the input and output devices transfer a 16-bit data word into or out of the ACC, respectively. We will base our design on the simplest I/O scheme, called *programmed I/O*.

The ALU and the CU together form the central processing unit (CPU), which we will also refer to as the processor. In the programmed I/O scheme, during the execution of the RWD instruction, the CPU commands the input device to send a data word and then waits. The input device gathers the data from the input medium and, when the data are ready in its data buffer, informs the CPU that the data are ready. The CPU then gates the data into the ACC over the DILs. During the WWD, the CPU gates the ACC content onto DOLs, commands the output device to accept the data, and waits. When the data are gated into its data buffer, the output device informs the CPU of the data acceptance. The CPU then proceeds to execute the next instruction in the sequence.

The sequence of operations described earlier is known as the data communication *protocol* (or "handshake") between the CPU and the peripheral device. A DATA flip-flop in the CU is used to facilitate the I/O handshake. The RWD and WWD protocols are described here in detail:

1. RWD
   a. CPU resets the DATA flip-flop.
   b. CPU sends a 1 on the INPUT control line, thus commanding the input device to send the data.
   c. CPU waits for the DATA flip-flop to be set by the input device.
   d. The input device gathers the data into its data buffer, gates it onto the DILs, and sets the DATA flip-flop.
   e. CPU now gates the DILs into the ACC, resets the INPUT control line, and resumes instruction execution.
2. WWD
   a. CPU resets the DATA flip-flop.
   b. CPU gates the ACC onto DOLs and sends a 1 on the OUTPUT control line, thus commanding the output device to accept the data.
   c. CPU waits for the DATA flip-flop to be set by the output device.
   d. The output device, when ready, gates DOL into its data buffer and sets the DATA flip-flop.
   e. CPU resets the OUTPUT control line, removes data from the DOLs, and resumes instruction execution.

As can be seen from the preceding protocols, the CPU controls the complete I/O process and waits for the input and output to occur. Since the I/O devices are much slower than the CPU and the CPU idles waiting for the data, the programmed I/O scheme is the slowest of the I/O schemes used in practice. But it is simple to design and generally has a low overhead in terms of the I/O protocols needed, especially when small amounts of data are to be transferred. Also note that this scheme is adequate in the environments in which the CPU cannot be allocated to any other task while the data transfer is taking place. Chapter 7 provides the details of the other popular I/O schemes and generalizes the ASC I/O scheme described in this chapter.

6.6 CONTROL UNIT

The CU is the most complex block of computer hardware from a designer's point of view. Its function is to generate control signals needed by other blocks of the machine in a predetermined sequence to bring about the sequence of actions called for by each instruction.

Figure 6.7 shows a block diagram of the ASC CU and lists all the external and internal inputs and the outputs (i.e., control signals produced).

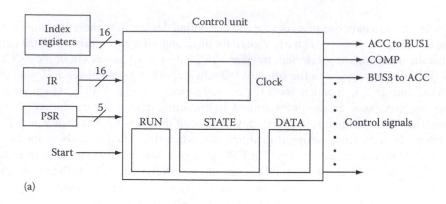

(a)

External inputs:	From registers	From console
	PSR contents	SWITCH BANK contents
	IR contents	START
	Index registers contents	MASTER CLEAR
	From I/O devices	
	Set DATA flip-flop	
Internal Inputs:	Contents of RUN and DATA flip-flops and	
	STATE register	
	CLOCK	

Signals generated for internal use:
State change signals
Fetch   Reset DATA flip-flop
Defer (indirect address)   Reset RUN flip-flop
Execute

Outputs:

To memory	To ALU	To I/O
READ	TRA1	INPUT
WRITE	TRA2	OUTPUT
	ADD	
	COMP	
	SHR	
	SHL	

To the bus structure			
ACC to BUS1	INDEX to BUS2	BUS3 to ACC	DIL to ACC
MAR to BUS1	MBR to BUS2	BUS3 to INDEX	ACC to DOL
IR$_{7-0}$ to BUS1	1 to BUS2	BUS3 to MAR	
PC to BUS1	SWITCHBANK to BUS2	BUS3 to MBR	
1 to BUS1		BUS3 to PC	
−1 to BUS1		BUS3 to MONITOR	
		BUS3 to IR	

(b)

Figure 6.7   Control unit. (a) Block diagram. (b) Signals. *Note:* CLOCK is connected to all the registers. SWITCH BANK, MONITOR, MASTER, CLEAR, and START are console facilities.

Inputs to the CU are

1. The opcode, indirect bit, and index flag from IR
2. Contents of the PSR
3. Index register bits 0–15, to test for the zero or nonzero index register in TIX and TDX instructions

In addition to these inputs, control signals generated by the CU are functions of the contents of the following:

1. DATA flip-flop: Used to facilitate the handshake between the CPU and I/O devices.
2. RUN flip-flop: Set by the START switch on the console (see Section 6.7) and indicates the RUN state of the machine. RUN flip-flop must be set for control signals to activate a microoperation. RUN flip-flop is reset by the HLT instruction.
3. STATE register: A 2-bit register used to distinguish between the three phases (states) of the instruction cycle. The CU is thus viewed as a three-state sequential circuit.

6.6.1 Types of Control Units

As mentioned earlier, the function of the CU is to generate the control signals in the appropriate sequence to bring about the instruction cycle that corresponds to each instruction in the program. In ASC, an instruction cycle consists of three phases. Each phase in the instruction cycle is composed of a sequence of *microoperations*. A microoperation is one of the following:

1. A simple *register transfer* operation, to transfer contents of one register to another register
2. A complex register transfer involving ALU, such as the transfer of the complement of the contents of a register and the sum of the contents of two registers to the destination register
3. A memory read or write operation

Thus, a machine instruction is composed of a sequence of microoperations (i.e., a *register transfer sequence*). We will use the terms *register transfer* and *microoperation* interchangeably.

There are two popular implementation methods for CUs:

1. Hardwired control unit (HCU): The output (i.e., control signals) of the CU is generated by the logic circuitry built of gates and flip-flops.
2. Microprogrammed control unit (MCU): The sequence of microoperations corresponding to each machine instruction is stored in a read-only memory (ROM) called *control ROM* (CROM). The sequence of microoperations is called the *microprogram*, and the microprogram consists of *microinstructions*. A microinstruction corresponds to one or more microoperations, depending on the CROM storage format. The control signals are generated by decoding the microinstructions.

The MCU scheme is more flexible than the HCU scheme because in it the meaning of an instruction can be changed by changing the microinstruction sequence corresponding to that instruction, and the instruction set can be extended simply by including a new ROM containing the corresponding microoperation sequences. Hardware changes to the CU thus are minimal in this implementation. In an HCU, any such change to the instruction set requires substantial changes to the hardwired logic. HCUs, however, are generally faster than MCUs and are used where the CU must be fast. Most of the recent machines have MCUs.

Among the machines that have an MCU, the degree to which the microprogram can be changed by the user varies from machine to machine. Some do not allow the user to change the microprogram, some allow partial changes and additions (e.g., machines with *writable control store*), and some do not have an instruction set of their own and allow the user to microprogram the complete instruction set suitable for his application. This latter type of machine is called a *soft machine*. We will design an HCU for ASC in this section. An MCU design is provided in Section 6.8.

6.6.2 Hardwired Control Unit for ASC

An HCU can either be synchronous or asynchronous. In a synchronous CU, each operation is controlled by a clock and the CU state can be easily determined knowing the state of the clock. In an asynchronous CU, completion of one operation triggers the next and hence no clock exists. Because of its nature, the design of an asynchronous CU is complex, but if it is designed properly, it can be made faster than a synchronous CU. In a synchronous CU, the clock frequency must be such that the time between two clock pulses is sufficient to allow the completion of the slowest microoperation. This characteristic makes a synchronous CU relatively slow. We will design a synchronous CU for ASC.

6.6.3 Memory versus Processor Speed

The memory hardware is usually slower than the CPU hardware, although the speed gap is narrowing with the advances in hardware technology. Some memory organizations that help reduce this speed gap are described in Chapter 9. We will assume a semiconductor RAM for ASC with an access time equal to two register transfer times. Thus, during a memory read, if the address is gated into the MAR along with the READ control signal, the data will be available in the MBR by the end of the next register transfer time. Similarly, if the data and address are provided in MBR and MAR, respectively, along with the WRITE control signal, the memory completes writing the data by the end of the second register transfer time. These characteristics are shown in Figure 6.8. Note that the contents of MAR cannot be altered until the read or write operation is completed.

6.6.4 Machine Cycles

In a synchronous CU, the time between two clock pulses (*register transfer time*) is determined by the slowest register transfer operation. In the case of ASC, the slowest register transfer is the one that involves the adder in the ALU. The register transfer time of a processor is known as the *processor cycle time*, or *minor cycle*. A *major cycle* of a processor consists of several minor cycles. Major cycles with either fixed or variable number of minor cycles have been used in practical machines. An instruction cycle typically consumes one or more major cycles.

To determine how long a major cycle needs to be, we will examine the fetch, address calculation, and execute phases in detail. The microoperations required during fetch phase can be allocated to minor cycles:

Minor Cycles	Microoperations	Comments
T_1	MAR ← PC, READ memory	
T_2	PC ← PC + 1	Memory read time
T_3	IR ← MBR	

Once the READ memory signal is issued at T_1, the instruction to be fetched will be available at the end of T_2, because the memory read operation requires two minor cycles. MAR cannot be altered during T_2, but the bus structure can be used during this time to perform other operations. Thus, we can increment the PC during this time slot, as required by the fetch phase. By the end of T_3, the instruction is available in IR.

The opcode is decoded by two 4 to 16 decoders (with active-high outputs) connected to the IR, as shown in Figure 6.9. We will make our HCU design specific to the 16 instructions in the instruction set. Therefore, only one of the two decoders is utilized in our design. The bottom decoder is for extended instruction set (i.e., $IR_{11} = 1$). When instruction set is extended, changes will be needed to more than one part of the HCU.

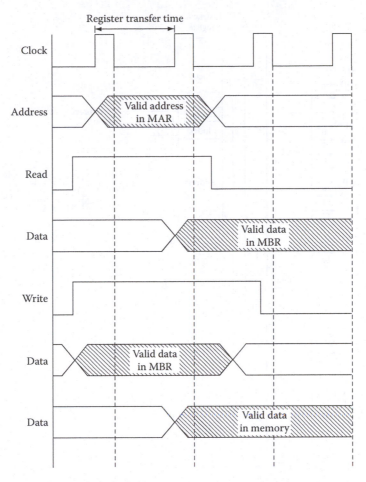

Figure 6.8 ASC memory timing characteristics.

If the instruction in IR is a zero-address instruction, we can proceed to the execute cycle. If not, effective address must be computed. To facilitate the address computation, a 2–4 decoder is connected to the index flag bits of the IR, as shown in Figure 6.9. The outputs of the decoder are used to connect the referenced index register to BUS2. Note that when the index flag is 00 (corresponding to no index register referencing), none of the three index registers are connected to BUS2; hence the bus lines are all at zero. Figure 6.10 shows the circuits needed to direct the data into the selected index register from BUS3 and the generation of ZERO-INDEX signal, which, if 0, indicates that the selected index register contains zero, as required during TIX and TDX instructions. In the following, INDEX refers to the index register selected by the circuit of Figure 6.10.

Thus, indexing can be performed during T_4:

$$T_4 : \text{MAR} \leftarrow 00000000 \not{c} \text{ IR}_{7-0} + \text{INDEX}$$

where $\not{c}$ is the concatenation operator.

Indexing in T_4 is not needed for zero-address instructions. If the assembler inserts 0s into the unused fields of zero-address instructions, the indexing operation during T_4 will not affect the execution of those instructions in any way. Also note that indexing is not allowed in LDX, STX, TIX, and TDX instructions. Microoperations in T_4 thus need to be altered for these instructions.

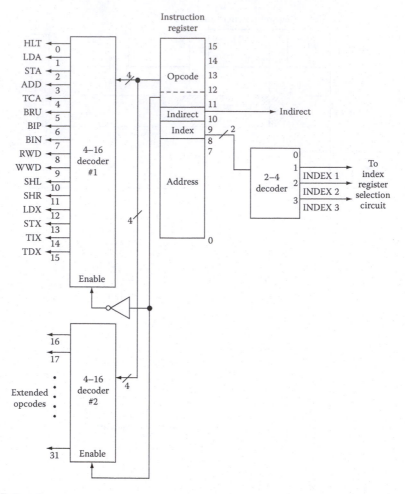

Figure 6.9 IR decoders.

From the opcode assignment, it is seen that indexing is not required for instructions with an opcode in the range of 10000 through 11110. Hence, microoperations in T_4 can use the MSB of opcode (IR_{15}) to inhibit indexing for index reference instructions. Thus,

$$T_4 : \text{IF } IR_{15} = 0 \text{ THEN MAR} \leftarrow IR_{7-0} + \text{INDEX}$$

$$\text{ELSE MAR} \leftarrow IR_{7-0}.$$

The fetch phase thus consists of four minor cycles and accomplishes direct and indexed (if called for) address computations. Effective address is in the MAR at the end of the fetch phase. The first three machine cycles of the fetch phase are same for all 16 instructions. The fourth minor cycle will differ since the zero-address instructions do not need an address computation.

When indirect address is called for, the machine enters the defer phase, where a memory read is initiated and the result of the read operation is transferred into MAR as the effective address. We will thus redefine the functions of the three phases of ASC instruction cycle as follows:

FETCH: Includes direct and indexed address calculation
DEFER: Entered only when indirect addressing is called for
EXECUTE: Unique to each instruction

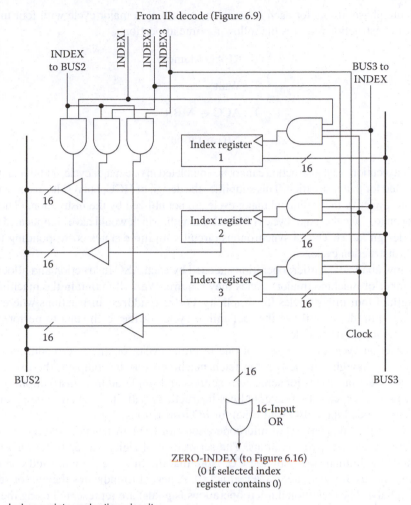

Figure 6.10   Index register selection circuit.

Since the address (indexed if needed) is in MAR at the end of fetch phase, the defer phase can use the following time allocation:

$$T_1 : \text{READ Memory}$$

$$T_2 : \text{Wait}$$

$$T_3 : \text{MAR} \leftarrow \text{MBR}$$

$$T_4 : \text{No operation}$$

Although only three minor cycles are needed for defer, we have made the defer phase four minor cycles long for simplicity since the fetch phase was four minor cycles long. This assumption results in some inefficiency. If T_4 of defer phase can be used to perform the microoperations needed during the execute phase of the instruction, the CU would be more efficient, but its complexity would increase.

The execute phase differs for each instruction. Assuming a major cycle with four minor cycles again, the LDA instruction requires the following time allocation:

T_1 : READ Memory

T_2 : Wait

T_3 : ACC ← MBR

T_4 : No operation

In ASC, if the execution of an instruction cannot be completed in one major cycle, additional major cycles must be allocated for that instruction. This simplifies the design of HCU, but results in some inefficiency if all the minor cycles of the additional major cycle are not utilized by the instruction. (The alternative would have been to make the major cycles of variable length, which would have complicated the design.)

We will design the HCU as a synchronous circuit with three states corresponding to the three phases of the instruction cycle.

We will now analyze the microoperations needed by each ASC instruction and allocate them to whatever number of machine (major) cycles they require. We will maintain the machine cycles of constant length of four minor cycles for simplicity. For zero-address instructions where no address calculation is required, we will use the four minor cycles of the fetch state to perform execution phase microoperations, if possible.

ASC instruction cycle thus consists of one machine cycle for those zero-address instructions in which there is enough time left in the fetch machine cycle to complete the execution of the instruction, two machine cycles for some zero-address and single-address instructions with no indirect addressing, three cycles for single-address instructions with indirect addressing, and multiple machine cycles (depending on I/O wait time) for I/O instructions.

Table 6.2 lists the complete set of microoperations for LDA instruction and the control signals needed to activate those microoperations. The set of control signals to activate a microoperation must be generated simultaneously by the CU. Note that the microoperations (and control signals) that are simultaneous are separated by a comma (,). A period (.) indicates the end of such a set of operations (signals). The conditional microoperations (signals) are represented using the notation:

IF condition THEN operation(s) ELSE operation(s).

The ELSE clause is optional, if an alternate set of operations is not required when the "condition" is not true (or 1).

The transitions between fetch (F), defer (D), and execute (E) states are allowed only in minor cycle 4 (CP_4) of each machine cycle, to retain the simplicity of design. If a complete machine cycle is not needed for a particular set of operations, these state transitions could occur earlier in the machine cycle, but the state transition circuit would be more complex. The state transitions are represented by the operations STATE ← E, STATE ← D, and STATE ← F. These are equivalent to transferring the codes corresponding to each state into the STATE register. We will use the following coding:

Code	State
00	F
01	D
10	E
11	Not used

Note that in CP_2 of the fetch cycle, a constant register containing a 1 is needed to facilitate increment PC operation. Such a constant register is connected to BUS2 by the signal 1 to BUS2. In CP_4 of the

Table 6.2 Microoperations for LDA

Machine	Minor Cycle	Microoperations	Control Signals
Cycle	(Clock pulse)		
FETCH	CP_1	MAR ← PC.	PC to BUS1, TRA1, BUS3 to MAR, READ.
		READ MEMORY.	
	CP_2	PC ← PC + 1.	PC to BUS1, 1 to BUS2, ADD, BUS3 to PC.
	CP_3	IR ← MBR.	MBR to BUS2, TRA2, BUS3 to IR.
	CP_4	IF IR_{15} = 0 THEN	IF IR_{15} = 0, THEN IR_{7-0} to BUS1, INDEX to
		MAR ← IR_{7-0} + INDEX	BUS2, ADD, BUS3 TO MAR
		ELSE	ELSE
		MAR ← IR_{7-0}.	IR_{7-0} TO BUS1, TRA1, BUS3 to MAR.
		IF IR_{10} = 1 THEN STATE ← D	IF IR_{10} = 1, THEN D to STATE
		ELSE STATE ← E.	ELSE E to STATE.
DEFER	CP_1	READ MEMORY.	READ.
	CP_2	WAIT.	—
	CP_3	MAR ← MBR.	MBR to BUS2, TRA2, BUS3 to MAR.
	CP_4	STATE ← E.	E to STATE.
EXECUTE	CP_1	READ MEMORY.	READ.
	CP_2	WAIT.	—
	CP_3	ACC ← MBR.	MBR to BUS2, TRA2, BUS3 to ACC.
	CP_4	STATE ← F.	F to STATE.

fetch cycle, indexing is controlled by IR_{15} and indirect address computation is controlled by IR_{10}. The state transition is either to defer (D) or to execute (E), depending on whether IR_{10} is 1 or 0, respectively.

We now analyze the remaining instructions to derive the complete set of control signals required for ASC.

6.6.5 One-Address Instructions

Fetch and defer states are identical to the ones shown in Table 6.2 for all one-address instructions. Table 6.3 lists the execute-phase microprograms for these instructions.

6.6.6 Zero-Address Instructions

The microoperations during the first three minor cycles of fetch cycle will be similar to that of LDA, for zero-address instructions also. Since there is no address computation, some of the execution-cycle operations can be performed during the fourth minor cycle and no execution cycle is needed. Microoperations for all zero-address instructions are listed in Table 6.4. TCA is performed in two steps and hence needs the execution cycle. SHR, SHL, and HLT are completed in the fetch cycle itself.

6.6.7 Input/Output Instructions

During RWD and WWD, the processor waits until the end of the execute cycle to check to see if the incoming data are ready (DATA = 1) or if they have been accepted (DATA = 1). The transition

Table 6.3 Microoperations for One-Address Instructions

Machine Cycle	Minor Cycle	STA	ADD	BRU	LDX	STX
EXECUTE	CP_1	—	READ.	PC ← MAR.	READ.	—
	CP_2	—	—	—	—	—
	CP_3	MBR ← ACC, WRITE.	ACC ← MBR +ACC.	—	INDEX ← MBR.	MBR ← INDEX, WRITE.
	CP_4	STATE ← F.	STATE ← F.	STATE ← F.	STATE ← F.	STATE ← F.

Machine Cycle	Minor Cycle	TIX	TDX	BIP	BIN
EXECUTE	CP_1	INDEX ← INDEX + 1.	INDEX ← INDEX − 1.	—	—
	CP_2	IF ZERO-INDEX = 0 THEN PC ← MAR.	IF ZERO-INDEX ≠ 0 THEN PC ← MAR.	—	—
	CP_3	—	—	—	—
	CP_4	STATE ← F.	STATE ← F.	IF ACCUMULATOR—POSITIVE = 1 THEN PC ← MAR. STATE ← F.	IF N = 1 THEN PC ← MAR. STATE ← F.

Table 6.4   Microoperations for Zero-Address Instructions

Cycle	Minor Cycle	RWD	WWD	TCA	SHR	SHL	HLT
FETCH	CP$_1$						
	CP$_2$	Same microoperations as those for LDA.					
	CP$_3$						
	CP$_4$	DATA ← 0, INPUT ← 1, STATE ← E.	DATA ← 0, OUTPUT ← 1, STATE ← E, DOL ← ACC.	STATE ← E.	ACC ← SHR (ACC).	ACC ← SHL (ACC).	RUN ← 0.
EXECUTE	CP$_1$	—	—	AC ← ACC'.			
	CP$_2$	—	—	ACC ← ACC + 1.	Execution cycle not needed; remains in F State.		
	CP$_3$	—	—	—			
	CP$_4$	IF DATA = 1 THEN ACC ← DIL, INPUT ← 0, STATE ← F.	IF DATA = 1 THEN OUTPUT ← 0.	STATE ← F.			

to fetch state occurs only if these conditions are satisfied. If not, the processor waits (loops) in the execute state until the conditions are met. Hence, the number of cycles needed for these instructions depends on the speeds of I/O devices.

Tables 6.5 and 6.6 list the control signals as implied by the microoperations in Tables 6.3 and 6.4, respectively. Logic diagrams that implement an HCU can be derived from the control signal information in these tables.

Figure 6.11 shows the implementation of the four-phase clock to generate CP$_1$, CP$_2$, CP$_3$, and CP$_4$. A 4-bit shift register is used in the implementation. The master clock is an oscillator that starts emitting clock pulses as soon as the power to the machine is turned on. When the START button on the console (discussed in the next section) is pushed, the RUN flip-flop and the MSB of the shift register are set to 1. The master clock pulse is used to circulate the 1 in the MSB of the shift register through the other bits on the right, to generate the four clock pulses. The sequence of these pulses continues as long as RUN flip-flop is set. The HLT instruction resets the RUN flip-flop, thus stopping the four-phase clock. A "master clear" button on the console clears all the flip-flops when the RUN is not on.

A 2 to 4 decoder is used to generate F, D, and E signals corresponding to the fetch (00), defer (01), and execute (10) states. Figure 6.12 shows the state change circuitry and its derivation, assuming D flip-flops. The CP$_4$ is used as the clock for the state register, which along with the transition circuits is shown in Figure 6.12.

Figure 6.13 shows the circuit needed to implement the first three minor cycles of the fetch. The fourth minor cycle of fetch is implemented by the circuit in Figure 6.14. Here, the RUN flip-flop is reset if the IR contains the HLT instruction. Indexing is performed only when IR$_{15}$ is a 0 and the instruction is not a TCA and not a HLT. Since the reset RUN flip-flop overrides any other operation, it is sufficient to implement the remaining condition (IR$_{15}$ = 0 AND NOT TCA) for indexing. This condition is handled by the NOR gate. Corresponding to the four index register reference instructions, only the address portion of IR is transferred (without indexing) to MAR. The control signals corresponding to the other four instructions are similarly implemented.

Figure 6.15 shows the control circuits for the defer cycle and Figure 6.16 for the execute cycle. The state transition signals of Figure 6.12 are not repeated in these circuit diagrams nor has the logic minimization been attempted in deriving these circuits.

Table 6.5 Control Signals for One-Address Instructions

Machine Cycle	Minor Cycle	STA	ADD	BRU	LDX	STX
EXECUTE	CP$_1$	—	READ.	MAR to BUS1, TRA1, BUS3 to PC.	READ.	—
	CP$_2$					
	CP$_3$	ACC to BUS1, TRA1, BUS3 to MBR,	ACC to BUS1, MBR to BUS2, ADD,	—	MAR to BUS2, TRA2, BUS3 to INDEX.	INDEX to BUS2, TRA2, BUS3 to MBR,
	CP$_4$	WRITE. F to STATE.	BUS3 to ACC. F to STATE.	F to STATE.	F to STATE.	WRITE. F to STATE.

Machine Cycle	Clock Pulse	TIX	TDX	BIP	BIN
EXECUTE	CP$_1$	1 to BUS1, INDEX to BUS 2, ADD, BUS3 to INDEX.	–1 TO BUS1, INDEX to BUS2, ADD, BUS3 to INDEX.	—	—
	CP$_2$	IF ZERO-INDEX = 0 THEN MAR to BUS1, TRA1, BUS3 to PC.	IF ZERO-INDEX ≠ 0 THEN MAR to BUS1, TRA1, BUS3 to PC.	—	—
	CP$_3$				
	CP$_4$	F to STATE.	F to STATE.	IF ACCUMULATOR POSITIVE = 1 THEN MAR to BUS1, TRA1, BUS3 to PC, F to STATE.	IF N = 1 THEN MAR to BUS1 TRA1, BUS3 to PC, F to STATE.

Table 6.6 Control Signals for Zero-Address Instructions

Machine Cycle	Clock Pulse	RWD	WWD	TCA	SHR	SHL	HLT
FETCH	CP$_1$						
	CP$_2$	Same as those for LDA in Table 6.2.					
	CP$_3$						
	CP$_4$	0 to DATA, 1 to INPUT, E to STATE.	0 to DATA, 1 to OUTPUT, E to STATE. ACC to DOL.	E to STATE.	ACC to BUS1, SHR, BUS3 to ACC.	ACC to BUS1, SHL, BUS3 to ACC.	0 to RUN.
EXECUTE	CP$_1$	—	—	ACC to BUS1, COMP, BUS3 to ACC.	Execution cycle not needed; remains in F state.		
	CP$_2$	—	—	ACC to BUS1, 1 to BUS2, ADD, BUS3 to ACC.			
	CP$_3$	—	—				
	CP$_4$	IF DATA = 1 THEN DIL to ACC, 0 to INPUT, F to STATE.	IF DATA = 1 THEN 0 to OUTPUT, F to STATE.	F to STATE.			

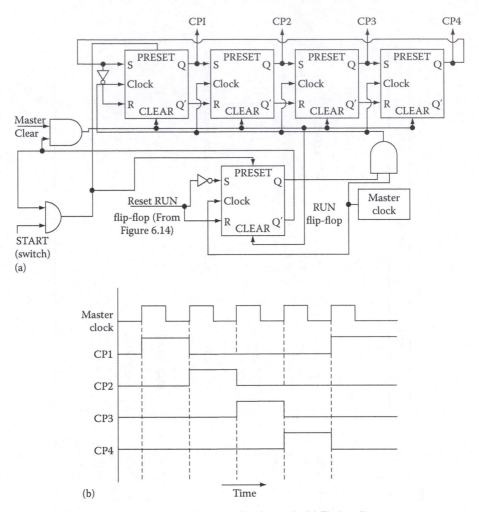

Figure 6.11 Four-phase clock using SR flip-flops. (a) Clock circuit. (b) Timing diagram.

6.7 CONSOLE

We have included the design of the console here for completeness. This section can be skipped without a loss of continuity. We will not consider console operation, however, in the MCU design in the next section.

Figure 6.17 shows the ASC console. The console (control panel) enables the operator to control the machine. It can be used for loading programs and data into memory, observing the contents of registers and memory locations, and starting and stopping the machine. There are 16 lights (monitors) on the console that can display the contents of a selected register or memory location. There is a *switch bank* consisting of 16 two-position switches that can be used for loading a 16-bit pattern into either PC or a selected memory location. There are two LOAD switches: LOAD PC and LOAD MEM. When LOAD PC is pushed, the bit pattern set on the switch bank is transferred to PC. When LOAD MEM is pushed, the contents of the switch bank are transferred to the memory location addressed by PC. There is a set of DISPLAY switches to enable the display of contents of ACC, PC, IR, index registers, and PSR, or the memory location addressed by PC. The DISPLAY and LOAD switches are push-button switches and are mechanically ganged together so that only one of these switches is pushed (operative) at any time. The switch that was previously pushed pops

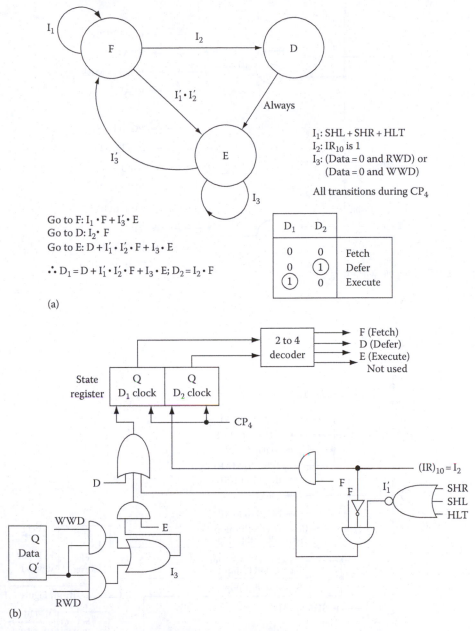

I_1: SHL + SHR + HLT
I_2: IR_{10} is 1
I_3: (Data = 0 and RWD) or
(Data = 0 and WWD)

All transitions during CP_4

Go to F: $I_1 \cdot F + I_3' \cdot E$
Go to D: $I_2 \cdot F$
Go to E: $D + I_1' \cdot I_2' \cdot F + I_3 \cdot E$

$\therefore D_1 = D + I_1' \cdot I_2' \cdot F + I_3 \cdot E;\ D_2 = I_2 \cdot F$

(a)

D_1	D_2	
0	0	Fetch
0	1	Defer
1	0	Execute

(b)

Figure 6.12 (a) State register and transitions. (b) Transition circuits.

out when a new switch is pushed. The MASTER CLEAR switch clears all ASC registers. The START switch sets the RUN flip-flop, which in turn starts the four-phase clock. There is also a POWER ON/OFF switch on the console.

Each switch on the console invokes a sequence of microoperations. Typical operations possible using the console along with the sequences of microoperations invoked are listed here:

Loading PC: Set the switch bank to the required 16-bit pattern.

$$\text{LOAD PC} \quad PC \leftarrow \text{Switch bank.}$$

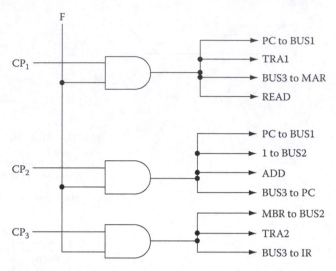

Figure 6.13 Fetch cycle (CP₁–CP₃).

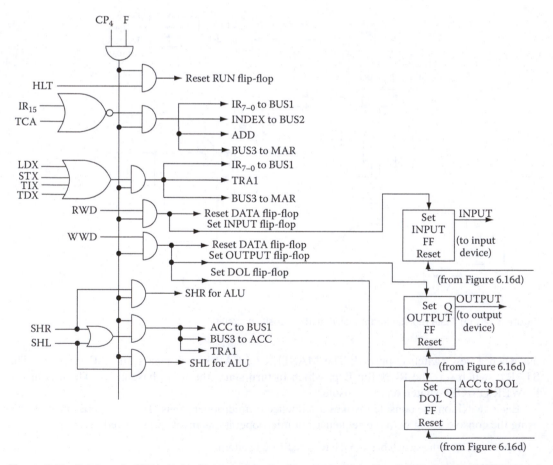

Figure 6.14 Fetch cycle (CP₄). *Note:* State change signals are shown in Figure 6.12. INPUT, OUTPUT, and ACC to DOL signals are set and remain set until reset in CP₄.

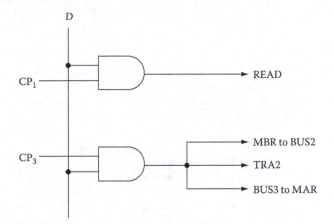

Figure 6.15 Defer cycle.

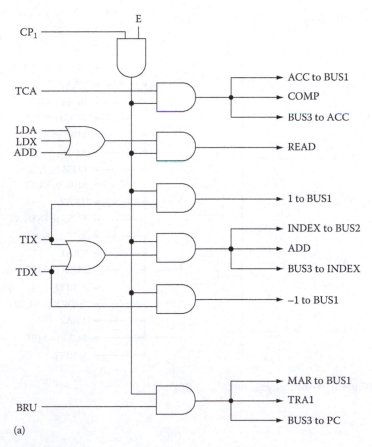

(a)

Figure 6.16 Execute cycle. (a) CP_1.

(continued)

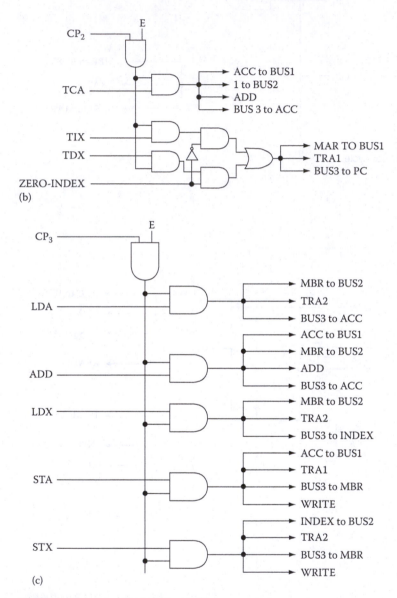

Figure 6.16 (continued) Execute cycle. (b) CP_2. (c) CP_3.

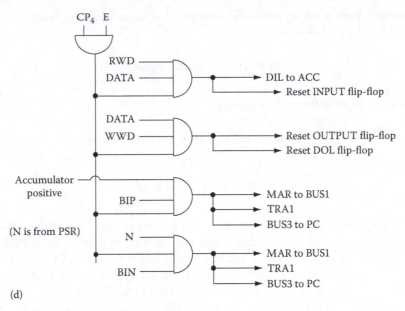

Figure 6.16 (continued) Execute cycle. (d) CP_4.

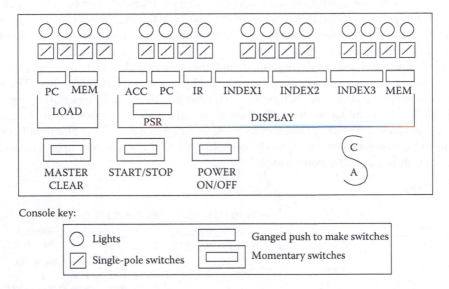

Figure 6.17 ASC console.

Loading memory: Load PC with the memory address. Set switch bank to the data to be loaded into the memory location.

LOAD MEM   MAR ← PC.

MBR ← Switch Bank.

WRITE.

Display registers: Each of the register contents can be displayed on monitors (lights) by pushing the corresponding display switch.

$$MONITOR \leftarrow Selected\ register$$

Display memory: Set PC to the memory address.

$$DISPLAY\ MEM \quad MAR \leftarrow PC, READ.$$

$$PC \leftarrow PC+1.$$

$$MONITOR \leftarrow MBR.$$

While loading and displaying memory, the PC value is incremented by 1 at the end of a load or a display. This enables easier loading and monitoring of the consecutive memory locations. To execute a program, the program and data are first loaded into the memory, PC is set to the address of the first executable instruction, and the execution is started by pushing the START switch. Since the content of any register or memory location can be placed on BUS3, a 16-bit *monitor* register is connected to BUS3. The lights on the console display the contents of this register. The switch bank is connected to BUS2.

Figure 6.18 shows the control circuitry needed to display memory. The console is active only when the RUN flip-flop is RESET. When one of the LOAD or DISPLAY switches is depressed, the *console active* flip-flop is set for a time period equal to three clock pulses (all the console functions can be completed in three-clock-pulse time).

The RUN flip-flop is also set during this period, thus enabling the clock. The clock circuitry is active long enough to give three pulses and deactivated by resetting the RUN flip-flop. The START/STOP switch complements the RUN flip-flop each time it is depressed, thus starting or stopping the clock. Note that except for the START/STOP switch, the console is inactive when the machine is running. Circuits to generate control signals corresponding to each LOAD and DISPLAY switch must be included to complete the design. The console designed here is very simple compared to the consoles of machines available commercially.

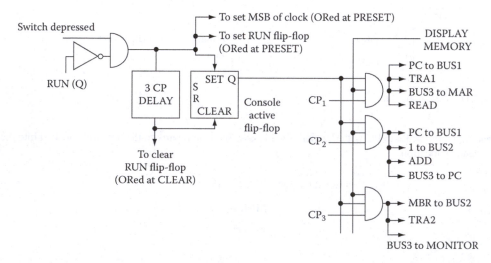

Figure 6.18 Console circuits for DISPLAY MEMORY.

6.8 MICROPROGRAMMED CONTROL UNIT (MCU)

An HCU requires extensive redesign of the hardware if the instruction set has to be expanded or if the function of an instruction has to be changed. In practice, a flexible CU is desired to enable tailoring the instruction set to the application environment. An MCU offers such flexibility. In an MCU, microprograms corresponding to each instruction in the instruction set are stored in a control ROM (CROM). A microcontrol unit (μCU) executes the appropriate microprogram based on the instruction in IR. The execution of a microinstruction is equivalent to the generation of control signals to bring out that microoperation. The μCU is usually hardwired and is comparatively simple to design since its function is only to execute microprograms in the CROM.

Figure 6.19 shows a block diagram of an MCU. Microprograms corresponding to fetch, defer, and execute cycles of each instruction are stored in the CROM. The beginning address of the fetch sequence is loaded into μMAR when the power is turned on. Then the CROM transfers the first microinstruction of fetch into μMBR. The μCU decodes this microinstruction to generate control signals required to bring about that microoperation. The μMAR is normally incremented by 1 at each clock pulse to execute the next microinstruction in sequence. This sequential execution is altered at the end of the fetch microprogram, since the execution of the microprogram corresponding to the execute cycle of the instruction now residing in IR must be started. Hence, the μMAR must be set to the CROM address where the appropriate execute microprogram begins. At the end of execution of each execute microprogram, control is transferred to the fetch sequence. The function of the μCU is thus to set the μMAR to the proper value, that is, the current value incremented by 1 or a jump address depending on the opcode and the status signals such as ZERO-INDEX, N, and Z.

For illustration purposes, a typical microinstruction might be PC ← PC + 1. When this microinstruction is brought into μMBR, the decoder circuits generate the following control signals: PC to BUS1, 1 to BUS2, ADD, and BUS3 to PC, and the μMAR is incremented by 1. We describe the design of an MCU for ASC in the following section.

6.8.1 Microprogrammed Control Unit for ASC

Table 6.7 shows the complete microprogram for ASC. The microprogram resembles any high-level language program and consists of "executable microinstructions" (which produce control signals as the result of execution) and "control microinstructions" (which change the sequence of

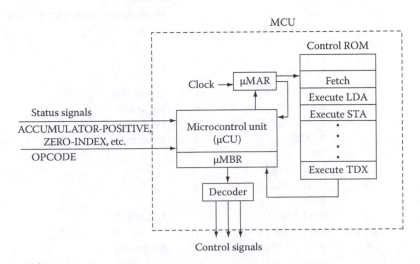

Figure 6.19 MCU model.

Table 6.7   ASC Microprogram

Address		Microinstruction	Comments
0		GO to 0.	Halt loop
1		GO TO LDA.	
2		GP TO STA.	
3		GO TO ADD.	
4		GO TO TCA.	
5		GO TO BRU.	Jump to
6		GO TO BIP.	appropriate
7		GO TO BIN.	execute
8		GO TO RWD.	routine
9		GO TO WWD.	
10		GO TO SHL.	
11		GO TO SHR.	
12		GO TO LDX.	
13		GO TO STX.	
14		GO TO TIX.	
15		GO TO TDX.	
16		—	
17		Reserved for other opcodes.	
18		—	
—		—	
—		—	
31		—	
32	FETCH	MAR ← PC, READ.	Fetch begins
33		PC ← PC + 1.	
34		IR ← MBR.	Fetch ends
35		GO TO* OPCODE.	Jump to 0 through 31 based on OPCODE in IR_{15-11}
36	LDA	MAR ← IR_{7-0} + INDEX.	Index
37		IF IR_{10} = 0 THEN GO TO M.	No indirect
38		READ.	Indirect
39		WAIT.	No operation, waiting for memory read
40		MAR ← MBR.	Execute
41	M	READ.	
42		WAIT.	
43		ACC ← MBR.	
44		GO TO FETCH.	End of LDA
45	STA	MAR ← IR_{7-0} + INDEX.	Begin STA
46	L1	IF IR_{10} = 0 THEN GO TO M1.	No indirect
47		READ.	Indirect
48		WAIT.	
49		MAR ← MBR.	
50	M1	MBR ← ACC, WRITE.	
51		WAIT.	
52		GO TO FETCH.	End of STA
53	ADD	MAR ← IR_{7-0} + INDEX.	
54	L2	IF IR_{10} = 0 THEN GO TO M2.	No indirect
55		READ.	Indirect

Table 6.7 (continued) ASC Microprogram

Address		Microinstruction	Comments
56		WAIT.	
57		MAR ← MBR.	
58	M2	READ.	
59		WAIT.	
60		ACC ← ACC + MBR.	
61		GO TO FETCH.	End of ADD
62	TCA	ACC ← NOT (ACC).	
63		ACC ← ACC + 1.	
64		GO TO FETCH.	End of TCA
65	BRU	MAR ← IR_{7-0} + INDEX.	
66		IF IR_{10} = 0 THEN GO TO M3.	No indirect
67		READ.	Indirect
68		WAIT.	
69		MAR ← MBR.	
70	M3	PC ← MAR.	Execute BRU
71		GO TO FETCH.	End BRU
72	BIP	IF ACC < = 0	If ACC is not positive,
		THEN GO TO FETCH.	go to fetch
73	BIN	IF ACC > = 0	If ACC is not
		THEN GO TO FETCH.	negative, go to fetch
75		GO TO BRU.	
76	RWD	DATA ← 0, INPUT ← 1.	Begin RWD
77	LL1	WAIT.	Wait until data are ready
78		IF DATA = 0 THEN GO TO LL1.	
79		ACC ← DIL, INPUT ← 0.	Data to ACC
80		GO TO FETCH.	End RWD
81	WWD	DATA ← 0, OUTPUT ← 1,	Begin WWD
		DOL ← ACC.	
82	LL2	WAIT.	Wait until data are accepted
83		IF DATA = 0 THEN GO TO LL2.	
84		OUTPUT ← 0.	
85		GO TO FETCH.	End WWD
86	SHL	ACC ← SHL (ACC)	Shift left
87		GO TO FETCH.	
88	SHR	ACC ← SHR (ACC)	Shift right
89		GO TO FETCH.	
90	LDX	MAR ← IR_{7-0}.	Begin LDX
91		IF IR_{10} = 0 THEN GO TO M6.	No indirect
92		READ.	Indirect
93		WAIT.	
94		MAR ← MBR.	
95	M6	READ.	Execute
96		WAIT.	
97		INDEX ← MBR.	
98		GO TO FETCH.	End LDX
99	STX	MAR ← IR_{7-0}.	Begin STX

(continued)

Table 6.7 (continued)   ASC Microprogram

Address		Microinstruction	Comments
100		IF IR_{10} = 0 THEN GO TO M7.	No indirect
101		READ.	Indirect
102		WAIT.	
103		MAR ← MBR.	
104	M7	MBR ← INDEX, WRITE.	Execute
105		WAIT.	
106		GO TO FETCH.	End STX
107	TIX	INDEX ← INDEX + 1.	Increment
108		IF ZERO-INDEX = 1 THEN	ZERO-INDEX
109	LL3	GO TO FETCH.	is nonzero
110		MAR ← IR_{7-0}.	
111		IF IR_{10} = 0 THEN GO TO M8.	No indirect
112		READ.	Indirect
113		WAIT.	
114	M8	MAR ← MBR.	
115		PC ← MAR.	Jump
116	TDX	GO TO FETCH.	
117		INDEX ← INDEX − 1.	Decrement
		IF ZERO-INDEX = 0 THEN	
118		GO TO FETCH.	
		GO TO LL3.	

execution of the microprogram based on some conditions). The first 32 microinstructions are merely jumps to the microinstruction sequences corresponding to the 32 possible instructions in ASC. Location 0 contains an infinite loop corresponding to the execution of HLT instruction. The START switch on the console takes ASC off this loop. That is, when the START switch is depressed, the microprogram begins execution at 32 where the fetch sequence begins. As soon as the instruction is fetched into IR, the microprogram jumps to one of the 32 locations (based on the opcode in IR) and in turn to the appropriate execution sequence.

The indexing and indirect address computation are each now part of the execution sequence of instructions that use those addressing modes. As in HCU, the index register is selected by the IR decoder circuit, and if the index flag corresponds to 00, no index register will be selected. The microinstruction sequence corresponding to indirect address computation (i.e., locations 38–40) is repeated for each instruction that needs it, for simplicity. (Alternatively, this sequence could have been written as a subprogram that could be called by each instruction sequence as needed.) At the end of each instruction execution, the microprogram returns to the fetch sequence.

To represent the microprogram in a ROM, the microprogram should be converted into binary. This process is very similar to an assembler producing the object code. There are two types of instructions in the microprogram, as mentioned earlier. To distinguish between the two types, we will use a microinstruction format with a 1-bit microopcode. A microopcode of 0 indicates type 0 instructions that produce the control signals, while microopcode 1 indicates type 1 instructions (jumps) that control the microprogram flow. The formats of the two types of microinstructions are shown later:

0		Control signals
1	Condition	Branch address

Each bit of the type 0 instruction can be used to represent a control signal. Thus, when this micro-instruction is brought to μMBR, each nonzero bit in the microinstruction would produce the corresponding control signal. This organization would not need any decoding of the microinstruction. The disadvantage is that the microinstruction word will be very long, thereby requiring a large CROM. One way to reduce the microinstruction word length is by encoding signals into several fields in the microinstruction, wherein a field represents control signals that are not required to be generated simultaneously. Each field is then decoded to generate the control signals.

An obvious signal encoding scheme for ASC is shown in Figure 6.20. Here, control signals are partitioned based on the buses they are associated with (i.e., ON BUS1, ON BUS2, and OFF BUS3) and the ALU. The control signals to all these four facilities need to be generated simultaneously. This listing shows that each of the fields requires three bits to represent the signals in that field. Figure 6.20b classifies the remaining control signals into four fields of 2 bits each. Thus, the ASC type 0 instruction would be 21 bits long.

An analysis of the microprogram in Table 6.7 shows that there are eight different branching conditions. These conditions are shown in Figure 6.21. A 3-bit field is required to represent these conditions. The coding of this 3-bit field is also shown in Figure 6.21. Since there are a total of 119 microinstructions, a 7-bit address is needed to completely address the CROM words. Thus, type 1 microinstructions can be represented in 11 bits. But to retain the uniformity, we will use a 21-bit format for type 1 instructions also. The formats are shown in Figure 6.21b.

Code	ON BUS1	ON BUS2	OFF BUS3	ALU
0 0 0	None	None	None	None
0 0 1	ACC	INDEX	ACC	TRA1
0 1 0	MAR	MBR	INDEX	TRA2
0 1 1	$(IR)_{ADRS}$	1	IR	ADD
1 0 0	PC	SWITCH BANK	MAR	COMP
1 0 1	1		MBR	SHL
1 1 0	−1		PC	SHR
1 1 1			MONITOR	

(a)

Code	FIELD1	FIELD2	FIELD3	MEMORY
0 0	None	None	None	None
0 1	Not used	DIL to ACC	0 to INPUT	READ
1 0	1 to INPUT	ACC to DOL	0 to OUTPUT	WRITE
1 1	1 to OUTPUT	INHIBIT CLOCK	0 to DATA	Not used

(b)

μ OPCODE	ON BUS1	ON BUS2	OFF BUS3	ALU	FIELD1	FIELD2	FIELD3	MEMORY
0								
1	3	3	3	3	2	2	2	2

(c)

Figure 6.20 Signal partitioning and encoding for microinstruction type 0. (a) 3-bit fields. (b) 2-bit fields. (c) Type 0 format.

Branching Condition	Condition Code	Comment
Unconditional branch	000	
GO TO* OPCODE	001	Branch based on opcode
Branch if $IR_{10} = 0$	010	Indirect
Branch if ACC < = 0	011	i.e., ACCUMULATOR-POSITIVE = 0
Branch if ACC > = 0	100	i.e., N = 0
Branch if DATA = 0	101	DATA flip-flop is reset
Branch if ZERO-INDEX = 1	110	Index register content nonzero.
Branch if ZERO-INDEX = 0	111	Index register content is zero.

(a)

1	Condition code	μ Address	Not used

(b)

Figure 6.21 Type 1 microinstruction encoding. (a) Type 1 microinstruction condition code field encoding. (b) Type 1 format.

μOPCODE	ACC to BUS1	NONE	BUS3 to MBR	TRA1	NONE	NONE	NONE	WRITE
0	001	000	101	001	00	00	00	10

MBR←ACC, WRITE.

(a)

μOPCODE	Condition code	μAddress	Not used
1	000	0100000	–

GO TO FETCH

(b)

Figure 6.22 Typical encoded microinstructions. (a) Type 0. (b) Type 1.

The microprogram of Table 6.7 must be encoded using the instruction formats shown in Figures 6.20 and 6.21. Figure 6.22 shows some examples of encoding.

The hardware structure to execute the microprogram is shown in Figure 6.23. The μMAR is first set to point to the beginning of the fetch sequence (i.e., address 32). At each clock pulse, the CROM transfers the microinstruction at the address pointed to by the μMAR into μMBR. If the first bit of μMBR is 0, the control decoders are activated to generate control signals and μMAR is incremented by 1. If the first bit of μMBR is a 1, the condition code decoders are activated. Based on the condition code, the μMAR is updated as follows:

Condition	μMAR Receives
START	32.
000	Address field of μMBR.
001	Opcode from IR.
010 through 111	Address field of μMBR if the corresponding condition is satisfied; otherwise, increment μMAR by 1.

Refer to Figure 6.21a for details on condition codes.

LET

$E_1 = (\mu opcode)'$

$E_2 = CC1$

$E_3 = START \cdot RUN'$

$E_4 = CC0 +$	Unconditional
$CC2 \cdot IR'_{10} +$	indirect
$CC3 \cdot (N + Z) +$	$ACC < = 0$
$CC4 \cdot (N' + Z) +$	$ACC > = 0$
$CC5 \cdot DATA' +$	DATA flip-flop reset
$CC6 \cdot ZERO\text{-}INDEX +$	$INDEX \neq 0$
$CC7 \cdot (ZERO\text{-}INDEX)'$	$INDEX = 0.$

where, CC0–CC7 are decoded condition codes.

Then

E_1: $\mu MAR \leftarrow \mu MAR + 1$

E_2: $\mu MAR \leftarrow 00\phi\ OPCODE$ (from IR)

E_3: $\mu MAR \leftarrow 32$

E_4: $\mu MAR \leftarrow$ Address portion of μMBR.

Figure 6.23 MCU hardware.

(continued)

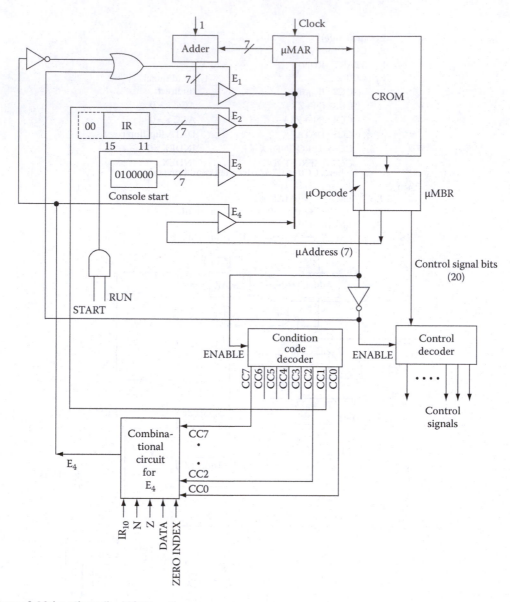

Figure 6.23 (continued) MCU hardware.

The console operations are also facilitated by the microcode. The console hardware should gate a 32 into μMAR and start the clock every time the START switch is activated and the RUN flip-flop is reset. Similarly, corresponding to the LOAD PC switch, the following microprogram is required:

120 LOAD PC PC ← SWITCH BANK.

121 GO TO 0.

Again, at each activation of LOAD PC switch, the address 120 should be gated into μMAR and clock started. At the end of LOAD PC, the machine goes into halt state. The detailed design of console circuits is let as an exercise.

In an MCU, the time required to execute an instruction is a function of the number of microinstructions in the corresponding sequence. The MCU returns to fetch the new instruction once the sequence is executed (the HCU waited until CP_4 for a state change). The overall operation is, however, slower than that of an HCU since each microinstruction is to be retrieved from the CROM and the time required between two clock pulses is thus equal to slowest register transfer time + CROM access time.

The function of an instruction is easily altered by changing the microprogram. This requires a new CROM. Other than this, no hardware changes are needed. This attribute of an MCU is used in the "emulation" of computers where an available machine (host) is made to execute another (target) machine's instructions by changing the control store of the host.

In practice, CROM size is a design factor. The length of the microprogram must be minimized to reduce CROM size, thereby reducing the cost of MCU. This requires that each microinstruction contains as many microoperations as possible, thereby increasing the CROM word size, which in turn increases the cost of MCU. A compromise is thus needed. As in ASC-MCU, each microinstruction has to be encoded and the number of each encoded field in an instruction has to be kept low to reduce the CROM width. A microoperation then will span more than one CROM word. Facilities are then needed to buffer the CROM words corresponding to a microoperation so that all required control signals are generated simultaneously. In a *horizontally microprogrammed* machine, each bit of the CROM word corresponds to a control signal. That is, there is no encoding of bits. Hence, the execution of the microprogram is fast since no decoding is necessary. But the CROM words are wide. In a *vertically microprogrammed* machine, the bits in the CROM word are encoded to represent control signals. This results in shorter words, but the execution of the microprograms becomes slow owing to decoding requirements.

We conclude this section with the following procedure for designing an MCU, once the bus structure of the machine is established:

1. Arrange the microinstruction sequences into one complete program (microprogram).
2. Determine the microinstruction format(s) needed. The generalized format is as follows:

CONDITION	BRANCH ADDRESS	CONTROL SIGNALS

This format might result in less efficient usage of the microword since BRANCH ADDRESS field is not used in those microinstructions that generate control signals and increment μMAR by 1 to point to the next microinstruction.
3. Count the total number of control signals to be generated. If this number is large, thus making the microinstruction word length too large, partition the control signals into distinct groups. Only one of the control signals from each partition is generated at a given time. Control signals that need to be generated simultaneously should be in separate groups. Assign fields in the control signal portion of the microinstruction format and encode them.
4. After the encoding in step 3, determine the μMAR size from the total number of microinstructions needed.
5. Load the microprogram into the CROM.
6. Design the circuit to initialize and update the contents of the μMAR.

6.9 SUMMARY

The detailed design of ASC provided here illustrates the sequence of steps in the design of a digital computer. In practice, the chronological order shown here cannot be strictly followed. Several iterations between the steps are needed in the design cycle before a complete design is obtained. Several architectural and performance issues arise at each step in the design process. We have ignored those issues in this chapter but address them in subsequent chapters of this book. This chapter introduced the design of MCUs. Almost all modern-day computers use MCUs, because of the flexibility needed to update processors quickly, to meet the market needs.

PROBLEMS

6.1 Rewrite the control sequence for conditional branch instructions of ASC so that the instruction cycle terminates in one major cycle if the condition is not satisfied. Discuss the effect of this modification on the hardware of the CU.

6.2 In a practical computer, the shift instructions accommodate multiple shifts. Use the address field of SHR and SHL to specify the number of shifts. How can you accommodate this modification by
   a. Changes to HCU
   b. Changes to microprogram
   c. Changes to assembler

6.3 Write register transfer sequences for the following new instructions on ASC: Subtract:

$$\text{SUB MEM} \quad \text{ACC} \leftarrow \text{ACC} - \text{M[MEM]}$$

Load immediate:

$$\text{LD1} \quad \text{data} \quad \text{ACC} \leftarrow \text{data}$$

Assume data are in bits 0 through 7 of the instruction.

6.4 Write register transfer sequences for the following new instructions:
   JSR: Subroutine jump; use memory location 0 for storing the return address
   RET: Return from subroutine

6.5 Write the register transfer sequence needed to implement a multiply instruction on ASC. That is,

$$\text{MPY MEM} \quad \text{ACC} \leftarrow \text{ACC} * \text{M[MEM]}.$$

Use the repeated addition of ACC to itself M[MEM] times. Assume that the product fits in one word. List the hardware changes needed (if any).

6.6 Write microprogram to implement the solutions for Problems 6.3 through 6.5.

6.7 Write the microprogram to implement the following move multiple words instruction:

$$\text{MOV A,B}$$

which moves N words from memory locations starting at A to those at B. The value of N is specified in the accumulator. Assume that A and B are represented as 4-bit fields in the 8-bit operand field of the instruction, and indexing and indirecting are not allowed. What is the limitation of this instruction?

6.8 Implement the move instruction of the previous problem, by changes to the HCU.

6.9 Use two words to represent the move instruction of Problem 6.7. The first word contains the opcode, address modifiers, and the address of A, and the second word contains the address modifiers and the address of B (the opcode field in this word is not used). Write the microprogram.

6.10 Discuss how you can accommodate more than one I/O device on ASC. Specify the hardware modification needed.

6.11 Rewrite the microprogram sequence for HLT instruction, if the JSR and RET instructions of Problem 6.4 are implemented by extending the HLT opcode. That is, use the unused bits of the HLT instruction to signify JSR and RET.

6.12 Design a single-bus structure for ASC.

6.13 Code the microprogram segments or LDA and TIX instructions in Table 6.7 in binary and hexadecimal.

6.14 We have assumed single level of indirect addressing for ASC. What changes are needed to extend this to multiple levels whereby the indirection is performed until the most significant bit of the memory word, addressed at the current level, is nonzero?

6.15 Change the ASC instruction format to accommodate both preindexed and postindexed indirect addressing. Rewrite the microcode to accommodate those operations.

6.16 Use a 16 bit up counter as the PC for ASC. What changes are needed for the CU hardware? Does it alter the operation of any of the instructions?

6.17 Extend the JSR and RET instructions of Problems 6.4 to allow nested subroutine calls. That is, a subroutine can call another subroutine. Then the returns from the subroutines will be in a last-in-first-out order. Use a stack in your design. (Refer to Appendix B for details of stack implementation.)

6.18 Design a paged-addressing mechanism for ASC. Assume that the two most significant bits of the 8-bit direct address are used to select one of the four 10-bit segment registers. The contents of a segment register concatenated with the least significant 6 bits of the address field form the 16-bit address. Show the hardware details. What is the effect of this addressing scheme on indirect and index address computations?

6.19 Do we still need the index and indirect addressing modes, after implementing the paged-addressing scheme of the previous problem? Justify your answer.

6.20 Assume that the ASC memory is organized as 8 bits per word. That means each single-address instruction now occupies two words and zero-address instructions occupy one word each. The ASC bus structure remains the same. MBR is now 8 bits long and is connected to the least significant 8 bit of the bus structure. Rewrite the fetch microprogram.

6.21 Include four general-purpose registers into ASC structure to replace the accumulator and index registers. Each register can be used either as an accumulator or as an index register. Expand ASC instruction set to utilize these registers. Assume the same set of operations as are now available on ASC. Is there any merit in including register-to-register operations in the instruction set? Design the new instruction format.

6.22 If you are restricted to using only eight instructions to build a new machine, which of the 16 instructions would you discard? Why?

6.23 How would you handle an invalid opcode in the MCU?

6.24 Study the architectures of Intel, Motorola, and MIPS processors along the features of ASC described in Chapters 5 and 6.

6.25 What are the differences in the operating characteristics of machines with fixed-length and variable-length instructions?

6.26 ASC memory now uses the word addressing mode. What changes are needed to ASC structure to accommodate addressing bytes of the memory? That is, each word of the memory is now considered as two bytes and we should be able to access either of the bytes.

6.27 In ASC, zero-address instructions now use only 5 bits of the word, while the other instructions use all the 16 bits. If the memory is byte addressable (as in Problem 6.26), we can use one byte to represent zero-address instructions (3 bits are still wasted) and two bytes to represent other instructions. What changes are needed for the CU to operate with such instructions of different length?

6.28 Assume that a machine has 32-bit instructions, memory addresses are 12 bits long, and there are 8 registers. We would like to have 8 instructions with two memory addresses and one register reference, 50 instructions with one memory address and one register reference, and 50 instructions with no memory addressing or register reference. Design a variable-length opcode. Do we have any room left for additional instructions?

6.29 Show the schematic of the CU to fetch and decode instructions in Problem 6.28.

BIBLIOGRAPHY

Baron, R.J. and Higbie, L., *Computer Architecture*, Reading, MA: Addison Wesley, 1992.

Patterson, D.A. and Hennessey, J.L., *Computer Organization and Design: The Hardware/Software Interface*, San Mateo, CA: Morgan Kaufmann, 2008.

Stallings, W., *Computer Organization and Architecture*, Upper Saddle River, NJ: Prentice Hall, 2006.

Stone, H.S., *High Performance Computer Architecture*, Reading, MA: Addison Wesley, 1993.

Tannenbaum, A., *Structured Computer Organization*, Upper Saddle River, NJ: Prentice Hall, 2005.

CHAPTER 7

Input/Output

In the design of a simple computer (ASC), we assumed one input device and one output device transferring data in and out of the accumulator using a programmed input/output (I/O) mode. An actual computer system, however, consists of several input and output devices or *peripherals*. Although the programmed I/O mode can be used in such an environment, it is slow and may not be suitable, especially when the machine is used as a real-time processor responding to irregular changes in the external environment. Consider the example of a processor used to monitor the condition of a patient in a hospital. Although the majority of its patient data gathering operations can be performed in a programmed I/O mode, alarming conditions such as abnormal blood pressure or temperature occur irregularly, and detection of such events requires that the processor be interrupted by the event from its regular activity. We discuss the general concept of interrupt processing and interrupt-driven I/O in this chapter.

The transfer of data between the processor and a peripheral consists of the following steps:

1. Selection of the device and checking the device for readiness
2. Transfer initiation, when the device is ready
3. Data transfer
4. Conclusion

These steps can be controlled by the processor or the peripheral or both. Contingent upon where the transfer control is located, three modes of I/O are possible. They are

1. Programmed I/O
2. Interrupt-mode I/O
3. Direct-memory access (DMA)

We will discuss each of these modes in detail following a discussion of the general I/O model. Pertinent details of I/O structures of some popular computer systems are provided as examples.

7.1 GENERAL I/O MODEL

The *I/O structure* of ASC with one input device and one output device is shown in Figure 7.1. ASC communicates with its peripherals through data-input lines (DILs) and data-output lines (DOLs). There are two *control lines*: INPUT and OUTPUT. The *data flip-flop* in the control unit is used to coordinate the I/O activities.

Figure 7.2 shows the generalization of ASC I/O structure to include multiple I/O devices. To address multiple devices, a device number (or address) is needed. This device address can be represented in the 8-bit operand field of RWD and WWD instructions. In fact, since the index and

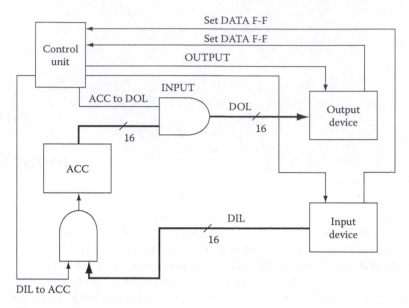

Figure 7.1   ASC I/O structure.

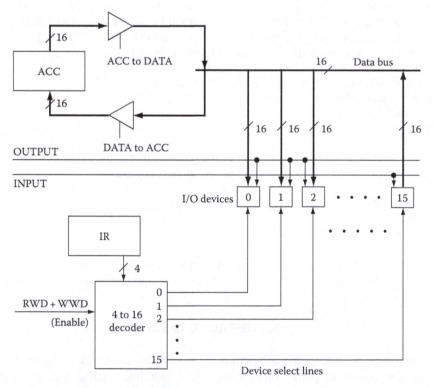

Figure 7.2   Interfacing multiple devices to ASC.

indirect fields of these instructions are not used, device addresses as large as 11 bits can be used. In Figure 7.2, it is assumed that only 4 bits are used to represent the device address. Thus, it is possible to connect 16 input and 16 output devices to ASC. A 4 to 16 decoder attached to device address bits is activated during RWD and WWD instruction cycles to select one of the 16 input or output devices, respectively. If the INPUT control line is active, the selected input device sends the data

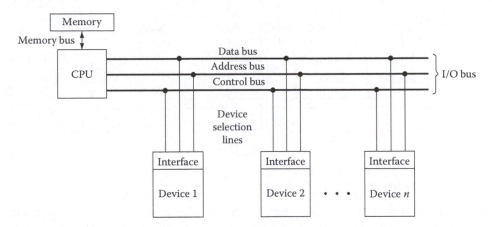

Figure 7.3   General I/O structure.

to the processor. If the OUTPUT control line is active, the selected output device receives the data from the processor. The DILs and DOLs are replaced by the bidirectional data bus.

Figure 7.3 shows the generalized I/O structure used in practice. Here, the device address is carried on the *Address bus* and is decoded by each device. Only the device whose address matches that on the address bus will participate in either input or output operation. The *Data* bus is bidirectional. The *Control bus* carries *control* signals (such as INPUT and OUTPUT). In addition, several *status signals* (such as DEVICE BUSY and ERROR) originating in the device interface also form part of the control bus.

In the structure shown in Figure 7.3, the memory is interfaced to the central processing unit (CPU) through a Memory bus (consisting of address, data, and control/status lines), and the peripherals communicate with the CPU over the I/O bus. Thus, there is a *memory address space* and a separate *I/O address space*. The system is said to use the *isolated I/O* mode. This mode of I/O requires a set of instructions dedicated for I/O operations. In some systems, both the memory and I/O devices are connected to the CPU through the same bus, as shown in Figure 7.4. In this structure, the device addresses are a part of the memory address space. Hence, the load and store instructions used with memory operands can be used as the read and write instructions with respect to those addresses configured for I/O devices. This mode of I/O is known as *memory-mapped I/O*. The advantages of memory-mapped I/O is that separate I/O instructions are not needed; the disadvantages are that some of the memory address space is used up by the I/O, and it can be difficult to distinguish between the memory and I/O-oriented operations in a program.

Although the earlier description implies two buses in the system structure for the isolated I/O mode, it need not be so in practice. It is possible to *multiplex* memory and I/O addresses on the same bus while using control lines to distinguish between the two operations.

Figure 7.5 shows the functional details of a *device interface*. A device interface is unique to a particular device since each device is unique with respect to its data representation and read/write (R/W)

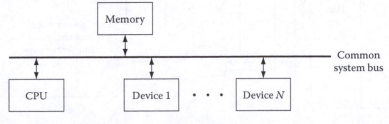

Figure 7.4   Common bus (for both memory and I/O).

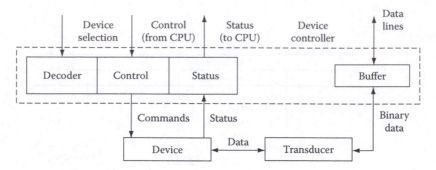

Figure 7.5 Device interface.

operational characteristics. The device interface consists of a controller that receives commands (i.e., control signals) from the CPU and reports the status of the device to the CPU. If the device is, for example, a tape reader, typical control signals are as follows: IS DEVICE BUSY?, ADVANCE TAPE, REWIND, and the like. Typical status signals are DEVICE BUSY, DATA READY, DEVICE NOT OPERATIONAL, etc. Device selection (address decoding) is also shown as a part of the controller. The *Transducer* converts the data represented on the I/O medium (tape, disk, etc.) into the binary format and stores it in the data *Buffer*, if the device is an input device. In the case of an output device, the CPU sends data into the buffer, and the transducer converts these binary data into a format suitable for output onto the external medium (e.g., 0/1 bit write format onto a magnetic tape or disk, ASCII patterns for a printer).

Figure 7.6 shows the functional details of an interface between the CPU and a magnetic tape unit. The interface is very much simplified and illustrates only the major functions. When the device

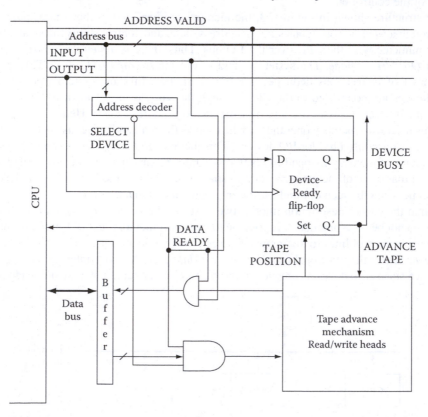

Figure 7.6 Tape interface.

address on the Address bus corresponds to that of the tape device, the SELECT DEVICE signal becomes active. The CPU outputs the ADDRESS VALID control signal a little after the decoder outputs are settled, which is used to clock the Device Ready flip-flop. The Device Ready flip-flop is cleared since the output of the decoder is zero. The Q' output of the Device Ready flip-flop then becomes the DEVICE BUSY status signal, while the tape is being advanced to the next character position. Once the position is attained, the tape mechanism generates the TAPE POSITION signal, which sets the Device Ready flip-flop (through its asynchronous set input) and in turn generates the DATA READY signal for the CPU. During the data read operation, the INPUT signal being active, the device reads the data from the tape, loads it into its Buffer, and gates it onto the Data bus. In response to the DATA READY, the CPU gates the Data bus into its accumulator.

During the data write, the CPU sends the data over the data bus into the device buffer, sets the OUTPUT control signal, sets the address bus with the device address, and outputs the ADDRESS VALID signal. The operation of the Device Ready flip-flop is similar to that during the data read operation. The data are written onto the tape when the DATA READY signal becomes active. The DATA READY also serves as the data accepted signal for the CPU, in response to which the CPU removes the data and address from respective buses. We will describe the CPU and I/O device handshake further in the next section.

7.2 I/O FUNCTION

The major functions of a device interface are

1. Timing
2. Control
3. Data conversion
4. Error detection and correction

The timing and control aspects correspond to the manipulation of control and status signals to bring about the data transfer. In addition, the operating speed difference between the CPU and the device must be compensated for by the interface. In general, data conversion from one code to the other is needed, since each device (or the medium on which data are represented) may use a different code to represent data. Errors occur during transmission and must be detected and if possible corrected by the interface. A discussion of each of these I/O functions follows.

7.2.1 Timing

So far in this chapter, we have assumed that the CPU controls the bus (i.e., it is *bus master*). In general, when several devices are connected to the bus, it is possible that a device other than the CPU can become bus master. Thus, among the two devices involved in the data transfer on the bus, one will be the bus *master* and the other will be the *slave*. The sequence of operations needed for a device to become the bus master is described later in this chapter.

The data transfer on a bus between two devices can be either *synchronous* or *asynchronous*. Figure 7.7 shows the timing diagram for a typical synchronous bus transfer. The clock serves as the timing reference for all the signals. Either the rising or falling edge can be used. During the read operation shown in Figure 7.7a, the bus master activates the READ signal and places the ADDRESS of the slave device (the device from which it is trying to read data) on the bus. All the devices on the bus decode the address, and the device whose address matches the address on the bus responds by placing the data on the bus several clock cycles later. The slave may also provide *status information* (such as error, no error) to the master. The number of clock

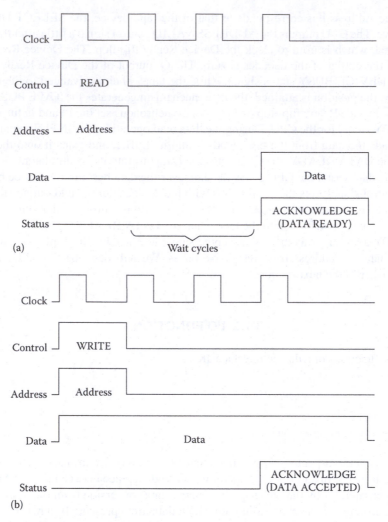

Figure 7.7 Synchronous transfer. (a) Read. (b) Write.

cycles (wait cycles) required for this response depends on the relative speeds of master and slave devices. If the waiting period is known, the master can gate the data from the data bus at the appropriate time. In general, to provide a flexible device interface, the slave device is required to provide a control signal—ACKNOWLEDGE (abbreviated ACK) or DATA READY—to indicate the validity of data on the bus. Upon sensing the ACK, the master gates the data into its internal registers.

Figure 7.7b shows the timing for a synchronous write operation. Here, the bus master activates the WRITE control signal and places the address of the slave device on the address lines of the bus. While the devices are decoding the address, the master also places the data on the data bus. After a wait period, the slave gates the data into its buffer and places the ACK (DATA ACCEPTED) on the bus, in response to which the master removes the data and WRITE control signal from the bus.

Note that all the operations described earlier are synchronized with the clock. In an *asynchronous* transfer mode, the sequence of operations is the same as the aforementioned, except that there will be no clock.

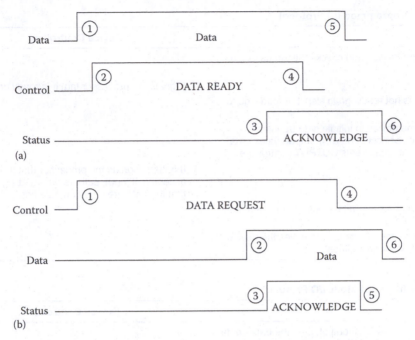

Figure 7.8 Asynchronous transfer. (a) Source initiated. (b) Destination initiated.

Figure 7.8 shows the timing diagrams for an asynchronous transfer between the source and destination devices. In Figure 7.8a, the source initiates the transfer by placing data on the bus and setting DATA READY. After accepting the data, the destination ACKs in response to which the DATA READY signal is removed, after which the ACK is removed. Note that the data are removed from the bus only after the ACK is received. In Figure 7.8b, the destination device initiates (i.e., requests) the transfer, in response to which the source puts the data on the bus. The ACK sequence is the same as in Figure 7.8a.

Peripheral devices usually operate in an asynchronous mode with respect to the CPU because they are not usually controlled by the same clock that controls the CPU. The sequence of events required to bring about the data transfer between two devices is called the *protocol* or the *handshake*.

7.2.2 Control

During the data-transfer handshake, some events are brought about by the bus master, and some are brought about by the slave device. The data transfer is completely controlled by the CPU in the programmed I/O mode. A typical protocol for this mode of I/O is shown in Table 7.1. We have combined the protocols for both input and output operations. A device will be either in the input or the output mode at any given time.

This sequence repeats for each transfer. The speed difference between the CPU and the device renders this mode of I/O inefficient.

An alternative is to distribute part of the control activities to the device controller. Now, the CPU sends a command to the device controller to input or output data and continues its processing activity. The controller collects the data from (or sends data to) the device and "interrupts" the CPU. The CPU disconnects the device after the data transfer is complete (i.e., the CPU services the interrupt) and returns to the mainline processing from which it was interrupted. A typical sequence of events during an *interrupt-mode I/O* is shown in Table 7.2.

Table 7.1   Programmed I/O Protocol

Processor	Device Controller
1. Selects the device and checks the status of the device	
	2. Signals the processor that it is ready or not ready
3. If device is not ready, go to step 1; if ready, go to step 4	
4. Signals the device to initiate data transfer (send data or accept data); if output, gates data onto data lines and sets OUTPUT control line	
	5. If output, signals the processor that the data are accepted; if input, gathers data and signals CPU that the data are ready on data lines
6. If input, accepts the data; if output, removes data from data lines	
7. Disconnects the device (i.e., removes the device address from address lines)	

Table 7.2   Interrupt-Mode I/O Protocol

Processor	Device Controller
1. Selects the device: if output, puts the data on data lines and sets OUTPUT control line; if input, sets INPUT control line	
2. Continues the processing activity	
	3. If output, collects data from data lines and transmits to the medium; if input, collects data from the medium into the data buffer
	4. Interrupts the processor
5. Processor recognizes the interrupt and saves its processing status; if output, removes data from data lines; if input, gates data into accumulator	
6. Disconnects the device	
7. Restores the processing status and returns to processing activity	

The protocol assumes that the CPU always initiates the data transfer. In practice, a peripheral device may first interrupt the CPU, and the type of transfer (input or output) is determined during the CPU–peripheral handshake. The data input need not be initiated only by the CPU. Interrupt-mode I/O reduces the CPU wait time (for the slow device) but requires a more complex device controller than that in the programmed I/O mode. The causes of interrupt and the popular interrupt structures are discussed in the next section.

In addition to the earlier protocols, other control and timing issues are introduced by the characteristics of the *link* (i.e., the data line or lines that connect the device and the CPU). The data link can be *simplex* (unidirectional), *half duplex* (either direction, one way at a time), *full duplex* (both directions simultaneously), *serial*, or *parallel*. Serial transmission requires that a constant clock rate be maintained throughout data transmission to avoid synchronization problems; in parallel transmission, care must be taken to avoid data "skewing" (i.e., data arriving at different times on the bus lines due to different electrical characteristics of individual bit lines).

Data transfer between peripheral devices located in the vicinity of the CPU is usually performed in parallel mode, while the remote devices communicate with the CPU in serial mode. We will assume parallel mode transfer in the following sections. Serial transfer mode is described further in Section 7.7.

7.2.3 Data Conversion

The data representation on the I/O medium is unique to each medium. For example, a magnetic tape uses either ASCII or EBCDIC code to represent data. Internally, the CPU might use a binary or BCD (decimal) representation. In addition, the interface link might be organized as serial by bit, serial by character (quasi-parallel), or serial by word (fully parallel). Thus, two levels of data conversion are to be accomplished by the interface: conversion from peripheral to link format and from link to CPU format.

7.2.4 Error Detection and Correction

Errors may occur whenever data are transmitted between two devices. One or more extra bits known as *parity bits* are used as part of the data representation to facilitate error detection and correction. Such parity bits, if not already present in the data, are included into the data stream by the interface before transmission and checked at the destination. Depending on the number of parity bits, various levels of error detection and correction are possible. Error detection and correction are particularly important in I/O because the peripheral devices exposed to the external environment are error prone. Errors may be due to mechanical wear, temperature and humidity variations, mismounted storage media, circuit drift, incorrect data-transfer sequences (protocols), and the like.

Figure 7.9a shows a parity bit included into a data stream of 8 bits. The parity bit P is 1 if *odd parity* is used (the total number of 1s is odd) and 0 if *even parity* is used (the total number of

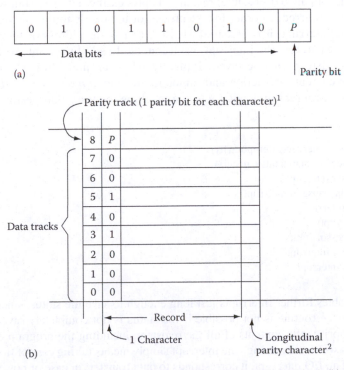

Figure 7.9 Parity bits for error detection on a magnetic tape. (a) Parity checking[1]. (b) Magnetic tape data representation. [1]$P = 1$ for odd parity (total number of 1s is odd); $P = 0$ for even parity (total number of 1s is even). [2]1 bit for each track; parity check for the complete record.

1s is even). When this 9-bit data word is transmitted, the receiver of the data computes the parity bit. If P matches the computed parity, there is no error in transmission; if an error is detected, the data can be retransmitted.

Figure 7.9b shows a parity scheme for a magnetic tape. An extra track (track 8) is added to the tape format. The parity track stores the parity bit for each character represented on the eight tracks of the tape. At the end of a record, a longitudinal parity character consisting of a parity bit for each track is added. More elaborate error detection and correction schemes are often used; references listed in the Bibliography section at the end of this chapter provide details of those schemes.

7.3 INTERRUPTS

In a number of conditions, the processor may be interrupted from its normal processing activity. Some of the conditions are

1. Power failure as detected by a sensor
2. Arithmetic conditions such as overflow and underflow
3. Illegal data or illegal instruction code
4. Errors in data transmission and storage
5. Software-generated interrupts (as intended by the user)
6. Normal completion of an asynchronous transfer

In each of these conditions, the processor must discontinue its processing activity, attend to the interrupting condition, and (if possible) resume the processing activity from where it had been when the interrupt occurred. In order for the processor to be able to resume normal processing after servicing the interrupt, it is essential to at least save the address of the instruction to be executed just before entering the interrupt service mode. In addition, contents of the accumulator and all other registers must be saved. Typically, when an interrupt is received, the processor completes the current instruction and jumps to an *interrupt service routine*. An interrupt service routine is a program preloaded into the machine memory that performs the following functions:

1. Disables further interrupts (temporarily)
2. Saves the processor status (all registers)
3. Enables further interrupts
4. Determines the cause of interrupt
5. Services the interrupt
6. Disables interrupts
7. Restores processor status
8. Enables further interrupts
9. Returns from interrupt

The processor disables further interrupts just long enough to save the status, since a proper return from interrupt service routine is not possible if the status is not completely saved. The processor status usually comprises the contents of all the registers, including the program counter (PC) and the program status word. "Servicing" the interrupt simply means taking care of the interrupt condition: in the case of an I/O interrupt, it corresponds to data transfer; in case of power failure, it is the saving of registers and status for normal resumption of processing when the power is back; during an arithmetic condition, it is checking the previous operation or simply setting a flag to indicate the arithmetic error.

Once the interrupt service is complete, the processor status is restored. That is, all the registers are loaded with the values saved during step 2. Interrupts are disabled during this restore period. This completes the interrupt service, and the processor returns to the normal processing mode.

7.3.1 Interrupt Mechanism for ASC

An interrupt may occur at any time. The processor recognizes interrupts only at the end of the execution of the current instruction. If the interrupt needs to be recognized earlier (say, at the end of a fetch, before execution), a more complex design is needed because the processor status has to be rolled back to the end of the execution of the previous instruction. Assuming the simpler case for ASC, the fetch microsequence must be altered to recognize the interrupt. Let us assume that there is an interrupt input (INT) into the control unit.

In the following interrupt scheme for ASC, we assume that there will be only one interrupt at a time, that is, no interrupts will occur until the current interrupt is completely serviced. (We will remove this restriction later.) We will reserve memory locations 0 through 5 for saving registers (PC, ACC, index registers, and PSR) before entering the interrupt service routine located in memory locations 6 and onward (see Figure 7.10). We also assume that the following new instructions are available:

1. SPS (store PSR in a memory location)
2. LPS (load PSR from a memory location)
3. ENABLE interrupt
4. DISABLE interrupt

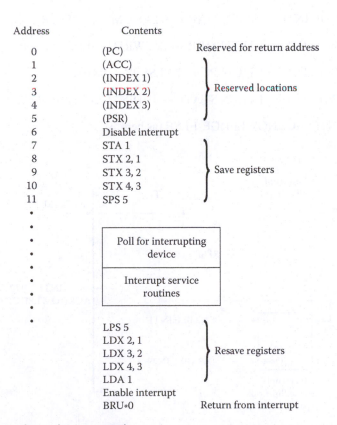

Figure 7.10 Interrupt software (memory map).

Recall that ASC has used only 16 of the 32 possible opcodes. Any of the remaining 16 opcodes can be used for these new instructions.

The fetch sequence now looks like the following:

T_1 : IF INT = 1 THEN MAR ← 0 ELSE MAR ← PC, READ.

T_2 : IF INT = 1 THEN MBR ← PC, WRITE ELSE PC ← PC + 1.

T_3 : IF INT = 1 THEN PC ← 6 ELSE IR ← MBR.

T_4 : IF INT = 1 THEN STATE ← F ELSE (as before).

If INT is high, one machine cycle is used for entering the interrupt service routine (i.e., stores PC in location 0; sets PC = 6). The first part of the service routine (in Figure 7.10) stores all the registers. The devices are then *polled*, as discussed later in this section, to find the interrupting device. The interrupt is serviced and the registers are restored before returning from the interrupt.

This interrupt handling scheme requires that the INT line be at 1 during the fetch cycle. That is, although the INT line can go to 1 any time, it has to stay at 1 until the end of the next fetch cycle in order to be recognized by the CPU. Further, it must go to 0 at the end of T_4. Otherwise, another fetch cycle in the interrupt mode is invoked. This timing requirement on the INT line can be simplified by including an *interrupt-enable* flip-flop and an *interrupt flip-flop* (INTF) into the control unit and gating the INT line into INTF at T_1, as shown in Figure 7.11. The fetch sequence to accommodate these changes is shown here:

T_1 : IF INTF = 1 THEN MAR ← 0 ELSE MAR ← PC, READ.

T_2 : IF INTF = 1 THEN MBR ← PC, WRITE ELSE PC ← PC + 1.

T_3 : IF INTF = 1 THEN PC ← 7 ELSE IR ← MBR.

T_4 : IF INTF = 1 THEN STATE ← F, DISABLE INTE, RESET.

INTF, ACKNOWLEDGE ELSE (as before).

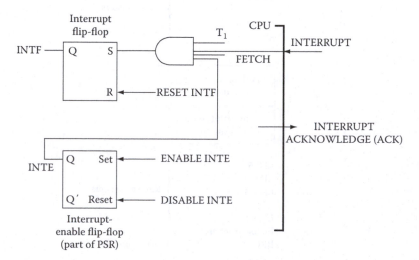

Figure 7.11 Interrupt hardware.

Note that the interrupt sequence now disables interrupts in T_4, thereby not requiring the DISABLE interrupt instruction at location 6 in Figure 7.10. The interrupt service routine execution then starts at location 7. An ENABLE interrupt instruction is required. This instruction sets the interrupt-enable flip-flop and allows further interrupt. Note also that an interrupt ACK signal is generated by the CPU during T_4 to indicate to the external devices that the interrupt has been recognized. In this scheme, the interrupt line must be held high until the next fetch cycle (i.e., one instruction cycle at the worst case) for the interrupt to be recognized. Once an interrupt is recognized, no further interrupts are allowed unless interrupts are enabled.

7.3.2 Multiple Interrupts

In the previous interrupt scheme, if another interrupt occurs during the interrupt service, it is completely ignored by the processor. In practice, other interrupts do occur during the interrupt service and should be serviced based on their priorities relative to the priority of the interrupt being serviced. The interrupt service routine of Figure 7.10 can be altered to accommodate this by inserting an ENABLE interrupt instruction in location 12 right after saving registers and a DISABLE interrupt instruction just before the block of instructions that resave the registers. Although this change recognizes interrupts during interrupt service, note that if an interrupt occurs during the service, memory locations 0–5 will be overwritten, thus corrupting the processor status information required for a normal return from the first interrupt.

If the processor status is saved on a stack rather than in dedicated memory locations, as done in the previous scheme, multiple interrupts can be serviced. At the first interrupt, status 1 is pushed onto the stack; at the second interrupt, status 2 is pushed onto the top of the stack. When the second interrupt service is complete, status 2 is popped from the stack, thus leaving status 1 on the stack intact for the return from the first interrupt. A stack thus allows the "nesting" of interrupts.

7.3.3 Polling

Once an interrupt is recognized, the CPU must invoke the appropriate interrupt service routine for each interrupt condition. In an interrupt-mode I/O scheme, CPU polls each device to identify the interrupting device. Polling can be implemented in either software or hardware. In the software implementation, the polling routine addresses each I/O device in turn and reads its status. If the status indicates an interrupting condition, the service routine corresponding to that device is executed. Polling thus incorporates a priority among devices since the highest priority device is addressed first followed by lower-priority devices.

Figure 7.12 shows a hardware polling scheme. The binary counter contains the address of the first device initially and counts up at each clock pulse if the CPU is in the polling mode. When the count reaches the address of the interrupting device, the interrupt request (IRQ) flip-flop is set, thereby preventing the clock from incrementing the binary counter. The address of the interrupting device is then in the binary counter.

In practice, a priority *encoder* is used to connect device interrupt lines to the CPU, thereby requiring only one clock pulse to detect the interrupting device. Examples of this scheme are given in Section 7.10.

7.3.4 Vectored Interrupts

An alternative to polling as a means of recognizing the interrupting device is to use a vectored interrupt structure like the one shown in Figure 7.13. Here, the CPU generates an ACK signal in response to an interrupt. If a device is interrupting (i.e., if its INTF is set), the ACK flip-flop in

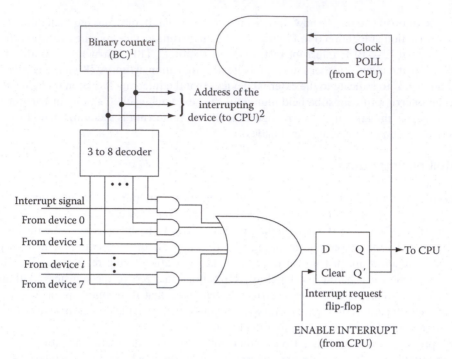

Figure 7.12   Polling hardware (eight devices). *Notes*: [1]BC starts counting when POLL = 1 and the interrupt is enabled. [2]Interrupting device (D_1) sets INT request flip-flop, and BC is stopped at i (i = 0 through 7). CPU receives i as an address.

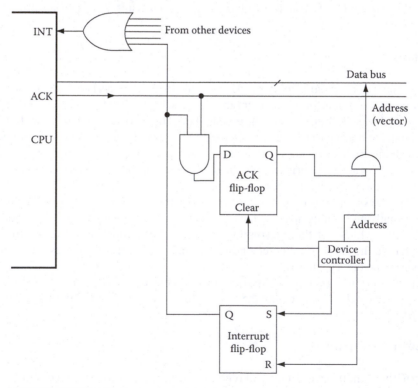

Figure 7.13   Vectored interrupt structure.

the device interface is set by the ACK signal and the device sends a vector onto the data bus. The CPU reads this vector to identify the interrupting device. The vector is either the device address or the address of the memory location where the interrupt service routine for that device begins. Once the device is identified, the CPU sends a second ACK addressed to that device, at which time the device resets its ACK and INTFs (circuitry for the second ACK is not shown in Figure 7.13). In the structure of Figure 7.13, all the interrupting devices send their vectors onto the data bus simultaneously in response to the first ACK. Thus, the CPU cannot distinguish between the devices. To prevent this and to isolate the vector of a single interrupting device, the I/O devices are connected in a *daisy-chain* structure in which the highest priority device receives the ACK first, followed by the lower-priority devices. The first interrupting device in the chain inhibits the further propagation of the ACK signal. Figure 7.14 shows a typical daisy-chain interface. Since the daisy chain is in the ACK path, this structure is called the *backward daisy chain*. In this scheme, the interrupt signals of all the I/O devices are ORed together and input to the CPU. Thus, any device can interrupt the CPU at any time. The CPU has to process the interrupt in order to determine the priority of the interrupting device compared to the interrupt it is processing at that time (if any). If it is processing a higher-priority interrupt, the CPU will not generate the ACK.

To eliminate the overhead of comparing the priorities of interrupts, when the CPU is processing an interrupt, a *forward daisy chain* can be used in which the higher-priority device prevents the interrupt signals of the lower-priority device from reaching the CPU. Figure 7.15 shows a forward daisy chain. Here, each device interface is assumed to have an interrupt status flip-flop in addition to the usual interrupt and ACK flip-flops. The interrupt status flip-flop will be set when the interrupt service for that device begins, and it is reset at the end of the interrupt service. Therefore, as long as the INTF of the device is set (i.e., as long as the device is interrupting) or the interrupt status flip-flop is set (i.e., the device is being serviced), the interrupt signal from the lower-priority device is prevented from reaching the CPU.

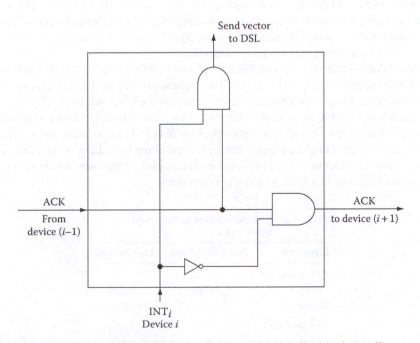

Figure 7.14 Daisy-chain interface. *Note:* Device (i – 1) has a higher priority than device (i).

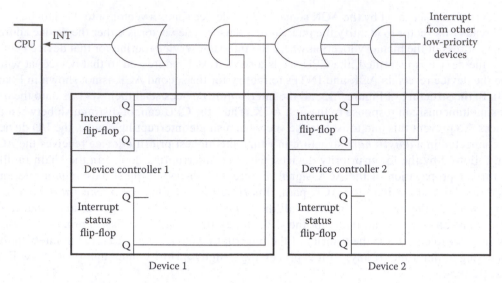

Figure 7.15   Forward daisy chain.

7.3.5 Types of Interrupt Structures

In the interrupt structure shown in Figure 7.13, interrupt lines of all the devices are ORed together, and hence, any device can interrupt the CPU. This is called a *single-priority* structure because all devices are equal (in importance) as far as interrupting the CPU is concerned. Single-priority structures can adopt either polling or vectoring to identify the interrupting device. A *single-priority polled* structure is the least complex interrupt structure, since polling is usually done by software, and the slowest because of polling. In a *single-priority vectored* structure, the CPU sends out an ACK signal in response to an interrupt, and the interrupting device returns a vector that is used by the CPU to execute the appropriate service routine. This structure operates faster than the polled structure but requires a more complex device controller.

The forward daisy-chain structure of Figure 7.15 is a *multipriority* structure, since interrupting the CPU depends on the priority of the device. The highest priority device can always interrupt the CPU in this structure. A higher-priority device can interrupt the CPU, while the CPU is servicing an interrupt from a lower-priority device. An interrupt from a lower-priority device is prohibited from reaching the CPU when it is servicing a higher-priority device. Once an interrupt is recognized by the CPU, the recognition of the interrupting device is done either by polling or by using vectors. The *multipriority vectored* structure is the fastest and most complex of the interrupt structures. Table 7.3 summarizes the characteristics of these interrupt structures.

Table 7.3   Characteristics of Interrupt Structures

Structure	Response Time	Complexity
Single priority		
Polled	Slowest	Lowest
Vectored	Fast	Medium
Multipriority		
Polled	Slow	Low
Vectored	Fastest	Highest

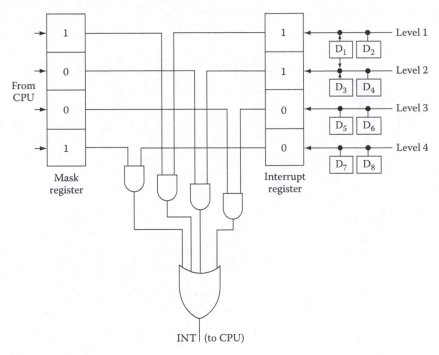

Figure 7.16 Masking interrupts.

In actual systems, more than one INT line will be provided at the CPU so that a hardware multilevel structure is possible. Within each level, devices can be daisy chained and each level assigned a priority. The priorities may also be dynamically changed (changed during the system operation). Figure 7.16 shows a masking scheme for such dynamic priority operations. The MASK register is set by the CPU to represent the levels that are permitted to interrupt (levels 2 and 3 are masked out and levels 1 and 4 are enabled). An INT signal is generated only if the enabled levels interrupt; that is, the levels that are masked out cannot interrupt the CPU. Note that in this scheme a device can be connected to more than one level (D_1 is connected to levels 1 and 2). Each level in Figure 7.16 receives its own ACK signal from the CPU. The ACK circuitry is not shown.

7.4 DIRECT-MEMORY ACCESS

The programmed and interrupt-mode I/O structures transfer data from the device into or out of a CPU register (accumulator in ASC). If the amount of data to be transferred is large, these schemes would overload the CPU. Data are normally required to be in the memory, especially when voluminous, and some complex computations are to be performed on them. A DMA scheme enables a device controller to transfer data directly into or from main memory. The majority of data-transfer control operations are now performed by a device controller. CPU initiates the transfer by commanding the DMA device to transfer the data and then continues with its processing activities. The DMA device performs the data transfer and interrupts the CPU only when it is completed.

Figure 7.17 shows a DMA transfer structure. The DMA device (either a DMA *controller* or a DMA *channel*) is a limited-capability processor. It will have a word-count register (WCR), an address register (AR), and a data buffer. To start a transfer, the CPU initializes the AR of the DMA channel with the memory address from (or to) which the data must be transferred and the WCR with the number of units of data (words or bytes) to be transferred. Note that the data bus is connected to these two registers. Usually, these registers are addressed by the CPU as output devices using

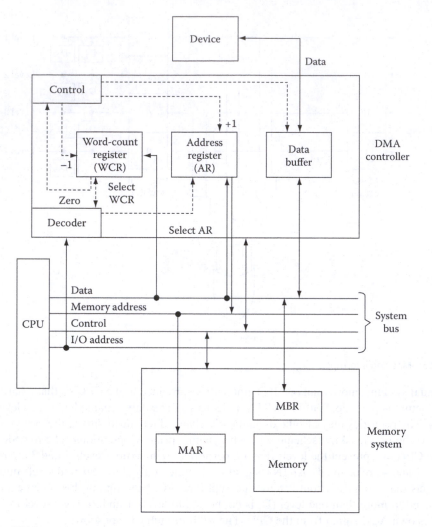

Figure 7.17   DMA transfer structure. *Note:* Memory and I/O ADDRESS buses are shown separately for clarity; a single-address bus with a control line indicating memory or I/O operation is an alternative.

the address bus; the initial values are transferred into them via the data bus. The DMA controller can decrement WCR and increment AR for each word transferred. Assuming an input transfer, the DMA controller starts the input device and acquires the data in its buffer register. This word is then transferred into the memory location addressed by AR; that is,

$$MAR \leftarrow AR.$$

$$MBR \leftarrow \text{Data buffer, WRITE MEMORY.}$$

These transfers are done using the address and data buses.

In this scheme, the CPU and the DMA device controller both try to access the memory through MAR and MBR. Since the memory cannot be simultaneously accessed by both the DMA and the CPU, a priority scheme is used to prohibit CPU from accessing the memory during a DMA operation. That is, a memory cycle is assigned to the DMA device for the transfer of data during which the CPU is prevented from accessing memory. This is called *cycle stealing*, since the DMA device

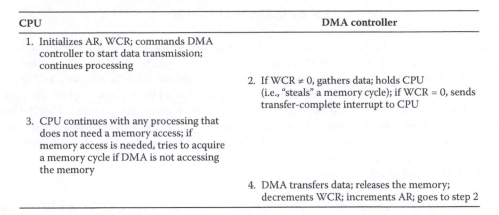

CPU	DMA controller
1. Initializes AR, WCR; commands DMA controller to start data transmission; continues processing	
	2. If WCR ≠ 0, gathers data; holds CPU (i.e., "steals" a memory cycle); if WCR = 0, sends transfer-complete interrupt to CPU
3. CPU continues with any processing that does not need a memory access; if memory access is needed, tries to acquire a memory cycle if DMA is not accessing the memory	
	4. DMA transfers data; releases the memory; decrements WCR; increments AR; goes to step 2

Figure 7.18   DMA transfer.

"steals" a memory cycle from the CPU when it is required to access the memory. Once the transfer is complete, the CPU can access the memory. The DMA controller decrements the WCR and increments AR in preparation for the next transfer. When WCR reaches 0, a transfer-complete interrupt is sent to the CPU by the DMA controller. Figure 7.18 shows the sequence of events during a DMA transfer.

DMA devices always have higher priority than the CPU for memory access because the data available in the device buffer may be lost if not transferred immediately. Hence, if an I/O device connected via DMA is fast enough, it can steal several consecutive memory cycles, thus "holding" back the CPU from accessing the memory for several cycles. If not, the CPU will access the memory in between each DMA transfer cycle.

DMA controllers can be either dedicated to one device or shared among several I/O devices. Figure 7.19 shows a bus structure that enables such a sharing. This bus structure is called *compatible I/O bus structure* because an I/O device is configured to perform both programmed and DMA transfers. DMA channels are shared by all the I/O devices. The I/O device may be transferring data at any time through either programmed or DMA paths. Some computer systems use a multiple-bus structure in which some I/O devices are connected to the DMA bus, while others communicate with the CPU in a programmed I/O mode.

Refer to Section 7.10 for a complete description of the I/O structure of a modern processor (Motorola 68000).

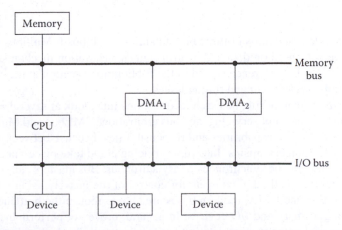

Figure 7.19   Compatible I/O bus structure.

7.5 BUS ARCHITECTURE

The importance of the bus architecture is as great as that of the processor, for few systems can operate without frequent data transfers along the buses in the system. As described earlier in this chapter, the system bus is responsible for interfacing the processor with the memory and disk systems, as well as other I/O devices. In addition, the bus system provides the system clock and handles arbitration for attached devices, allowing configuration of more advanced I/O transfer mechanisms. Several standard bus architectures have evolved over the years. We will now generalize the bus control and arbitration concepts and provide relevant details of three standard bus architectures.

7.5.1 Bus Control (Arbitration)

When two or more devices capable of becoming bus masters share a bus, a bus controller is needed to resolve the contention for the bus. Typically, in a CPU and I/O device structure, the CPU handles the bus control function. In modern computer system structures, it is common to see more than one processor connected to the bus. (A DMA structure is one such example.) One of these processors can handle the bus control function, or a separate bus controller may be included in the system structure.

Figure 7.20 shows three common *bus arbitration* techniques. Here, the bus busy signal indicates that the bus has already been assigned to one of the devices as a master. The bus controller assigns the bus to a new requesting device, if the bus is not busy and if the priority of the requesting device is higher than that of the current bus master or other requesting devices. Bus request and bus grant control lines are used for this purpose. These signals are analogous to interrupt and ACK signals, respectively.

In Figure 7.20a, the devices are daisy chained along the bus grant path, with the highest priority device being Device 1 and the lowest priority device being Device N. Any of the devices can send a bus request to the controller. When the bus is not busy, the controller sends a grant along the daisy chain. The highest priority device among the requesting devices sets the bus busy, stops the bus grant signal from going further in the daisy chain, and becomes the bus master. In Figure 7.20b, in response to the bus request from one or more devices, the controller polls them (in a predesigned priority order) and selects the highest priority device among them and grants the bus to it. Only one bus grant line is shown. But only the selected device will be activated as bus master (i.e., accepts the bus grant). All the other devices will ignore it. Alternatively, we could have independent bus grant lines as in Figure 7.20c. In Figure 7.20c, there are independent bus request and grant lines. The controller resolves the priorities among the requesting devices and sends a grant to the highest priority device among them.

7.5.2 Bus Standards

This section provides the details of three bus standards: Multibus I, Multibus II, and VMEbus. The description here is focused on data and control signals, arbitration mechanisms, and interrupt generation and handling. The references listed in the Bibliography section at the end of the chapter provide further details on electrical and timing issues.

When examining and comparing bus architectures, one must look at several attributes, such as the transfer mechanism, interrupt servicing, and bus arbitration. VMEbus and Multibus I and II are excellent buses to demonstrate capabilities and responsibilities, because between them they implement a variety of different mechanisms. Interrupt servicing could take place via direct or vectored techniques. Bus transfers can be synchronous or asynchronous. Bus masters can implement a serial (daisy chained) or parallel method to compete for control of the bus. By using techniques such as asynchronous arbitration and DMA, modern buses are able to reduce the bottleneck the processor faces when accessing memory and I/O devices. In general, processor performance advances more quickly than the bus performance, necessitating faster bus standards. As faster standardized buses

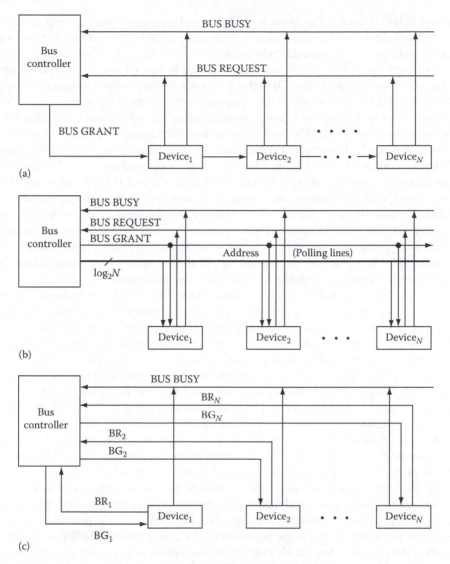

Figure 7.20 Bus arbitration techniques. (a) Daisy chain. (b) Polling. (c) Independent request/grant.

become prevalent, this gap may eventually disappear or at least be reduced to a level where zero wait state execution becomes a reality.

7.5.2.1 Multibus I

Although an older standard (Institute of Electrical and Electronics Engineers [IEEE] Standard 796), Multibus I has proved that a well-designed architecture can provide a viable system for many years. Systems utilizing this bus have existed since about 1976, but some implementations have been used in more modern systems. The lengthy existence of the multibus architecture was a direct result of the design goals: simplicity, processor and system flexibility, ease of upgrading, and suitability for harsh environments.

A Multibus I system comprises one or more boards connected via a passive backplane. The architecture is processor agnostic, so much so that systems utilizing the bus have been designed around processors ranging from the Intel 8080 to the 80386 and other families such as the Z80 and Motorola 68030. Multibus I supports both single- and multiprocessor architectures. The IEEE Standard 796 provides for a wide variation in implementations, and this section attempts to describe the bus in terms of its overall capability while noting relevant allowed variations.

Multibus I control signals are active low and terminated by pull-up resistors. Two clock signals are present. The bus clock (BCLK) runs at 10 MHz and is used to synchronize bus contention operations. The CCLK is also 10 MHz but is routed to the masters and slaves to use as the master can use to initiate a read or write to memory or I/O space. During a write operation, an active signal denotes that the address carried on the address lines is valid. During a read operation, the transfer from active to inactive indicates that the master has received the requested data from the slave. Slaves raise the transfer acknowledge (XACK) signal to indicate to a master that it has completed a requested operation. The initialize signal (INIT) can be generated to reset the system to its initial state. The lock (LOCK) signal may be used to lock the bus. This is explained further later, during the discussion of multiprocessing features.

Multibus I supports up to 24 address lines (ADR0–ADR23). Memory sizes greater than 16 MB (the maximum allowed with 24 address lines) could be handled by a memory management unit. Both 8- and 16-bit processors are supported, and in the case of an 8-bit system, both even and odd bytes are accessible via a swapping mechanism. In I/O access, either 8 or 16 address lines are used, giving an I/O address space of up to 64K separate from the data space.

There are 16 data lines (DAT0–DAT15), although for 8-bit systems only the first 8 are valid. DAT0 is the least significant bit. The data lines are shared between memory and I/O devices. The use of both 8- and 16-bit systems is permitted by the byte high enable signal (BHEN), which is used on 16-bit systems to signify the validity of the other eight lines. Two inhibit lines (INH1 and INH2) can be asserted by a slave to inhibit another slave's bus activity during a memory read or write.

Eight interrupt lines (INT0–INT7) are available and can be configured to work in either a direct or bus-vectored interrupt scheme. The interrupt acknowledge (IACK) signal (INTA) is used in the bus-vectored mechanism by a master to freeze the interrupt status and request the interrupting device to place its vector address onto the bus data lines.

Bus arbitration is handled via five bus exchange lines. The bus busy (BUSY) line is driven by the master currently in ownership of the bus to signify the state of the bus. It is a bidirectional signal driven by an open-collector gate and synchronized by BCLK. In either serial or parallel-priority schemes, other bus masters use the bus priority in (BPRN) to indicate a request for bus control along the chain or parallel circuitry. In serial mechanism, the bus priority out signal (BPRO) is passed on to the next higher bus master to propagate a request for bus control. In the parallel-priority scheme, each bus master requesting access to the bus raises the bus request signal (BREQ). The parallel circuitry resolves the priorities, and enables BPRN for the highest priority master requesting control. An optional signal, the common bus request (CBREQ) can be used by any master to signal a request for bus control. This allows a master of lower priority to request bus control.

Multibus I data transfer is asynchronous and supports DMA transfer. Devices in the system may be either masters or slaves, with slaves such as memory unable to initiate a transfer. Multibus I supports up to 16 bus masters, using either serial or parallel-priority schemes. Data transfer takes place in the following steps:

1. Bus master places the memory or I/O address on the address lines.
2. Bus master generates the appropriate command signal.
3. Slave either accepts the data in a write or places the data on the data lines for a read.
4. Slave sends the transfer ACK signal back to the master.
5. Bus master removes the signal on the command lines and then clears the address and data lines.

Since Multibus I data transfers are asynchronous, it is possible that due to error, a transfer could extend indefinitely. To prevent this, a bus time-out can be implemented to terminate the cycle after a preset interval of at least 1 ms. For any memory transfer, a slave can assert the inhibit lines to inhibit the transfer of another slave. This operation has been implemented for use in diagnostic applications and devices and is not widely used in normal operation. When transferring between 8- and 16-bit devices, the BHEN and the least significant address line (ADR0) are used to define whether the even or odd byte will be transferred. In an even-byte transfer, both are inactive. For the odd-byte transfer, BHEN is inactive and ADR0 is active. When transferring between two 16-bit devices, both signals are active.

Multibus I supports two methods of interrupts: direct (nonbus vectored) and bus vectored. Direct interrupts are handled by the master without the need for the device address being placed on the address lines. Bus-vectored interrupts are handled by the master interrogating the interrupting slave instead of determining the vector address. When the IRQ occurs, the bus master interrupts its processor, which generates the INT command and freezes the state of the interrupt logic so that the priority of the request can be analyzed. After the INT command, the bus master determines the address of the highest priority request and places the address on the bus address lines. Depending on the size of the address of the interrupt vector, either one or two more INTA commands must be generated. The second one causes the slave to transmit the low-order (or only) byte of its interrupt vector on the data lines. If necessary (for 16-bit addresses), the third INTA causes the high-order byte to be placed on the data lines. The bus master then uses the high-order byte to be placed on the data lines. The bus master then uses this address to service the interrupt.

Since Multibus I can accommodate several bus masters, there must be a mechanism for the masters to negotiate for control of the bus. This can take place through either a serial or a parallel-priority negotiation. In the serial negotiation, a daisy-chain technique is used. The priority in and out of each master is connected to each other in order of their priority. When a bus master requests control of the bus, it generates its BPRO signal and blocks its BPRN signal, thus locking out the requests of all the lower-priority masters. In the parallel mechanism, a bus arbiter receives the BPRN and BREQ signals of each master. The bus arbiter then determines the priority of the request and performs the request. Bus arbiters are usually not designed into the backplane.

When the I/O of the processor is bottlenecked by the bus, Multibus I allows the use of bus extensions, which take over high-throughput transfers such as DMA. In a standard Multibus I system, the maximum throughput is limited to 10 MB/s. Processors later in the life of the Multibus I architecture were capable of a much higher rate; the 20 MHz Intel 80386 chip can transfer at 40 MB/s. Multibus I may implement both the iLBX and iSBX extensions. ILBX provides a fast memory-mapped interface in an expansion board with the same form factor as the Multibus I standard. A maximum of two masters can share the bus, which limits the need for complex bus arbitration. The arbitration that does take place has been modified to be asynchronous to data transfers. ILBX slaves are available to the system as byte-addressable memory resources controlled directly from the iLBX bus lines. For 16-bit transfers, these improvements allow a maximum throughput of 19 MB/s. The iSBX expansion board implements its own I/O and DMA transfers, taking over much of the function of the Multibus I architecture. The implementation of the iSBX bus extension marked the beginning of an evolution toward Multibus II.

7.5.2.2 Multibus II

Multibus I was introduced in 1974 and it is fundamentally CPU and memory bus. Then it evolved to a multiple-master shared-memory bus capable of solving most real-time applications of that time. But in the late 1980s, the users demanded a new standard bus. Therefore, Intel set up a consortium with 18 industry leaders to define the next generation of bus—Multibus II.

The consortium decides that no single bus can be used to satisfy all user needs; therefore, a multiple-bus structure consisting of four subsystems was defined. The four subsystems are the iSBX bus for incremental I/O expansions, a local CPU and memory expansion bus, and two system buses—one serial and one parallel.

Consider a local area network or LAN. This system is functionally partitioned—each node of the LAN is independent of others and optimized for a part of overall problem. This solution gives the system architect freedom to choose the hardware and software for each node that best fits the subtask. Moreover, each of the systems can be upgraded individually. The Multibus II allows creating a "very local" network within a single chassis. Dividing a multi-CPU application into a set of networked subsystems allows optimization of each subsystem for the subtask it works on. If a subsystem is complex, the resources may be spread over multiple boards and communicate via local expansion bus.

A double Eurocard format, the IEEE 1101 Standard, with dual 96 pin DIN connectors was chosen for Multibus II standard. A U-shaped front panel, licensed from Siemens, Germany, was chosen for its enhanced electromagnetic interference (EMI) and radio-frequency interference (RFI) shielding properties.

The popularity of Multibus I products encouraged adoption of iSBX (IEEE 894) for incremental I/O bus. The IEEE/ANSI 1296 specification does not define the exact bus for local expansion. The bus varies depending on performance required in subsystem design. Intel initiated iLBX II standard optimized for 12 MHz Intel 80286 processor. For high-performance local expansion, buses (Futurebus) can be used.

Because the CPU–memory bus on most buses is not adequate for system-level requirements, *system space* was defined in Multibus II specification. It consists of two parts: interconnect space and message space. Interconnect space is used for initialization, self-diagnostics, and configuration requirements. In order to maintain compatibility with existing buses, the traditional CPU–memory space is retained.

Intel implemented the Multibus II parallel system bus in a single VLSI chip called the message-passing coprocessor (MPC). This chip consists of 70,000 transistors and contains almost all logic needed to interface processor to the system bus. The parallel system bus is defined as a 32 bit bus clocking at 10 MHz, thus allowing data transfers up to 40 Mbit/s.

The serial system bus (SSB) is defined as a 2 Mbit/s serial bus but not implemented. The software interface to an SSB must be identical to that for a parallel bus.

Being a major part of IEEE/ANSI 1296 Multibus II specification, interconnect address space addresses board identification, initialization, configuration, and diagnostics requirements. Interconnect space is implemented as an ordered set of 8-bit registers on long-word (32 bit) boundaries—in this way, a small Endian microprocessor such as the 8086 family and a big Endian microprocessor such as Motorola 68000 family access the information in an identical manner. The software can use the interconnect address space to get information about the environment it operates in, the functionality of board, and the slot in which the board operates.

The identification registers contain information about board type, its manufacturer, and installed components. They are read-only registers. Configuration registers are R/W locations that can be set and changed by system software. Diagnostics registers are used for self-diagnosis of the board.

Interconnect space is based on the idea that it is possible to locate boards within the backplane by their physical slot positions. This principle, called geographical addressing, is used for system-wide initialization. Each board has firmware with a standardized 32-byte header format containing a 2-byte vendor ID, a 10-byte board name, and other vendor-defined information. At boot time, the system software scans each board to locate its resources and then loads appropriate drivers. This method eliminates the need for reconfiguration each time a new board is added to the system. Each board in the system performs its own initialization and testing using the firmware and passes all information needed to the operating system, which, in turn, generates a resource-location map to be used as a basis for message-passing addresses, thus achieving slot independence. In general, a board manufacturer also supplies other function records to make additional functionality of the board accessible through interconnect space. Types of function record for common functions such as memory configuration and serial I/O are defined.

Each Multibus II board should be capable of self-testing and reporting the status in interconnection space. The self-testing can be invoked during power-on initialization or explicitly from the console. If a hardware failure is detected, a yellow LED on the front panel will illuminate, helping the operating easily define and replace the board.

The high performance of Multibus II is achieved by decoupling activities between the CPU–memory local bus and the system bus. This approach gives two advantages: the parallelism of operations is increased and one bus bandwidth does not limit transfer rate of another. The local bus and the system bus work independently and in parallel. This is achieved by using nine 32-byte first in, first out (FIFO) buffers integrated into the MPC. Five of them are used for interrupts (one sends and four receive) and four for data transfer (two to send, two to receive).

The Multibus II specification introduces a hardware-recognized data type called a *packet*. The packets are moved on subsequent clock edges of the 10 MHz synchronous bus; thus, a single packet can occupy the bus for no longer than 1 μs. An address is assigned to each board in the system. This address is used in source and destination fields. Seven different types are defined by the standard. The packets are divided into two groups: unsolicited and solicited. Unsolicited packets are used for interrupts, while solicited packets are used for data transfers. The data fields are user defined and may be from 0 to 32 (28 for unsolicited packets) bytes long in 4-byte increments.

Unsolicited packets are always a "surprise" for the destination-bus MPC. These packets are similar to interrupts in shared-memory systems with additional feature to carry up to 28 bytes of information. General interrupt packets can be sent between any two boards. Broadcast interrupts are sent to all boards in the system. Three other types (buffer request, reject, and grant) are used to initiate large data transfers. Unlike unsolicited packets, solicited packets are not unpredictable to the destination MPC. These packets are used for large data transfers between boards. Up to 16 MB of data can be transferred by solicited packets. All operations on packets such as creation, bus arbitration, error checking, and correction are done by MPC and transparent to the local processor.

The multibus II system bus utilizes a distributed arbitration scheme. Each board has daisy-chain circuitry. The system bus is continually monitored by each of MPCs. The scheme supports two arbitration algorithms: fairness and high priority. In the fairness mode, if the bus is being used, the MPC waits before requesting the bus; when the bus is not busy, the MPC makes the request and waits for a bus grant; once the MPC uses the bus, it does not request the bus until all other requesters have used it. If no bus requests have arrived since the last usage of the bus, the MPC accesses the bus without performing an arbitration cycle. This mechanism prevents the bus from being monopolized by a single board. Since each MPC uses the bus for a maximum of 1 μs and arbitration is resolved in parallel, no clock cycles are wasted and all transfers operate back to back.

The high-priority mode is used for interrupts. The requesting MPC bus controller is guaranteed the next access to the bus. Usually interrupt packets are sent in high-priority mode, and this means that most interrupt packets have a maximum latency of 1 μs, although in very rare instances, when N boards initiate interrupt packets within the same 1 μs window, packets may have up to N—1 μs latency.

The parallel system bus is implemented on a single 96 pin connector. The signals are divided into five groups: central control, address/data, system control, arbitration, and power. The parallel system bus is synchronous and great care is taken to maintain a clear 10 MHz system clock. The IEEE/ANSI 1296 specification details precisely what happens upon each of the synchronous clock edges so there is no ambiguity. Numerous state machines that track bus activity are defined to guarantee compatibility.

The board in slot 0, called the central service module (CSM), generates all of the central control signals. It can be implemented on a CPU board, on a dedicated board, or on the backplane itself. The module drives reset (RST*)—active-low signals are denoted by a*—to initialize the system; combinations of DCLOW* and PROT* are used to distinguish between cold start, warm start, and power-failure recovery. Two system clocks (BCKL* at 10 MHz and CCLK* at 20 MHz) are generated.

The IEEE/ANSI 1296 specification defines the parallel system bus as a full 32 bit (AD0*–AD31*) with parity control (PAR0*–PAR3*). Because it is defined as a multiplexed address/data bus, the system control lines are used to distinguish between data and addresses. Since all transfers are checked for parity error, in case of parity failure, the MPC bus controller retries the operation. If the error is not recovered within 16 tries, MPC interrupts its host processor and asks for assistance.

Table 7.4 System Control Lines

Signal	Function during Request Phase	Function during Reply Phase
SC0*	Request phase	Reply phase
SC1*	Lock	Lock
SC2*	Data width	End of transfer
SC3*	Data width	Bus owner read
SC4*	Address space	Replier ready
SC5*	Address space	Agent status
SC6*	Read or write	Agent status
SC7*	Reserved	Agent status
SC8*	Even parity on SC (7–4)*	Even parity on SC (7–4)*
SC9*	Even parity on SC (3–0)*	Even parity on SC (3–0)*

There are 10 system control lines (SC0*–SC9*), and their functions are also multiplexed. SC0* defines the current state of the bus cycle (request or reply) and how SC1–SC7 lines should be interpreted. SC8 provides even parity over SC4–SC7, and SC9 provides even parity over SC0–SC3. Table 7.4 taken from the handbook by DiGiacomo (1990) summarizes functions of the system control lines.

There is a CBREQ line BREQ*. The specification defines that the distributed arbitration scheme grants the bus to the numerically higher requesting board, as identified on lines ARB0–ARB5. As mentioned earlier, the scheme supports two arbitration modes: fairness and high priority.

The parallel system bus is particularly easy to interface to. An I/O replier needs to implement only a single *replying agent* state machine as shown in Figure 7.21. In the following example, the assumption is made that the requestor makes only valid requests.

The replying agent is a bus monitor. State transitions occur on the falling clock edge. The replier remains in the "wait for request" state until the start of request cycle is detected (SC0* is low). If the request is addressed to the replier (ADDR is high), then there is a state transition to a new state controlled by a local ready signal (REPRDY). If REPRDY is low, then it waits until the device is ready. When ready, it waits for requestor to be ready (SC3* is low) and performs the data transfer. Then it checks if it is a multibus transfer (SC2 is high); if it is, the state machine decides to accept or ignore the data in the remainder of the cycle. If additional data cannot be handled, then the replier sends the continuation error and waits for the requestor to terminate the cycle. If the additional data can be handled, the replier oscillates between the replier wait state and replier handshake state until the last packet (SC2* is low) is received. If it is not a multibus transfer, the replier returns to the wait state.

Because of the simple standardized interface and processor independence, the Multibus II became very popular on the market. Many vendors produced Multibus II-compatible boards with many different functions. Later, the IEEE/ANSI 1296.2 Standard was adopted expanding the Multibus II with "live insertion" capabilities.

7.5.2.3 *VMEbus*

The VMEbus is a standard backplane interface that simplifies integration of data processing, data storage, and peripheral control devices in a tightly coupled hardware configuration. The VMEbus interfacing system is defined by the VMEbus specification. This system has been designed to do the following:

1. Allow communication between devices without disturbing the internal activities of other devices interfaced to the VMEbus.
2. Specify the electrical and mechanical system characteristics required to design devices that will reliably and unambiguously communicate with other devices interfaced to the VMEbus.

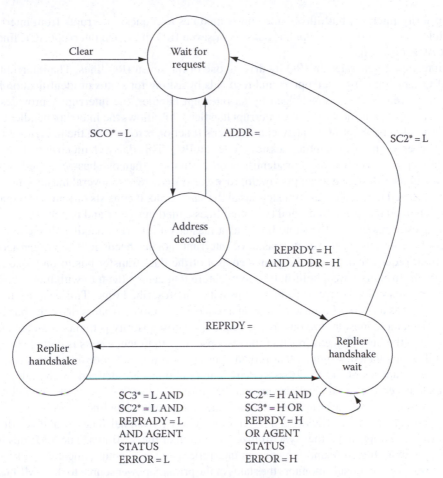

Figure 7.21 Multibus II state diagram. (Courtesy of Intel Corporation.)

3. Specify protocols that precisely define the interaction between devices.
4. Provide terminology and definitions that describe system protocol.
5. Allow a broad range of design latitude so that the designer can optimize cost and/or performance without affecting system compatibility.
6. Provide a system where performance is primarily device limited, rather than system interface limited.

The VMEbus functional structure consists of backplane interface logic, four groups of signal lines called buses, and a collection of functional modules. The lowest layer, called the backplane assess layer, is composed of the backplane interface logic, the utility bus modules, and the arbitration bus modules. The VMEbus data-transfer layer is composed of the data-transfer bus and priority interrupt bus modules.

The data-transfer bus allows bus masters to direct the transfer of binary data between themselves and slaves. The data-transfer bus consists of 32 data lines, 32 address lines, 6 address modifier (AM) lines, and 5 control lines. There are nine basic types of data-transfer bus cycles: read, write, unaligned write, block read, block write, read–modify–write, address-only, and interrupt ACK cycle.

The slave detects data-transfer bus cycle initiated by a master and, when those cycles specify its participation, transfers data between itself and the master.

The priority interrupt bus allows interrupter modules to request interrupts from interrupt handler modules. The priority interrupt bus consists of seven IRQ lines, one interrupt ACK line, and an interrupt ACK daisy chain.

The interrupter generates an IRQ by driving one of the seven IRQ lines. The interrupt handler detects IRQs generated by interrupters and responds by asking for status or identification information. When its request is acknowledged by an interrupt handler, the interrupter provides 1, 2, or 4 bytes of status or identification to the interrupt handler. This allows the interrupt handler to service the interrupt. The interrupt ACK daisy-chain driver's function is to activate the interrupt ACK daisy chain whenever an interrupt handler acknowledges an IRQ. This daisy chain ensures that only one interrupter responds with its status or identification when more than one has generated an IRQ.

The VMEbus is designed to support multiprocessor systems where several masters and interrupt handlers may need to use the data-transfer bus at the same time. It consists of four bus request lines, four daisy-chained bus grant lines, and two other lines called bus clear and bus busy.

The requester resides on the same board as a master on interrupt handler. It requests the use of the data-transfer bus whenever its master or interrupt handler needs it. The arbiter accepts bus requests from requester modules and grants control of the data-transfer bus to only one requester at a time. Some arbiters have a built-in time-out feature that causes them to withdraw a bus grant if the requesting board does not start using the bus within a prescribed time. This ensures that the bus is not locked up as a result of transient edge on a request line. Other arbiters drive the bus clear line when they detect a request for the bus from a requester whose priority is higher than the one that is currently using the bus. This ensures that the response time to urgent events is bounded.

The utility bus includes signals that provide periodic timing and coordinate the power-up and power-down sequences of the VMEbus system. Three modules are defined by the utility bus: the system clock driver, the serial clock driver, and the power monitor. It consists of a two-clock line, a system-reset line, a power fail line, a system fail line, and a serial data line.

The system clock driver provides a fixed-frequency 16 MHz signal. The serial clock driver provides a periodic timing signal that synchronizes operation of the VMEbus. The VMEbus is part of the VME system architecture and provides an interprocessor serial communication path.

The power monitor module monitors the status of the primary power source to the VMEbus system. When power strays outside the limits required for reliable system operation, it uses the power fail lines to broadcast a warning to all boards on the VMEbus system in time to affect graceful shutdown.

The system controller board resides in slot 1 of the VMEbus backplane and includes all the one-of-a-kind functions that have been defined by the VMEbus. These functions include the system clock driver, the arbiter, the interrupt ACK daisy-chain driver, and the bus timer.

Two signaling protocols are used on the VMEbus: closed-loop protocols and open-loop protocols. Closed-loop protocols are interlocked bus signals, while open-loop protocols use broadcast bus signals. The address strobe (AS) and data strobes are interlocked signals that are especially important. They are interlocked with the data ACK or bus error signals and coordinate the transfer of addresses and data. There is no protocol for acknowledging a broadcast signal; instead, the broadcast is maintained long enough to ensure that all appropriate modules detect the signal. Broadcast signals may be activated at any time.

The smallest addressable unit of storage on the VMEbus is the byte. The masters broadcast the address over the data-transfer bus at the beginning of each cycle. These addresses may consist of 16, 24, or 32 bits. The 16-bit addresses are called short addresses, the 24-bit addresses are called standard addresses, and the 32-bit addresses are called extended addresses. The master broadcasts a 6 bit AM code along with each address to tell slaves whether the address is short, standard, or extended.

There are four basic data-transfer capabilities associated with the data-transfer bus: D08 (EO) (even and odd byte), D08 (O) (odd-byte only), D16, and D32.

Five basic types of data-transfer cycles are defined by the VMEbus specification. These cycle types include the read cycle, the write cycle, the block read cycle, the block write cycle, and the read–modify–write cycle. Two more types of cycles are defined: the address-only cycle and the interrupt ACK cycle. With the exception of the address-only cycle, which does not transfer data, all these cycles can be used to transfer 8, 16, or 32 bits of data.

Read and write cycle can be used to read or write 1, 2, 3, or 4 bytes of data. The cycle begins when the master broadcasts an address and an AM mode. Block transfer cycles are used to read or write a block of 1–256 bytes of data. The VMEbus specification limits the maximum length of block transfers to 256 bytes. Read–modify–write cycles are used both to read from and write to a slave location in an indivisible manner, without permitting another master to access that location. The VMEbus protocol allows a master to broadcast the address for the next cycle, while the data transfer for the previous cycle is still in progress. The VMEbus provides two ways that processors in a multiprocessing system can communicate with each other: by using IRQ lines and by using location monitors.

The location monitor functional module is intended for use in multiple-processor systems. It monitors all data-transfer cycles over the VMEbus and activates an onboard signal whenever an access is done to any of the locations that it is assigned to watch. In multiple-processor systems, events that have global importance need to be broadcast to some or all of the processors. This can be accomplished if each processor board includes a location monitor.

The VMEbus provides both the performance and versatility needed to appeal to a wide range of users. Its rapid rise in popularity has made it the most popular 32-bit bus. VMEbus systems accommodate the inevitable changes easily and without making existing equipment obsolete.

Refer to Section 7.10 for details on another popular standard—the Peripheral Component Interface (PCI) standard.

The Universal Serial Bus (USB) is a more recent standard interface bus. A group of major companies like Intel, Microsoft, and others developed this standard. The standard defines the cables, connectors, and protocols for communication between computers and external devices. Usually the bus is used for two-way data transfer and can also supply power to external electronic devices. The USB 1.0, 2.0, and 3.0 are currently predominant connection technologies for a wide range of I/O and external memory devices. The USB technology was developed to standardize the connection capabilities of personal computers with computer peripherals (including keyboards, pointing devices, digital cameras, printers, portable media players, disk drives, and network adaptors), for both communication and supplying electrical power. USB 1.0 was introduced in 1996 with data rates of 1.5 Mbit/s (low bandwidth) and 12 Mbit/s (full bandwidth). In 2000, USB 2.0 was released that provided the data transmission rates up to 480 Mbits/s. It also brought significant improvements to features like battery charging specifications, link power management, and physical dimensions of the connectors. USB connectors and cables including Mini-A/B and Micro-USB were then standardized in 2000 and 2007, respectively. USB 3.0 with speeds up to 5 Gbits/s was released in Nov 2008. Apart from higher bandwidth, the major features of this version included the reduction in power consumption while retaining backward compatibility with USB 2.0. The USB group plans to further increase the data rates of the USB standard in the near future.

7.6 CHANNELS

A channel is a more sophisticated I/O controller than a DMA device but performs I/O transfers in a DMA mode. It is a limited-capability processor that can perform all operations a DMA system can. In addition, a channel is configured to interface several I/O devices to the memory, while a DMA controller is usually connected to one device. A channel performs extensive error detection and correction, data formatting, and code conversion. Unlike DMA, the channel can interrupt CPU under any error condition. There are two types of channels: *multiplexer* and *selector*.

A *multiplexer channel* is connected to several low- and medium-speed devices (card readers, paper tape readers, etc.). The channel scans these devices in turn and collects data into a buffer.

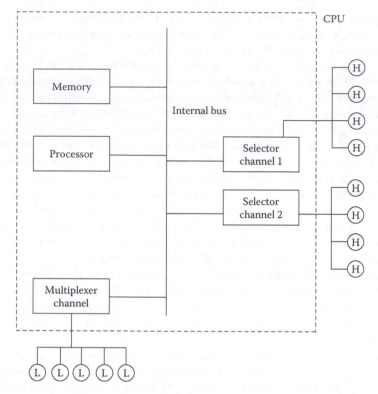

Figure 7.22 Typical computer system. *Note:* H, high-speed device; L, low-speed device.

Each unit of data may be tagged by the channel to indicate the device it came from (in an input mode). When these data are transferred into the memory, the tags are used to identify the memory buffer areas reserved for each device. A multiplexer channel handles all the operations needed for such transfers from multiple devices after it has been initialized by the CPU. It interrupts the CPU when the transfer is complete. Two types of multiplexer channels are common: (1) character multiplexers, which transfer one character (usually one byte) from each device, and (2) block multiplexers, which transfer a block of data from each device connected to them.

A *selector channel* interfaces high-speed devices such as magnetic tapes and disks to the memory. These devices can keep a channel busy because of their high data-transfer rates. Although several devices are connected to each selector channel, the channel stays with one device until the data transfer from that device is complete.

Figure 7.22 shows a typical computer system structure with several channels. Each device is assigned to one channel. It is possible to connect a device to more than one channel through a multichannel switching interface. Channels are normally treated as a part of the CPU in conventional computer architecture; that is, channels are CPU-resident I/O processors.

7.7 I/O PROCESSORS

Channels and interrupt structures perform the majority of I/O operations and control, thus freeing the central processor for internal data processing. This enhances the throughput of the computer system. A further step in this direction of distributing the I/O processing functions to peripherals is to make channels more versatile, like full-fledged processors. Such I/O processors are called *peripheral* or *front-end processors* (FEPs). With the advent of microprocessors and the availability of less-expensive hardware devices, it is now possible to make the FEP versatile enough while keeping the cost low. A large-scale computer system uses several minicomputers as FEPs, while a minicomputer might use another mini- or a microcomputer for the FEP. Since FEPs are programmable, they serve as flexible I/O devices to the CPU. They also perform as much processing on the data as possible (source processing) before data are transferred into the memory. If the FEP has a writable control store (i.e., the control ROM is field programmable), the microprogram can be changed to reflect the device interface needed.

The coupling between the FEP and the central processor is either through a disk system or through the shared memory (Figure 7.23). In a *disk-coupled* system, the FEP stores data on the disk unit, which in turn are processed by the central processor. During the output, the central processor stores data on the disk and provides the required control information to the FEP to enable data output. This system is easier than the shared-memory system to implement even when the two processors (FEP and CPU) are not identical because timing and control aspects of the processor–disk interface are essentially independent.

In the shared-memory system, each processor acts as a DMA device with respect to the shared memory. Hence, a complex handshake is needed, especially when the two processors are not identical. This system will generally be faster, however, since an intermediate direct-access device is not used. Figure 7.23 shows the FEP–CPU coupling schemes.

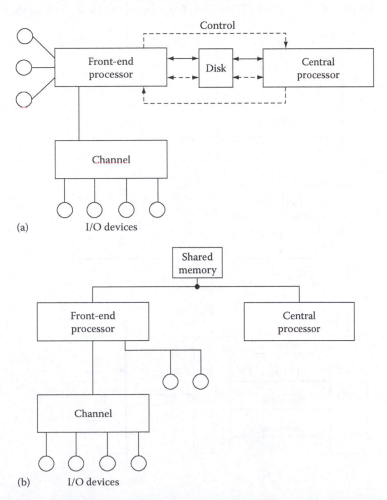

Figure 7.23 CPU/IOP interconnection. (a) Disk-coupled system. (b) Shared-memory system.

7.7.1 IOP Organization

Figure 7.24 shows the organization of a typical shared-memory CPU–IOP interface. The shared main memory stores the CPU and IOP programs and contains a CPU/IOP communication area. This communication area is used for passing information between the two processors in the form of messages. The CPU places I/O-oriented information in the communication area. This information consists of the device addresses, the memory buffer addresses for data transfers, types and modes of transfer, address of the IOP program, etc. This information is placed into the communication area by the CPU using a set of *command words* (CWs). In addition to these CWs, the communication area contains the space for IOP STATUS.

While initiating an I/O transfer, the CPU first checks the status of the IOP to make sure that the IOP is available. It then places the CW into the communication area and commands the IOP to start through the START I/O signal. The IOP gathers the I/O parameters, executes the appropriate IOP program, and transfers the data between the devices. If the data transfer involves the memory, a DMA mode of transfer is used by acquiring the memory bus using the bus arbitration protocol. Once the transfer is complete, the IOP sends a transfer-complete interrupt to the CPU through the I/O DONE line.

The CPU typically has three I/O-oriented instructions to handle the IOP: TEST I/O, START I/O, and STOP I/O, each of which will be a CW. The instruction set of an IOP consists of data-transfer instructions of the type: READ (or WRITE) *n* units from (or to) device X to (or from) memory buffer

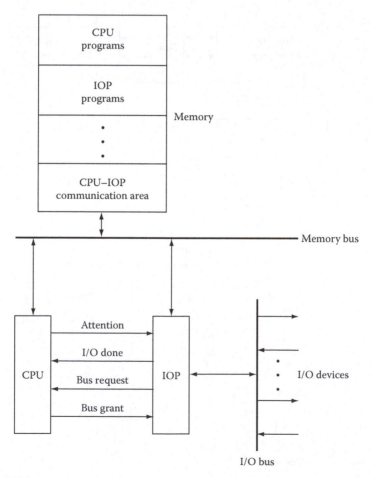

Figure 7.24   Typical CPU–IOP interface.

starting at location Z. In addition, the IOP instruction set may contain address manipulation instructions and a limited set of IOP program control instructions. Depending on the devices handled, there may be device-specific instructions such as rewind tape, print line, and seek disk address.

7.8 SERIAL I/O

We assumed parallel buses transferring data (word or byte) between the CPU, memory, and I/O devices in the earlier sections of this chapter. Devices such as low-data-rate terminals use *serial buses* (i.e., serial mode of data transfer containing a single line in either direction of transfer). The transmission usually is in the form of an asynchronous stream of characters. That is, there is no fixed time interval between the two adjacent characters. Figure 7.25 shows the format of an 8-bit character for asynchronous serial transmission. The transmission line is assumed to stay at 1 during the idle state, in which there is no character being transmitted. A transition from 1 to 0 indicates the beginning of transmission (*start* bit). The start bit is followed by 8 data bits, and at least 2 *stop* bits terminate the character. Each bit is allotted a fixed time interval, and hence the receiver can decode the character pattern into the proper 8-bit character code.

Figure 7.26 shows a serial transmission controller. The CPU transfers data to be output into the interface buffer. The interface controller generates the start bit, shifts the data in the buffer 1 bit at a time onto output lines, and terminates with 2 stop bits. During an input, the start and stop bits are removed from the input stream and the data bits are shifted into the buffer and in turn transferred into the CPU. The electrical characteristics of this asynchronous serial interface have been standardized by the Electronic Industries Association, and it is called EIA RS232C.

The data-transfer rate is measured in bits per second. For a 10 character/s transmission (i.e., 10 character/s × 11 bits per character, assuming 2 stop bits per character), the time interval for each bit in the character is 9.09 ms.

Digital pulses or signal levels are used in the communication of data between local devices and the CPU. As the distance between the communicating devices increases, digital pulses get distorted due to the line capacitance, noise, and other phenomena, and after a certain distance cannot be recognized as 0s and 1s. Telephone lines are usually used for transmission of data over long distances. In this transmission, a signal of appropriate frequency is chosen as the *carrier*. The carrier is *modulated* to produce two distinct signals corresponding to the binary values 0 and 1. Figure 7.27 shows a typical communication structure. Here, the digital signals produced by the source computer are *frequency modulated* by the modulator–demodulator (modem) into analog signals consisting of two frequencies. These analog signals are converted into their digital counterparts by the modem at the

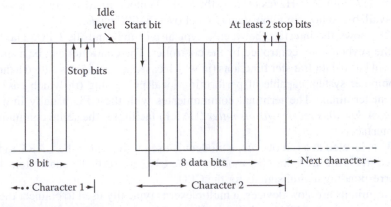

Figure 7.25   Serial data format.

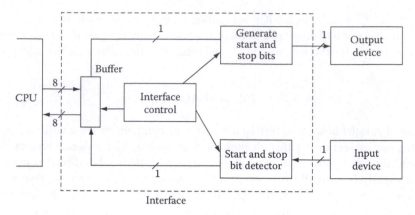

Figure 7.26 Serial data controller.

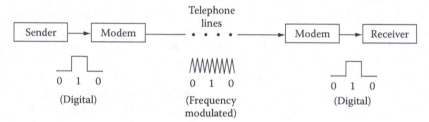

Figure 7.27 Long-distance transmission of digital signals.

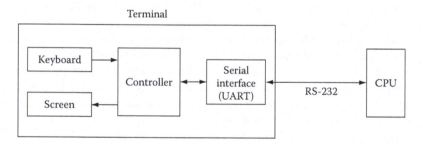

Figure 7.28 Typical terminal–CPU interface.

destination. If an ordinary "voice-grade" telephone line is used as the transmission medium, the two frequencies are 1170 and 2125 Hz (as used by the Bell 108 modem). More sophisticated modulation schemes are available today that allow data rates of over 20,000 bits/s.

Figure 7.28 shows the interface between a typical terminal and the CPU. The video display (screen) and the keyboard are controlled by the controller. The controller can be a simple interface that can perform basic data-transfer functions (in which case the terminal is called a dumb terminal) or a microcomputer system capable of performing local processing (in which case the terminal is an intelligent terminal). The terminal communicates with the CPU usually in a serial mode. A *universal asynchronous receiver/transmitter* (UART) facilitates the serial communication using the RS-232 interface.

If the CPU is located at a remote site, a modem is used to interface the UART to the telephone lines, as shown in Figure 7.29. Another modem at the CPU converts the analog telephone line signals to the corresponding digital signals for the CPU.

Since the terminals are slow devices, a multiplexer is typically used to connect the set of terminals at the remote site to the CPU, as shown in Figure 7.30.

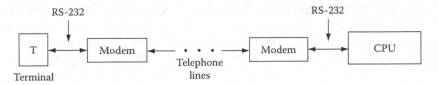

Figure 7.29 Remote terminal interface to CPU.

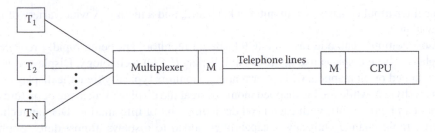

Figure 7.30 Multiplexed transmission. *Notes*: T, terminal; M, modem.

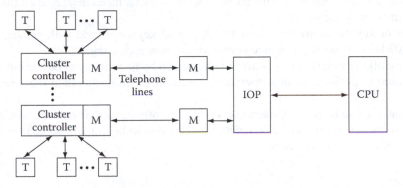

Figure 7.31 Terminal-based computer network. *Notes:* T, terminal; M, modem.

Figure 7.31 shows the structure of a system with several remote terminal sties. The terminal multiplexer of Figure 7.30 at each site is now replaced by a cluster controller. The cluster controller is capable of performing some processing and other control tasks, such as priority allocation. The cluster controllers in turn are interfaced to the CPU through an FEP.

The communication between the terminals and the CPU in networks of the type shown in Figure 7.31 follows one of the several standard protocols such as BSC, synchronous data-link control (SDLC), and *high-level data-link control* (HDLC). These protocols provide the rules for initiating and terminating transmission, handling error conditions, and data framing (rules for identifying serial bit patterns into characters, character streams into messages, etc.).

7.9 COMMON I/O DEVICES

A variety of I/O devices are used to communicate with computers. They can be broadly classified into the following categories:

1. Online devices such as terminals that communicate with the processor in an interactive mode
2. Off-line devices such as printers that communicate with the processor in a noninteractive mode
3. Devices that help in real-time data acquisition and transmission (analog-to-digital (A/D) and digital-to-analog (D/A) converters)
4. Storage devices such as tapes and disks that can also be classified as I/O devices

We will provide very brief descriptions of selected devices in this section. This area of computer systems technology also changes very rapidly and newer and more versatile devices are announced on a daily basis. As such, the most up-to-date information on their characteristics is only available from vendor's literature and magazines, such as the ones listed in the Bibliography section of this chapter.

7.9.1 Terminals

A typical terminal consists of a monitor, a keyboard, and a mouse. A wide variety of terminals are now available.

The most common cathode ray tube (CRT)-based monitors are being rapidly replaced by flat panel displays. These displays use liquid crystal display (LCD) technology. Displays can be either character mapped or bit mapped. Character-mapped monitors typically treat the display as a matrix of characters (bytes), while the bit-mapped monitors treat the display as an array of picture elements (pixels) that can be on or off, with each pixel depicting 1 bit of information. Display technology is experiencing rapid change with newer capabilities added to displays almost daily. Sophisticated alphanumeric and graphic displays are now commonly available. Touch screen capability as an input mechanism is now common. This allows selection from a menu displayed on the screen, just by touching an item on the menu.

A variety of keyboards are now available. A typical keyboard used with personal computers consists of 102 keys. Various ergonomic keyboards are now appearing.

The mouse allows pointing to any area on the screen by its movement on the mouse pad. The buttons are used to perform various operations based on the information at the selected spot on the screen.

In addition to these three components of a typical terminal, various devices such as light pen, joystick, microphones (for direct audio input), speakers (for audio output), and cameras (for video input) are commonly used I/O devices.

7.9.2 Mouse

The mouse is the most common input device today. It typically has two buttons and a scroll wheel. The scroll wheel allows the scrolling of the image on the screen, and the buttons allow the selection of a particular position on the screen, as the mouse is moved on the mouse pad. The left button selects the cursor position, and the right button typically provides the menu of possible operations at the selected cursor position. Earlier mice used the position of a ball underneath the mouse to track its motion mechanically. It is now common to see optical mice.

Apple Computer's wireless Mighty Mouse's tracking engine is based on laser technology that delivers 20 times the performance of standard optical tracking, giving more accuracy and responsiveness on more surfaces. It offers 360° scrolling capability, perfectly positioned to roll smoothly under just one finger. Touch-sensitive technology employed under its seamless top shell detects where we are clicking. The force-sensing buttons on either side of Mighty Mouse let us squeeze the mouse to activate a whole host of customizable features instantly. It is available in wired and wireless versions. Table 7.5 provides additional details.

7.9.3 Printers

Printers have come a long way from the noisy daisy-wheel and dot-matrix printers of the 1980s. Today, we can print any document with crisp, realistic colors and sharp text in essentially any font we can imagine. Whether we want to print photos, family projects, or documents on the go, there is a specially designed printer for the job. The most common type of printer

Table 7.5 Apple Mighty Mouse Specifications

	Wireless (MA272LL/A)	Wired (MA086LL/A)
Power source	AA battery	USB
Cables	None	USB 1.1 (compatible with either USB 1.1 or USB 2.0 ports)
Scroll ball	360° scroll ball with adjustable scrolling	
Buttons	Up to four programmable buttons: full-body button with touch-sensitive technology beneath for left, right, and scroll ball clicking. Force-sensing side buttons	
Tracking	Laser	Optical
System requirements	Mac OS X v10.4.6 or later	Mac OS X (programmability requires Mac OS X v10.4.2 Tiger or later) or Windows 2000 or Windows XP

found in homes today is the *inkjet* printer. This printer works by spraying ionized ink onto the paper with magnetized plates directing the ink to the desired shape. Inkjet printers are capable of producing high-quality text and images in black and white or color, approaching the quality that is produced by more costly laser printers. Many inkjet printers today are capable of printing photo-quality images. *Laser* printers provide the highest quality text and images. They operate by using a laser beam to produce an electrically charged image on a drum, which is then rolled through a reservoir of toner. The toner is picked up by the electrically charged portions of the drum and transferred to the paper through a combination of heat and pressure. While full-color laser printers are available, they tend to be much more expensive than black and white versions and require a great deal of printer memory to produce high-resolution images. A *portable* printer is a compact mobile inkjet printer that can fit in a briefcase, weigh very little, and run on battery power. Infrared-compatible (wireless) printers, available in both inkjet and LaserJet models, allow us printing from a handheld device, laptop computer, or digital camera. The wireless short-range radio technology that allows this to happen is called Bluetooth. *All-in-one* devices (inkjet or laser based) that combine the functions of printer, scanner, copier, and fax into one machine are now available.

As an example, the HP Color LaserJet 1600 family is a basic color laser printer designed for light printing needs. It has a 264 MHz processor and 16 MB of memory. Table 7.6 provides its specifications.

7.9.4 Scanners

Scanners are used to transfer the content of a hard copy document to the machine memory. The image can be retained as is and transmitted as needed. It can also be processed (reduced, enlarge, rotate, etc.). If the image is to be treated as a word-processable document, optical character recognition is first performed after scanning it. Scanners of various capabilities and price ranges are now available.

The Microtek ScanMaker i900 is a flatbed scanner that can handle legal-size originals and features a dual-scanning bed that produces the best film scans. It is connected to the processor via USB or FireWire. It employs Digital ICE Photo Print technology to correct dust and scratches and ColorRescue for color-balance corrections to film and reflective scans. In addition to the typical flatbed glass plate, i900 has glassless film scanner with the glass scan surface underneath. By eliminating the glass, the way a stand-alone film scanner does, the i900 can capture more tonal information from film—from light shades to deep shadows. The scanner's optical resolution is 3200 by 6400 dots per inch (dpi).

Table 7.6   HP LaserJet 1600 Specifications

Property	Specification
Technology	Laser
Printer type	Personal printer laser
Max media size	8.5 in. × 14 in. (216 × 356 mm): letter, legal, executive, envelopes (No. 10 Monarch)
Printer output type	Color
Max resolution (black and white)	600 × 600 dpi
Printer speed	8 pages/min
Printer output	Black draft A4 (8.25 × 11.7 in.), 8 pages/min
	Black normal A4 (8.25 × 11.7 in.), 8 pages/min
	Color best A4 (8.25 × 11.7 in.), 8 pages/min
	Color draft A4 (8.25 × 11.7 in.), 8 pages/min
	Color normal A4 (8.25 × 11.7 in.)
Printer output quality	Black best, black draft, black normal, color best, color draft, color normal
Connectivity technology	Wired technology
Ports required	1 hi-speed USB 2.0 4 pin USB type B
Operational power consumption	190 W
Operational power consumption (standby)	13 W

7.9.5 A/D and D/A Converters

Figure 7.32 shows a digital processor controlling an analog device. Real-time and process-control environments, monitoring laboratory instruments, fall into the application mode shown in the figure. Here, the analog signal produced by the device is converted into a digital bit pattern by the A/D converter. The processor outputs the data in digital form, which is converted into analog form by the D/A converter. D/A converters are resistor–ladder networks that convert the input n-bit digital information into the corresponding analog voltage level.

A/D converters normally use a counter along with a D/A converter. The contents of the counter are incremented, converted into analog signals, and compared to the incoming analog signal. When the counter value corresponds to a voltage equivalent to the input analog voltage, the counter is stopped from incrementing further. The contents of the counter, then, correspond to the equivalent digital signal.

7.9.6 Tapes and Disks

Storage devices such as magnetic tapes (reel-to-reel, cassette, and streaming cartridge) and disks (hard and floppy, magnetic and optical) described in Chapter 9 are also used as I/O devices.

Table 7.7 lists typical data-transfer rates offered by some I/O devices, mainly to show the relative speeds of these devices. The speed and capabilities of these devices change so rapidly that such tables become outdated even before they are published. The description provided in this section is

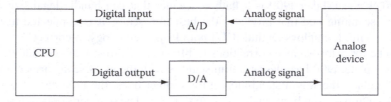

Figure 7.32 Interfacing an analog device.

Table 7.7   Data-Transfer Rates of I/O Devices

Device	Transfer Rate
Display terminals	10–240 characters/s
Impact printers	100–3000 lines/min
Nonimpact printers	100–40,000 lines/min
A/D and D/A converters	15–300×10^6 samples/s
Magnetic tape	15,000–30,000 characters/s
Cassette tape	10–400 characters/s
Hard disk	10–$25 \ 10^9 \times$ bit/s
Floppy disk	25,000 characters/s

intentionally very brief. The reader should refer to vendors' literature for current information on these and other devices.

The compact disks have been a common storage device for decades. Their evolution has now reached a threshold, with their capacity being increasing to nearly 100 GB on a single disk with multiple layers, using technologies such as DVD, HD disk, and Blu-ray disk. Recent laptops are increasingly produced with no disk drives in them, making such compact disks obsolete.

7.9.7 Touchpads and Trackpads

Touchpads and trackpads are devices used to translate the movement of the user's fingers on a specialized surface into the movement of the mouse pointer on the screen. They use surfaces with special sensors known as tactile sensors. The most popular method of operation is using capacitive sensing. In this method, a series of conductors are arranged in parallel with insulating material in between, to form a grid. An alternating current is passed through these conductors. When a finger is placed over the grid, it touches the intersection and creates a virtual ground discharging the stored charge. This leads to the change in the apparent capacitance at the location. This is sensed as a signal to control the pointer on the screen. By 1982, they were first introduced in Apollo desktop computers, located on the right side of the keyboard. Later with the advent of laptops, they are shifted besides the keyboard closer to the users. Improvements in touch and track technologies now provide the capabilities like multitouch and taps being read as the right/left click of a traditional mouse. Such features are presently used for operations like selecting an area and scaling the selected area on the screen.

7.9.8 Touch Screens

Touch screens recognize the touch of a finger to select data or interact with a program displayed on a screen. Recent touch screens support the capability to process input from multiple fingers at the same time. Also known as gestures, these multitouch inputs are typically used to scale pictures, to select an area on the screen for copy/cut operations, and other editing options. Touch screens have been the prominent I/O mechanism for smartphones, tablets, and laptops. Given that it is more intuitive to directly touch the area of the screen that needs to be accessed, recent operating systems are designed to operate using this new I/O facility.

7.9.9 Gaming Console

The gaming console is an interactive entertainment computer or customized computer system. Such systems have been evolving since the mid-1970s. Today, they are composed of handheld controllers as input devices and the screen as the output device. The handheld controllers take inputs through the motion of the user's hands as well as tactile buttons for gaming activates. The inputs received from these devices are used to progress further in a game. Several kinds of handheld devices are connected wirelessly to the gaming systems to give increased mobility and a better gaming experience.

7.10 EXAMPLES

Computer systems are generally configured around a single- or multiple-bus structure. There is no general baseline structure for commercial systems; each system is unique. But most of the modern systems utilize one or more standard bus architectures to allow easier interface of devices from disparate vendors. This section provides brief descriptions of three I/O structures:

1. Motorola 68000 (a complete I/O structure example)
2. Intel 21285 (a versatile interconnect processor)
3. Control Data 6600 (an I/O structure of historical interest)

All of these systems have now been replaced by higher-performance, more versatile versions. Nevertheless, these structures depict the pertinent characteristics more simply. Refer to manufacturers' manuals for details on the latest versions.

7.10.1 Motorola 68000 (MC68000)

Although MC68000 has been replaced by higher-performance counterparts, it is included here to provide a complete view of the I/O structure of a commercially available processor system. The MC68000 is a popular microprocessor with a 32-bit internal architecture that is capable of interfacing with both 8- and 16-bit-oriented peripherals. Figure 7.33 shows the functional grouping of signals available on the processor. A brief description of these signals follows.

The MC68000 memory space is byte addressed. Address bus lines (A1–A23) always carry an even address corresponding to a 16-bit (2 byte) word. The upper data strobe (UDS) and lower data strobe (LDS), if active, refer to the upper or lower byte respectively of the memory word selected by A1–A23. D0–D15 constitutes the 16-bit bidirectional data bus. The direction of the data transfer is indicated by the R/W control signal. The AS signal, if active, indicates that the address (as determined by the address bus, UDS, and LDS) is valid. DTACK is used to indicate that the data have been accepted (during a write operation) or that the data are ready on the data bus (during a read). The processor waits for the DTACK during the I/O operations with variable speed peripherals. There are three bus arbitration signals: BR, BG, and bus grant acknowledge (BGAC). There are three interrupt lines (IPL0–IPL2) that allow an interrupting device to indicate one of the seven levels of interrupts possible: level 1 being the lowest priority, level 7 the highest, and level 0 meaning no interrupt. The processor communicates with 8-bit-oriented peripherals of MC6800 through the ENABLE (E), VALID MEMORY ADDRESS (VMA), and VALID PERIPHERAL ADDRESS (VPA) control signals. The communication through these lines is synchronous. The FUNCTION CODE (FC0–FC2) output signals indicate the type of bus activity currently undertaken (such as interrupt ACK and supervisor/user program/data memory access) by the processor. There are three system control signals: BUS ERROR (BERR), RESET, and HALT.

7.10.1.1 Programmed I/O on MC68000

The MC68000 uses memory-mapped I/O, since no dedicated I/O instructions are available. The I/O devices can be interfaced either to the asynchronous I/O lines or to the MC68000-oriented synchronous I/O lines. We will now illustrate interfacing an input and an output port to MC68000 to operate in the programmed I/O mode through the asynchronous control lines, using the parallel interface/timer chip MC68230.

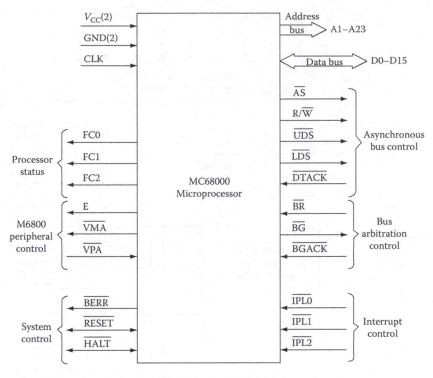

Figure 7.33 Motorola 68000 processor. (Courtesy of Motorola Inc.)

The MC 68230 (see Figure 7.34) is a peripheral interface/timer (PI/T) consisting of two independent sections: the ports and the timer. In the port section, there are two 8-bit ports (PA0–7 and PB0–7), four handshake signals (H1–H4), two general I/O pins (I/O), and six dual-function pins. The dual-function pins can individually work either as a third port (C) or an alternate function related to either port A or B the timer. Pins H1–H4 are used in various modes: to control data transfer to and from ports, as general-purpose I/O pins or as interrupt-generating inputs with corresponding vectors.

The timer consists of a 24-bit counter and a prescaler. The timer I/O pins (TIN, TOUT, TIACK) also serve as port C pins.

The system data bus is connected to pins D0–D7. Asynchronous transfer between the MC68000 and PI/T is facilitated by data-transfer acknowledge (DTACK), register selects (RS1–RS5), timer interrupt acknowledge (TIACK), (R/W), and port interrupt acknowledge (PIACK).

We will restrict this example to the port section. The PI/T has 23 internal registers, each of which can be addressed using RS1–RS5. Associated with each of the three ports are a data register, a control register, and a data-direction register (DDR). Register 0 is the port general control register (PGCR) that controls all the ports. Bits 6 and 7 of this register are used to configure the ports in one of the four possible modes, as shown in the following table. The other bits of the PGCR are used for I/O handshaking operations.

Bit 7	Bit 6	Mode
0	0	Unidirectional (8 bit)
0	1	Unidirectional (16 bit)
1	0	Bidirectional (8 bit)
1	1	Bidirectional (16 bit)

When the ports are configured in unidirectional mode, the corresponding control registers are used to further define the submode of operation. For example, in the unidirectional 8-bit mode, bit 7 of a control

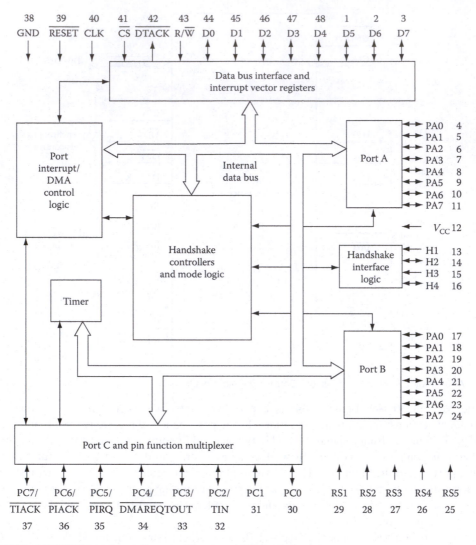

Figure 7.34 MC68230 functional diagram. (Courtesy of Motorola Inc.)

register being 1 signifies a bit-oriented I/O in the sense that each bit of the corresponding port can be programmed independently. To configure a bit as an output or an input bit, the corresponding DDR bit is set to 1 or 0, respectively. DDR settings are ignored in the bidirectional transfer mode.

Figure 7.35 shows the MC68000 interface with MC68230 in which the following addresses are used:

Register Address (Hex)
PGCR 100001
PADDR 100003
PBDDR 100007
PACR 10000D
PBCR 10000F
PADR 100011
PBDR 100013

Figure 7.36 shows an assembly language program that reads an 8-bit input from port A and writes the data to port B.

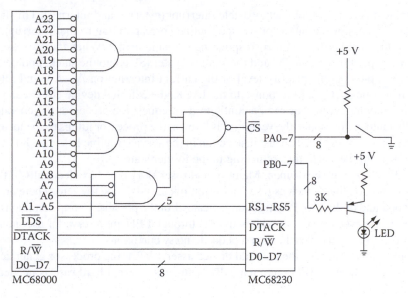

Figure 7.35 Interfacing I/O ports to MC68000.

```
; ASSIGN ADDRESSES
PGCR        EQU      $100001   ; $ indicate hex
PADDR       EQU      $100003
PBDDR       EQU      $100007
PACR        EQU      $10000D
PBCR        EQU      $10000F
PADR        EQU      $100011
PBDR        EQU      $100013
; INITIALIZE ALL REGISTERS
; MOVE B Source Destination Moves bytes
; # indicate immediate address mode
            MOVE. B    #$00, PGCR ; Select unidirectional
                                  ; 8-bit mode
            MOVE. B    #$FF, PACR ; Bit input/output mode for port A
            MOVE. B    #$FF, PBCR ; Bit input/output mode for port B
            MOVE. B    #$00, PADDR; All bits of A are input
            MOVE. B    #$FF, PBDDR; All bits of B are output
; READ AND WRITE
START:      MOVE. B    PADR.D1    ; Input from A into processor
                                  ; register Di
            MOVE. B    D1. PBDR   ; Output to B
            JMP        START      ; Repeat
```

Figure 7.36 MC68000 assembly language program for I/O.

7.10.1.2 MC68000 interrupt system

The MC68000 interrupt system is divided into two types: *internal* and *external*. The internal interrupts are called *exceptions* and correspond to conditions such as divide by zero, illegal instruction, and user-defined interrupts by the use of trap instructions. The external interrupts correspond to the seven levels of interrupts brought through the $\overline{IPL0} - \overline{IPL2}$ lines, as mentioned earlier, level 0 indicates no interrupt, level 7 is the highest priority interrupt and is nonmaskable by the processor, and the other levels are recognized by the processor if the processor is in a lower-priority processing mode.

A 4-byte field is reserved for each possible interrupt (external and internal). This field contains the beginning address of the interrupt service routine corresponding to that interrupt. The beginning address of the field is the vector corresponding to that interrupt. When the processor is ready to service an interrupt, it pushes the PC and the status register (SR) onto the stack, updates the priority mask bits in SR, puts out the priority level of the current interrupt on A1–A3, and sets FC0–FC2 to 111 to indicate the (IACK). In response to the IACK, the external device either can send an 8 bit vector number (nonautovector) or can be configured to request the processor to generate the vector automatically (autovector). Once the vector (v) is known, the processor jumps to the location pointed to by the memory address (4v). The last instruction in the service routine has to be RTE, which pops the PC and SR from the stack, thus returning to the former state.

Figure 7.37 shows the vector map. Memory addresses 00H (i.e., hexadecimal 00) through 2FH contain vectors for conditions such as reset, bus error, trace, and divide the zero. There are seven autovectors. The operand of the TRAP instruction indicates 1 of the possible 15 trap conditions, thus generating 1 of the 15 trap vectors. Vector addresses 40H through FFH are reserve for user nonautovectors. Spurious interrupt vector handles the interrupt due to noisy conditions.

In response to the IACK, if the external device asserts $\overline{VPA}$, the processor generates one of the seven vectors (19H through 1FH) automatically. Thus, no external hardware is needed to provide the interrupt vector.

For nonautovector mode, the interrupting device places a vector number (40H through FFH) on the data bus lines D0–D7 and asserts $\overline{DTACK}$. The processor reads the vector and jumps to the appropriate service routine.

Due to system noise, it is possible for the processor to be interrupted. In that case, on the receipt of IACK, an external timer can activate the $\overline{BERR}$ (after a certain time). When the processor receives the BERR in response to IACK, it generates the spurious interrupt vector (18H).

Figure 7.38 shows the hardware needed to interface two external devices to the MC68000. The priority encoder (74LS148) generates the 3-bit interrupt signal for the processor based on the relative priorities of the devices connected to it. When the processor recognizes the interrupt, it puts the

Vector address		Vector number
00H to 2FH	Reset, bus error, etc.	00H to 16H
30H to 5CH	Unassigned	0CH to 17H
60H, 62H	Spurious interrupt	18H
64H, 66H	Autovector 1	19H
6BH, 6AH	Autovector 2	1AH
6CH, 6EH	Autovector 3	1BH
70H, 72H	Autovector 4	1CH
74H, 76H	Autovector 5	1DH
78H, 7AH	Autovector 6	1EH
7CH, 7EH	Autovector 7	1FH
80H to BCH	TRAP instructions	20H to 2FH
COH	Unassigned	30H
:		:
FCH		3FH
100H	User interrupts (nonautovector)	40H
:		:
.		.
3FCH		FFH

Figure 7.37   MC68000 vector map.

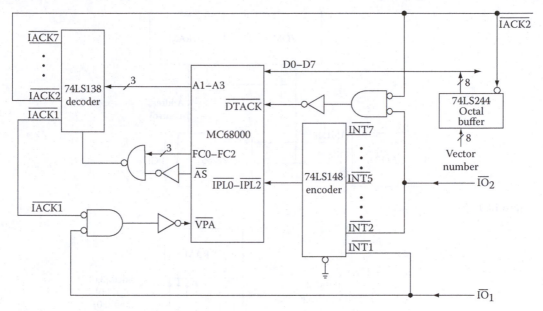

Figure 7.38 Vectored interrupts on MC68000.

priority level on A1–A3, sets F0–F2 to 111, and activates $\overline{AS}$. The 3–8 decoder (74LS138) thus generates the appropriate IACK. When device 1 is interrupting, the $\overline{VPA}$ line is activated in response to $\overline{IACK1}$, thus activating an autovector mode. For device 2, $\overline{IACK2}$ is used to gate its vector onto data lines D0–D7 and to generate $\overline{DTACK}$.

7.10.1.3 MC68000 DMA

To perform a DMA, the external device requests the bus by activating $\overline{BR}$. One clock period after receiving $\overline{BR}$, MC68000 enables the bus grant line $\overline{BG}$ and relinquishes (tristates) the bus after completing the current instruction cycle (as indicated by the $\overline{AS}$ going high). The external device then enables $\overline{BGACK}$, indicating that the bus is in use. The processor waits for the $\overline{BGACK}$ to go high in order to use the bus.

7.10.2 Intel 21285

This section is extracted from Intel Corporation's *21285 Core Logic for the SA-110 Microprocessor Datasheet*, September 1998.

Figure 7.39 shows a block diagram of a system with an Intel 21285 I/O Processor connected to Intel SA-110 (StrongARM) microprocessor. SA-110 is optimized for embedded applications such as intelligent adapter cards, switches, routers, printers, scanners, RAID controllers, process-control applications, and set-top boxes.

The Intel 21285 I/O Processor consists of the following components: synchronous dynamic random access memory (SDRAM) interface, ROM interface, peripheral component interface (PCI), DMA controllers, interrupt controllers, programmable timers, X-Bus interface, serial port, bus arbiter, joint test architecture group (JTAG) interface, dual address cycle (DAC) support, and power management support.

The SDRAM controller controls from 1 to 4 arrays of synchronous DRAMs (SDRAMs) consisting of 8, 16, and 64 MB parts. All SDRAMs share command and address bits but have separate block and chip select bits, as shown in Figure 7.40.

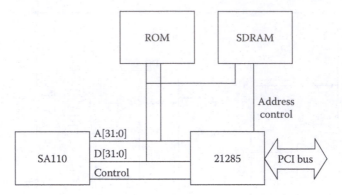

Figure 7.39 System diagram.

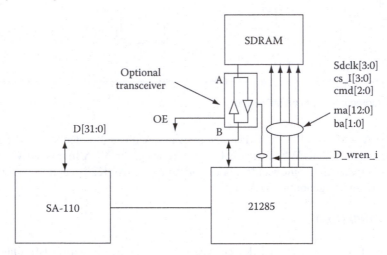

Figure 7.40 SDRAM configuration. *Note:* When SA-110 reads or writes SDRAM, the data do not pass through 21285.

SDRAM operations performed by the 21285 are refresh, read, write, and mode register set. Reads and writes are generated by either the SA-110, PCI bus masters (including intelligent I/O (I²O) accesses), or DMA channels. The SA-110 is stalled, while the selected SDRAM bank is addressed. It is unstalled after the data have been latched into the SDRAM from D bus or when the first SDRAM data have been driven to the D bus. PCI memory write to SDRAM occurs if the PCI address matches the SDRAM base AR or the configuration space register (CSR) base AR, and the PCI command is either a memory write or a memory write and invalidate. PCI memory read from SDRAM occurs if the PCI address matches the SDRAM base AR or the CSR base AR, and the command is either a memory read, memory read line, or memory read multiple.

There are four registers controlling arrays 0 through 3. These four registers define each of the four SDRAM arrays' start address, size, and address multiplexing. Software must ensure that the arrays of SDRAM are mapped so there is no overlap of addresses. The arrays do not need to all be the same size; however, the start address of each array must be naturally aligned to the size of the array. The arrays do not need to form a contiguous address space, but to do so with different size arrays, place the largest array at the lowest address, next largest array above, etc.

Figure 7.41 shows the ROM configuration. The ROM output enable and write enable are connected to address bits [30:31], respectively. The ROM address is connected to address bits [24:2].

The ROM can always be addressed by the SA-110 at 41000000h through 41FFFFFFh. After reset, the ROM is also aliased at every 16 MB throughout memory space, blocking access to

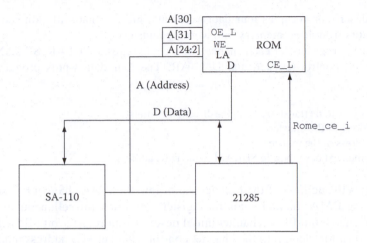

Figure 7.41 ROM configuration.

SDRAM. This allows the SA-110 to boot from ROM at address 0. After any SA-110 write, the alias address range is disabled.

The SA-110 is stalled, while the ROM is read. Each data word may require one, two, or four ROM reads, depending on the ROM width. The data are collected and packed into data words to be driven onto D [31:0]. When the ROM read completes, the 21285 unstalls the SA-110. The SA-110 is stalled, while the ROM is written. The ROM write data must be placed on the proper byte lanes by software running on the SA-110, that is, the data are not aligned in hardware by the 21285. Only one write is done, regardless of the ROM width. When the ROM write completes, the 21285 unstalls the SA-110. PCI memory write to ROM occurs when the PCI address matches the expansion ROM base AR, bit [0] of the expansion ROM base AR is a 1, and the PCI command is either a memory write or a memory write and invalidate. The PCI memory write address and data are collected in the inbound FIFO to be written to ROM at a later time. The 21285 target disconnects after one data phase. PCI memory read to ROM occurs when the PCI address matches the expansion ROM base AR, bit [0] of the expansion ROM base AR is a 1, and the PCI command is either a memory read, memory read line, or memory read multiple. Timing during ROM accesses can be controlled by values in the SA-110 control register. The ROM access time, burst time, and tristate time can be specified.

The 21285 is a programmable, two-way DMA channel that can move blocks of data from SDRAM to PCI or PCI to SDRAM. The DMA channels read parameters from a list of descriptors in memory, perform the data movement, and stop when the list is exhausted. For DMA operations, the SA-110 sets up the descriptors in SDRAM. Figure 7.42 shows DMA descriptors in local

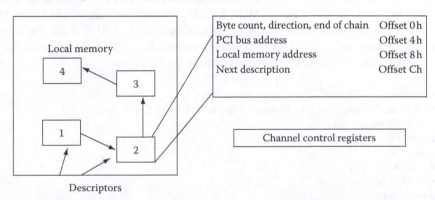

Figure 7.42 Local memory descriptors.

memory. Each descriptor occupies four data words and must be naturally aligned. The channels read the descriptors from local memory into working registers.

There are several registers for each channel: byte count register, PCI AR, SDRAM AR, descriptor pointer register, control register, and DAC AR. The four data words provide the following information:

1. Number of bytes to be transferred and the direction of transfer
2. PCI bus address of the transfer
3. SDRAM address of the transfer
4. Address of the next descriptor in SDRAM or the DAC address

The SA-110 writes the address of the first descriptor into the DMA channel n descriptor pointer register, writes the DMA channel n control register with other miscellaneous parameters, and sets the channel enable bit. If the channel initial descriptor in register bit [4] is clear, the channel reads the descriptor block into the channel control, channel PCI address, channel SDRAM address, and channel descriptor point register. The channel transfers the data until the byte count is exhausted and then sets the channel transfer done bit [2] in the DMA channel n control register. If the end of chain bit [31] in the DMA channel n-byte count register (which is in bit [31] of the first word of the descriptor) is clear, the channel reads the next descriptor and transfers the data. If it is set, the channel sets the chain done bit [7] in the DMA channel n control register and then stops.

The PCI arbiter receives requests from five potential bus masters (four external and the 21285), and a grant is made to the device with the highest priority. The main register for the bus arbiter is X-Bus cycle/arbiter register. Its offset is 148 h, and its function is used to control either the parallel port (X-Bus) or the internal PCI arbiter.

There are two levels of priority groups with the low-priority groups as one entry in the high-priority group. Priority rotates evenly among the low-priority groups. Each device, including the 21285, can appear either in the low-priority group or in the high-priority group, according to the value of the corresponding priority bit in the arbiter control register. The master is in the high-priority group if the bit is 1 and low-priority group if the bit is a 0. Priorities are reevaluated every time frame_1 is asserted, at the start of each new transaction on the PCI bus. The arbiter grants the bus to the higher-priority device in the next clock style if it interrupts when a lower-priority device is using the bus. The master that initiated the last transaction has the lowest priority in the group.

PCI is a local bus standard (a data bus that connects directly to the microprocessor) developed by Intel Corporation. Most modern PCs include a PCI bus in addition to a more general Industry Standard Architecture (ISA) expansion bus. Many analysts believe that PCI will eventually supplant ISA entirely. PCI is a 64-bit bus, though it is usually implemented as a 32-bit bus. It can run at clock speeds of 33 or 66 MHz. At 32 bit and 33 MHz, it yields a throughput rate of 133 Mbps. PCI is not tied to any particular family of microprocessors. It acts like a tiny "LAN" inside the computer, in which multiple devices can each talk to each other, sharing a communication channel that is managed by the chipset. The PCI target transactions include the following:

1. Memory write to SDRAM
2. Memory read, memory read line, memory read multiple to SDRAM
3. Type 0 configuration write
4. Type 0 configuration read
5. Write to CSR
6. Read to CSR
7. Write to I^2O address
8. Read to I^2O address
9. Memory read to ROM

The PCI master transactions include the following:

1. DAC support from SA-110 or DMA
2. Memory write, memory write, and invalidate from SA-110 or DMA
3. Selecting PCI command for writes
4. Memory write, memory write, and invalidate from SA-110
5. Selecting PCI command for writes
6. Memory read, memory read line, memory read multiple from SA-110 or DMA
7. I/O write
8. I/O read
9. Configuration write
10. Special cycle
11. IACK read
12. PCI request operation
13. Master latency timer

The message unit provides a standardized message-passing mechanism between a host and a local processor. It provides a way for the host to read and write lists over the PCI bus at offsets of 40 and 44 h from the first base address. The function of the FIFOs is to hold message frame addresses, which are offsets to the message frames. The I²O message units support four logical FIFOs:

Name	Function
Inbound free_list	Manage I/O requests from the host processor to SA-110
Inbound post_list	
Outbound free_list	Manages messages that are replies from
Outbound post_list	SA-110 to the host processor

The I²O inbound FIFOs are initiated by the SA-110. The host sends messages to the local processor. The local processor operates on the messages. The SA-110 allocates memory space for both the inbound free_list and post_list FIFOs and initializes the inbound pointers. The SA-110 initializes the inbound free_list by writing valid memory frame address (MFA) values to all entries and increments the number of entries in the inbound free_list count register. The I²O outbound FIFOs are initialized by the SA-110. The SA-110 allocates memory space for both outbound free_list and outbound post_list FIFOs. The host processor initializes the free_list FIFO by writing valid MFAs to all entries (see Figures 7.43 and 7.44).

The 21285 contains four 24-bit timers. There are control register and SR for each timer. A timer has two modes of operation: free run and periodic. The timers and their interrupt can be individually enabled/disabled. The interrupts can be enabled/disabled by setting the appropriate

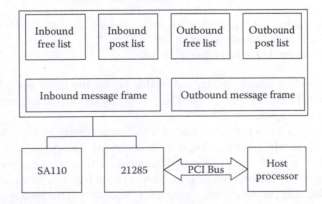

Figure 7.43 I²O message units.

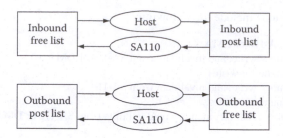

Figure 7.44   Configuration of MFAs.

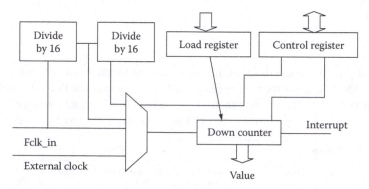

Figure 7.45   Timer block diagram.

bits in the IRQ enable set/fast interrupt request (FIQ) enable set registers. Timer 4 can be used as a watchdog timer but requires that the watchdog enable bit be set in the SA-110 control register. Figure 7.45 shows a block diagram of a typical timer. The control, status, and data registers for this capability are accessible only from the SA-110 interface (all offsets are from 4200 0000H).

There are three registers that govern the operations of a timer. In the periodic mode, the load register contains a value that is loaded onto a down counter and decremented to zero. The load register is not used in the free-run mode. The control register allows the user to select a clock rate that will be used to decrement the counter, the mode of operation, and whether the timer is enabled. The clear register resets the timer interrupt.

The 21285 has only one register that might be considered an SR, and that is the value register. It can be read to obtain the 24-bit value in the down counter.

In terms of the software interface, the user needs to determine

1. Clock speed that will be used to decrement the down counter
2. Whether free-run or periodic mode will be used
3. Whether to generate the IRQ or FIQ interrupt or both

Once determined, the clock speed, mode, and enable bit for the timer must be written into the timer's control register. The interrupt type must be enabled by setting the appropriate bits in the IRQ enable set/FIQ enable set registers.

In the free-run mode, a down counter is loaded with the maximum 24-bit value and decremented until it reaches zero (0), at which time an interrupt is generated. The periodic mode loads the down counter with the value in the load register and decrements until it reaches zero (0), at which time an interrupt is generated. Upon reaching zero (0), timers in the free-run mode are reloaded with the maximum 24-bit value, and periodic mode timers are reloaded from the load register. In both cases, once reloaded, the counters decrement to zero (0) to repeat the cycle.

The 21285 has a UART that can support bit rates from approximately 225 bps to approximately 200 kbps. The UART contains separate transmit and receive FIFOs that can hold 16 data entries each. The UART generates receive and transmit interrupts that can be enabled/disabled by setting the appropriate bits in the IRQ enable set/FIQ enable set registers. The control, status, and data registers for this capability are accessible only from the SA-110 interface (all offsets are from 4200 0000H).

There are four registers that govern the operations of the UART. The H_UBRLCR register allows the user to set the data length, enable parity, select odd or even parity, select the number of stop bits, enable/disable the FIFOs, and generate a break signal (the tx pin is held low). The M_UBRLCR and L_UBRLCR registers, together, contain the 12-bit baud rate divisor (BRD) value. The UARTCON register contains the UART enable bit along with bits for using the SIREN HP SIR protocol and the infrared data (IrDa) encoding method.

The UART contains a single data register. It allows the transmit and receive FIFOs to be accessed. Data to and from the UART can be accessed either via FIFOs or as a single data word. The access method is determined by the enable FIFO bit in the H_UBRLCR control register.

The 21285 has two SRs: RXSTAT indicates any framing, parity, or overrun errors associated with the received data. A read of RXSTAT must follow a read of UARTDR because RXSTAT provides the status associated with the last piece of data read from UARTDR. This order cannot be reversed. UARTFLG contains transmit and receive FIFO status and an indication whether the transmitter is busy.

When the UART receives a frame of data (start bit, data bits, stop bits, and parity), the data bits are stripped from the frame and put in the receive FIFO. When the FIFO is more than half full, a receive interrupt is generated. If the FIFO is full, an overrun error will be generated. The framing data are examined, and if incorrect, a framing error is generated. Parity is checked and the parity error bit is set accordingly. The data words are accessed by reading UARTDR and any errors associated can be read from RXSTAT.

When the UART transmits data, a data word is taken from the transmit FIFO and the framing data are added (start bit, stop bits, and parity). The frame of data is loaded into a shift register and clocked out. When the FIFO is more than half empty, a transmit interrupt is generated.

The 21285 is compliant with the IEEE 1149.1 "IEEE Standard Test Access Port and Boundary-Scan Architecture." It provides five pins to allow external hardware and software to control the test access port (TAP).

The 21285 provides an nIRQ and nFIQ interrupt to the SA110 processor. These interrupts are the logical NOR of the bits in the IRQ status and FIQ status, respectively. These registers collect the interrupt status for 27 interrupt sources. All the interrupts have the equal priority. The control, status, and data registers for this capability are accessible only from the SA-110 interface. Figure 7.46 is a block diagram of a typical interrupt.

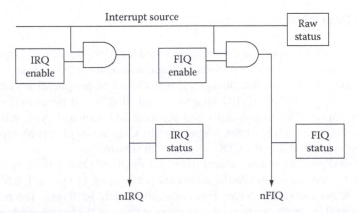

Figure 7.46 Interrupt block diagram.

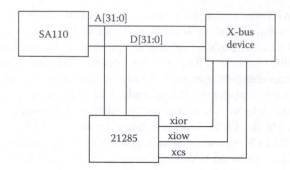

Figure 7.47 Bus configuration block diagram.

The following control registers deal with interrupts. The IRQ enable set register allows individual interrupts to be enabled. The IRQ enable clear register disables an individual interrupt. The IRQ soft register allows the software to generate an interrupt. The following SRs too deal with interrupts. The IRQ enable register indicates which interrupts are being used by the system. The IRQ raw SR indicates which interrupts are active. The IRQ SR is the logical AND of the IRQ enable register and the IRQ raw SR.

When an interrupting device activates its interrupt, the information is loaded into the IRQ raw SR. These data are ANDed with the IRQ enable register to set bits in the IRQ SR. The bits in the IRQ SR are NORed together to generate the nIRQ interrupt to the SA110. The SA110 determines what caused the interrupt by reading the IRQ SR. An interrupt is cleared by resetting the appropriate bit in the IRQ enable clear register.

The 21285 provides 8-, 16-, and 32-bit parallel interfaces with the SA110. The duration and timing relationship between the address, data, and control signals is accomplished via control register settings. This provides the 21285 a great deal of flexibility to accommodate a wide variety of I/O devices. The control, status, and data registers for this capability are accessible only from the SA-110 interface (all offsets are from 4200 0000H). Figure 7.47 shows a block diagram of the X-bus interface.

The 21285 provides the following control and SRs to deal with the X-bus: The X-bus cycle register allows the user to set the length of an R/W cycle, apply a divisor to the clock used during R/W cycles, choose a chip select, and determine if the X-bus or PCI bus is being used and the X-bus and PCI interrupt levels. The X-bus I/O strobe register allows the user to control when the xior and xiow signals go low and how long they stay low within the programmed R/W cycle length.

7.10.3 Control Data 6600

The CDC 6600 first announced in 1964 was designed for two types of use: large-scale scientific processing and time sharing of smaller problems. To accommodate large-scale scientific processing, a high-speed, floating-point, multifunctional CPU was used. The peripheral activity was separated from CPU activity by providing 12 I/O channels controlled by 10 peripheral processors. This architecture with multiple functional units and separate I/O structure with multiple peripheral processors operating in parallel has been adopted by the Cray series of supercomputers (described in Chapter 10). Figure 7.48 shows the CDC 6600 system structure.

The 10 peripheral processors have access to central memory. One of these processors acts as a control processor for the system, while the others are performing I/O tasks. Each of these processors has its own memory, used for storing programs and for data buffering. The peripheral processors access the central memory in a time-shared manner through the barrel mechanism. The barrel

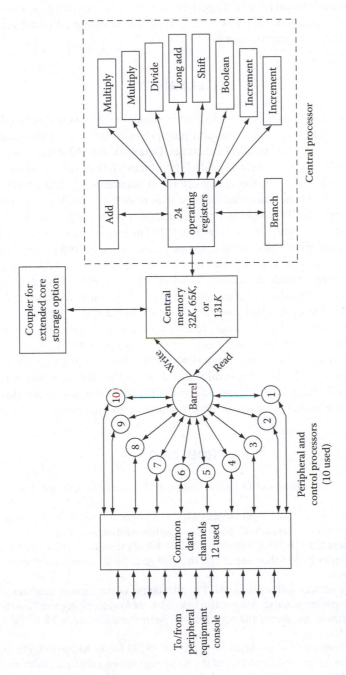

Figure 7.48 CDC 6600 system structure. (Courtesy of Control Data Corporation.)

mechanism has a 100 ns cycle. Each peripheral processor is connected to the central memory for 100 ns in each cycle of the barrel. The peripheral processor provides the memory-accessing information to the barrel during its 100 ns time slot. Actual memory access takes place during the remaining 900 ns of the barrel cycle, and the peripheral processor can start a new access during its next time slot. Thus, the barrel mechanism handles 10 peripheral processor requests in an overlapped manner. I/O channels are 12-bit bidirectional paths. Thus, one 12-bit word can be transferred into or out of the memory every 1000 ns by each channel.

7.11 SUMMARY

Various modes of data transfer between I/O devices and the CPU were discussed in this chapter. The advances in hardware technology and dropping prices of hardware have made it possible to implement cost-effective and versatile I/O structures. Details of the I/O structures of representative commercially available machines were provided. Interfacing I/O devices to the CPU is generally considered a major task. However, the recent efforts in standardizing I/O transfer protocols and the emergence of standard buses have reduced the tedium of this task. It is now possible to easily interface an I/O device or a CPU made by one manufacturer with compatible devices from another manufacturer. The trend in delegating the I/O tasks to I/O processors has continued. In fact, modern computer structures typically contain more than one general-purpose processor, some of which can be dedicated to I/O tasks, as the need arises. This chapter has provided details of representative I/O devices. Newer and more versatile devices are being introduced daily. Any list of speed and performance characteristics of these devices provided in a book would quickly become obsolete. Refer to the magazines listed in the Bibliography section for up-to-date details.

Several techniques are used in practical architectures to enhance the I/O system performance. Each system is unique in its architectural features. The concepts introduced in this chapter form the baseline features that help to evaluate practical architectures. In practice, the system structure of a family of machines evolves over several years. Tracing the family of machines to gather the changing architectural characteristics and to distinguish the influence of advances in hardware and software technologies on the system structure would be an interesting and instructive project.

PROBLEMS

7.1   What is the maximum number of I/O devices that can be interfaced to ASC, given its current instruction format?

7.2   It is required to interface five input devices (device address 0 through 4) and seven output devices (device address 5 through 11) to ASC. Show the complete hardware needed.

7.3   A computer system has a 16-bit address bus. The first 4 K of the memory must be configured with ROM and the remaining as RAM. Show the circuit that will generate an error if an attempt is made to write into the ROM area.

7.4   Design a priority encoder with eight inputs. The 3-bit output of the encoder indicates the number of the highest priority input that is active. Assume that input 0 has the lowest priority and 7 has the highest priority.

7.5   What is the maximum number of I/O ports that can be implemented using MC68230s on a system bus with 16 address lines?

7.6   Design an 8-bit input port and an output port using MC68230 for the MC68000. The input port receives a 4-bit BCD value on its low-order bits, and the output port drives a seven-segment display from its bits. Show the assembly language program needed.

7.7   Rewrite the fetch microinstruction sequence for ASC, assuming that the processor status is saved on a stack before entering interrupt service. Include a stack pointer register.

7.8 Rewrite the sequence in Problem 7.7, assuming a vectored interrupt mode. Assume that the device sends its address as the vector.

7.9 Write microprograms to describe the handshake between the peripheral and CPU during (a) interrupt-mode I/O and (b) DMA. Use the HDL structure common to ASC microprograms to describe these asynchronous operations.

7.10 Develop a complete hardware block diagram of a multiplexer channel. List the microoperations (μops) of each block in the diagram.

7.11 Repeat Problem 7.10 for a selector channel.

7.12 Develop a schematic for a printer controller. Assume that one character (8 bit) is transferred to the printer at a time, in a programmed I/O mode.

7.13 Repeat Problem 7.12, assuming that the printer has a buffer that can hold 80 characters. Printing is activated only when the buffer is full.

7.14 Design a bus controller to resolve the memory-access requests by the CPU and a DMA device.

7.15 The structure of Figure 7.12 polls the peripheral devices in the order of the device numbers. Design a structure in which the order of polling (i.e., priorities) can be specified. (Hint: Use a buffer to store the priorities.)

7.16 Design a set of instructions to program the DMA controller. Specify the instruction formats, assuming ASC as the CPU.

7.17 A cassette recorder–player needs to be interfaced to ASC. Assume that the cassette unit has a buffer to store 1 character and the cassette logic can read or record a character into or out of this buffer. What other control lines are needed for the interface? Generate the interface protocol.

7.18 Compare the timing characteristics of programmed input mode with that of the vector-mode, interrupt-driven input scheme, on ASC. Use the major cycles needed to perform the input as the unit of time.

7.19 Assume that there is a serial output line connected to the LSB of the accumulator. The contents of the accumulator are to be output on this line. It is required that a start bit of 0 and 2 stop bits (1) delimit the data bits. Write an assembly language program on ASC to accomplish the data output. What changes in hardware are needed to perform this transfer?

7.20 Look up the details of the power-failure detection circuitry of a computer system you have access to, to determine how a power-failure interrupt is generated and handled.

7.21 There are memory-port controllers that allow interfacing multiple memories to computer systems. What is the difference between a memory-port controller and a bus controller?

7.22 Suppose a processor receives multiple IRQs at about the same time, it should service all of them before returning control to the interrupted program. What mechanism is ideal for holding these multiple requests: stack, queue, or something else? Why?

7.23 A processor executes 1000 K instructions per second. The bus system allows a bandwidth of 5 MB/s. Assume that each installation requires on average 8 byte of information for the instruction itself and the operands. What bandwidth is available for the DMA controller?

7.24 If the processor of Problem 7.23 performs programmed I/O and each byte of I/O requires two instructions, what bandwidth is available for the I/O?

7.25 In this chapter, we assumed that the interrupt service always starts at the beginning of the instruction cycle. It may not be desirable to wait until the end of current instruction cycle during complex instructions (such as multiple moves). What are the implications of interrupting an instruction cycle in progress? How would you restart the interrupted instruction?

7.26 Look up the details of the following:
 a. Small computer system interface (SCSI)
 b. FireWire (IEEE 1394)
 c. USB 1.0 and 2.0
 d. High-performance peripheral interface (HiPPI)

7.27 Which mode of I/O is most suitable for interfacing the following devices?
 a. Keyboard
 b. Mouse
 c. Serial printer
 d. Modem
 e. Flash drive (memory stick)

7.28   Amdahl's law for the overall speedup of a computer system states that

$$S = \frac{1}{(1-f) + f/k},$$

where

  S is the overall system speedup
  f is the fraction of the work performed by the faster component
  k is the speedup of the new component

Discuss the significance of this law. How is it useful in determining system (processor, I/O) upgrades?

7.29   Investigate the interactions between the CPU and a graphic processing unit (GPU) in a system you have access to. What interfaces are used?

7.30   Investigate the protocols and interfaces used by a tablet PC to connect to the Internet through Wi-Fi.

7.31   How does the interface scheme of a networked printer differ from that of a dedicated printer?

BIBLIOGRAPHY

Amdahl, G.M., Validity of the single processor approach to achieving large scale Computing capabilities, *Proceedings of AFIPS Spring Joint Computer Conference*, Atlantic City, NJ, April 1967, pp. 483–485.

Apple Computers, http://www.apple.com/mightymouse/index.html *21285 Core Logic for the SA-110 Microprocessor Data Sheet*, Santa Clara, CA: Intel, 1998.

Computer, Los Alamitos, CA: IEEE Computer Society (monthly).

Cramer, W. and Kane, G., *68000 Microprocessor Handbook*, Berkeley, CA: Osborne McGraw Hill, 1986.

DiGiacomo, J., *Digital Bus Handbook*, New York: McGraw Hill, 1990.

Goldner, J.S., The Emergence of iSCSI, *ACM Queue*, June 2003, pp. 44–53.

Harris, D. and Harris, S., *Digital Design and Computer Architecture*, San Francisco, CA: Morgan Kaufmann, 2012.

Hewlett Packard, http://h71036.www7.hp.com/hho/cache/774–0–0–39–121.html?jumpid=reg_R1002_CAEN

HP Printers, http://www.printerworks.com/DataSheets/CLJ1600datasheet.pdf

IEE/ANSI 796 Standard, New York: IEEE, 1983.

IEE/ANSI 1296 Standard, New York: IEEE, 1988.

MC68000 Users Manual, Austin, TX: Motorola, 1982.

Microtek, http://reviews.cnet.com/Microtek_ScanMaker_i900/4505-3136_7-30822156.html

Patterson, D.A. and Hennessey, J.L., *Computer Architecture: A Quantitative Approach*, San Mateo, CA: Morgan Kaufmann, 2011.

Shiva, S.G., *Advanced Computer Architectures*, Boca Raton, FL: Taylor & Francis, 2006.

Simple Multibus II I/O Replier, Mountain view, CA: PLX Technology, 1989.

Stallings, W., *Computer Organization and Architecture*, Saddle River, NJ: Prentice Hall, 2012.

CHAPTER 8

Processor and Instruction-Set Architectures

The main objective of Chapter 6 was to illustrate the design of a simple but complete computer (ASC). Simplicity was our main consideration, and so architectural alternatives possible at each stage in the design were not presented. Chapter 7 extended the input/output (I/O) subsystem of ASC in light of I/O structures found in practical machines. This chapter describes selected architectural features of some popular processors as enhancements of ASC architecture. We will concentrate on the details of instruction set, addressing modes, and register and processor structures. Architectural enhancements to the memory, control unit, and ALU are presented in subsequent chapters. For the sake of completeness, we will also provide in this chapter a brief description of a family of micro-processors from Intel Corporation. Subsequent chapters provide the details of several advanced architectural features employed in contemporary systems. We will first distinguish between the four popular types of computer systems now available.

8.1 TYPES OF COMPUTER SYSTEMS

A modern-day computer system can be classified as a *supercomputer,* a *large-scale machine* (or a *mainframe*), a *minicomputer,* or a *microcomputer.* Other combinations of these four major categories (mini-microsystem, micro-minisystem, etc.) have also been used in practice. We will include the currently popular laptop and tablet systems under the microcomputer family. It is difficult to produce sets of characteristics that would definitively place a system in one of these categories today; original distinctions are becoming blurred with advances in hardware and software technologies. A desktop microcomputer system of today, for example, can provide roughly the same processing capability as that of a large-scale computer system of the 1990s. Nevertheless, Table 8.1 lists some of the characteristics of the four classes of computer systems. As can be noted, there is a considerable amount of overlap in their characteristics. A supercomputer is defined as the most powerful computer system available to date.

The physical size of a "minimum configuration" system is probably still a distinguishing feature: A supercomputer or a large-scale system would require a wheelbarrow to move it from one place to another, while a minicomputer can be carried without any mechanical help; a microcomputer can fit easily on an 8- by 12-inches board (or even on one IC) and can be carried in one hand.

With the advances in hardware and software technologies, architectural features found in supercomputers and large-scale machines eventually appear in mini- and microcomputer systems. The tablet PCs and smartphones of today have features of minicomputers a few years ago. As such, the features discussed in this and subsequent chapters of this book are assumed to apply equally well to all classes of computer systems.

Table 8.1   Characteristics of Contemporary Computers

Characteristic	Supercomputer	Large-Scale Computer	Minicomputer	Microcomputer
CPU size	Not easily portable	Not easily portable	2×2 ft (60×60 cm) rack	Small circuit card (or one IC)
Relative cost of minimum system	1000s	100s	10s	1
Typical word size (bits)	64	32–64	16–64	4–32
Typical processor cycle time (μs)	0.01	0.5–0.75	0.2–1	1
Typical application	High-volume scientific	General purpose	Dedicated or general purpose	Dedicated or general purpose

8.2 OPERAND (DATA) TYPES AND FORMATS

The selection of processor word size is influenced by the types and magnitudes of data (or operands) that are expected in the application for the processor. The instruction set is also influenced by the types of operands allowed. Data representation differs from machine to machine. The data format depends on the word size, code used (ASCII or EBCDIC), and the arithmetic (1s or 2s complement) mode employed by the machine. The most common operand or data types are fixed-point (or integer) BCD, floating-point (or real) binary, and character strings. Chapter 2 provided the details of these data types and their representations. Typical representation formats for each of these types are summarized in the succeeding text.

8.2.1 Fixed Point

The fixed-point binary number representation consists of a sign bit (0 for positive, 1 for negative) and $(n - 1)$ magnitude bits for an n-bit machine. Negative numbers are represented either in 2s complement or in 1s complement form, the 2s complement form being the most common. The binary point is assumed to be at the right end of the representation, since the number is an integer.

8.2.2 Decimal Data

Machines that allow decimal (BCD) arithmetic mode use 4 bits (*nibble*) per decimal digit and pack as many digits as possible into the machine word. In some machines, a separate set of instructions is used to operate on such data. In others, the arithmetic mode is changed from BCD to binary (and vice versa) by an instruction provided for that purpose. Thus, the same arithmetic instructions can be used to operate on either type of data. A BCD digit occupies 4 bits, and hence two BCD digits can be packed into a byte.

For instance, IBM 370 allows the decimal numbers to be of variable length (from 1 to 16 digits). The length is specified (or implied) as part of the instruction. Two digits are packed into each byte. Zeros are padded on the most significant end if needed. A typical data format is shown in Figure 8.1. In the packed format shown in Figure 8.1a, the least significant nibble represents the sign, and each magnitude digit is represented by a nibble. In the unpacked form shown in Figure 8.1b, IBM 370 uses 1 byte for each decimal digit: the upper nibble is the zone field and contains 1111, and the lower nibble has the decimal digit. All the arithmetic operations are on packed numbers; the unpacked format is useful for I/O since the IBM 370 uses EBCDIC code (1 byte per character).

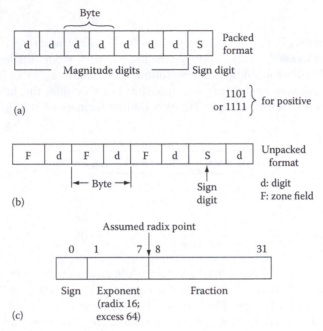

Figure 8.1   IBM 370 data representation. (a) Packed format. (b) Unpacked format. (c) Floating point.

8.2.3 Character Strings

These are represented typically 1 byte per character, using either ASCII or EBCDIC code. The maximum length of the character string is a design parameter of the particular machine.

8.2.4 Floating-Point Numbers

As discussed in Chapter 2, the representation consists of the sign of the number, an exponent field, and a fraction field. In the most commonly used IEEE standard representation, the fraction is represented using either 24 bits (single precision) or 56 bits (double precision).

Figure 8.1c shows the IBM 370 representation where the exponent is a radix-16 exponent and is expressed as an excess-64 number (i.e., a 64 is added to the true exponent so that the exponent representation ranges from 0 through 127 rather than −64 to 63).

Example 8.1

The number $(23.5)_{10}$ is shown in the IBM representation as follows:

$$(23.5)_{10} = (10111.1)_2 = (0.101111) \times (2)^5$$

$$= (0001\ 0111.1)_2$$

$$= 0.(0001\ 0111\ 1)_2 \times (16)^2.$$

0	1000010	0001 0111 1000 0000 0000 0000
Sign	Exponent	Fraction

Note that the exponent is $2 + 64 = 66$; the normalization is done with 4-bit shifts (base-16 exponent) rather than single-bit shifts used in IEEE standard.

8.2.5 Endian

As mentioned earlier in the book, there are two ways of representing a multiple-byte data element in a byte-addressable architecture. In the *little Endian* mode, the least significant byte of the data is stored in the low addressed byte followed by remaining bytes in higher addressed locations. The *big Endian* mode does the opposite. For example, the hexadecimal number 56789ABC requires 4 byte for storage. The two Endian formats of storage are shown in the following:

Byte Address	Little Endian	Big Endian
0	BC	56
1	9A	78
2	78	9A
3	56	BC

It is important to keep the Endian notation intact while accessing data. For instance, if we want to add two 4-byte numbers, the order of addition will be bytes 0, 1, 2, and 3 in the little Endian and bytes 3, 2, 1, and 0 in the big Endian. Thus, little Endian representation is more convenient for addition compared with that of big Endian. On the other hand, since the most significant byte of data (where the sign bit is normally located) is in the first byte in the big Endian representation, it will be convenient to check if the number is positive or negative.

The Intel series of processors use little Endian, while the Motorola processor family uses big Endian. Some Intel processors have instructions that reverse the order of data bytes within a register.

In general, byte ordering is very important when reading data from or writing data to a file. If an application writes data files in little Endian format, a big Endian machine has to reverse the byte order appropriately when utilizing that data, and vice versa. Some popular applications (Adobe Photoshop, JPEG, MacPaint, etc.) use big Endian and others (Windows BMP, GIF, etc.) use little Endian. Applications such as Microsoft WAV and TIF support both formats.

8.2.6 Register versus Memory Storage

ASC is an *accumulator architecture* since the arithmetic and logic instructions implied that one of the operands is in the accumulator. This resulted in shorter instructions and also reduced the complexity of the instruction cycle. But, since the accumulator is the only temporary storage of operands within the processor, memory traffic becomes heavy. Accessing operands from the memory is more time consuming than accessing them from the accumulator. As the hardware technology progressed to VLSI era, it became cost effective to include multiple registers into the processor. In addition, these registers were designated general purpose, allowing them to be used in all operations of the processor as needed, rather than dedicating them as accumulator, index register, etc. Thus, the GPR *architectures* allowed faster access to data stored within the CPU. Also, the register addressing requires much fewer bits compared with memory addresses, making the instructions shorter. With the availability of a large number of registers, some processors have arranged these registers into a stack. As we see later in this section, *stack architectures* allow accessing the data with instructions that contain only an opcode—the operands are implied to be on the top two levels of the stack. Some architectures use register-based stacks, some use memory-based stacks, and some use a combination of the two. In general, it is best to retain as much of the data as possible in CPU registers to allow easy and fast access, since accessing data from the memory is slower.

8.3 REGISTERS

Each of the registers in ASC is designed to be a special-purpose register (accumulator, index register, program counter, etc.). The assignment of functions limits the utility of these registers, but such machines are simpler to design. As the hardware technology has advanced, yielding lower cost hardware, the number of registers in the CPU has increased. The majority of registers are now designated as GPRs and hence can be used as accumulators, index registers, or pointer registers (i.e., registers whose content is an address pointing to a memory location). The *processor status register* is variously referred to as the status register, condition code register, program status word, and so on.

Figure 8.2 shows the register structures of MC6800 and MC68000 series processors. The MC68000 series of processors operate in either *user* or *supervisory* mode. Only user-mode registers common to all the members in the processor series are shown. In addition, these processors have several supervisor mode registers. The type and number of supervisor mode registers vary among the individual processors in the series. For example, the MC68000 has a system SP and uses 5 bits in the most significant byte of the status register, while the MC68020 has seven additional registers.

Figure 8.3 shows the register structures of INTEL 8080, 8086, and 80386 processors. Note that the same set of GPRs (A, B, C, and D) is maintained, except that their width has increased from 8 bits in the 8080 to 32 bits in the 80386. These registers also handle special functions for certain instructions. For instance, C is used as a counter for loop control and to maintain the number of shifts during shift operations, and B is used as a base register. The *source* (SI) and *destination* (DI) index registers are used in string manipulation.

INTEL 8086 series views the memory as consisting of several *segments*. There are four segment registers. The code-segment (CS) register points to the beginning of code (i.e., program) segment, the *data-segment* (DS) register points to the beginning of DS, the *stack segment* (SS) points to the beginning of SS, and the *extra segment* (ES) register points to that of the extra (data) segment. The *instruction pointer* (IP) is the program counter, and its value is a displacement from the address pointed to by CS. We will discuss the base–displacement addressing mode later in this chapter. Similarly, the SP contains a displacement value from the SS contents and points to the top level of the stack. The base pointer (BP) is used to access stack frames that are nontop stack levels. Refer to Section 8.8 for details on the Intel Pentium series of processors.

DEC PDP-11, a popular minicomputer family of the 1960s, could operate in three modes: user, supervisory, and kernel. All instructions were valid in kernel mode, but certain instructions (such as HALT) were not allowed in the other two modes. There were two sets of six registers each (R0-R5): set 0 is for the user mode and set 1 is for the other modes. There were three SPs, one for each mode (R6) and one program counter. All registers could be used in operand (contain data), pointer (contain address), and indexed modes. The processor status word format is shown in Figure 8.4.

DEC VAX-11, a popular minicomputer family of the late 1900s, maintained the general register structure of PDP-11 and contained sixteen 32-bit GPRs, a 32-bit processor status register, four 32-bit registers used with system console operations, four 32-bit clock function and timing registers, and a 32-bit floating-point accelerator control register.

IBM 370 is a 32-bit machine. There are 16 GPRs, each 32 bits long. They can be used as base registers, index registers, or operand registers. There are four 64-bit floating-point registers. The program status word is shown in Figure 8.5.

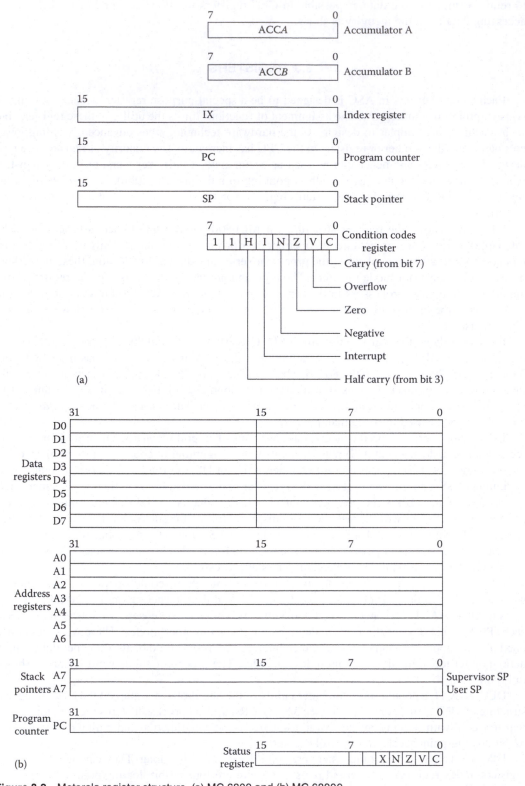

Figure 8.2   Motorola register structure. (a) MC 6800 and (b) MC 68000.

Register pairs B, C, and D, E—each register 8 bits long; 8-bit operands utilize individual register and 16-bit operands utilize register pairs.

Register pair H, L—usually used for address storage.

Stack pointer—16 bits.

Program counter—16 bits.

Accumulator—8 bits.

Butter registers to store ALU operands—two, 8 bits long.

Status flag register—5 bits: Z is zero, C is carry, S is sign, P is parity, and AC is auxiliary carry (for decimal arithmetic). (Also, see Figure 8.12)

(a)

8-bit data registers (AH, AL; BH, BL; CH, CL; DH, and DL). Each pair can be used as a 16 bit register. The pairs are designated AX, BX, CX, and DX.

16-bit registers

SP: Stack pointer BP: Base pointer IP: Instruction pointer (PC)

SI: Source index DI: Destination index

Segment registers: Code (CS), Data (DS), Stack (SS), Extra (ES)

Status register: Containing Overflow, Direction, Interrupt, Trap, Sign, Zero, Auxiliary carry, Parity, Carry flags. (Also, see Figure 8.14)

(b)

32-bit registers

General data and address registers: EAX, EBX, ECX, EDX, ESI, EDI, EBP, and ESP

Instruction pointer: EIP

Flag register: EFLAGS

(All are extended versions of corresponding 8066 registers)

16-bit registers

Segment registers: CS, SS, DS, ES, FS, and GS (the last four are used in data addressing)

(c)

Figure 8.3 Register structures of INTEL microprocessors. (a) 8080. (b) 8086. (c) 80386.

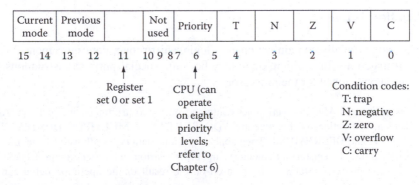

Figure 8.4 PDP-11 processor status word.

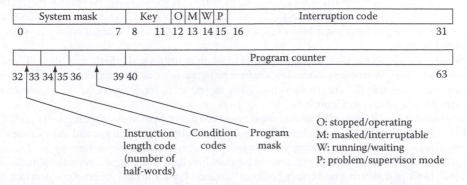

Figure 8.5 IBM 370 processor status word.

8.4 INSTRUCTION SET

The selection of a set of instructions for a machine is influenced heavily by the application for which the machine is intended. If the machine is for general-purpose data processing, then the basic arithmetic and logic operations must be included in the instruction set. Some machines have separate instructions for binary and decimal (BCD) arithmetic, making them suitable for both scientific and business-oriented processing, respectively. Some processors can operate in either binary or decimal mode; that is, the same instruction operates on either type of data. The logical operations AND, OR, NOT, and EXCLUSIVE-OR and the SHIFT and CIRCULATE operations are needed to access the data at bit and byte levels and to implement other arithmetic operations such as multiply and divide and floating-point arithmetic. Control instructions for branching (conditional and unconditional), halt, and subroutine call and return are also required. I/O can be performed using dedicated instructions for I/O or in a memory-mapped I/O mode (where the I/O devices are treated as memory locations and hence all the operations using memory are also applicable to I/O).

In general, the processors can be classified according to the type of operand used by each instruction. The operand is located either in a register or in a memory location. The typical architectures are as follows:

1. Memory to memory: All the operands are in the memory. The operation is performed on these operands, and results left as memory operands, without the need for a register.
2. Register to memory: At least one of the operands is in a register and the rest in the memory.
3. Load/store: The operand has to be loaded into a register before an operation can be performed on it. The result in the register is stored in memory as needed.

8.4.1 Instruction Types

Instruction sets typically contain arithmetic, logic, shift/rotate, data movement, I/O, and program control instructions. In addition, there may be some special-purpose instructions depending on the application envisioned for the processor.

Arithmetic instructions: ASC instruction set contained only two arithmetic instructions (ADD and TCA). Typical instructions of this type are ADD, SUBTRACT, MULTIPLY, DIVIDE, NEGATE, INCREMENT, and DECREMENT. They apply to various data types allowed by the architecture, access data from CPU registers and memory, and utilize various addressing modes. These instructions affect the processor status register based on the result of the operation being zero, carry, overflow, etc.

Logic instructions: These instructions are similar to arithmetic instructions, except that they perform Boolean logic operations such as AND, OR, NOT, EXCLUSIVE-OR, COMPARE, and TEST. They affect processor status register bits. These instructions are used in bit manipulation (i.e., set, clear, complement bits).

Shift/rotate instructions: ASC instruction set had two shift instructions (SHR and SHL). These are *arithmetic shift* instructions since they conformed to the rules of 2s complement system. Typical instruction sets will have instructions that allow logical shift, rotate with carry bit, rotate without carry bit, and so on (see Figure 8.6).

Data movement instructions: ASC instruction set contained four data movement instructions (LDA, STA, LDX, and STX) that moved data between memory and registers (accumulator and index). Instruction sets contain instructions for moving various types and sizes of data between memory locations, registers to memory, and between registers. Some machines have block move instructions that move a whole block of data from one memory buffer to the other. These instructions are equivalent to a move instruction set in a loop. Such high-level instructions are called *macroinstructions*.

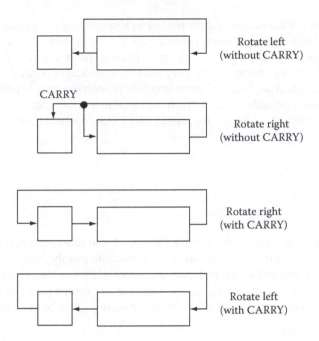

Figure 8.6 Rotate operations.

I/O instructions: ASC provided two I/O instructions (RWD, WWD). In general, an input instruction transfers data from a device (or a port) to a register or memory, and an output instruction transfers data from a register or memory to a device (or a port). The memory-mapped I/O scheme allows the use of all data movement, arithmetic and logic instructions operating with memory operands as I/O instructions. Machines using the isolated I/O scheme have architecture-specific I/O instructions. Block I/O instructions are common to handle array and string type of data.

Control instructions: ASC program control instructions are BRU, BIP, BIN, HLT, TDX, and TIX. Control instructions alter the sequence of program execution through branches (conditional and unconditional), subroutine calls and returns, and halts.

Other instructions: Some instructions do not fit well into the aforementioned types. For instance, an NOP instruction allows the programmer to hold the place in the program. It can be replaced with another instruction later. Although an NOP does not perform any useful function, it consumes time. NOPs are used to adjust the execution time of the program to accommodate timing requirements. There might be instructions specific to the application at hand–string or image processing, HLL support, and so on.

8.4.2 Instruction Length

The length of an instruction is a function of the number of operands in the instruction. Typically, a certain number of bits are needed for the opcode; register references do not require a large number of bits, while memory references consume the major portion of an instruction. Hence, memory-reference and memory-to-memory instructions will be longer than the other types of instructions. A variable-length format can be used with these to conserve the amount of memory occupied by the program, but this increases the complexity of the control unit.

The number of memory addresses in an instruction dictates the speed of execution of instruction, in addition to increasing the instruction length. One memory access is needed to fetch the instruction from the memory if the complete instruction is in one memory word. If the instruction is longer than one word, multiple memory accesses are needed to fetch it unless the memory architecture provides for the access of multiple words simultaneously (as described in Chapter 9). Index and indirect address computations are needed for each memory operand in the instruction, thereby adding to the instruction-processing time. During the instruction execution phase, corresponding to each memory operand, a memory read or write access is needed. Since the memory access is slower than a register transfer, the instruction-processing time is considerably reduced if the number of memory operands in the instruction is kept to a minimum.

Based on the number of addresses (operands), the following instruction organizations can be envisioned:

1. Three address
2. Two address
3. One address
4. Zero address

We now compare the earlier organizations using the typical instruction set required to accomplish the four basic arithmetic operations. In this comparison, we assume that the instruction fetch requires one memory access in all cases and all the addresses are direct addresses (i.e., no indexing or indirecting needed). Hence, the address computation does not require any memory accesses. Therefore, we will compare only the memory accesses needed during the execution phase of the instruction:

Three-address machine:

ADD	A, B, C	$M[C] \leftarrow M[A] + M[B]$
SUB	A, B, C	$M[C] \leftarrow M[A] - M[B]$
MPY	A, B, C	$M[C] \leftarrow M[A] \cdot M[B]$
DIV	A, B, C	$M[C] \leftarrow M[A] / M[B]$

A, B, and C are memory locations. Each of the earlier instructions requires three memory accesses during execution. In practice, the majority of operations are based on two operands with the result occupying the position of one of the operands. Thus, the instruction length and address computation time can be reduced by using a two-address format, although three memory accesses are still needed during the execution:

Two-address machine:

ADD	A, B	$M[A] \leftarrow M[A] + M[B]$
SUB	A, B	$M[A] \leftarrow M[A] - M[B]$
MPY	A, B	$M[A] \leftarrow M[A] \cdot M[B]$
DIV	A, B	$M[A] \leftarrow M[A] / M[B]$

The first operand is lost after the operation. If one of the operands can be retained in a register, the execution speeds of the earlier instructions can be increased. Further, if the operand register is implied by the opcode, a second operand field is not required in the instruction, thus reducing the instruction length:

One-address machine:

ADD	A	$ACC \leftarrow ACC + M[A]$
SUB	A	$ACC \leftarrow ACC - M[A]$
MPY	A	$ACC \leftarrow ACC \cdot M[A]$
DIV	A	$ACC \leftarrow ACC / M[A]$
LOAD	A	$ACC \leftarrow M[A]$
STORE	A	$M[A] \leftarrow ACC$

Here, ACC is an accumulator or any other register implied by the instruction. Load and store instructions are needed. If both the operands can be held in registers, the execution time is decreased. Further, if the opcode implies the two registers, instructions can be of zero-address type:

Zero-address machine:

ADD		SL ← SL + TL, POP
SUB		SL ← SL − TL, POP
MPY		SL ← SL · TL, POP
DIV		SL ← SL / TL, POP
LOAD	A	PUSH, TL ← M[A]
STORE	A	M[A] ← TL, POP

Here, SL and TL are, respectively, the second and top levels of a last-in, first-out stack. In a zero-address machine, all the arithmetic operations are performed on the top two levels of a stack, so no explicit addresses are needed as part of the instruction. However, some memory-reference (one-address) instructions such as LOAD and STORE (or MOVE) to move the data between the stack and the memory are required.

Appendix B describes the two most popular implementations of stack. If the zero-address machine uses a hardwired stack, then the previously mentioned arithmetic instructions do not require any memory access during execution; if a RAM-based stack is used, each arithmetic instruction requires three memory accesses.

Assuming n bits for an address representation and m bits for the opcode, the instruction lengths in the previous four organizations are as follows:

Three address: $m + 3n$ bits
Two address: $m + 2n$ bits
One address: $m + n$ bits
Zero address: m bits

Table 8.2 offers programs to compute the function $F = A \cdot B + C \cdot D$ using each of the previously mentioned instruction sets. Here, A, B, C, D, and F are memory locations. Contents of A, B, C, and D are assumed to be integer values, and the results are assumed to fit in one memory word. Program sizes can be easily computed from benchmark programs of this type, using the typical set of operations performed in an application environment. The number of memory accesses needed for each program is also shown in the figure. Although these numbers provide a measure of relative execution times, the time required for other (nonmemory access) operations should be added to this time to complete the execution time analysis. Results of such benchmark studies are used in the selection of instruction sets and instruction formats and comparison of processor architectures.

For example, in IBM 370, the instructions are 2, 4, or 6 bytes long: DEC PDP-11 employs an innovative addressing scheme to represent both single and double operand instructions in 2 bytes. An instruction in INTEL 8080 is 1, 2, or 3 bytes long. The instruction formats are discussed later in this section.

8.4.3 Opcode Selection

Assignment of opcode for instructions in an instruction set can significantly influence the efficiency of the decoding process at the execute phase of the instruction cycle. (ASC opcodes have been arbitrarily assigned.) Two opcode assignments are generally followed:

1. Reserved opcode method
2. Class code method

Table 8.2   Programs to Compute F = A · B + C · D

Program	Program Length (bits)	Number of Memory Accesses	
		FETCH	EXECUTE
Three-address			
MPY A, B, A			
MPY C, D, C	$3(m + 3n)$	3	$3 \cdot 3 = 9$
ADD A, C, F			
Two-address			
MPY A, B			
MPY C, D	$5(m + 2n)$	5	$5 \cdot 3 = 15$
ADD A, C			
SUB F, F			
ADD F, A			
One-address			
LOAD A			
MPY B			
STORE F			
LOAD C	$7(m + n)$	7	$7 \cdot 1 = 7$
MPY D			
ADD F			
STORE F			
Zero-address			
LOAD A			
LOAD B			
MPY			
LOAD C	$3m + 5(m + n)$	8	$5 \cdot 1 = 5$
LOAD D			(Assuming hardware stack)
MPY			
ADD			
STORE F			

In the reserved opcode method, each instruction would have its own opcode. This method is suitable for instruction sets with fewer instructions. In the class code method, the opcode consists of two parts: a class code part and an operation part. The class code identifies the type or class of instruction, and the remaining bits in the opcode (operation part) identify a particular operation in that class. This method is suitable for larger instruction sets and for instruction sets with variable instruction lengths. Class codes provide a convenient means of distinguishing between various classes of instructions in the instruction set. The two opcode assignment modes are illustrated here:

Reserved opcode instruction:

Opcode	Address(es)

Class code instruction:

Class code	Operation	Address(es)

In practice, it may not be possible to identify a class code pattern in an instruction set. When the instructions are of fixed length and the bits of the opcode are always completely decoded, there may not be any advantage to assigning opcodes in a class code form. For example, the INTEL 8080 instruction set does not exhibit a class code form; in IBM 370, the first 2 bits of the opcode distinguish between 2-, 4-, and 6-byte instructions. Mostek 6502, an 8-bit microprocessor, used a class code in the sense that part of the opcode distinguishes between the allowed addressing modes of the same instruction. For example, the "add memory to accumulator with carry (ACC)" instruction has the following opcode variations:

Addressing Mode	Opcode (H)
Immediate mode	69
Zero page	65
Zero page, index X	75
Nonzero page	6D
Nonzero page, index X	7D
Nonzero page, index Y	79
Preindexed indirect	61
Postindexed indirect	71

We will describe the paged addressing mode later in this chapter.

8.4.4 Instruction Formats

The majority of machines use a fixed-field format within the instruction while varying the length of the instruction to accommodate varying number of addresses. Some machines vary even the field format within an instruction. In such cases, a specific field of the instruction defines the instruction format.

Figure 8.7 shows the instruction formats of the Motorola 6800 and 68000 series of machines. The MC6800 is an 8-bit machine and can directly address 64 kB of memory. Instructions are 1, 2, or 3 bytes long. Members of the MC68000 series of processors have an internal 32-bit (data and address) architecture. The MC68000 provides a 16-bit data bus and a 24-bit address bus, while the top-of-the-line MC68020 provides 32-bit data and 32-bit address buses. The instruction format varies from register-mode instructions that are one (16-bit) word long to memory-reference instructions that can be two to five words long.

Figure 8.8 shows the instruction formats for DEC PDP-11 and VAX-11 series. PDP-11 was a 16-bit processor series. In its instruction format, R indicates one of the eight GPRs, I is the direct/indirect flag, and M indicates the mode in which the register is used. The registers can be used in operand, autoincrement pointer, autodecrement pointer, or index modes. We will describe these modes later in this chapter. The VAX-11 series of 32-bit processors have a much generalized instruction format. An instruction consists of the opcode (1 or 2 bytes long) followed by from 0 to 6 operand specifiers. An operand specifier is between 1 and 8 bytes long. Thus, a VAX-11 instruction can be from 1 to 50 bytes long.

Figure 8.9 shows the instruction formats of IBM 370, a 32-bit processor whose instructions can be from 2 to 6 bytes long. Here, R1, R2, and R3 refer to one of the 16 (GPRs); B1 and B2 are base-register designations (one of the GPRs); DATA indicates immediate data; D1 and D2 refer to a 12-bit displacement; and L1 and L2 indicate the length of data in bytes. The base–displacement addressing mode is described later in this chapter.

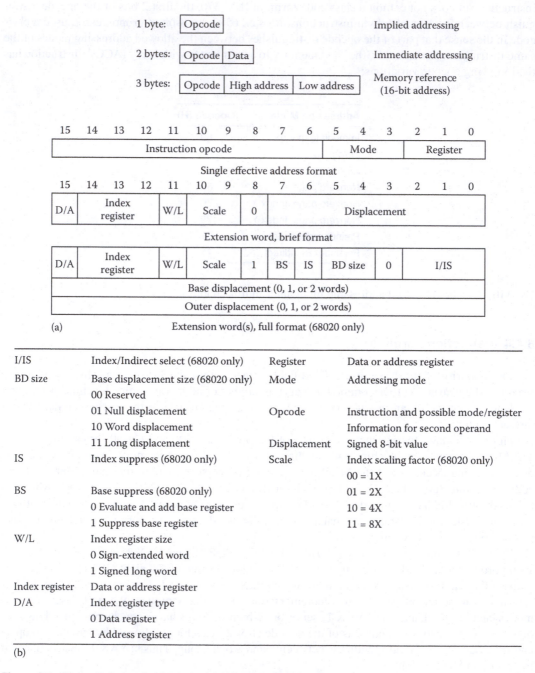

Figure 8.7 Motorola instruction formats. (a) MC 6800. (b) MC 88000.

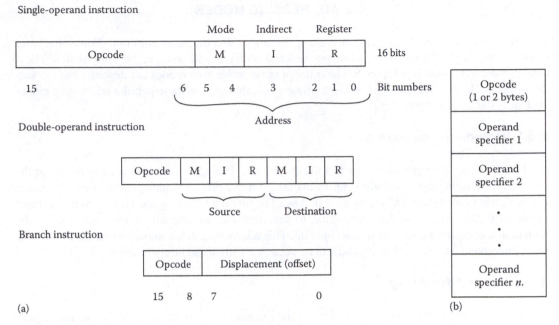

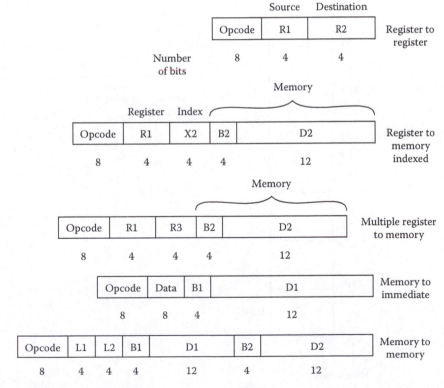

Figure 8.8 DEC instruction formats. (a) PDP-11. (b) VAX-11.

Figure 8.9 IBM 370 instruction formats.

8.5 ADDRESSING MODES

The direct, indirect, and indexed addressing modes are the most common. Some machines allow preindexed indirect, some allow postindexed indirect, and some allow both modes. These modes were described in Chapter 5. The other popular addressing modes are described in the succeeding text. In practice, processors adopt whatever combination of these popular addressing modes that suits the architecture.

8.5.1 Immediate Addressing

In this mode, the operand is a part of the instruction. The address field is used to represent the operand itself rather than the address of the operand. From a programmer's point of view, this mode is equivalent to the literal addressing used in ASC. The literal addressing was converted into a direct address by the ASC assembler. In practice, to accommodate an immediate addressing mode, the instruction contains a data field and an opcode. This addressing mode makes operations with constant operands faster since the operand can be accessed without additional memory fetch.

8.5.2 Paged Addressing

When this mode of addressing is allowed, the memory is assumed to consist of several equal-sized blocks, or *pages*. The memory address can then be treated as having a page number and a location address (offset) within the page. Figure 8.10 shows a paged memory containing 256 pages,

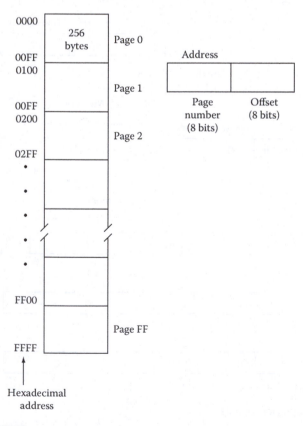

Figure 8.10 Paged memory.

with each page 256 bytes long. The instruction format in such an environment will have two fields to represent a memory address: a *page field* (8 bits long in this case) and an *offset* (with enough bits to address all the locations within a page—8 bits in this case). Address modifiers (to indicate indexing, indirect, etc.) are also included in the instruction format.

If the pages are large enough, the majority of the memory references by the program will be within the same page, and those locations can be addressed with just the offset field, thus saving the number of bits in the address field of the instruction, compared to the case of the direct addressing scheme. If the referenced location is beyond the page, page number is needed to access it.

Note that the page number can be maintained in a register; thus, the instruction contains only the offset part of the address and a reference to the register containing the page number. The segment registers of INTEL 8086 essentially serve this purpose. The base-register addressing mode described next is a variation of paged addressing mode.

Usually, the "zero page" of the memory is used for the storage of the most often used data and pointers. A page bit in the instruction can be used to specify whether the address corresponds to the zero page or the "current page" (i.e., the page in which the instruction is located). Alternatively, the opcode might imply a zero-page operand. Mostek 6502 uses such an addressing scheme. The 16-bit address is divided into an 8-bit page address and an 8-bit offset within the page (of 256 bytes). Further, the processor has unique opcodes for zero-page mode instructions. These instructions are assembled into 2 bytes (implying that the higher byte of the address is zero), while the nonzero-page instructions need 3 bytes to represent the opcode and a 16-bit address.

Paged memory schemes are useful in organizing virtual memory systems, as described in Chapter 9.

8.5.3 Base-Register Addressing

In some machines, one of the CPU registers is used as a *base register*. The beginning address of the program is loaded into this register as a first step in program execution. Each address referenced in the program is an offset (displacement) with respect to the contents of the base register. Only the base-register identification and the displacement are represented in the instruction format, thus conserving bits. Since the set of instructions to load the base register is part of the program, the relocation of the programs is automatic. IBM 370 series machines use this scheme; any of the GPRs can be designated as a base register by the programmer. The segment mode of addressing used by INTEL 8086 series is another example of base-register addressing.

8.5.4 Relative Addressing

In this mode of addressing, an offset (or displacement) is provided as part of the instruction. This offset is added to the current value of PC during execution to find an effective address. The offset can be either positive or negative. Such addressing is usually used for branch instructions. Since the jumps are usually within a few locations of the current address, the offset can be a small number compared with those of the actual jump address, thus reducing the bits in the instruction. DEC PDP-11 and Mostek 6502 use such an addressing scheme.

8.5.5 Implied (Implicit) Addressing

In this mode of addressing, the opcode implies the operand(s). For instance, in zero-address instructions, the opcode implies that the operands are on the top two levels of the stack; in one-address instructions, one of the operands is implied to be in a register similar to the accumulator.

The implementation of the addressing modes varies from processor to processor. Addressing modes of some practical machines are listed in the succeeding text.

M6502: M6502 allows direct, preindexed-indirect, and postindexed-indirect addressing. Zero-page addressing is employed to reduce the instruction length by 1 byte. The addresses are 2 bytes long. A typical absolute address instruction uses 3 bytes. A zero-page address instruction uses only 2 bytes.

MC6800: The addressing modes on MC6800 are similar to those of M6502 except that the indirect addressing mode is not allowed.

MC68000: This processor allows 14 fundamental addressing modes that can be generalized into the six categories shown in Table 8.3.

INTEL 8080: The majority of the memory references for INTEL 8080 are based on the contents of register pair H, L. These registers are loaded, incremented, and decremented under program control. The instructions are thus only 1 byte long and imply indirect addressing through the H, L register pair. Some instructions have the 2-byte address as part of them (direct addressing). There are also single-byte instructions operating in an implied addressing mode.

INTEL 8086: The segment-register-based addressing scheme of this series of processors is a major architectural change from the 8080 series. This addressing scheme makes the architecture flexible, especially for the development of compilers and operating systems.

DEC PDP-11: The two instruction formats of PDP-11 shown earlier in this chapter provide a versatile addressing capability. Mode (M) bits allow the contents of the referenced register to be (1) an operand, (2) a pointer to a memory location that is incremented (autoincrement) or decremented (autodecrement) automatically after accessing that memory location, or (3) an index. In the index mode, the instructions are two words long. The content of the second word is an address and is indexed by the referenced register. The indirect bit (I) allows direct and indirect addressing in all

Table 8.3   MC68000 Addressing Modes

Mode	Effective Address
Register direct addressing	
Address register direct	EA = An
Data register direct	EA = Dn
Absolute data addressing	
Absolute short	EA = {Next word}
Absolute long	EA = {Long words}
Program counter relative addressing	
Relative with offset	EA = {PC} + d16
Relative with index and offset	EA = {PC} + {Rn} + d8
Register indirect addressing	
Register indirect	EA = {An}
Postincrement register indirect	EA − {An}. An ← An + N
Predecrement register indirect	An ← An = N, EA = {An}
Register indirect with offset	EA = {An} + d16
Indexed register indirect with offset	EA = {An} + {Rn} + d8
Immediate data addressing	
Immediate	Data = Next word(s)
Quick immediate	Inherent data
Implicit addressing	
Implicit register	EA = SR, USP, SP, PC

Notes: EA, effective address; {}, contents of; Dn, data register; An, address register; dn, n-bit offset; Rn, data or address register.

Table 8.4 PDP-11 Instruction Examples

Mnemonic	Octal Opcode	Comments
CLR	0050 *nn*	Clear word, *nn* is the register reference
CLRB	1050 *nn*	Clear byte
ADD	06 *nn mm*	Add, *nn* = source, *mm* = destination
ADD R2, R4	06 02 04	R4 ← R2 + R4
CLR (R5)	0050 25	Autoincrement pointer, clear R5 ← 0, R5 ← R5 + 1
ADD @X(R2), R1	06 72 01 Address	@ indicates indirect; X indicates indexing R1 ← M[M[Address + R2]] + R1 immediate
ADD #10, R0	062700 000010	R0 ← 10
INC Offset	005267 Offset	M[PC + Offset] ← M[PC + Offset] + 1

the four register modes. In addition, PDP-11 allows PC-relative addressing, in which an offset is provided as part of the instruction and added to the current PC value to find the effective address of the operand. Some examples are shown in Table 8.4.

DEC VAX-11: The basic addressing philosophy of VAX-11 follows that of PDP-11 except that the addressing modes are much more generalized, as implied by the instruction format shown earlier in this chapter.

IBM 370: Any of the 16 GPRs can be used as an index register, an operand register, or a base register. The 12-bit displacement field allows 4 kB of displacement from the contents of the base register. A new base register is needed if the reference exceeds the 4 kB range. Immediate addressing is allowed. In decimal arithmetic where the memory-to-memory operations are used (6-byte instructions), the lengths of the operands are also specified, thus allowing variable-length operands. Indirect addressing is not allowed.

8.6 INSTRUCTION-SET ORTHOGONALITY

Instruction-set orthogonality is defined by two characteristics: *independence* and *consistency*. An independent instruction set does not contain any redundant instructions. That is, each instruction performs a unique function and does not duplicate the function of another instruction. Also, the opcode/operand relationship is independent and consistent in the sense that any operand can be used with any opcode. Ideally, all operands can equally well be utilized with all the opcodes, and all addressing modes can be consistently used with all operands. Basically, the uniformity offered by an orthogonal instruction set makes the task of compiler development easier. The instruction set should be complete while maintaining a high degree of orthogonality.

8.7 RISC VERSUS CISC

The advent of VLSI provided the capability to fabricate a complete processor on an IC chip. In such ICs, the control unit typically occupied a large portion (of the order of 50%–60%) of the chip area, thereby restricting the number of processor functions that could be implemented in hardware. A solution to this problem was the design of processors with simpler control units. Reduced instruction-set computers (RISCs) enable the use of a simple control unit since their instruction sets tend to be small. Simplification of the control unit was the main aim of the RISC designs first initiated at IBM in 1975 and later, in 1980, at the University of California, Berkeley. The characteristics and the definition of RISC have changed considerably since those early designs. There is currently no single accepted definition of RISC.

One of the motivations for the creation of RISC was the criticism of certain problems inherent in the design of the established complex instruction-set computers (CISCs). A CISC has a relatively large number of complicated, time-consuming instructions. It also has a relatively large number of addressing modes and different instruction formats. These in turn result in the necessity of having a complex control unit to decode and execute these instructions. Some of the instructions may be so complex that they are not necessarily faster than a sequence of several RISC instructions that could replace them and perform the same function.

The complexity of the CISC control unit hardware implies a longer design time. It also increases the probability of a larger number of design errors. And those errors are subsequently difficult, time consuming, and costly to locate and correct.

To provide a perspective, let us look at an example of a CISC. One such system is the DEC VAX-11/780 with its 304 instructions and 16 addressing modes. It has sixteen 32-bit registers. The VAX supports a considerable number of data types, among which are six types of integers, four types of floating points, packed decimal strings, character strings, variable-length bit fields, numeric strings, and more. An instruction can be as short as 2 bytes and as long as 14 bytes. There may be up to six operand specifiers. As another example, Motorola MC68020 a 32-bit micropro-cessor is a CISC. It also has 16 general-purpose CPU registers, recognizes seven data types, and implements 18 addressing modes. The MC68020 instructions can be from one word (16 bits) to 11 words in length. These and many other CISCs have a great variety of data and instruction formats, addressing modes, and instructions. A direct consequence of this variety is the highly complex control unit. For instance, even in the less sophisticated MC68000, the control unit takes up over 60% of the chip area.

Naturally, an RISC-type system is expected to have fewer than 304 instructions. There is no exact consensus about the number of instructions that an RISC system should have. The Berkeley RISC I had 31, the Stanford MIPS had over 60, and the IBM 801 had over 100. Although the RISC design is supposed to minimize the instruction count, that in itself is not the definitive characteristic of an RISC. The instruction environment is also simplified, by reducing the number of addressing modes, by reducing the number of instruction formats, and by simplifying the design of the control unit. On the basis of the RISC designs reported so far, we can look at the following list of RISC attributes as an informal definition of an RISC:

1. A relatively low number of instructions.
2. A low number of addressing modes.
3. A low number of instruction formats.
4. Single-cycle execution of all instructions.
5. Memory access performed by load/store instructions only.
6. The CPU has a relatively large register set. Ideally, most operations can be done register to register. Memory access operations are minimized.
7. An HCU (may be changed to microprogrammed as the technology develops).
8. An effort is made to support (HLL) operations inherently in the machine design by using a judicious choice of instructions and optimized (with respect to the large CPU register set) compilers.

It should be stressed that these attributes are to be considered as a flexible framework for a definition of an RISC machine, rather than a list of design attributes common to most of the

RISC systems. The boundaries between RISC and CISC have not been rigidly fixed. Still, these attributes give at least an idea of what to expect in an RISC system.

RISC machines are used for many high-performance applications. Embedded controllers are a good example. They can also be used as building blocks for more complex multiprocessing systems. It is precisely its speed and simplicity that make an RISC appropriate for such applications.

A large instruction set presents too large a selection of choices for the compiler of any HLL. This in turn makes it more difficult to design the optimizing stage of a CISC compiler. Furthermore, the results of this "optimization" may not always yield the most efficient and fastest machine language code.

Some CISC instruction sets contain a number of instructions that are particularly specialized to fit certain HLL instructions. However, a machine language instruction that fits one HLL may be redundant for another and would require excessive effort on the part of the designer. Such a machine may have a relatively low cost–benefit factor.

Since an RISC has relatively few instructions, few addressing modes, and a few instruction formats, a relatively small and simple (compared with CISC) decode and execute hardware sub-system of the control unit is required. The chip area of the control unit is considerably reduced. For example, the control area of RISC I constitutes 6% of the chip area; on RISC II, 10%; and the Motorola MC68020, 68%. In general, the control area for CISCs takes over 50% of the chip area. Therefore, on an RISC VLSI chip, there is more area available for other features. As a result of the considerable reduction of the control area, the RISC designer can fit a large number of CPU registers on the chip. This in turn enhances the throughput and the HLL support.

The control unit of an RISC is simpler, occupies a smaller chip area, and provides faster execution of instructions. A small instruction set and a small number of addressing modes and instruction formats imply a faster decoding process. A large number of CPU registers permit programming that reduces memory accesses. Since CPU register-to-register operations are much faster than memory accesses, the overall speed is increased. Since all instructions have the same length and all execute in one cycle, we obtain a streamlined instruction handling that is particularly well suited to pipe-lined implementation.

Since the total number of instructions in an RISC system is small, a compiler (for any HLL), while attempting to realize a certain operation in machine language, usually has only a single choice, as opposed to a possibility of several choices in a CISC. This makes that part of the compiler shorter and simpler for an RISC.

The availability of a relatively large number of CPU registers in an RISC permits a more efficient code optimization stage in a compiler by maximizing the number of faster register-to-register operations and minimizing the number of slower memory accesses.

All in all, an RISC instruction set presents a reduced burden on the compiler writer. This tends to reduce the time of design of RISC compilers and their costs. Since an RISC has a small number of instructions, a number of functions performed by some instructions of a CISC may need two, three, or more instructions of an RISC. This causes the RISC code to be longer and constitutes an extra burden on the machine and assembly language programmer. The longer RISC programs consequently require more memory locations for their storage.

It is argued that RISC performance is due primarily to extensive register sets and not to the RISC principle in general. Again, from the VLSI standpoint, it is precisely due to the RISC principles, which permitted a significant reduction of the control area, that the RISC designers were able to fit so many CPU registers on a chip in the first place.

It has been argued that RISCs may not be efficient with respect to operating systems and other auxiliary operations. The efficiency of RISCs with respect to certain compilers and benchmarks has already been established. There is no substantiated reason to doubt that efficient operating systems and utility routines can be generated for an RISC.

A number of RISC processors have been introduced with various performance rating, register and bus structures, and instruction-set characteristics. The reader is referred to the magazines listed in the reference section for further details.

8.8 EXAMPLE SYSTEMS

Practical CPUs are organized around either a single-bus or multiple-bus structure. As discussed earlier, the single-bus structure offers simplicity of hardware and uniform interconnections at the expense of speed, while the multiple-bus structures allow simultaneous operations on the structure but require complex hardware. Selection of the bus structure is thus a compromise between hardware complexity and speed. The majority of modern-day microprocessors are organized around single-bus structures since implementation of a bus consumes a great deal of the silicon available on an IC.

The majority of CPUs use parallel buses. If faster operating speeds are not required, serial buses can be employed, as is usually the case with the processor ICs used to build calculators. The design goal in fabricating a calculator IC is to pack as many functions as possible into one IC. By using serial buses, silicon area on the IC is conserved and used for implementing other complex functions.

Figure 8.11 shows a comparison of three-, two-, and single-bus schemes for ASC. Note that the ADD operation can be performed in one cycle using the three-bus structure. In the two-bus structure, the contents of the ACC are first brought into the BUFFER register in the ALU in one cycle, and the results are gated into ACC in the second cycle, thus requiring two cycles for addition. In the single-bus structure, the operands are transferred into ALU buffer registers BUFFER1 and BUFFER2 one at a time and the result is then transferred into the ACC during the third cycle.

We will now provide brief descriptions of basic architectural features of representative processors from the Intel family (8080, 8086, and Pentium). We have not yet described the advanced architectural features used by these machines. As such, this section needs a revisit after learning those concepts from subsequent chapters. We will also describe more recent processors of Intel and other families in subsequent chapters of this book. The following material is extracted from Intel hardware and software developer manuals.

The 4004 microprocessor was the first designed by Intel in 1969, followed by the 8080 and the 8085. The Pentium II architecture is based on 8086 introduced in 1978. The 8086 had a 20-bit address and could reference 1 MB of memory, 16-bit registers, and a 16-bit external data bus. The sister processor, the 8088, had an 8-bit bus. Memory was divided into 64 kB segments, and four 16-bit segment registers were used as pointers to addresses of the active segments in the so-called real mode.

8.8.1 Intel 8080

The 8080 uses an 8-bit internal data bus for all register transfers (refer to Figure 8.12). The arithmetic unit operates on two 8-bit operands residing in the temporary register (TEMP REG) and the accumulator latch. The CPU provides an 8-bit bidirectional data bus, a 16-bit address bus, and a set of control signals. These are used in building a complete microcomputer system. This is single-bus architecture from the point of view of its ALU. The temporary register and the accumulator latch (internal to the ALU) are used for storing the second operand during operations requiring two operands. These internal buffers are transparent to programmers.

The 8080 has a synchronous HCU operating with a major cycle (or *machine cycle* in Intel terminology) consisting of three, four, or five minor cycles (or *state* in Intel terminology). An instruction cycle consists on one to five major cycles, depending on the type of instruction. The machine cycles are controlled by a two-phase nonoverlapping clock. "SYNC" identifies the beginning of a machine cycle. There are 10 types of machine cycles: fetch, memory read, memory write, stack read, stack write, input, output, interrupt acknowledge, halt acknowledge, and interrupt acknowledge while halt.

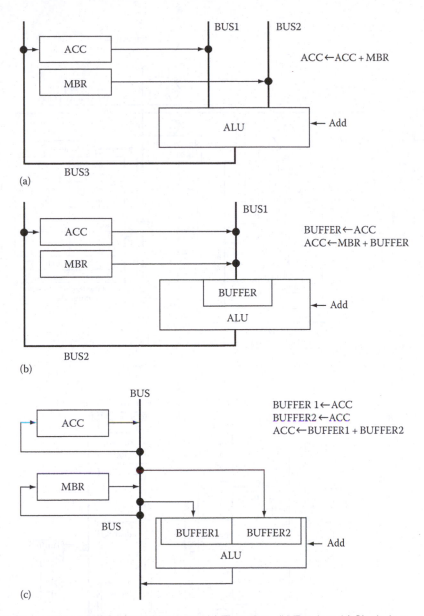

Figure 8.11 Comparison of ASC bus architectures. (a) Three-bus. (b) Two-bus. (c) Single-bus.

Instructions in the 8080 contain 1–3 bytes. Each instruction requires from one to five machine or memory cycles for fetching and execution. Machine cycles are called M1, M2, M3, M4, and M5. Each machine cycle requires from three to five states, T1, T2, T3, T4, and T5, for its completion. Each state has the duration of one clock period (0.5 μs). There are three other states (WAIT, HOLD, and HALT), which last from one to an indefinite number of clock periods, as controlled by external signals. Machine cycle M1 is always operation code fetch cycle and lasts four or five clock periods. Machine cycles M2, M3, M4, and M5 normally last three clock periods each.

To further understand the basic timing operation of the INTEL 8080, refer to the instruction cycle shown in Figure 8.13. During T1, the content of the program counter is sent to the address bus, SYNC is true, and the data bus contains the status information pertaining to the cycle that

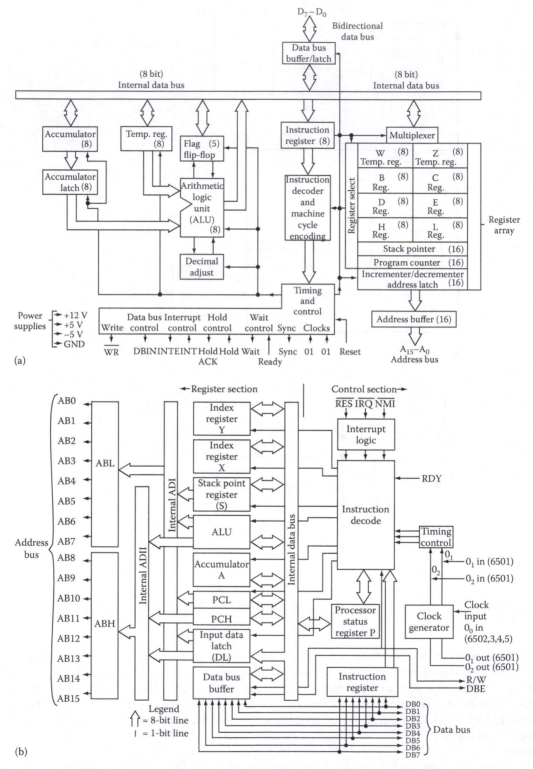

Figure 8.12 INTEL 8080 processor. (a) Functional Block diagram. (b) Internal architecture. (Reprinted with permission from *Intel 8080 Microcomputer Systems Users Manual*, Santa Clara, CA, Intel Corporation, 1977. Copyright 1977. All mnemonics copyright Intel Corporation 1977.)

is currently being initiated. T1 is always followed by another state, T2, during which the condition of the READ, HOLD, and HALT acknowledge signals is tested. If READY is true, T3 can be entered; otherwise, the CPU will go into the WAIT state (TW) and stays there for as long as READY is false. READY thus allows the CPU speed to be synchronized to a memory with any access time or to any input device. The user, by properly controlling the READY line, can single-step through a program.

During T3, the data coming from memory are available on the data bus and are transferred into the instruction register (during M1 only). The instruction decoder and control section then generate the basic signals to control the internal data transfer, the timing, and the machine cycle requirements of the instruction.

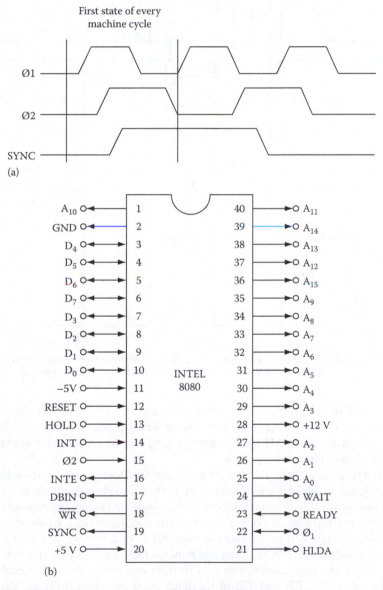

Figure 8.13 INTEL 8080. (a) Nonoverlapping clock. (b) Pinout diagram.

(continued)

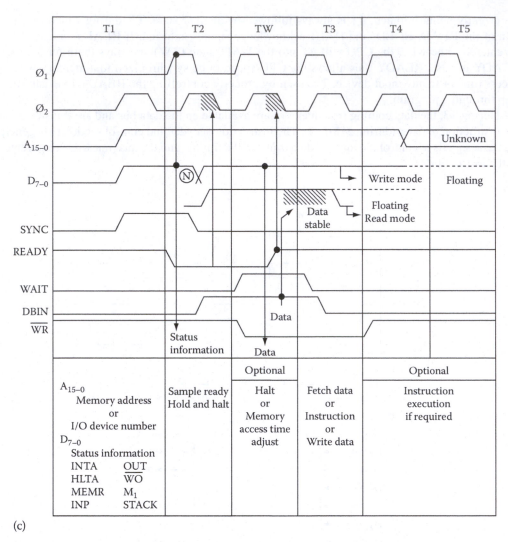

(c)

Figure 8.13 (continued)   INTEL 8080. (c) Instruction cycle. (Reprinted with permission from Intel Corporation. Copyright 1981. All mnemonics copyright Intel Corporation 1981.)

At the end of T4 (if the cycle is complete) or at the end of T5 (if it is not), the 8080 goes back to T1 and enters machine cycle M2, unless the instruction requires only one machine cycle for its execution. In such cases, a new M1 cycle is entered. The loop is repeated for as many cycles and states as may be required by the instruction.

Instruction-state requirements range from a minimum of four states for nonmemory referencing instructions (such as register and accumulator arithmetic instructions) up to 18 states for the most complex instructions (such as instructions to exchange the contents of registers H and L with the contents of the two top locations of the stack). At the maximum clock frequency of 2 MHz, this means that all instructions will be executed in intervals from 2 to 9 μs. If a HALT instruction is executed, the processor enters a WAIT state and remains there until an interrupt is received.

Table 8.5 shows the μop sequence of two instructions (one column for each instruction). The first three states (T1, T2, and T3) of the fetch machine cycle (M1) are the same for all instructions.

Table 8.5   INTEL 8080 Microinstructions

Mnemonic		JMP adrs	CALL adrs
Opcode ($D_7 - D_0$)		1100 0011	1100 1101
M1	T4	X	SP ← SP − 1
	T5		
M2	T1	PC OUT, STATUS	PC OUT, STATUS
	T2	PC ← PC + 1	PC ← PC + 1
	T3	Z ← B2	Z ← B2
M3	T1	PC OUT, STATUS	PC OUT, STATUS
	T2	PC ← PC + 1	PC ← PC + 1
	T3	W ← B3	W ← B3
M4	T1		SP OUT, STATUS
	T2		Data bus ← PCH
	T3		SP ← SP − 1
M5	T1		SP OUT, STATUS
	T2		Data bus ← PCL
	T3		
	T4		
	T5		
During the fetch of next instruction		WZ OUT, STATUS PC ← WZ + 1	WZ OUT, STATUS PC ← WZ + 1

Source: Reprinted with permission from Intel Corporation. Copyright 1981. All mnemonics copyright Intel Corporation 1981.

8.8.2 Intel 8086

The 8086 processor structure shown in Figure 8.14 consists of two parts: the execution unit (EU) and the bus interface unit (BIU). The EU is configured around a 16-bit ALU data bus. The BIU interfaces the processor with the external memory and peripheral devices via a 20-bit address bus and a 16-bit data bus. The EU and the BIU are connected via the 16-bit internal data bus and a 6-bit control bus (Q bus). EU and BIU are two independent units that work concurrently, thus increasing the instruction-processing speed. The BIU generates the instruction address and transfers instructions from the memory into an instruction buffer (instruction queue). This buffer can hold up to six instructions. The EU fetches instructions from the buffer and executes them. As long as the instructions are fetched from sequential locations, the instruction queue remains filled and the EU and BIU operations remain smooth. When the EU refers to an instruction address beyond the range of the instruction queue (as may happen during a jump instruction), the instruction queue needs to be refilled from the new address. The EU has to wait until the instruction is brought into the queue. Instruction buffers are further described in Chapter 9.

The Intel 80286 introduced the protected mode, using the segment register as a pointer into address tables. The address table provided 24-bit addresses so that 16 MB of physical memory could be addressed and also provided support for virtual memory.

The Intel 80386 included 32-bit registers to be used for operands and for addressing. The lower 16 bits of the new registers were used for compatibility with software written for earlier versions of the processor. The 32-bit addressing and 32-bit address bus allowed each segment to be as large as 4 GB. 32-bit operand and addressing instructions, bit manipulation instructions, and paging were introduced in the 80386. The 80386 had six parallel stages for the bus interface, code prefetching,

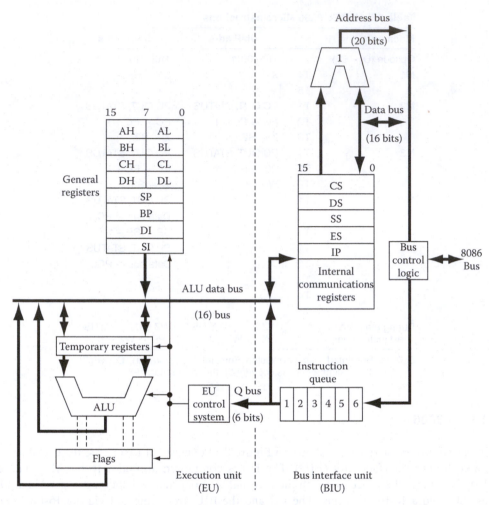

Figure 8.14 INTEL 8086 elementary block diagram. (Reprinted with permission from Intel Corporation. Copyright 1981. All mnemonics copyright Intel Corporation 1981.)

instruction decoding, execution, segment addressing, and memory paging. The 80486 processor added more parallel execution capability by expanding the 80386 processor's instruction decode and EUs into five pipelined stages. Chapter 10 introduces pipelining. Each stage could do its work on one instruction in one clock executing one instruction per CPU clock. An 8 kB on-chip L1 cache was added to the 80486 processor to increase the proportion of instructions that could execute at the rate of one per clock cycle. Chapter 9 introduces cache memory concepts. The 80486 included an on-chip floating-point unit (FPU) to increase the performance.

8.8.3 Intel Pentium

The Intel Pentium processor (or the P6), introduced in 1993, added a second execution pipeline to achieve performance of executing two instructions per clock cycle. The L1 cache was increased to 8 kB devoted to instructions and another 8 kB for data. Write-back and write-through modes were included for efficiency. Branch prediction with an on-chip branch table was added to increase pipeline performance for looping constructs.

The Intel Pentium® Pro processor, introduced in 1995, included an additional pipeline so that the processor could execute three instructions per CPU clock. The Pentium Pro processor provides microarchitecture for flow analysis, out-of-order execution, branch prediction, and speculative execution to enhance pipeline performance. A 256 kB L2 cache that supports up to four concurrent accesses was included. The Pentium Pro processor also has a 36-bit address bus, giving a maximum physical address space of 64 GB.

The Intel Pentium II processor is an enhancement of the Pentium Pro architecture that combines functions of the P6 microarchitecture and included support for the Intel MMM single instruction, multiple data (SIMD) processing technique and eight 46-bit integer MMX™ instruction registers for use in multimedia and communication processing. The Pentium II was available with clock speeds from 233 to 450 MHz. Chapters 10 through 12 introduce these advanced concepts.

The Pentium II processor can directly address $2^{32} - 1$ or 4 GB of memory on its address bus. Memory is organized into 8-bit bytes. Each byte is assigned a physical memory location, called a physical address. To provide access into memory, the Pentium II provides three modes of addressing: flat, segmented, and real-addressing modes. The *flat addressing mode* provides a linear address space. Program instructions, data, and stack contents are present in sequential byte order. Only 4 GB of byte-ordered memory is supported in flat addressing. The *segmented memory* model provides independent memory address spaces that are typically used for instructions, data, and stack implementations. The program issues a logical address, which is a combination of the segment selector and address offset, which identifies a particular byte in the segment of interest. $2^{16} - 1$ or 16, 232 segments of a maximum of $2^{32} - 1$ or 4 GB are addressable by the Pentium II providing a total of approximately 2^{48} or 64 terabytes (TB) addressable memory. The *real-addressing* mode is provided to maintain the Intel 8086-based memory organization. This is provided for backward compatibility for programs developed for the earlier architectures. Memory is divided into segments of 64 kB. The maximum size of the linear address space in real-address mode is 2^{20} bytes.

The Pentium II has three separate processing modes: protected, real-address, and system management. In *protected* mode, the processor can use any of the preceding memory models, flat, segmented, or real. In *real-address* mode, the processor can only support the real-addressing mode. In the *system management* mode, the processor uses an address space from the system management RAM (SMRAM). The memory model used in the SMRAM is similar to the real-addressing mode.

The Pentium II provides 16 registers for system processing and application programming. The first eight registers are the GPRs and are used for logical and arithmetic operands, address calculations, and memory pointers. The registers are shown in Figure 8.15. The EAX register is used as an accumulator for operands and the results of calculations. EBX is used as a pointer into the DS segment memory, ECX for loop and string operations, EDX for an I/O pointer, ESI for a pointer to data in the segment pointed to by the DS register, source pointer for string operations, EDI for a pointer to data (or destination) in the segment pointed to by the ES register, and destination pointer for string operations.

ESP is used for an SP (in the SS segment), and the EBP register is used as a pointer to data on the stack (in the SS segment). The lower 16 bits of the registers correspond to those in the 8086

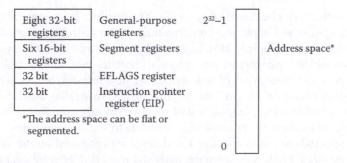

Figure 8.15 Pentium® Pro processor basic execution environment.

architecture and 80286 and can be referenced by programs written for those processors. Six segment registers hold a 16-bit address for segment addresses. The CS register handles CS selectors where instructions are loaded from memory for execution. The CS register is used together with the EIP register that contains the linear address of the next instruction to be executed. The DS, ES, FS, and GS registers are DS selectors that point to four separate DSs that the application program can access. The SS register is the SS selector for all stack operations. The processor contains a 32-bit EFLAGS register that contains processor status, control, and system flags that may be set by the programmer using special-purpose instructions.

The extended instruction pointer (EIP) register contains the address in the current CS for the next instruction to be executed. It is incremented to the next instruction or it is altered to move ahead or backward by a number of instructions when executing jumps, call interrupts, and return instructions. The EIP register cannot be accessed directly by programs; it is controlled implicitly by control-transfer instructions, interrupts, and exceptions. The EIP register controls program flow and is compatible with all Intel microprocessors, regardless of prefetching operations.

There are many different data types that the Pentium II processor supports. The basic data types are of 8-bit bytes, 2 byte or 16-bit words, double words of 4 bytes (32 bits), and quad words that are 8 bytes or 64 bits in length. Additional data types are provided for direct manipulation of signed and unsigned integers, BCD, pointers, bit fields, strings, floating-point data types (to be used with the FPU), and special 64-bit MMX data types.

A number of operands (zero or more) are allowed by the Intel architecture. The addressing modes fall into four categories: immediate, register, memory pointer, and I/O pointer.

Immediate operands are allowed for all arithmetic instructions except division and must be smaller than the maximum value of an unsigned double word, 2^{32}.

In register addressing mode, the source and destination operands can be given as any one of the 32-bit GPRs (and their 16-bit subsets), the 8-bit GPRs, the EFLAGS register, and some system registers. The system registers are usually manipulated by implied means from the system instruction.

Source and destination operands in memory locations are given as a combination of the segment selector and the address offset within that segment. Most applications of this process are to load the segment register with the proper segment and then include that register as part of the instruction. The offset value is the effective address of the memory location that is to be referenced. This value may be direct or combinations of a base address, displacement, index, and scale into a memory segment.

The processor supports an I/O address space that contains up to 65,536 8-bit I/O ports. Ports that are 16-bit and 32-bit may also be defined in the I/O address space. An I/O port can be addressed with either an immediate operand or a value in the DX register.

8.8.3.1 Instruction Set

The Intel architectures can be classified as CISCs. The Pentium II has several types of powerful instructions for the system and application programmer. These instructions are divided into three groups: integer instructions (to include MMX), floating-point instructions, and system instructions.

Intel processors include instructions for integer arithmetic, program control, and logic functions. The subcategories of these types of instructions are data transfer, binary arithmetic, decimal arithmetic, logic, shift and rotate, bit and byte, control transfer, string, flag control, segment register, and miscellaneous. The following paragraphs will cover the mostly used instructions.

The move instructions allow the processor to move data to and from registers and memory locations provided as operands to the instruction. Conditional moves based on the value, comparison, or status bits are provided. Exchange, compare, push and pop stack operations, port transfers, and data-type conversions are included in this class.

The binary arithmetic instruction class includes integer add, add with carry, subtract, subtract with borrow, signed and unsigned multiply, and signed and unsigned divide, and instructions to increment, decrement, negate, and compare integers are provided decimal arithmetic instructions. This class of instructions deals with the manipulation of decimal and ASCII values and adjusting the values after add, subtract, multiply, or divide functions. Logical AND, OR, XOR (EXCLUSIVE-OR), and NOT instructions are available for integer values. Shift arithmetic left and right include logical shift and rotate left and right, with and without carry manipulate integer operands of single- and double-word length.

The Pentium allows the testing and setting of bits in registers and other operands. The bit and byte functions are as follows: bit test, bit test and set, bit test and reset, bit test and complement, bit scan forward, bit scan reverse, set byte if equal/set byte if zero, and set byte if not equal/set byte if not zero.

The control instructions allow the programmer to define jump, loop, call, return, and interrupt; check for out-of-range conditions; and enter and leave high-level procedure calls. The jump functions include instructions for testing zero, not zero, carry, parity, and less or greater than conditions. Loop instructions use the ECX register to test conditions for loop constructs.

The string instructions allow the programmer to manipulate strings by providing move, compare, scan, load, store, repeat, input, and output for string operands. A separate instruction is included for byte, word, and double-word data types.

The MMX instructions execute on packed-byte, packed-word, packed-double-word, and quad-word operands. MMX instructions are divided into subgroups of data transfer, conversion, packed arithmetic, comparison, logical shift and rotate, and state management. The data transfer instructions are MOVD and MOVQ for movement of double and quad words. Figure 8.16 shows the mapping from the special MMX registers to the mantissa of the FPU. The Pentium II correlates the MMX registers and the FPU registers so that either the MMX instructions or the FPU instructions can manipulate values. Conversion instructions deal mainly with packing and unpacking of word and double-word operands. The MMX arithmetic instructions add, subtract, and multiply packed bytes, words, and double words.

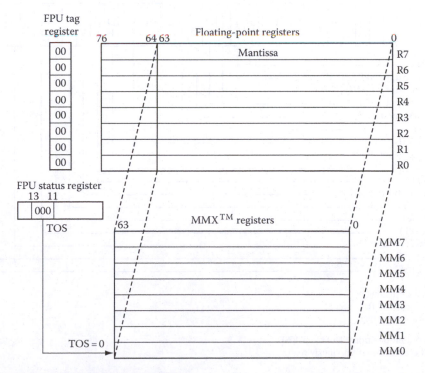

Figure 8.16   Mapping of MMX registers to FPU.

The floating-point instructions are executed by the processor's FPU. These instructions operate on floating-point (real), extended integer, and BCD operands. Floating-point instructions are divided into categories similar to the arithmetic instructions: data transfer, basic arithmetic, comparison, transcendental (trigonometry functions), load constants, and FPU control.

System instructions are used to control the processor and to support operating systems and executive. These instructions include loading and storing of the global descriptor table (GDT) and the local descriptor table (LDT), task, machine status, cache, table look-aside buffer manipulation, bus lock, and halt and providing performance information. The EFLAGS instructions allow the state of selected flags in the EFLAGS register to be read or modified. Figure 8.17 shows the EFLAGS register.

8.8.3.2 *Hardware Architecture*

The Pentium II is manufactured using Intel's 0.25 μm manufacturing process and contains over 7.5 million transistors and runs at clock speeds from 233 to 450 MHz. It contains a 32 kB L1 cache that is separated into a 16 kB instruction cache and a 16 kB data cache and a 512 kB L2 cache that operates on a dedicated 64-bit cache bus. It supports memory cacheability for up to 4 GB of addressable memory space. It uses a dual independent bus (DIB) architecture for increased bandwidth and performance. The system bus speed is increased from 66 to 100 MHz. It contains the MMX media enhancement technology for improved multimedia performance. It uses a pipelined (FPU) that supports IEEE Standard 754, 32 and 64-bit formats and an 80-bit format. It is packaged in the new single edge contact cartridge to allow higher frequencies and more handling protection.

The Pentium II is a three-way *superscalar* architecture, meaning it can fetch, decode, and execute up to three instructions per clock cycle. Program execution in the Pentium II is carried out

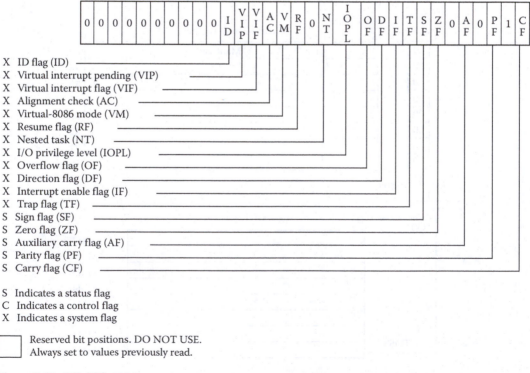

Figure 8.17 EFLAGS register.

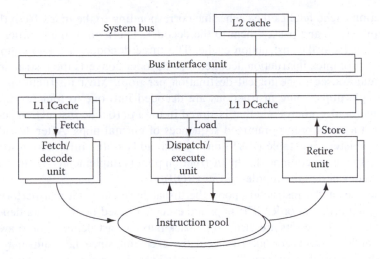

Figure 8.18 Instruction execution stages. (From Intel's *P6 Family of Processors Hardware Developer's Manual*.)

by a 12-stage (fine-grain) *pipeline* consisting of the following stages: instruction prefetch, length decode, instruction decode, rename/resource allocation, µop scheduling/dispatch, execution, write back, and retirement. The pipeline can be broken down into three major stages with each of the twelve stages either supporting or contributing to these three stages. The three stages are fetch/ decode, dispatch/execute, and retire. Figure 8.18 shows each of these stages, their interface through the instruction pool, connections to the L1 cache, and the bus interface.

The fetch/decode unit fetches the instructions in their original program order from the instruction cache. It then decodes these instructions into a series of µops that represent the dataflow of that instruction. It also performs a speculative prefetch by also fetching the next instruction after the current one from the instruction cache. Figure 8.19 shows the fetch/decode unit.

The L1 instruction cache is a local instruction cache. The <DISP>Next_IP</DISP> unit provides the L1 instruction cache index, based on inputs from the branch-target buffer (BTB).

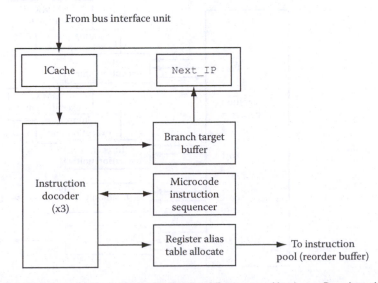

Figure 8.19 Fetch/decode unit. (From Intel's *P6 Family of Processors Hardware Developer's Manual*.)

The L1 instruction cache fetches the cache line corresponding to the index from the NexIP, and the next line (prefetch), and passes them to the decoder. There are three parallel decoders that accept this stream from the instruction cache. The three decoders are two simple instruction decoders and one complex instruction decoder. The decoder converts the instruction into triadic µops (two logical sources, one logical destination per µop). Most instructions are converted directly into single µops, some instructions are decoded into one to four µops, and the complex instructions require microcode that is stored in the microcode instruction sequencer. This microcode is just a set of preprogrammed sequences of normal µops. After decoding, the µops are sent to the register alias table (RAT) unit, which adds status information to the µops and enters them into the instruction pool. The instruction pool is implemented as an array of content addressable memory called the reorder buffer (ROB).

Through the use of the instruction pool, the dispatch/execute unit can perform instructions out of their original program order. The µops are executed based on their data dependencies and resource availability. The results are then stored back into the instruction pool to await retirement based on program flow. This is considered speculative execution since the results may or may not be used based on the flow of the program. This is done to keep the processor as busy as possible at all times. The processor can schedule at most five µops per clock cycle (3 µops is typical), one for each resource port. Figure 8.20 shows the dispatch/execute unit.

The dispatch unit selects µops from the instruction pool depending on their status. If the status indicates that a µop has all of its operands, then the dispatch unit checks to see if the execution resource needed by that µop is also available. If both are true, the reservation station removes that µop and sends it to the resource where it is executed. The results of the µop are later returned to the pool. If there are more µops available than can be executed, then the µops are executed in an FIFO order. The core is always looking ahead for other instructions that could be speculatively executed and is typically looking 20–30 instructions in front of the IP.

The retire unit is responsible for ensuring that the instructions are completed in their original program order. Completed means that the temporary results of the dispatch/execute stage are permanently committed to memory. The combination of the retire unit and the instruction pool allows instructions to be started in any order but always be completed in the original program order. Figure 8.21 shows the retire unit.

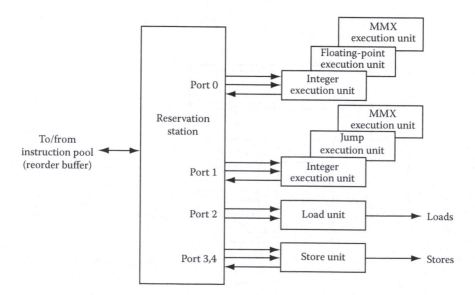

Figure 8.20 Dispatch/execute unit. (From Intel's *P6 Family of Processors Hardware Developer's Manual*.)

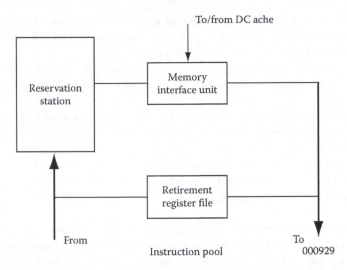

Figure 8.21 Retire unit. (From Intel's *P6 Family of Processors Hardware Developer's Manual.*)

In every clock cycle, the retire unit checks the status of μops in the instruction pool. It is looking for μops that have executed and can be removed from the pool. Once removed, the original target of the μops is written based on the original instruction. The retire unit not only must notice which μops are complete but also must reimpose the original program order on them. After determining which μops can be retired, the retire unit writes the results of this cycle's retirements to the retirement register file (RRF). The retire unit is capable of retiring 3 μops per clock.

As shown previously, the instruction pool removes the constraint of linear instruction sequencing between the traditional fetch and execute phases.

The BIU is responsible for connecting the three internal units (fetch/decode, dispatch/execute, and retire) to the rest of the system. The bus interface communicates directly with the L2 cache bus and the system bus. Figure 8.22 shows the BIU.

The memory order buffer (MOB) allows to pass loads and stores by acting like a reservation station and ROB. It holds suspended loads and stores and redispatches them when a blocking condition (dependency or resource) disappears. Loads are encoded into a single μop since they only need to specify the memory address to be accessed, the width of the data being retrieved, and the destination register. Stores need to provide a memory address, a data width, and the data to be written. Stores therefore require two μops, one to generate the address and one to generate the data. These μops recombine for the store to complete. Stores are also never reordered among themselves. A store is dispatched only when both the address and the data are available, and there are no older stores awaiting dispatch.

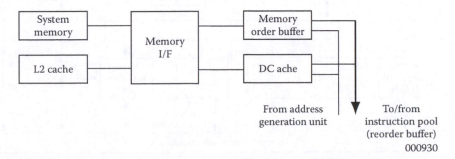

Figure 8.22 BIU. (From Intel's *P6 Family of Processors Hardware Developer's Manual.*)

A combination of three processing techniques enables the processor to be more efficient by manipulating data rather than processing instructions sequentially. The three techniques are multiple branch prediction, data-flow analysis, and speculative execution.

Multiple branch prediction uses algorithms to predict the flow of the program through several branches. While the processor is fetching instructions, it is also looking at instructions further ahead in the program.

Data-flow analysis is the process performed by the dispatch/execute unit. By analyzing data dependencies between instructions, it schedules instructions to be executed in an optimal sequence, independent of the original program order.

Speculative execution is the process of looking ahead of the program counter and executing instructions that are likely to be needed. The results are stored in a special register and only used if needed by the actual program path. This enables the processor to stay busy at all times, thus increasing the performance.

The bus structure of the Pentium II is referred to as a DIB architecture. The DIB is used to aid processor bus bandwidth. By having two independent buses, the processor can access data from either bus simultaneously and in parallel. The two buses are the L2 cache bus and the system bus.

The cache bus refers to the interface between the processor and the L2 cache, which for the Pentium II is mounted on the substrate with the core. The L2 cache bus is 64 bits wide and runs at half the processor core frequency.

The system bus refers to the interface between the processor, system core logic, and other bus agents. The system bus does not connect to the cache bus. The system bus is also 64 bits wide and runs at about 66 MHz and is pipelined to allow simultaneous transactions. Figure 8.23 shows a block diagram of the Pentium II processor.

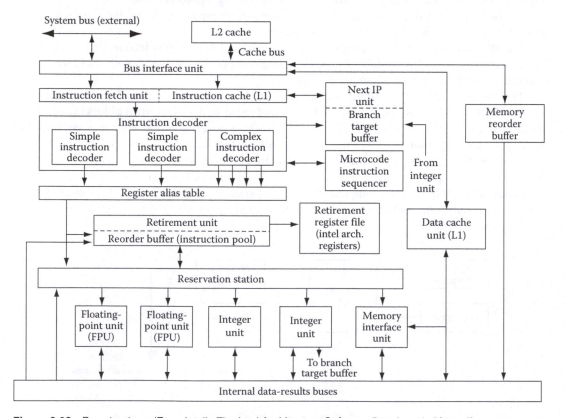

Figure 8.23 Bus structure. (From Intel's *The Intel Architecture Software Developer's Manual.*)

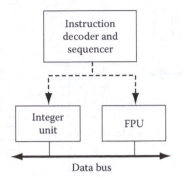

Figure 8.24 Integer and FPUs. (From Intel's *The Intel Architecture Software Developer's Manual*.)

The Pentium II contains two integer and FPUs that can all operate in parallel, with all sharing the same instruction decoder, sequencer, and system bus.

The FPU supports real, integer, and BCD-integer data types and the floating-point processing algorithms defined in the IEEE 754 and 854 Standards for floating-point arithmetic. The FPU uses eight 80-bit data registers for storage of values. Figure 8.24 shows the relationship between the integer and FPUs.

The Pentium II uses I/O ports to transfer data. I/O ports are created in system hardware by circuitry that decodes the control, data, and address pins on the processor. The I/O port can be an input port, an output port, or a bidirectional port. The Pentium II allows I/O ports to be accessed in two ways: as a separate I/O address space and memory-mapped I/O. I/O addressing is handled through the processor's address lines. The Pentium II uses a special memory–I/O transaction on the system bus to indicate whether the address lines are being driven with a memory address or an I/O address.

Accessing I/O ports through the I/O address space is handled through a set of I/O instructions and a special I/O protection mechanism. This guarantees that writes to I/O ports will be completed before the next instruction in the instruction stream is executed. Accessing I/O ports through memory-mapped I/O is handled with the processor's general-purpose move and string instructions, with protection provided through segmentation or paging.

The Pentium II has two levels of cache, the L1 cache and the L2 cache. Memory is cacheable for up to 4 GB of addressable memory space.

The L1 cache is 32 kB that is divided into two 16 kB units to form the instruction cache and the data cache. The instruction cache is four-way set associative, and the data cache is two-way set associative each with a 32-byte cache line size. The L1 cache operates at the same frequency as the processor and provides the fastest access to the most frequently used information. If there is an L1 cache miss, then the L2 cache is searched for the data.

The Pentium II supports between 256 kB and 1 MB of L2 cache with 521 kB most common. The L2 is a four-way set associative with a 32-byte cache line size. The L2 cache uses a dedicated 64-bit bus to transfer data between the processor and the cache. Cache coherency is maintained through the modified, exclusive, shared, invalid (MESI) snooping protocol. The L2 cache can support up to four concurrent cache accesses as long as they are two different banks. Figure 8.25 shows the cache architecture of the Pentium II.

8.8.3.3 MMX Technology

Considered the most significant enhancement to the Intel architecture in the last 10 years, this technology provides the improved video compression/decompression, image manipulation, encryption, and I/O processing needed for multimedia applications. These improvements are achieved

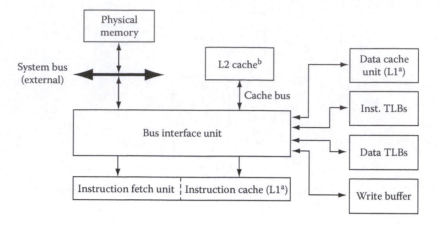

^aFor the Intel486 processor, the L1 cache is a unified
  instruction and data cache.

^bFor the Pentium ⊕and Intel486 processors, the L2 Cache
  is external to the processor package and there is
  no cache bus (i.e., the L2 cache interface with
  the system bus).

Figure 8.25 Cache architecture. (From Intel's *The Intel Architecture Software Developer's Manual*.)

through the use of the following: SIMD technique, 57 new instructions, eight 64-bit wide MMX registers, and four new data types.

The SIMD technique allows for the parallel processing of multiple data elements by a single instruction. This is accomplished by performing an MMX instruction on one of the MMX packed data types. For example, an add instruction performed on two packed bytes would add eight different values with the single add instruction.

The 57 new instructions cover the following areas: basic arithmetic comparison operations, conversion instructions, logical operations, shift operations, and data transfer instructions. These are general-purpose instructions that fit easily into the parallel pipelines of the processor and are designed to support the MMX packed data types.

MMX contains eight 64-bit registers (MM0–MM7) that are accessed directly by the MMX instructions. These registers are used to perform calculations only on the MMX data types. The

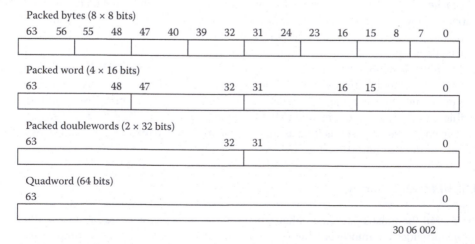

Figure 8.26 MMX data types. (From Intel's *The Intel Architecture Software Developer's Manual*.)

registers have two data access modes: 64-bit and 32-bit. The 64-bit mode is used for transfers between MMX registers, and the 32-bit mode is for transfers between integer registers and MMX registers.

The four new data types included in MMX are packed byte, packed word, packed double word, and quad word (see Figure 8.26). Each of these data types is a grouping of signed and unsigned fixed-point integers, bytes, words, double words, and quad words into a single 64-bit quantity. These are then stored in the 64-bit MMX registers. Then the MMX instruction executes on all of the values in the register at once.

The single edge connect cartridge (SEC) is a metal and plastic cartridge that completely encloses the processor core and the L2 cache. To enable high-frequency operations, the core and L2 cache are surface mounted directly to a substrate inside the cartridge. The cartridge then connects to the motherboard via a single edge connector. The SEC allows the use of high-performance BSRAMs (widely available and cheaper) for the L2 cache. The SEC also provides better handling protection for the processor.

8.9 SUMMARY

This chapter provided the details of instruction-set architectures through examples of commercially available machines. Processor structures including the memory and register architectures were also discussed. Advanced instruction-processing structures such as pipelining and parallel (superscalar) execution will be covered in subsequent chapters of the book. The reader needs to revisit the details in this chapter after learning these advanced concepts. We have traced the development of a family of Intel microprocessors in this chapter. We will continue this trace in subsequent chapters with the newer members of the family. The processor architectures change rapidly. The only way to keep up with them is to refer to the trade magazines listed in the reference section of this chapter and the manufacturer's manuals.

PROBLEMS

8.1 Study the processor and system characteristics of a large-scale computer, a minicomputer, and a microprocessor system you have access to, with reference to the features described in this chapter.

8.2 Assume that the processor status register of an 8-bit machine contains the following flags: CARRY, ZERO, OVERFLOW, EVEN PARITY, and NEGATIVE. Determine the value of each of these flags after an ADD operation on two operands A and B for the various values of A and B shown in the following:

A	B
+1	−1
12	−6
31	31
12	−13
127	−127

8.3 Represent the following numbers in IBM and IEEE standard floating-point notation:
(a) 46.24×10^{-3} and (b) 2.46×10^{2}

8.4 A 16-bit machine has a byte-addressable memory. Bytes 0 and 1 of the memory contain decimal values 31 and 241, respectively. If the machine uses 2s complement system, what is the value in the 16-bit word made up of the two bytes 0 and 1, if the memory is (a) big Endian and (b) little Endian?

8.5 Determine the μops for the ADD instruction, for the machine of Problem 8.4 assuming the memory is (a) big Endian and (b) little Endian.

8.6 Design a 4-bit per level, eight-level stack using shift registers. In addition to allowing PUSH and POP operations, the stack must generate an "overflow" signal when an attempt to PUSH into a full stack is made and an "underflow" signal when an empty stack is popped.

8.7 Design the stack of Problem 8.4 using an RAM.

8.8 Assume that ASC has an SP register that is initiated to 0 when power is turned on. Assume also two instructions LSP and SSP for loading and storing the SP from and into a memory location, respectively. The following operations are required:

a. PUSH	TL ← ACC
b. POP	ACC ← TL
c. ADD	SL ← TL + SL; POP

Write subroutines for these operations using the ASC instruction set including the two new instructions LSP and SSP.

8.9 Write the ASC microinstruction sequences for PUSH, POP, and ADD instruction in Problem 8.6.

8.10 In a machine such as IBM 370 that does not allow indirect addressing mode, how can an operand whose address is in a memory location be accessed?

8.11 Assume that a branch instruction with a PC-relative mode of addressing is located at $X1$. If the branch is made to location $X2$, what is the value of the address field of the instruction? If the address field is 10 bits long, what is the range of branches possible?

8.12 Using the notation of Chapter 5, express the effective address of the operand for each of the addressing modes described in this chapter.

8.13 Write programs for zero-, one-, two-, and three-address machines to evaluate $ax^2 + bx + c$, given the values for x, a, b, and c.

8.14 Given that

$$F = M + N + P - (Q - R - S)/T + V$$

where F, M,...., V each represent a memory location, generate code for
a. A stack machine.
b. A one-address machine.
c. A two-address machine.
d. Assume 8 bits of opcode and 16 bits for an address and compute the length of the program in each of the aforementioned cases.
e. How many memory cycles are needed to execute the aforementioned programs?
f. Assume that the stack is ten levels deep. What is the effect on the program in (a) if the stack is restricted to four levels?

8.15 Answer Problem 8.14(e) assuming that the memory of each of the machines is organized as eight bits/word and only one memory word can be accessed per memory cycle.

8.16 An example of macroinstruction is a TRANSLATE instruction. The instruction replaces an operand with a translated version of itself, the translated version being derived from a table stored in a specified memory location. Assume that in ASC the operand is located in the accumulator and the beginning address of the table is located in the second word of the instruction (i.e., the instruction is two words long). The accumulator content thus provides the offset into the table. Write a microprogram for TRANSLATE.

8.17 Assuming that a machine using base–displacement addressing mode has N-bit base registers, investigate how the compilers for that machine access data and instructions in blocks that are larger than 2N words.

8.18 A machine to compute the roots of the quadratic $ax^2 + bx + c = 0$ is needed. Develop the instruction set needed for zero-, one-, two-, and three-address architectures.

8.19 The instruction set of a machine has the following number and types of instructions:

Ten 3-address instructions
Thirty-six 2-address instructions (including 10 memory-reference instructions)
Fifty 1-address instructions
Thirty 0-address instructions

The machine has eight 16-bit registers. All the instructions are 16-bit register to register, except for the memory-reference, load, store, and branch instructions. Develop an instruction format and an appropriate opcode structure.

8.20 Develop the subroutine call and return instructions for all the three types of machines. Assume that a stack is available.

8.21 Discuss the relative merits and effects on the instruction cycle implementation of the following parameter-passing techniques:

    a. Pass the parameters in registers.

    b. Pass the parameters on stack.

    c. Pass the parameter addresses in registers.

    d. Pass the parameter addresses on stack.

8.22 In a four-address machine, the fourth address in the instruction corresponds to the address of the next instruction to be fetched. This means that the PC is not needed. What types of instructions would not utilize the fourth address?

8.23 A 32-bit machine needs 250 instructions, each 32 bits long. All instructions are two-address instructions. Show the instruction format assuming all addresses are direct addresses. What is the range of memory that can be addressed? How would you change the instruction format to accommodate addressing a 64 K word memory?

8.24 There are three classes of instructions in an instruction set (two address, one address, and zero address). All instructions must be 32 bits long. The machine should be able to address 64 K, 32-bit words of memory. Use a class code method for assigning opcodes and design the instruction formats. What is the maximum number of each class of instructions the machine can have?

8.25 Select two contemporary machines (one RISC and one CISC) and compare their instruction sets with respect to various features discussed in this chapter.

8.26 Processor manufacturers have recently introduced multicore architectures. Select a multicore architecture and study its features in terms of concepts described in this chapter.

8.27 One common term appearing in manufacturer's manuals is "SoC." What is its significance?

8.28 What should be the significant architectural features for a processor to be suitable for a smartphone?

BIBLIOGRAPHY

Brey, B., *Intel Microprocessors: Architecture, Programming, and Interfacing*, Englewood Cliffs, NJ: Prentice Hall, 2003.

Computer, Los Alamitos, CA: IEEE Computer Society, Published monthly, http://www.computer.org.

Electronic Design, Cleveland, OH: Penton Media, Published twice monthly, http://www.electronicdesign.com.

Embedded Systems Design, San Francisco, CA: CMP Media, Published monthly, http://www.embedded.com.

IBM System 370 Principles of Operation, GA22–7000. Poughkeepsie, New York: IBM Corporation, 1976.

Intel 8080 Microcomputer Systems Users Manual, Santa Clara, CA: Intel Corporation, 1977.

Intel Architecture Software Developer's Manuals—Volume 1: Basic Architecture (24319001. pdf); Volume 2: Instruction Set Reference (24319101.pdf) and Volume 3: System Programming Guide (24319201.pdf) at http://www.intel.com/design/PentiumII/manuals, Santa Clara, CA: Intel Corporation.

M6502 Hardware Manual, Norristown, PA: MOS Technology Inc., 1976.

MC68000 Technical Features, Phoenix, AZ: Motorola Semiconductor Products, 1982.

MC68000 Users Manual, Austin, TX: Motorola, 1982.

MCS-80 Users Manual, Santa Clara, CA: Intel Corporation, 1977.

Messmer, H., *The Indispensable PC Hardware Book*, Reading, MA: Addison-Wesley, 1997.

P6 Family of Processors Hardware Developer's Manual, Santa Clara, CA: Intel Corporation, http://www.intel.com/design/PentiumII/manuals/24400101.pdf.

Patterson, D.A. and Hennessey, J.L., *Computer Organization and Design: The Hardware/Software Interface*, San Mateo, CA: Morgan Kaufmann, 2011.

PDP-11 Processor Handbook, Maynard, MA: Digital Equipment Corporation, 1978.

Pentium II Processor Technology, ECG Technology Communications, ECG046/0897, Houston, TX: Compaq Computer Corporation, August, 1997.

Pentium II Xeon Processor Technology Brief, Order Number: 243792–002, Santa Clara, CA: Intel Corporation, 1998.

Shiva, S.G., *Advanced Computer Architectures*, Boca Raton, FL: Taylor & Francis, 2006.

VAX Hardware Handbook, Maynard, MA: Digital Equipment Corporation, 1979.

CHAPTER 9

Memory and Storage

We have demonstrated the use of flip-flops in storing binary information. Several flip-flops put together form a register. A register is used either to store data temporarily or to manipulate data stored in it using the logic circuitry around it. The *memory* subsystem of a digital computer is functionally a set of such registers where data and programs are stored. The instructions from the programs stored in memory are retrieved by the control unit of the machine (digital computer system) and are decoded to perform the appropriate operation on the data stored either in memory or in a set of registers in the processing unit.

For optimum operation of the machine, it is required that programs and data be accessible by control and processing units as quickly as possible. The *main memory* (primary memory) allows such a fast access. This fast-access requirement adds a considerable amount of hardware to the main memory and thus makes it expensive. Chapter 5 provided the details of the random-access memory (RAM) used with a simple computer (ASC). To reduce memory cost, data and programs not immediately needed by the machine are normally stored in a less-expensive *secondary memory* subsystem (ASC does not have a secondary memory). They are brought into the main memory as the processing unit needs them. The larger the main memory, the more information it can store and hence the faster the processing, since most of the information required is immediately available. But because main-memory hardware is expensive, a speed–cost trade-off is needed to decide on the amounts of main and secondary storage needed.

This chapter provides models of operation for the most commonly used types of memories and a brief description of memory devices and organization, followed by the description of *virtual* and *cache* memory schemes. We will provide the models of operation for the four most commonly used types of memories in the next section. Section 9.2 lists the parameters used in evaluating memory systems, and Section 9.3 describes the *memory hierarchy* in computer systems. Section 9.4 describes the popular semiconductor memory devices and memory organizations. Representative memory integrated circuits (ICs) are briefly described in Appendix A, and the design of primary memory using these ICs is described in Section 9.5. Memory speed enhancement concepts are introduced in Section 9.6, and size enhancement is discussed in Section 9.7. Section 9.8 provides address extension concepts. The chapter ends with example memory systems in Section 9.9.

9.1 TYPES OF MEMORY

Depending on the mechanism used to store and retrieve data, a memory system can be classified as one of the following four types:

1. RAM
   a. Read/write memory (RWM)
   b. Read-only memory (ROM)

2. Content-addressable memory (CAM) or associative memory (AM)
3. Sequential-access memory (SAM)
4. Direct-access memory (DAM)

Primary memory is of the RAM type. CAMs are used in special applications in which rapid data search and retrieval are needed. SAM and DAM are used as secondary memory devices.

9.1.1 RAM

As shown in Chapter 5, in a RAM, any addressable location in the memory can be accessed in a random manner. That is, the process of reading from and writing into a location in a RAM is the same and consumes an equal amount of time no matter where the location is physically in the memory. The two types of RAM available are RWM and ROMs:

RWM: The most common type of main memory is the RWM, whose model is shown in Figure 9.1. In an RWM, each memory register or memory location has an "address" associated with it. Data are input into (written into) and output from (read from) a memory location by accessing the location using its "address." The memory address register (MAR) of Figure 9.1 stores such an address. With n bits in the MAR, 2^n locations can be addressed, and they are numbered from 0 to $2^n - 1$.

Transfer of data in and out of memory is usually in terms of a set of bits known as a *memory word*. Each of the 2^n words in the memory of Figure 9.1 has m bits. Thus, this is a $(2^n \times m)$-bit memory. This is a common notation used to describe RAMs. In general, an $(N \times M)$-unit memory contains N words of M units each. A "unit" is a bit, a byte (8 bits), or a word of certain number of bits. A memory buffer register (MBR) is used to store the data to be written into or read from a memory word. To read the memory, the address of the memory word to be read from is provided in MAR, and the read signal is set to 1. A copy of the contents of the addressed memory word is then brought by the memory logic into the MBR. The content of the memory word is thus not altered

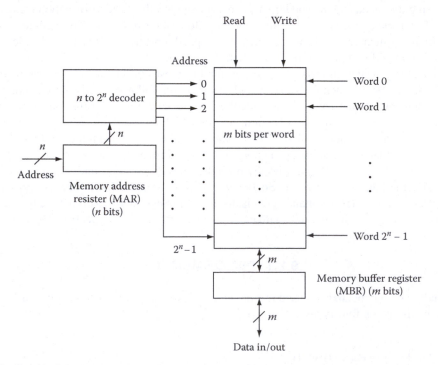

Figure 9.1   Read/write memory.

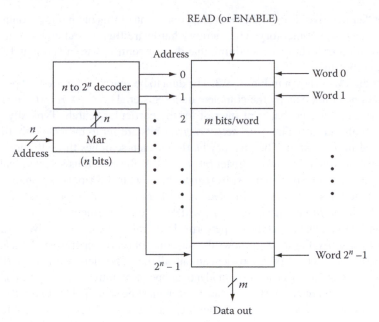

Figure 9.2 Read-only memory.

by a read operation. To write a word into the memory, the data to be written are placed in MBR by external logic; the address of the location into which the data are to be written is placed in MAR; and the write signal is set to 1. The memory logic then transfers the MBR content into the addressed memory location. The content of the memory word is thus altered during a write operation.

A memory word is defined as the most often accessed unit of data. The typical word sizes used in memory organizations of commercially available machines are 6, 16, 32, 36, and 64 bits. In addition to addressing a memory word, it is possible to address a portion of it (e.g., half-word, quarter-word) or a multiple of it (e.g., double word, quad word), depending on the memory organization. In a "byte-addressable" memory, for example, an address is associated with each byte (usually 8 bits/byte) in the memory, and a memory word consists of one or more bytes.

ROM: The literature routinely uses the acronym RAM to mean RWM. We will follow this popular practice and use RWM only when the context requires us to be more specific. We have included MAR and MBR as components of the memory system in this model. In practice, these registers may not be located in the memory subsystem, but other registers in the system may serve the functions of these registers.

ROM is also a RAM, except that data can only be read from it. Data are usually written into a ROM either by the memory manufacturer or by the user in an off-line mode; that is, by special devices that can write (burn) the data pattern into the ROM. A model of ROM is shown in Figure 9.2. A ROM is also used as main memory and contains data and programs that are not usually altered in real time during the system operation. The MBR is not shown in Figure 9.2. In general, we assume that the data on output lines are available as long as the memory enable signal is on and it is latched into an external buffer register. A buffer is provided as part of the memory system, in some technologies.

9.1.2 Content-Addressable Memory

In this type of memory, the concept of address is not usually present: rather, the memory logic searches for the locations containing a specific pattern, and hence the descriptor "content addressable" or "associative" is used. In the typical operation of this memory, the data to be searched for are first provided to the memory. The memory hardware then searches for a match and either identifies the location or locations containing that data or returns with a "no match" if none of the locations contain the data.

A model for an *AM* is shown in Figure 9.3. The data to be searched for are first placed in the *data register*. The data need not occupy the complete data register. The *mask register* is used to identify the region of the data register that is of interest for the particular search. Typically, corresponding mask register bits are set to 1. The *word-select register* bits are set to indicate only those words that are to be involved in the search. The memory hardware then searches through those words in only those bit positions in which the mask register bits are set. If the data thus selected match the content of the data register, the corresponding *results register* bit is set to 1. Depending on the application, all words responding to the search may be involved in further processing, or a subset of the respondents may be selected. The *multiple-match resolver* (MMR) circuit implements this selection process.

Note that the data-matching operation is performed in parallel. Hence, extensive hardware is needed to implement this memory. Figure 9.3b shows the sequence of search operation. A "select first" MMR circuit is used here to select the first respondent among all the respondent words for further processing.

Associative memories are useful when an identical operation must be performed on several pieces of data simultaneously or when a particular data pattern must be searched for in parallel. For example, if each memory word is a record in a personnel file, the records corresponding to the set of female employees 25 years old can be searched for by setting the data and mask register bits appropriately. If sufficient logic is provided, all records responding to the search can also be updated simultaneously.

In practice, CAMs are built out of RAM components and as such have the same addressing capability. In fact, the MMR returns the address of the responding word or words in response to a search. The major application of CAM is for storing data on which rapid search and update operations are performed. The virtual memory scheme described later in this chapter shows an application for CAMs. We will return to the description of CAMs in Section 9.4.2, where the detailed design of a CAM system is given.

9.1.3 Sequential-Access Memory

A serial-input/serial-output shift register is the simplest model of sequential memory. In the right-shift register of Figure 9.4a, data enter from the left input and leave the shift register from the right output. Because these are the only input and output available on the device, the data must be written into and read from the device in the sequence in which they are stored in the register. That is, every data item in sequence (from the first data item until the desired item) must be accessed in order to retrieve the required data. Thus, it is a SAM. In particular, this model corresponds to a *first-in/first-out* (FIFO) SAM, since the data items are retrieved in the order in which they were entered. This organization of a SAM is also called a *queue*. Note that in addition to the right-shift register, mechanisms for input and output of data (similar to MAR and MBR in the RAM model) are needed to build a FIFO memory.

Figure 9.4b shows the model for a *last-in/first-out* (LIFO) SAM. Here, a shift register that can shift both right and left is used. Data always enter through the left input and leave the register through the left output. To write, the data are placed on the input and the register is shifted right. While reading, the data on the output are read and the register is shifted left, thereby moving each item in the register left and presenting the next data item at the output. Note that the data are accessed from this device in a LIFO manner. This organization of a SAM is also called a *stack*. The data input operation is described as PUSHing the data into the stack, and data are retrieved by POPing the stack.

Figure 9.5 shows FIFO and LIFO organizations for 4 bit data words using 8 bit shift registers. Each of these SAM devices thus can store eight 4 bit words.

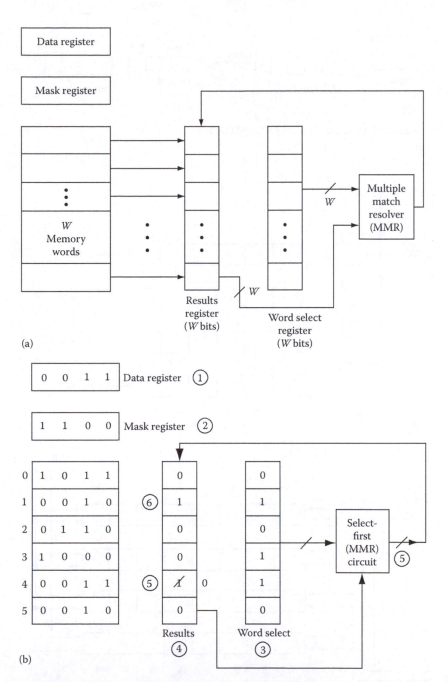

Figure 9.3 AM. (a) Model. (b) Operation. (1) The data being searched for is 0011. (2) The most significant 2 bits of mask register are 1s. Hence, only the corresponding 2 bits of data are compared with those of memory words 0 through 5. (3) The word select register bit setting indicates that only words 1, 3, and 4 are to be involved in the search process. (4) The results register indicates that words 1 and 4 have the needed data. (5) The select-first (MMR) circuit resets the results register bit, corresponding to word 4. (6) Word 1 is the final respondent. This information can be used for updating word 1 contents. *Note:* Comparison of the data register with memory words is done in parallel. The addresses shown (0 through 5) are for reference only.

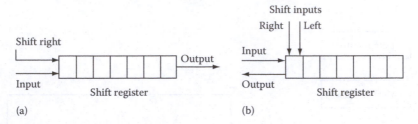

Figure 9.4   Sequential-access device. (a) FIFO. (b) LIFO.

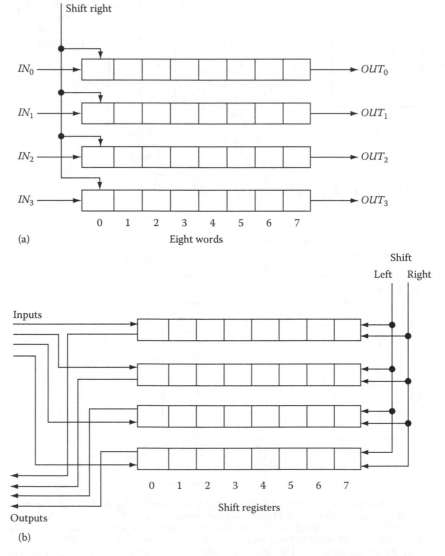

Figure 9.5   Eight 4-bit word sequential-access device using shift registers. (a) FIFO. (b) LIFO.

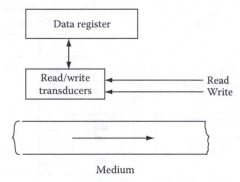

Figure 9.6   Sequential-access storage.

Figure 9.6 shows the generalized model of a sequential-access storage system. Here, the read/write transducers read data from or write data onto the data storage medium at its current position. The medium is then moved to the next position. Thus, each data item must be examined in sequence to retrieve the desired data. This model is applicable to secondary storage devices as magnetic tape.

9.1.4 Direct-Access Memory

Figure 9.7 shows the model of a DAM device (a magnetic disk) in which data are accessed in two steps:

1. The transducers move to a particular position determined by the addressing mechanism (cylinder, track).
2. The data on the selected track are accessed sequentially until the desired data are found.

This type of memory is used for secondary storage. It is also called *semi*-RAM, since the positioning of read/write transducers to the selected cylinder is random and only the accessing of data within the selected track is sequential.

9.2 MEMORY SYSTEM PARAMETERS

The most important characteristics of any memory system are its capacity, data-access time, the data-transfer rate, the frequency at which memory can be accessed (the cycle time), and cost.

The *capacity* of the storage system is the maximum number of units (bits, bytes, or words) of data it can store. The capacity of a RAM, for instance, is the product of the number of memory words and the word size. A $2k \times 4$ memory, for example, can store $2k$ ($k = 1024 = 2^{10}$) words, each containing 4 bits, or a total of $2 \times 1024 \times 4$ bits.

The *access time* is the time taken by the memory module to access the data after an address is provided to the module. The data appear in the MBR at the end of this time in a RAM. The access time in a non-RAM is a function of the location of the data on the medium with reference to the position of read/write transducers.

The *cycle time* is a measure of how often the memory can be accessed. The cycle time is equal to the access time in *nondestructive* readout memories in which the data can be read without being destroyed. In some storage systems, data are destroyed during a read operation (destructive readout). A rewrite operation is necessary to restore the data. The cycle time in such devices

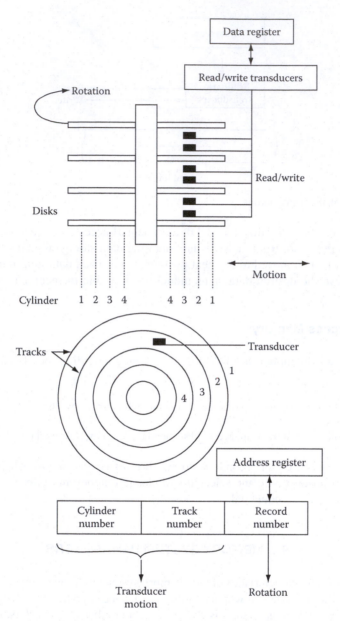

Figure 9.7 Direct-access storage.

is defined as the time it takes to read and restore data, since a new read operation cannot be performed until the rewrite has been completed.

The *data-transfer rate* is the number of bps at which the data can be read out of the memory. This rate is the product of the reciprocal of access time and the number of bits in the unit of data (data word) being read. This parameter is of more significance in non-RAM systems than in RAMs.

The *cost* is the product of capacity and the price of memory device per bit. RAMs are usually more costly than other memory devices.

Some of the other parameters of interest are fault tolerance, radiation hardness, weight, and data compression and integrity depending on the application the memory is used.

9.3 MEMORY HIERARCHY

The *primary memory* of a computer system is always built out of RAM devices, thereby allowing the processing unit to access data and instructions in the memory as quickly as possible. It is necessary that the program or data be in the primary memory when the processing unit needs them. This would call for a large primary memory when programs and data blocks are large, thereby increasing the memory cost. In practice, it is not really necessary to store the complete program or data in the primary memory as long as the portion of the program or data needed by the processing unit is in the primary memory.

A *secondary memory* built out of direct- or serial-access devices is then used to store programs and data not immediately needed by the processing unit. Since random-access devices are more expensive than secondary memory devices, a cost-effective memory system results when the primary memory capacity is minimized. But this organization introduces an overhead into the memory operation, since mechanisms to bring the required portion of the programs and data into primary memory as needed will have to be devised. These mechanisms form what is called a *virtual memory* scheme.

In a virtual memory scheme, the user assumes that the total memory capacity (primary plus secondary) is available for programming. The operating system manages the moving in and out of portions (segments or pages) of program and data into and out of the primary memory.

Even with current technologies, the primary memory hardware is slow compared with the processing unit hardware. To reduce this speed gap, a small but faster memory is usually introduced between the main memory and the processing unit. This memory block is called *cache memory* and is usually 10–100 times faster than the primary memory. A virtual memory mechanism similar to that between primary and secondary memories is then needed to manage operations between main memory and cache. The set of instructions and data that are immediately needed by the processing unit are brought from the primary memory into cache and retained there. A parallel fetch operation is possible in that while the cache unit is being filled from the main memory, the processing unit can fetch from the cache, thus narrowing the memory-to-processor speed gap.

Note that the registers in the processing unit are temporary storage devices. They are the fastest components of the computer system memory.

Thus, in a general-purpose computer system, there is a memory hierarchy in which the highest speed memory is closest to the processing unit and is most expensive. The least-expensive and slowest memory devices are farthest from the processing unit. Figure 9.8 shows the memory hierarchy.

Thus, the memory system hierarchy common to most modern-day computer systems consists of the following levels:

1. Central processing unit (CPU) registers
2. Cache memory, a small, fast, RAM block
3. Primary (main) memory, the RAM from which the processor accesses all programs and data (via the cache memory)
4. Secondary (mass) memory consisting of semi-random-access and SAM elements such as magnetic disks and tapes

The fastest memory is at level 1, and the speed decreases as we move toward higher levels. The cost per bit is highest at level 1 and lowest at level 4. The major aim of the hierarchical memory design is to enable a speed–cost trade-off that will provide a memory system of the desired capacity with the highest speed possible at the lowest cost.

Let us consider a memory system with an n-level hierarchy. Let C_i ($i = 1$ through n) be the cost per bit of the ith level in the previous hierarchy, while S_i is the capacity (i.e., total number

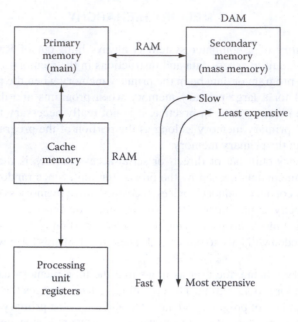

Figure 9.8 Memory hierarchy.

of bits) at the ith level. Then the average cost per bit (C_a) of the memory system is given by the following equation:

$$C_a = \frac{\sum_{i=1}^{n} C_i S_i}{\sum_{i=1}^{n} S_i}. \tag{9.1}$$

Typically, we would like to have C_a as close to C_n as possible. Since $C_i >>> C_{i+1}$, in order to minimize the cost, it requires that $S_i <<< S_{i+1}$.

As seen by the example systems described in earlier chapters, additional levels have been added to the previous hierarchy in recent computer systems. In particular, it is common to see two levels of cache (between the processor and the main memory), typically called level 1 and level 2 or on- and off-chip cache, depending on the system structure. In addition, a disk cache could be present (between the primary and secondary memory), to enable faster disk access.

9.4 MEMORY DEVICES AND ORGANIZATIONS

The basic property that a memory device should possess is that it must have two well-defined states that can be used for the storage of binary information. In addition, the ability to switch from one state to another (i.e., reading and writing a 0 or 1) is required, and the switching time must be small in order to make the memory system fast. Further, the cost per bit of storage should be as low as possible.

The address-decoding mechanism and its implementation distinguish RAM from non-RAM. Since RAM needs to be fast, the address decoding is done all electronically, thus involving no physical movement of the storage media. In a non-RAM, either the storage medium or the read/write mechanism (transducers) is usually moved until the appropriate address (or data) is found. This sharing of the addressing mechanism makes non-RAM less expensive than RAM, while the mechanical movement makes it slower than RAM in terms of data-access times. In addition to the memory device characteristics, decoding of the external address and read/write circuitry affect the speed and cost of the storage system.

Semiconductor and magnetic technologies have been the popular primary memory device technologies. Magnetic core memories were used extensively as primary memories during the 1970s. They are now obsolete, since the semiconductor memories have the advantages of lower cost and higher speed. One advantage of magnetic core memories is that they are *nonvolatile*. That is, the data are retained by the memory even after the power is turned off. Semiconductor memories, on the other hand, are *volatile*. Either a backup power source must be used to retain the memory contents when the power is turned off or the memory contents are dumped to a secondary memory and restored when needed to circumvent the volatility of these memories.

The most popular secondary storage devices have been magnetic tape and disk. Optical disks are now becoming cost-effective with the introduction of compact disk (CD) ROMs (CD-ROM), write-once read-many-times (or read-mostly) (WORM) disks, and erasable disks.

In each technology, memory devices can be organized in various configurations with varying cost and speed characteristics. We will now examine representative devices and organizations of semiconductor memory technology.

9.4.1 RAM Devices

Two types of semiconductor RAMs are now available: *static* and *dynamic*. In a static RAM, each memory cell (MC) is built out of a flip-flop. Thus, the content of the MC (either 1 or 0) remains intact as long as the power is on. Hence, the memory device is *static*. A dynamic MC, however, is built out of a capacitor. The charge level of the capacitor determines the 1 or 0 state of the cell. Because the charge decays with time, these MCs must be *refreshed* (i.e., recharged) every so often to retain the memory content. Dynamic memories require complex refresh circuits, and because of the refresh time needed, they are slower than static memories. But more dynamic MCs can be fabricated on the same area of silicon than static MCs can. Thus, when large memories are needed and speed is not a critical design parameter, dynamic memories are used; static memories are used in speed-critical applications.

9.4.1.1 Static RAM

The major components of RAM are the address-decoding circuit, the read/write circuit, and the set of memory devices organized into several words. The memory device that can store a bit and has the appropriate hardware to support decoding and read/write operations is called an *MC*.

Flip-flops are used in forming static RAM cells. Figure 9.9a shows an MC built out of a JK flip-flop. The flip-flop is not clocked. When the enable signal is 1, either the input signal enters the flip-flop or the contents of the flip-flop are seen on the output based on the value of the read/write signal. If the enable is 0, the cell outputs 0 and also makes $J = K = 0$, leaving the contents of the flip-flop unchanged. The read/write signal is 1 for reading (i.e., output $\leftarrow Q$) and 0 for writing (i.e., $Q \leftarrow$ input). A symbol for this MC is shown in Figure 9.9b.

A (4×3) bit RAM, built out of such MCs, is shown in Figure 9.10. The 2 bit address in the MAR is decoded by a 2–4 decoder, to select one of the four memory words. For the memory to be active, the memory enable line must be 1. If not, none of the words are selected (i.e., all outputs of the decoder are 0). When the memory enable line is 1 and the $R/\overline{W}$ line is 1, the outputs of MCs enabled by the selected word line will be input into the set of OR gates whose outputs are connected to the output lines. Output lines receive signals from MCs that are in the enabled word only, since all other MC outputs in each bit position are 0. If the memory is enabled and the $R/\overline{W}$ line is 0, only the selected word will receive the INPUT information.

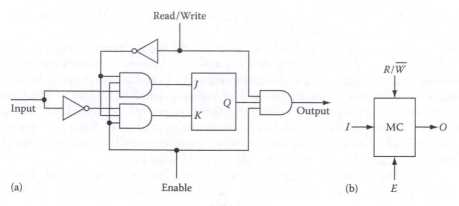

Figure 9.9   A semiconductor MC: (a) Circuit. (b) Symbol.

If the number of words is large, as in any practical semiconductor RAM, the OR gates shown in Figure 9.10 become impractical. To eliminate these gates, the MCs are fabricated with either open-collector or tristate outputs. If open-collector outputs are provided, outputs of MCs in each bit position are tied together to form a wired OR, thus eliminating an OR gate. However, the pull-up resistors required and the current dissipation by the gates limit the number of the outputs of gates that can be wire-ORed. MCs with tristate outputs can be used in such limiting cases. Outputs of MCs in each bit position are then tied together to form an output line.

When the number of words in the memory is large, the linear decoding technique of Figure 9.10 results in complex decoding circuitry. To reduce the complexity, coincident decoding schemes are used. Figure 9.11 shows such a scheme. Here, the address is divided into two parts: X and Y. The low-order bits of the address (Y) select a column in the MC matrix, and the high-order bits of the address (X) select a row. The MC selected is the one at the intersection of the selected row and column. When the data word consists of more than one bit, several columns of the selected row are selected by this coincident decoding technique. The enable input of the MC is now obtained by ANDing the X and Y selection lines.

Some commercial memory ICs provide more than one enable signal input on each chip. These multiple enable inputs are useful in building large memory systems employing coincident memory decoding schemes. The outputs of these ICs will also be either open collector or tristate, to enable easier interconnection.

9.4.1.2 Dynamic Memory

Several dynamic MC (DMC) configurations have been used. Figure 9.12 shows the most common DMC, built from a single MOS transistor and a capacitor. Read and write control is achieved by using two MOS transistors.

Consider the n-channel MOS (NMOS) transistor Q_1 in Figure 9.12a. The transistor has three terminals: drain, source, and gate. When the voltage on the gate is positive (and exceeds certain threshold value), the transistor conducts, thus connecting the drain to the source. If the gate voltage is negative, the transistor is off, thus isolating the drain and the source.

When the write control is high, Q_1 is on and D_{in} is transferred to the gate of Q_2, across the capacitor. If D_{in} is high, the capacitor is charged; otherwise, the capacitor is discharged through the gate-to-source resistance of Q_2, while the write is active. A discharged capacitor corresponds to storing a low in the MC, and thus, a stored low can be maintained indefinitely. A charged capacitor corresponds to storing a high. The high value can be maintained only as long as the capacitor remains charged. To fabricate denser memories, the capacitor is made very small and hence the capacitance will be quite

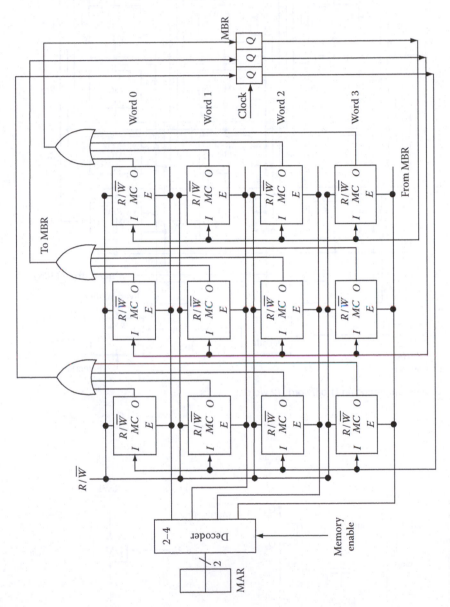

Figure 9.10 A 4-word, 3-bit semiconductor memory.

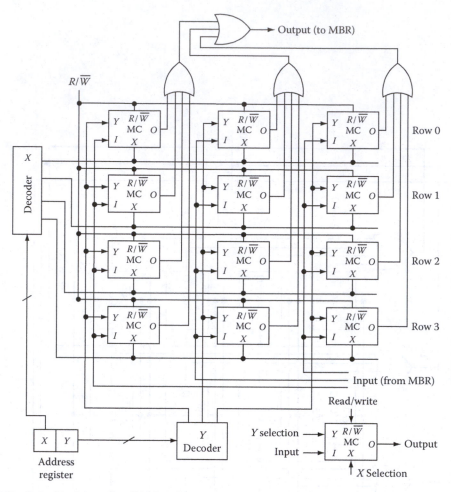

Figure 9.11   Coincident decoding (1 bit/word; 12 words).

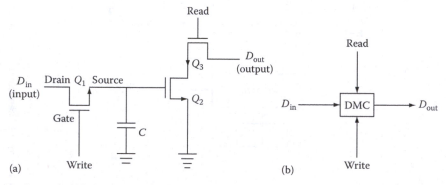

Figure 9.12   Dynamic MC: (a) Circuit. (b) Symbol.

small (on the order of a fraction of a picofarad). The capacitor thus discharges in several hundred milliseconds, requiring that the charge be restored or "refreshed," approximately every 2 ms.

When the read control goes high, Q_3 is on and the drain of Q_2 is connected to D_{out}. Since the stored data are impressed on the gate of Q_2 and the output is from the drain of Q_2, D_{out} will be the complement of the data stored in the cell. The output data are usually inverted by the external circuitry. Note that the read operation is destructive since the capacitor is discharged when the data are read. Thus, the data must be refreshed.

Figure 9.13 shows a (16 × 1) bit dynamic memory using the DMC of Figure 9.12. The DMCs are internally organized in a 4 × 4 matrix. The high-order two bits of the address select one of the rows. When the read is on, the data from the selected row are transferred to sense amplifiers. Since the capacitance of output lines is much higher than the capacitor in the DMC, output voltage is very low; consequently, sense amplifiers are required to detect the data value in the presence of noise. These amplifiers are also used to refresh the memory. The low-order two bits of the address are used to select one of the four sense amplifiers for the 1 bit data output. The data in the sense amplifiers

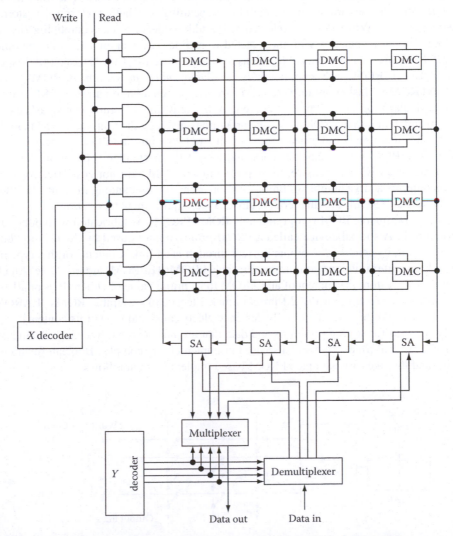

Figure 9.13   (16 × 1) Bit dynamic memory. *Note:* SA, sense amplifiers; DMC, dynamic MC.

are rewritten into the row of DMCs. To write 1 bit data, the selected row is first read, and the data in the selected sense amplifier are changed to the new data value just before the rewrite operation.

The need for the refresh results in the requirement of complex refresh circuitry and also reduces the speed of dynamic memory devices. But because of the small size of DMCs, it is possible to fabricate very dense memories on a chip. Usually the refresh operation is made transparent by performing a refresh when the memory is not being otherwise used by the system, thereby gaining some speed. When the memory capacity required is small, the refresh circuitry can be built on the memory chip itself. Such memory ICs are called *integrated dynamic RAMs* (iRAMs). Dynamic memory controller ICs that handle refresh and generate all the control signals for dynamic memory are available. These are used in building large dynamic memory systems.

9.4.1.3 Read-Only Memory

ROM is a RAM with data permanently stored in it. When the *n*-bit address is input to the ROM, the data stored at the addressed location are output on its output lines. ROM is basically a combinational logic device. Figure 9.14 shows a four-word ROM with 3 bits/word. The links at junctions of word lines and bit lines are either open or close depending on whether a 0 or a 1 is stored at the junction, respectively. When a word is selected by the address decoder, each output line (i.e., bit line) with a closed link at its junction with the selected word line will contain a 1, while the other lines contain a 0. In Figure 9.14, contents of locations 0 through 3 are 101, 010, 111, and 001, respectively.

Two types of ROMs are commercially available: *mask-programmed* ROMs and *user-programmed* ROMs. Mask-programmed ROMs are used when a large number of ROM units containing a particular program and/or data are required. The IC manufacturer can be asked to "burn" the program and data into the ROM unit. The program is given by the user and the IC manufacturer prepares a mask and uses it to fabricate the program and data into the ROM as the last step in the fabrication. The ROM is thus custom fabricated to suit the particular application. Since custom manufacturing of an IC is expensive, mask-programmed ROMs are not cost-effective unless the application requires a large number of units, thus spreading the cost among the units. Further, since the contents of these ROMs are unalterable, any change requires new fabrication.

A user-programmable ROM (programmable ROM [PROM]) is fabricated with either all 1s or all 0s stored in it. A special device called a *PROM programmer* is used by the user to "burn" the required program, by sending the proper current through each link. Contents of this type of ROM cannot be altered after initial programming. Erasable PROMs (EPROM) are available. An ultraviolet light is used to restore the content of an EPROM to its initial value of either all 0s or all 1s. It can then be reprogrammed using a PROM programmer. Electrically alterable ROMs (EAROMs) are another kind of ROM that uses a specially designed electrical signal to alter its contents.

ROMs are used for storing programs and data that are not expected to change during program execution (i.e., in real time). They are also used in implementing complex Boolean functions, code converters, and the like. An example of ROM-based implementation follows.

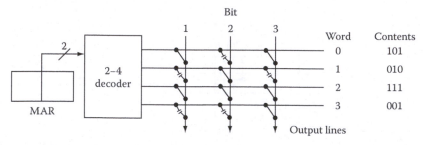

Figure 9.14 A 4-word, 3-bit/word ROM.

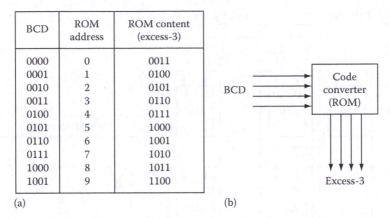

BCD	ROM address	ROM content (excess-3)
0000	0	0011
0001	1	0100
0010	2	0101
0011	3	0110
0100	4	0111
0101	5	1000
0110	6	1001
0111	7	1010
1000	8	1011
1001	9	1100

(a)   (b)

Figure 9.15 ROM-based implementation of code converter: (a) Code conversion table. (b) Code converter.

Example 9.1

Implement a binary-coded decimal (BCD)-to-excess-3 decoder using a ROM.

Figure 9.15 shows the BCD-to-excess-3 conversion. Since there are 10 input (BCD) combinations, a ROM with 16 words ($2^3 < 10 < 2^4$) must be used. The first 10 words of the ROM will contain the 10 excess-3 code words. Each word is 4 bits long. The BCD input appears on the four-address input lines of the ROM. The content of the addressed word is output on the output lines. This output is the required excess-3 code.

9.4.2 Associative Memory

A typical associative MC (AMC) built from a JK flip-flop is shown in Figure 9.16. The response is 1 when either the data bit (D) and the memory bit (Q) match while the mask bit (M) is 1 or when the mask bit is 0 (corresponding to a "do not compare"). A truth table and a block diagram for the simplified cell are also shown in the figure. In addition to the response circuitry, an AMC will have read, write, and enable circuits similar to the RAM cell shown in Figure 9.9.

A four-word, 3 bits/word AM built out of the previous cells is shown in Figure 9.17. The data and mask registers are each 3 bits long. The word-select register of Figure 9.3 is neglected here. Hence, all memory words are selected for comparison with the data register. Response outputs of all cells in a word are ANDed together to form the word response signal. Thus, a word is a respondent if the

Data (D)	Mask (M)	Q	Response
0	0	0	1
0	0	1	1
0	1	0	1
0	1	1	0
1	0	0	1
1	0	1	1
1	1	0	0
1	1	1	1

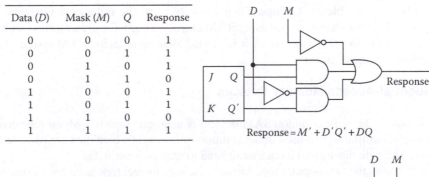

$Response = M' + D'Q' + DQ$

Figure 9.16 A simplified AMC.

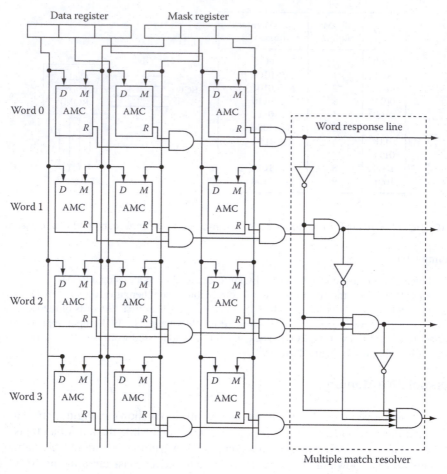

Figure 9.17 A 3-bit, 4-word AM.

response output of each and every cell in the word is a 1. The MMR circuit shown selects the first respondent. The first respondent then drives to 0 the response outputs of other words following it.

Input (or write) and output (or read) circuitry is also needed. It is similar to that of the RWM system shown in Figure 9.10 and hence is not shown in Figure 9.17.

Small AMs are available as IC chips. Their capacity is on the order of 8 bits/word by eight words. Larger AM systems are designed using RAM chips. As such, these memories can be used in both RAM and CAM modes. Because of their increased logic complexity, CAM systems cost much more than RWMs of equal capacity.

9.4.3 Sequential-Access Memory Devices

Magnetic tape is the most common SAM device. A magnetic tape is a Mylar tape coated with magnetic material (similar to that used in home music systems) on which data are recorded as magnetic patterns. The tape moves past a read/write head to read or write data.

Figure 9.18 shows the two popular tape formats: the reel-to-reel tape used for storing large volumes of data (usually with large-scale and minicomputer systems) and the cassette tape used for small data volumes (usually with microcomputer systems). Data are recorded on tracks. A track on a magnetic tape runs along the length of the tape and occupies a width just sufficient to store a bit. On a nine-track tape, for example, the width of the tape is divided into nine tracks, and each character

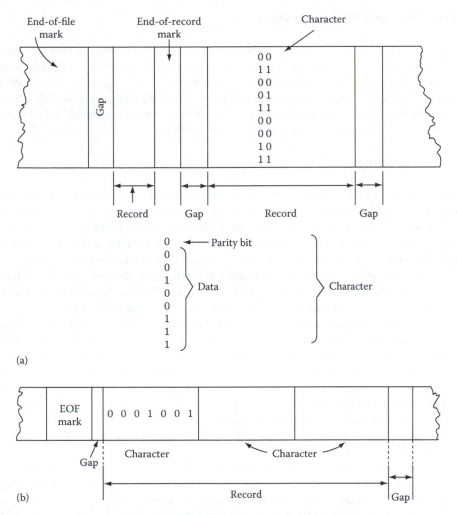

Figure 9.18 Magnetic tape formats: (a) Nine-track reel to reel. (b) Cassette tape.

of data is represented with 9 bits (1 bit on each track). One or more of these bits are usually a parity bit, which facilitates error detection and correction, as shown in Figure 9.18a.

Several characters grouped together form a *record*. The records are separated by an inter-record gap (about 3/4 in.) and an end-of-record mark (which is a special character). A set of records forms a *file*. The files are separated by an end-of-file mark and a gap (about 3 in.). On cassette tapes, the data are recorded in a serial mode on one track as shown in Figure 9.18b.

Recording, or *writing*, on magnetic devices is the process of creating magnetic flux patterns on the device; the sensing of the flux pattern when the medium moves past the read/write head constitutes the *reading* of the data. In reel-to-reel tapes, the data are recorded along the tracks digitally. In a cassette tape, each bit is converted into an audio frequency and is recorded. Digital cassette recording techniques are also becoming popular. Note that the information on a magnetic tape is nonvolatile.

Magnetic tapes permit recording of vast amounts of data at a very low cost. However, the access time, being a function of the position of the data on the tape with respect to the read/write head position along the length of the tape, can be very long. Sequential-access devices thus form low-cost secondary memory devices, which are primarily used for storing the data that do not have to be used frequently, such as system backup, archival storage, and transporting data to other sites.

9.4.4 Direct-Access Storage Devices

Magnetic and optical disks are the most popular direct- or semi-random-access storage devices. Accessing data in these devices requires two steps: random or direct movement of read/write heads to the vicinity of data, followed by a sequential access. These mass-memory devices are used as secondary storage devices in a computer system for storing data and programs. Solid-state drives (SSDs) that use the faster flash-based memory, instead of magnetic or optical disks to store data, have also begun to emerge as a newer kind of storage medium.

9.4.4.1 Magnetic Disk

A magnetic disk (see Figure 9.7) is a flat circular surface coated with a magnetic material, much like a phonograph record. Several such disks are mounted on a rotating spindle. Each surface will have a read/write head. Each surface is divided into several concentric circles (*tracks*), with track 0 being the outermost. By first positioning read/write heads to the proper track, the data on the track can be accessed sequentially. A track is normally divided into several *sectors*, each of which corresponds to a data word. The address of a data word on the disk thus corresponds to a track number and a sector number. All the tracks under the read/write heads at a given time form a *cylinder*. The time taken by the read/write heads to position themselves on a cylinder is called *seek time*. Once the read/write heads are positioned, the time taken for the data to appear below them is the *rotational delay*. The *access time* is the sum of seek time and rotational delay. *Latency* is a function of the disk rotation speed. It is the average time taken for the desired sector to appear under the read/write head once the disk arm is positioned at the track.

The *disk directory*, maintained on the disk drive, maps logical file information to physical address consisting of a cylinder number, surface number, and sector number. At the beginning of each read and write operation, the disk directory is read. As such, the track on which the directory is placed plays an important role in improving the access time.

Earlier disk drives came with removable disks called disk packs. All the disk drives today come as sealed units (Winchester drives).

Magnetic disks are available as either *hard disks* or *floppy disks*. The data storage and access formats for both types of disks are the same. Floppy disks have a flexible disk surface and were very popular storage devices, especially for microcomputer systems, although their data-transfer rate is slower than that of hard disk devices. Hard disks have remained the main media for data storage and backup. Floppy disks have almost become obsolete, yielding to *flash memory devices* and *thumb drives*.

9.4.4.2 SSDs

SSDs are a new kind of mass storage medium that uses ICs to store data persistently, instead of magnetic disk platters as used by conventional hard disk drives (HDDs). Since they use ICs for memory, these drives do not contain any moveable mechanical parts. These drives typically use NAND-based flash memory for data storage.

Flash memory is a kind of nonvolatile computer storage chip that can be erased and reprogrammed electrically. It stores information in an array of MCs made from floating-gate transistors. They are developed from electrically erasable programmable ROM (EEPROM) modules and are required to be erased in large blocks before new data can be rewritten on them.

This is similar to the kind of memory used in Universal Serial Bus (USB) flash drives (also known as thumb drives); however, it is much slower in such USB drives. USB drives use the slower USB as the interface for accessing the flash memory when compared to the faster Serial AT Attachment (Serial ATA or SATA) used by SSDs. SSDs also tend to use more complex memory controllers as these drives are generally used as fixed disks, instead of removable ones like USB drives.

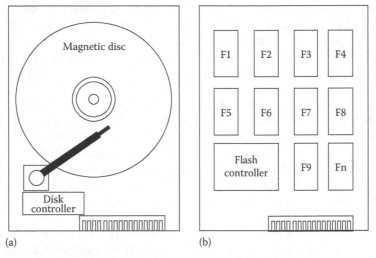

Figure 9.19 Mass storage devices. (a) Conventional HDD. (b) SSD. In SSD, the data is stored on Flash memory modules F1, F2…Fn using the Flash Controller. In HDD, the Disk Controller uses the mechanical head to write and read data on the circular magnetic discs.

Figure 9.19 shows images of a conventional HDD (a) and an SSD (b). Magnetic disk platters used in conventional hard drives make them sensitive to physical shocks and vibrations, since they are required to perform high-precision movement during their operation. Moreover, their data throughput speed is limited by the maximum rotating speed of the disk spindle on which these drives are mounted.

Since SSDs use flash storage instead of such disk platters, they are more robust and resistant to physical shocks. Due to the high speed of NAND memory, these drives are typically in the order of four or more times faster than conventional hard drives and use less power as well. These drives are also silent in operation and cause no vibrations since they have no moving parts.

Data are accessed on SSDs with the help of a controller chip that reads and writes by sending electrical signals to the NAND-based flash memory. This is contrary to HDDs where mechanical heads are used to read and write data on the spinning magnetic disks. In particular, to read and write, the mechanical head detects and modifies the magnetization of the track on the spinning disk directly below it.

Typically SSDs use the same conventional I/O interfaces like SATA, which are used by HDDs and hence can be easily replaced in current generation systems.

The current generation of SATA interface (SATA revision 3.0) offers data throughput speeds of up to 6 GB/s. Due to the use of high-speed flash storage in SDDs, these drives are able to better utilize the higher bandwidth provided by SATA 3.0 when compared to HDDs.

As with most new technologies, SSDs have a drawback of higher cost per gigabit when compared to HDDs. However, the prices of such drives have continued to decline.

A new category of disk drives known as "hybrid drives" is also available that use both SSDs and HDDs in a single unit. Typically, these drives use the SSD component as a cache to store frequently accessed data and use the HDD component for everything else. Thus, these drives are able to offer better performance than conventional HDDs, but lower than that of SDDs, while maintaining a price in between HDDs and SSDs.

9.4.4.2.1 Major Components of SSD

An SSD consists of two major components: controller and memory. The memory component is used to store data that are operated on and controlled by the controller.

The **memory** component consists of either DRAM volatile memory or NAND flash nonvolatile memory. DRAM volatile memory is faster than the NAND counterpart; however it loses data once the power supplied to it is interrupted. Hence, these are used in SSDs in situations where speed is important and data persistence is not. To provide some form of data persistence, these drives are usually connected to external power supplies, to prevent data loss in case of unexpected power failure.

Most SSDs currently use NAND flash nonvolatile memory that is cheaper and slower than DRAM, but offers persistent storage. NAND flash memory can be further categorized into single-level cell (SLC) and multilevel cell (MLC) memory. SLC NAND memory consists of a single level per cell that is used to store a bit of information, compared to MLC that consists of multiple levels per cell, each capable of storing bits of information. Since MLC can store more bits per cell, they are cheaper than SLC flash; however, this also increases the chances of data errors in the drive. MLC-based memory is thus used for consumer grade SSD, and SLC is used for enterprise grade.

The **controller**, also known as the flash memory controller (FMC), acts as the interface between the memory and the host computer. It consists of an embedded processor that runs firmware-level code to operate and maintain the flash memory and the data stored on it. The functions of FMC include (Bechtolsheim 2008, Werner 2010) the following:

Error correction: The controller uses error-correcting code (ECC) to fix any potential errors while storing or retrieving data.
Wear leveling: Cells in flash storage have a limit to the number of times it can be erased and rewritten. Wear leveling performed by FMC includes arranging data in memory such that all cells are used evenly. This process is used to prolong the service life of the drive.
Bad block mapping: The controller maps bad sectors (or blocks) so that they can be avoided during future SSD operations.
Scrubbing: It is a background process that regularly inspects stored data and corrects any errors using techniques like ECC. This helps in correcting errors sooner and prevents accumulation of errors that may lead to situations that cannot be repaired.
Write and read caching: The controller caches recently and frequently used data to improve the performance of the drive.
Encryption: The controller offers the capability to encrypt user data on the drive for security.

Figure 9.20 is the block diagram of the SanDisk SSD P4. It shows the arrangement of the FMC and the NAND memory in the drive. In this figure, the FMC (NAND flash SSD controller) is connected to NAND flash memory (NAND flash array – 0 to 1) on its right side. These two together represent

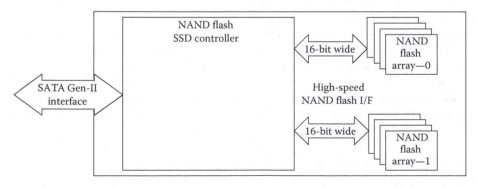

Figure 9.20 Block diagram of SanDisk SSD P4. (From SanDisk Product Manual, Rev 1.1.)

the two major components of the SSD. The FMC is connected to the SATA I/O interface (SATA Gen-II interface) on its left side that is then eventually connected to the host computer.

9.4.4.3 Redundant Array of Independent Disks

Even the smallest of the computer system nowadays uses a disk drive for storing data and programs. A disk head crash at the end of a data gathering session or a business day results in expensive data regathering and loss of productivity. One solution to this problem is to duplicate data on several disks. This redundancy provides the fault tolerance (fault handling) when one of the disks fails. Further, if each disk has its own controller and cache, more than one of these disks can be accessed simultaneously, thereby increasing the throughput of the system. Traditionally, large computer systems used single large expensive disk (SLED) drives.

In the late 1980s small disk drives were available for microcomputer systems. RAID systems were formed by utilizing an array of inexpensive disk units. The word "inexpensive" is now replaced by "independent." RAID provides the advantages of both fault tolerance and high performance and hence is used in critical applications employing large file servers, transaction application servers, and desktop systems for applications such as computer-aided design (CAD), multimedia editing, and playback, where high-transfer rates are needed.

A RAID system consists of a set of matched hard drives and a RAID controller. The array management software provides the logical to physical mapping of data and is part of the RAID controller. Descriptions of several popular configurations (levels) of RAID are given later. The following material is extracted from http://www.acnc.com/raid.html:

> *RAID level 0: disk striping without fault tolerance*: In this mode, the data are divided into blocks and these blocks are placed on drives in the array in an interleaved fashion. Figure 9.21 shows the data representation of a data file with blocks K, L, M, N, O, etc. RAID level 0 requires at least two drives. If independent controllers are used, one drive can be writing or reading a block, while the other is seeking the next block.
>
> This mode does not offer any redundancy and does not use parity overhead. It utilizes the array capacity fully. For example, if there are two 120 GB drives in the array, the total capacity of the system would be 240 GB. It is not a true RAID because it is not redundant and it is not fault tolerant. If one of the drives fails, then all the contents of that drive are not accessible, making the data in the system not usable.
>
> *RAID level 1: mirroring and duplexing*: RAID level 1 requires at least two drives. As shown in Figure 9.22, each block of data is stored at least in two drives (mirroring) so as to provide fault tolerance. Even when one drive fails, the system can continue to operate by using the data from the mirrored drive.

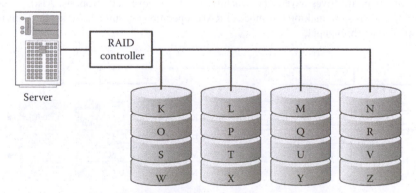

Figure 9.21   RAID level 0: Striped disk array without fault tolerance.

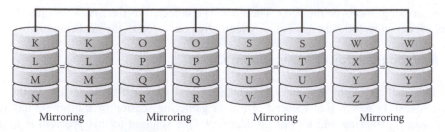

Figure 9.22 RAID level 1: Mirroring and duplexing.

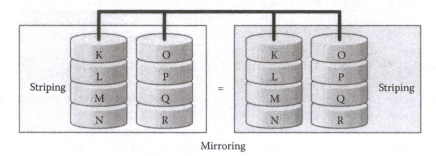

Figure 9.23 RAID level 0 + 1: High data-transfer performance.

This mode does not offer any improvement in data-access speed. The capacity of the disk system is half the sum of individual drive capacities. It is used in applications such as accounting, payroll, and other financial environments that require very high availability.

RAID level 0 + 1: high data-transfer performance: RAID level 0 + 1 requires a minimum of four drives and is the combination of RAID 0 and RAID 1. It is created by first creating two RAID 0 sets and mirroring them, as shown in Figure 9.23.

Although this mode offers high fault tolerance and I/O rates, it is very expensive due to its 100% overhead. It is used in applications such as imaging and general file servers, where high performance is needed.

RAID level 2: Hamming code: Instead of writing data blocks of arbitrary size, RAID 2 writes data 1 bit/strip, as shown in Figure 9.24. Thus, to accommodate one ASCII character, we will need eight drives. Additional drives are used to hold the Hamming code used to detect 2 bit errors and correct 1 bit errors. When any of the data drives fail, the Hamming code can be used to reconstruct the data on the failed drive. Since 1 bit of the data is written on each drive along with the corresponding Hamming code, all drives have to be synchronized to retrieve data properly. Also, generation of Hamming code is slow, making this mode of RAID operation not suitable for applications requiring high-speed data throughput.

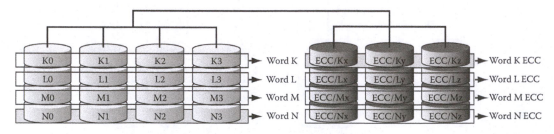

Figure 9.24 RAID level 2: Hamming code ECC.

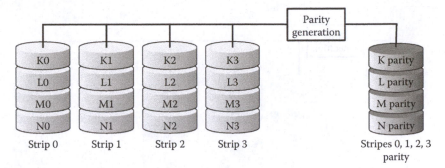

Figure 9.25 RAID level 3: Parallel transfer with parity.

RAID level 3: parallel transfer with parity: In this RAID level shown in Figure 9.25, parity is used instead of mirroring to protect against disk failure. It requires a minimum of three drives to implement. One drive is designated as parity drive and it contains the parity information computed from the other drives. A simple parity bit is calculated by exclusively ORing the corresponding bits in each stripe. This provides the error detection capability. Additional parity bits are needed to provide the error correction.

This mode offers reduced costs, since fewer drives are needed to implement redundant storage. However, performance is degraded by the parity computation and bottlenecks are caused because of single dedicated parity drive.

RAID level 4: independent data disks with shared parity: RAID 4 shown in Figure 9.26 is RAID 0 with parity. The data are written as blocks (strips) of equal size on each drive creating a stripe across all the drives followed by a parity drive that contains the corresponding parity strip. The parity drive becomes the bottleneck since all data read/write operations require access to it. In practice, not all applications provide for uniform data block sizes. Storing varying size data blocks becomes impractical in this mode of RAID.

RAID level 5: independent data disks with distributed parity blocks: RAID 5 may be the most popular and powerful RAID configuration. As shown in Figure 9.27, it provides striping of data as well as striping of parity information for error recovery. The parity block is distributed among the drives of array. It requires a minimum of three drives to implement. Bottlenecks induced by the parity drive are eliminated as well as cost is reduced.

This mode offers high read data-transaction rates and medium write data-transaction rates. Disk failure has a medium impact on throughput. If a disk failure occurs, it is very difficult to rebuild RAID 5 compared with RAID level 1. Typical applications have been in file and application servers, database servers, web, e-mail, and news servers.

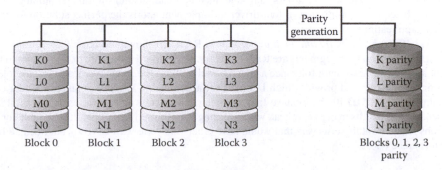

Figure 9.26 RAID level 4: Independent data disks with shared parity disk.

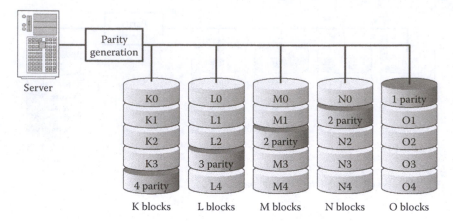

Figure 9.27   RAID level 5: Independent data disks with distributed parity blocks.

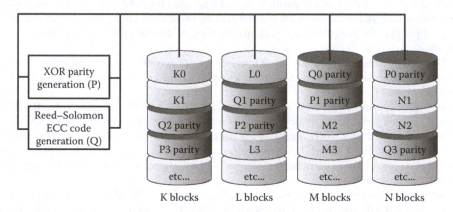

Figure 9.28   RAID level 6: Independent data disks with two independent distributed parity schemes.

RAID level 6: independent data disks with two independent distributed parity schemes: RAID 6 shown in Figure 9.28 is essentially an extension of RAID 5 that allows for additional fault tolerance by using a second independent distributed parity scheme (dual parity). Data are striped on a block level across a set of drives, just like in RAID 5, and a second set of parity is calculated and written across all the drives. Two independent parity computations are used in order to provide protection against double disk failure. Two different algorithms are employed to achieve this. RAID 6 provides for an extremely high data fault tolerance and can sustain multiple simultaneous drive failures. It requires a minimum of four drives to implement and is the perfect solution for mission critical applications.

RAID level 10: high reliability and high performance: RAID 10 shown in Figure 9.29 is implemented as a striped array whose segments are RAID 1 arrays. It requires a minimum of four drives to implement and has the same fault tolerance as RAID level 1. It has the same overhead for fault tolerance as mirroring alone and provides high I/O rates through striping RAID 1 segments. Under certain circumstances, RAID 10 array can sustain multiple simultaneous drive failures. However, it is very expensive and results in high overhead while offering very limited scalability. It is suitable for applications such as database servers that would have otherwise gone with RAID 1 but need additional performance boost.

As can be expected, a large system might use more than one level of RAID depending on the cost, speed, and fault tolerance requirements.

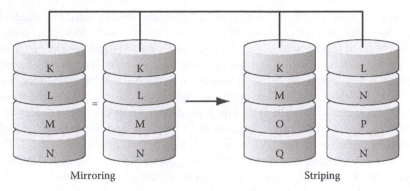

Figure 9.29 RAID level 10: Reliability combined with high performance.

9.4.4.4 Optical Disks

Four types of optical disks are available. CD-ROMs are similar to mask-programmed ROMs in which the data are stored on the disk during the last stage of disk fabrication. The data, once stored, cannot be altered. CD-recordable (CD-R) and WORM optical disks allow writing the data once. The portions of the disk that are written once cannot be altered. CD-rewritable (CD-RW) or *erasable* optical disks are similar to magnetic disks that allow repeated erasing and storing of data. *Digital video disks* (DVDs) allow much higher density storage and offer higher speeds than CDs.

Manufacturing of CD-ROMs is similar to that of phonograph records except that digital data are recorded by burning pit and no-pit patterns on the plastic substrate on the disk with a laser beam. The substrate is then metallized and sealed. While reading the recorded data, 1s and 0s are distinguished by the differing reflectivity of an incident laser beam. Since the data once recorded cannot be altered, the application for CD-ROMs has been in storing the data that do not change. The advantage of CD-ROMs over magnetic disks is their low cost, high density, and non-erasability.

Erasable optical disks use a coating of a magneto-optic material on the disk surface. To record data on these disks, a laser beam is used to heat the magneto-optic material in the presence of a bias field applied by a bias coil. The bit positions that are heated take on the magnetic polarization of the bias field. This polarization is retained when the surface is cooled. If the bias field is reversed and the surface is heated, the data at the corresponding bit position are erased. Thus, changing the data on the disk requires a two-step operation. All the bits in a track are first erased and new data are then written onto the track. During reading, the polarization of the read laser beam is rotated by the magnetic field. Thus, the polarization of the laser beam striking the written bit positions is different from that of the rest of the media.

WORM devices are similar to erasable disks except that the portions of the disk that are written once cannot be erased and rewritten.

Optical disk technology offers densities of about 50,000 bits and 20,000 tracks/in., resulting in a capacity of about 600 MB/3.5 in. disk. Corresponding numbers for the magnetic disk technology are 150,000 bits and 2000 tracks/in., or 200 MB/3.5 in. disk. Thus, optical storage offers a 3-to-1 advantage over magnetic storage in terms of capacity. It also offers a better per-unit storage cost. The storage densities of magnetic disks are also rapidly increasing, especially since the advent of vertical recording formats.

The data-transfer rates of optical disks are much lower than magnetic disks owing to the following factors: It takes two revolutions to alter the data on the track; the rotation speed needs to be about half that of magnetic disks to allow time for heating and changing of bit positions; and, since the optical read/write heads are bulkier than their magnetic counterparts, the seek times are higher.

DVD also comes in recordable and nonrecordable forms. Digital versatile disks, as they are called now, are essentially quad-density CDs that rotate about three times faster than CDs and use about twice the pit density. DVDs are configured in single- and double-sided and single- and double-layer versions and offer storage of about 20 GB of data, music, and video. The progress in laser technology has resulted in the use of blue-violet lasers utilizing a 450 nm wavelength. Two competing formats designed to supersede the DVD format are now on the market. The Blu-ray format developed by a consortium of nine consumer electronics manufacturers (such as Sony, Pioneer, and Samsung) offers disks up to 25 GB/layer. There are up to quadruple layer Blu-ray disks available in the market that offers a total capacity of 128 GB. The high-definition/density DVD (HD-DVD) format developed by NEC and Toshiba offers disks up to 15 GB/layer and is available in single-layer and dual-layer formats. There are other DVD formats in the horizon offering 30–35 GB DVDs. These include HD-DVD, advanced optical disk (AOD), and holographic versatile disk (HVD).

The storage density and the speed of access of disks improve so rapidly that a listing of such characteristics soon becomes outdated. Refer to the magazines listed in the Bibliography section of this chapter for such details.

9.5 MEMORY SYSTEM DESIGN USING ICs

Refer to Appendix A for details of representative memory ICs. Memory system designers use such commercially available memory ICs to design memory systems of required size and other characteristics. The major steps in such memory designs are the following:

1. Based on speed and cost parameters, determining the type of memory ICs (static or dynamic) to be used in the design.
2. Selecting an available IC of the type selected previously, based on access time requirements and other physical parameters, such as the restriction on the number of chips that can be used and the power requirements. It is generally better to select an IC with the largest capacity in order to reduce the number of ICs in the system.
3. Determining the number of ICs needed N = (total memory capacity)/(chip capacity).
4. Arranging the previous N ICs in a P × Q matrix, where Q = (number of bits per word in memory system)/(number of bits per word in the IC) and P = N/Q.
5. Designing the decoding circuitry to select a unique word corresponding to each address.

We have not addressed the issue of memory control in this design procedure. The control unit of the computer system, of which the memory is a part, should produce control signals to strobe the address into the MAR, enable read/write, and gate the data in and out of MBR at appropriate times.

The following example illustrates the design.

Example 9.2

Design a $4k \times 8$ memory, using Intel 2114 RAM chips.

1. Number of chips needed $= \dfrac{\text{Total memory capacity}}{\text{Chip capacity}} = \dfrac{4k \times 8}{1k \times 4} = 8.$
2. The memory system MAR will have 12 bits, since $4k = 4 \times 1024 = 2^{12}$; the MBR will have 8 bits.
3. Since 2114s are organized with 4 bits/word, two chips are used in forming a memory word of 8 bits. Thus, the eight 2114s are arranged in four rows, with two chips per row.
4. The 2114 has 10 address lines. The least significant 10 bits of the memory system MAR are connected to the 10 address lines of each 2114. A 2–4 decoder is used to decode the most significant 2 bits of the MAR and to select one of the four rows of 2114 chips through the $\overline{CS}$ signal on each 2114 chip.
5. I/O lines of chips in each row are connected to the MBR. Note that these I/O lines are configured as tristate. The $\overline{WE}$ lines of all the 2114 chips are tied together to form the system $\overline{WE}$.

The memory system is shown in Figure 9.30. Note that the number of bits in the memory word can be increased in multiples of 4 simply by including additional columns of chips. If the number of words needs to be extended beyond $4k$, additional decoding circuitry will be needed.

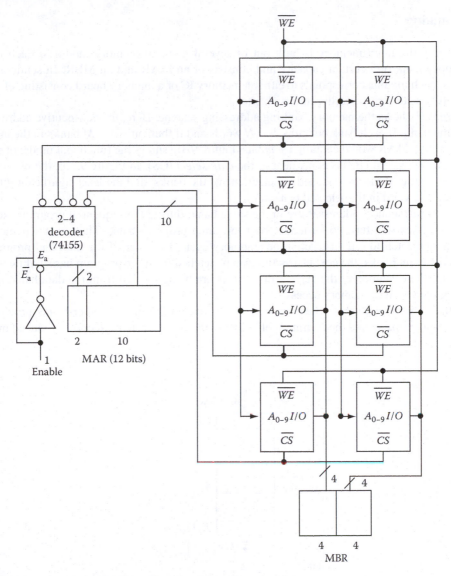

Figure 9.30 A 4k × 8 memory. *Note:* Power and ground are not shown.

9.6 SPEED ENHANCEMENT

Traditionally, memory cycle times have been much longer than processor cycle times; this speed gap between the memory and the processor means that the processor must wait for memory to respond to an access request. With the advances to hardware technology, faster semiconductor memories are now available and have replaced core memories as primary memory devices. But the processor–memory speed gap still exists, since the processor hardware speeds have also increased. Several techniques have been used to reduce this speed gap and optimize the cost of the memory system.

The obvious method of increasing the speed of the memory system is by using a higher-speed memory technology. Once the technology is selected, the access speeds can be increased further by judicious use of address-decoding and access techniques. Six such techniques are described in the following sections.

9.6.1 Banking

Typically, the main memory is built out of several physical memory modules. Each module is a memory bank of a certain capacity and consists of an MAR and an MBR. In semiconductor memories, each module corresponds to either a memory IC or a memory board consisting of several memory ICs, as described earlier in this chapter.

Figure 9.31 shows the memory banking addressing scheme. Here, the consecutive addresses lie in the same bank. If each bank contains $2^n = N$ words and if there are $2^m = M$ banks in the memory, then the system MAR would contain $n + m$ bits. Figure 9.31a shows the functional model of a bank. The *bank-select* signal (BS) is equivalent to the *chip select* (CS). In Figure 9.31b, the most significant m bits of the MAR are decoded to select one of the banks, and the least significant n bits are used to select a word in the selected bank.

Since the subsequent addresses are in the same bank, during the sequential program execution process, accesses will have to be made from the same program bank. Thus, this scheme limits the instruction fetch to one instruction per memory cycle. However, if the data and program are stored in different banks, the next instruction can be fetched from the program bank, while the data required for the execution of the current instruction are being fetched from the data bank, thereby increasing the speed of memory access.

Another advantage of this scheme is that even if one bank fails, the other banks provide continuous memory space, and the operation of the machine is unaffected (except for reduced memory capacity).

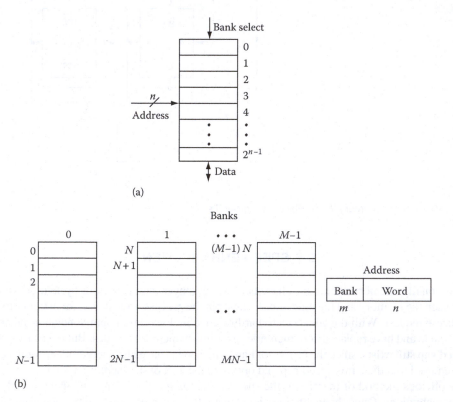

Figure 9.31 Memory banking. *Note:* $2^m = M$, $2^n = N$: (a) A bank. (b) Banking scheme.

9.6.2 Interleaving

Interleaving the memory banks is a technique to spread the subsequent addresses to separate physical banks in the memory system. This is done by using the low-order bits of the address to select the bank, as shown in Figure 9.32.

The advantage of the interleaved memory organization is that the access request for the next word in the memory can be initiated while the current word is being accessed, in an overlapping manner. This mode of access increases the overall speed of memory access. The disadvantage of this scheme is that if one of the memory banks fails, the complete memory system becomes inoperative.

An alternative organization is to implement the memory as several subsystems, each subsystem consisting of several interleaved banks. Figure 9.33 shows such an organization.

There is no general guideline to select one addressing scheme over the other among the three described earlier. Each computer system uses its own scheme. The basic aim of these schemes is to spread subsequent memory references over several physical banks so that faster accessing is possible.

9.6.3 Multiport Memories

Multiple-port (multiport) memories are available in which each port corresponds to an MAR and an MBR. Independent access to the memory can be made from each port. The memory system resolves the conflicts between the ports on a priority basis. Multiport memories are useful in an environment where more than one device accesses the memory. Examples of such systems are a

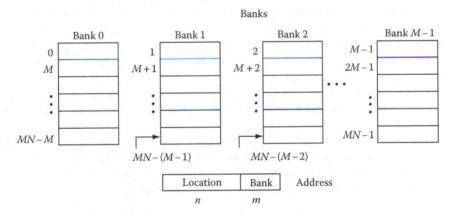

Figure 9.32 Interleaving. *Note:* $2^m = M$, $2^n = N$.

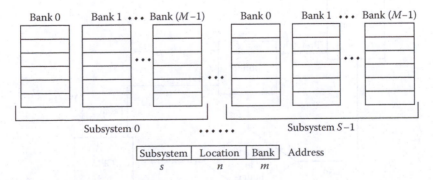

Figure 9.33 Interleaved subsystems. *Note:* $2^m = M$, $2^n = N$, $2^s = S$.

single-processor system with a direct-memory access (DMA), I/O controller (see Chapter 7), and a multiprocessor system with more than one CPU (see Chapter 12).

9.6.4 Wider-Word Fetch

Consider the interleaved memory in Figure 2.30. If M MBRs are employed one for each block and if all the banks are activated (i.e., bank-select portion of the address is not used), then we can fetch M words in one cycle into the MBRs. This is the concept of wider-word fetch. For instance, the IBM 370 fetches 64 bits (two words) in each memory access. This enhances execution speed, because the second word fetched most likely contains the next instruction to be executed, thus saving a "wait" for the fetch. If the second word does not contain the required instruction (e.g., during a jump), a new fetch is required.

9.6.5 Instruction Buffer

The M MBRs in the previous scheme can be considered as an instruction buffer. Providing a first-in, first-out (FIFO) buffer (or queue) between the CPU and the primary memory enhances the instruction fetch speed. Instructions from the primary memory are fed into the buffer at one end, and the CPU fetches the instructions from the other end, as shown in Figure 9.34. As long as the instruction execution is sequential, the two operations of filling the buffer (prefetching) and fetching from the buffer into CPU can go on simultaneously. But when a jump (conditional or unconditional) instruction is executed, the next instruction to be executed may or may not be in the buffer. If the required instruction is not in the buffer, the fetch operation must be directed to the primary memory, and the buffer is refilled from the new memory address. If the buffer is large enough, the complete range of the jump (or loop) may be accommodated in the buffer. In such cases, the CPU can signal the buffer to FREEZE, thereby stopping the prefetch operation. Once the loop or jump is satisfied, both FETCH operations can continue normally.

The buffer management requires hardware components to manage the queue (e.g., check for queue full or queue empty) and mechanisms to identify the address range in the buffer and to freeze and unfreeze the buffer.

Intel 8086 processor uses a 6 byte-long instruction buffer organized as a queue. CDC 6600 uses an instruction buffer that can store eight 60 bit words, that is, 16–32 instructions, since the instructions are either 15 or 30 bits long. Figure 9.35 shows the instruction buffer organization. Instructions from main memory are brought into the buffer through a buffer register. The lowest level of the

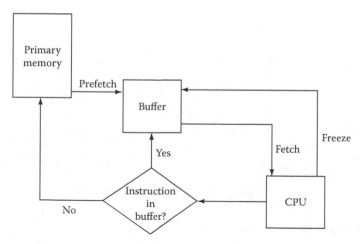

Figure 9.34   Instruction buffer.

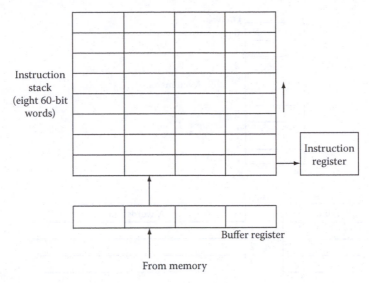

Figure 9.35 CDC 6600 instruction buffer.

buffer is transferred to the instruction register for execution, while the contents of the buffer move up one position. A new set of instructions (60 bits) enters the buffer through the buffer register. When a branch instruction is encountered, if the address of the branch is within the range of the buffer, the next instruction is retrieved from it. If not, instructions are fetched from the new memory address.

9.6.6 Cache Memory

Analyses of memory reference characteristics of programs have shown that typical programs spend most of their execution times in a few main modules (or routines) and tight loops. Therefore, the addresses generated by the processor (during instruction and data accessing) over short time periods tend to cluster around small regions in the main memory. This property, known as the *program locality principle*, is utilized in the design of cache memory scheme. The cache memory is a small (of the order of 1/2k to 2k words) but fast (of the order of 5–10 times the speed of main memory) memory module inserted between the processor and the main memory. It contains the information most frequently needed by the processor, and the processor accesses the cache rather than the main memory, thereby increasing the access speed.

Figure 9.36 shows a cache memory organization. Here, the primary memory and cache are each divided into blocks of $2^n = N$ words. The block size depends on the reference characteristics implied by the program locality and the main-memory organization. For instance, if the main memory is organized in an N-way interleaved memory, it is convenient to make each block N words long, since all N words can be retrieved from the main memory in one cycle. We will assume that the cache capacity is $2^b = B$ blocks. In the following discussion, we will refer to the block in the main memory as a main-memory *frame* and a block in the cache memory as a *block* or a *cache line*.

In addition to the *data* area (with a total capacity of $B \times N$ words), the cache also has a *tag* area consisting of B tags. Each tag in the tag area identifies the address range of the N words in the corresponding block in the cache. If the primary memory address A contains p bits and the least significant n bits are used to represent the N words in each frame, the remaining $(p - n)$ bits of the address form the tag for the block, whereby the tag is the beginning address of the block of N words.

When the processor references a primary memory address A, the cache mechanism compares the tag portion of A to each tag in the tag area. If there is a matching tag (i.e., a cache *hit*), the corresponding cache block is retrieved from the cache, and the least significant n bits of A are used to

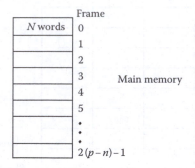

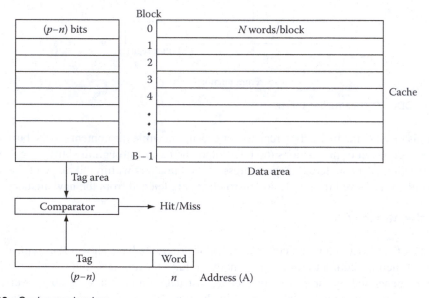

Figure 9.36 Cache mechanism.

select the appropriate word from the block, to be transmitted to the processor. If there is no matching tag (i.e., a cache *miss*), first the frame corresponding to A is brought into the cache and then the required word is transferred to the processor.

Example 9.3

Figure 9.37 shows the cache mechanism for a system with 64 kB of primary memory and a 1 kB cache. There are 8 bytes/block.

In general, a cache mechanism consists of the following three functions:

1. *Address translation function*: Determines if the referenced block is in the cache or not and handles the movement of blocks between the primary memory and the cache.
2. *Address mapping function*: Determines where the blocks are to be located in the cache.
3. *Replacement algorithm*: Determines which of the blocks in the cache can be replaced, when space is needed for new blocks during a cache miss. These functions are described next.

9.6.6.1 *Address Translation*

The address translation function of Figure 9.36 is simple to implement but enforces an N-word boundary to each frame. Thus, even if the referenced address A corresponds to the last word in the frame, all the words of the frame are transferred to the cache. If the translation function were to

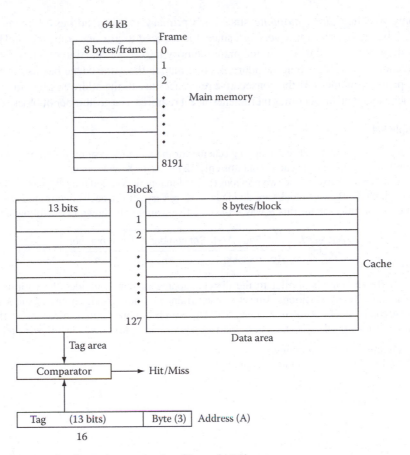

Figure 9.37   An example of a cache mechanism (Example 9.3).

be general, in the sense that an N-word frame starting from an arbitrary address A is transferred to the cache, then each tag would have to be p bits long, rather than $(p - n)$ bits. In addition to this overhead, in this general scheme it will not always be possible to retrieve N words from A in one memory cycle, even if the primary memory is N-way interleaved. Because of these disadvantages of the general addressing scheme, the scheme shown in Figure 9.36 is the most popular one.

9.6.6.2 Address Mapping Function

In the address mapping scheme of Figure 9.36, a main-memory frame can occupy any of the cache blocks. Therefore, this scheme is called a *fully associative* mapping scheme. The disadvantage with this scheme is that all the B tags must be searched in order to determine a hit or miss. If B is large, this search can be time consuming. If the tag area is implemented using a RAM, an average of $B/2$ RAM cycles are needed to complete the search. An alternative is to implement the tag area of the cache using an AM. Then the search through B tags requires just one AM (compare) cycle.

One way of increasing the tag search speed is to use the least significant b bits of the tag to indicate the cache blocks in which a frame can reside. Then, the tags would be only $(p–n–b)$ bits long, as shown in the following:

Address A (p bits)	Tag	Block number	Word number
	$p–n–b$	b	n

This mapping is called *direct mapping* since each primary memory address A maps to a unique cache block. The tag search is now very fast since it is reduced to just one comparison. However, this mechanism divides the 2^p addresses in the main memory into $B = 2^b$ partitions. Thus, 2^{p-b} addresses map to each cache block. Note that the addresses that map to the same cache blocks are $B \times N$ words apart in the primary memory. If the consecutive references are to the addresses that map to the same block, the cache mechanism becomes inefficient since it requires a large number of block replacements.

Example 9.4

Figure 9.38 shows the direct-mapping scheme for the memory system of Example 9.1. Note that now the data area of the cache contains only 32 blocks of 8 bytes each.

A compromise between the two previous types of mapping is called the *set-associative* mapping, in which the B cache blocks are divided into $2^k = K$ partitions, each containing B/K blocks. The address partitioning for this K-way set-associative mapping is shown in the following:

Address A (p bits)	Tag	Set number	Word number
	$(p-n-b+k)$	$(b-k)$	n

Note that this scheme is similar to the direct mapping in that it divides the main-memory addresses into $2(b-k)$ partitions. However, each frame can now reside in one of the K corresponding cache blocks, known as a *set*. The tag search is now limited to K tags in the set. Figure 9.39 shows the details of a set-associative cache scheme. Also note the following relations:

$K = 0$ implies direct mapping.
$K = b$ implies fully associative mapping.

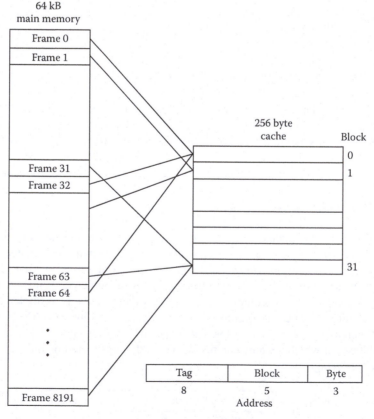

Figure 9.38 Direct mapping (Example 9.4) (*Note:* Tag area is not shown).

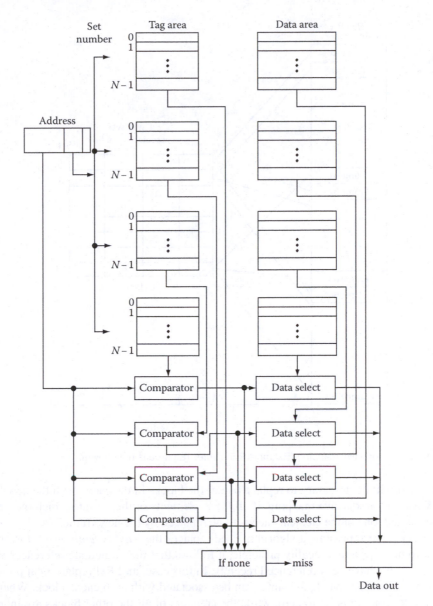

Figure 9.39 The structure of a four-way set-associative cache with *N* sets.

Example 9.5

Figure 9.40 shows the four-way set-associative scheme for the memory system of Example 9.1 Note that now the data area of the cache contains a total of 128 blocks, or 1 kB.

9.6.6.3 *Replacement Algorithm*

When there is a cache miss, the frame corresponding to the referenced address is brought into the cache. The placement of this new block in the cache depends on the mapping scheme used by the cache mechanism. If the direct mapping is used, there is only one possible cache block the frame can occupy. If that cache block is already occupied by another frame, it is replaced by the new frame. In fully associative mapping, the new frame may occupy any vacant block in the cache. If there are no

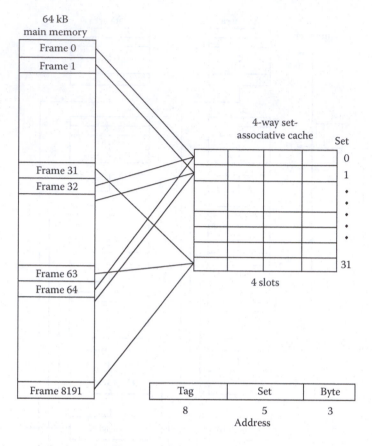

Figure 9.40 Set-associative mapping (Example 9.5) (*Note:* Tag area is not shown).

vacant blocks, it will be necessary to replace one of the blocks in the cache with the new frame. In case of a *K*-way set-associative mapping, if all the *K* elements of the set into which the new frame maps are occupied, one of the elements needs to be replaced by the new frame.

The most popular replacement algorithm used replaces the *least recently used (LRU)* block in the cache. From the program locality principle, it follows that the immediate references will be to those addresses that have been referenced recently. In this case, an LRU replacement policy works well. To identify the LRU block, a counter can be associated with each cache block. When a block is referenced, its counter is set to zero, while the counters of all the other blocks are incremented by 1. At any time, the LRU block is the one whose counter has the highest value. These counters are usually called *aging counters* since they indicate the age of the cache blocks.

The *FIFO* replacement policy has also been used. Here, the block that is in the cache longest is replaced. To determine the block to be replaced, each time a frame is brought into the cache, its identification number is loaded into a queue. The output of the queue thus always contains the identification of the frame that entered the cache first. Although this mechanism is easy to implement, it has the disadvantage that under certain conditions, blocks are replaced too frequently.

Other possible policies are (1) the *least frequently used (LFU)* policy, in which the block that has experienced the least number of references in a given time period is replaced, and (2) the *random* policy, in which a block among the candidate blocks is randomly picked for replacement. Simulation has shown that the performance of the random policy (which is not based on the usage characteristics of the cache blocks) is only slightly inferior to that of the policies based on usage characteristics.

9.6.6.4 *Write Operations*

The description mentioned previously assumed only read operations from the cache and primary memory. The processor can also access data from the cache and update it. Consequently, the data in the primary memory must also be correspondingly updated. Two mechanisms are used to maintain this data *consistency: write back* and *write through*. In the *write-back* (WB) mechanism, when the processor writes something into a cache block, that cache block is tagged as a *dirty block*, using a 1 bit tag. Before a dirty block is replaced by a new frame, the dirty block is copied into the primary memory. In the *write-through* scheme, when the processor writes into a cache block, the corresponding frame in the primary memory is also written with the new data.

Obviously, the write-through mechanism has higher overhead on memory operations than the WB mechanism. But it easily guarantees the data consistency, an important feature in computer systems in which more than one processor accesses the primary memory. Consider, for instance, a system with an I/O processor. Since the I/O processor accesses the memory in a DMA mode (not using the cache) and the central processor accesses the memory through the cache, it is necessary that the data values be consistent.

In multiple processor systems with a common memory bus, a *write-once* mechanism is used. Here, the first time a processor writes to its cache block, the corresponding frame in the primary memory is also written, thus updating the primary memory. All the other cache controllers invalidate the corresponding block in their cache. Subsequent updates to the cache affect only (local) cache blocks. The dirty blocks are copied to the main memory when they are replaced.

The write-through policy is the more common since typically the write requests are of the order of 15%–20% of the total memory requests, and hence, the overhead due to write through is not significant.

The tag portion of the cache usually contains a *valid bit* corresponding to each tag. These bits are reset to zero when the system power is turned on, indicating the invalidity of the data in all the cache blocks. As and when a frame is brought into the cache, the corresponding valid bit is set to 1.

9.6.6.5 *Performance*

The average access time, T_a, of memory system with cache is given by the following equation:

$$T_a = hT_c + (1-h)T_m, \tag{9.2}$$

where
T_c and T_m are average access times of cache and primary memory, respectively
h is the *hit ratio*
$(1 - h)$ is the *miss ratio*

We would like T_a to be as close to T_c as possible. Since $T_m \ggg T_c$, h should be as close to 1 as possible. From Equation 9.2, it follows that

$$\frac{T_c}{T_a} = \frac{1}{h+(1-h)T_m / T_c}. \tag{9.3}$$

Typically, T_m is about 5–10 times T_c. Thus, we would need a hit ratio of 0.75 or better to achieve reasonable speedup. A hit ratio of 0.9 is not uncommon in contemporary computer systems.

Progress in hardware technology has enabled very cost-effective configuration of large-capacity cache memories. In addition, as seen with the Intel Pentium architecture described earlier, the processor chip itself contains the first-level cache, and the second-level cache is configured externally to the processor chip. Some processor architectures contain separate data and instruction caches. Brief descriptions of representative cache structures are provided later in this chapter.

9.7 SIZE ENHANCEMENT

The main memory of a machine is not usually large enough to serve as the sole storage medium for all programs and data. Secondary storage devices such as disks are used to increase capacity. However, the program and data must be in the main memory during the program execution. Mechanisms to transfer programs and data to the main memory from the secondary storage as needed during the execution of the program are thus required. Typically, the capacity of the secondary storage is much higher than that of the main memory. But the user assumes that the complete secondary storage space (i.e., *virtual address space*) is available for programs and data, although the processor can access only the main-memory space (i.e., *real or physical address space*). Hence the name *virtual storage*.

In early days, when programmers developed large programs that did not fit into main-memory space, they would divide the programs into independent partitions known as *overlays*. These overlays were then brought into the main memory as and when needed for execution. Although this process appears simple to implement, it increases the program development and execution complexity and is visible to the user. Virtual memory mechanisms handle overlays in an automatic manner transparent to the user.

Virtual memory mechanisms can be configured in one of the following ways:

1. Paging (PG) systems
2. Segmentation systems
3. Paged-segmentation systems

In PG systems, both real and virtual address spaces are divided into small, equal-sized partitions called *pages*. Segmentation systems use memory *segments*, which are unequal-sized blocks. A segment is typically equivalent to an overlay described earlier. In paged-segmentation systems, segments are divided into pages, each segment usually containing a different number of pages. We will now describe PG systems that are simpler to implement than the other two systems. Refer to the books on operating systems listed in the Bibliography section of this chapter for further details.

Consider the memory system shown in Figure 9.41. The main memory is 2^p pages long. That is, there are 2^p *real pages* and we will denote each page slot in the main memory as a *frame*. The secondary storage consists of 2^q pages (i.e., there are 2^q *virtual pages*). Both real and virtual pages are 2^d words long. Also $q \gg p$. Since the user assumes that he has the complete virtual space to program in, the virtual address A_v (i.e., the address produced by the program in referencing the memory system) consists of $(q + d)$ bits. However, the main memory or physical address A_p consists of $(p + d)$ bits.

When a page is brought from the secondary storage into a main-memory frame, the *page table* is updated to reflect the location of the page in the main memory. Thus, when the processor refers to a memory address, the $(q + d)$-bit address is transformed into the $(p + d)$-bit primary address, using the page-table information. If the referenced virtual page is not in the main memory, Table 2 (disk directory) is searched to locate the page on the disk. That is, Table 2 transforms A_v into the physical address A_{vp} on the disk, which will be in the form of drive number, head number, track number, and displacement within the track. The page at A_{vp} is then transferred to an available real page location from which the processor accesses the data.

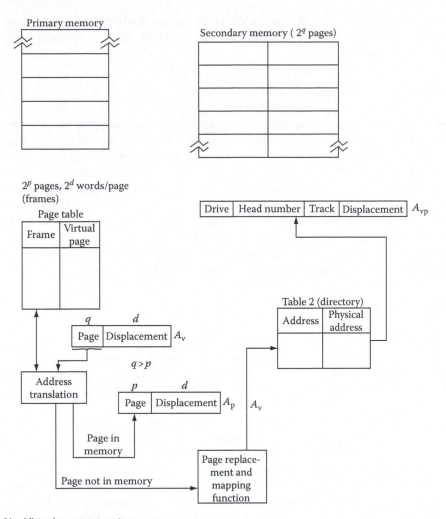

Figure 9.41 Virtual memory system.

The operation of virtual memory is similar to that of a cache. As such, the following mechanisms are required:

1. *Address translation*: Determines whether the referenced page is in the main memory or not and keeps track of the page movements in both memories.
2. *Mapping function*: Determines where the pages are to be located in the main memory. The direct, fully associative, and set-associative mapping schemes are used.
3. *Page replacement policy*: Decides which of the real pages are to be replaced. The LRU and FIFO policies are commonly used.

Although functionally similar, the cache and virtual memory systems differ in several characteristics. These differences are described as follows.

9.7.1 Page Size

The page size (PS) in virtual memory schemes is of the order of 512 to 4 kB when compared with 8–16 byte block sizes in cache mechanisms. The PS is determined by monitoring the address reference patterns of application program, as exhibited by the program locality principle. In addition, the access parameters (such as block size) of the secondary storage devices influence the PS selection.

9.7.2 Speed

When a cache miss occurs, the processor waits for the corresponding block from the main memory to arrive in the cache. To minimize the wait time, the cache mechanisms are implemented totally in hardware and operate at the maximum speed possible. A miss in virtual memory mechanism is termed a *page fault*. A page fault is usually treated as an operating system call in the virtual memory mechanisms. Thus, the processor is switched to perform some other task, while the memory mechanism brings the corresponding page into the main memory. The processor is then switched back to that task. Since the speed is not the main criterion, virtual memory mechanisms are not completely implemented using hardware. With the popularity of microprocessors, special hardware units called *memory management units* (MMUs) that handle all the virtual memory functions are now available. One such device is described later in this chapter.

9.7.3 Address Translation

The number of entries in the tag area of the cache mechanism is very small compared with the entries in the page table of a virtual memory mechanism because of the sizes of the main memory and secondary storage. As a result, maintenance of the page table and minimizing the page stable search time are important considerations. We will describe the popular address translation schemes next.

Figure 9.42 shows a page-table structure for the virtual memory scheme of Figure 9.41. The page table contains 2^q slots. The page field of the virtual address forms the index to the page table. When a secondary page is moved into the main memory, the main-memory frame number is entered into the corresponding slot in the page table. The *residence bit* is set to 1 to indicate that the main-memory frame contains a valid page. The residence bit being 0 indicates that the corresponding main-memory frame is empty. The page-table search requires only one access to the main memory, in which the page table is usually maintained. But in this scheme, since there are only 2^p frames in the main memory, $2^q - 2^p$ slots in the page table are always empty. Since q is large, the page table wastes the main-memory space.

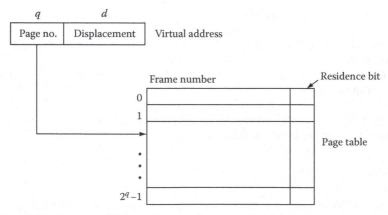

Figure 9.42 Page-table structure with 2^q entries.

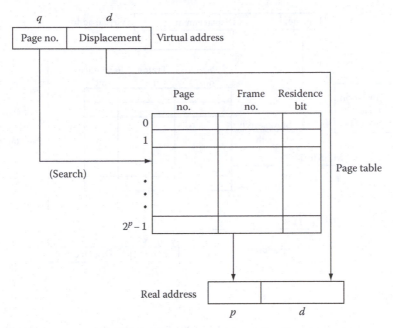

Figure 9.43 Page-table structure with 2^p entries.

Example 9.6

Consider a memory system with 64k of secondary memory and 8k of main memory and 1k pages. Then

$$d = 10 \text{ bits}$$
$$(q + d) = 16 \text{ bits}$$
$$(p + d) = 13 \text{ bits}$$

(i.e., $q = 6$ and $p = 3$). There are eight frames in the main memory. The secondary storage contains 64 pages. The page table consists of 64 entries. Each entry is 4 bits long (one 3 bit frame number and 1 residence bit).

Note that in order to retrieve data from the main memory, two accesses are required (the first to access the page table and the second to retrieve the data) when there is no page fault. If there is a page fault, then the time required to move the page into the main memory needs to be added to the access time.

Figure 9.43 shows a page-table structure containing 2^p slots. Each slot contains the virtual page number and the corresponding main-memory frame number in which the virtual page is located. In this structure, the page table is first searched to locate the page number in the virtual address reference. If a matching number is found, the corresponding frame number replaces the q-bit virtual page number. This process requires, on average, $2^p/2$ searches through the page table.

Example 9.7

For the memory system of Example 9.6, a page table with only eight entries is needed in this scheme. Each entry consists of 10 bits (a residence bit, 3 bit main-memory frame number, and a 6 bit secondary storage page number).

Figure 9.44 shows a scheme to enhance the speed of page-table search. Here, the most recently referenced entries in the page table are maintained in a fast memory (usually an AM) called the *translation lookaside buffer* (TLB). The search is initiated in both TLB and the page table simultaneously. If a match is found in the TLB, the search in the page table is terminated. If not, the page-table search continues.

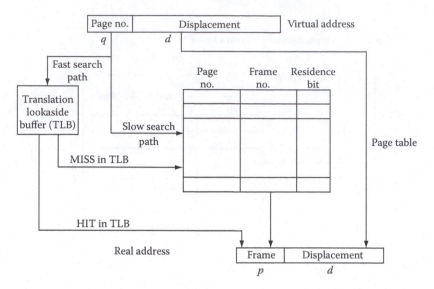

Figure 9.44 TLB.

Consider a 64 kB main memory with 1 kB PS. That is, $d = 10$ and $p = 6$, since the main-memory address is 16 bits long. The page table contains only $2^6 = 64$ entries and can easily be maintained in the main memory. If the main-memory capacity is increased to 1 MB, then the page table would contain $1k$ entries, which is a significant portion of the main memory. Thus, as the main-memory size grows, the size of the page table poses a problem. In such cases, the page table could be maintained in the secondary storage. The page table is then divided into several pages. It is brought into the main memory one page at a time to perform the comparison.

Figure 9.45 shows the *paged-segmentation* scheme of virtual memory management. In this scheme, the virtual address is divided into a segment number, a page number, and a displacement. The segment number is an index into a segment table whose entries point to the base address of the corresponding page tables. Thus, there is a page table for each segment. The page table is searched in the usual manner to arrive at the main-memory frame number.

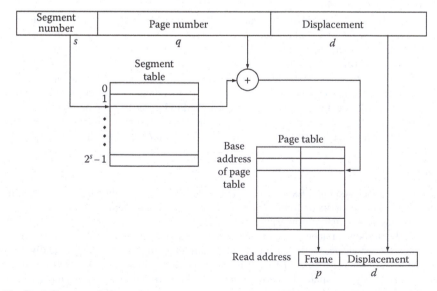

Figure 9.45 Paged-segmentation scheme.

9.8 ADDRESS EXTENSION

In our description of the virtual memory schemes, we have assumed that the capacity of the secondary storage is much higher than that of the main memory. In addition, we assumed that all memory references by the CPU are in terms of virtual addresses. In practice, the addressing range of the CPU is limited, and the direct representation of the virtual address as part of the instruction is not possible.

Consider the instruction format of ASC, for instance. Here, the direct addressing range is only the first 256 words of the memory. This range was extended to $64k$ words by indexing and indirect addressing. Suppose the secondary storage is $256k$ words long, thus requiring 18 bits in the virtual address. Since ASC can only represent a 16 bit address, the remaining 2 address bits must be generated by some other mechanism. One possibility is to output the two most significant address bits into an external register using the WWD command and the lower 16 bits of the address on the address bus, thus enabling the secondary storage mechanism to reconstruct the required 18 bit address, as shown in Figure 9.46. To make this scheme work, the assembler should be modified to assemble all programs in an 18 bit address space where the lower 16 bits are treated as displacement addresses from a 2 bit page (or bank) address maintained as a constant with respect to each program segment. During execution, the CPU executes a WWD to output the page address into the external register as the first instruction in the program segment.

The address extension concept is illustrated in Figure 9.46 usually termed *bank switching*. In fact, this scheme can be generalized to extend the address further. As shown in Figure 9.47, the most significant m bits of the address can be used to select one of the 2^m n-bit registers. If the address bus is p bits wide, then the virtual address will be $(n + p)$ bits long. In this description, we have assumed that the data bus is used to output the m-bit address portion. In practice, it need not be so, and the address bus can be time-multiplexed to transfer the two address portions. The DEC PDP-11/23 system uses this address extension scheme. As can be seen in Figure 9.48, processors that use either paged or base–displacement addressing modes accommodate the address extension naturally.

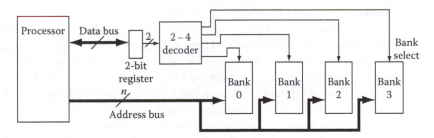

Figure 9.46 Bank switching.

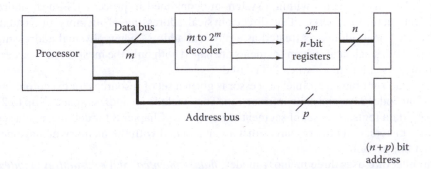

Figure 9.47 Address extension.

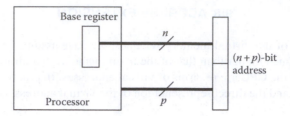

Figure 9.48   Address extension with base register.

9.9 EXAMPLE SYSTEMS

A variety of structures have been used in the implementation of memory hierarchies in modern computer systems. In early systems, it was common to see the cache mechanism implemented completely using hardware, while the virtual memory mechanism was software intensive. With advances in hardware technology, MMU became common. Some MMUs manage only the cache level of the hierarchy, while the others manage both the cache and virtual levels. It is now common to see most of the memory management hardware implemented on the processor chip itself. In this section we will provide a brief description of three memory systems. We first expand on the description of the Intel Pentium processor structure of Chapter 8 to show its memory management details. This is followed by the hardware structure details of the Motorola 68020 processor. The final example describes the memory management aspects of the Sun Microsystems' System-3.

9.9.1 Memory Management in Intel Processors

This section is extracted from the Intel Architecture Software Developer's Manuals listed in the Bibliography section of this chapter. An operating system designed to work with a processor uses the processor's memory management facilities to access memory. These facilities provide features such as segmentation and PG, which allow memory to be managed efficiently and reliably. Programs usually do not address physical memory directly when they employ the processor's memory management facilities. The processor's addressable memory space is called *linear address space*. Segmentation provides a mechanism for dividing the linear address space into smaller protected address spaces called *segments*. Segments in Intel processors are used to hold the code, data, and stack for a program or to hold system data structures. Each program has its own set of segments if more than one program (or task) is running on a processor. The processor enforces the boundaries between these segments and ensures that one program does not interfere with the execution of another program by writing into the other program's segments. The segmentation also allows typing of segments so that the operations that are performed on a particular type of segment can be restricted. All of the segments within a system are contained in processor's linear address space. The Intel architecture supports either direct physical addressing of memory or virtual memory through PG. A linear address is treated as a physical address when physical addressing is used. All the code, data, stack, and system segments are paged with only the most recently accessed pages being held in physical memory when PG is used.

Any program running on an Intel processor is given a set of resources for executing instructions and for storing code, data, and state information. They include an address space of up to 2^{32} bytes, a set of general data registers, a set of segment registers (see Chapter 8 for details), and a set of status and control registers. Only the resources that are associated with the memory management will be described in this section.

Intel architecture uses three memory modes: *flat*, *segmented*, and *real-address modes*. The linear address space appears as a single and continuous address space to a program with the flat

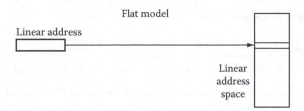

Figure 9.49 Flat model.

memory model as shown in Figure 9.49. The code, data, and the procedure stacks are all contained in this address space. The linear address space for this model is byte addressable. The size of the linear address space is 4 GB.

With the segmented memory model, the linear address space appears to be a group of independent segments, as shown Figure 9.50. The code, data, and stacks are stored in separate segments when this model is used. The program also needs to issue a logical address to address a byte in a segment when this model is used. The segment selector is used here to identify the segment to be accessed, and the offset is used to identify a byte in the address space of the segment. A processor can address up to 16,383 segments of different sizes and types. Each segment can be as large as 2^{32} bytes.

The real-address mode (Figure 9.51) uses a specific implementation of segmented memory in which the linear address space for the program and the operating system consists of an array of segments of up to 64 kB each. The maximum size of the linear address space in real-address mode is 2^{20} bytes.

A processor can operate in three operating modes: *protected*, *real-address*, and *system management*. The processor can use any of the memory models when in the protected mode. The use of memory models depends on the design of the operating system. Different memory models are used when multitasking is implemented. The processor only supports the real-address mode memory model when the real-address mode is used. The processor switches to a separate address space called

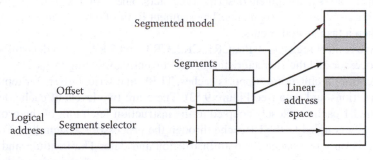

Figure 9.50 Segmented model.

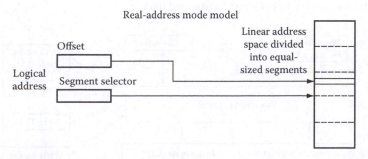

Figure 9.51 Real-address model. *Note:* The linear address space can be paged when using the flat or segmented model.

System table registers

	47	16	15	0
GDTR	32-bit linear base address		16-bit table limit	
IDTR	32-bit linear base address		16-bit table limit	

System segment registers

System description registers (automatically loaded)

	15	0				Attributes	
Task register	Segment selector		32-bit linear base address	Segment limit			
LDTR	Segment selector		32-bit linear base address	Segment limit			

Figure 9.52 Memory management registers.

the system management RAM (SMRAM) when the system management mode is used. The memory model used to address bytes in this address space is similar to the real-address mode model.

The Intel processor provides four memory management registers: the global descriptor table register (GDTR), the local descriptor table register (LDTR), the interrupt descriptor table register (IDTR), and the task register (TR). These registers specify the locations of the data structures that control segmented memory management. Figure 9.52 shows the details.

The GDTR is 48 bits long and is divided into two parts: a 16 bit table limit and a 32 bit linear base address. The 16 bit table limit specifies the number of bytes in the table, and the 32 bit base address specifies the linear address of byte 0 of global descriptor table (GDT). There are two instructions that are used to load (LGDT) and store (SGDT) the GDTR. The IDTR is also 48 bits long and is divided into two parts: 16 bit table limit and 32 bit linear base address. The 16 bit table limit specifies the number of bytes in the table, and the 32 bit base address specifies the linear address of byte 0 of IDT. LIDT and SIDT are used to load and store this register. The LDTR has system segment registers and segment descriptor registers. The system segment registers are 16 bits long, and they are used to select segments. The segment limit of TR is automatically loaded along with the TR when a task switch occurs.

There are five control registers: CR0, CR1, CR2, CR3, and CR4, which determine the operating mode of the processor and the characteristics of the currently executing tasks.

The Intel Pentium architecture supports caches, TLBs, and write buffers for temporary on-chip storage of instructions and data (see Figure 9.53). There are two levels of cache: level 1 (L1) and level 2 (L2). The L1 cache is closely coupled to the instruction fetch unit of the processor, and the L2 cache is closely coupled to the L1 cache through the processor's system bus. For the Intel 486 processor, there is only one cache for both instruction and data. The Pentium and its subsequent processor series have two separate caches: one for instruction and the other for data. The Intel Pentium processor's L2 cache is external to the processor package, and it is optional. The L2 cache is a unified cache for storage of both instruction and data.

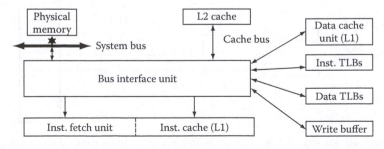

Figure 9.53 Intel cache architecture.

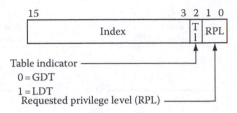

Figure 9.54 Segment selector.

A linear address contains two items: a segment selector and an offset. A segment selector is a 16 bit identifier for a segment. It usually points to the segment descriptor that defines the segment. A segment selector contains three items: index, T1 (table indicator) flag, and requested privilege level (RPL). The index is used to select one of 8192 (2^{32}) descriptors in the GDT or LDT. The T1 flag is used to select which descriptor table to use. The RPL is used to specify the privilege level of the selector. Figure 9.54 shows how these items are arranged in a segment selector.

For the processor to access a segment, the segment must be loaded into the appropriate segment register. The processor provides six registers for holding up to six segment selectors. Each of these segment registers supports a specific kind of memory reference (code, stack, or data). The code-segment (CS), data-segment (DS), and stack-segment (SS) registers must be loaded with valid segment selectors for any kind of program execution to take place. However, the processor also provides three additional DS registers: ES, FS, and GS. These registers can be used to make additional data segments available to the programs.

A segment register has two parts: a visible part and a hidden part. The processor loads the base address, segment limit, and access information into the hidden part and segment selector into the visible part, as shown in Figure 9.55.

A segment descriptor is an entry in a GDT or LDT. It provides the processor with the size and location of a segment as well as access control and status information. In the Intel system, compilers, linkers, loaders, or the operating system typically creates the segment descriptors. Figure 9.56 shows the general descriptor format for all types of Intel segment descriptors.

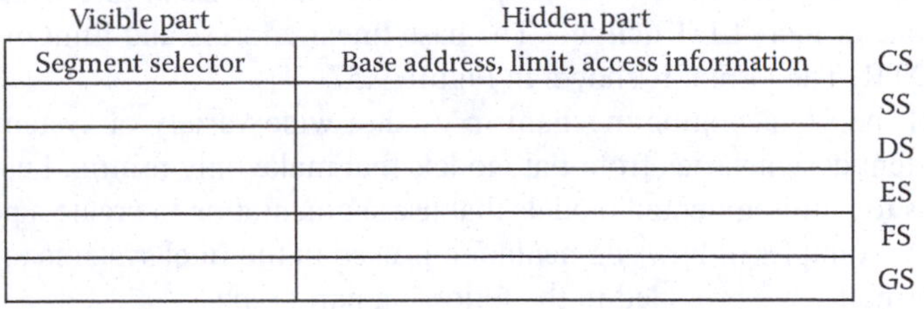

Visible part	Hidden part	
Segment selector	Base address, limit, access information	CS
		SS
		DS
		ES
		FS
		GS

Figure 9.55 Segment register.

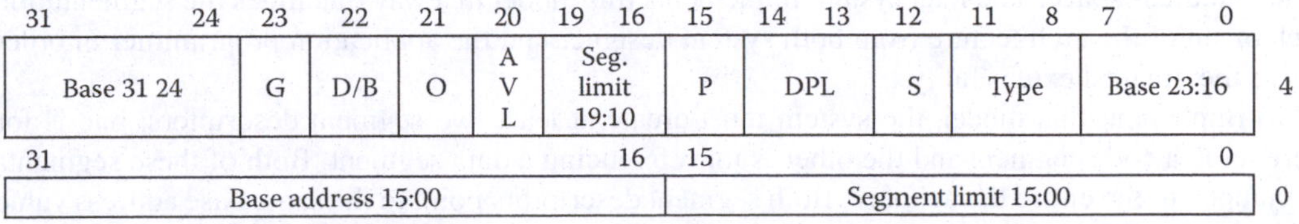

31	24	23	22	21	20	19 16	15	14 13	12	11 8	7 0	
Base 31 24		G	D/B	O	A V L	Seg. limit 19:10	P	DPL	S	Type	Base 23:16	4

31	16	15	0	
Base address 15:00		Segment limit 15:00		0

Figure 9.56 Segment descriptor. *Notes*: AVL, available for use by system software; BASE, segment base address; D/B, default operation size (0 = 16 bit segment; 1 = 32 bit segment); DPL, descriptor privilege level; G, granularity; LIMIT, segment limit; P, segment present; S, descriptor type (0 = system: 1 = code or data); TYPE, segment type.

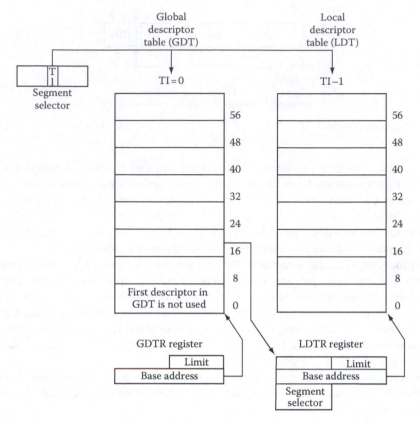

Figure 9.57 Global and LDTs.

A segment descriptor table is a data structure in the linear address space and contains an array of segment descriptors. It can contain up to 8192 (2^{32}) 8 byte descriptors. There are two kinds of descriptor tables: the GDT and the local descriptor tables (LDTs). Each Intel system must have one GDT defined and one or more LDT defined. The base linear address and limit of the GDT must be loaded into the GDTR. The format is shown in Figure 9.57.

On the basis of the segmentation mechanism used, a wide variety of system designs are possible. The Intel system design varies from flat models that make only minimal use of segmentation to protect programs to multisegmented models that use segmentation to create a powerful operating environment. A few examples of how segmentation is used in the Intel systems to improve memory management performance are provided in the following paragraphs.

The basic flat model is the simplest memory model for a system. The detail of this model is shown in Figure 9.58. It gives the operating system and programs access to a continuous unsegmented address space. The Intel system implements this model in a way that hides the segmentation mechanism of the architecture from both system designer and the application programmer in order to give the greatest extensibility.

To implement this model, the system must create at least two segment descriptors: one is for referencing a code segment and the other is for referencing a data segment. Both of these segments are mapped to the entire linear space. Both segment descriptors point to the same base address value of 0 and have the same segment limit of 4 GB.

The protected flat model is similar to the basic flat model. The only difference is in the segment limits. In this model, the segment limits are set to include only the range of addresses for which

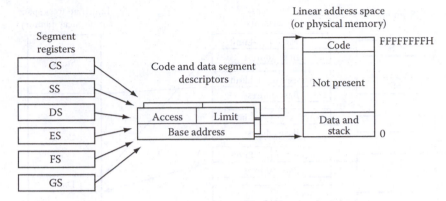

Figure 9.58 Basic flat model.

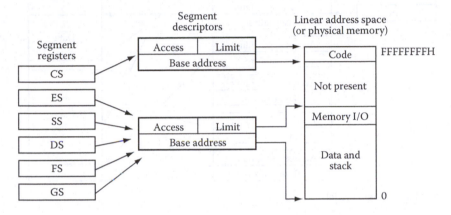

Figure 9.59 Protected flat model.

physical memory actually exists. This model is shown in Figure 9.59. This model can provide better protection by adding a simple PG structure.

The multisegmented model uses the full capabilities of the segmentation mechanism to provide the protection of the code, data structures, and tasks. The model is shown in Figure 9.60. In this model, each program is given its own table of segment descriptors and its own segments. The segments are completely private to their assigned programs or shared among programs. As shown in the figure, each segment register has a separate segment descriptor associated with it, and each segment descriptor points to a separate segment. Access to all segments and to the execution environments of individual programs running on the Intel system is controlled by hardware.

Four control registers facilitate PG options. CR0 contains system control flags that control flats that control the operating mode and states of the processor. Bit 31 of CR0 is the PG flag. The PG flag enables the page-translation mechanism. This flag must be set if the processor is implementing a demand-paged virtual memory system or if the operating system is designed to run more than one program (or task) in virtual 8086 mode. CR2 contains the linear address that causes a page fault. CR3 contains the physical address of the base of the page directory. It is referred to as the page-directory base register (PDBR). Only the 20 most significant bits of the PDBR are specified—the lower 12 bits are assumed to be 0. Bit 4 of CR4 is the PS extension (PSE) flag. The PSE flag enables a larger PS of 4 MB. When the PSE flag is not set, the more common page length of 4 kB is used. Figure 9.61 shows the various configurations that can be achieved by setting the PG and PSE flags as well as the PS flag.

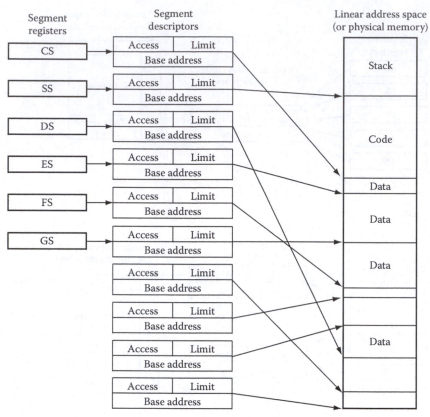

Figure 9.60 Multisegment model.

PG flag. CR0	PSE flag. CR4	PS flag. PDE	Page size	Physical address size
0	X	X	—	Paging disabled
1	0	X	4 kB	32 bits
1	1	0	4 kB	32 bits
1	1	1	4 MB	32 bits

Figure 9.61 Page and physical address sizes.

When PG is enabled, the processor uses the information contained in three data structures to translate linear address into physical addresses. The first data structure is a page directory. This is an array of 32 bit page-directory entries (PDEs) contained in a 4 kB page. Up to 1024 PDEs can be held in a page directory. The second data structure is the page table. This is an array of 32 bit page-table entries (PTEs) contained in a 4 kB page. Up to 1024 PTEs can be held in a page table. The third data structure is the page. This is either a 4 kB or 4 MB flat address space depending on the setting of the PS flag. Figures 9.62 and 9.63 show the formats of page-directory and PTEs when 4 kB pages and 4 MB pages are used.

The only difference in these formats is that the 4 MB format does not utilize page tables. As a result, there are no PTEs. The functions of the flags and fields in the previous entries are as follows:

• Page base address (bits 12–31/22–31): For 4 kB pages, this specifies the physical address of the first byte of a 4 kB page in the PTE and the physical address of the first byte of a page table in the PDE. For 4 MB pages, this specifies the physical address of the first byte of a 4 MB page in the page directory. Only bits 22–31 are used. The rest of the bits is reserved and must be set to 0.

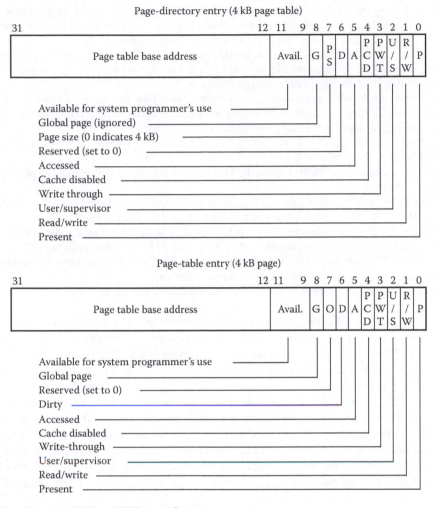

Figure 9.62 Format of PDE and PTE for 4 kB pages.

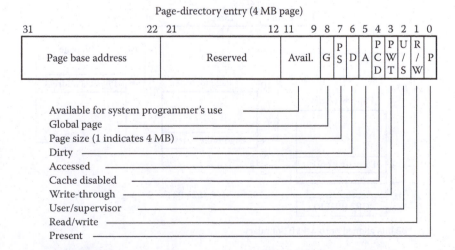

Figure 9.63 Format of PDE for 4 MB pages.

- Present (P) flag (bit 0): This indicates whether the page or the page table being pointed to by the entry is currently loaded in physical memory. When this flag is clear, the page is not in memory. If the processor attempts to access the page, it generates a page-fault exception.
- Read/write (R/W) flat (bit 1): This specifies the read/write privileges for a page or group of pages (in the case of a PDE that points to a page table). When the flag is clear, the page is read only, and when the flag is set, the page can be read or written to.
- User/supervisor (U/S flat (bit 2): This specifies the user/supervisor privileges for a page or a group of pages. When this flag is clear, the page is assigned the supervisor privilege level, and when the flag is set, the page is assigned the user privilege level.
- Page-level write-through (PWT) flat (bit 3): This controls the write-through or WB caching policy of individual pages or page tables.
- Page-level cache disable (PCD) flat (bit 4): This controls whether or not individual pages or page tables can be cached.
- Accessed (A) flat (bit 5): This indicates whether or not a page or page table has been accessed.
- Dirty (D) flag (bit 6): This indicates whether or not a page has been written to.
- PS flag (bit 7): This determines the PS. When this flag is clear, the PS is 4 kB and the PDE points to a page table. When the flag is set, the PS is 4 MB for normal 32 bit addressing and the PDE points to a page. If the PDE points to a page table, all the pages associated with that page table will be 4 kB long.
- Global (G) flag (bit 8): This indicates a global page when set.
- Available-to-software bits (bits 9–11): These bits are available for use by software.

TLBs are on-chip caches that the processor utilizes to store the most recently used page directory and PTEs. The Pentium processor has separate TLBs for data and instruction caches as well as for 4 kB and 4 MB PSs. PG is most performed using the contents of the TLBs. If the TLBs do not contain the translation information for a requested page, then bus cycles to the page directory and page tables in memory are performed.

Figure 9.64 shows the page-directory and page-table hierarchy when mapping linear addresses to 4 kB pages. The entries in the page directory point to page tables, and the entries in a page table point to pages in physical memory. This PG method can be used to address up to 2^{20} pages, which spans a linear address space of 2^{32} bytes (4 GB). To select the various table entries, the linear address is divided into three sections. Bits 22–31 provide an offset to an entry in the page directory. The selected entry provides the base physical address of a page table. Bits 12–21 provide an offset to an

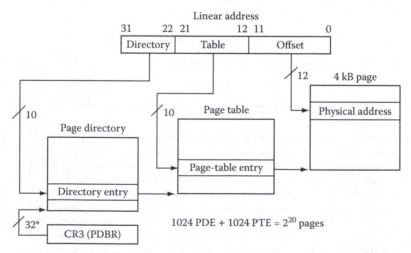

Figure 9.64 Linear address translation (4 kB pages).

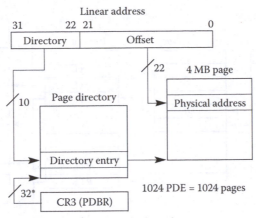

Figure 9.65 Linear address translation (4 MB pages).

entry in the selected page table. This entry provides the base physical address of a page in physical memory. Bits 0–11 provide an offset to a physical address in the page.

Figure 9.65 shows the use of a page directory to map linear addresses to 4 MB pages. The entries in the page directory point to 4 MB pages in physical memory. Page tables are not used for 4 MB pages. These PSs are mapped directly from one or more PDEs.

The PDBR points to the base of the page directory. Bits 22–31 provide an offset to an entry in the page directory. The selected entry provides the base physical address of a 4 MB page. Bits 0–21 provide an offset to a physical address in the page.

As explained previously, when the PSE flag in CR4 is set, both 4 MB pages and page tables for 4 kB pages can be accessed from the same page directory. Mixing 4 kB and 4 MB pages can be useful. For example, the operating system or executive's kernel can be placed in a large page to reduce TLB misses and thus improve overall system performance. The processor to maintain 4 MB page entries and 4 kB page entries uses separate TLB. Because of this, placing often used code such as the kernel in a large page frees up 4 kB page TLB entries for application programs and tasks.

The Intel architecture supports a wide variety of approaches to memory management using the PG and segmentation mechanisms. There is no forced correspondence between the boundaries of pages and segments. A page can contain the end of one segment and the beginning of another. Likewise, a segment can contain the end of one page and the beginning of another.

Segments can be mapped to pages in several ways. One such implementation is demonstrated in Figure 9.66. This approach gives each segment its own page table, by giving each segment a single entry in the page directory that provides the access control information for PG the entire segment.

9.9.2 Intel Sandy Bridge Cache Hierarchy

In this section we discuss the cache hierarchy used by Intel in their processors based on Intel microarchitecture code named Sandy Bridge. It is extracted from the Intel 64 and IA-32 Architectures Optimization Reference Manual (http://www.intel.com/content/www/us/en/architecture-and-technology/64-ia-32-architectures-optimization-manual.html).

The Intel cache hierarchy contains a first-level instruction cache, a first-level data (L1D) cache, and a second-level (L2) cache in each core. The L1D cache may be shared by two logical processors if the processors support Intel HyperThreading Technology. The L2 cache is shared by instructions and data. All cores in a physical processor package connect to a shared last-level cache (LLC) via a ring connection.

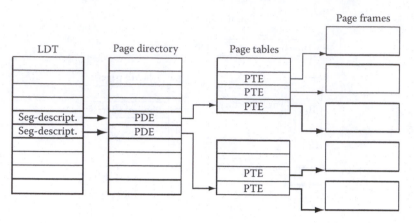

Figure 9.66 Memory management convention that assigns a page table to each segment.

Table 9.1   Cache Parameters

Level	Capacity	Associativity (Ways)	Line Size (Bytes)	Write Update Policy	Inclusive
L1 data	32 kB	8	64	Write back	—
Instruction	32 kB	8	N/A	N/A	—
L2 (unified)	256 kB	8	64	Write back	No
Third level (LLC)	Varies, query CPUID leaf 4	Varies with cache size	64	Write back	Yes

The caches use the services of the instruction TLB (ITLB), data TLB (DTLB), and shared TLB (STLB) to translate linear addresses to physical address. Table 9.1 shows the cache parameters. Data coherency in all cache levels is maintained using the modified, exclusive, shared, invalid (MESI) protocol, which is also known as the Illinois protocol.

9.9.2.1 Loads

When an instruction reads data from a memory location that has WB type, the processor looks for it in the caches and memory. Table 9.2 shows the access lookup order and best-case latency. The actual latency can vary depending on the cache queue occupancy, LLC ring occupancy, memory components, and their parameters. The latency of L3 (third level or LLC), as shown in Table 9.2, varies with product segment and stock-keeping unit (SKU). The values apply to second-generation Intel Core processor families.

The LLC is inclusive of all cache levels above it—data contained in the core caches must also reside in the LLC. Each cache line in the LLC holds an indication of the cores that may have this

Table 9.2   Lookup Order and Load Latency

Level	Latency (Cycles)	Bandwidth (per Core per Cycle)
L1 data	4	2 × 16 bytes
L2 (unified)	12	1 × 32 bytes
Third level (LLC)	26–31	1 × 32 bytes
L2 and L1D cache in other cores if applicable	43—clean hit 60—dirty hit	

line in their L2 and L1 caches. If there is an indication in the LLC that other cores may hold the line of interest and its state might have to modify, there is a lookup into the L1D cache and L2 of these cores too. The lookup is called "clean" if it does not require fetching data from the other core caches. The lookup is called "dirty" if modified data have to be fetched from the other core caches and transferred to the loading core.

The latencies shown in Table 9.2 are the best-case scenarios. Sometimes a modified cache line has to be evicted to make space for a new cache line. The modified cache line is evicted in parallel to bringing the new data and does not require additional latency.

However, when data are written back to memory, the eviction uses cache bandwidth and possibly memory bandwidth as well. Therefore, when multiple cache misses require the eviction of modified lines within a short time, there is an overall degradation in cache response time. Memory access latencies vary based on occupancy of the memory controller queues, DRAM configuration, data-direction register (DDR) parameters, and DDR PG behavior (if the requested page is a page hit, page miss, or page empty).

9.9.2.2 Stores

When an instruction writes data to a memory location that has a write-back memory type, the processor first ensures that it has the line containing this memory location in its L1D cache, in exclusive or modified MESI state. If the cache line is not there, in the right state, the processor fetches it from the next levels of the memory hierarchy using a read for ownership request. The processor looks for the cache line in the following locations, in the specified order:

1. L1D cache
2. L2 cache
3. Last-level cache
4. L2 and L1D cache in other cores, if applicable
5. Memory

Once the cache line is in the L1D cache, the new data are written to it, and the line is marked as modified.

Reading for ownership and storing the data happens after instruction retirement and follows the order of store instruction retirement. Therefore, the store latency usually does not affect the store instruction itself. However, several sequential stores that miss the L1D cache may have cumulative latency that can affect performance. As long as the store does not complete, its entry remains occupied in the store buffer. When the store buffer becomes full, new micro-ops cannot enter the execution pipe and execution might stall.

9.9.2.3 L1D Cache

The L1D cache is the first-level data cache. It manages all load and store requests from all types through its internal data structures. The L1D cache

- Enables loads and stores to issue speculatively and out of order
- Ensures that retired loads and stores have the correct data upon retirement
- Ensures that loads and stores follow the memory ordering rules of the IA-32 and Intel 64 instruction set architecture

Table 9.3 shows the L1D cache components. The data cache unit (DCU) is organized as 32 kB, eight-way set associative. Cache line size is 64 bytes arranged in eight banks. Internally, accesses are up to 16 bytes, with 256 bit Intel Advanced Vector Extensions (AVX) instructions utilizing two 16 byte accesses. Two load operations and one store operation can be handled each cycle. The L1D cache maintains requests that cannot be serviced immediately to completion. Some reasons

Table 9.3   L1 Data Cache Components

Component	Intel Microarchitecture Code Name Sandy Bridge	Intel Microarchitecture Code Name Nehalem
DCU	32 kB, 8 ways	32 kB, 8 ways
Load buffers	64 entries	48 entries
Store buffers	36 entries	32 entries
LFBs	10 entries	10 entries

for requests that are delayed are as follows: cache misses, unaligned access that splits across cache lines, data not ready to be forwarded from a preceding store, loads experiencing bank collisions, and load block due to cache line replacement. The L1D cache can maintain up to 64 load micro-ops from allocation until retirement. It can maintain up to 36 store operations from allocation until the store value is committed to the cache or written to the line fill buffers (LFBs) in the case of nontemporal stores. The L1D cache can handle multiple outstanding cache misses and continues to service incoming stores and loads. Up to 10 requests of missing cache lines can be managed simultaneously using the LFB. The L1D cache is a WB write-allocate cache. Stores that hit in the DCU do not update the lower levels of the memory hierarchy. Stores that miss the DCU allocate a cache line.

9.9.2.4 Loads

The L1D cache architecture can service two loads per cycle, each of which can be up to 16 bytes. Up to 32 loads can be maintained at different stages of progress, from their allocation in the out of order engine until the loaded value is returned to the execution core.

Loads can

- Read data before preceding stores when the load address and store address ranges are known not to conflict
- Be carried out speculatively, before preceding branches are resolved
- Take cache misses out of order and in an overlapped manner

Loads cannot

- Speculatively take any sort of fault or trap
- Speculatively access uncacheable memory

The common load latency is five cycles. When using a simple addressing mode, base plus offset that is smaller than 2048, the load latency can be four cycles. This technique is especially useful for pointer-chasing code. However, overall latency varies depending on the target register data type due to stack bypass.

Table 9.4 lists overall load latencies. These latencies assume the common case of flat segment, that is, segment base address is zero. If segment base is not zero, load latency increases.

Table 9.4   Effect of Addressing Modes on Load Latency

Data Type/Addressing Mode	Base + Offset > 2048; Base + Index [+ Offset]	Base + Offset < 2048
Integer	5	4
MMX, SSE, 128-bit AVX	6	5
X87	7	6
256-bit AVX	7	7

9.9.2.5 Stores

Stores to memory are executed in two phases:

- Execution phase. Fills the store buffers with linear and physical address and data. Once store address and data are known, the store data can be forwarded to the following load operations that need it.
- Completion phase. After the store retires, the L1D cache moves its data from the store buffers to the DCU, up to 16 bytes/cycle.

9.9.2.6 Address Translation

The DTLB can perform three linear to physical address translations every cycle, two for load addresses, and one for a store address. If the address is missing in the DTLB, the processor looks for it in the STLB, which holds data and instruction address translations. The penalty of a DTLB miss that hits the STLB is seven cycles. Large page support includes 1GB pages, in addition to 4k and 2M/4M pages.

The DTLB and STLB are four-way set associative. Table 9.5 specifies the number of entries in the DTLB and STLB.

9.9.2.7 Store Forwarding

If a load follows a store and reloads the data that the store writes to memory, the data can forward directly from the store operation to the load. This process, called store to load forwarding, saves cycles by enabling the load to obtain the data directly from the store operation instead of through memory. Store forwarding allows quick move of complex structures without losing the ability to forward the subfields.

The memory control unit can handle store-forwarding situations with less restrictions compared to previous microarchitectures.

The following rules must be met to enable store to load forwarding:

- The store must be the last store to that address, prior to the load.
- The store must contain all data being loaded.
- The load is from a WB memory type and neither the load nor the store is nontemporal accesses.

Stores cannot forward to loads in the following cases:

- There are 4 and 8 byte loads that cross 8 byte boundary, relative to the preceding 16 or 32 byte store.
- Any load that crosses a 16 byte boundary of a 32 byte store.

Table 9.6 shows the store to load-forwarding behavior. For a given store size, all the loads that may overlap are shown and specified by "F." Cases that cannot forward are shown as "N."

Table 9.5   DTLB and STLB Parameters

TLB	PS	Entries
DTLB	4 kB	64
	2 MB/4 MB	32
	1 GB	4
STLB	4 kB	512

Table 9.6　Store-Forwarding Conditions (4–16 byte Stores)

Store Size	Load Size	Load Alignment															
		0	1	2	3	4	5	6	7	8	9	10	11	12	13	14	15
4	1	F	F	F	F												
	2	F	F	F	N												
	4	F	N	N	N												
8	1	F	F	F	F	F	F	F	F								
	2	F	F	F	F	F	F	F	N								
	4	F	F	F	F	F	N	N	N								
	8	F	N	N	N	N	N	N	N								
16	1	F	F	F	F	F	F	F	F	F	F	F	F	F	F	F	F
	2	F	F	F	F	F	F	F	F	F	F	F	F	F	F	F	N
	4	F	F	F	F	F	N	N	N	F	F	F	F	F	N	N	N
	8	F	N	N	N	N	N	N	N	F	N	N	N	N	N	N	N
	16	F	N	N	N	N	N	N	N	N	N	N	N	N	N	N	N

9.9.2.8 *Memory Disambiguation*

A load operation may depend on a preceding store. Many microarchitectures block loads until all preceding store addresses are known. The memory disambiguator predicts which loads will not depend on any previous stores. When the disambiguator predicts that a load does not have such a dependency, the load takes its data from the L1 data cache even when the store address is unknown. This hides the load latency. Eventually, the prediction is verified. If an actual conflict is detected, the load and all succeeding instructions are re-executed.

The following loads are not disambiguated. The execution of these loads is stalled until addresses of all previous stores are known:

- Loads that cross the 16 byte boundary.
- There are 32 byte Intel AVX loads that are not 32 byte aligned.

The memory disambiguator always assumes dependency between loads and earlier stores that have the same address bits 0:11.

9.9.2.9 *Bank Conflict*

Since 16 byte loads can cover up to three banks and two loads can happen every cycle, it is possible that six of the eight banks may be accessed per cycle, for loads. A bank conflict happens when two load accesses need the same bank (their address has the same 2–4 bit value) in different sets, at the same time. When a bank conflict occurs, one of the load accesses is recycled internally.

In many cases two loads access exactly the same bank in the same cache line, as may happen when popping operands off the stack, or any sequential accesses. In these cases, conflict does not occur and the loads are serviced simultaneously.

9.9.3 Motorola 68020 Memory Management

Figure 9.67 shows a block diagram of MC68020 instruction cache. Data are not cached in this system. The cache contains 64 entries. Each entry consists of a 26 bit tag field and a 32 bit data (instruction) field. The tag field contains the most significant 24 bits (A8–A31) of the memory address, a valid bit, and the function code bit FC2. FC2 = 1 indicates a supervisory mode, and FC2 = 0 indicates a

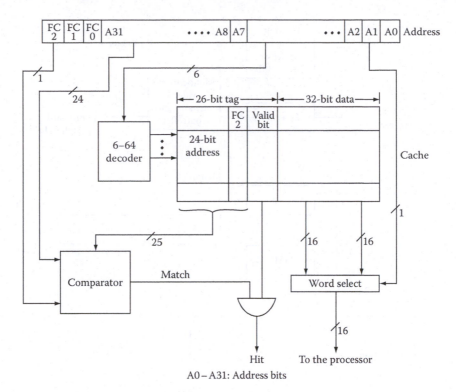

Figure 9.67 MC68020 cache organization.

user mode of operation. Address bits A2 through A7 are used to select one of the 64 entries in the cache. Thus, the cache is direct mapped. The data field can hold two 16 bit instruction words.

During the cache access, the tag field from the selected entry is first matched with A8–A31. If there is a match and the valid hit is 1, A1 selects the upper or lower 16 bit word from the data area of the selected entry. If there is no match or if the valid bit is 0, a miss occurs. Then, the 32 bit instruction from the referenced address is brought into the cache, and the valid bit is set to 1.

In this system, the on-chip cache access requires two cycles, while the off-chip memory access requires three cycles. Thus, during hits, the memory access time is reduced considerably. In addition, the data can be accessed simultaneously during the instruction fetch.

9.9.4 Sun-3 System Memory Management

Figure 9.68 shows the structure of the Sun-3 workstation, a microcomputer system based on MC68020. In addition to the processor, the system uses a floating-point coprocessor (MC68881), an MMU, a cache unit, 4 or 8 MB of memory, an Ethernet interface for networking, a high-resolution monitor with a 1 MB display memory, keyboard and mouse ports, two serial ports, bootstrap EPROM, ID PROM, configuration EEROM, time-of-day clock, and memory bus interface. External devices are interfaced to 3/100 and 3/200 systems through a VMEbus interface. We will restrict this description to the memory management aspects of the system.

Although MC68020 has a memory management coprocessor (MC68851), Sun Microsystems chose to use an MMU of their own design. This MMU supports mapping of up to 32 MB of physical memory at a time out of a total virtual address space of 4 GB (Figure 9.69).

The virtual address space is divided into four physical (i.e., real) address spaces: one physical memory and three I/O device space. The physical address space is further divided into 2048 segments, with each segment containing up to 16 pages. The PS is 8 kB. The segment and page tables are stored

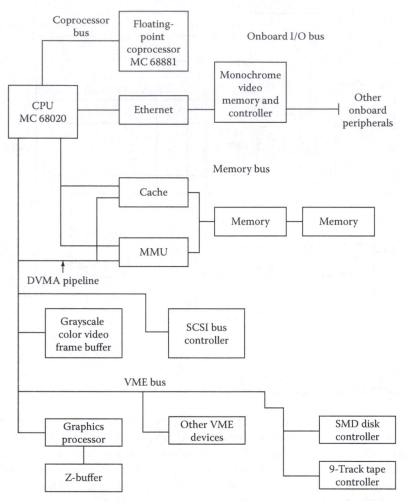

Figure 9.68 Sun-3 system block diagram. (Courtesy of Sun Microsystems Inc.)

in a high-speed memory attached to the MMU rather than in the main memory; this speeds up the address translation since page faults do not occur during the segment and page-table accesses.

The MMU contains a 3 bit *context register*. The address translation mechanism is shown in Figure 9.69. The contents of the context register are concatenated with the 11 bit segment number. These 14 bits are used as an index to the segment table to fetch a page-table pointer. The page-table pointer is then concatenated to the 4 bit virtual page number and used to access a page descriptor from the page table. The page table yields a 19 bit physical page number. This is concatenated to the 13 bit displacement from the virtual address to drive the 32 bit physical address that allows an addressing range of 4 GB in each of the four-address spaces.

In addition to the page number, the PTEs contain protection, statistics, and page-type fields. The protection field contains the following bits:

Valid bit: This bit indicates that the page is valid. If it is not set, any access to the corresponding page results in a page fault.
Write enable bit: When set, this bit indicates that a write operation is allowed on the page.
Supervisor bit: When set, this bit indicates that the page may be accessed only when the processor is in supervisory mode; user mode accesses are not allowed.

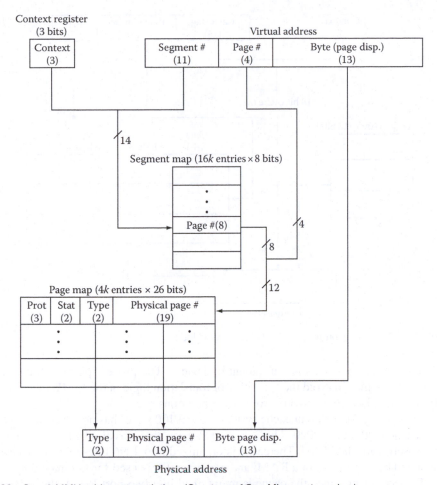

Figure 9.69 Sun-3 MMU address translation. (Courtesy of Sun Microsystems Inc.)

Statistics field: These bits identify the most recently used pages. When a page is successfully accessed, the MMU sets the "accessed" bit for that page. If the access is a write, the "modified" bit for that page is also set. These bits are used during page replacement. Only the "modified" pages are transferred to the backing store during page replacement.

Type bits: These bits indicate which of the four physical spaces the page resides in. On the basis of these bits, the memory access request is routed to main memory, the onboard I/O bus, or the VMEbus. The memory may reside on the onboard I/O bus or VMEbus, and such memory can be mapped into user programs but cannot be used as the main memory.

The MMU overlaps the address translation with physical memory access. When the address translation is started, the 13 bit displacement value is sent to the memory as a row address. This activates all the MC in every page that corresponds to that row and sets them to fetch their contents. When the translation is completed, the translated part of the address is sent as a column address to the memory. By this time, the memory cycle is almost complete, and the translated address selects the memory word at the appropriate row. Because of this overlapped operation, the translation time is effectively zero.

The MMU controls all accesses to the memory including DMA by devices. In the so-called *direct virtual memory access* (DVMA) mode, the devices are given virtual addresses by their drivers. The MMU interposes itself between devices and the memory and translates the virtual

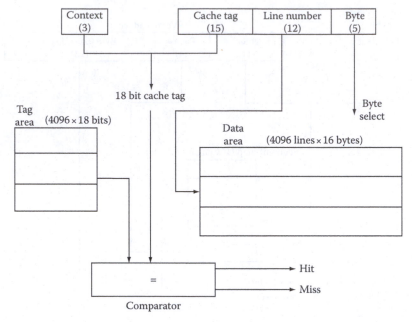

Figure 9.70 Sun-3 cache. (Courtesy of Sun Microsystems Inc.)

addresses to real addresses. This is transparent to devices. The process places a high load on the address translation logic. To avoid the saturation of translation logic in the MMU due to this load, a separate translation pipeline is provided for DMA operations.

Sun-3 systems use two different cache systems. The 3/100 series has no external cache but uses the on-chip cache of the MC68020. The 3/200 series uses an external 64 kB cache (Figure 9.70) that works closely with the MMU. The cache is organized into 4096 lines of 16 bytes each and is direct mapped. The tag area is in a RAM, and a comparator is used for tag matching. The cache uses the WB mechanism. Since the cache is maintained in a two-port RAM, the WB operations can be performed in parallel with the cache access by the processor. The cache works with virtual rather than real addresses. The address translation is started on each memory reference in parallel with the cache search. If there is a cache hit, the translation is aborted. The reader is referred to the manufacturer's manual for further details.

9.10 SUMMARY

The four basic types of memories (RAM, CAM, SAM, and DAM) were introduced in this chapter. Design of typical MCs and large memory systems with representative ICs was described. The main memory of any computer system is now built out of semiconductor RAM devices, since advances in semiconductor technology have made them cost-effective. Magnetic and optical disks are the current popular secondary memory devices.

Speed–cost trade-off is the basic parameter in designing memory systems. Although large-capacity semiconductor RAMs are now available and used in system design, memory hierarchies consisting of a combination of random- and direct-access devices are still seen even on smaller computer systems.

The memory subsystem of all modern-day computer systems consists of a memory hierarchy ranging from the fast and most expensive processor registers to the slow and least-expensive secondary storage devices. The purpose of the hierarchy is the capacity, speed, and cost trade-off to achieve the desired capacity and speed at the lowest possible cost. This chapter described the popular schemes for enhancing the speed and capacity of the memory system. The virtual memory schemes are now common even in microcomputer system. The modern-day microprocessors offer on-chip caches and memory management hardware to enable building of large and fast memory systems. As the memory capacity increases, the addressing range of the processor becomes a limitation. The popular schemes to extend the addressing range were also described in this chapter.

The speed and cost characteristics of memory devices keep improving almost daily. As a result, we have avoided listing those characteristics of the devices currently available. Refer to the manufacturers' manuals and magazines listed in the Bibliography section of this chapter for such information.

PROBLEMS

9.1 Complete the following table:

Memory System Capacity	Number of Bits in MAR	Number of Bits in MBR	Number of Chips Needed if Chip Capacity Is		
			1k × 4	2k × 1	1k × 8
64k × 4					
64k × 8					
32k × 4					
32k × 16					
32k × 32					
10k × 8					
10k × 10					

9.2 What is the storage capacity and maximum data-transfer rate of
   a. A magnetic tape, 800 bits/in., 2400 ft long, 10 in./s?
   b. A magnetic disk with 50 tracks and 4k bits/track, rotating at 3600 rpm?
   c. What is the maximum access time assuming a track-to-track move time of 1 ms?

9.3 Redesign the MC of Figure 9.9 to make it suitable for coincident decoding (i.e., two enable signals).

9.4 Design the decoding circuitry for a 2k × 4 memory IC, using linear and coincident decoding schemes. Make the MC array as square as possible in the latter case. Compare the complexities of these designs.

9.5 Design a 16k × 8 memory using the following ICs:
   a. 1024 × 1
   b. 2k × 4
   c. 1k × 8
   Each IC has on-chip decoding, tristate outputs, and an enable pin.

9.6 Arrange the 16 chips needed in the design of Problem 9.5b as a 4 × 4 array and design the decoding circuitry.

9.7 You are given a 16k × 32 RAM unit. Convert it into a 64k × 8 RAM. Treat the 16k × 32 RAM as one unit that you cannot alter. Only include logic external to the RAM unit.

9.8 A processor has a memory addressing range of 64k with 8 bits/word. The lower and upper 4k of the memory must be ROM, and the rest of the memory must be RAM. Use 1k × 8 ROM and 4k × 4 RAM chips, and design the memory system. Design a circuit to generate an error signal if an attempt is made to write into a ROM area.

9.9 An 8k memory is divided into 32 equal-size blocks (or pages) of 256 words each. The address bits are then grouped into two fields: the page number and the number of the memory word within a page. Draw the memory, MAR, and MBR configurations.

9.10 Four of the 32 pages of the memory of Problem 9.9 must be accessible at any time. Four auxiliary 5 bit registers, each containing a page address, are used for this purpose. The processor outputs a 10 bit address, the most significant 2 bits of which select one of the auxiliary registers and the least significant 8 bits select a word within a page, the page number obtained from the selected auxiliary register. Design the circuit to convert the 10 bit address output by the processor into the 13 bit address required.

9.11 Assume that a dynamic RAM controller is available for an 8k RAM with multiplexed addresses. Draw the schematic diagram of the controller showing the address (multiplexed) and the data and control signals. Describe its operation.

9.12 The computations in a particular computer system with an AM require that the contents of any field in the AM would be incremented by 1.
    a. Design a circuit to perform this increment operation on a selected field in all memory words that respond to a search.
    b. Extend the circuit to accommodate the simultaneous addition of a constant value greater than 1 to the selected field of all respondents.

9.13 Using appropriate shift registers, draw the schematic for a FIFO memory (queue) with eight words and 4 bits/word. It is required that the memory provides the queue full and queue empty conditions as the status signals. Derive the conditions under which these status signals are valid. Design the circuit.

9.14 Develop a circuit similar to the one in the previous problem for an 8 × 4 LIFO (stack) memory.

9.15 In practice, a RAM is used to simulate the operation of a stack by manipulating the address value in the MAR. Here, the data values in the stack do not actually shift during PUSH and POP operations. Rather, the address of the RAM location corresponding to the data input and output is changed appropriately. Determine the sequence of operations needed for PUSH and POP.

9.16 Develop the schematic for a two-port memory. Each port in the memory corresponds to an MAR and an MBR. Data can be simultaneously accessed by these ports. In other words
    a. Data can be read from both ports simultaneously
    b. Data can be written into two different locations simultaneously
    c. If both ports address the same location during a write operation, the data from PORT 1 should be written. That is, PORT 1 has higher priority

9.17 Study the memory system characteristics of computer systems you have access to with reference to the features described in this chapter.

9.18 Show the schematic diagram of a 64 kB memory system, using 8 kB memory blocks, assuming (a) interleaving (b) blocking, and (c) memory system to consist of four subsystems, with interleaving in the subsystem and blocking across the subsystems.

9.19 In each of the designs in Problem 9.18, show the physical location of the following addresses: 0, 48, 356, and 8192.

9.20 Assume that ASC memory is built by (two-way) interleaving two 32k words blocks. What is the effect of this memory organization on the microinstruction sequences for LDA and TCA instruction? Rewrite these sequences to efficiently utilize the interleaved memory.

9.21 Assume that ASC memory is built using eight interleaved blocks of 8k words each. Include an instruction buffer 8 words long into the ASC control unit. Develop the microoperations needed to manage the buffer. Assume that the buffer will be refilled (as a block) when the last instruction is fetched or when a jump is to an address beyond the address range in the buffer.

9.22 Rework Problem 9.21 assuming that the eight memory blocks are banked rather than interleaved.

9.23 Generalize Equations 9.2 and 9.3 to n-level memory hierarchies.

9.24 The characteristics of a four-level memory hierarchy are shown as follows:

Level	Memory Type	Access Time	Hit Ratio
1	Cache	100 ns	0.8
2	Main	1000 ns	0.85
3	Disk	1 ms	0.9
4	Tape	50 ms	1.0

What is the average access time?

9.25 Given that, in a virtual memory system, if the probability of a page fault is p, the main-memory access time is t_m and the time required to move a secondary page into the main memory is t_s, derive an expression for the average access time t_a.

9.26 A computer system has a 64 kB main memory and a 4 kB (data area only) cache. There are 8 bytes/cache line. Determine (1) the number of comparators needed and (2) the size of the tag field, for each of the following mapping schemes:
   a. Fully associative
   b. Direct
   c. Four-way set associative

9.27 Show the schematic diagrams of the cache memory in Problem 9.26 assuming that the data and tag areas of the cache are built out of 128 byte RAM ICs.

9.28 In Problem 9.26, assume that the access time of the main memory is 10 times that of the cache, and the cache hit ratio is 0.85. If the across efficiency is defined as the ratio of the average access time with a cache to that without a cache, determine the access efficiency.

9.29 In Problem 9.28, if the cache access time is 100 ns, what hit ratio would be required to achieve an average access time of 500 ns?

9.30 A computer system has 128 kB of secondary storage and an 8 kB main memory. The PS is 512 bytes. Design a virtual memory scheme using (a) direct and (b) fully associative mapping.

9.31 Determine the minimum and maximum page-table sizes and the corresponding number of accesses needed to search the page table, for each scheme in Problem 9.30.

9.32 A memory system has the following characteristics access times: cache, 100 ns; main memory, 1000 ns; TLB, 40 ns; and secondary storage, 3000 ns. The page table is in main memory, and the page-table search requires only one access to the main memory. Determine the minimum and maximum access times for the memory system.

9.33 A processor has a direct addressing range of 64 kB. Show two schematics to extend the address range to 512 kB.

9.34 In a machine with both cache and virtual memories, is the cache tag compared with the virtual address or the physical address? Justify your answer.

9.35 Assume that a system has both main-memory and disk caches. Discuss how write-through and WB mechanisms work on this machine. What are the advantages and disadvantages of each mechanism?

9.36 What is the purpose of L1D cache in the Intel cache hierarchy? List its components for Intel microarchitecture code named Sandy Bridge and Nehalem.

9.37 What is software-based RAID? Find some examples on how it is being implemented today?

9.38 You have 12 drives, each of size 1 TB to setup in RAID. List the space available and space used for protection in RAID 0, 6, and 10.

9.39 Look up the term JBOD. How is it different from SLED?

BIBLIOGRAPHY

Belady, L.A., A study of replacement algorithms for a virtual storage computer, *IBM Systems Journal*, 5(2), 78–101, 1966.

Bell, J.D., Casanet, D., and Bell, C.G., An investigation of alternative cache organization, *IEEE Transactions on Computers*, C-23, 346–351, April 1974.

Bechtolsheim, A. *The Solid State Storage Revolution, Storage Developer Conference,* Santa Clara, CA, 2008.

Bipolar Memory Data Manual, Sunnyvale, CA: Signetics, 1987.

Bipolar Microcomputer Components Data Book, Dallas, TX: Texas Instruments, 1987.

Computer, Los Alamitos, CA: IEEE Computer Society, Published Monthly, http://www.computer.org.

EE Times, Manhasset, NY: CMP Media, Published Weekly, www.eetimes.com.

Electronic Design, Cleveland, OH: Penton Media, Published twice monthly, http://www.electronicdesign.com.

Gallant, J., Cache coherency protocols, *EDN*, 2(4), 41–50, 1991.

Intel Architecture Software Developer's Manuals—Volume 1: Basic Architecture (24319001.pdf): Volume 2: Instruction Set Reference (24319101.pdf) and Volume 3: System Programming Guide (24319201.pdf) at http://download.intel.com/design/PentiumII/manuals, Santa Clara, CA: Intel Corp.

MC68851 Paged Memory Management Unit User's Manual, Englewood Cliffs, NJ: Prentice Hall, 1986.

Patterson, D.A., Gibson, G., and Katz, R., A case for redundant arrays of inexpensive disks (RAID), *Proceedings of the ACM SIGMOD Conference on Management of Data*, Chicago, IL, June 1998, pp. 109–116.

RAID Architectures, AC & NC, http://www.acnc.com/raid.html.

Shiva, S.G., *Introduction to Logic Design*, New York: Marcel Dekker, 1998.

Shiva, S.G., *Advanced Computer Architectures*, Boca Raton, FL: Taylor & Francis, 2006.

Sun-3 Architecture: A Sun Technical Report. Mountain View, CA: Sun Microsystems Inc., 1986.

Tannenbaum, A.S., Woodhull, A.S., and Woodhull, A., *Operating Systems: Design and Implementation*, Englewood Cliffs, NJ: Prentice Hall, 2006.

(TTR 2012) The Tech Report: SSD prices in steady, substantial decline, http://techreport.com/review/23149/ssd-prices-in-steady-substantial-decline, retrieved on October 23, 2012.

VAX Hardware Handbook, Maynard, MA: Digital Equipment Corporation, 1979.

Werner, J., A Look Under the Hood at Some Unique, SSD Features, SandForce, Inc., Flash Memory Summit 2010, Santa Clara, CA.

CHAPTER **10**

Arithmetic/Logic Unit Enhancement

A simple computer (ASC) arithmetic/logic unit (ALU) was designed to perform addition, complementation, and single-bit shift operations on 16-bit binary data, in parallel. The four basic arithmetic operations (addition, subtraction, multiplication, and division) can be performed using this ALU and appropriate software routines, as described in Chapter 5. Advances in integrated circuit (IC) technology have made it practical to implement most of the ALU functions required in hardware, thereby minimizing the software implementation and increasing the speed. In this chapter, we will describe a number of enhancements that make the ASC ALU resemble a practical ALU.

The majority of modern-day processors use *bit-parallel* ALUs. It is possible to use a bit-serial ALU where slow operations are acceptable. For example, in a calculator IC, serial arithmetic may be adequate due to the slow nature of human interaction with the calculator. In building processors on single ICs, a considerable amount of silicon "real estate" can be conserved by using a serial ALU. The silicon area thus saved can be used in implementing other complex operations that may be required.

As discussed in Chapter 8, the most common operands an ALU deals with are binary (fixed- and floating-point), decimal, and character strings. We will describe the ALU enhancements that facilitate fixed-point binary and logical data manipulation in the next section. Section 10.2 deals with decimal arithmetic. Section 10.3 describes the pipelining concept as applied to ALUs. It is now common to see either multiple functional units within an ALU or multiple ALUs within the central processing unit (CPU). Section 10.4 outlines the use of multiple units to enable the parallelism of operations, thereby achieving higher speeds. Section 10.5 provides some details of several commercially available systems.

10.1 LOGICAL AND FIXED-POINT BINARY OPERATIONS

ASC ALU is designed to perform a single-bit right or left shift. Other types of shift operations are useful in practice. The parallel binary adder used in ASC ALU is slow, since the carry has to ripple from the least significant bit (LSB) to the most significant bit (MSB) before the addition is complete. Several algorithms for fast addition have been devised. We will describe one such algorithm in this section. Various multiplication and division schemes have been devised; we will describe the popular ones.

10.1.1 Logic Operations

The single-bit shifts performed by the ASC ALU are called *arithmetic shift* operations since the results of shift conformed to the 2s complement arithmetic used by the ALU.

ASC ALU is designed for 1-bit shift at a time. It is desirable to have the capability of performing multiple shifts using one instruction. Since the address field in the ASC instruction format for shift instructions is not used, it may represent the number of shifts, as in

$$\text{SHR } n \quad \text{and} \quad \text{SHL } n,$$

where the operand n specifies the number of shifts. The ALU hardware is designed to perform arbitrary number of shifts in one cycle or in multiple cycles by using the single-shift operation repeatedly. The control unit must be changed to recognize the shift count in the address field and perform the appropriate number of shifts.

In practical ALUs, it is common to see rotate operations (rotate right/left, with/without carry, etc.) as described earlier (see Figure 8.6).

Logic operations such as AND, OR, EXCLUSIVE OR, and NOT are also useful ALU functions. The operands for these operations could be the contents of registers and/or memory locations. For instance, in ASC,

$$\text{AND Z}$$

might imply bit-by-bit ANDing of the accumulator with the contents of memory location Z, and

$$\text{NOT}$$

might imply the bit-by-bit complementation of the accumulator.

The *logical shift* operations are useful in manipulating logical data (and characters strings). During a logical shift, the bits are shifted left or right, and zeros are typically inserted into vacant positions. Sign copying is not performed as in arithmetic shifting.

10.1.2 Addition and Subtraction

The pseudoparallel adder used in the ASC ALU is slow; the addition is complete only after the CARRY has rippled through the MSB. There are several techniques to increase the speed of addition. We will discuss one such technique (CARRY LOOK-AHEAD) as follows.

Consider the ith stage of the pseudoparallel adder with inputs a_i, b_i, and C_{i-1} (CARRY IN) and the outputs S_i and C_i (CARRY OUT). We define two functions:

$$\text{Generate: } G_i = a_i \cdot b_i. \tag{10.1}$$

$$\text{Propagate: } P_i = a_i \oplus b_i. \tag{10.2}$$

These functions imply that the state i generates a CARRY if $(a_i \cdot b_i) = 1$ and the CARRY C_{i-1} is propagated to C_i if $(a_i \oplus b_i) = 1$. Substituting G_i and P_i into the equations for the SUM and CARRY functions of the full adder, we get

$$\text{SUM: } S_i = a_i \oplus b_i \oplus C_{i-1},$$

$$= P_i \oplus C_{i-1}. \tag{10.3}$$

$$\text{CARRY: } C_i = a_i \cdot b_i \cdot C'_{i-1} + a'_i \cdot b_i \cdot C_{i-1} + a_i \cdot b'_i \cdot C_{i-1} + a_i \cdot b_i \cdot C_{i-1},$$

$$= a_i b_i + (a'_i b_i + a_i b'_i) C_{i-1},$$

$$= a_i b_i + (a_i \oplus b_i) C_{i-1},$$

$$= G_i + P_i C_{i-1}. \tag{10.4}$$

G_i and P_i can be generated simultaneously since a_i and b_i are available. Equation 10.3 implies that S_i can be simultaneously generated if all CARRY IN (C_{i-1}) signals are available. From Equation 10.4,

$$C_0 = G_0 + P_0 C_{-1} \quad (C_{-1} \text{ is CARRY IN to the right-most bit})$$

$$C_1 = G_1 + C_0 P_1$$

$$= G_1 + G_0 P_1 + C_{-1} P_0 P_1$$

$$\vdots$$

$$C_i = G_i + G_{i-1} P_i + G_{i-2} P_{i-1} P_i + \cdots$$

$$+ G_0 P_1 P_2 \cdots P_i + C_{-1} P_0 P_1 \cdots P_i. \tag{10.5}$$

From Equation 10.5 it is seen that C_i is a function of only Gs and Ps of the ith and earlier stages. All C_is can be simultaneously generated because Gs and Ps are available. Figure 10.1 shows a 4-bit carry look-ahead adder (CLA) schematic and the detailed circuitry. This adder is faster than the ripple-carry adder since the delay is independent of the number of stages in the adder and equal to the delay of the three stages of circuitry in Figure 10.1 (i.e., about six gate delays). There are 4- and 8-bit off-the-shelf CLA units available. It is possible to connect several such units to form larger adders.

Refer to Section 10.5 for details of an off-the-shelf ALU.

10.1.3 Multiplication

Multiplication can be performed by repeated addition of the multiplicand to itself, multiplier number of times. In the binary system, multiplication by a power of 2 corresponds to shifting the multiplicand left by one position, and hence multiplication can be performed by a series of shift and add operations.

Example 10.1

Consider the multiplication of two 4-bit numbers:

Multiplicand (D)	1011		
	×1101	Multiplier (R)	$D = d_{n-1}, \ldots, d_1 d_0$
	1011		
	0000		
	1011	Partial products	$R = r_{n-1}, \ldots, r_1 r_0$
	1011		
	10001111	Product	

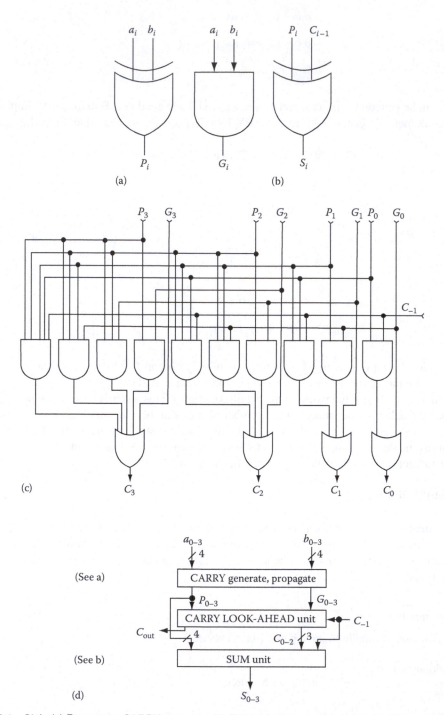

Figure 10.1 CLA. (a) Propagate, CARRY generate. (b) SUM. (c) CLA circuitry (4 bit). (d) CLA schematic.

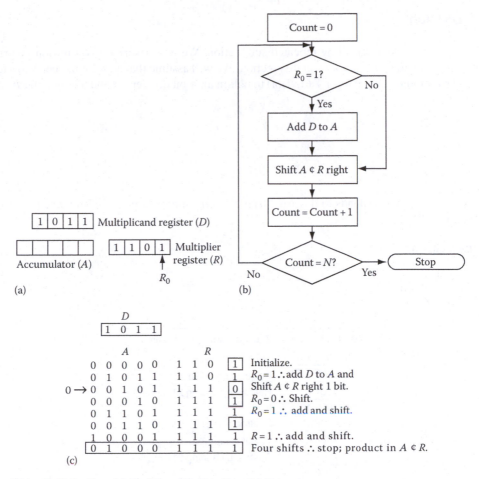

Figure 10.2 Multiplication. (a) Registers. (b) Algorithm. (c) Example.

Note that each nonzero partial product previously shown is the multiplicand shifted left an appropriate number of bits. Since the product of two n-bit numbers is $2n$ bits long, we can start off with a $2n$-bit accumulator containing all 0s and obtain the product by the following algorithm:

1. Perform the following step n times ($i = 0$ through $n - 1$).
2. Test the multiplier bit r_i.

If $r_i = 0$, shift accumulator right 1 bit.

If $r_i = 1$, add multiplicand (D) to the most significant end of the accumulator and shift the accumulator right 1 bit.

Figure 10.2a shows a set of registers that can be used for multiplying two 4-bit numbers. In general, for multiplying two n-bit numbers, R and D registers are each n bits long, and the accumulator (A) is ($n + 1$) bits long. The concatenation of registers A and R (i.e., $A \not\subset R$) will be used to store the product. The extra bit in A can be used for representing the sign of the product. We will see from the multiplication example shown in Figure 10.2c that the extra bit is needed to store the carry during partial product computation. The multiplication algorithm is shown in Figure 10.2b.

Various multiplication algorithms that perform multiplication faster than that in the previous example are available. Hardware multipliers are also available as off-the-shelf units. One such multiplier is described in Section 10.5. The book by Hwang (1979) listed as a reference in the Bibliography section at the end of this chapter provides details of multiplication algorithms and hardware.

10.1.4 Division

Division can be performed by repeated subtraction. We will describe two division algorithms that utilize shift and add operations in this section. We will assume that an n-bit integer X (dividend) is divided by another n-bit integer Y (divisor) to obtain an n-bit quotient Q and a remainder R, where

$$\frac{X}{Y} = Q + \frac{R}{Y} \tag{10.6}$$

and

$$0 \leq R < Y.$$

The first algorithm corresponds to the usual trial-and-error procedure for division and is illustrated with Example 10.2.

Example 10.2

Let
$X = 1011$ (i.e., $n = 4$)
and $Y = 0011$,
$Q = q_3 q_2 q_1 q_0$.

$0011\,\lfloor 000\ 1011$	Expand X with $(n-1)$ zeros.
$-001\ 1$	Subtract $2^{n-1}\,y$.
$-000\ 1101$	Result is negative $\therefore q_{n-1} = 0$.
$+001\ 1$	Restore by adding $2^{n-1}\,y$ back.
$000\ 1011$	
$-\ 00\ 11$	Subtract $2^{n-2}\,y$.
$-000\ 0001$	Result is negative $\therefore q_{n-2} = 0$.
$+\ 00\ 11$	Restore by adding $2^{n-2}\,y$.
$000\ 1011$	
$-\ \ 0\ 011$	Subtract $2^{n-3}\,y$.
$000\ 0101$	Result is positive $\therefore q_{n-3} = 1$.
$-\ \ \ \ \ 0011$	Subtracting $2^{n-4}\,y$.
$000\ 0010$	Result is positive $\therefore q_{n-4} = 1$.
	Stop, after n steps.
$\therefore$ Remainder $= (0010)$, Quotient $= (0011)$.	

Note that in the subtractions performed in this example, the two numbers are first compared, the smaller number is subtracted from the larger, and the result has the sign of the larger number. For example, in the first step, 0001011 is smaller than 0011000; hence, 0011000 − 0001011 = 0001101, and the result is negative.

This is called a *restoring division* algorithm, since the dividend is restored to its previous value if the result of the subtraction at any step is negative. If numbers are expressed in complement form, subtraction can be replaced by addition. The algorithms for an n-bit division are generalized as follows:

1. Assume the initial value of dividend D is $(n-1)$ zeros concatenated with X.
2. Perform the following for $i = (n-1)$ through 0: subtract $2^i \cdot y$ from D. If the result is negative, $q_i = 0$ and restore D by adding $2^i \cdot y$; if the result is positive, $q_i = 1$.
3. Collect q_i to form the quotient Q; the value of D after the last step is the remainder.

The second division algorithm is the *nonrestoring division* method. Example 10.3 illustrates this method.

Example 10.3

Let $X = 1011$, $Y = 0110$.

$0011 \lfloor 000\ 1011$	Trial dividend.
$-001\ 0$	Subtract 2^{n-1}. Y.
$-010\ 0101$	Negative result $\therefore q_3 = 0$.
$+\ \ 01\ 10$	Add $2^{n-2}\ y$. Y.
$-000\ 1101$	Negative result $\therefore q_2 = 0$.
$+\ \ \ 0\ 110$	Add 2^{n-3}. Y.
$-000\ 0001$	Negative result $\therefore q_1 = 0$.
$+\ \ \ \ \ 0110$	Add 2^{n-4}. Y.
$-000\ 1101$	Subtract $2^{n-3}\ y$.
$-\ \ \ \ \ 0011$	
$+000\ 0101$	Positive result $\therefore q_0 = 1$.

$\therefore$ Quotient $= 0001$, Remainder $= 0101$.

This method can be generalized into the following steps:

1. Assume initial value of dividend D is $(n-1)$ 0s concatenated with X; set $i = n - 1$.
2. Subtract $2^i Y$ from D.
3. If the result is negative, $q_i = 0$; if the result is positive, $q_i = 1$, $i = i - 1$; go to step 4.
4. If the result is negative, add $2^i \cdot Y$ to the result; otherwise, subtract $2^i \cdot Y$ from the result to form new D. If $i = 0$, go to step 5; otherwise, go to step 3.
5. If the final result is negative, $q_0 = 0$; otherwise $q_0 = 1$. If the final result is negative, correct the remainder by adding $2^0 \cdot Y$ to it.

The multiplication and division algorithms previously discussed assume the operands are positive, and no provision is made for a sign bit, although the MSB of the result in division algorithms can be treated as a sign bit. Direct methods for multiplication and division of numbers represented in 1s and 2s complement forms are available.

The hardware implementation of multiply and divide is usually an optional feature on smaller machines. Multiply and divide algorithms can be implemented either in hardware or in firmware by developing microprograms to implement these algorithms or by using software routines, as in Chapter 5. Figure 10.3 shows the register usage for binary multiply–divide operations on the IBM 370.

10.1.5 Stack-Based ALU

As described in Chapter 8, a zero-address machine uses the top two levels of the stack as operands for the majority of its operations. The two operands are discarded after the operation, and the result of the operation is pushed onto the stack. Stack-based ALU architectures have the following advantages:

1. The majority of operations are on the top two levels of the stack; hence, faster execution times are possible because address decoding and data fetch are not excessive.
2. Intermediate results in a computation usually can be left on the stack, thereby reducing the memory access needed and increasing the throughput.
3. Program lengths are reduced, since zero-address instructions are shorter compared to one-, two-, and three-address instructions.

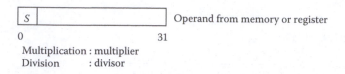

0   31
Multiplication : multiplier
Division   : divisor

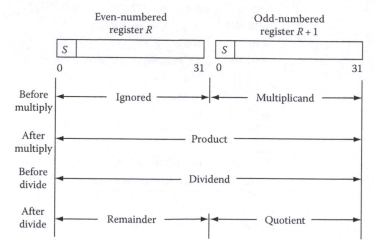

Figure 10.3   Register usage for fixed-point multiply and divide on IBM 370.

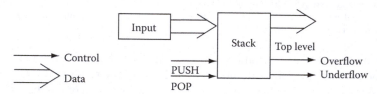

Figure 10.4   Stack model.

Figure 10.4 shows the model of a stack. The data from the input register are pushed onto the stack when the PUSH control signal is activated. Contents of the top level of the stack are always available at the top-level output. POP when activated pops the stack, thereby moving the second level to the top level. The underflow error is generated when an attempt is made to pop the empty stack, and the overflow error condition occurs if an attempt is made to push the data into a full stack.

Appendix B provides the details of the two popular implementations of stacks. The hardware (i.e., shift register-based) implementation is used in ALU designs rather than the random-access memory (RAM)-based implementation since fast stack operations are desired. Some machines use a combination of the shift register- and RAM-based implementations in implementing stacks for the ALU. Here, a set of ALU registers are organized as the top few levels of the stack, and the subsequent levels are implemented in the RAM area. Thus, the levels needed most often can be accessed fast.

Figure 10.5 shows the details of a stack-based ALU. The functional unit is a combinational circuit that can perform operations such as addition and subtraction on its input operands X and Y and produce the result Z. The function of this module depends on the control signals from the

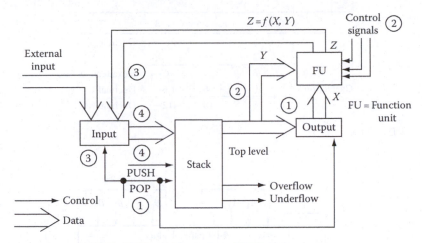

Figure 10.5 Stack ALU. *Note:* Numbers (①, ②, ...) indicate the sequence of operations.

control unit. In order to add two numbers on the top two levels of the stack, the following sequence of micro-operations is needed:

	Micro-Operation	Comments
①	POP	OUTPUT ← top level (i.e., operand X is now available at the functional unit); second and subsequent levels of the stack move up. The second operand is available on Y input.
②	ADD	Activate ADD control signal; the functional unit produces the sum of X and Y at its output Z.
③	POP	Pop the second operand from the stack and at the same time gate Z into the input register.
④	PUSH	Push the result from the input register onto the stack.

The control unit should produce such micro-operation sequences for each of the operations allowed in the ALU. Note in the previous sequence that we assumed that the data are on the top two levels of the stack as a result of previous ALU operations.

10.2 DECIMAL ARITHMETIC

Some ALUs allow decimal arithmetic. In this mode, 4 bits are used to represent a binary-coded decimal (BCD) digit, and the arithmetic can be performed in two modes: *bit serial* and *digit serial*.

Example 10.4

Figure 10.6 shows the addition of two decimal digits. In case 1, the sum results in a valid digit. In cases 2 and 3, the sum exceeds the highest valid digit 9, and hence a 6 is added to the sum to bring it to the proper decimal value. Thus, the "ADD 6" correction is needed when the sum of the digits is between $(A)_{16}$ and $(F)_{16}$.

Figure 10.7a shows a digit serial (bit parallel) decimal adder. A bit-serial adder is shown in Figure 10.7b. The bit-serial adder is similar to the serial adder discussed in Chapter 3, except for the ADD 6 correction circuit. The sum bits enter the B register from the left-hand input. At BIT 4,

	Case 1	Case 2	Case 3	
	5	6	7	Decimal
	+3	4	5	
	8	A	C←	Hexadecimal
Correction	0	+ 6	+ 6←	Decimal
	8	10	12←	Decimal

Figure 10.6   BCD addition.

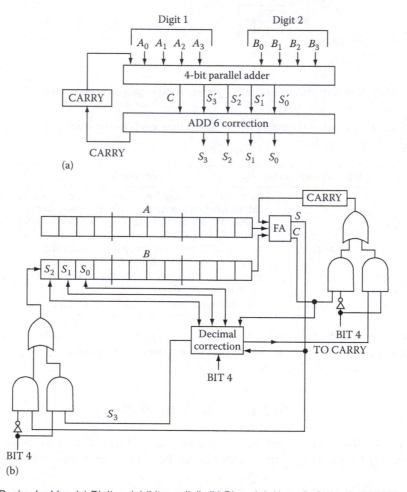

(a)

(b)

Figure 10.7   Decimal adder. (a) Digit serial (bit parallel). (b) Bit serial. *Note: S,* SUM; *C,* CARRY.

the decimal correction circuit examines sum bits S_0, S_1, S_2, and S_3 and returns the corrected sum to the B register while generating an appropriate carry for the next digit.

Some processors use separate instructions for decimal arithmetic, while others use special instructions to switch the ALU between decimal and binary arithmetic modes. In the latter case, the same set of instructions operates both on decimal and binary data. The Mostek 6502 uses a set decimal instruction to enter decimal mode and clear decimal instruction to return to binary mode. The arithmetic is performed in digit serial mode. Hewlett–Packard 35 system uses a serial arithmetic on 13-digit (52-bit) floating-point decimal numbers.

10.3 PIPELINING

Example 10.5

Consider the addition of two floating-point numbers shown as follows:

$$(0.5 \times 10^{-3}) + (0.75 \times 10^{-2})$$

$$= 0.05 \times 10^{-2} + 0.75 \times 10^{-2} \quad \text{Equalize exponents.}$$

$$= 0.80 \times 10^{-2} \quad \text{Add mantissa.}$$

If the mantissa of the sum is greater than 1, the exponent needs to be adjusted to make the mantissa less than 1 as in the following:

$$0.5 \times 10^{-3} + 0.75 \times 10^{-3}$$

$$= 0.5 \times 10^{-3} + 0.75 \times 10^{-3} \quad \text{Equalize exponents.}$$

$$= 1.25 \times 10^{-3} \quad \text{Add mantissa.}$$

$$= 0.125 \times 10^{-2} \quad \text{Normalize mantissa.}$$

The three steps previously involved in the floating-point addition can be used to design a three-stage adder shown in Figure 10.8. The first stage performs equalization of exponents and passes the two modified operands to the second stage. The second stage performs addition and passes the result to the third stage. The third stage performs normalization. Each stage receives its own control signals. R_1 and R_2 are holding registers to hold the data between stages. This implementation is called a *pipeline*.

Note that the three stages work independently on the data provided to them. Thus, when the operands move to the second stage from the first stage, a new set of operands can be inserted into the first stage. Similarly, when the second stage passes the results to the third, it receives the next set of operands from the second stage. This operation continues and resembles that of a factory assembly line.

If we assume that each stage takes 1 time unit to complete its operation, the total time to compute the sum is 3 units. However, due to the pipeline mode of operation, we see one result emerging from stage three every time unit, after the first 3 time units (i.e., pipeline is full) from the beginning (when pipeline is empty). Thus, to add N sets of operands, the pipeline will need $3 + (N - 1)$ time units, rather than the $3N$ time units needed with a nonpipeline implementation. The *speedup* offered by this pipeline is

$$\frac{3N}{(3 + N - 1)}.$$

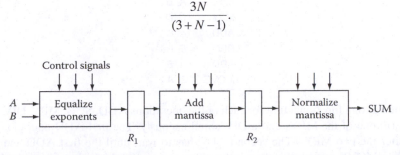

Figure 10.8 Floating-point add pipeline.

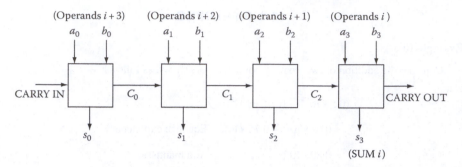

Figure 10.9   Floating-point binary adder. *Note:* SUM i complete at i, but the SUM bits were generated during $i, i - 1, i - 2$, and $i - 3$.

For large values of N, the speedup tends to 3. In general, an M-stage pipeline offers a speedup of M. One requirement is that the operands need to be aligned to keep the pipeline busy. Also, since each stage of the pipeline is required to operate independently, a complex multistage control unit design is needed. We will provide further details on pipelines in Chapter 11.

Example 10.6

Figure 10.9 shows a pipeline scheme to add two 4-bit binary numbers. Here, the carries propagate along the pipeline, while the data bits to be added are input through the lateral inputs of the stages. Note that once the pipeline is full, each stage will be performing an addition. At any time, four sets of operands are being processed. The addition of the operands entering the pipeline at time t will be complete at time $(t + 4\Delta)$, where Δ is the processing time of each stage. It is necessary to store the sum bits produced during four time slots to get the sum of a set of operands.

10.4 ALU WITH MULTIPLE FUNCTIONAL UNITS

An obvious method of increasing the throughput of the ALU is to employ multiple functional units in the design. Each functional unit can be either a general-purpose unit capable of performing all the ALU operations or a dedicated unit that can perform only certain operations, as many of these functional units as needed are activated by the control unit simultaneously, thereby achieving a high throughput. The control unit design becomes very complex since it is required to maintain the status of all the functional units and schedule the processing functions in an optimum manner, to take advantage of the possible parallelism.

Consider the computation: $Z = A*B + C*D + E/F$. A program to perform this on a two-address machine would be

MPY	A, B
MPY	C, D
ADD	A, C
DIV	E, F
ADD	A, E
MOVE	Z, A

If the machine has one functional unit for each of the operations ADD, SUB, MPY, DIV, and MOVE, we can perform one of the MPYs and a DIV simultaneously. The first ADD operation can only be performed after the two MPYs. The second ADD has to wait until the first ADD and the DIV are completed. The MOVE can be performed only after the second ADD. It is thus possible to invoke more than one operation simultaneously as long as we obey the precedence constraints imposed by the computation desired. If the machine has two MPY units, both the MPYs can be done simultaneously.

To make the previous operation efficient, we will need a complex instruction-processing mechanism that retrieves the instructions from the program and issues them to the appropriate functional unit, based on the availability of functional units. Processors with such mechanisms that can invoke more than one instruction simultaneously are known as *superscalar* architectures. In this description, we have assumed that each functional unit is capable of performing one function. This need not be the case. In practice, these units could be full-fledged processors that can handle integer, floating point, and other operations, and there may also be multiples of each type. These units are often pipelined to offer better performance. These architectures also use sophisticated compilers that can schedule operations appropriately to make best use of the computing resources. Compilers generate approximate schedules for instructions and the instruction-processing hardware resolves the dependencies among the instructions.

The Intel Pentium described in Chapter 8 is a superscalar architecture with multiple integer, floating-point, and MMX units (see Figures 8.22 through 8.29). It had three instruction-processing stages (fetch/decode, dispatch/execute, and retire) retrieving instructions from the instruction pool and reordering and reserving them to run as many of the functional units as possible simultaneously.

In very long instruction word (*VLIW*) architectures, each instruction word is capable of holding multiple instructions. Once the instruction word is fetched, all the component instructions can be invoked assuming that the functional units are available to handle them. This requires compilers and assemblers that can pack multiple instructions into the instruction word. If we assume that a VLIW has five slots, one for each type of functional unit to perform the operations ADD, SUB, MPY, DIV, and MOVE, we can rewrite the previous program as

```
MPY A, B       DIV E, F
MPY C, D
ADD A, C
ADD A, E
MOVE Z, A
```

If the machine had two MPY units, we could have packed two MPYs into the first instruction word.

In practice, VLIW architectures pack 4–8 instructions per instruction word. They use sophisticated compilers that handle the dependency among the instructions and package them appropriately. Since the instructions are scheduled at compile time, if there is any change in hardware environment during run time that makes it impractical to obey that schedule, a recompilation of the code would be necessary.

The Intel Itanium (IA-64) described in Chapter 11 is an example of VLIW architecture.

10.5 EXAMPLE SYSTEMS

In this section, we provide brief descriptions of the following commercially available systems, ranging in complexity from an ALU IC to a supercomputer system:

1. SN74181, an ALU IC capable of performing 16 common arithmetic/logic functions.
2. Texas Instruments' MSP430, a hardware multiplier.
3. Motorola 68881 floating-point coprocessor that supports an integer processor.
4. Control Data Corporation's CDC 6600, an early system with multiple functional units, included here for its historical interest. The system formed the basis for the architecture of the Cray series of supercomputers.
5. Cray X-MP, an early supercomputer system, with multiple pipelined functional units.

10.5.1 Multifunction ALU IC (74181)

An ALU can be built using off-the-shelf ICs. For example, the SN 74181 shown in Figure 10.10 is a multifunction ALU. It can perform 16 binary arithmetic operations on two 4-bit operands A and B. The function to be performed is selected by pins S_0, S_1, S_2, and S_3 and includes addition, subtraction, decrement, and straight transfer, as shown in Figure 10.10b. C_n is the CARRY INPUT and C_{n+4} is the CARRY OUTPUT if the IC is used as a 4-bit ripple-carry adder. The IC can also be used with a look-ahead carry generator (74,182) to build high-speed adders of multiple stages forming 8-, 12-, and 16-bit adders.

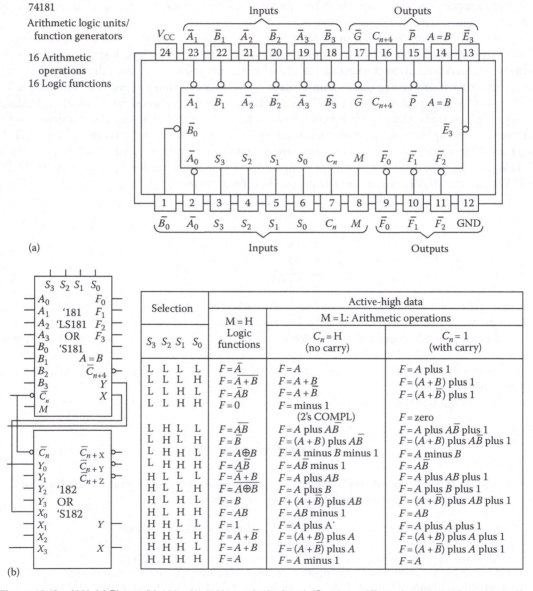

Figure 10.10 ALU. (a) Pinout. (b) 4 bit with ALU carry look-ahead. (Courtesy of Texas Instruments Incorporated.)

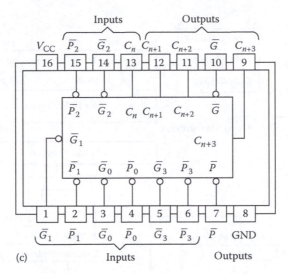

Figure 10.10 (continued)   ALU. (c) Look-ahead carry generator (74182). (Courtesy of Texas Instruments Incorporated.)

10.5.2 Texas Instruments' MSP430 Hardware Multiplier

This section is extracted from *The MSP430 Hardware Multiplier: Functions and Applications*, SLAA042, April 1999, Texas Instruments Inc.

The MSP430 hardware multiplier allows three different multiply operations (modes):

1. Multiplication of unsigned 8- and 16-bit operands (MPY)
2. Multiplication of signed 8- and 16-bit operands (MPYS)
3. Multiply-and-accumulate (MAC) function using unsigned 8- and 16-bit operands

Any mixture of operand lengths (8 and 16 bits) is possible. Additional operations are possible when supplemental software is used, such as signed MAC.

Figure 10.11 shows the hardware modules that comprise the MSP430 multiplier. The accessible registers are explained in the following paragraphs. Figure 10.11 is not intended to depict the physical reality; it illustrates the hardware multiplier from the programmer's point of view. Figure 10.12 shows the system structure where the MSP430 is a peripheral.

The hardware multiplier is not part of the MSP430 CPU—it is a peripheral that does not interfere with the CPU activities. The multiplier uses normal peripheral registers that are loaded and read using CPU instructions. The programmer-accessible registers are explained next.

Operand 1 registers: The operational mode of the MSP430 hardware multiplier is determined by the address where operand 1 is written:

1. *Address 130h*: executes unsigned multiplication (MPY)
2. *Address 132h*: executes signed multiplication (MPYS)
3. *Address 134h*: executes unsigned MAC

The address of operand 1 alone determines the operation to be performed by the multiplier (after modification of operand 2). No operation is started by modification of operand 1 alone.

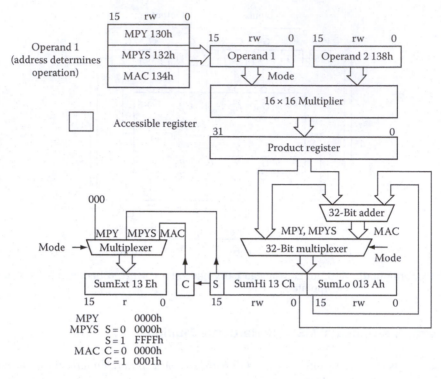

Figure 10.11   Hardware multiplier block diagram.

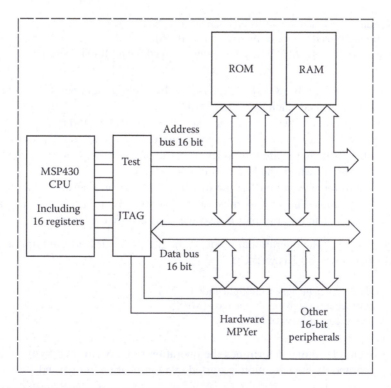

Figure 10.12   Hardware multiplier internal connections.

Example 10.7

A multiply unsigned (MPY) operation is defined and started. The two operands reside in $R14$ and $R15$.

```
MOV R15, &130h      ; Define MPY operation
MOV R14, &138h      ; Start MPY with operand 2
...                 ; Product in SumHi I SumLo
```

Operand 2 register: The operand 2 register (at address 138h) is common to all three multiplier modes. Modification of this register (normally with a MOV instruction) starts the selected multiplication of the two operands contained in operand 1 and operand 2 registers. The result is written immediately into the three hardware registers SumExt, SumHi, and SumLo. The result can be accessed with the next instruction, unless indirect addressing modes are used for source addressing.

SumLo register: This 16-bit register contains the lower 16 bit of the calculated product or sum. All instructions may be used to access or modify the SumLo register. The high byte cannot be accessed with byte instructions.

SumHi register: The contents of this 16-bit register, which depend on the previously executed operation, are as follows:

1. *MPY*: The most significant word of the calculated product.
2. *MPYS*: The most significant word, including the sign of the calculated product. 2s complement notation is used for the product.
3. *MAC*: The most significant word of the calculated sum.

All instructions may be used with the SumHi register. The high byte cannot be accessed using byte instructions.

SumExt register: The sum extension register SumExt allows calculations with results exceeding the 32-bit range. This read-only register holds the most significant part of the result (bits 32 and higher). The content of SumExt is different for each multiplication mode:

1. *MPY*: SumExt always contains zero, with no carry possible. The largest possible result is 0FFFFh × 0FFFFh = 0FFFE0001h.
2. *MPYS*: SumExt contains the extended sign of the 32-bit result (bit 31). If the result of the multiplication is negative (MSB = 1), SumExt contains 0FFFFh. If the result is positive (MSB = 0), SumExt contains 0000h.
3. *MAC*: SumExt contains the carry of the accumulate operation. SumExt contains 0001 if a carry occurred during the accumulation of the new product to the old one and zero otherwise.

SumExt register simplifies multiple word operations—straightforward additions are performed without conditional jumps, saving time, and ROM space.

Example 10.8

The new product of an MPYS operation (operands in $R14$ and $R15$) is added to a signed 64-bit result located in RAM word RESULT through RESULT + 6:

```
MOV     R15, &MPYS        ; First operand
MOV     R14, &OP2         ; Start MPYS with operand 2
ADD     SumLo, RESULT     ; Lower 16 bits of result
ADDC    SumHi, RESULT+2   ; Upper 16 bits
ADDC    SumExt, RESULT+4  ; Result bits 32 to 47
ADDC    SumExt, RESULT+6  ; Result bits 48 to 63
```

Hardware multiplier rules: The hardware multiplier is essentially a word module. The hardware registers can be addressed in word mode or in byte mode, but the byte mode can only address the lower bytes. The upper byte cannot be addressed.

The operand registers of the hardware multiplier (addresses 0130h, 0132h, 0134h, and 0138h) behave like the CPU's working registers $R0$–$R15$ when modified in byte mode: the upper byte is cleared in this case. This allows for any combination of 8- and 16-bit multiplications.

Multiplication modes: The three different multiplication modes available are explained in the following sections.

Unsigned multiply: The two operands written to operand registers 1 and 2 are treated as unsigned numbers with

00000h being the smallest number
0FFFFFh being the largest number

The maximum possible result is obtained with input operands 0FFFFh and 0FFFFh:

$$0FFFFh \times 0FFFFh = 0FFFE0001h$$

No carry is possible, the SumExt register always contains zero. Table 10.1 gives the products for some selected operands.

Signed multiply: The two operands written to operand registers 1 and 2 are treated as signed 2s complement numbers with

08000h being the most negative number (–32768)
07FFFh being the most positive number (+32767)

The SumExt register contains the extended sign of the calculated result:

SumExt = 00000h: The result is positive.
SumExt = 0FFFFh: The result is negative.

Table 10.2 gives the signed-multiply products for some selected operands.

MAC: The two operands written to operand registers 1 and 2 are treated as unsigned numbers (0h to 0FFFFh). The maximum possible results are obtained with input operands 0FFFFh and 0FFFFh:

$$0FFFFh \times 0FFFFh = 0FFFE0001h$$

Table 10.1   Results with Unsigned-Multiply Mode

Operands	SumExt	SumHi	SumLo
0000 × 0000	0000	0000	0000
0001 × 0001	0000	0000	0001
7FFF × 7FFF	0000	3FFF	0001
FFFF × FFFF	0000	FFFF	0001
7FFF × FFFF	0000	7FFF	8001
8000 × 7FFF	0000	3FFF	8000
8000 × FFFF	0000	7FFF	8000
8000 × 8000	0000	4000	0000

Table 10.2   Results with Signed-Multiply Mode

Operands	SumExt	SumHi	SumLo
0000 × 0000	0000	0000	0000
0001 × 0001	0000	0000	0001
7FFF × 7FFF	0000	3FFF	0001
FFFF × FFFF	0000	0000	0001
7FFF × FFFF	FFFF	FFFF	8001
8000 × 7FFF	FFFF	C000	8000
8000 × FFFF	0000	0000	8000
8000 × 8000	0000	4000	0000

Table 10.3   Results with Unsigned MAC Mode

Operands	SumExt	SumHi	SumLo
0000 × 0000	0000	C000	0000
0001 × 0001	0000	C000	0001
7FFF × 7FFF	0000	FFFF	0001
FFFF × FFFF	0001	BFFE	0001
7FFF × FFFF	0001	3FFE	8001
8000 × 7FFF	0000	FFFF	8000
8000 × FFFF	0001	3FFF	8000
8000 × 8000	0001	0000	0000

This result is added to the previous content of the two sum registers (SumLo and SumHi). If a carry occurs during this operation, the SumExt register contains 1 and is cleared otherwise:

SumExt = 00000h: No carry occurred during the accumulation.
SumExt = 00001h: A carry occurred during the accumulation.

The results of Table 10.3 assume that SumHi and SumLo contain the accumulated content C000,0000 before the execution of each example. See Table 10.1 for the results of an MPY without accumulation.

Multiplication word lengths: The MSP430 hardware multiplier allows all possible combinations of 8- and 16-bit operands. Notice that input registers operand 1 and operand 2 behave like CPU registers, where the high-register byte is cleared if the register is modified by a byte instruction. This simplifies the use of 8-bit operands. The following are examples of 8-bit operand use for all three modes of the hardware multiplier.

```
; Use the 8-bit operand in R5 for an unsigned multiply.
; MOV.B R5, &MPY   ; The high byte is cleared
;
; Use an 8-bit operand for a signed multiply.
MOV.B R5, &MPYS   ; The high byte is cleared
SXT &MPYs ; Extend sign to high byte
;
; Use an 8-bit operand for a multiply-and-accumulate.
; MOV.B R5, &MAC   ; The high byte is cleared
```

Operand 2 is loaded in a similar fashion. This allows all four possible combinations of input operands:

16×16   8×16   16×8   8×8

Table 10.4   CPU Cycles Needed with Different Multiplication Modes

Operation	Software Loop	Hardware MPYer	Speed Increase
Unsigned multiply MPY	139...171	8	17.4...21.4
Unsigned MAC	137...169	8	17.1...21.1
Signed multiply MPYS	145...179	8	18.1...22.4
Signed MAC	143...177	17	8.4...10.4

Speed comparison with software multiplication: Table 10.4 shows the speed increase for the four different 16 × 16 bit multiplication modes. The software loop cycles include the subroutine call (CALL #MULxx), the multiplication subroutine itself, and the RET instruction. Only CPU registers are used for the multiplication. See the *Metering Application Report* for details on the four multiplication subroutines.

The cycles given for the hardware multiplier include the loading of the multiplier operands operand 1 and operand 2 from CPU registers and—in the case of the signed MAC operation—the accumulation of the 48-bit result into three CPU registers.

Refer to the application report cited previously for further details on programming the MSP430 and other application details.

10.5.3 Motorola 68881 Coprocessor

The MC68881 floating-point coprocessor provides (Institute of Electrical and Electronics Engineers) IEEE Standard 754 floating-point operations capability as a logical extension to the integer data processing capabilities of MC68000 series of microprocessors (MPUs). The major features of the MC68881 are the following:

1. Eight general-purpose floating-point data registers
2. 67-bit arithmetic unit
3. 67-bit barrel shifter
4. 46 instructions
5. Full conformance to ANSI/IEEE 754-1985 standard
6. Supports additional functions not on IEEE standard
7. Seven data formats
8. Twenty-two constants available in on-chip ROM (π, e, etc.)
9. Virtual memory/machine operations
10. Efficient mechanisms for exception processing
11. Variable size data bus compatibility

The coprocessor interface mechanism attempts to provide the programmer with an execution model based on sequential instruction execution in the MPU and the coprocessor. Floating-point instructions are executed concurrently with MPU integer instructions. The coprocessor and the MPU communicate via standard bus protocols.

Figure 10.13 shows the MC68881 simplified block diagram. The MC68881 is logically divided into two parts:

1. Bus interface unit (BIU)
2. Arithmetic processing unit (APU)

The task of the BIU is to communicate with the MPU. It monitors the state of the APU even though it operates independently of the APU. The BIU contains the coprocessor interface registers, communication status flags, and the timing control logic.

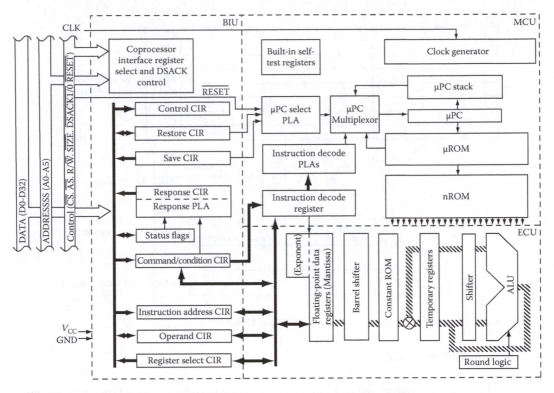

Figure 10.13 MC68881 simplified block diagram. (Courtesy of Motorola Inc.)

The task of the APU is to execute the command words and operands sent from the BIU. It must report its internal status to the BIU. The eight floating-point data registers, control, status, and instruction address registers are located in the APU. The high-speed arithmetic unit and the barrel shifter are also located in the APU. Figure 10.14 presents the programming model for the coprocessor.

The MC68881 floating-point coprocessor supports seven data formats:

1. Byte integer
2. Word integer
3. Long-word integer
4. Single-precision real
5. Double-precision real
6. Extended-precision real
7. Packed decimal string real

The integer data formats consist of standard 2s complement format. Both single-precision and double-precision floating-point formats are implemented as specified in the IEEE standard. Mixed mode arithmetic is accomplished by converting integers to extended-precision floating-point numbers before being used in the computation.

The MC68881 floating-point coprocessor provides six classes of instructions:

1. Moves between coprocessor and memory
2. Move multiple registers
3. Monadic operations
4. Dyadic operations
5. Branch, set, or trap conditionally
6. Miscellaneous

These classes provide 46 instructions, which include 35 arithmetic operations.

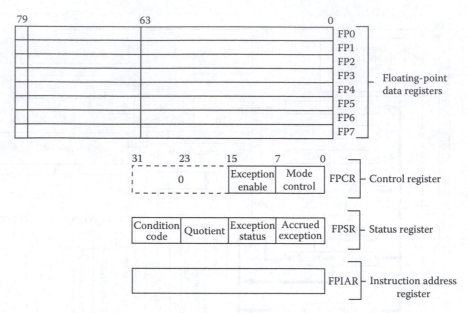

Figure 10.14   MC68881 programming model. (Courtesy of Motorola Inc.)

10.5.4 Control Data 6600

The CDC 6600 first introduced in 1964 was designed for two types of use: large-scale scientific processing and time sharing of smaller problems. To accommodate large-scale scientific processing, a high-speed, floating-point CPU employing multiple functional units was used. Figure 10.15 shows the structure of the system. The peripheral activity was separated from the CPU activity by using 12 input/output (I/O) channels controlled by 10 peripheral processors. Here, we will concentrate on the operation of the CPU.

As seen in Figure 10.15, the CPU comprises 10 functional units: 1 add, 1 double-precision add, 2 multiply, 1 divide, 2 increment, 1 shift, 1 Boolean, and 1 branch units. The CPU obtains its

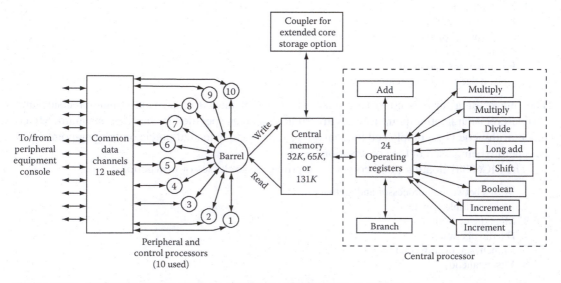

Figure 10.15   CDC 6600 system structure. (Courtesy of Control Data Corporation.)

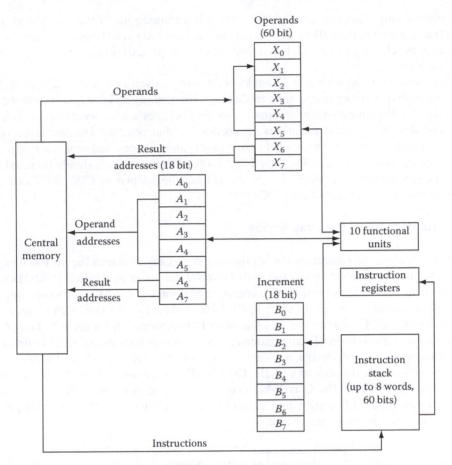

Figure 10.16 Central processor operating registers of CDC 6600. (Courtesy of Control Data Corporation.)

programs and data from the central memory. It can be interrupted by a peripheral processor. The 10 functional units in the central processor can operate in parallel on one or two 60-bit operands to produce a 60-bit result. The operands and results are provided in the operating registers shown in Figure 10.16. A functional unit is activated by the control unit as soon as the operands are available in the operating registers. Since the functional units work concurrently, a number of arithmetic operations can be performed in parallel.

As an example, the computation of $Z = A \cdot B + C \cdot D$, where Z, A, B, C, and D are memory operands, progresses as follows: first, the contents of A, B, C, and D are loaded into a set of CPU registers (say R_1, R_2, R_3, and R_4, respectively). Then, R_1 and R_2 are assigned as inputs to a multiply unit with another register, R_5, allocated to receive its output. Simultaneously, R_3 and R_4 are assigned as inputs to another multiply unit with R_6 as its output. R_5 and R_6 are assigned as inputs to the add unit with R_7 as its output. As soon as the multiply units provide their results into R_5 and R_6, they are added by the add unit and the result from R_7 is stored into Z. There is a queue associated with each functional unit. The CPU simply loads these queues with the operand and results register information. As and when a functional unit is free, it retrieves this information from its queue and operates on it, thus providing a very high parallelism.

There are 24 operating registers: eight 18-bit index registers, eight 18-bit address registers, and eight 60-bit floating-point registers. Figure 10.14 shows the data and address paths. Instructions are either 15 or 30 bits long. An instruction stack capable of holding 32 instructions enhances instruction execution speed.

The control unit maintains a *scoreboard*, which is a running file of the status of all registers and functional units and their allocation. As new instructions are fetched, resources are allocated to execute them by referring to the scoreboard. Instructions are queued for later processing if resources cannot be allocated.

Central memory is organized in 32 banks of 4K words. Each consecutive address calls for a different bank. Five memory trunks are provided between memory and five floating-point registers. An instruction calling for an address register implicitly initiates a memory reference on its trunk. An overlapped memory access and arithmetic operation is thus possible. The concurrent operation of functional units, high transfer rates between registers and memory, and separation of peripheral activity from the processing activity make CDC 6600 a fast machine. It should be noted that the CRAY supercomputer family designed by Seymour Cray, the designer of CDC 6600 series, retains the basic architectural features of the CDC 6600.

10.5.5 Architecture of the Cray Series

The Cray-1, a second-generation vector processor from Cray Research Incorporated (now Cray Corporation), has been described as the most powerful computer of the late 1970s. Benchmark studies show that it is capable of sustaining computational rates of 138 MFLOPS over long periods of time and attaining speeds of up to 250 MFLOPS in short bursts. This performance is about 5 times that of the CDC 7600 or 15 times that of an IBM System/370 Model 168. Thus, Cray-1 is uniquely suited to the solution of computationally intensive problems encountered in fields such as weather forecasting, aircraft design, nuclear research, geophysical research, and seismic analysis.

Figure 10.17 shows the structure of the Cray X-MP (successor to Cray-1). A four-processor system (X-MP/4) is shown. The Cray X-MP consists of four sections: multiple Cray-1-like CPUs, the memory system, the I/O system, and the processor interconnection. The following paragraphs provide a brief description of each section.

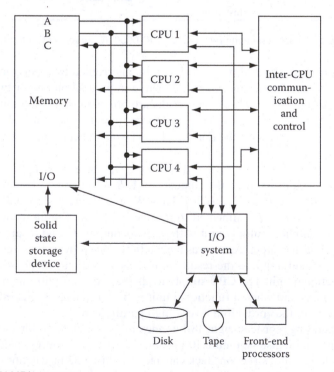

Figure 10.17   Cray X-MP/4 structure.

10.5.5.1 Memory

The memory system is built out of several sections, each divided into banks. Addressing is interleaved across the sections and within sections across the banks. The total memory capacity can be up to 16 megawords with 64 bits per word. Associated with each memory word, there is an 8-bit field dedicated to single error correction/double error detection (SECDED). The memory system offers a bandwidth of 25–100 Gbps. It is multiported, with each CPU connected to four ports (two for reading, one for writing, and one for independent I/O). Accessing a port ties it up for one clock cycle, and a bank access takes four clock cycles.

Memory contention can occur several ways: a bank conflict occurs when a bank is accessed while it is still processing a previous access, a simultaneous conflict occurs if a bank is referenced simultaneously on independent lines from different CPUs, and a line conflict occurs when two or more of the data paths make a memory request to the same memory section during the same clock cycle. Memory conflict resolution may require wait states to be inserted. Because memory conflict resolution occurs element by element during vector references, it is possible that the arithmetic pipelines being fed by these vector references may experience clock cycles with no input. This can produce a degradation in the arithmetic performance attained by the pipelined functional units. Memory performance is typically degraded by 3%–7% on average due to memory contention and in particularly bad cases by 20%–33%.

The secondary memory, known as the solid-state device (SSD), is used as an exceptionally fast-access disk device although it is built out of MOS random access memory ICs. The access time of SSD is 40 μs, compared to the 16 ms access time of the fastest disk drive from Cray Research Inc. The SSD is used for the storage of large-scale scientific programs that would otherwise exceed main memory capacity and reduce bottlenecks in I/O-bound applications. The central memory is connected to the SSD through either one or two 1000 MB/s channels. The I/O subsystem is directly connected to the SSD, thereby allowing prefetching of large data sets from the disk system to the faster SSD.

10.5.5.2 Processor Interconnection

The interconnection of the CPUs assumes a coarse-grained multiprocessing environment. That is, each processor (ideally) executes a task almost independently, requiring communication with other processors once every few million or billion instructions.

Processor interconnection comprises the clustered share registers. The processor may access any cluster that has been allocated to it in either user or system monitor mode. A processor in monitor mode has the processor ability to interrupt any other processor and cause it to go into monitor mode.

10.5.5.3 Central Processor

Each CPU is composed of low-density ECL logic with 16 gates per chip. Wire lengths are minimized to cut the propagation delay of signals to about 650 ps. Each CPU is a register-oriented vector processor (Figure 10.18) with various sets of registers that supply arguments to and receive results from several pipelined, independent functional units. There are eight 24-bit address registers $(A_0–A_{70})$, eight 64-bit scalar registers $(S_0–S_7)$, and eight vector registers $(V_0–V_7)$. Each vector register can hold up to sixty-four 64-bit elements. The number of elements present in a vector register for a given operation can be contained in a 7-bit vector length (VL) register. A 64-bit vector mask (VM) register allows masking of the elements of a vector register prior to an operation. Sixty-four 24-bit address save registers $(B_0–B_{63})$ and 64 scalar save registers $(T_0–T_{63})$ are used as buffer storage areas for the address and scalar registers, respectively.

The address registers support an integer add and an integer multiply functional unit. The scalar and vector registers each support integer add, shift, logical, and population count functional units.

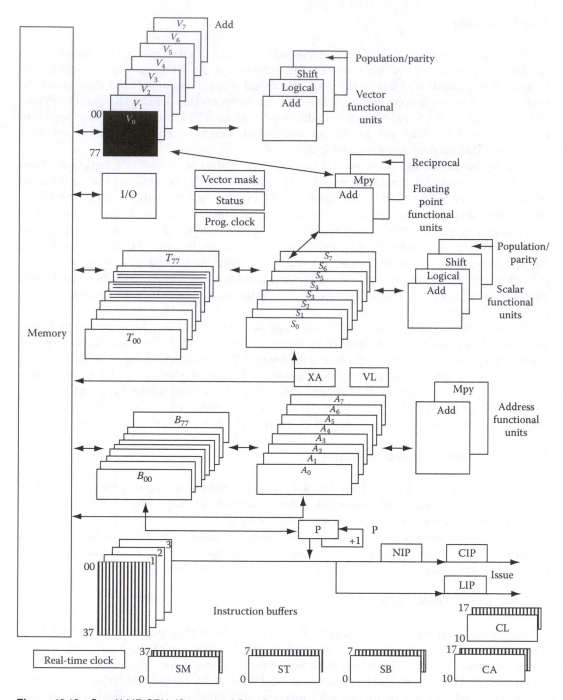

Figure 10.18   Cray X-MP CPU. (Courtesy of Cray Research Inc.)

Three floating-point arithmetic functional units (add, multiply, and reciprocal approximation) take their arguments from either the vector or the scalar registers.

The result of the vector operation is either returned to another vector register or may replace one of the operands to the same functional unit (i.e., "written back") provided there is no recursion.

An 8-bit status register contains such flags as processor number, program status, cluster number, interrupt, and error detection enables. This register can be accessed through an *S* register.

The exchange address register is an 8-bit register maintained by the operating system. This register is used to point to the current position of the exchange routine in memory. In addition, there is program clock used for accurately measuring duration intervals.

As mentioned earlier, each CPU is provided with four ports to the memory: with one port reserved for the input/output subsystem (IOS) and the other three, labeled *A*, *B*, and *C*, supporting data paths to the registers. All the data paths can be active simultaneously, as long as there are no memory access conflicts.

Data transfer between scalar and address registers and the memory occurs directly (i.e., as individual elements into and out of referenced registers). Alternatively, block transfers can occur between the buffer registers and the memory. The transfer between scalar and address registers and the corresponding buffers is done directly. Transfers between the memory and the vector registers are done only directly.

Block transfer instructions are available for loading to and storing from *B* (using port *A*) and *T* (using port *B*) buffer registers. Block stores from the *B* and *T* registers to memory use port *C*. Loads and stores directly to the address and scalar registers use port *C* at a maximum data rate of 1 word every 2 clock cycles.

Transfers between the *B* and *T* registers and address and scalar registers occur at the rate of 1 word per clock cycle, and data can be moved between memory and the *B* and *T* registers at the same rate of 1 word per clock cycle. Using the three separate memory ports, data can be moved between common memory and the buffer registers at a combined rate of 3 words per clock cycle, 1 word into *B* and *T* and 1 word from one of them.

The functional units are fully segmented (i.e., pipelined), which means that a new set of arguments may enter a unit every clock period (8.5 ns). The segmentation is performed by holding the operands entering the unit and the partial results at the end of every segment until a flag allowing them to proceed is set. The number of segments in a unit determines the start-up time for that unit. Table 10.5 shows the functional unit characteristics.

Table 10.5 Functional Unit Characteristics (Cray X-MP)

Type	Operation	Registers Used	Number of Bits	Unit Time (Clock Periods)
Address	Integer add	A	24	2
	Integer multiply	A	24	4
Scalar	Integer add	S	64	3
	Shift	S	64	2 or 3
	Logical	S	64	1
	Population	S	64	3 or 4
	Parity and leading zero	S and A	64	3 or 4
Vector	Integer add	V		
	Shift	V and A	64	3
	Logical	V	64	3 or 4
	Second logical	V	64	2
	Population and parity	V	64	3
Floating point	Add	S or V	64	6
	Multiply	S or V	64	
	Reciprocal	S or V	64	
Memory transfer	Scalar load	S	64	
	Vector load (64 elements)	V	64	

Example 10.9

Consider again the vector addition:

$$C[i] = A[i] + B[i] \quad 1 \le i \le N.$$

Assume that N is 64, A and B are loaded into two vector registers, and the result vector is stored in another vector register. The unit time for floating-point addition in six clock periods, including one clock period for transferring data from vector registers to the add unit and one clock to store the result into another vector register, would take ($64 \times 8 = 512$) clock periods to execute in scalar mode.

This vector operation performed in the pipeline mode is shown in Figure 10.19. Here, the first element of the result will be stored into the vector register after 8 clock periods. Afterward there will be one result every clock period. Therefore, the total execution time is ($8 + 63 = 71$) clock periods.

If $N < 64$, the previous execution times are reduced correspondingly. If $N > 64$, the computation is performed on units of 64 elements at a time. For example, if N is 300, the computation is performed on 4 sets of 64 elements each, followed by the final set with the remaining 44 elements.

The vector length register contains the number of elements (n) to be operated upon at each computation. If M is the length of vector registers in the machine, the following program can be used to execute the previous vector operation for an arbitrary value of N.

```
begin = 1
n = (N mod M)
for i = 0, (N/M)
  for j = begin, begin + n - 1
  C[j] = A[j] + B[j]
  endfor
begin = begin + n
n = M.
endfor
```

Here, first the (N mod M) elements are operated upon, followed by N/M sets of M elements.

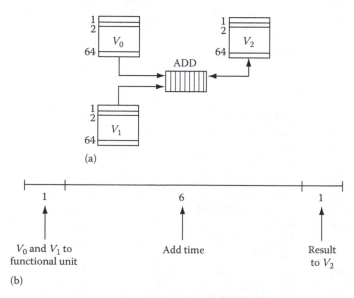

(a)

(b)

Figure 10.19 Vector pipelining on Cray X-MP. (a) The pipeline. (b) Timing.

In practice, the vector length will not be known at compile time. The compiler generates the code similar to the one presented previously, such that the vector operation is performed in sets of length less than or equal to *M*. This is known as *strip mining*. Strip-mining overhead must also be included in start-up time computations of the pipeline.

In order for multiple functional units to be active simultaneously, intermediate results must be stored in the CPU registers. When properly programmed, the Cray architecture can arrange CPU registers such that the results of one functional unit can be input to another independent functional unit. Thus, in addition to the pipelining within the functional units, it is possible to pipeline arithmetic operations between the functional units. This is called *chaining*.

Chaining of vector functional units and their overlapped, concurrent operation are important characteristics of this architecture, which is about a vast speedup in the execution times. Example 10.10 shows a loop where overlapping would occur.

Example 10.10

Consider the following loop:

```
For J = 1,64
C(J) = A(J) + B(J)
D(J) = E(J)*F(J)
Endfor
```

Here, the addition and multiplication can be done in parallel because the functional units are totally independent.

Example 10.11 illustrates chaining.

Example 10.11

```
For J = 1,64
C(J) = A(J) + B(J)
D(J) = C(J)*E(J)
Endfor
```

Here, the output of the add functional unit is an input operand to the multiplication functional unit. With chaining, we do not have to wait for the entire array *C* to be computed before beginning the multiplication. As soon as $C(1)$ is computed, it can be used by the multiply functional unit concurrently with the computation of $C(2)$. That is, the two functional units form the stages of a pipeline, as shown in Figure 10.20.

Assuming that all the operands are in vector registers, this computation is done without vectorization (i.e., no pipelining or chaining) and requires ($64 \times 8 = 512$ plus $64 \times 9 = 576$) 1088 clock periods. It can be completed in (chain start-up time of $17 + 63$ more computations) 80 clock periods if vectorization (pipelining and chaining) is employed.

Note that the effect of chaining is to increase the depth of the pipeline and hence the start-up overheads. If the operands are not already in vector registers, they need to be loaded first and the result stored in the memory. The two load paths and the path that stores data to memory can be considered as functional units in chaining. The start-up time for a load vector instruction is 17 cycles, and thereafter 1 value per cycle may be fetched; then any operation using this data may access 1 value per cycle after 18 cycles. Figure 10.21 shows an example of this. Here, port A is used to read in V_0 and port *B* to read in V_1. This occurs in parallel. As soon as each vector register has its first operand, the floating-point add may begin processing, and as soon as the first operand is placed in V_2, port *C* may be used to store it back to memory.

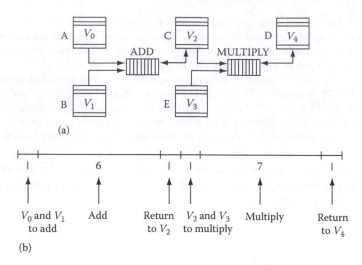

(a)

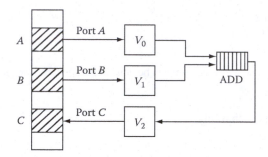

(b)

Figure 10.20   Chaining in Cray X-MP. (a) Chain. (b) Timing.

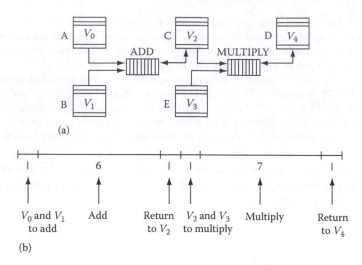

Figure 10.21   Memory and functional unit chaining on Cray X-MP.

In a chain, a functional unit can only appear once. Two fetches and one store are possible in each chain. This is because Cray systems supply only one of each of the previous types of functional units. This demands that if two floating-point adds are to be executed, they must occur sequentially. Because there are two ports for fetching vectors and one port for storing vectors, the user may view the system as having two load functional units and a store functional unit on Cray X-MP. On Cray-1, there is only one memory functional unit.

An operand can only serve as input to one arithmetic functional unit in a chain. An operand can, however, be input to both inputs of a functional unit requiring two operands. This is because a vector register is tied to a functional unit during a vector instruction. When a vector instruction is issued, the functional unit and registers used in the instruction are reserved for the duration of the vector instruction.

Cray has coined the term "chime" (chained vector time) to describe the timing of vector operations. A chime is not a specific amount of time, but rather a timing concept representing the number of clock periods required to complete one vector operation. This equates to length of a vector register plus a few clock periods for chaining. For Cray systems, a chime is equal to 64 clock periods plus a few more. A chime is thus a measure that allows the user to estimate the speedup available from pipelining, chaining, and overlapping instructions.

The number of chimes needed to complete a sequence of vector instructions is dependent on several factors. Since there are three memory functional units, two fetches and one store operation may appear in the same chime. A functional unit may be used only once within a chime. An operand may appear as input to only one functional unit in a chime. A store operation may chain onto any previous operation.

Example 10.12

Figure 10.22 shows the number of chimes required to perform the following code on Cray X-MP, Cray-1, and the Cray-2 (Levesque and Williamson 1989) systems:

```
For I = 1 to 64
  A(I) = 3.0* A(I) + (2.0 + B(I)* C(I)
Endfor
```

The Cray X-MP requires only two chimes, while the Cray-1 requires four, and the Cray-2, which does not allow chaining, requires six chimes to execute the code. Clock cycles of the Cray-1, the Cray X-MP, and the Cray-2 are 12.5, 8.5, and 4.1 ns, respectively. If a chime is taken to be 64 clock cycles, then the time for each machine to complete the code is

Cray-1	4 chimes × 64 × 12.5 ns = 3200 ns
Cray X-MP	2 chimes × 64 × 8.5 ns = 1088 ns
Cray-2	6 chimes × 64 × 4.1 ns = 1574 ns

Thus, for some instruction sequences, the Cray X-MP with the help of chaining can actually outperform Cray-2, which has a faster clock. Since Cray-2 does allow overlapping, the actual gain of Cray X-MP may not be as large for very large array dimensions.

During vector operations, up to 64 target addresses could be generated by 1 instruction. If a cache were to be used as intermediate memory, the overhead to search for 64 addresses would be prohibitive. Use of individually addressed registers eliminates this overhead. One disadvantage of not using a cache is that the programmer (or the compiler) must generate all the references to the individual registers. This adds to the complexity of code (or compiler) development.

```
                                     CHIME 1 : A → V0

CHIME 1 : A → V0                     CHIME 2 :   B → V1
            B → V1                             2.0 + V1 → V3
         2.0 + V1 → V3                         3.0 * V0 → V4
         3.0 * V0 → V4
                                     CHIME 3 :       C → V5
CHIME 2 :       C → V5                         V3 * V5 → V6
            V3 * V5 → V6                        V4 + V6 → V7
            V4 + V6 → V7
                 V7 → A              CHIME 4 :              V7 → A

   (a)                                  (b)

                     CHIME 1 :   A → V0

                     CHIME 2 :   B → V1
                               3.0 * V0 → V4

                     CHIME 3 :       C → V5
                               2.0 + V1 → V3

                     CHIME 4 :            V3 * V5 → V6

                     CHIME 5 :            V4 + V6 → V7

                     CHIME 6 :                 V7 → A
                        (c)
```

Figure 10.22 Chime characteristics. (a) Cray X-MP. (b) Cray-1. (c) Cray-2.

10.5.5.4 Instruction Fetch

Control registers are part of the special-purpose register set and are used to control the flow of instructions. Four instruction butters, each containing 32 words (128 parcels, 16 bits each), are used to hold instructions fetched from memory. Each instruction buffer has its own instruction buffer address register (IBAR). The IBAR serves to indicate what instructions are currently in the buffer. The contents of IBAR are the high-order 17 bit of the words in the buffer. The instruction buffer is always loaded on a 32-word boundary. The P register is the 24-bit address of the next parcel to be executed. The current instruction parcel (CIP) contains the instruction waiting to issue, and the next instruction parcel (NIP) contains the NIPs to issue after the parcel in the CIP. Also, there is a last instruction parcel (LIP) that is used to provide the second parcel to a 32-bit instruction without using an extra clock period.

The P register contains the address of the next instruction to be decoded. Each buffer is checked to see if the instruction is located in the buffers. If the address is found, the instruction sequence continues. However, if the address is not found, the instruction must be fetched from memory after the parcels in the CIP and NIP have been issued. The least recently filled buffer is selected to be overwritten, so that the current instruction is among the first eight words to be read. The rest of the buffer is then filled in a circular fashion until the buffer is full. It will take three clock pulses to complete the filling of the buffer. Any branch to an out-of-buffer address causes a delay in processing of 16 clock pulses.

Some buffers are shared between all of the processors in the system. One of these is the real-time clock. Other registers of this type include a cluster consisting of 48 registers. Each cluster contains 32 (1 bit) semaphore or synchronization registers and eight 64-bit shared-T (ST) registers and eight 24-bit or shared-B (SB) registers. A system with two processors will contain three clusters, while a four-processor system will contain five clusters.

10.5.5.5 I/O System

The input and output of the X-MP are handled by the IOS. The IOS is made of 2–4 interconnected I/O processors. The IOS receives data from four 100 MB/s channels connected directly to the main memory of the X-MP. Also, four 6 MB/s channels are provided to furnish control between the CPU and the IOS. Each processor has its own local memory and shares a common buffer. The IOS supports a variety of front-end processors and peripherals such as disk drives and drives.

To support the IOS, each CPU has two types of I/O control registers: current address and channel limit registers. The current address registers point to the current word being transferred. The channel limit registers contain the address of the last word to be transferred.

10.5.5.6 Other Systems in the Series

Since the first Cray-1 supercomputer was powered up back in 1975, Cray systems went through generations of developments and enhancements. Cray systems received their first official customer order by the National Center for Atmospheric Research in 1976 and made their first international shipment to the European Centre for Medium-Range Weather Forecasts. In the early 1980s Cray systems introduced their Cray-2 and Cray X-MP that was the first multiprocessor system at that time. Cray Y-MP introduced in the late 1980s was the first to sustain one gigaflop computation rate. Cray Y-MP 2E was the first air-cooled supercomputer.

In the 1990s Cray systems introduced Cray XMS and Cray EL systems although the Cray Y-MP continued to be produced and enhanced to exceed the 1 gigaflop. Cray Y-MP was further enhanced into Cray Y-MP 8, Cray Y-MP EL, and Cray Y-MP 4E systems. The Cray T family of wireless systems was introduced in the mid-1990s with Cray T90. It was followed by the first Cray T3E that sustained one teraflop performance, followed by Cray T94 system.

During 2004–2009 Cray systems introduced Cray XT series XT, XT4, XT5, and XT9. Cray CX1000 a rack-mounted supercomputer was introduced in 2010 followed by XE6 the next-generation massive parallel processing system. Cray Y-MP extends the X-MP's 24-bit addressing scheme to 32 bits, thus allowing an address space of 32 million 64-bit words. It runs on a 6 ns clock and uses eight processors, thus doubling the processing speed.

Cray XT4 utilizes from 548 to 30,508 processing elements. Each processing element is built using a 2.6 MHz AMD dual-core processor. The system memory ranges from 4.3 to 239 TB. The system offers from 5.6 to 318 teraflops of peak.

In 2011, Cray systems introduced Cray XK family, the successor to the XE family. Cray XK6 was upgradeable to more than 50 petaflops, followed by the current Cray XK7 in 2012. Chapter 12 provides further details on Cray XK7.

10.6 SUMMARY

The spectrum of enhancements possible for ALUs spanning the range of employing faster hardware and algorithms to multiple functional units were described in this chapter. Several algorithms have been proposed over the years to increase the speed of the ALU functions. Only representative algorithms were described in this chapter. The reader is referred to the books listed in the Bibliography section for further details. The advances in hardware technology have resulted in several fast, off-the-shelf ALU building blocks. Some examples of such ICs were given. Examples of pipelined ALU architectures were provided. The trend in ALU design is to employ multiple processing units and activate them in parallel so that very high speeds can be achieved through superscalar and VLIW architectures described in this chapter.

PROBLEMS

10.1 List the architectural features of the ALUs of a mini-, micro-, a large-scale, and a supercomputer system you have access to with reference to the characteristics described in this chapter.

10.2 Look up the TTL data book to find a CLA IC. Design a 16-bit CLA using an appropriate number of these ICs.

10.3 Perform the multiplication of the following operands using the shift and add algorithm. Use the minimum number of bits in the representation of each number and the results.

   a. 24×8

   b. 2×24

   c. 17×64

10.4 Perform the following division operations using the restoring division algorithm. Use the minimum number of bits required.

   a. $24/8$

   b. $43/15$

   c. $129/23$

10.5 Solve Problem 10.4 using nonrestoring division algorithm.

10.6 For the stack-based ALU shown in Figure 10.5, derive the micro-operation sequences needed for the following operations:

   a. SHR: Shift the contents of top level right, once.

   b. COMP: Replace the top level with its 2s complement. (Assume that the functional unit can perform the 2s complement of operand X.)

   c. SUB: $SL \leftarrow SL - TL$, POP.

10.7   A hardware stack is used to evaluate arithmetic expressions. The expressions can contain both REAL and INTEGER values. The data representation contains a TAG bit with each operand. The TAG bit is 1 for REAL and 0 for INTEGER data. List the control signals needed for the stack-based ALU of Figure 10.5 to perform addition and subtraction of the operands in the top two levels of the stack. Include any additional hardware needed to accommodate this new data representation.

10.8   Give an algorithm to generate the sum of two numbers in excess-3 representation (i.e., each digit corresponds to 4 bits, in excess-3 format). Design a circuit for the adder similar to that in Figure 10.7.

10.9   Develop the detailed block diagrams for each of the stages in the floating-point add pipeline of Figure 10.8. Assume the IEEE standard representation for the floating-point numbers.

10.10   Develop the stages needed for a floating-point multiplier, assuming that the numbers are represented in the IEEE standard form.

10.11   A popular method of detecting the OVERFLOW and UNDERFLOW conditions in a shift register-based stack is by using an additional shift register with number of bits equal to the number of levels in the stack and shifting a 1 through its bits. Design the complete circuit for an N-level stack.

10.12   Design the UNDERFLOW and OVERFLOW detection circuits for a RAM-based stack.

10.13   It is required to compute $C_i = A_i + B_i$ ($i = 1 - 50$), where A, B, and C are arrays of floating-point numbers, using a six-stage add pipeline. Assume that each stage requires 15 ns. What is the total computation time?

10.14   Derive the formula for the total execution time of Problem 10.13 using a pipeline of M stages, with each stage requiring T ns. Assume that the arrays have N elements each.

10.15   Describe the conditions under which an n-stage pipeline is n times faster than a serial machine.

10.16   Differentiate and contrast between MSP430 and MSP430X multipliers in terms of instruction set and bits of address used.

10.17   Discuss the limitations of the superscalar techniques associated with ALU with multiple functional units.

BIBLIOGRAPHY

August, M.C. et al., Cray X-MP: The birth of a supercomputer, *IEEE Computer*, 22, 45–52, January 1989.

Cheung, T. and Smith, J.E., An analysis of the Cray X-MP memory system, *Proceedings of IEEE International Conference on Parallel Processing*, Bellaire, MI, 1984, pp. 499–505.

Cray Research Inc., Mendota Heights, MN, http://www.craysupercomputers.com/downloads/CrayXT4/CrayXT4_Datasheet.pdf

Cray Research Inc., *Cray X-MP and Cray Y-MP Computer Systems* (Training Workbook), Mendota Heights, MN, 1988.

Culler, D.E., Singh, J., and Gupta, A., *Parallel Computer Architecture: A Hardware/Software Approach*, San Francisco, CA: Morgan Kaufmann, 1998.

Hill, M.D., Hockney, R.W., and Jesshope, C.R., *Parallel Computers 2*, Philadelphia, PA: Adam Hilger, 1988.

Hwang, K., *Computer Arithmetic: Principles, Architecture and Design*, New York: John Wiley, 1979.

IBM Corporation, *IBM System 370 Principles of Operation, GA22-7000*, IBM Corporation, Poughkeepsie, NY, 1973.

Jouppi, N.P. and Sohi, G.S., *Readings in Computer Architecture*, San Francisco, CA: Morgan Kaufmann, 1999.

Kogge, P., *The Architecture of Pipelined Computers*, New York: McGraw-Hill, 1981.

Lazou, C., *Supercomputers and Their Use*, New York: Oxford University Press, 1988.

Levesque, J.M. and Williamson, J.L., *A Guidebook to Fortran on Supercomputers*, San Diego, CA: Academic Press Inc., 1989.

MC6888/MC68882, *Floating-Point Coprocessor User's Manual*, Englewood Cliffs, NJ: Prentice Hall, 1987.

Ortega, J.M., *Introduction to Parallel and Vector Solution of Linear Systems*, New York: Plenum Press, 1988.

Patterson, D.A. and Hennessey, J.L., *Computer Architecture: A Quantitative Approach*, San Mateo, CA: Morgan Kaufmann, 1996.

Shiva, S.G., *Advanced Computer Architectures*, Boca Raton, FL: Taylor & Francis, 2006.

Stone, H.S., *High Performance Computer Architecture*, Reading, MA: Addison Wesley, 1993.

The Bipolar Microcomputer Components Data Book, Dallas, TX: Texas Instruments Inc., 1978.

The MSP430 Hardware Multiplier: Function and Applications, Dallas, TX: Texas Instruments Inc., April 1999.

The TTL Data Book, Dallas, TX: Texas Instruments Inc., 1986.

CHAPTER 11

Control Unit Enhancement

Two popular implementations of the control unit (hardwired control unit [HCU] and microprogrammed control unit [MCU]) were described in Chapter 6. The subsequent chapters provided the details of architectural features found in modern-day machines. Inclusion of any new feature into the architecture requires a corresponding enhancement to the control unit. The following parameters are usually considered in the design of a control unit:

Speed: The control unit should generate control signals fast enough to utilize the processor bus structure most efficiently and minimize the instruction and program execution times.

Cost and complexity: The control unit is the most complex subsystem of a processor. The complexity should be reduced as much as possible to make the maintenance easier and the cost low. In general, random logic implementations of the control unit (i.e., HCU) tend to be the most complex, while read-only memory (ROM)-based designs (i.e., MCU) tend to be the least complex.

Flexibility: HCUs are inflexible in terms of adding new architectural features to the processor since they require a redesign of the hardware. MCUs offer a very high flexibility since microprograms can be easily updated without a substantial redesign of the hardware involved.

With the advances in hardware technology, faster and more versatile processors are introduced to the market very rapidly. This requires that the design cycle time for newer processors must be as small as possible. Since the design costs must be recovered over a short life span of the new processor, they must be minimized. MCUs offer such flexibility and low-cost redesign capabilities, although they are inherently slow compared with HCUs. This speed differential between the two designs is getting smaller, since in the current technology, the MCU is fabricated on the same integrated circuit (IC) (i.e., with the same technology) as the rest of the processor. We will concentrate on the popular speed-enhancement techniques used in contemporary machines in this chapter.

11.1 SPEED ENHANCEMENT

In a simple computer (ASC), the control unit fetches an instruction, decodes it, and executes it, before fetching the next instruction as dictated by the program logic. That is, the control unit brings about the instruction cycles one after the other. With this *serial* instruction execution mode, the only way to increase the speed of execution of the overall program is to minimize the instruction cycle time of the individual instructions. This concept is called the *instruction cycle speedup*. The program execution time can be reduced further, if the instruction cycles can be overlapped, that is, if the next instruction can be fetched and/or decoded during the current instruction cycle. This overlapped operation mode is termed *instruction execution overlap* or more commonly *pipelining*.

Another obvious technique would be to bring about the instruction cycle of more than one instruction simultaneously. This is the *parallel* mode of instruction execution. We will now describe these speed-enhancement techniques.

11.1.1 Instruction Cycle Speedup

Recall that the *processor cycle time* (i.e., major cycle time) depends on the register-transfer time of the processor bus structure. If the processor structure consists of multiple buses, it is possible to perform several register transfers simultaneously. This requires that the control unit produce the appropriate control signals simultaneously.

In a *synchronous* HCU, the processor cycle time is fixed by the slowest register transfer. Thus, even the fastest register-transfer operation consumes a complete processor cycle. In an *asynchronous* HCU, the completion of one register transfer triggers the next; therefore, if properly designed, the asynchronous HCU would be faster than the synchronous HCU. Since the design and maintenance of an asynchronous HCU is difficult, the majority of the practical processors have synchronous HCUs. An MCU is slower than an HCU since the microinstruction execution time is the sum of processor cycle time and the control ROM (CROM) access time.

The HCU of ASC has the simplest configuration possible. Each instruction cycle is divided into one or more phases (state or *major cycles*), each phase consisting of four processor cycles (i.e., *minor cycles*). A majority of actual control units are synchronous control units that are, in essence, enhanced versions of the ASC control unit. The only optimization performed in the ASC control unit was to reduce the number of major cycles needed to execute certain instructions (SHR, SHL) by not entering an execute cycle, since all the required microoperations (μops) to implement those instructions could be completed in one major cycle. Further optimization is possible. It is not necessary to use up a complete major cycle if the μops corresponding to an instruction execution (or fetch or defer) can be completed in a part of the major cycle. For example, the μops for the execution of each branch instruction (BRU, BIP, BIN) could all be completed in one minor cycle rather than in a complete major cycle as they are currently implemented in ASC control unit. Thus, three minor cycles could be saved in the execution of branch instructions by returning to fetch cycle after the first minor cycle in the execute cycle. When such enhancements are implemented, the state-change circuitry of the control unit becomes more complex, but the execution speed increases.

Note that in the case of an MCU, the concept of the major cycle is not present, and the basic unit we work with is the minor cycle (i.e., processor cycle time + CROM access time). The lengths of microprograms corresponding to each instruction are different. Each microinstruction is executed in one minor cycle, and the microprograms do not include any idle cycles. In developing the microprogram for ASC, the μop sequences from the HCU design were reorganized to make them as short as possible.

Section 8.6 provided the instruction cycle details of Intel 8080, illustrating the instruction cycle speedup concept. Although this is an obsolete processor, it was selected for its simplicity. The more recent Intel processors described in Section 11.4 adopt these techniques very extensively.

11.1.2 Instruction Execution Overlap

Even in early processors such as Intel 8080, for instructions such as ADD r, once the memory operand is fetched into the central processing unit (CPU), addition is performed while the CPU is fetching the next instruction in sequence from the memory. This overlap of instruction fetch and execute phases increases the program execution speed.

In general, the control unit can be envisioned as a device with three subfunctions: fetch, decode (or address computation), and execute. If the control unit is designed in a modular form with one module for each of these functions, it is possible to overlap the instruction-processing functions.

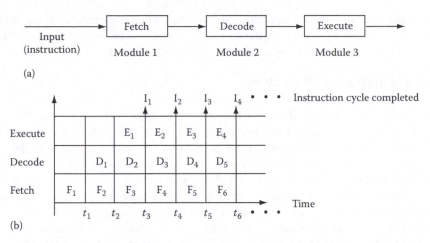

Figure 11.1 Pipelined instruction processing. (a) Pipelined control unit. (b) Instruction execution overlap.

The overlapped processing is brought about by a *pipeline*. As described in previous chapter, a *pipeline* is a structure that, like an automobile assembly line, consists of several stations, each of which is capable of performing a certain subtask. The work flows from one station to the next. As the work leaves a station, the subsequent unit of the work is picked up by that station. When the work leaves the last station in the pipeline, the task is complete. If the pipeline has N stations and the work stays at each station for T seconds, the complete processing time for a task is $(N \times T)$ seconds. However, since all the N stations are working in an overlapped manner (on various tasks), the pipeline outputs one completed task every T seconds (after the initial period in which the pipeline is being filled).

Figure 11.1 introduces the concept of an instruction-processing *pipeline*. The control unit has three modules. The processing sequence is shown in Figure 11.1b. Any time after t_2, the first module will be fetching instruction $(I + 1)$, and the second module will be decoding instruction I, while the last module will be executing instruction $(I - 1)$. From t_3 onward, the pipeline flows full and the throughput is one instruction per time slot.

For simplicity, we have assumed that each module in this pipeline consumes the same amount of processing time. If such equal time partitioning of the processing task cannot be made, intermediate registers to hold the result and flags that indicate the completion of one task and beginning of the next task are needed.

We have assumed that the instructions are always executed in the sequence they appear in the program. This assumption is valid as long as the program does not contain a branch instruction. When a branch instruction is encountered, the next instruction is to be fetched from the target address of the branch instruction. If it is a conditional branch, the target address would not be known until the instruction reaches the execute stage. If a branch is indeed taken, then the instructions following the branch instruction that are already in the pipeline need to be discarded, and the pipeline needs to be filled from the target address. Another approach would be to stop fetching subsequent instructions into the pipeline, once the branch instruction is fetched, until the target address is known. The former approach is preferred for handling conditional branches since there is a good chance that the branch might not occur; in that case, the pipeline would flow full. The pipeline flow suffers only when a branch occurs. For unconditional branches, the latter approach can be used. We will return to a detailed treatment of pipeline concepts later in this chapter.

To implement the control unit as a pipeline, each stage should be designed to operate independently, performing its own function while sharing resources such as the processor bus structure and the main memory system. Such designs become very complex.

The pipeline concept is now used very extensively in all modern processors. It is typical for the processors today to have four or five stages in their pipelines. As the hardware technology progresses, processors with deeper pipelines (i.e., pipelines with larger number of stages) have been introduced. These processors belong to the so-called *superpipelined* processor class. Section 11.4 and subsequent chapters provide some examples of this class of machines.

11.1.3 Parallel Instruction Execution

As seen by the description in Chapters 8 through 10, the Intel Pentium series of processors contain two execution units: the integer unit and the floating-point (FP) unit. The control unit of these processors fetches instructions from the same instruction stream (i.e., program), decodes them, and delivers them to the appropriate execution unit. Thus, the execution units would be working in parallel, each executing its own instruction. The control unit must now be capable of creating these parallel steams of execution and synchronizing those streams appropriately, based on the precedence constraints imposed by the program. That is, the result of the computation must be the same, whether the program is executed by serial or parallel execution of instructions. This class of architectures, where more than one instruction steam is processed simultaneously, is known as *superscalar* architectures introduced in the previous chapter of this book. Section 11.4 and subsequent chapters provide some examples.

11.1.4 Instruction Buffer and Cache

The instruction buffers and instruction cache (ICache) architectures introduced earlier in the book also bring about instruction-processing overlap, although at the complete instruction level. That is, the operation of fetching instructions into the buffer or cache (prefetching) is overlapped with the operation of retrieving and executing instructions that are in the buffer.

11.2 HARDWIRED VERSUS MICROPROGRAMMED CONTROL UNITS

All the speedup techniques described in the previous section have been adopted by HCUs of practical machines. As mentioned earlier, the main advantage of the HCU is its speed, while the disadvantage is its inflexibility. Although asynchronous HCUs offer a higher speed capability than synchronous HCUs, the majority of practical machines have synchronous HCUs because they have the simpler design of the two.

In the current very large scale integrated (VLSI) era, complete processing systems are fabricated on a single IC. Because the control unit is the most complex unit, it occupies a large percentage of the chip "real estate." Its complexity increases as the number of instruction in the instruction set of the machine increases. The *reduced instructions set computers* (RISCs) introduced earlier in the book address this complexity problem.

In an MCU, the execution time for an instruction is proportional to the number of μops required and hence the length of microprogram for the instruction. Since an MCU starts fetching the next instruction once the last μop of the current instruction microprogram is executed, MCU can be treated as an asynchronous control unit. An MCU is slower than the HCU because of the addition of CROM access time to the register-transfer time. But it is more flexible than HCU and requires minimum changes in the hardware if the instruction set is to be modified or enhanced.

All the speedup techniques described in Section 11.1 are used in practical MCUs. Another level of pipelining, shown in Figure 11.2, is possible in an MCU. Here, the fetching of the next microinstruction is overlapped with the execution of the current microinstruction.

The CROM word size is one of the design parameters of an MCU. Although the price of ROMs is decreasing, the cost of data path circuits required within the control unit increases as the CROM

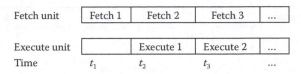

Figure 11.2 Pipelining in an MCU.

word size increases. Therefore, the word size should be reduced to reduce the cost of the MCU. We will now examine the microinstruction format used by practical machines with respect to their cost-effectiveness.

The most common format for a microinstruction is shown in Figure 11.3a. The "instruction" portion of the microinstruction is used in generating the control signals, and the "address" portion indicates the address of the next microinstruction. The execution of such a microinstruction corresponds to the generation of control signals and transferring the address portion to the µMAR to retrieve the next microinstruction. The advantage of this format is that very little external circuitry is needed to generate the next microinstruction address, while the disadvantage is that the conditional branches in the microprogram cannot be easily coded. The format shown in Figure 11.3b allows for conditional branching. It is now assumed that when the condition is not satisfied, the µMAR is simply incremented to point to the next microinstruction in sequence. However, this requires additional µMAR circuitry. The microinstruction format shown in Figure 11.3c explicitly codes the jump addresses corresponding to both the outcomes of the test condition, thus reducing the external µMAR circuitry. The CROM word size will be large in all these formats since they contain one or more address fields. It is possible to reduce the CROM word size if the address representation can be made implicit. The format shown in Figure 11.3d is similar to the one used by the MCU of ASC. This format distinguishes between the two types of microinstructions through the µop code: Type 1 microinstruction produces control signals, while type 0 manages the microprogram flow. Incrementing of µMAR after the execution of a type 1 microinstruction is implicit in this format. As seen in Chapter 6, this format of microinstruction requires fairly complex circuitry to manage the µMAR.

Figure 11.4 shows the two popular forms of encoding the "instructions" portion of a microinstruction. In the *horizontal* (or *unpacked*) microinstruction, each bit in the instruction represents a control signal. Hence, all the control signals can be generated simultaneously, and no external decoding is needed to generate the control signal, thus making the MCU fast. The disadvantage is that this requires larger CROM words. Also, instruction encoding is cumbersome because a thorough familiarity with the processor hardware structure is needed to prevent the generation of control signals that cause conflicting operations in the processor hardware.

In the *vertical* (or *packed*) microinstruction, the instruction is divided into several fields, each field corresponding to either a resource or a function in the processing hardware. (In the design

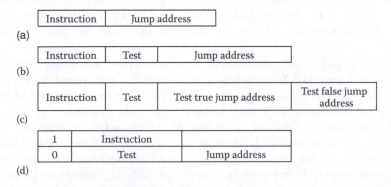

Figure 11.3 Examples of microinstruction formats. (a) Common format. (b) With test condition and single branch address. (c) With test condition and two branch addresses. (d) Two instruction types.

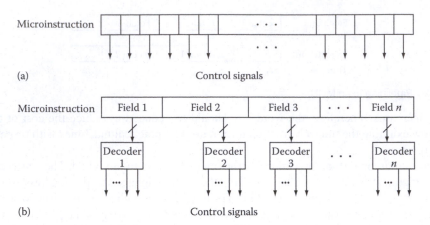

Figure 11.4 Popular microinstruction-encoding formats. (a) Horizontal. (b) Vertical. *Note:* Only the instruction field of the microinstruction is shown; the address and test fields are not shown.

of ASC MCU, each field corresponded to a resource such as arithmetic and logic unit (ALU) and BUS1). Vertical microinstruction reduces the CROM word size, but the decoders needed to generate control signals from each field of the instruction contribute to the delays in control signals. Encoding for vertical microinstruction is easier than that for horizontal microinstruction because of the former's function/resource partitioning.

In the foregoing discussion, we have assumed that all the control signals implied by the microinstruction are generated simultaneously and the next clock pulse fetches a new microinstruction. This type of microinstruction encoding is called *monophase encoding*. It is also possible to associate each field or each bit in the microinstruction with a time value. That is, the execution of each microinstruction now requires more than one clock pulse. This type of microinstruction encoding is called *polyphase encoding*. Figure 11.5a shows the monophase encoding, where all the control signals are generated simultaneously at the clock pulse. Figure 11.5b shows an *n*-phase encoding where μops $M1$ through Mn are associated with times value $t1$ through tn, respectively.

11.3 PIPELINE PERFORMANCE ISSUES

Figure 11.6 shows an instruction-processing pipeline consisting of six stages. The first stage fetches instructions from the memory, one instruction at a time. At the end of each fetch, this stage also updates the program counter to point to the next instruction in sequence. The decode stage decodes the instruction, and the next stage computes the effective address of the operand, followed by a stage that fetches the operand from the memory. The operation called for by the instruction is then performed by the execute stage and the results are stored in the memory by the next stage.

The operation of this pipeline is depicted by the modified time–space diagram called a *reservation table* shown in Figure 11.7. In a reservation table, each row corresponds to a stage in the pipeline and each column corresponds to a pipeline cycle. An "X" at the intersection of the *i*th row and the *j*th column indicates that stage *i* would be busy performing a subtask at cycle *j*, where cycle 1 corresponds to the initiation of the task in the pipeline. That is, stage *i* is "reserved" (and hence not available for any other task) at cycle *j*. The number of columns in the reservation table for a given task is determined by the sequence in which the subtasks corresponding to that task flow in the pipeline. The reservation table in Figure 11.7 shows that each stage completes its task in one cycle time and hence an instruction cycle requires six cycles to be completed, although one instruction is completed every cycle (once the pipeline is full).

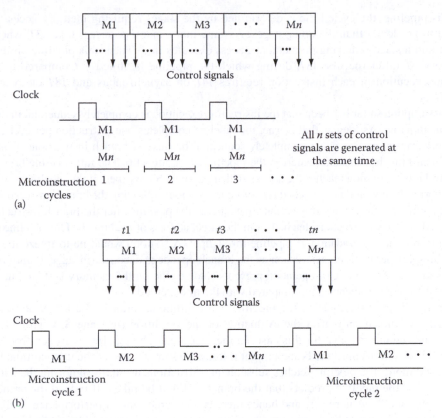

Figure 11.5 Phase encoding of microinstructions. (a) Monophase. (b) Polyphase. *Notes:* (1) M1–M*n* are the *n* sets of control signals implied by the microinstruction. (2) *t1–tn* are microcontrol unit clock pulses (phases).

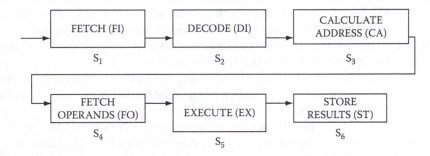

Figure 11.6 An instruction pipeline.

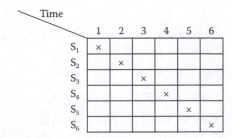

Figure 11.7 Reservation table.

In this pipeline, the cycle time is determined by the stages requiring memory access because they tend to be slower than the other stages. Assume that a memory access takes $3T$, where T is a time unit and a stage requiring no memory access executes in T. Then, this pipeline produces one result every $3T$ (after the first $18T$, during which the pipeline is "filled"). Compared to this, the sequential execution of each instruction requires $14T$ on asynchronous and $18T$ on synchronous control units.

The assumption so far has been that the instruction execution is completely sequential in the operation of this pipeline. As long as that is true, the pipeline completes one instruction per cycle. In practice, the program execution is not completely sequential because of branch instructions. Consider an unconditional branch instruction entering the pipeline of Figure 11.6. The target of the branch is not known until the instruction reaches the address calculate stage S3. By then, if the pipeline is allowed to function normally, it would have fetched two more instructions following the branch instruction. When the target address is known, the instructions that entered the pipeline after the branch instruction must be discarded and new instructions fetched from the target address of the branch. This pipeline draining operation results in a degradation of pipeline throughput. A solution would be to freeze the pipeline from fetching further instructions as soon as the branch opcode is decoded (in stage 2) until the target address is known. This mode of operation prevents some traffic on the memory system, but does not increase the pipeline efficiency any, compared with the first method.

When the instruction entering the pipeline is a conditional branch, the target address of the branch will be known only after the evaluation of the condition (in stage S5). Three modes of pipeline handling are possible for this case. In the first mode, the pipeline is frozen from fetching subsequent instructions until the branch target is known, as in the case of the unconditional branch. In the second mode, the pipeline fetches subsequent instructions normally, ignoring the conditional branch. That is, the pipeline predicts that the branch will not be taken. If indeed the branch is not taken, the pipeline flows normally and hence there is no degradation of performance. If the branch is taken, the pipeline must be drained and restarted at the target address. The second mode is preferred, since the pipeline functions normally about 50% of the time on an average. The third mode would be to start fetching the target instruction sequence into a buffer (as soon as the target address is computed in S3), while the nonbranch sequence is being fed into the pipeline. If the branch is not taken, the pipeline continues with its normal operation and the contents of the buffer are ignored. If the branch is taken, the instructions already in the pipeline are discarded (i.e., the pipeline is flushed) and target instruction is fetched from the buffer. The advantage here is that fetching instructions from the buffer is faster than that from the memory.

In the last two modes of operation, all the activity of the pipeline with respect to instructions entering the pipeline following the conditional branch must be marked *temporary* and made permanent only if the branch is not taken.

These problems introduced into the pipeline by branch instructions are called *control hazards*. We will describe mechanisms that reduce the effect of control hazards on pipeline performance in this section.

Example 11.1

Consider now the instruction sequence given in the following, to be executed on a processor that utilizes the pipeline of Figure 11.6:

LOAD R1, MEM1	R1 ← (MEM1)
LOAD R2, MEM2	R2 ← (MEM2)
MPY R3, R1	R3 ← (R3) * (R1)
ADD R4, R2	R4 ← (R4) + (R2).

Cycles

	1	2	3	4	5	6	7	8	9
Load R1, Mem1	FI	DI	CA	FO Read Mem1	EX Write R1	ST			
Load R2, Mem2		FI	DI	CA	FO Read Mem 2	EX Write R2	ST		
Mpy R3, R1			FI	DI	CA	FO Read R1, R3	EX Write R3	ST	
Add R4, R2				FI	DI	CA	FO Read R2, R4	EX Write R4	ST

Figure 11.8   Pipeline operation and resource requirements.

where R1, R2, R3, and R4 are registers; MEM1 and MEM2 are memory addresses; "()" denotes "contents of," and "←" indicates a data transfer.

Figure 11.8 depicts the operation of the pipeline. During cycles 1 and 2, only one memory access is needed. In cycle 3, two simultaneous read accesses to memory are needed—one due to CA and the other due to FI. In cycle 4, three memory read accesses (FO, CA, and FI) are needed. In cycle 5, two memory read (FO, CA) and one register write (EX) are required. Cycle 6 requires one access each for memory read and write and register read and write. Cycles 7–9 do not require memory access, but they do require one memory write. Cycle 7 requires one register read and one register write and cycle 8 needs one register write. The memory and register accesses have been summarized in Figure 11.9.

As seen from Figure 11.9, the memory system must accommodate three accesses per cycle for this pipeline to operate properly. If we assume that the machine has separate data and ICaches, then two simultaneous accesses can be handled. This solves the problem in cycles 5 and 6 (assuming that the machine accesses ICache during CA). But, during cycle 4, two accesses (FI, CA) to data cache would be needed. One way to solve this problem is to *stall* the ADD instruction (i.e., initiate ADD instruction later than cycle 4) until cycle 6 as shown in Figure 11.10. The stalling process results in a degradation of pipeline performance. The memory and register accesses for Figure 11.10 have been summarized in Figure 11.11.

Note that the pipeline controller must evaluate the resource requirements of each instruction before the instruction enters the pipeline, so that the previously mentioned *structural hazards* are eliminated.

In addition to the collision problems in pipelines due to improper initiation rate, data interlocks occur because of shared data between the stages of the pipeline, and conditional branch instructions degrade the performance of an instruction pipeline, as mentioned earlier. These problems and some common solutions are described in this section.

	Memory Read	Memory Write	Register Read	Register Write
Cycle 1	1			
Cycle 2	1			
Cycle 3	2			
Cycle 4	3			
Cycle 5	2			1
Cycle 6	1		1	1
Cycle 7			1	1
Cycle 8				1
Cycle 9				

Figure 11.9   Memory and register access (read and write).

Cycles

	1	2	3	4	5	6	7	8	9
Load R1, Mem1	FI	DI	CA	FO Read Mem1	EX Write R1	ST			
Load R2, Mem2		FI	DI	CA	FO Read Mem2	EX Write R2	ST		
Mpy R3, R1			FI	DI	CA	FO Read R1, R3	EX Write R3	ST	
Add R4, R2				Stall	Stall	FI	DI	CA	FO Read R2, R4

Figure 11.10 Pipeline operation with stalls.

	Memory Read	Memory Write	Register Read	Register Write
Cycle 1	1			
Cycle 2	1			
Cycle 3	2			
Cycle 4	2			
Cycle 5	2			1
Cycle 6	1		1	1
Cycle 7				1
Cycle 8	1			
Cycle 9			1	

Figure 11.11 Memory and register access (read and write for pipeline operation with stalls).

11.3.1 Data Interlocks

An instruction-processing pipeline is most efficient when instructions flow through its stages in a smooth manner. In practice, this is not always possible because of the interinstruction dependencies. These interinstruction dependencies are due to the sharing of resources such as a memory location or a register by the instructions in the pipeline. In such sharing environments, the computation cannot proceed if one of the stages is operating on the resource, while the other has to wait for the completion of that operation.

Example 11.2

Consider the following instruction sequence:

$$LOAD\ R1,\ MEM1 \quad R1 \leftarrow (MEM1)$$

$$LOAD\ R2,\ MEM2 \quad R2 \leftarrow (MEM2)$$

$$MPY\ R1,\ R2 \quad R1 \leftarrow (R2) * (R1)$$

$$ADD\ R1,\ R2 \quad R1 \leftarrow (R1) + (R2).$$

Figure 11.12 shows the operation of the pipeline of Figure 11.6 for this sequence. Note that as a result of the second LOAD instruction, R2 is loaded with the data from MEM2 during cycle 6.

	Cycles								
	1	2	3	4	5	6	7	8	9
Load R1, Mem1	FI	DI	CA Mem1	FO Read Mem1	EX Write R1	ST			
Load R2, Mem2		FI	DI	CA Mem2	FO Read Mem2	EX Write R1	ST		
Mpy R3, R1			FI	DI	CA	FO Read R1, R2	EX Write R1	ST	
Add R4, R2				FI	DI	C A	FO Read R2, R2	EX Write R1	ST

Figure 11.12 Data hazards.

But the MPY instruction reads R2 during cycle 6 also. In general, R2 cannot be guaranteed to contain the proper data until the end of cycle 6, and hence MPY instruction would operate with erroneous data. Similar data hazard occurs in cycle 7. For the results to be correct, we must ensure that R2 and R1 are read after they have been written into by the previous instruction, in each of these cycles. One possible solution is to forward the data to where it is needed in the pipeline as early as possible. For instance, since the ALU requires contents of R2 in cycle 6, the memory read mechanism can simply forward the data to ALU while it is being written into R2, thus accomplishing the write and read simultaneously. The concept of internal forwarding is described further, later in this section.

In general, the following scenarios are possible among instructions *I* and *J* where *J* follows *I* in the program:

1. Instruction *I* produces a result, which is required by J. Then, *J* has to be delayed until *I* produces the result.
2. Both *I* and *J* are required to write into a common memory location or a register, but the order of writing might get reversed because of the operation of the pipeline.
3. J writes into a register whose previous contents must be read by *I*. Then, *J* must be delayed until the register contents are read by *I*.

If the order of operations is reversed by the pipeline from what was implied by the instruction sequence in the program, then the result will be erroneous. Since, an instruction either READs from a resource or WRITEs into it, there are four possible orders of operations by two instructions that are sharing that resource. They are

1. READ/READ (READ after READ)
2. READ/WRITE (READ after WRITE)
3. WRITE/READ (WRITE after READ)
4. WRITE/WRITE (WRITE after WRITE)

In each case, the first operation is from the earlier instruction *I* and the second operation is from the later instruction J. If the orders are reversed, a conflict occurs. That is, a READ/WRITE conflict occurs if the WRITE operation is performed by *J* before the resource has been READ by *I*, and so on.

Reversing the order of READ/READ is not detrimental since data are not changed by either instructions and hence it is not considered a conflict. After a WRITE/WRITE conflict, the result in the shared resource is the wrong one for subsequent read operations. If it can be established that there are no READs in between the two WRITEs, the pipeline can allow the WRITE from *J* and disable the WRITE from *I* when it occurs. If the order of either READ/WRITE or WRITE/READ is reversed, the instruction reading the data gets an erroneous value. These conflicts must be detected by the pipeline mechanism to make sure that results of instruction execution remain as specified by the program.

There are in general two approaches to resolve conflicts. The first one is to compare the resources required by the instruction entering the pipeline with those of the instructions that are already in the pipeline and stall (i.e., delay the initiation) the entering instruction if a conflict is expected. That is, in the instruction sequence $[I, I + 1,..., J, J + 1,...]$, if a conflict is discovered between the instruction J entering the pipeline and instruction I in the pipeline, then the execution of instructions J, $J + 1,...$ is stopped until I passes the conflict point. The second approach is to allow the instructions $J, J + 1,...$ to enter the pipeline and handle the conflict resolution at each potential stage where the conflict might occur. That is, suspend only instruction J and allow $J + 1, J + 2,...$ to continue. Of course, suspending J and allowing subsequent instructions to continue might result in further conflicts. Thus, a multilevel conflict resolution mechanism may be needed making the pipeline control very complex. The second approach, known as instruction deferral, may offer better performance although it requires more complex hardware and independent functional units. Section 11.3.4 describes instruction deferral further.

One approach to avoid WRITE/READ conflicts is data forwarding, in which the instruction that WRITEs the data also forwards a copy of the data to those instructions waiting for it. A generalization of this technique is the concept of internal forwarding, which is described next.

Internal forwarding: Internal forwarding is a technique to replace unnecessary memory accesses by register-to-register transfers, during a sequence of read–operate–write operations on the data in the memory. This results in a higher throughput since slow memory accesses are replaced by faster register-to-register operations. This scheme also resolves some data interlocks between the pipeline stages.

Consider the memory location M with which registers r1 and r2 exchange data. There are three possibilities of interest:

1. Write–read forwarding: The following sequence of two operations,

$$M \leftarrow (r1)$$

$$r2 \leftarrow (M),$$

where "$\leftarrow$" designates a data transfer and "()" designates the "contents of," can be replaced by

$$M \leftarrow (r1)$$

$$r2 \leftarrow (r1),$$

thus saving one memory access.

2. Read–read forwarding: The following sequence of two operations,

$$r1 \leftarrow (M)$$

$$r2 \leftarrow (M),$$

can be replaced by

$$r1 \leftarrow (M)$$

$$r2 \leftarrow (r1),$$

thus saving one memory access.

3. Write–write forwarding (overwriting): The following sequence of two operations,

$$M \leftarrow (r1)$$

$$M \leftarrow (r2),$$

can be replaced by

$$M \leftarrow (r2),$$

thus saving one memory access.

The internal forwarding technique can be applied to a sequence of operations as shown by the following example.

Example 11.3

Consider the operation $P = (A^*B) + (C^*D)$ where P, A, B, C, and D are memory operands. This can be performed by the following sequence:

$$R1 \leftarrow (A)$$

$$R2 \leftarrow (R1) * (B)$$

$$R3 \leftarrow (C)$$

$$R4 \leftarrow (R3) * (D)$$

$$P \leftarrow (R4) + (R2).$$

The data-flow sequence for these operations is shown in Figure 11.13a. By internal forwarding, the data-flow sequence can be altered to that in Figure 11.13b. Here, A and C are forwarded to the corresponding multiply units, eliminating register R1 and R3, respectively. The results from these multiply units are forwarded to the adder, eliminating the transfers to R2 and R4.

In this example, a generalized architecture that allows operations between memory and register operands (as in second and fourth instructions) and register operands (as in the last instruction) was assumed. There are two other possibilities: load/store and memory/memory architectures. In load/store architecture, the arithmetic operations are always between two register operands, and memory is used for load and store only. In memory/memory architectures, operations can be performed on two memory operands directly. The assessment of the performance of the earlier instruction sequence on these architectures is left as an exercise.

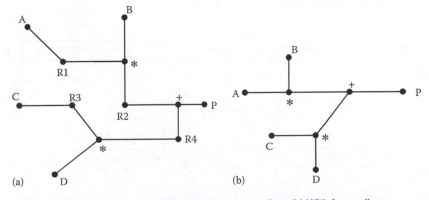

Figure 11.13 Example of internal forwarding. (a) Without forwarding. (b) With forwarding.

The internal forwarding technique as described earlier is not restricted to pipelines alone but is applicable to general multiple processor architectures. In particular, internal forwarding in a pipeline is a mechanism to supply data produced by one stage, to another stage that needs them, directly (i.e., without storing them in and reading them from the memory). The following example illustrates the technique for a pipeline.

Example 11.4

Consider the computation of the sum of the elements of an array A. That is,

$$SUM = \sum_{i=1}^{n} A_i.$$

Figure 11.14a shows a pipeline for computing this SUM. It is assumed that the array elements and the SUM are located in memory. Assuming that each stage in the pipeline completes its task in one pipeline cycle time, the computation of the first accumulation would be possible. Then onward, the fetch sum unit has to wait for two cycles before it can obtain the proper value for the SUM. This data interlock results in the degradation of the throughput of the pipeline to one output every three cycles.

One obvious thing to do is to feed the output of the add unit back to its input, as shown in Figure 11.14b. This requires the buffering of the intermediate sum value in the add unit. Once all the elements are accumulated, the value of SUM can be stored into the memory.

If the add unit requires more than one cycle to compute the sum, the solution again results in the degradation of throughput since the adder becomes the bottleneck.

Kogge (1981) provided a solution for such problems by rewriting the summing as

$$SUM_i = SUM_{i-d} + A_i,$$

where
    i is the current iteration index
    SUM_{i-d} is the intermediate SUM, d iterations ago
    d is the number of cycles required by the add unit

Because SUM_{i-d} is available at the ith iteration, this computation can proceed every iteration. But the computation results in d partial sums, each accumulating elements of A, d apart. At the end, these d partial sums should be added to obtain the final SUM. Thus, the pipeline can work efficiently at least during the computation of partial sums.

Thus, in the pipeline (1) earlier, if $d = 2$, we will require storing of two partial sums, SUM-1 and SUM-2 obtained one and two cycles ago, respectively, from the current iteration. The buffer holding these partial sums should be arranged as a first-in-first-out (FIFO) buffer so that the fetch sum unit fetches the appropriate value for each iteration.

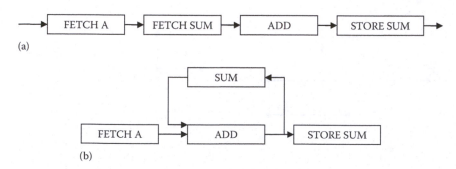

Figure 11.14 Pipeline for computing array sum. (a) Without forwarding. (b) With forwarding.

This type of solution is practical when changing the order of computation does not matter as in associative and commutative operations such as addition and multiplication. Even in these, changing the order might result in unexpected errors. For instance, in many numeric analysis applications, the relative magnitudes of the numbers to be added are arranged to be similar. If the order of addition is changed, this structure would change, resulting in a large number added to a small number, thereby altering the error characteristics of the computation.

11.3.2 Conditional Branches

As described earlier in this chapter, conditional branches degrade the performance of an instruction pipeline. The hardware mechanisms described earlier to minimize branch penalty were static in nature in the sense that they did not take into consideration the dynamic behavior of the branch instruction during program execution. Two compiler-based static schemes and two hardware-based dynamic schemes are described in the succeeding text.

11.3.2.1 Branch Prediction

The approaches described earlier are in a way branch-prediction techniques, in the sense that one of them predicted that the branch will be taken and the other predicted that the branch will not be taken. The run-time characteristics of the program can also be utilized in predicting the target of the branch. For instance, if the conditional branch corresponds to the end-of-do-loop test in a Fortran program, it is safe to predict that the branch is to the beginning of the loop. This prediction would be correct every time through the loop except for the last iteration. Obviously, such predictions are performed by the compiler, which generates appropriate flags to aid prediction during program execution.

In general, the target of the branch is guessed, and the execution continues along that path while marking all the results as tentative until the actual target is known. Once the target is known, the tentative results are either made permanent or discarded. Branch prediction is very effective as long as the guesses are correct. Refer to Section 11.4 for a description of the branch-prediction mechanisms of several Intel processors.

11.3.2.2 Delayed Branching

The delayed branching technique is widely used in the design of MCUs. Consider the two-stage pipeline for the execution of instructions shown in Figure 11.15a. The first stage fetches the instruction and the second executes it. The effective throughput of the pipeline is one instruction per cycle, for sequential code. When a conditional branch is encountered, the pipeline operation suffers while the pipeline fetches the target instruction, as shown in Figure 11.15b. But, if the branch instruction is interpreted as "execute the next instruction and then branch conditionally," then the pipeline can be kept busy while the target instruction is being fetched, as shown in Figure 11.15c. This mode of operation where the pipeline executes one or more instructions following a branch before executing the target of the branch is called delayed branching.

If n instructions enter the pipeline after a branch instruction and before the target is known, then the branch-delay slot is of length n as shown in the following:

> Branch instruction
> Successor instruction 1
> Successor instruction 2
> Successor instruction 3
> ⋮
> Successor instruction n
> Branch-target instruction

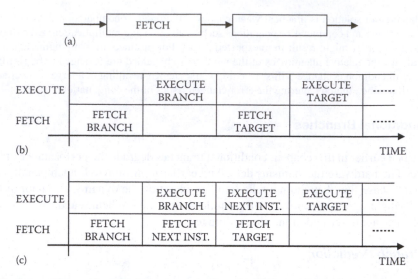

Figure 11.15 Delayed branching. (a) Two-stage pipeline. (b) Without delayed branching. (c) With delayed branching.

The compiler rearranges the program such that the branch instruction is moved n instructions prior to where it normally occurs. That is, the branch slot is filled with instructions that need to be executed prior to branch and the branch executes in a delayed manner. Obviously, the instructions in the branch-delay slot should not affect the condition for branch.

The length of the branch slot is the number of pipeline cycles needed to evaluate the branch condition and the target address (after the branch instruction enters the pipeline). The earlier these can be evaluated in the pipeline, the shorter is the branch slot.

As can be guessed, this technique is fairly easy to adopt for architectures that offer single-cycle execution of instructions utilizing a two-stage pipeline. In these pipelines, the branch slot can hold only one instruction. As the length of the branch slot increases, it becomes difficult to sequence instructions such that they can be properly executed, while the pipeline is resolving a conditional branch.

Modern-day RISC architectures adopt pipelines with small number (typically 2–5) of stages for instruction processing. They utilize delayed branching technique extensively.

The rearrangement of instructions to accommodate delayed branching is done by the compiler and is usually transparent to the programmer. Since the compiled code is dependent on the pipeline architecture, it is not easily portable to other processors. As such, delayed branching is not considered a good architectural feature, especially in aggressive designs that use complex pipeline structures.

All these techniques are static in nature in the sense that the predictions are made at compile time and are not changed during program execution. The following techniques utilize hardware mechanisms to dynamically predict the branch target. That is, the prediction changes if the branch changes its behavior during program execution.

11.3.2.3 Branch-Prediction Buffer

A branch-prediction buffer is a small memory buffer indexed by the branch instruction address. It contains 1 bit per instruction that indicates whether the branch was taken or not. The pipeline fetches subsequent instruction based on this prediction bit. If the prediction is wrong, the prediction bit is inverted. Ideally, the branch-prediction buffer must be large enough to contain 1 bit for each branch instruction in the program, or the bit is attached to each instruction and fetched during the instruction fetch. But that increases the complexity. Typically, a small buffer indexed by several low-order bits of the branch instruction address is used to reduce the complexity. In such cases, more than one instruction maps to each bit in the buffer, and hence the prediction may not be correct with respect to the branch instruction at hand, since some other instruction could have altered the prediction bit. Nevertheless, the prediction is assumed to be correct. Losq et al. (1984) named the branch-prediction buffer a decode history table.

The disadvantage with this technique is that by the time the instruction is decoded to detect that it is a conditional branch, other instructions would have entered the pipeline. If the branch is successful, the pipeline needs to be refilled from the target address. To minimize this effect, if the decode history table also contains the target instruction, in addition to the information as to the branch was taken or not, that instruction can enter the pipeline immediately, if the branch is a success.

11.3.2.4 Branch History

The branch history technique (Sussenguth, 1971) uses a branch history table, which stores for each branch, the most probable target address. This target could very well be the target it reached last time during the program execution.

Figure 11.16 shows a typical branch history table. It consists of two entries for each instruction: the instruction address and the corresponding branch address. It is stored in a cache-like memory. As soon as the instruction is fetched, its address is compared with the first field of the table. If there is a match, the corresponding branch address is immediately known. The execution continues with this assumed branch, as in branch-prediction technique. The branch address field in the table is continually updated, as and when the branch targets are resolved.

As can be guessed, the implementation of the branch history table requires an excessive number of accesses to the table, thereby creating a bottleneck at the cache containing the table.

11.3.2.5 Multiple Instruction Buffers

We described the use of multiple instruction buffers to reduce the effect of conditional branching on the performance of pipeline, earlier. The details of a practical architecture utilizing that feature are provided here.

Figure 11.17 shows the structure of IBM 360/91 instruction-processing pipeline, where the instruction fetch stage consists of two buffers: the S-buffer is the sequential instruction prefetch buffer and the T-buffer is for the prefetch of target instruction sequence. This stage is followed by decode and other stages similar to the other pipelines described in this chapter.

The contents of the S-buffer are invalidated when a branch is successful and the contents of T-buffer are invalidated when the branch is not successful. The decode unit fetches instructions from the appropriate buffer.

When the decoder issues a request for the next instruction, the S-buffer is looked up (i.e., a single-level, nonsequential search) for sequential instructions or the T-buffer is looked up if a conditional

Instruction address	Branch address

Figure 11.16 Branch history table.

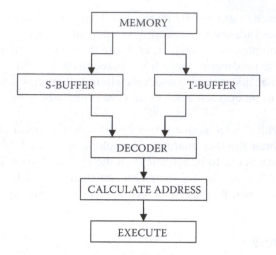

Figure 11.17 Instruction pipeline with multiple instruction buffers.

branch has been just successful. If the instruction is available in either buffer, it is brought into the decode stage without any delay. If not, the instruction needs to be fetched from the memory, which will incur a memory access delay.

All nonbranch instructions enter the remaining stages of the pipeline after the decode is complete. For unconditional branch instruction, the instruction at its target address is immediately requested by the decode unit, and no further decoding is performed until that instruction arrives from the memory. For conditional branch instruction, the sequential prefetching is suspended, while the instruction is traversing the remaining stages of the pipeline. Instructions are prefetched from the target address simultaneously, until the branch is resolved. If the branch is successful, further instructions are fetched from the T-buffer. If the branch is not successful, normal fetching from the S-buffer continues.

11.3.3 Interrupts

A complex hardware/software support is needed to handle interrupts on a pipelined processor since many instructions are executed in an overlapped manner and any of those instructions can generate an interrupt. Ideally, if instruction I generates an interrupt, the execution of instructions $I + 1$, $I + 2$, etc., in the pipeline should be postponed until the interrupt is serviced. However, the instructions $I - 1$, $I - 2$, etc., that have already entered the pipeline must be completed before the interrupt service is started. Such an ideal interrupt scheme is known as precise interrupt scheme.

Note also that instructions are usually not processed in order (i.e., the order in which they appear in the program). When delayed branching is used, the instructions in branch-delay slot are not sequentially related. If an instruction in the branch-delay slot generates the interrupt and the branch is taken, the instructions in the branch-delay slot and the branch-target instruction must be restarted after interrupt is processed. This necessitates multiple program counters since these instructions are not sequentially related.

Several instructions can generate interrupts simultaneously and these interrupts can be out of order. That is, the interrupt due to instruction I must be handled prior to that due to $I + 1$. To ensure that, a status vector is attached to each instruction as it traverses the pipeline. The machine state changes implied by the instruction are marked temporary until the last stage. At the last stage, if the vector indicates that there has been no interrupt, the machine state is changed. If an interrupt is indicated, an in-order processing of interrupts is performed before state changes are committed.

11.3.4 Instruction Deferral

Instruction deferral is used to resolve the data interlock conflicts in the pipeline. The concept is to process as much of an instruction as possible at the current time and defer the completion until the data conflicts are resolved, thus obtaining a better overall performance than stalling the pipeline completely until data conflicts are resolved.

11.3.4.1 CDC 6600 Scoreboard

The processing unit of CDC 6600 shown in Figure 11.18 consists of 16 independent functional units (5 for memory access, 4 for FP operations, and 7 for integer operations). The input operands to each functional unit come from a pair of registers and the output is designated to one of the registers. Thus, three registers are allocated to each functional unit corresponding to each arithmetic instruction. The control unit contains a scoreboard, which maintains the status of all the registers, the status of functional units, and the register-functional unit associations. It contains an instruction queue called reservation station. Instructions first enter into the reservation station. Each arithmetic instruction corresponds to a 3-tuple consisting of two source register and one destination register designations. The load and store instructions consist of a 2-tuple corresponding to the memory address and register designation. The source operands in each instruction have a tag associated with them. The tag provides an indication of the availability of the operand in the register. If the operand is not available, it indicates the functional unit from which the operand is expected (Thornton, 1970).

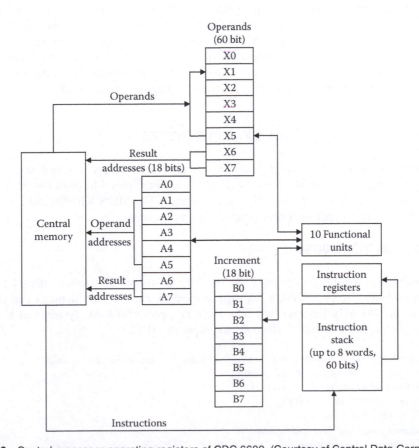

Figure 11.18 Central processor-operating registers of CDC 6600. (Courtesy of Control Data Corporation.)

For each instruction, the scoreboard determines whether the instruction can be executed immediately or not based on the analysis of data dependencies. If the instruction cannot be executed immediately, it is placed in the reservation station and the scoreboard monitors its operand requirements and decides when all the operands are available. The scoreboard also controls when a functional unit writes its result into the destination register. Thus, the scoreboard resolves all conflicts. The scoreboard activity with respect to each instruction can be summarized into the following steps:

1. If a functional unit is not available, the instruction is stalled. When the functional unit is free, the scoreboard allocates the instruction to it, if no other active functional unit has the same destination register. This resolves structural hazards and write/write conflicts.
2. A source operand is said to be available if the register containing the operand is being written by an active functional unit or if no active functional unit is going to write it. When the source operands are available, the scoreboard allocates that instruction to the functional unit for execution. The functional unit then reads the operands and executes the instruction. Write/read conflicts are resolved by this step. Note that the instructions may be allocated to functional units out of order (i.e., not in the order specified in the program).
3. When a functional unit completes the execution of an instruction, it informs the scoreboard. The scoreboard decides when to write the results into the destination register making sure that read/write (R/W) conflicts are resolved. That is, the scoreboard does not allow writing of the results. If there is an active instruction whose operand is the destination register of the functional unit wishing to write its result or if the active instruction has not read its operands yet, the corresponding operand of the active instruction was produced by an earlier instruction. Writing of the result is stalled until R/W conflict clears.

Because all functional units can be active at any time, an elaborate bus structure is needed to connect registers and funs. The 16 functional units of the CDC 6600 were grouped into four groups with a set of buses (data trunks) for each group. Only one functional unit in each group could be active at any time. Note also that all results are written to the register file and the subsequent instruction has to wait until such a write takes place. That is, there is no forwarding of the results. Thus, as long as the write/write conflicts are infrequent, the scoreboard performs well.

11.4 EXAMPLE SYSTEMS

Most of the concepts introduced in this chapter were used by the processors described in earlier chapters of this book. In this section we provide additional examples. Motorola 88000 series of processors are selected to represent early RISC architectures. The MIPS R10000 and the Intel series represent contemporary superpipelined, superscalar architectures.

11.4.1 Motorola MC88100/88200

The MC88100 is the first processor in the MC8800 family of RISC processors. The MC88000 family also includes the MC88200, a high-speed memory-caching and demand-paged memory-management unit (MMU). Together they offer the user a powerful RISC system that is claimed to become the new standard in high-performance computing (HPC):

1. The MC88100 instruction set contains 51 instructions. These instructions include:
   • Integer add, subtract, multiply, and divide
   • FP add, subtract, multiply, and divide
   • Logical AND, OR, and XOR
   • Bit-field instructions
   • Memory-manipulation instructions
   • Flow-control instructions

2. There are a small number of addressing modes: three addressing modes for data memory, four addressing modes for instruction memory, and three register addressing modes.
3. A fixed-length instruction format is implemented. All instructions are 32 bits long.
4. All instructions are either executed in one processor cycle or dispatched to pipelined execution units that produce results every processor cycle.
5. Memory access is performed by load/store instructions.
6. The CPU contains 32, 32-bit user registers as well as numerous registers used to control pipelines and save the context of the processor during exception processing.
7. The control unit is hardwired.
8. High-level language support exists in the form of procedure parameter registers and a large register set in general.

The most important design feature that contributes to the single-cycle execution of all instructions is the multiple-execution units (see Figure 11.19): the instruction unit, the data unit, the FP unit, and the integer unit. The integer unit and the FP unit execute all data manipulation instructions. Data memory accesses are performed by the data unit, and instruction fetches are performed by the instruction unit. They operate both independently and concurrently. These execution units allow

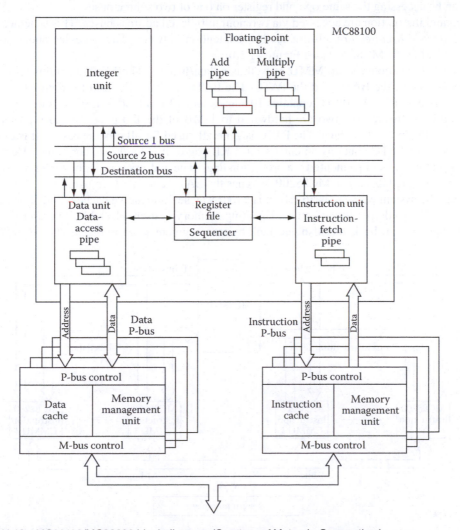

Figure 11.19   MC88100/MC88200 block diagram. (Courtesy of Motorola Corporation.)

the MC88100 to perform up to five operations in parallel. It can access program memory; execute an arithmetic, logical, or bit-field instruction access data memory; execute an FP or integer-divide instruction; or execute an FP or integer-multiply instruction.

Three of the execution units are pipelined: the data unit and the FP unit. The FP unit actually has two pipelines, the add pipe and the multiply pipe.

Data interlocks are avoided within the pipelines by using a scoreboard register. Each time an instruction enters the pipe, the bit corresponding to the destination register is set in the scoreboard register. The subsequent instruction entering the pipe checks to see if the bit corresponding to its source registers is set. If so, it waits. Upon completion of the instruction, the pipeline mechanism clears the destination register's bit, freeing it to be used as source operand.

The MC88100 incorporates delayed branching to reduce pipeline penalties associated with changes in program flow due to conditional branch instructions. This technique enables the instruction fetched after the branch instruction to be optionally executed. Therefore, the pipeline flow is not broken.

There are also three internal buses within the MC88100, two source operand buses and a destination bus. These buses enable the pipeline to access operand registers concurrently. Two different pipes can be accessing the same operand register on one of two source buses.

Data and instruction are accessed via two nonmultiplexed address buses. This scheme, known as the *Harvard architecture*, eliminates bus contention between data accesses and instruction fetches by using the MC88200 (see Figure 11.20).

The MC88200 contains an MMU as well as a data/ICache. The MMU contains two address translation caches, the BATC and the PATC. The BATC is generally used for high-bit addresses. It is fully associative. The BATC contains 10 entries for 512 KB blocks. Eight of these entries are controlled by software, and two are hardwired to 1 MB of the I/O page. Memory protection is implemented by protection flags. The PATC is generally used for all other accesses. It is also fully associative. The PATC contains 56 entries for 4 KB pages. It is hardware controlled. Here again, memory protection is implemented via protection flags. The address translation tables are standard two-level mapping tables. The MC88200 searches the BATC and PATC in parallel.

The cache system capacity is 16 KB. It is a four-way set-associative cache with 256 sets. There are four 32-bit words per line of cache. Updating memory is user selectable. The user can either write through or copy back. It is also selectable by area, segment, page, or block. The MC88200 uses

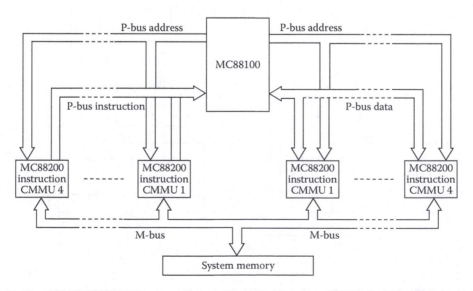

Figure 11.20   MC88100/MC88200 system diagram example. (Courtesy of Motorola Corporation.)

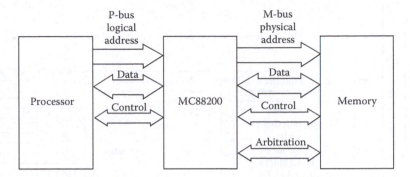

Figure 11.21 MC88200 logical block diagram. (Courtesy of Motorola Corporation.)

the least recently used (LRU) replacement algorithm. Lines can be disabled by line for fault-tolerant operation. Additionally, the MC88200 provides a "snoop" capability for coherency.

The MC88200 interfaces the P-bus with the M-bus. The P-bus is a synchronous bus with dedicated address and data lines. It has an 80/MB/speak throughput at 20 MHz. Checker mode is provided to allow shadowing. The M-bus is a synchronous 32-bit bus. It has multiplexed address and data lines (Figure 11.21).

11.4.2 MIPS R10000 Architecture*

The R10000 is a single-chip superscalar RISC microprocessor that is a follow-on to the MIPS RISC processor family that includes chronologically the R2000, R3000, R6000, R4400, and R8000. The integer and FP performance of the R1000 makes it ideal for applications such as engineering workstations, scientific computing, 3D graphic workstation, database servers, and multiuse systems. The R10000 uses the MIPS architecture with nonsequential dynamic execution scheduling (ANDES), which supports two-integer and two FP execute instructions plus one load/store instruction per cycle. The R10000 has the following major features:

- 64-bit MIPS IV ISA
- Decoding four instructions each pipeline cycle, appending them to one of three instruction queues
- Five execution pipelines connected to separate internal integer and FP execution (or functional) units
- Dynamic instruction scheduling and out-of-order execution
- Speculative instruction issue (also termed "speculative branching")
- Precise exception mode (exceptions can be traced back to the instruction that caused them)
- Nonblocking caches
- Separate on-chip 32 KB primary instruction and data caches
- Individually optimized secondary cache and system interface ports
- Internal controller for the external secondary cache
- Internal system interface controller with multiprocessor support

A block diagram of the processor and its interfaces is shown in Figure 11.22.

11.4.2.1 Instruction Set (MIPS IV)

The R10000 implements the MIPS IV ISA. MIPS IV is a superset of the MIPS III ISA and is backward compatible. At a frequency of 200 MHz, the R10000 delivers a peak performance for

---

* This section is extracted from *MIPS R10000 User's Manual*, Version 2.0, MIPS Technologies Inc., 1997; *MIPS R10000 Technical Brief*, Version 2.0, MIPS Technologies Inc., 1997; and *ZDNET review of R10000*, December 1997.

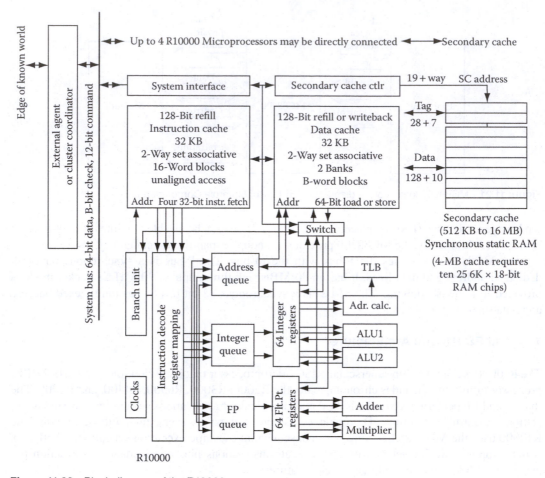

Figure 11.22 Block diagram of the R10000.

800 MIPS with a peak data transfer rate of 3.2 GB/s to secondary cache. MIPS has defined an ISA, implemented in the following sets of CPU designs:

MIPS I, implemented in the R2000 and R3000
MIPS II, implemented in the R6000
MIPS III, implemented in the R4400
MIPS IV, implemented in the R8000 and R10000 (added prefetch, conditional move I/FP, index load/store FP, etc.)

The original MIPS I CPU ISA has been extended forward three times. Each extension is backward compatible. The ISA extensions are inclusive in the sense that each new architecture level (or version) includes the former levels. The result is that a processor implementing MIPS IV is also able to run MIPS I, MIPS II, or MIPS III binary programs without change.

11.4.2.2 Superscalar Pipeline

A superscalar processor is one that can fetch, execute, and complete more than one instruction in parallel. The R10000 is a four-way superscalar architecture, which fetches and decodes four instructions per cycle. Each decoded instruction is appended to one of three instruction queues. Each queue can perform dynamic scheduling of instructions. The queues determine the execution

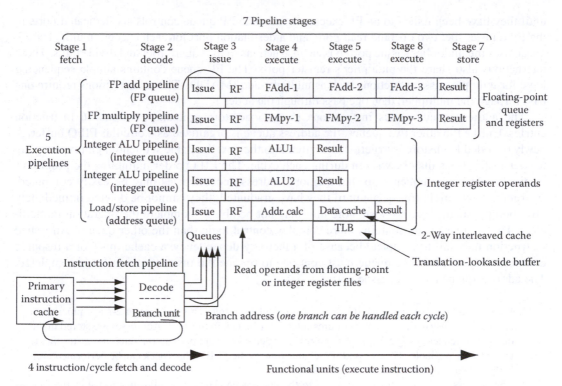

Figure 11.23 Superscalar pipeline architecture in the R10000.

order based on the availability of the required execution units. Instructions are initially fetched and decoded in order but can be executed and completed out of order, allowing the processor to have up to 32 instructions in various stages of execution. Instructions are processed in six partially independent pipelines, as shown in Figure 11.23. The fetch pipeline reads instructions from the ICache, decodes them, renames their registers, and places them in three instruction queues.

The instruction queues contain integer, address calculate, and FP instructions. From these queues, instructions are dynamically issued to the five pipelined execution units. Each pipeline in R10000 includes stages for fetching (stage 1 in Figure 11.23), decoding (stage 2), issuing instructions (stage 3), reading register operands (stage 3), executing instructions (stages 4 through 6), and storing result (stage 7).

The processor keeps the decoded instructions in three instruction queues: integer queue, address queue, and FP queue. These queues allow the processor to fetch each instruction at its maximum rate without stalling, because of instruction conflicts or dependencies. Each queue uses instruction tags to keep track of the instruction in each execution pipeline stage. These tags set a "done" bit in the active list as each instruction is completed.

The integer queue issues instructions to the two-integer arithmetic units: ALU1 and ALU2. The integer queue contains 16 instruction entries. Up to four instructions may be written during each cycle; newly decoded integer instructions are written into empty entries in no particular order. Instructions remain in this queue only until they have been issued to an ALU. Branch and shift instructions can be issued only to ALU1. Integer-multiply and integer-divide instructions can be issued only to ALU2. Other integer instructions can be issued to either ALU. The integer queue controls six dedicated pots to the integer register file: two operand read ports and a destination write port for each ALU.

The FP queue issues instructions to the FP multiplier and the FP adder. The FP queue contains 16 instruction entries. Up to four instructions may be written during each cycle; newly decoded FP instructions are written into empty entries in random order. Instructions remain in this queue only until they have been issued to an FP execution unit. The FP queue controls six dedicated ports to the FP register file: two operand read ports and a destination port for each execution unit. The FP queue uses the multiplier's issue port to issue instructions to the square-root and divide units. These instructions also share the multiplier's register ports. The FP queue contains simple sequencing logic for multiple-pass instructions such as multiply–add (MADD). These instructions require one pass through the multiplier, then one pass through the adder.

The address queue issues instructions to the load/store unit and contains 16 instruction entries. Unlike the other two queues, the address queue is organized as a circular FIFO buffer. A newly decoded load/store instruction is written into the next available sequential empty entry; up to four instructions may be written during each cycle. The FIFO order maintains the program's original instruction sequence so that memory address dependencies may be easily computed. Instructions remain in this queue until they have graduated; they cannot be deleted immediately after being issued, since the load/store unit may not be able to complete the operation immediately. The address queue contains more complex control logic than the other queues. An issued instruction may fail to complete because of a memory dependency, a cache miss, or a resource conflict; in these cases, the queue must continue to reissue the instruction until it is completed. The address queue has three issue ports:

1. It issues each instruction once to the address calculation unit. This unit uses a 2-stage pipeline to compute the instruction's memory address and to translate it in the TLB. Addresses are stored in the queue's dependency logic. This port controls two dedicated read ports to the integer register file. If the cache is available, it is accessed at the same time as the TLB. A tag check can be performed even if the data array is busy.
2. The address queue can reissue accesses to the data cache. The queue allocates usage of the four sections of the cache, which consist of the tag and data sections of the two cache banks. Load and store instructions begin with a tag check cycle, which checks to see if the desired address is already in caches. If it is not, a refill operation is initiated, and this instruction waits until it has completed. Load instructions also read and align a double-word value from the data array. This access may be either concurrent to or subsequent to the tag check. If the data are present and no dependencies exist, the instruction is marked done in the queue.
3. The address queue can issue store instructions to the data cache. A store instruction may not modify the data cache until it graduates. Only one store can graduate per cycle, but it may be anywhere within the four oldest instructions, if all previous instructions are already completed.

The access and store ports share four register file ports (integer read and write, FP read and write). These shared parts are also used for jump and link and jump register instructions and for move instructions between the integer and register files.

The three instruction queues can issue one new instruction per cycle to each of the five execution pipelines:

1. Integer queue issues instructions to the two-integer ALU pipelines.
2. Address queue issues one instruction to the load/store unit pipeline.
3. FP queue issues instructions to the FP adder and multiplier pipelines.
4. A sixth pipeline, the fetch pipeline, reads and decodes instructions from the ICache.

The 64-bit integer pipeline has the following characteristics:

1. 16-entry integer instruction queue that dynamically issues instructions
2. 64-bit 64-location integer physical register file, with seven read and three write ports
3. 64-bit arithmetic logic units
   a. Arithmetic logic unit, shifter, and integer branch comparator
   b. Arithmetic logic unit, integer multiplier, and divider

The load/store pipeline has the following characteristics:

1. 16-entry address queue that dynamically issues instructions and uses the integer register file for base and index registers.
2. 16-entry address stack for use by nonblocking loads and stores.
3. 44-bit virtual address calculation unit.
4. 64-entry fully associative TLB, which converts virtual addresses to physical addressing, using a 40-bit physical address. Each entry maps two pages, with sizes ranging from 4 KB to 16 MB, in powers of 4.

The 64-bit FP pipeline has the following characteristics:

1. 16-entry instruction queue, with dynamic issue
2. 64-bit 64-location FP physical register file, with five read and three wire ports (32 logical registers)
3. 64-bit parallel multiply unit (3-cycle pipeline with 2-cycle latency), which also performs move instructions
4. 64-bit add unit (3-cycle pipeline with 2-cycle latency), which handles addition, subtraction, and miscellaneous FP operations
5. Separate 64-bit divide and square-root units, which can operate concurrently (these units share their issue and completion logic with the FP multiplier)

11.4.2.3 Functional Units

The five execution pipelines allow overlapped instruction by issuing instructions to the following five functional units: two-integer ALUs (ALU1 and ALU2), load/store unit (address calculate), FP adder, and FP multiplier. There are also three "iterative" units to compute more complex results.

The integer-multiply and integer-divide operations are performed by an integer-multiply/integer-divide execution unit; these instructions are issued to ALU2. ALU2 remains busy for the duration of the divide. FP divides are performed by the divide execution unit; these instructions are issued to the FP multiplier. Floating-point square root is performed by the square-root execution unit; these instructions are issued to the FP multiplier.

The instruction decode and rename unit has the following characteristics:

1. Processes four instructions in parallel
2. Replaces logical register numbers with physical register numbers (register renaming)
3. Maps integer registers into 33-word-by-6-bit mapping table that has 4 write and 12 read ports
4. Maps FP registers into a 32-word-by-6-bit mapping table that has 4 write and 16 read ports
5. Has a 32-entry active list of all instructions within the pipeline

The branch unit has the following characteristics:

1. Allows one branch per cycle.
2. Conditional branches can be executed speculatively, up to 4 deep.
3. 44-bit adder to compute branch addresses.
4. The branch return cache contains four instructions following a subroutine call, for rapid use when returning from leaf subroutines.
5. The program trace RAM stores the program counter for each instruction in the pipeline.

11.4.2.4 Pipeline Stages

In stage 1, the processor fetches four instructions each cycle, independent of their alignment in the ICache—except that the processor cannot fetch across a 16-word cache block boundary. These words are then aligned in the four-word instruction register. If any instructions were left from the previous decode cycle, they are merged with new words from the ICache to fill the instruction register.

In stage 2, the four instructions in the instruction register are decoded and renamed. (Renaming determines any dependencies between instructions and provides precise exception handling.) When renamed, the logical registers referenced in an instruction are mapped to physical registers. Integer and FP registers are renamed independently. A logical register is mapped to a new physical register whenever that logical register is the destination of an instruction.

Thus, when an instruction places a new value in a logical register, that logical register is renamed (mapped) to a new physical register, while its previous value is retained in the old physical register.

As each instruction is renamed, its logical register numbers are compared to determine if the dependencies exist between the four instructions decoded during this cycle. After the physical register numbers become known, the physical register busy table indicates whether or not each operand is valid. The renamed instructions are loaded into integer or FP instruction queues.

Only one branch instruction can be executed during stage 2. If the instruction register contains a second branch instruction, this branch is not decoded until the next cycle. The branch unit determines the next address for the program counter; if a branch is taken and then reversed, the branch resume cache provides the instructions to be decoded during the next cycle.

In stage 3, decoded instructions are written into the queues. Stage 3 is also the start of each of the five execution pipelines.

In stages 4–6, instructions are executed in the various functional units. These units and their execution process are described in the following:

FP *multiplier (three-stage pipeline)*: Single- or double-precision multiply and conditional move operations are executed in this unit with a one-cycle latency and a one-cycle repeat rate. The multiplication is completed during the first two cycles; the third cycle is used to pack and transfer the result.

FP *divide and square-root units*: Single- or double-precision division and square-root operations can be executed in parallel by separate units. These units share their issue and completion logic with the FP multiplier.

FP *adder (three-stage pipeline)*: Single- or double-precision add, subtract, compare, or convert operations are executed with a two-cycle latency and a one-cycle repeat rate. Although a final result is not calculated until the third pipeline state, internal bypass paths set a two-cycle latency for dependent add or multiply instructions.

Integer ALU1 (one-stage pipeline): Integer add, subtract, shift, and logic operations are executed with a one-cycle latency and a one-cycle repeat rate. This ALU also verifies predictions made for branches that are conditional on integer register values.

Integer ALU2 (one-stage pipeline): Integer add, subtract, and logic operations are executed with a one-cycle latency and a one-cycle repeat rate. Integer-multiply and integer-divide operations take more than one cycle.

A single memory address can be calculated every cycle for use by either an integer or an FP load or store instruction. Address and load operations can be calculated. Address is translated from a 44-bit virtual address into a 40-bit physical address using a TLB. The TLB contains 64 entries, each of which can translate two pages. Each entry can select a page size ranging from 4 KB to 16 MB, inclusive, in powers of 4, as shown in Figure 11.24.

Load instructions have a two-cycle latency if the addressed data are already within the data cache. Store instructions do not modify the data cache or memory until they graduate.

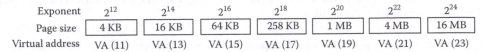

Exponent	2^{12}	2^{14}	2^{16}	2^{18}	2^{20}	2^{22}	2^{24}
Page size	4 KB	16 KB	64 KB	258 KB	1 MB	4 MB	16 MB
Virtual address	VA (11)	VA (13)	VA (15)	VA (17)	VA (19)	VA (21)	VA (23)

Figure 11.24 TLB page sizes.

11.4.2.5 *Cache*

The R10000 contains an on-chip cache consisting of 32 KB of primary data cache and a separate 32 KB ICache. This is considered a large on-chip cache. This processor controls an off-chip secondary cache that can range in size from 512 KB to 16 MB.

The primary data cache consists of two equal 16 KB banks. The data cache is two-way interleaved and each of the two banks is two-way set associative using an LRU replacement algorithm. The data cache line size is 32 bytes and the data cache contains an affixed block size of eight words.

The interleaved data cache design was used because the access times of most currently available RAM are long relative to processor cycle time. The interleaved data cache allows the memory requests to be overlapped, which in turn allows the ability to hide more of the access and recovery times of each bank. To speed access delays, the processor can do the following:

1. Execute up to 16 load and store instructions speculatively and out of order, using nonblocking primary and secondary caches, meaning it looks ahead in its instruction system to find load and store instructions that can be executed early; if the addressed data block is not in the primary cache, the processor initiates cache refills as soon as possible.
2. If a speculatively executed load initiates a cache refill, the refill is completed even if the load instruction is aborted.
3. The external interface gives priority to its refill and interrogate operations. When the primary cache is refilled, any required data can be streamed directly to waiting load instructions.
4. The external interface can handle up to four nonblocking memory accesses to secondary cache and main memory.

Because of an interleaved data cache, increased complexity is required to support them. This complexity begins with each cache bank containing independent tag and data arrays. These four sections can be allocated separately to achieve high utilization. Five separate circuits compete for cache bandwidth. These circuits are the address calculated, tag check, load unit, store unit, and external interface.

The data cache is nonblocking. This allows the cache accesses to continue even though a cache miss has occurred. This is important to the cache performance because it gives the data cache the ability to stack memory references by queuing up multiple cache misses and servicing them simultaneously.

The data cache uses a write-back protocol, which means a cache store writes data into the cache instead of writing it directly to memory. When data are written to the data cache, it is tagged directly to memory. When data are written to the data cache, it is tagged as a dirty block. Prior to this dirty block being replaced by a new frame, it is written back to the off-chip secondary cache. The secondary cache writes back to the main memory. This protocol is used to maintain data consistency. Note that the data cache is written back prior to the secondary cache writing back to the main memory.

Because the data cache is a subset of the secondary cache, the data cache is said to be inconsistent when it has been modified from the corresponding data in the secondary cache. The data cache can be in only one of the following four states at any given time: *invalid, clean exclusive, dirty exclusive,* and *shared.*

The processor requires the cache blocks to have as a single owner at all times. Thus, the processor adheres to certain ownership rules:

The processor assumes ownership of a cache block if the state of the block becomes dirty exclusive.
For a processor upgrade request, the processor assumes ownership of the block after receiving an external ACK completion response.
The processor gives up ownership of a cache block if the state of the cache block changes to invalid, clean exclusive, or shared.
Clean exclusive and shared cache blocks are always considered to be owned by memory.

The events that trigger a change in the state of a data cache block are included in the following events:

1. Primary data cache R/W miss
2. Primary data cache hit
3. Subset enforcement
4. A cache instruction
5. External intervention shared request
6. Intervention exclusive request

The secondary cache is located off-chip and can range in size from 512 KB to 16 MB. It is interfaced with the R10000 by a 128-bit data bus, which can operate at a maximum of 200 MHz, yielding a maximum transfer rate of 3.2 GB/s. This dedicated cache bus cannot be interrupted by any other bus traffic in the system. Thus, when a cache miss occurs, the access to the secondary cache is immediate. Therefore, it is true to say that the secondary cache interface approaches zero wait state performance. This means that when the cache receives a data request from the processor, it will always be able to return data in the following clock cycle. There is no external interface circuitry required for the secondary cache system. The secondary cache maintains many of the same features as the primary data cache. It is two-set associative and uses an LRU replacement algorithm. It also uses the writes back protocol and can be in one of the following four states: *invalid*, *clean exclusive*, *dirty exclusive*, and *shared*.

The events that trigger the changing of a secondary cache block are the same as the primary data cache except for events (2)–(4) in the following:

1. Primary data cache R/W miss
2. Data cache write hit to a *shared* or *clean exclusive* block
3. Secondary cache read miss
4. Secondary cache write hit to a *shared* or *clean exclusive* block
5. A cache instruction
6. External intervention shared request
7. Intervention exclusive request
8. Invalidate request

The R10000 has 32 KB on-chip ICache that is two-way set associative. It has a fixed block size of 16 words and a line size of 64 bytes and uses the LRU replacement algorithm. At any given time, the ICache may be in one of two states, valid or invalid. The following are the events that cause the ICache block to change states:

1. A primary ICache read miss
2. Subset property enforcement
3. Any of various cache instructions
4. External intervention exclusive and invalidate requests

The behavior of the processor when executing load and store instructions is determined by the cache algorithm specified for the accessed address. The processor supports five different cache algorithms:

1. Uncached
2. Cacheable noncoherent
3. Cacheable coherent exclusive
4. Cacheable coherent exclusive on write
5. Uncached accelerated

The loads and stores under the uncached cache algorithm bypass the primary and secondary caches. Under the cacheable noncoherent cache algorithm, load and store secondary cache misses result in processor noncoherent block read requests. Under the cacheable coherent exclusive cache algorithm, load and store secondary cache misses result in processor coherent block read exclusive requests. Such processor requests indicate to external agents containing caches that a coherency check must be performed and that the cache block must be returned in an exclusive state. The cacheable coherent exclusive on wire cache algorithm is similar to the cacheable coherent exclusive cache algorithm except that load secondary cache misses result in processor coherent block read shared requests. The R10000 implements a new cache algorithm, uncached accelerated. This allows the kernel to mark the TLB entries for regions of the physical address space, or certain blocks of data, as uncached. After signaling the hardware, it may now gather a number of uncached writes together as a series of writes to the same address or sequential writes to all addresses in the block. These are put to an uncached accelerated buffer and then issued to the system interface as processor block write requests. The uncached accelerated algorithm differs from the uncached algorithm in that block write gathering is not performed.

11.4.2.6 Processor-Operating Modes

The three operating modes are listed in order of decreasing system privilege:

1. *Kernel mode* (highest system privilege): can access and change any register. The innermost core of the operating system runs in kernel mode.
2. *Supervisor mode*: has fewer privileges and is used for less-critical sections of the operating system.
3. *User mode* (lowest system privilege): prevents users from interfering with one another.

The selection between the three modes can be made by the operating system (when in kernel mode) by writing into the status register's KSU field. The processor is forced into kernel mode when the processor is handling an error (the ERL bit is set) or an exception (EXL bit is set). Figure 11.25 shows the selection of operating modes with respect to the KSU, EXU, and ERL bits. It also shows different instruction sets, and addressing modes are enabled by the status register's XX, UX, SX, and KX bits. In kernel mode, the KX bit allows 64-bit addressing; all instructions are always valid. In supervisor mode, the SX bit allows 64-bit addressing and the MIPS III instruction. The MIPS IV ISA is enabled all the time in supervisor mode. In user mode, the UX bit allows 64-bit addressing and the MIPS III instructions; the XX bit allows the new MIPS IV instruction.

The processor uses either 32-bit or 64-bit address spaces, depending on the operating and addressing modes set by the status register. The processor uses the following addresses: virtual address VA and region bits VA. If a region is mapped, virtual addresses are translated in the TLB. Bits VS are not translated in the TLB and are sign extensions of bit VA.

In both 32-bit and 64-bit address modes, the memory address space is divided into many regions as shown in Figure 11.26. Each has specific characteristics and uses. The user can access only the useg region in 32-bit mode and xuseg in 64-bit mode as shown in the table. The supervisor can

XX 31	KX 7	SX 6	UX 5	KSU 4:3	ERL 2	EXL 1	Description	ISA III	ISA IV	Addressing Mode 32 Bit 64 Bit
0	—	—	0	10	0	0	User mode	No	No	32
1	—	—	0	10	0	0		No	Yes	32
0	—	—	1	10	0	0		Yes	No	64
1	—	—	1	10	0	0		Yes	Yes	64
—	—	0	—	01	0	0	Supervisor mode	No	Yes	32
—	—	1	—	01	0	0		Yes	Yes	64
—	0	—	—	00	0	0	Kernel mode	Yes	Yes	32
—	1	—	—	00	0	0		Yes	Yes	64
—	0	—	—	—	0	1	Exception level	Yes	Yes	32
—	1	—	—	—	0	1		Yes	Yes	64
—	0	—	—	—	1	X	Error level	Yes	Yes	32
—	1	—	—	—	1	X		Yes	Yes	64

Figure 11.25 Operating mode versus addressing modes. *Note:* No means the ISA is disabled; Yes means the ISA is enabled. Dashes (–) indicate "don't care".

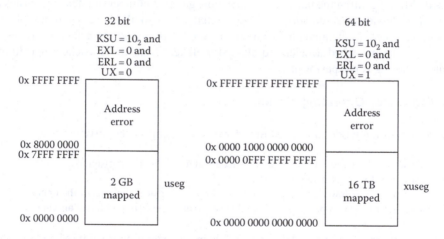

Figure 11.26 User-mode virtual address space.

access user region as well as sseg (in 32-bit mode) or xsseg and csseg (in 64-bit mode) shown in Figure 11.26. The kernel can access all regions except those restricted because bits VA are not implemented in the TLB.

In user mode, a single uniform virtual address space-labeled user segment is available; its size is 2 GB in 32-bit mode (useg) and 16 TB in 64-bit mode (xuseg). When $UX = 0$ in the status register, user-mode addressing is compatible with the 32-bit addressing modes, and a 2 GB user address space is available labeled useg. All valid user-mode virtual addresses have their most significant bit cleared to 0. Any attempt to reference an address with the most significant bit set while in user mode causes an address error exception. When UX = 1 in the status register, user-mode addressing is extended to the 64-bit model shown in Figure 11.26. In 64-bit user mode, the processor provides a single, uniform virtual address space for 2^{44} bytes labeled xuseg. All valid user-mode virtual addresses have bits 63:44 equal to 0; an attempt to reference an address with bits 63:44 not equal to 0 causes an address error exception (Figure 11.27).

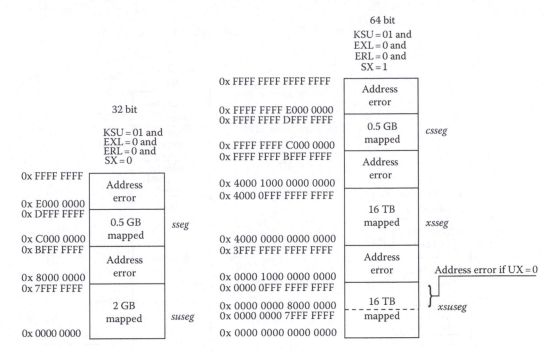

Figure 11.27 Supervisor mode address space.

A supervisor mode is designed for layered operating systems in which a true kernel runs in processor kernel mode and the rest of the operating system runs in supervisor mode (Figure 11.28). In this mode, when $SX = 0$ in the status register and the most significant bit of the 32-bit virtual address is set to 0, the suseg virtual address space is selected; it covers the full 231 bytes of the current user address space. The virtual address is extended with the contents of the 8-bit field to form a unique virtual address. When $SX = 0$ in the status register and the three most significant bits of the 32-bit virtual address are 110_2, the sseg virtual address space is selected; it covers 2^{29} bytes of the current supervisor address space (Figure 11.27).

The processor operates in kernel mode when the status register contains the kernel mode bit values as shown in Figure 11.25. The kernel mode virtual address space is divided into regions differentiated by the high-order bits of the virtual address. In this mode, when $KX = 0$, in the status register and the most significant bit of the virtual address, A31, is cleared, the 32-bit kuseg virtual address space is selected. It covers the full 2 GB of the current user address space. When $KX = 0$ in the status register and the most significant three bits of the virtual address are 100_2, 32-bit kseg() virtual address space is selected. References to kseg() are not mapped through the TLB. When $KX = 0$ in the status register and the most significant three bits of the 32-bit virtual address are 10_2, the ksseg virtual address space is selected; it is the current 512 MB supervisor virtual space. When $KX = 1$ in the status register and bits 63:62 of the 64-bit virtual address are 10_2, the xkphys virtual address space is selected; it is a set of eight kernel physical spaces. Each kernel physical space contains either one of four 2^{40}-byte physical pages. References to this space are not mapped; the physical address selected is taken directly from bits 39:0 of the virtual address. Bits 61:59 of the virtual address specify the cache algorithm. If the cache algorithm is either uncached or uncached accelerated, the space contains four physical pages; access to addresses whose bits 56:40 are not equal to 0 causes an address error exception.

Virtual to physical address translations are maintained by the operating system, using page tables in memory. A subset of these translations is loaded into a hardware buffer (TLB). The TLB contains 64 entries, each of which maps a pair of virtual pages. Formats of TLB entries are shown

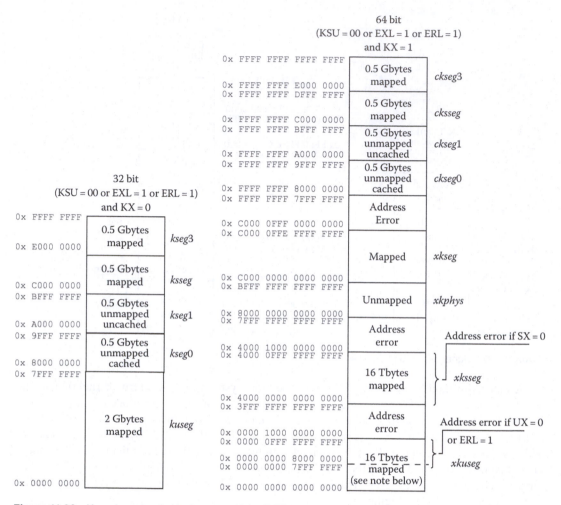

Figure 11.28 Kernel mode physical space. *Note:* If ERL = 1, the selected 2 GB space becomes uncached and unmapped.

in Figure 11.29. The cache algorithm fields of the TLB, entrylo0, and entrylo1, and Config registers indicate how data are cached.

Figure 11.29 shows the TLB entry formats for both 32- and 64-bit modes. Each field of an entry has corresponding field in the entryhi, Entrylo0, entrylo1, or PageMask registers.

Because a 64 bit is unnecessarily large, only the low 44 address bits are translated. The high tow virtual address bits (bits 63:62) select between user, supervisor, and kernel address spaces. The intermediate address bits (61:44) must either be all zeros or all ones depending on the address region. For data cache accesses, the joint TLB translates addresses from the address calculate unit. For instruction accesses, the JTLB translates the PC address if it misses in the instruction TLB. That entry is copied into the ITLB for subsequent accesses.

Each independent task or process has a separate address space, assigned a unique 8-bit address space identifier (ASID). This identifier is stored with each TLB entry to distinguish between entries loaded for different processes. The ASID allows the processor to move from one process to another (called a context switch) without having to invalidate TLB entries.

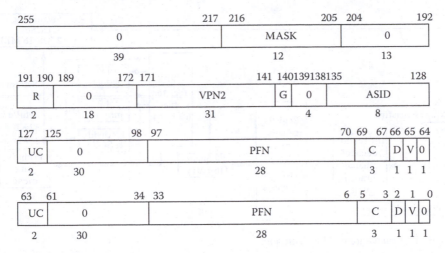

Figure 11.29 TLB entries.

11.4.2.7 *Floating-Point Units*

The R10000 contains two primary FP units. The adder unit handles add operations, and the multiply unit handles multiply operations. In addition, two secondary FP units exist, which handle long-latency operations such as divide and square root.

Addition, subtraction, and conversion instructions have a two-cycle latency and a one-cycle repeat rate and are handled within the adder unit. Instructions that convert integer values to single-precision FP values have a four-cycle latency as they must pass through the adder twice. The adder is busy during the second cycle after the instruction is issued.

All FP multiply operations execute with a two-cycle latency and a one-cycle repeat rate and are handled within the multiplier unit. The multiplier performs multiply operations. The FP divide and square-root units perform calculations using iterative algorithms. These units are not pipelined and cannot begin another operation until the current operation is completed. Thus, the repeat rate approximately equals the latency. The ports of the multiplier are shared with the divide and square-root units. A cycle is lost at the beginning of the operation (to fetch the operand) and at the end (to store the result).

The FP MADD operation, which occurs frequently, is computed using separate multiply and add operations. The MADD instruction has a four-cycle latency and a one-cycle repeat rate. The combined instruction improves performance by eliminating the fetching and decoding of an extra instruction. The divide and square-root units use separate circuitry and can be operated simultaneously. However, the FP queue cannot issue both instructions during the same cycle.

The FP add, multiply, divide, and square-root units read their operands and store their results in the FP register file. Values are loaded to or stored from the register file by the load/store and move units.

A logic diagram of FP operations is shown in Figure 11.30 in which data and instructions are read from the secondary cache into the primary caches and then into the processor. There they are decoded and appended to the FP queue and passed into the FP register file where each is dynamically issued to the appropriate functional unit. After execution in the functional unit, results are stored, through the register file, in the primary data cache.

The FP queue can issue one instruction to the adder unit and one instruction to the multiplier unit. The adder and multiplier each has two dedicated read ports and a dedicated write ports in the

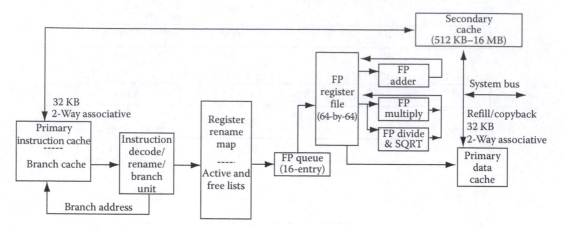

Figure 11.30 Logical diagram of FP operations.

FP register file. Because of their low repeat rates, the divide and square-root units do not have their own issue port. Instead, they decode instructions issued to the multiplier unit, using its operand registers and bypass logic. They appropriate a second cycle alter for storing their result.

When an instruction is issued, up to two operands are read from dedicate read ports in the FP register file. After the operation has been completed, the result can be written back into the register file using a dedicated write port. For the add and multiply units, this write occurs four cycles after its operands were read.

The control of FP execution is shared by the following units:

1. The FP queue determines operand dependencies and dynamically issues instructions to be the execution units. It also controls the destination registers and register bypass.
2. The execution units control the arithmetic operations and generate status.
3. The graduate unit saves the status until the instructions graduate, and then it updates the FP *status* register.

The FP unit is the hardware implementation of coprocessor 1 in the MIPS IV ISA. The MIPS IV ISA defines 32 logical floating-point general registers (FGRs), as shown in Figure 11.31. Each FGR is 64 bits wide and can hold either 32-bit single-precision or 64-bit double-precision values. The hardware actually contains 64 physical 64-bit registers in the FP register file, from which the 32 logical registers are taken.

An FP instruction uses a 5-bit logical number to select an individual FGR. These logical numbers are mapped to physical registers by the rename unit (in pipeline stage 2), before the FP unit executes them. Physical registers are selected using 6-bit addresses.

The FR bit (26) in the status register determines the number of logical FP registers. When FR = 1, FP load and stores operate as follows:

1. Single-precision operands are read from the low half of a register, leaving the upper half ignored. Single-precision results are written into the low half of the register. The high half of the result register is architecturally undefined; in the R10000 implementation, it is set to zero.
2. Double-precision arithmetic operations use the entire 64-bit contents of each operand or result register.

Because of register renaming, every new result is written into a temporary register, and conditional move instructions select between a new operand and the previous old value. The high half of the destination register of a single-precision conditional move instruction is undefined, even if no move occurs.

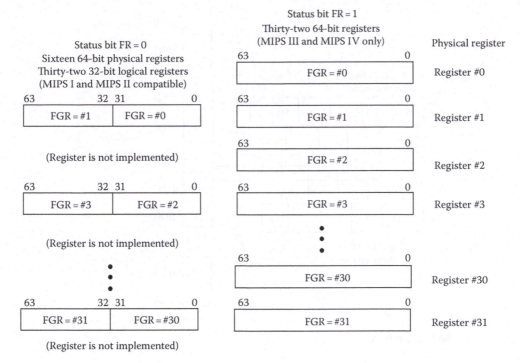

Figure 11.31   FP registers in 16-bit mode.

The load/store unit consists of the address queue, address calculation unit, TLB, address stack, store buffer, and primary data cache. The load/store unit performs load, store, prefetch, and cache instructions. All load or store instructions begin with a three-cycle sequence, which issues the instruction, calculates its virtual address, and translates the virtual address to physical. The address is translated only once during the operation. The data cache is accessed and the required data transfer is completed provided there was a primary data cache hit. If there is a cache miss or if the necessary shared register ports are busy, the data cache and data cache tag access must be repeated after the data are obtained from either the secondary cache or main memory.

The TLB contains 64 entries and translates virtual addresses to physical addresses. The virtual address can originate from either the adder calculation unit or the program counter (PC).

11.4.2.8 Interrupts

Figure 11.32 shows the interrupt structure of the R10000.

Several successive derivatives were developed by extending the R10000. These included R12000, R12000A, R14000, R14000A, R16000, and R16000A. The clock frequency of models after R12000 was kept as low as possible to maintain the power dissipation in the range of 15–20 W. This allowed them to be packaged in Silicon Graphics, Inc.'s (SGI) HPC systems.

The most recent derivative R16000A refers to the R16000 microprocessors that had clock rates higher than 700 MHz. The first version in this series was introduced in February 2004 and it operated at 800 MHz. Later, versions with 900 MHz and 1 GHz were introduced and shipped to customers. The users of R16000 included HP and SGI. SGI used the microprocessor in their workstations known as Fuel and Tezro and in their servers and supercomputers known as the Origin 3000. HP used the R16000A in their fault-tolerant servers named NonStop Himalaya S-Series.

In December 2012, Imagination Technologies bought the processing technology firm MIPS.

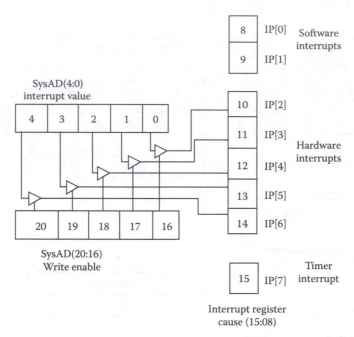

Figure 11.32   Interrupts in R10000.

11.4.3 Intel Corporation's Itanium

The Itanium processor (with 733–800 MHz speed) is the first in a family of 64-bit processors from Intel released in 2001, and the Itanium 2 processor (900 MHz–1 GHz) was launched in 2002. Figure 11.33 shows the block diagram of Itanium processor. The architectural components of the processor include

1. Functional unit
2. Cache (three levels: L1, L2, and L3)
3. Register stack engine (RSE)
4. Bus

The processor consists of four FP units, four integer units, four MMX units, and an IA-32 decode and control unit. Out of the four FP units, two work at 82-bit precision, while the other two are used for 32-bit precision operations. All four can perform fused multiply accumulate (FMAC) operations along with two-operand addition and subtraction, as well as FP and integer multiplication. They also execute several single-operand instructions, such as conversions between FP and integers, precision conversion, negation, and absolute value. Integer units are used for arithmetic and other integer or character manipulations and MMX units to accommodate instructions for multimedia operations. IA-32 decode and control unit provides compatibility with the other families.

Figure 11.34 shows the cache hierarchy. It consists of three caches: L1, L2, and L3. L1 has separate data (L1D) and ICache (L1I).

L1 instruction (L1I) cache is a dual-ported, 16 KB cache. One port is used for instruction fetch and the other is used for prefetches. It is a 4-way set-associative cache memory. It is fully pipelined and is physically indexed and tagged.

L1 data (L1D) cache is four ported, 16 KB in size. It supports two concurrent loads and two stores. It is used for caching only the integer data. It is physically indexed and tagged for loads and stores.

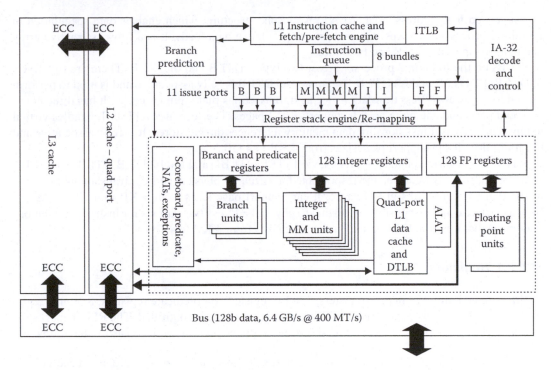

Figure 11.33 Intel Itanium architecture. (Reproduced from *Intel Itanium 2 Architecture, Hardware Developer's Manual*, July 2002, http://www.intel.com/design/itanium2/manuals/25110901.pdf. With permission.)

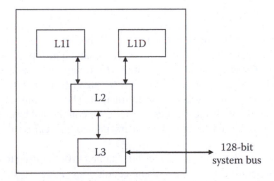

Figure 11.34 Intel Itanium cache. (Reproduced from *Intel Itanium 2 Architecture, Hardware Developer's Manual*, July 2002, http://www.intel.com/design/itanium2/manuals/25110901.pdf. With permission.)

L2 cache is 256 KB, four ported that supports up to four concurrent accesses. It is 8-way set-associative and is physically indexed and tagged. L2 cache is used to handle the L1I and L1D cache misses and accesses up to four FP data units. When data are requested, either one, two, three, or all the four ports are used, but when instructions are accessed in L2 cache, all the ports are used. If a cache miss occurs in L2 cache, the request is forwarded to L3 cache.

L3 cache is either 1.5 or 3 MB in size, fully pipelined, and single ported. It is 12-way set-associative and can support eight requests. It is physically indexed and tagged and handles all the requests caused by L2 cache miss.

Advanced load address table (ALAT) is a cache structure, which enables data speculation in Itanium 2 processor. It keeps information on speculative data loads. It is a fully associative array and handles up to two loads and two stores.

TLBs on the Itanium 2 processor are of two types: DTLB and the ITLB. There are two levels of DTLB: L1 DTLB and L2 DTLB. The first-level DTLB is fully associative and is used to perform virtual to physical address translations for load transactions that hit in L1 cache. It has three ports: two read ports and one write port. It supports 4 KB pages. The second-level DTLB handles virtual to physical address translations for data memory references during stores. It is fully associative and has four ports. It can support page sizes from 4 KB to 4 GB.

ITLB is categorized into two levels: level 1 ITLB (ITLB1) and level 2 ITLB (ITLB2). ITLB1 is dual ported, fully associative, and responsible for virtual to physical address translations to enable instruction truncation hits in L1I cache. It can support page sizes from 4 KB to 4 GB. ITLB2 is fully associative and responsible for virtual to physical address translations for instruction memory references that miss the ITLB1. It supports page sizes from 4 KB to 4 GB.

11.4.3.1 Registers

Itanium contains 128 integer registers, 128 FP registers, 64 predicate registers, 8 branch registers, and a 128-register "application store" where machine state is controlled. The RSE manages the data in the integer registers. Figure 11.35 shows the complete register set. It consists of the following:

1. General-purpose registers: These are a set of 128, 64-bit general-purpose registers, named gr0–gr127. These registers are partitioned into static general registers, which are 0 through 31 and stacked general registers that are 32 through 127. gr0 always reads 0 when sourced as an operand, and illegal operation fault occurs if an attempt to write to gr0 occurs.
2. FP registers: FP computation uses a set of 128, 82-bit floating registers, named fr0–fr127. These registers are divided into subsets: static FP registers that include fr0 through fr31 and rotating FP registers (fr32–fr127). fr0 always reads +0.0 when sourced; fr1 always reads +1.0 when sourced. A fault occurs if either of the fr0 or fr1 is used as the destination.
3. Predicate registers: These are a set of 64, 1-bit registers named pr0–pr63. They hold the results of compare instructions. These registers are partitioned into subsets: static predicate registers that include pr0–pr15 and rotating predicate registers that extend from pr16 to pr63. pr0 always reads 1 and the result is discarded if it is used as a destination. The rotating registers support software pipeline loops, while static predicate registers are used for conditional branching.
4. Branch registers: There are 8, 64-bit branch registers, named br0–br7 used to hold indirect branching information.
5. Kernel registers: There are 8, 64-bit kernel registers, named kr0–kr7 used to communicate information from the kernel (operating system) to an application.
6. Current frame marker (CFM): It describes the state of the general register stack. It is a 64-bit register used for stack-frame operations.
7. Instruction pointer (IP): It is a 64-bit pointer and holds pointer to current 16-byte aligned bundle in IA-64 mode or offset to 1-byte aligned instruction in IA-32 mode.
8. Performance monitor data registers (PMD): They are the data registers for performance monitor hardware.
9. User mask (UM): It is a set of single-bit values used to monitor FP register usage, performance monitor.
10. Processor identifiers (CPUID): They describe processor implementation-dependent features.
11. There are several other 64-bit registers with operating system-specific, hardware-specific, or application-specific uses covering hardware control and system configuration.

RSE is used to remove the latency (delay) caused by the necessary saving and restoring of data processing registers when entering and leaving a procedure. The stack provides for fast procedure calls by passing arguments in registers as opposed to the stack.

When a procedure is called, a new frame of registers is made available to the called procedure without the need for an explicit save of the caller's registers. The old registers remain in the large on-chip physical register file as long as there is enough physical capacity. When the number of registers needed overflows the available physical capacity, a state machine called the RSE saves the registers to memory to free up the necessary registers needed for the upcoming call. The RSE maintains the illusion of infinite number of registers.

On a call return, the base register is restored to the value that the caller was using to access registers prior to the call. Often a return is encountered even before these registers need to be saved, making it unnecessary to restore them. In cases where the RSE has saved some of the callee's registers, the processor stalls on the return until the RSE can restore the appropriate number of the callee's registers.

The bus in Itanium processor is 128 bits wide and operates at a clock frequency of 400 MHz, transferring up to 6.4 GB/s.

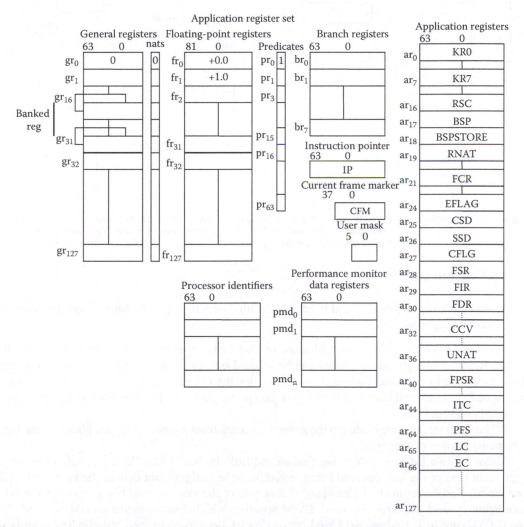

Figure 11.35 Itanium register set. (Reproduced from *Intel Itanium Architecture Software Developer's Manual*, Vol. 1: Application Architecture, October 2002, http://developer.intel.com/design/itanium/manuals/245317.pdf. With permission.)

(continued)

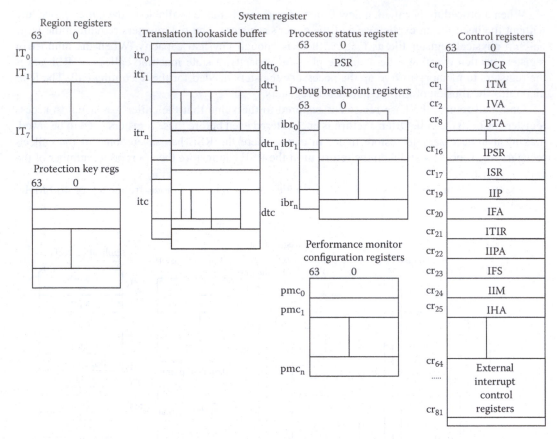

Figure 11.35 (continued)   Itanium register set. (Reproduced from *Intel Itanium Architecture Software Developer's Manual*, Vol. 1: Application Architecture, October 2002, http://developer.intel.com/design/itanium/manuals/245317.pdf. With permission.)

11.4.3.2 Memory

Memory is byte addressable and is accessed with 64-bit pointers. 32-Bit pointers are manipulated in 64-bit registers.

Addressable unit and alignment: Memory in IA-64 can be addressed in units of 1, 2, 4, 8, 10, and 16 bytes. Although data on IA-64 can be aligned on any boundary, IA-64 recommends that items be aligned on naturally aligned boundaries for the object size. For example, words should be aligned on word boundaries. There is one exception: 10-byte FP values should be aligned on 16-byte boundaries.

When the quantities are loaded to the general registers from memory, they are placed in the least significant portion of the register.

There are two Endian models: big Endian and little Endian. Little Endian results in the least significant byte of the load operand being stored in the least significant byte of the target operand, and the big Endian results in the most significant byte of the load operand being stored in the least significant byte of the target operand. IA-64 specifies which Endian model should be used. All IA-32 CPUs are little Endian. All IA-64 instruction fetches are performed little Endian regardless of current Endian mode. The instruction appears in the reverse order if the instruction data are read using big Endian.

Table 11.1   Types of Instructions

Instruction Type	Description
A	Integer ALU
I	Non-ALU integer
M	Memory
F	FP
B	Branch
L + X	Extended

Source: Reproduced from *Intel Itanium Architecture Software Developer's Manual*, Volume 1: Application Architecture, October 2002.

11.4.3.3 Instruction Package

There are six types of instructions shown in Table 11.1.

Three 41-bit instructions are grouped together into 128-bit sized and aligned containers called bundles. The "pointer" (0–4 bits) indicates the kinds of instructions that are packed. Itanium architecture allows issuing of independent instructions in these bundles for parallel execution. Out of the 32 possible kinds of packaging, 8 are not used, thus reducing to 24. In little-Endian format, a bundle appears in Figure 11.36.

The instruction from the bundle is executed in the order described in the following:

1. Ordering of bundles is done from lowest to highest memory address. Instructions present in the lower memory address precede the instructions that are in the higher memory addresses.
2. Instructions within a bundle are ordered from instructions 1–3.

11.4.3.4 Instruction Set Transition Model

There are two operating environments supported by Itanium architecture:

1. IA-32 system environment: supports IA-32, which is a 32-bit operating system
2. Itanium system environment: supports Itanium-based operating systems (IA-64)

The processor can execute either IA-32 or Itanium instructions at any time. Intel Architecture (IA-64) is compatible with the 32-bit software (IA-32). The software can be run in real mode (16 bits), protected mode (32 bits), and virtual mode 86 (16 bits). Thus, the CPU will be able to operate in both IA-64 and IA-32 modes. There are special instructions to go from one mode to the other, as is shown in Figure 11.37.

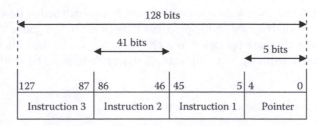

Figure 11.36   Instruction bundle.

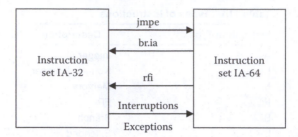

Figure 11.37 Model of instruction set transition. (Reproduced from *Intel Itanium Architecture, Software Developer's Manual*, December 2001, http://devresource.hp.com/drc/STK/docs/refs/24531703s.pdf. With permission.)

There are three instructions and interruptions that make the transition between the IA-32 and Itanium instruction sets (IA-64). They are

1. JMPE (IA-32 instruction): jumps to a 64-bit instruction and changes to IA-64 mode.
2. br.ia (IA-64 instruction): moves to a 32-bit instruction and changes to IA-32 mode.
3. rfi (IA-64 instruction): it is the return of the interruption; the return happens both to an IA-32 situation and to an IA-64, depending on the situation present at the moment when the interruption is invoked.

Itanium processor is based on explicit parallel instruction computing (EPIC) technology, which provides pipelining and is capable of doing parallel execution of up to six instructions. Some of the EPIC technology features, provided by Itanium processor, are as follows:

Predication: EPIC uses predicated execution of branches to reduce the branch penalty. It can be illustrated by an example—consider the C source code:

$$\text{if}(x == 7)z = 2 \quad \text{else} \quad z = 7.$$

The instruction flow would generally be written as

1. Compare x to 7
2. If not equal goto line 5
3. $z = 2$
4. goto line 6
5. $z = 7$
6. //Program continues from here

In the code, line 2 or 4 causes at least one break (goto) in the instruction flow irrespective of the value of x. This causes interruption in the program flow twice in one pass. Itanium architecture assigns the result of the compare operation to a predicate bit, which then allows or disallows the output to be committed to memory. Thus, the same C source code could be written using this IA-64 instruction flow:

Compare x to 7 and store result in a predicate bit (let it be "P")
If $P == 1, z = 2$.
If $P == 0, z = 7$.

In this code, if the value of the predicate bit (P) is equal to 1 (which indicates the $x == 7$), z is assigned 2 otherwise 7. Once the P is set, it can be tested again and again in subsequent code.

Speculation: Along with predication, EPIC supports control and data speculation to handle branches.

Control speculation: Control speculation is the execution of an operation before the branch that guards it. Consider the code sequence:

$$if(ab)\text{load}(\text{Id\_addr1}, \text{target1})$$

$$else \text{ load } (\text{Id\_addr2}, \text{target2}).$$

If the operation load (ld _ addr1, target1) were to be performed prior to the determination of ($a > b$), then the operation would be control speculative with respect to the controlling condition ($a > b$). Under normal execution, the operation load (ld _ addr1, target1) may or may not execute. If the new control speculative load causes an exception, then the exception should only be serviced if ($a > b$) is true. When the compiler uses control speculation, it leaves a check operation at the original location. The check verifies whether an exception has occurred and if so it branches to recovery code.

Data speculation: This is also known as "advance loading." It is the execution of a memory load prior to a store that preceded it. Consider the code sequence in the following:

$$\text{store}(\text{st\_addr, data})$$

$$\text{load}(\text{Id\_addr, target})$$

$$\text{use(target)}.$$

In this example, if ld _ addr and st _ addr cannot be disambiguated (the process of determining at compile time the relationship between memory addresses) and if load were to be performed prior to the store, then the load would be data speculative with respect to the store. If memory addresses overlap during the execution, a data-speculative load issued before the store might return a different value than a regular load issued after the store. When the compiler data speculates a load, it leaves a check instruction at the original location of the load. The check verifies whether an overlap has occurred and if so it branches to recovery code.

Register rotation: EPIC supports register renaming by using register rotation technique, that is, the name of the register is generated dynamically. The rotating register base (RRB) is present in a rotating register file, which is added with register number given in an instruction and modulo the number of registers in rotating register file. This generated number is actually used as register address.

11.4.4 Other Intel Processors

In this section, we discuss the following Intel microarchitectures in detail, namely, NetBurst, Core, Atom, Nehalem, Sandy Bridge, and Ivy Bridge. We also explain the terms ring interconnect, last level cache (LLC), and data prefetching and provide branch-prediction examples for a few of these microarchitectures. The contents of this section is extracted from *Intel® 64 and IA-32 Architectures Optimization Reference Manual* Volume A and *Intel® 64 and IA-32 Architectures Software Developer's Manual* Volume 1.

11.4.4.1 Intel NetBurst Microarchitecture

The Intel NetBurst microarchitecture provides the following:

- The rapid execution engine.
- ALUs run at twice the processor frequency.
- Basic integer operations can dispatch in 1/2 processor clock tick.
- Hyper-pipelined technology.
- Deep pipeline to enable industry-leading clock rates for desktop PCs and servers.
- Frequency headroom and scalability to continue leadership into the future.
- Advanced dynamic execution.
- Deep, out-of-order, speculative execution engine:
  - Up to 126 instructions in flight.
  - Up to 48 loads and 24 stores in pipeline.*
- Enhanced branch-prediction capability (refer to Section 11.4.8.3):
  - Reduces the misprediction penalty associated with deeper pipelines.
  - Advanced branch-prediction algorithm.
  - 4K-entry branch-target array.
- New cache subsystem.
- First-level caches:
  - Advanced Execution Trace Cache stores decoded instructions.
  - Execution Trace Cache removes decoder latency from main execution loops.
  - Execution Trace Cache integrates path of program execution flow into a single line.
  - Low-latency data cache.
  - Second-level cache.
  - Full-speed, unified 8-way level 2 on-die Advanced Transfer Cache.
  - Bandwidth and performance increases with processor frequency.
- High-performance, quad-pumped bus interface to the Intel NetBurst microarchitecture system bus:
  - Supports quad-pumped, scalable bus clock to achieve up to 4X effective speed.
  - Capable of delivering up to 8.5 GB of bandwidth per second.
- Superscalar issue to enable parallelism.
- Expanded hardware registers with renaming to avoid register name space limitations.
- 64-byte cache line size (transfers data up to two lines per sector).

Figure 11.38 is an overview of the Intel NetBurst microarchitecture.

This microarchitecture pipeline is made up of three sections: (1) the front-end pipeline, (2) the out-of-order execution core, and (3) the retirement unit.

11.4.4.1.1 Front-End Pipeline

The front end supplies instructions in program order to the out-of-order execution core. It performs a number of functions:

- Prefetches instructions that are likely to be executed
- Fetches instructions that have not already been prefetched
- Decodes instructions into μops
- Generates microcode for complex instructions and special-purpose code
- Delivers decoded instructions from the execution trace cache
- Predicts branches using highly advanced algorithm

---

* Intel 64 and IA-32 processors based on the Intel NetBurst microarchitecture at 90 nm process can handle more than 24 stores in flight.

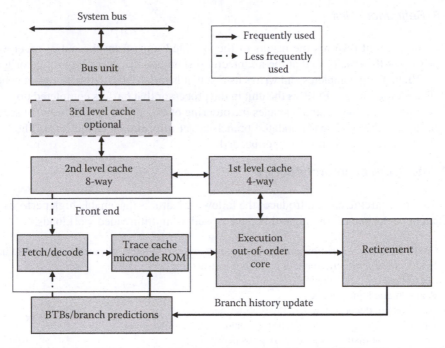

Figure 11.38 The Intel NetBurst microarchitecture. (From *Intel® 64 and IA-32 Architectures Optimization Reference Manual Volume A and Intel® 64.*)

The pipeline is designed to address common problems in high-speed, pipelined microprocessors. Two of these problems contribute to major sources of delays:

- Time to decode instructions fetched from the target
- Wasted decode bandwidth due to branches or branch target in the middle of cache lines

The operation of the pipeline's trace cache addresses these issues. Instructions are constantly being fetched and decoded by the translation engine (part of the fetch/decode logic) and built into sequences of μops called traces. At any time, multiple traces (representing prefetched branches) are being stored in the trace cache. The trace cache is searched for the instruction that follows the active branch. If the instruction also appears as the first instruction in a prefetched branch, the fetch and decode of instructions from the memory hierarchy cease and the prefetched branch becomes the new source of instructions (see Figure 11.38).

The trace cache and the translation engine have cooperating branch-prediction hardware. Branch targets are predicted based on their linear addresses using branch target buffers (BTBs) and fetched as soon as possible.

11.4.4.1.2 Out-of-Order Execution Core

The out-of-order execution core's ability to execute instructions out of order is a key factor in enabling parallelism. This feature enables the processor to reorder instructions so that if one μop is delayed, other μops may proceed around it. The processor employs several buffers to smooth the flow of μops. The core is designed to facilitate parallel execution. It can dispatch up to six μops per cycle (this exceeds trace cache and retirement μop bandwidth). Most pipelines can start executing a new μop every cycle, so several instructions can be in flight at a time for each pipeline. A number of ALU instructions can start at two per cycle; many FP instructions can start once every two cycles.

11.4.4.1.3 Retirement Unit

The retirement unit receives the results of the executed μops from the out-of-order execution core and processes the results so that the architectural state updates according to the original program order. When a μop completes and writes its result, it is retired. Up to three μops may be retired per cycle. The reorder buffer (ROB) is the unit in the processor that buffers completed μops, updates the architectural state in order, and manages the ordering of exceptions. The retirement section also keeps track of branches and sends updated branch-target information to the BTB. The BTB then purges prefetched traces that are no longer needed.

11.4.4.2 Intel Core Microarchitecture

Intel Core microarchitecture introduces the following features that enable high performance and power-efficient performance for single-threaded as well as multithreaded workloads:

- *Intel Wide Dynamic Execution* enables each processor core to fetch, dispatch, and execute in high bandwidths to support retirement of up to four instructions per cycle.
- Fourteen-stage efficient pipeline.
- Three arithmetic logical units.
- Four decoders to decode up to five instructions per cycle.
- Macrofusion and microfusion to improve front-end throughput.
- Peak issue rate of dispatching up to six μops per cycle.
- Peak retirement bandwidth of up to 4 μops per cycle.
- Advanced branch prediction (refer to Section 11.4.8.2).
- Stack pointer tracker to improve efficiency of executing function/procedure entries and exits.
- *Intel Advanced Smart Cache* delivers higher bandwidth from the second-level cache to the core and optimal performance and flexibility for single-threaded and multithreaded applications.
- Large second-level cache up to 4 MB and 16-way associativity.
- Optimized for multicore and single-threaded execution environments.
- 256-bit internal data path to improve bandwidth from L2 to first-level data cache.
- *Intel Smart Memory* Access prefetches data from memory in response to data access patterns and reduces cache-miss exposure of out-of-order execution.
- Hardware prefetchers to reduce effective latency of second-level cache misses.
- Hardware prefetchers to reduce effective latency of first-level data cache misses.
- Memory disambiguation to improve efficiency of speculative execution engine.
- *Intel Advanced Digital Media Boost* improves most 128-bit SIMD instruction with single-cycle throughput and FP operations.
- Single-cycle throughput of most 128-bit SIMD instructions.
- Up to eight FP operations per cycle.
- Three issue ports available to dispatching SIMD instructions for execution.

Intel Core 2 Extreme, Intel Core 2 Duo processors, and Intel Xeon processor 5100 series implement two processor cores based on the Intel Core microarchitecture; the functionality of the subsystems in each core are depicted in Figure 11.39.

11.4.4.2.1 Front End

The front end of Intel Core microarchitecture provides several enhancements to feed the Intel Wide Dynamic Execution engine:

- Instruction fetch unit prefetches instructions into an instruction queue to maintain steady supply of instruction to the decode units.
- Four-wide decode unit can decode four instructions per cycle or five instructions per cycle with macrofusion.

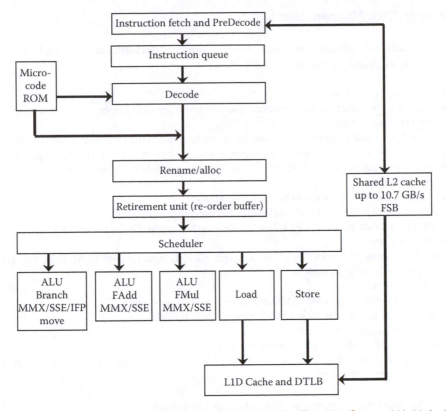

Figure 11.39 The Intel Core microarchitecture pipeline functionality. (From *Intel® 64 and IA-32 Architectures Optimization Reference Manual Volume A and Intel® 64.*)

- Macrofusion fuses common sequence of two instructions as one decoded instruction (µops) to increase decoding throughput.
- Microfusion fuses common sequence of two µops as one µop to improve retirement throughput.
- Instruction queue provides caching of short loops to improve efficiency.
- Stack pointer tracker improves efficiency of executing procedure/function entries and exits.
- Branch-prediction unit (BPU) employs dedicated hardware to handle different types of branches for improved branch prediction.
- Advanced branch-prediction algorithm directs instruction fetch unit to fetch instructions likely in the architectural code path for decoding.

11.4.4.2.2 Execution Core

The execution core of the Intel Core microarchitecture is superscalar and can process instructions out of order to increase the overall rate of instructions executed per cycle (IPC). The execution core employs the following feature to improve execution throughput and efficiency:

- Up to six µops can be dispatched to execute per cycle.
- Up to four instructions can be retired per cycle.
- Three full arithmetic logical units.
- SIMD instructions can be dispatched through three issue ports.
- Most SIMD instructions have 1-cycle throughput (including 128-bit SIMD instructions).
- Up to eight FP operations per cycle.
- Many long-latency computation operations are pipelined in hardware to increase overall throughput.
- Reduced exposure to data access delays using Intel Smart Memory Access.

11.4.4.3 Intel Atom Microarchitecture

Intel Atom microarchitecture maximizes power-efficient performance for single-threaded and multithreaded workloads by providing the following:

- Advanced µops execution
- Single-µop instruction execution from decode to retirement, including instructions with register-only, load, and store semantics
- Sixteen-stage, in-order pipeline optimized for throughput and reduced power consumption
- Dual pipelines to enable decode, issue, execution, and retirement of two instructions per cycle
- Advanced stack pointer to improve efficiency of executing function entry/returns
- Intel Smart Cache
- Second-level cache is 512 KB and 8-way associativity
- Optimized for multithreaded and single-threaded execution environments
- 256-bit internal data path between L2 and L1 data cache improves high bandwidth
- Efficient memory access
- Efficient hardware prefetchers to L1 and L2, speculatively loading data likely to be requested by processor to reduce cache-miss impact
- Intel Digital Media Boost
- Two issue ports for dispatching SIMD instructions to execution units
- Single-cycle throughput for most 128-bit integer SIMD instructions
- Up to six FP operations per cycle
- Up to two 128-bit SIMD integer operations per cycle
- Safe instruction recognition (SIR) to allow long-latency FP operations to retire out of order with respect to integer instructions

11.4.4.4 Intel Microarchitecture Code Name Nehalem

Intel microarchitecture code name Nehalem provides the foundation for many innovative features of Intel Core i7 processors. It builds on the success of 45 nm Intel Core microarchitecture and provides the following feature enhancements:

- Enhanced processor core
  - Improved branch prediction and recovery from misprediction
  - Enhanced loop streaming to improve front-end performance and reduce power consumption
  - Deeper buffering in out-of-order engine to extract parallelism
  - Enhanced execution units to provide acceleration in CRC, string/text processing, and data shuffling
- Smart memory access
  - Integrated memory controller provides low-latency access to system memory and scalable memory bandwidth
  - New cache hierarchy organization with shared, inclusive L3 to reduce snoop traffic
  - Two level TLBs and increased TLB size
  - Fast unaligned memory access
- Hyperthreading (HT) technology
  - Provides two hardware threads (logical processors) per core
  - Takes advantage of 4-wide execution engine, large L3, and massive memory bandwidth
- Dedicated power management innovations
  - Integrated microcontroller with optimized embedded firmware to manage power consumption
  - Embedded real-time sensors for temperature, current, and power
  - Integrated power gate to turn off/on per core power consumption
  - Versatility to reduce power consumption of memory, link subsystems

11.4.5 Intel Microarchitecture Code Name Sandy Bridge

Intel microarchitecture code name Sandy Bridge builds on the successes of Intel Core microarchitecture and Intel microarchitecture code name Nehalem. It offers the following innovative features:

- Intel Advanced Vector Extensions (Intel AVX).
- 256-bit FP instruction set extensions to the 128-bit Intel Streaming SIMD Extensions (SSE), providing up to 2X performance benefits relative to 128-bit code.
- Nondestructive destination encoding offers more flexible coding techniques.
- Supports flexible migration and coexistence between 256-bit AVX code, 128-bit AVX code, and legacy 128-bit SSE code.
- Enhanced front-end and execution engine.
- New Decoded ICache component that improves front-end bandwidth and reduces branch misprediction penalty.
- Advanced branch prediction.
- Additional macrofusion support.
- Larger dynamic execution window.
- Multiprecision integer arithmetic enhancements (ADC/SBB, MUL/IMUL).
- LEA bandwidth improvement.
- Reduction of general execution stalls (read ports, write-back conflicts, bypass latency, partial stalls).
- Fast FP exception handling.
- XSAVE/XRSTORE performance improvements and XSAVEOPT new instruction.
- Cache hierarchy improvements for wider data path.
- Doubling of bandwidth enabled by two symmetric ports for memory operation.
- Simultaneous handling of more in-flight loads and stores enabled by increased buffers.
- Internal bandwidth of two loads and one store each cycle.
- Improved prefetching.
- High-bandwidth low-latency LLC architecture.
- High-bandwidth ring architecture of on-die interconnect.
- System-on-a-chip support.
- Integrated graphics and media engine in second-generation Intel Core processors.
- Integrated PCIE controller.
- Integrated memory controller.
- Next-generation Intel Turbo Boost Technology.
- Leverage TDP headroom to boost performance of CPU cores and integrated graphic unit (GT).

11.4.5.1 Intel Microarchitecture Code Name Sandy Bridge Pipeline Overview

Figure 11.40 depicts the pipeline and major components of a processor core that is based on Intel microarchitecture code name Sandy Bridge. The pipeline consists of the following:

- An in-order issue front end that fetches instructions and decodes them into μops. The front end feeds the next pipeline stages with a continuous stream of μops from the most likely path that the program will execute.
- An out-of-order, superscalar execution engine that dispatches up to six μops to execution, per cycle. The allocate/rename block reorders μops to "dataflow" order so they can execute as soon as their sources are ready and execution resources are available.
- An in-order retirement unit that ensures that the results of execution of the μops, including any exceptions they may have encountered, are visible according to the original program order.

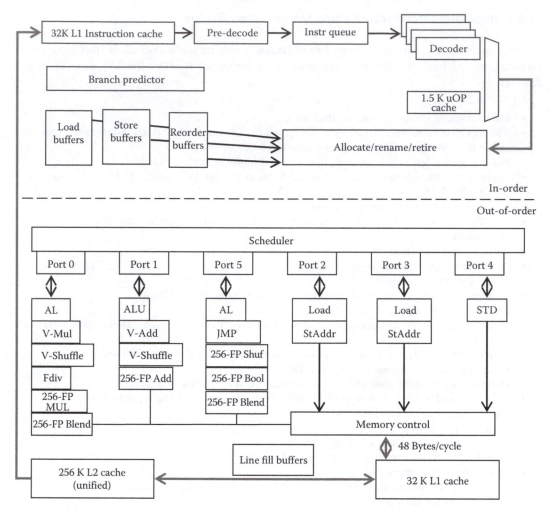

Figure 11.40 Intel microarchitecture code name Sandy Bridge pipeline functionality. (From *Intel® 64 and IA-32 Architectures Optimization Reference Manual Volume A and Intel® 64*.)

The flow of an instruction in the pipeline can be summarized in the following progression:

1. The BPU (refer to Section 11.4.8.1) chooses the next block of code to execute from the program. The processor searches for the code in the following resources, in this order:
    a. Decoded ICache
    b. ICache, via activating the legacy decode pipeline
    c. L2 cache, LLC, and memory, as necessary
2. The μops corresponding to this code are sent to the rename/retirement block. They enter into the scheduler in program order but execute and are deallocated from the scheduler according to data-flow order. For simultaneously ready μops, FIFO ordering is nearly always maintained.

    The μop execution is executed using execution resources arranged in three stacks. The execution units in each stack are associated with the data type of the instruction.

    Branch mispredictions are signaled at branch execution. It resteers the front end that delivers μops from the correct path. The processor can overlap work preceding the branch misprediction with work from the following corrected path.

3. Memory operations are managed and reordered to achieve parallelism and maximum performance. Misses to the L1 data cache go to the L2 cache. The data cache is nonblocking and can handle multiple simultaneous misses.

4. Exceptions (faults, traps) are signaled at retirement (or attempted retirement) of the faulting instruction.

Each processor core based on Intel microarchitecture code name Sandy Bridge can support two logical processors if Intel HT Technology is enabled.

11.4.5.2 Front End

This section describes the key characteristics of the front end. Table 11.2 lists the components of the front end, their functions, and the problems they address.

11.4.5.2.1 Legacy Decode Pipeline

The legacy decode pipeline comprises the ITLB, the ICache, the instruction predecode, and the instruction decode units.

ICache and ITLB: An instruction fetch is a 16-byte aligned lookup through the ITLB and into the ICache. The ICache can deliver every cycle 16 bytes to the instruction predecoder. Table 11.3 compares the ICache and ITLB with prior generation.

Upon ITLB miss, there is a lookup to the second-level TLB (STLB) that is common to the DTLB and the ITLB. The penalty of an ITLB miss and a STLB hit is seven cycles.

Instruction predecode: The predecode unit accepts the 16 bytes from the ICache and determines the length of the instructions.

The following length changing prefixes (LCPs) imply instruction length that is different from the default length of instructions. Therefore, they cause an additional penalty of three cycles per

Table 11.2 Components of the Front End of Intel Microarchitecture Code Name Sandy Bridge

Component	Functions	Performance Challenges
ICache	32 KB backing store of instruction bytes.	Fast access to hot code instruction bytes.
Legacy decode pipeline	Decode instructions to µops, delivered to the µop queue and decoded ICache.	Provides the same decode latency and bandwidth as prior Intel processors. Decoded ICache warm-up.
Decoded ICache	Provide stream of µops to the µop queue.	Provides higher µop bandwidth at lower latency and lower power than the legacy decode pipeline.
MSROM	Complex instruction µop flow store, accessible from both legacy decode pipeline and Decoded ICache.	
BPU	Determine next block of code to be executed and drive lookup of Decoded ICache and legacy decode pipelines.	Improves performance and energy efficiency through reduced branch mispredictions.
µop queue	Queues µops from the Decoded ICache and the legacy decode pipeline.	Hide front-end bubbles; provides execution µops at a constant rate.

Table 11.3 ICache and ITLB of Intel Microarchitecture Code Name Sandy Bridge

Component	Intel Microarchitecture Code Name Sandy Bridge	Intel Microarchitecture Code Name Nehalem
ICache size	32 byte	32 byte
ICache ways	8	4
ITLB 4K page entries	128	128
ITLB large page (2M or 4M) entries	8	7

LCP during length decoding. The previous processors incur a six-cycle penalty for each 16-byte chunk that has one or more LCPs in it. Since usually there is no more than one LCP in a 16-byte chunk, in most cases, Intel microarchitecture code name Sandy Bridge introduces an improvement over previous processors:

- Operand-size override (66H) preceding an instruction with a word/double immediate data. This prefix might appear when the code uses 16-bit data types, unicode processing, and image processing.
- Address-size override (67H) preceding an instruction with a modr/m in real, big real, 16-bit protected or 32-bit protected modes. This prefix may appear in boot code sequences.
- The REX prefix (4xh) in the Intel 64 instruction set can change the size of two classes of instructions: MOV offset and MOV immediate. Despite this capability, it does not cause an LCP penalty and hence is not considered an LCP.

Instruction decode: There are four decoding units that decode instruction into μops. The first can decode all IA-32 and Intel 64 instructions up to four μops in size. The remaining three decoding units handle single-μop instructions. All four decoding units support the common cases of single μop flows including microfusion and macrofusion.

The μops emitted by the decoders are directed to the μop queue and to the Decoded ICache. Instructions longer than four μops generate their μops from the MSROM. The MSROM bandwidth is four μops per cycle. Instructions whose μops come from the MSROM can start either from the legacy decode pipeline or from the Decoded ICache.

Microfusion: Microfusion fuses multiple μops from the same instruction into a single complex μop. The complex μop is dispatched in the out-of-order execution core as many times as it would if it were not microfused.

Microfusion enables you to use memory-to-register operations, also known as the complex instruction set computer (CISC) instruction set, to express the actual program operation without worrying about a loss of decode bandwidth.

Microfusion improves instruction bandwidth delivered from decode to retirement and saves power. Coding an instruction sequence by using single-μop instructions will increase the code size, which can decrease fetch bandwidth from the legacy pipeline.

The following are examples of microfused μops that can be handled by all decoders:

- All stores to memory, including store immediate. Stores execute internally as two separate functions, store address and store data.
- All instructions that combine load and computation operations (load+op), for example,
  - ADDPS XMM9, OWORD PTR [RSP+40]
  - FADD DOUBLE PTR [RDI+RSI*8]
  - XOR RAX, QWORD PTR [RBP+32]
- All instructions of the form "load and jump," for example,
  - JMP [RDI+200]
  - RET
- CMP and TEST with immediate operand and memory.
- An instruction with RIP-relative addressing is not microfused in the following cases.

- An additional immediate is needed, for example,
  - CMP [RIP+400], 27
  - MOV [RIP+3000], 142
- The instruction is a control flow instruction with an indirect target specified using RIP-relative addressing, for example,
  - JMP [RIP+5000000]

In these cases, an instruction that cannot be microfused will require decoder 0 to issue two μops, resulting in a slight loss of decode bandwidth.

In a 64-bit code, the usage of RIP-relative addressing is common for global data. Since there is no microfusion in these cases, performance may be reduced when porting a 32-bit code to 64-bit code.

Macrofusion: Macrofusion merges two instructions into a single μop. In Intel Core microarchitecture, this hardware optimization is limited to conditions specific to the first and second of the macrofusable instruction pair:

- The first instruction of the macrofused pair modifies the flags. The following instructions can be macrofused:
- In Intel microarchitecture code name Nehalem: CMP, TEST
- In Intel microarchitecture code name Sandy Bridge: CMP, TEST, ADD, SUB, AND, INC, DEC
- These instructions can fuse if
  - The first source/destination operand is a register
  - The second source operand (if exists) is one of the immediate, register, or non-RIP-relative memories
- The second instruction of the macrofusable pair is a conditional branch.

Macrofusion does not happen if the first instruction ends on byte 63 of a cache line, and the second instruction is a conditional branch that starts at byte 0 of the next cache line.

Since these pairs are common in many types of applications, macrofusion improves performance even on nonrecompiled binaries.

Each macrofused instruction executes with a single dispatch. This reduces latency and frees execution resources. You also gain increased rename and retire bandwidth, increased virtual storage, and power savings from representing more work in fewer bits.

11.4.5.2.2 Decoded ICache

The Decoded ICache is essentially an accelerator of the legacy decode pipeline. By storing decoded instructions, the Decoded ICache enables the following features:

- Reduced latency on branch mispredictions
- Increased μop delivery bandwidth to the out-of-order engine
- Reduced front-end power consumption

The Decoded ICache caches the output of the instruction decoder. The next time the μops are consumed for execution, the decoded μops are taken from the Decoded ICache. This enables skipping the fetch and decode stages for these μops and reduces power and latency of the front end. The Decoded ICache provides average hit rates of above 80% of the μops; furthermore, "hot spots" typically have hit rates close to 100%.

Typical integer programs average less than four bytes per instruction, and the front end is able to race ahead of the back end, filling in a large window for the scheduler to find instruction-level parallelism. However, for a high-performance code with a basic block consisting of many instructions, for example, Intel SSE media algorithms or excessively unrolled loops, the 16 instruction bytes per cycle is occasionally a limitation. The 32-byte orientation of the Decoded ICache helps such code to avoid this limitation.

The Decoded ICache automatically improves performance of programs with temporal and spatial locality. However, to fully utilize the Decoded ICache potential, you might need to understand its internal organization.

The Decoded ICache consists of 32 sets. Each set contains eight ways. Each way can hold up to six μops. The Decoded ICache can ideally hold up to 1536 μops.

The following are some of the rules how the Decoded ICache is filled with μops:

- All μops in a way represent instructions that are statically contiguous in the code and have their EIPs within the same aligned 32-byte region.
- Up to three ways may be dedicated to the same 32-byte aligned chunk, allowing a total of 18 μops to be cached per 32-byte region of the original IA program.
- A multi-μop instruction cannot be split across ways.
- Up to two branches are allowed per way.
- An instruction that turns on the MSROM consumes an entire way.
- A nonconditional branch is the last μop in a way.
- Microfused μops (load+op and stores) are kept as one μop.
- A pair of macrofused instructions is kept as one μop.
- Instructions with 64-bit immediate require two slots to hold the immediate.

When μops cannot be stored in the Decoded ICache due to these restrictions, they are delivered from the legacy decode pipeline. Once μops are delivered from the legacy pipeline, fetching μops from the Decoded ICache can resume only after the next branch μop. Frequent switches can incur a penalty.

The Decoded ICache is virtually included in the ICache and ITLB. That is, any instruction with μops in the Decoded ICache has its original instruction bytes present in the ICache. ICache evictions must also be evicted from the Decoded ICache, which evicts only the necessary lines.

There are cases where the entire Decoded ICache is flushed. One reason for this can be an ITLB entry eviction. Other reasons are not usually visible to the application programmer, as they occur when important controls are changed, for example, mapping in CR3 or feature and mode enabling in CR0 and CR4. There are also cases where the Decoded ICache is disabled, for instance, when the CS base address is not set to zero.

11.4.5.2.3 Microoperation Queue and the Loop Stream Detector

The μop queue decouples the front end and the out-of-order engine. It stays between the μop generation and the renamer as shown in Figure 11.40. This queue helps to hide bubbles that are introduced between the various sources of μops in the front end and ensures that four μops are delivered for execution, each cycle.

The μop queue provides postdecode functionality for certain instruction types. In particular, loads combined with computational operations and all stores, when used with indexed addressing, are represented as a single μop in the decoder or Decoded ICache. In the μop queue, they are fragmented into two μops through a process called unlamination: one does the load and the other does the operation. A typical example is the following "load plus operation" instruction:

$$ADD \quad RAX, [RBP + RSI]; rax := rax + LD(RBP + RSI)$$

Similarly, the following store instruction has three register sources and is broken into "generate store address" and "generate store data" subcomponents:

$$MOV \quad [ESP + ECX * 4 + 12345678], AL$$

The additional μops generated by unlamination use the rename and retirement bandwidth. However, it has an overall power benefit. For code that is dominated by indexed addressing (as often happens with array processing), recoding algorithms to use base (or base+displacement) addressing can sometimes improve performance by keeping the load plus operation and store instructions fused.

Loop stream detector (LSD): The LSD was introduced in Intel Core microarchitectures. The LSD detects small loops that fit in the μop queue and locks them down. The loop streams from the μop queue, with no more fetching, decoding, or reading μops from any of the caches, until a branch miss-prediction inevitably ends it.

The loops with the following attributes qualify for LSD/μop queue replay:

- Up to eight chunk fetches of 32-instruction bytes.
- Up to 28 μops (~28 instructions).
- All μops are also resident in the Decoded ICache.
- Can contain no more than eight taken branches. and none of them can be a CALL or RET.
- Cannot have mismatched stack operations, for example, more PUSH than POP instructions.

Many calculation-intensive loops, searches, and software string moves match these characteristics. Use the loop cache functionality opportunistically. For high-performance code, loop unrolling is generally preferable for performance even when it overflows the LSD capability.

11.4.5.3 Out-of-Order Engine

The out-of-order engine provides improved performance over prior generations with excellent power characteristics. It detects dependency chains and sends them to execution out-of-order while maintaining the correct dataflow. When a dependency chain is waiting for a resource, such as a second-level data cache line, it sends μops from another chain to the execution core. This increases the overall rate of IPC.

The out-of-order engine consists of two blocks, shown in Figure 2.1: core functional diagram, the rename/retirement block, and the scheduler.

The out-of-order engine contains the following major components:

- *Renamer.* The renamer component moves μops from the front end to the execution core. It eliminates false dependencies among μops, thereby enabling out-of-order execution of μops.
- *Scheduler.* The scheduler component queues μops until all source operands are ready. Schedules and dispatches ready μops to the available execution units in as close to an FIFO order as possible.
- *Retirement.* The retirement component retires instructions and μops in order and handles faults and exceptions.

Renamer: The renamer is the bridge between the in-order part in Figure 2.1 and the dataflow world of the scheduler. It moves up to four μops every cycle from the μop queue to the out-of-order engine. Although the renamer can send up to 4 μops (unfused, microfused, or macrofused) per cycle, this is equivalent to the issue port that can dispatch six μops per cycle. In this process, the out-of-order core carries out the following steps:

- Renames architectural sources and destinations of the μops to microarchitectural sources and destinations
- Allocates resources to the μops, for example, load or store buffers
- Binds the μop to an appropriate dispatch port

Some μops can execute to completion during rename and are removed from the pipeline at that point, effectively costing no execution bandwidth. These include

- Zero idioms (dependency breaking idioms)
- NOP
- VZEROUPPER
- FXCHG

The renamer can allocate two branches each cycle, compared to one branch each cycle in the previous microarchitecture. This can eliminate some bubbles in execution.

Microfused load and store operations that use an index register are decomposed to two μops; hence consume two out of the four slots the renamer can use every cycle.

Dependency breaking idioms: Instruction parallelism can be improved by using common instructions to clear register contents to zero. The renamer can detect them on the zero evaluation of the destination register.

Use one of these dependency breaking idioms to clear a register when possible:

- XOR REG,REG
- SUB REG,REG
- PXOR/VPXOR XMMREG,XMMREG
- PSUBB/W/D/Q XMMREG,XMMREG
- VPSUBB/W/D/Q XMMREG,XMMREG
- XORPS/PD XMMREG,XMMREG
- VXORPS/PD YMMREG, YMMREG

Since zero idioms are detected and removed by the renamer, they have no execution latency. There is another dependency breaking idiom—the "ones idiom":

- CMPEQ   XMM1, XMM1; "ones idiom" set all elements to all "ones"

In this case, the μop must execute, however, since it is known that regardless of the input data, the output data are always "all ones"; the μop dependency upon its sources does not exist as with the zero idiom, and it can execute as soon as it finds a free execution port.

Scheduler: The scheduler controls the dispatch of μops onto their execution ports. In order to do this, it must identify which μops are ready and where its sources come from: a register file entry or a bypass directly from an execution unit. Depending on the availability of dispatch ports and write-back buses and the priority of ready μops, the scheduler selects which μops are dispatched every cycle.

11.4.5.4 Execution Core

The execution core is superscalar and can process instructions out of order. The execution core optimizes overall performance by handling the most common operations efficiently while minimizing potential delays.

The out-of-order execution core improves execution unit organization over prior generation in the following ways:

- Reduction in read port stalls
- Reduction in write-back conflicts and delays
- Reduction in power
- Reduction of SIMD FP assists dealing with denormal inputs and underflowed outputs

Some high-precision FP algorithms need to operate with FTZ = 0 and DAZ = 0, that is, permitting underflowed intermediate results and denormal inputs to achieve higher numeric precision at the expense of reduced performance on prior generation microarchitectures due to SIMD FP assists. The reduction of SIMD FP assists in Intel microarchitecture code name Sandy Bridge applies to the following SSE instructions (and AVX variants): ADDPD/ADDPS, MULPD/MULPS, DIVPD/DIVPS, and CVTPD2PS.

The out-of-order core consists of three execution stacks, where each stack encapsulates a certain type of data. The execution core contains the following execution stacks:

- General-purpose integer
- SIMD integer and FP
- X87

The execution core also contains connections to and from the cache hierarchy. The loaded data are fetched from the caches and written back into one of the stacks.

The scheduler can dispatch up to six μops every cycle, one on each port. Table 11.4 summarizes which operations can be dispatched on which port.

After execution, the data are written back on a write-back bus corresponding to the dispatch port and the data type of the result. The μops that are dispatched on the same port but have different latencies may need the write-back bus at the same cycle. In these cases, the execution of one of the μops is delayed until the write-back bus is available. For example, MULPS (five cycles) and BLENDPS (one cycle) may collide if both are ready for execution on port 0: first the MULPS and four cycles later the BLENDPS. Intel microarchitecture code name Sandy Bridge eliminates such collisions as long as the μops write the results to different stacks. For example, integer ADD (one cycle) can be dispatched four cycles after MULPS (five cycles) since the integer ADD uses the integer stack, while the MULPS uses the FP stack.

When a source of a μop executed in one stack comes from a μop executed in another stack, a one- or two-cycle delay can occur. The delay occurs also for transitions between Intel SSE integer and Intel SSE FP operations. In some of the cases, the data transition is done using a μop that is added to the instruction flow. Table 11.5 describes how data, written back after execution, can bypass to μop execution in the following cycles.

Table 11.4 Dispatch Port and Execution Stacks

	Port 0	Port 1	Port 2	Port 3	Port 4	Port 5
Integer	ALU, shift	ALU, Fast LEA, Slow LEA, MUL	Load_Ad dr, Store_ad dr	Load_Ad dr Store_ad dr	Store_data	ALU, Shift, Branch, Fast LEA
SSE-Int, AVX-Int, MMX	Mul, Shift, STTNI, Int-Div, 128b-Mov	ALU, Shuf, Blend, 128b-Mov			Store_data	ALU, Shuf, Shift, Blend, 128-Mov
SSE-FP, AVX-FP_ low	Mul, Div, Blend, 256b-Mov	Add, CVT			Store_data	Shuf, Blend, 256b-Mov
X87, AVX-FP_ High	Mul, Div, Blend, 256b-Mov	Add, CVT			Store_data	Shuf, Blend, 256b-Mov

Table 11.5   Execution Core Write-Back Latency (Cycles)

	Integer	SSE-Int, AVX-Int, MMX	SSE-FP, AVX-FP_low	X87, AVX-FP_High
Integer	0	µop (port 0)	µop (port 0)	µop (port 0) + 1 cycle
SSE-Int, AVX-Int, MMX	µop (port 5) or µop (port 5) + 1 cycle	0	1-cycle delay	0
SSE-FP, AVC-FP_low	µop (port 5) or µop (port 5_ + 1 cycle)	1-Cycle delay	0	µop (port 5) + 1 cycle
X87, AVX-FP_High	µop (port 5) + 1 cycle	0	µop (port 5) + 1 cycle	0
Load	0	1-Cycle delay	1-cycle delay	2-Cycle delay

11.4.5.5 System Agent

The system agent implemented in the second-generation Intel Core processor family contains the following components:

- An arbiter that handles all accesses from the ring domain and from I/O (PCIe* and DMI) and routes the accesses to the right place.
- PCIe controllers connect to external PCIe devices. The PCIe controllers have different configuration possibilities that vary with product segment specifics: x16 + x4, x8 + x8 + x4, x8 + x4 + x4 + x4.
- DMI controller connects to the PCH chipset.
- Integrated display engine, flexible display interconnect, and display port, for the internal graphic operations.
- Memory controller.

All main memory traffic is routed from the arbiter to the memory controller. The memory controller in the second-generation Intel Core processor 2xxx series supports two channels of DDR, with data rates of 1066, 1333, and 1600 MHz and 8 bytes per cycle, depending on the unit type, system configuration, and DRAMs. Addresses are distributed between memory channels based on a local hash function that attempts to balance the load between the channels in order to achieve maximum bandwidth and minimum hot spot collisions.

For best performance, populate both channels with equal amounts of memory, preferably the exact same types of DIMMs. In addition, using more ranks for the same amount of memory results in somewhat better memory bandwidth, since more DRAM pages can be open simultaneously. For best performance, populate the system with the highest supported speed DRAM (1333 or 1600 MHz data rates, depending on the max supported frequency) with the best DRAM timings. The two channels have separate resources and handle memory requests independently. The memory controller contains a high-performance out-of-order scheduler that attempts to maximize memory bandwidth while minimizing latency. Each memory channel contains a 32 cache line write-data buffer. Writes to the memory controller are considered completed when they are written to the write-data buffer. The write-data buffer is flushed out to main memory at a later time, not impacting write latency.

Partial writes are not handled efficiently on the memory controller and may result in read–modify–write operations on the DDR channel if the partial writes do not complete a full cache line in time. Software should avoid creating partial write transactions whenever possible and consider alternative, such as buffering the partial writes into full cache line writes.

The memory controller also supports high-priority isochronous requests (such as USB isochronous and display isochronous requests). A high bandwidth of memory requests from the integrated display engine takes up some of the memory bandwidth and impacts core access latency to some degree.

11.4.5.6 *Intel Microarchitecture Code Name Ivy Bridge*

The third-generation Intel Core processors are based on Intel microarchitecture code name Ivy Bridge. Most of the features described in Sections 2.1.1 through 2.1.6 also apply to Intel microarchitecture code name Ivy Bridge. This section covers feature differences in microarchitecture that can affect coding and performance.

The support for new instructions enabling include

- Numeric conversion to and from half-precision FP values
- Hardware-based random number generator compliant to NIST SP 800-90A
- Reading and writing to FS/GS base registers in any ring to improve user-mode threading support

A small number of microarchitectural enhancements that can be beneficial to software is as follows:

- Hardware prefetch enhancement: A next-page prefetcher (NPP) is added in Intel microarchitecture code name Ivy Bridge. The NPP is triggered by sequential accesses to cache lines approaching the page boundary, either upward or downward.
- Zero-latency register move operation: A subset of register-to-register MOV instructions is executed at the front end, conserving scheduling and execution resource in the out-of-order engine.
- Front-end enhancement: In Intel microarchitecture code name Sandy Bridge, the μop queue is statically partitioned to provide 28 entries for each logical processor, irrespective of software executing in single thread or multiple threads. If one logical processor is not active in Intel microarchitecture code name Ivy Bridge, then a single thread executing on that processor core can use the 56 entries in the μop queue. In this case, the LSD can handle larger loop structure that would require more than 28 entries.
- The latency and throughput of some instructions have been improved over those of Intel microarchitecture code name Sandy Bridge. For example, 256-bit packed FP divide and square-root operations are faster; ROL and ROR instructions are also improved.

11.4.6 Ring Interconnect and Last Level Cache

The system-on-a-chip design provides a high-bandwidth bidirectional ring bus to connect between the IA cores and various subsystems in the uncore. In the second-generation Intel Core processor 2xxx series, the uncore subsystem includes a system agent, the GT, and the LLC.

The LLC consists of multiple cache slices. The number of slices is equal to the number of IA cores. Each slice has logic portion and data array portion. The logic portion handles data coherency, memory ordering, access to the data array portion, LLC misses and write back to memory, and more. The data array portion stores cache lines. Each slice contains a full cache port that can supply 32 bytes/cycle.

The physical addresses of data kept in the LLC data arrays are distributed among the cache slices by a hash function, such that addresses are uniformly distributed. The data array in a cache block may have 4/8/12/16 ways corresponding to 0.5M/1M/1.5M/2M block size.

However, due to the address distribution among the cache blocks from the software point of view, this does not appear as a normal N-way cache.

From the processor cores and the GT view, the LLC acts as one shared cache with multiple ports and bandwidth that scales with the number of cores. The LLC hit latency, ranging between 26 and 31 cycles, depends on the core location relative to the LLC block and how far the request needs to travel on the ring.

The number of cache slices increases with the number of cores; therefore, the ring and LLC are not likely to be a bandwidth limiter to core operation.

The GT sits on the same ring interconnect and uses the LLC for its data operations as well. In this respect, it is very similar to an IA core. Therefore, high-bandwidth graphic applications using cache bandwidth and significant cache footprint can interfere, to some extent, with core operations.

All the traffic that cannot be satisfied by the LLC, such as LLC misses, dirty line write back, noncacheable operations, and MMIO/IO operations, still travels through the cache-slice logic portion and the ring, to the system agent.

In the Intel Xeon Processor E5 Family, the uncore subsystem does not include the GT. Instead, the uncore subsystem contains many more components, including an LLC with larger capacity and snooping capabilities to support multiple processors, Intel QuickPath Interconnect interfaces that can support multisocket platforms, power management control hardware, and a system agent capable of supporting high-bandwidth traffic from memory and I/O devices.

11.4.7 Data Prefetching

Data can be speculatively loaded to the L1 DCache using software prefetching, hardware prefetching, or any combination of the two.

You can use the four SSE prefetch instructions to enable software-controlled prefetching. These instructions are hints to bring a cache line of data into the desired levels of the cache hierarchy. The software-controlled prefetch is intended for prefetching data but not for prefetching code.

The rest of this section describes the various hardware prefetching mechanisms provided by Intel microarchitecture code name Sandy Bridge and their improvement over previous processors. The goal of the prefetchers is to automatically predict which data the program is about to consume. If these data are not close by to the execution core or inner cache, the prefetchers bring it from the next levels of cache hierarchy and memory. Prefetching has the following effects:

- Improves performance if data are arranged sequentially in the order used in the program.
- May cause slight performance degradation due to bandwidth issues, if access patterns are sparse instead of local.
- On rare occasions, if the algorithm's working set is tuned to occupy most of the cache and unneeded prefetches evict lines required by the program, hardware prefetcher may cause severe performance degradation due to cache capacity of L1.

11.4.7.1 Data Prefetch to L1 Data Cache

Data prefetching is triggered by load operations when the following conditions are met:

- Load is from write-back memory type.
- The prefetched data are within the same 4K-byte page as the load instruction that triggered it.
- No fence is in progress in the pipeline.
- Not many other load misses are in progress.
- There is not a continuous stream of stores.

Two hardware prefetchers load data to the L1 DCache:

- *DCU prefetcher.* This prefetcher, also known as the streaming prefetcher, is triggered by an ascending access to very recently loaded data. The processor assumes that this access is part of a streaming algorithm and automatically fetches the next line.
- *IP-based stride prefetcher.* This prefetcher keeps track of individual load instructions. If a load instruction is detected to have a regular stride, then a prefetch is sent to the next address, which is the sum of the current address and the stride. This prefetcher can prefetch forward or backward and can detect strides of up to 2K bytes.

11.4.7.2 Data Prefetch to L2 and Last Level Cache

The following two hardware prefetchers fetched data from memory to the L2 cache and LLC:

Spatial prefetcher: This prefetcher strives to complete every cache line fetched to the L2 cache with the pair line that completes it to a 128-byte aligned chunk.

Streamer: This prefetcher monitors read requests from the L1 cache for ascending and descending sequences of addresses. Monitored read requests include L1 DCache requests initiated by load and store operations and by the hardware prefetchers and L1 ICache requests for code fetch. When a forward or backward stream of requests is detected, the anticipated cache lines are prefetched. Prefetched cache lines must be in the same 4K page.

The streamer and spatial prefetcher prefetch the data to the LLC. Typically, data are brought also to the L2 unless the L2 cache is heavily loaded with missing demand requests.

Enhancement to the streamer includes the following features:

- The streamer may issue two prefetch requests on every L2 lookup. The streamer can run up to 20 lines ahead of the load request.
- Adjusts dynamically to the number of outstanding requests per core. If there are not many outstanding requests, the streamer prefetches further ahead. If there are many outstanding requests, it prefetches to the LLC only and less far ahead.
- When cache lines are far ahead, it prefetches to the LLC only and not to the L2. This method avoids the replacement of useful cache lines in the L2 cache.
- Detects and maintains up to 32 streams of data accesses. For each 4K-byte page, you can maintain one forward and one backward stream.

11.4.8 Branch-Prediction Examples

Branch prediction is important to the performance of a deeply pipelined processor. It enables the processor to begin executing instructions long before the branch outcome is certain. Branch prediction predicts the branch target and enables the processor to begin executing instructions long before the branch true execution path is known. All branches utilize the BPU for prediction.

11.4.8.1 Intel Microarchitecture Code Name Sandy Bridge

This unit predicts the target address not only based on the EIP of the branch but also based on the execution path through which execution reached this EIP. The BPU can efficiently predict the following branch types:

- Conditional branches
- Direct calls and jumps
- Indirect calls and jumps
- Returns

11.4.8.2 Intel Core Microarchitecture

The BPU contains the following features:

- 16-Entry return stack buffer (RSB). It enables the BPU to accurately predict RET instructions.
- Front-end queuing of BPU lookups. The BPU makes branch predictions for 32 bytes at a time, twice the width of the fetch engine. This enables taken branches to be predicted with no penalty. Even though this BPU mechanism generally eliminates the penalty for taken branches, software should still regard taken branches as consuming more resources than do not-taken branches.

The BPU makes the following types of predictions:

- Direct calls and jumps. Targets are read as a target array, without regarding the taken or not-taken prediction.
- Indirect calls and jumps. These may either be predicted as having a monotonic target or as having targets that vary in accordance with recent program behavior.
- Conditional branches. Predicts the branch target and whether or not the branch will be taken.

11.4.8.3 Intel NetBurst Microarchitecture

Branch delay is the penalty that is incurred in the absence of correct prediction. For Pentium 4 and Intel Xeon processors, the branch delay for a correctly predicted instruction can be as few as zero clock cycles. The branch delay for a mispredicted branch can be many cycles, usually equivalent to the pipeline depth.

Branch prediction in the Intel NetBurst microarchitecture predicts near branches (conditional calls, unconditional calls, returns, and indirect branches). It does not predict far transfers (far calls, irets, and software interrupts).

Mechanisms have been implemented to aid in predicting branches accurately and to reduce the cost of taken branches. These include the following:

- Ability to dynamically predict the direction and target of branches based on an instruction's linear address, using the BTB.
- If no dynamic prediction is available or if it is invalid, the ability to statically predict the outcome based on the offset of the target: a backward branch is predicted to be taken, a forward branch is predicted to be not taken.
- Ability to predict return addresses using the 16-entry return address stack.
- Ability to build a trace of instructions across predicted taken branches to avoid branch penalties.

Static predictor: Once a branch instruction is decoded, the direction of the branch (forward or backward) is known. If there was no valid entry in the BTB for the branch, the static predictor makes a prediction based on the direction of the branch. The static prediction mechanism predicts backward conditional branches (those with negative displacement, such as loop-closing branches) as taken. Forward branches are predicted not taken.

To take advantage of the forward-not-taken and backward-taken static predictions, code should be arranged so that the likely target of the branch immediately follows forward branches.

Branch-target buffer: Once branch history is available, the Pentium 4 processor can predict the branch outcome even before the branch instruction is decoded. The processor uses a branch history table and a BTB to predict the direction and target of branches based on an instruction's linear address. Once the branch is retired, the BTB is updated with the target address.

Return stack: Returns are always taken, but since a procedure may be invoked from several call sites, a single predicted target does not suffice. The Pentium 4 processor has a return stack that can predict return addresses for a series of procedure calls. This increases the benefit of unrolling loops containing function calls. It also mitigates the need to put certain procedures in-line since the return penalty portion of the procedure call overhead is reduced.

Even if the direction and target address of the branch are correctly predicted, a taken branch may reduce available parallelism in a typical processor (since the decode bandwidth is wasted for instructions, which immediately follow the branch and precede the target, if the branch does not end the line and target does not begin the line). The branch predictor allows a branch and its target to coexist in a single trace cache line, maximizing instruction delivery from the front end.

11.5 SUMMARY

The major parameters of concern in the design of control units are speed, cost, complexity, and flexibility. HCUs offer higher speeds than MCUs, while the MCUs offer better flexibility. Speed-enhancement techniques applicable to both types of control units were discussed in this chapter. The commonly used microinstruction formats were introduced along with their speed-cost trade-offs.

Pipelining techniques have been adopted extensively to enhance the performance of serial processing structures. This chapter covers the basic principles and design methodologies for pipelines. Although the throughput of the machine is improved by employing pipeline techniques, the basic cycle time of instructions remains as in nonpipelined implementations. In earlier machines with little or no pipelining, the average clock per instruction (CPI) was 5–10. Modern RISC architectures have achieved a CPI value close to 1. Pipelines exploit instruction-level parallelism in the program whereby instructions that are not dependent on each other are executed in an overlapped manner. If the instruction-level parallelism is sufficient to keep the pipeline flowing full (with independent instruction), we achieve the ideal machine with a CPI of 1.

Three approaches have been tried to improve the performance beyond the ideal CPI case: superpipelining (described in this chapter) and superscalar and VLIW architectures (described in Chapter 10). A superpipeline uses deeper pipelines and control mechanisms to keep many stages of the pipeline busy concurrently. Note that as the latency of a stage gets longer, the instruction issue rate will drop, and also there is a higher potential for data interlocks. A superscalar machine allows the issue of more than one instruction per clock into the pipeline, thus achieving an instruction rate higher than the clock rate. In these architectures, the hardware evaluates the dependency among instructions and dynamically schedules a packet of independent instruction for issue at each cycle. In a VLIW architecture, a machine instruction corresponds to a packet of independent instructions, created by the compiler. No dynamic hardware scheduling is used.

Pipeline techniques are extensively used in all architectures today, from microprocessors to supercomputers. This chapter provided some details of four pipelined machines.

PROBLEMS

11.1 Study the following architectural features of a micro-, a mini-, a large-scale, and a supercomputer system with an HCU you have access to:
> Processor cycle time
> Number of clock phase and clock frequency
> Minor and major cycle details
> Speedup techniques used

11.2 Study the following architectural features of a micro-, a mini-, a large-scale, and a supercomputer system with an MCU you have access to:
> Processor cycle time
> CROM size, access time, and organization
> Microinstruction format
> Speedup techniques used

11.3 It is required that the LDA instruction cycle of ASC use a maximum of two major cycles. Derive the microinstruction sequence and list any changes needed to the bus structure.

11.4 Write the microprogram for the LDA instruction in Problem 11.3. List any changes needed for the microinstruction format.

11.5 Assume that an HCU needs to be designed for LDA, STA, TCA, and ADD instructions only of ASC. Examine the µop sequences for these instructions to determine the optimum length for the major cycle if (1) only equal-length major cycles are allowed and (2) major cycles can be of varying length.

11.6 In the microprogram for ASC (Chapter 6), the sequence for indirect address computations is repeated for each ASC instruction requiring that computation. Suppose that the indirect address computation is made into a subroutine called by ASC instructions as required, rewrite the microprogram for LDA. What is the effect on the execution time of LDA if indirect address is (1) called for? (2) not called for?

11.7 List the changes needed to convert ASC HCU to include three modules—fetch, defer, and execute—working in a pipeline mode. Assume that only LDA, STA, SHR, TCA, LDX, and HLT instructions are needed.

11.8 Repeat Problem 11.7 listing changes needed for ASC MCU.

11.9 ASC MCU currently employs two types of microinstructions. Determine the microinstruction format needed if only a single type of microinstruction is allowed.

11.10 Repeat Problem 11.9 assuming that only horizontal microinstruction is allowed.

11.11 A program consists of N instructions. Assume that all the N instruction cycles require the fetch, decode, and execute phases. If each phase consumes one time unit, compute the serial execution time of the program. What is the speedup, if a pipelined control unit with three stages is used?

11.12 Assume that ASC uses a three-stage pipelined control unit. Show the (timeline) trace of the following program:

LDA Y
ADD Y
STA Z
HLT

11.13 Repeat Problem 11.12 assuming a control unit with the following stages: fetch, decode, compute address, fetch operands, execute, and store results.

11.14 Compare the instruction sets of Intel 8080 and Intel Pentium II with respect to instruction execution speeds and modes of execution.

11.15 Compare the instruction sets of Intel Itanium and MIPS R10000 with respect to instruction set, instruction execution speed, and modes of execution.

11.16 Describe the conditions under which an n-stage pipeline is n times faster than a serial machine.

11.17 Design a pipeline ALU that can add two 8-bit integers. Assume that each stage can add two 2-bit integers and employ sufficient number of stages. Show the operation of the pipeline using a reservation table.

11.18 Design a pipeline for multiplying two FP numbers represented in IEEE standard format. Assume that addition requires T seconds and multiplication requires $2T$ s and shift takes $T/2$ s. Use as many stages as needed and distribute the functions such that each stage approximately consumes the same amount of time.

11.19 Compute the minimum number of pipeline cycles required to multiply 50 FP numbers in the pipeline of Problem 11.18, assuming that the output from the last stage can be fed back to any of the stages that require it, with the appropriate delay.

11.20 Compare the relative merits of "superpipelined" and "superscalar" architectures. To what application environment each technique is suitable? What is an "underpipelined" architecture?

11.21 List the components of the front end of Intel microarchitecture code name Sandy Bridge and their functions.

11.22 Investigate the areas of limitation for a superscalar CPU architecture.

11.23 What are the motivations for using branch-prediction schemes? What are the four simple compile-time prediction schemes? Explain each of them and how they differ.

11.24 Explain the steps for the 5-stage MIPS with BTB.

BIBLIOGRAPHY

Computer, New York: IEEE Computer Society, Published monthly.

Dual-Core Intel Itanium 2 Processor: Reference Manual Update, For Software Development and Optimization, Revision 0.9, January 2006. http://www.intel.com/content/www/us/en/processors/itanium/dual-core-update-itanium-2-processor-manual.html

Electronic Design, Cleveland, OH: Penton Media, Published twice monthly, http://www.electronicdesign.com.

EE Times, Manhasset, NY:CMP Media, Published weekly, www.eetimes.com.

Intel Itanium Architecture Software Developer's Manual, Volume 1: Application Architecture, Revision 2.3, May 2010. http://www.intel.com/content/www/us/en/processors/itanium/itanium-architecture-software-developer-rev-2-3-vol-1-manual.html

Jouppi, N.P. and Wall, D.W., Available instruction-level parallelism for superscalar and superpipelined machines, *Proceedings of the IEEE/ACM Conference on Architectural Support for Programming Languages and Operating Systems,* April 1989, pp. 272–282.

Kogge, P.M., *The Architecture of Pipelined Computers,* New York: McGraw-Hill, 1981.

Losq, J.J., Rao, G.S., and Sachar, H.E., Decode history table for conditional branch instructions, U.S. Patent No. 4,477,872, October, 1984.

MIPS R10000 Microprocessor User's Manual, Version 2.0, October 10, 1996, http://cch.loria.fr/documentation/docSGI/R10K/t5.ver.2.0.book.pdf.

MIPS R10000 Microprocessor User's Manual, Version 2.0, January 29, 1997, http://techpubs.sgi.com/library/manuals/2000/007–2490–001/pdf/007–2490–001.pdf.

MIPS Technologies, http://www.mips.com.

Shar, L.E. and Davidson, E.S., A multiminiprocessor system implemented through pipelining. *IEEE Computer,* 7(2), 42–51, February 1974.

Shiva, S.G., *Advanced Computer Architectures,* Boca Raton, FL: Taylor & Francis, 2006.

Smith, J.E. and Plezkun, A.R., Implementing precise interrupts in pipelined processors, *IEEE Transactions on Computers,* 37(5), 562–573, May 1988.

Stone, H.S., *High Performance Computer Architecture,* Reading, MA: Addison-Wesley, 1987.

Sussenguth, E., Instruction sequence control. U.S. Patent No. 3,559,183, January 26, 1971.

Thornton, J.E., *Design of a Computer: The Control Data 6600,* Glenview, IL: Scott, Foresman, 1970.

CHAPTER 12

Advanced Architectures

Up to this point, we have considered processors that have a single *instruction stream* (i.e., a single control unit bringing about a serial instruction execution) operating upon a single *data stream* (i.e., a single unit of data is being operated upon at any given time). We have also described various enhancements to each of the subsystems (memory, ALU, control, and input/output [I/O]) that increase the processing speed of the system. In addition, components from the fastest hardware technology available are used in the implementation of these subsystems to achieve higher speeds.

There are applications, especially in real-time processing, that make the machine throughput inadequate even with all these enhancements. It has thus been necessary to develop newer architectures with higher degrees of parallelism than these enhancements provide in order to circumvent the limits of technology and achieve faster processing speeds.

Suppose that the application allows the development of processing algorithms with a degree of parallelism A, the language used to code the algorithm allows a degree of parallelism of L, the compilers retain a degree of parallelism of C, and the hardware structure of the machine has a degree of parallelism of H. Then, for the processing to be most efficient, the following relation must be satisfied:

$$H \geq C \geq L \geq A. \tag{12.1}$$

Development of algorithms with a high degree of parallelism is application dependent. A great deal of research has been devoted to developing languages that allow parallel processing constructs and compilers that either extract the parallelism from a sequential program or retain the parallelism of the source code through the compilation to produce parallel code. In this chapter, we will describe the popular hardware structures used in executing parallel code.

Several hardware structures have evolved over the years, providing for the execution of programs with various degrees of parallelism at various levels. Flynn divided computer architectures into four main classes (Flynn's taxonomy) based on the number of instruction and data streams:

1. Single instruction stream, single data stream (SISD) machines, which are *uniprocessors* such as ASC and Intel 8080.
2. Single instruction stream, multiple data stream (SIMD) architectures, which are systems with multiple arithmetic–logic processors and a control processor (CP). Each arithmetic–logic processor processes a data stream of its own, as directed by the single CP. They are also called *array processors* or *vector processors*.
3. Multiple instruction stream, single data stream (MISD) machines, in which the single data stream is simultaneously acted upon by a multiple instruction stream. A system with a pipelined ALU can be considered an MISD, although that extends the definition of a data stream somewhat.
4. Multiple instruction stream, multiple data stream (MIMD) machines, which contain multiple processors, each executing its own instruction stream to process the data stream allocated to it. A computer system with one processor and an I/O channel working in parallel is the simplest example of an MIMD (i.e., a *multiprocessor* system).

We will provide the details of the last three of the earlier-listed classifications, in Sections 12.1 through 12.3. Section 12.4 addresses the cache coherency problem. Section 12.5 provides a brief description of dataflow architectures, an experimental architecture type that provides the finest granularity of parallelism. Section 12.6 describes systolic architectures. Section 12.7 provides brief descriptions of two Intel technologies that provide enhanced processor throughputs and two Cray supercomputer systems. This chapter concentrates on the so-called *tightly coupled* multiple processor (computer) systems. *Loosely coupled* multiple processor (computer) systems forming computer networks and distributed systems are described in Chapter 15.

12.1 MISD

Figure 12.1 shows the organization of an MISD machine. There are *n* processors (or processing stations) arranged in a pipeline. The data stream from the memory enters the pipeline at processor 1 and moves from station to station through processor *n*, and the resulting data stream enters the memory. The control unit of the machine is shown to have *n* subunits, one for each processor. Thus, there will be at any time *n* independent instruction streams.

Note that the concept of data stream in this model is somewhat broad. The pipeline will be processing several independent data units at a given time. However, for the purposes of this model, all those data units put together is considered a single data stream. As such, this classification is considered an anomaly. It is included here for the sake of completeness.

12.2 SIMD

Figure 12.2 shows the structure of an SIMD. There are *n* arithmetic–logic processors (P_1–P_n), each with its own memory block (M_1–M_n). The individual memory blocks combined constitute the system memory. The memory bus is used to transfer instructions and data to the CP. The CP decodes instructions and sends control signals to processors P_1–P_n. The processor interconnection network (IN) enables data exchange between the processors.

The CP, in practice, is a full-fledged uniprocessor. It retrieves instructions from memory, sends arithmetic–logic instructions to processors, and executes control (branch, stop, etc.) instructions itself. Processors P_1–P_n execute the same instruction at any time, each on its own data stream. On the basis of the arithmetic–logic conditions, some of the processors may be deactivated during certain operations. Such activation and deactivation of processors is handled by the CP.

The following example compares the operation of an SIMD with that of an SISD system.

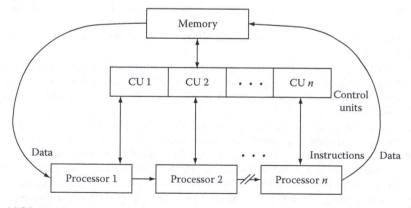

Figure 12.1   MISD structure.

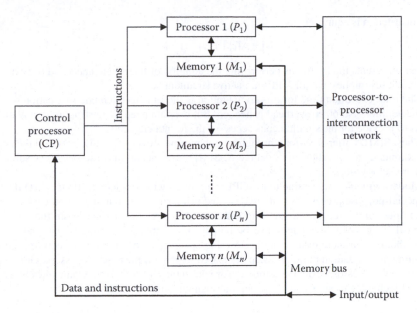

Figure 12.2 SIMD structure.

Example 12.1

Consider the addition of two N-element vectors A and B element-by-element to create the sum vector C. That is,

$$C[i] = A[i] + B[i] \quad 1 \le i \le N.$$

This computation requires N add times plus the loop control overhead on an SISD. Also, the SISD processor has to fetch the instructions corresponding to this program from the memory each time through the loop.

Figure 12.3 shows the SIMD implementation of this computation using N processing elements (PEs). This is identical to the SIMD model of Figure 12.2 and consists of multiple PEs, one CP, and the memory system. The processor IN of Figure 12.2 is not needed for this computation. The elements of arrays A and B are distributed over N memory blocks and hence, each PE has access to one pair of operands to be added. Thus, the program for the SIMD consists of one instruction: C = A + B.

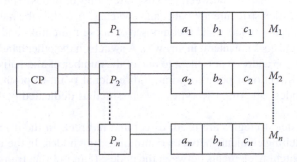

Figure 12.3 SIMD processing model.

This instruction is equivalent to

$$C[i] = A[i] + B[i] \quad (1 \le i \le N),$$

where i represents the PE that is performing the addition of the ith elements and the expression in parentheses implies that all N PEs are active simultaneously.

The total execution time for this computation is the time to fetch one instruction plus time for one addition. No other overhead is needed. But the data need to be structured in N memory blocks to provide for the simultaneous access of the N data elements.

Thus, SIMDs offer an N-fold throughput enhancement over the SISD, provided the application exhibits a data parallelism of degree N. SIMDs are special-purpose machines suitable for array or vector processing.

Modern graphics processing units (GPUs) are often wide implementations of SIMD. GPUs are special-purpose processors that are designed primarily for simultaneous processing of data by multiple cores executing a single instruction. They are used to accelerate the building of graphical images that are required to be output on a display. Their highly parallel structure makes them much more efficient in performing special-purpose tasks that require processing large amounts of data in parallel, when compared to general-purpose processors. GPUs can be present on video cards, or on motherboards, or within the CPU die itself. Companies like Nvidia and ATI technologies are some of the popular manufacturers of GPUs.

12.2.1 Interconnection Networks for SIMD Architectures

PEs and memory blocks in an SIMD architecture are interconnected by either an n-to-n or an n-by-n switch. Also, an n-by-n switch between the PEs is needed in general to allow fast exchange of data as the computation proceeds. Here onward, the acronym IN will be used to refer to the interconnection hardware rather than the term "switch." The term "switch" is more generally used to refer to a component of an IN (as discussed later in this section).

The PE-to-PE IN for an SIMD depends on the dataflow requirements of the application. In fact, depending on the application, most of the computation can be accomplished by the IN itself if the network is chosen appropriately, thus reducing the complexity of PEs drastically.

Several INs have been proposed and built over the last few years. This section introduces the terminology and performance measures associated with INs and describes common INs, as applied to SIMD systems. The next section extends this description to MIMD architectures.

12.2.1.1 Terminology and Performance Measures

An IN consists of several nodes interconnected by links. A node in general is either a PE or a memory block or a complete computer system consisting of PEs, memory blocks, and I/O devices. A link is the hardware interconnect between two nodes. The IN facilitates the transmission of messages (data or control information) between processes residing in nodes. The two functional entities that form the interconnection structure are paths and switches. A path is the medium by which a message is transmitted between two nodes and is composed of one or more links and switches. The link just transmits the message and does not alter it in any way. A switch on the other hand may alter the message (i.e., change the destination address) or route it to one of the number of alternative paths available.

A path can be unidirectional point-to-point, bidirectional point-to-point, or bidirectional, and visit more than two nodes. The first two types are classified as dedicated paths and the last type is the shared path.

Two message transfer strategies are used: direct and indirect. In the direct strategy, there will be no intervening switching elements between communicating nodes. In the indirect strategy, there will be one or more switching elements between the nodes. If an indirect transfer strategy is chosen, either a centralized or a decentralized transfer control strategy can be adopted. In a central strategy, all switching is done by a single entity (called the switch controller). In a decentralized strategy on the other hand, the switch control is distributed among a number of switching elements.

For instance, in a ring network, there is a path from each node to every other node. The path to the neighboring node from any node is of length 1, while the path between non-neighboring nodes is of length equal to the number of links that need to be traversed (i.e., number of hops needed) to transmit messages between those nodes. If a decentralized control strategy is used, each intermediate node in the path serves as a switch that decodes the destination address and transmits the message not addressed to it to the next node. In the examples provided earlier in this chapter, the CP was the controller of the ring and issues multiple shift instructions depending on the distance (see later) between the source and destination nodes.

In general, an IN should be able to connect all the nodes in the system to one another, transfer maximum number of messages per second reliably, and offer minimum cost. Various performance measures have been used to evaluate INs. They are described below.

Connectivity (or degree of the node) is the number of nodes that are immediate neighbors of a node, that is, the number of nodes that can be reached from the node in one hop. For instance, in a unidirectional ring, each node is of degree 1 since it is connected to only one neighboring node. In a bidirectional ring each node is of degree 2.

Bandwidth is the total number of messages the network can deliver in unit time. A message is simply a bit pattern of certain length consisting of data and/or control information.

Latency is a measure of the overhead involved in transmitting a message over the network from the source node to the destination node. It can be defined as the time required to transmit a zero-length message.

Average distance: The "distance" between two nodes is the number of links in the shortest path between those nodes in the network. The average distance is given by

$$d_{avg} = \frac{\sum_{d=1}^{r} dN_d}{N-1},$$ (12.2)

where
N_d is the number of nodes at distance d apart
r is the diameter (i.e., the maximum of the minimum distance between all pairs of nodes) of the network
N is the total number of nodes

It is desirable to have a low average distance. A network with low average distance would result in nodes of higher degree, that is, larger number of communication ports from each node, which may be expensive to implement. Thus, a normalized average distance can be defined as

$$d_{avg(normal)} = d_{avg} \times P,$$ (12.3)

where P is the number of communication ports per node.

Hardware *complexity* of the network is proportional to the total number of links and switches in the network. As such, it is desirable to minimize the number of these elements.

Cost is usually measured as the network hardware cost as a fraction of the total system hardware cost. The "incremental hardware cost" (i.e., cost modularity) in terms of how much additional hardware and redesign is needed to expand the network to include additional nodes is also important.

Place modularity is a measure of expandability, in terms of how easily the network structure can be expanded by utilizing additional modules.

Regularity: If there is a regular structure to the network, the same pattern can be repeated to form a larger network. This property is especially useful in the implementation of the network using VLSI circuits.

Reliability and fault tolerance is a measure of the redundancy in the network to allow the communication to continue, in case of the failure of one or more links.

Additional *functionality* is a measure of other functions (such as computations, message combining, and arbitration) offered by the network in addition to the standard message transmission function.

A complete IN (i.e., an *n*-by-*n* network in which there is a link from each node to every other node) is the ideal network since it would satisfy the minimum latency, minimum average distance, maximum bandwidth, and simple routing criteria. But, the complexity of this network becomes prohibitively large, as the number of nodes increases. Expandability also comes at a very high cost. As such, the complete interconnection scheme is not used in networks with large number of nodes. There are other topologies that provide a better cost/performance ratio.

The topologies can further be classified as either *static* or *dynamic*. In a static topology, the links between two nodes are passive, dedicated paths. They cannot be changed to establish a direct connection to other nodes. The interconnections are usually derived based on a complete analysis of the communication patterns of the application. The topology does not change as the computation progresses. In dynamic topologies, the interconnection pattern changes as the computation progresses. This changing pattern is brought about by setting the network's active elements (i.e., switches).

The mode of operation of a network can be *synchronous*, *asynchronous*, or combined as dictated by the data manipulation characteristics of the application. The ring network used in the examples earlier in this chapter is an example of a synchronous network. In these networks, communication paths are established and message transfer occurs synchronously. In asynchronous networks, connection requests are issued dynamically as the transmission of the message progresses. That means the message flow in an asynchronous network is less orderly compared with that in a synchronous network. A combined network exhibits both modes of operation.

12.2.1.2 Routing Protocols

Routing is the mechanism that establishes the path between two nodes for transmission of messages. It should be as simple as possible and should not result in a high overhead. It should preferably be dependent on the state of each node, rather than the state of the whole network. Three basic routing protocols (or switching mechanisms) have been adopted over the years: circuit switching, packet switching, and wormhole switching.

In *circuit switching*, a path is first established between the source and the destination nodes. This path is dedicated for the transmission of the complete message. That is, there is a dedicated hardware path between the two nodes that cannot be utilized for any other message until the transmission is complete. This mode is ideal when large messages are to be transmitted.

In *packet switching*, the message is broken into small units called packets. Each packet has a destination address that is used to route the packet to the destination through the network nodes, in a store and forward manner. That is, a packet travels from the source to the destination, one link at a time. At each (intermediate) node, the packet is usually stored (or buffered). It is then forwarded to the appropriate link based on its destination address. Thus, each packet might follow a different route to the destination and the packets may arrive at the destination in any order. The packets are reassembled at the destination to form the complete message. Packet switching is efficient for short messages and more frequent transmissions. It increases the hardware complexity of the switches, because of the buffering requirement.

The packet switching mode is analogous to the working of the postal system in which each letter is a packet. Unlike messages in an IN, the letters from a source to a destination follow the same route and they arrive at the destination usually in the same order they are sent. The circuit switching mode is analogous to the telephone network in which a dedicated path is first established and maintained throughout the conversation. The underlying hardware in the telephone network uses packet switching. But what the user sees is a (virtual) dedicated connection.

The *wormhole switching* (or cut-through routing) is a combination of these two methods. Here, a message is broken into small units (called flow control digits or flits), as in packet switching. But, all flits follow the same route to the destination, unlike packets. Since the leading flit sets the switches in the path and the others follow, the store and forward buffering overhead is reduced.

Routing mechanisms can be either static (or deterministic) or dynamic (or adaptive). Static schemes determine a unique path for the message from the source to the destination, based solely on the topology. Since they do not take into consideration the state of the network, there is the potential of uneven usage of the network resources and congestion. The dynamic routing, on the other hand, utilizes the state of the network in determining the path for a message and hence can avoid congestion by routing messages around the heavily used links or nodes.

The routing is realized by setting switching elements in the network appropriately. As mentioned earlier, the switch setting (or control) function can either be managed by a centralized controller or can be distributed to each switching element. Further detail on routing mechanisms is provided in the following sections, along with the description of various topologies.

12.2.1.3 Static Topologies

Figures 12.4 through 12.7 show some of the many static topologies that have been proposed, representing a wide range of connectivity. The simplest among these is the linear array in which the first and the last nodes are each connected to one neighboring node and each interior node is connected to two neighboring nodes. The most complex is the complete IN in which each node is connected to every other node. A brief description of each of these topologies follows.

12.2.1.3.1 Linear Array and Ring (LOOP)

In the one-dimensional mesh, or linear array (Figure 12.4a), each node is connected to its neighboring node, each interior node is connected to two of its neighboring nodes, while the boundary nodes are connected to only one neighbor. The links can either be unidirectional or bidirectional.

A ring or loop network is formed by connecting the two boundary nodes of the linear array (Figure 12.4b). In this structure, each node is connected to two neighboring nodes. The loop can

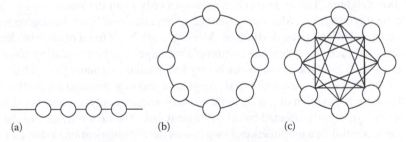

(a)						(b)						(c)

Figure 12.4 One-dimensional networks. (a) Linear array. (b) Ring. (c) Chordal ring.

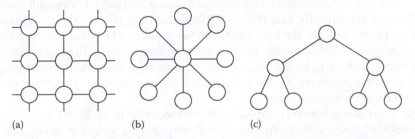

(a)						(b)						(c)

Figure 12.5 Two-dimensional networks. (a) Near neighbor mesh. (b) Star. (c) Binary tree.

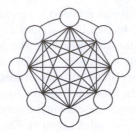

Figure 12.6 Completely connected network.

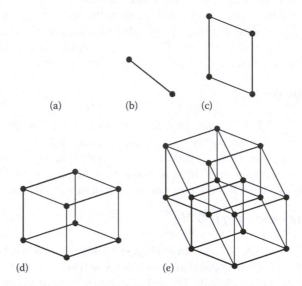

(a)   (b)   (c)

(d)   (e)

Figure 12.7 Hypercube. (a) Zero-cube. (b) One-cube. (c) Two-cube. (d) Three-cube. (e) Four-cube.

either be unidirectional or bidirectional. In a unidirectional loop, each node has a source neighbor and a destination neighbor. The node receives messages only from the source and sends messages only to the destination neighbor. Messages circulate around the loop from the source with intermediate nodes acting as buffers to the destination. Messages can be of fixed or variable length and the loop can be designed such that either one or multiple messages can be circulating simultaneously.

The logical complexity of the loop network is very low. Each node should be capable of originating a message destined for a single destination and recognize a message destined for itself and relay messages not destined for it. Addition of a node to the network requires only one additional link and flow of messages is not significantly affected by this additional link. The fault tolerance of the loop is very low. The failure of one link in a unidirectional loop causes the communication to stop (at least between the nodes connected by that link). If the loop is bidirectional, the failure of one link does not break the loop. Two link failure partitions the loop into two disconnected parts. Loops with redundant paths have been designed to provide the fault tolerance. The chordal ring (Figure 12.4c) in which the degree of each node is 3 is an example. The bandwidth of the loop can be a bottleneck as the communication requirements increase. It is also possible that one node can saturate the entire bandwidth of the loop.

Loop network evolved from the data communication environments in which geographically dispersed nodes were connected for file transfers and resource sharing. They have used bit-serial data links as the communication paths.

Note that when a loop network is used to interconnect PEs in an SIMD system, the message transfer between PEs is simultaneous since the PEs work in a lock-step mode. The transfer is typically controlled by the CP through SHIFT or ROTATE instructions issued once (for transfer between neighboring PEs) or multiple times (for transfers between remote PEs). The links typically carry data in parallel, rather than in bit-serial fashion.

12.2.1.3.2 Two-dimensional Mesh

A popular IN is the two-dimensional mesh (nearest neighbor) shown in Figure 12.5a. Here, the nodes are arranged in a two-dimensional matrix form and each node is connected to four of its neighbors (north, south, east, and west).

The connectivity of boundary nodes depends on the application. The IBM wire routing machine (WRM) uses a "pure mesh" in which boundary nodes have degree 3 and corners have degree 2. In the mesh network of ILLIAC-IV, the bottom node in a column is connected to the top node in the same column and the rightmost node in a row is connected to the leftmost node in the next row. This network is called a torus.

The cost of two-dimensional network is proportional to N, the number of nodes. The network latency is N. The message transmission delay depends on the distance of the destination node to the source and hence, there is a wide range of delays. The maximum delay increases with N.

A higher-dimensional mesh can be constructed analogous to the 1D and 2D meshes earlier. A k-dimensional mesh is constructed by arranging its N nodes in the form of a k-dimensional array and connecting each node to its (2^k) neighbors by dedicated, bidirectional links. The diameter of such a network is $^k\sqrt{N}$.

The routing algorithm commonly employed in mesh networks follows a traversal of one dimension at a time. For instance, in a 4D mesh, to find a path from the node labeled (a,b,c,d) to the node labeled (p,q,r,s), we first traverse along the first dimension to the node (p,b,c,d), then along the next dimension to (p,q,c,d), then along the third dimension to (p,q,r,d), and finally along the remaining dimension to (p,q,r,s).

It is not possible to add a single node to the structure in general. The number of nodes and the additional hardware needed depend on the mesh dimensions. As such, the cost and place modularities are poor. Because of the regular structure, the routing is simpler. The fault tolerance can be high, since the failure of a node or a link can be compensated by other nodes and other alternative paths possible.

12.2.1.3.3 Star

In the star interconnection scheme (Figure 12.5b), each node is connected to a central switch through a bidirectional link. The switch forms the apparent source and destination for all messages. The switch maintains the physical state of the system and routes messages appropriately.

The routing algorithm is trivial. The source node directs the message to the central switch, which in turn directs it to the appropriate destination on one of the dedicated links. Thus, if either of the nodes involved in message transmission is the central switch, the message path is just the link connecting them. If not, the path consists of two links.

The cost and place modularity of this scheme are good with respect to PEs but poor with respect to the switch. The major problems are the switch bottleneck and the catastrophic effect on the system in the case of the switch failure. Each additional node added to the system requires a bidirectional link to the switch and the extension of switch facilities to accommodate the new node.

Note that the central switch basically interconnects the nodes in the network. That is, the interconnection hardware is centralized at the switch, often allowing centralized instead of distributed control over routing. The interconnection structure within the switch could be of any topology.

12.2.1.3.4 Binary Trees

In binary tree networks (Figure 12.5c), each interior node has a degree 3 (two children and one parent), leaves have degree 1 (parent), and the root has degree 2 (two children).

A simple routing algorithm can be used in tree networks. To reach node Y from node X, traverse up in the tree from X until an ancestor of Y is reached and then traverse down to Y. To find the shortest path, we need to first find the ancestor of Y at the lowest level in the tree while ascending from X and then decide whether to follow right or left link while descending toward Y.

The nodes of the tree are typically numbered consecutively, starting with the root node at 1. The node numbers of the left and right children of a node z are $2z$ and $2z + 1$, respectively. Further, the node numbers at level i (with the root level being 1) will be i-bits long. Thus, the numbers for the left and right children of a node can be obtained by appending a 0 (left child) or a 1 (right child) to the parent's number. With this numbering scheme, in order to find a path from node X and to node Y, we first extract the longest common bit pattern from the numbers of X and Y. This is the node number of their common ancestor A. The difference in the lengths of numbers of X and A is the number of levels to be traversed up from X. To reach Y from A, we first remove the common (most significant) bits in numbers A and Y. We then traverse down based on the remaining bits of Y, going left on 0 and right on 1, until all the bits are exhausted (Almasi and Gottlieb 1989).

For a tree network with N nodes, the latency is $\log_2 N$, the cost is proportional to N, and the degree of nodes is 3 and is independent of N.

12.2.1.3.5 Complete Interconnection

Figure 12.6 shows the complete IN, in which each node is connected to every other node with a dedicated path. Thus, in a network with N nodes, all nodes are of degree $N - 1$ and there are $N (N - 1)/2$ links. Since all nodes are connected, the minimal length path between any two nodes is simply the link connecting them.

Routing algorithm is trivial. The source node selects the path to the destination node among the $(N - 1)$ alternative paths available and all nodes must be equipped to receive messages on a multiplicity of paths.

This network has poor cost modularity, since the addition of a node to an N node network requires N extra links and all nodes must be equipped with additional ports to receive the message from the new node. Thus, the complexity of the network grows very fast as the number of nodes is increased. Hence, complete INs are used in environments where only a small number of nodes (4–16) are interconnected. The place modularity is also poor for the same reasons as for the cost modularity.

This network provides a very high bandwidth and its fault tolerance characteristics are good. Failure of a link does not make the network inoperable, since alternative paths are readily available (although the routing scheme gets more complex).

The Intel Paragon and the MIT Alewife use 2D networks. The MasPar MP-1 and MP-2 use a 2D torus network where each node is connected to its eight nearest neighbors by the use of shared X connections. The Fujitsu AP3000 distributed-memory multicomputer has a 2D torus network, while the CRAY T3D and T3E both use the 3D torus topology.

12.2.1.3.6 Hypercube

The hypercube is a multidimensional near-neighbor network. A k-dimensional hypercube (or a k-cube) contains 2^k nodes, each of degree k. The label for each node is a binary number with k bits. The labels of neighboring nodes differ in only one bit position.

Figure 12.7 shows the hypercubes for various values of k. A 0-cube has only one node as shown in Figure 12.7a. A 1-cube connects two nodes labeled 0 and 1 as shown in Figure 12.7b. A 2-cube of Figure 12.7c connects four nodes labeled: 00, 01, 10, and 11. Each node is of degree 2 and the labels of neighboring nodes differ in only one bit position. A 3-cube shown in Figure 12.7d has eight nodes with labels ranging from 000 through 111 (or decimal 0 through 7). Node (000), for example, is connected directly to nodes (001), (010), and (100), and the message transmission between these neighboring nodes requires only one hop. To transmit a message from node (000) to node (011), two routes are possible: 000 to 001 to 011 and 000 to 010 to 011. Both routes require two hops. Note that the source and destination labels differ in two bits, implying the need for two hops. To generate these routes, a simple strategy is used. First, the message is at 000. The label 000 is compared with the destination address 011. Since they differ in bit positions 2 and 3, the message is routed to one of the corresponding neighboring nodes 010 or 001. If the message is at 010, this label is again compared with 011, noting the difference in position 3, implying that it be forwarded to 011. A similar process is used to route from 001 to 011.

Thus, the routing algorithm for the hypercube is simple. For a k-dimensional hypercube, the routing algorithm uses at most k steps. During step i, the messages are routed to the adjacent node in dimension i if the ith bit of X is 1; otherwise, the messages remain where they are.

Hypercube networks reduce network latency by increasing the degree of each node (i.e., connecting each of the N nodes to $\log_2 N$ neighbors). The cost is of the order of $N\log_2 N$ and the latency is $\log_2 N$.

Several commercial architectures using the hypercube topology (from Intel Corporation, NCUBE Corporation, and Thinking Machines Corporation) are now available. One major disadvantage of the hypercube topology is that the number of nodes should always be a power of two. Thus, the numbers of nodes need to double, every time a single node is required to be added to the network.

12.2.1.4 Dynamic Topologies

A parallel computer system with static interconnection topology can be expected to do well on applications that can be partitioned into processes with predictable communication patterns consisting mostly of exchanges among neighboring PEs. Some examples of such application domains are the analysis of events in space, vision, image processing, weather modeling, and VLSI design. If the application does not exhibit such predictable communication pattern, machines with static topologies become inefficient since the message transmission between non-neighboring nodes results in excessive transmission delays. Hence, computer systems using static INs tend to be more special-purpose compared with those using dynamic INs. Both SIMD and MIMD architectures have used static topologies.

Several dynamic INs have been proposed and built over the last few years, with a wide range of performance/cost characteristics. They can be classified under the following categories:

1. Bus networks
2. Crossbar networks
3. Switching networks

Bus networks: These are simple to build. Several standard bus configurations have evolved with data path widths as high as 64 bits. A bus network provides the least cost among the three types of dynamic networks and also has the lowest performance. Bus networks are not suitable for PE interconnection in an SIMD system. The next section provides further details on bus networks in the context of MIMD architectures.

12.2.1.4.1 Crossbar Networks

A crossbar network is the highest performance, highest cost alternative among the three dynamic network types. It allows any PE to be connected to any other nonbusy PE at any time.

Figure 12.8 shows an $N \times N$ crossbar, connecting N PEs to N memory elements. The number of PEs and memory elements need not be equal, although both are usually powers of 2. The number of memory elements is usually a small multiple of the number of PEs. There are N^2 crosspoints in the crossbar, one at each row–column intersection. If the PEs produce a 16-bit address and work with 16-bit data units, each crosspoint in the crossbar corresponds to the intersection of 32 lines plus some control lines. Assuming four control lines, to build a 16×16 crossbar, we would need at least $(16 \times 16 \times 36)$ switching devices. To add one more PE to the crossbar, only one extra row of

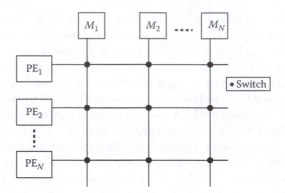

Figure 12.8 $N \times N$ crossbar.

crosspoints is needed. Thus, although the wire cost grows as the number of processors N, the switch cost grows as N^2, which is the major disadvantage of this network.

In the crossbar network of Figure 12.8, any PE can be connected to any memory and each of the N PEs can be connected to a distinct memory, simultaneously. To establish a connection between a PE and a memory block, the switch settings at only one crosspoint need to be changed. Since there is only one set of switches in any path, the crossbar offers uniform latency. If two PEs try to access the same memory, then there is a contention and one of them needs to wait. Such contention problems can be minimized by appropriate memory organizations.

The operation of the MIMD requires that the processor–memory interconnections be changed in a dynamic fashion. With high-speed switching needed for such operation, high-frequency capacitive and inductive effects result in noise problems that dominate crossbar design. Because of the noise problems and high cost, large crossbar networks are not practical.

The complete connectivity offered by crossbars may not be needed always. Depending on the application, a "sparse" crossbar network in which only certain crosspoints have switches may be sufficient, as long as it satisfies the bandwidth and connectivity requirements.

The aforementioned description uses memory blocks as one set of nodes connected to a set of PEs. In general, each node could be either a PE or a memory block or a complete computer system with PE, memory, and I/O components.

In an SIMD architecture using crossbar IN for PEs, the switch settings are changed according to the connectivity requirements of the application at hand. This is done either by the CP or by a dedicated switch controller. Alternatively, a decentralized control strategy can be used where the switches at each crosspoint forward the message toward the destination node.

12.2.1.4.2 Switching Networks

Single and multistage switching networks offer a cost/performance compromise between the two extremes of bus and crossbar networks. A majority of switching networks proposed are based on an interconnection scheme known as perfect shuffle. The following paragraphs illustrate the single- and multistage network concepts as applied to SIMD architectures, through the perfect shuffle IN. The discussion of other switching networks is deferred to the next section.

12.2.1.4.3 Perfect Shuffle (Stone 1971)

This network derives its name from its property of rearranging the data elements in the order of the perfect shuffle of a deck of cards. That is, the card deck is cut exactly in half and the two halves are merged such that the cards in similar position in each half are brought adjacent to each other.

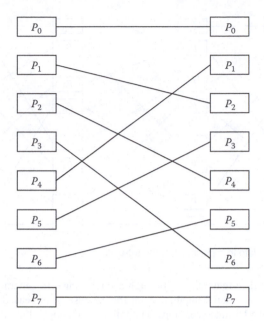

Figure 12.9 Perfect shuffle network.

Example 12.2

Figure 12.9 shows the shuffle network for eight PEs. Two sets of PEs are shown here for clarity. Actually, there are only eight PEs. The shuffle network first partitions the PEs into two groups (the first containing 0, 1, 2, and 3 and the second containing 4, 5, 6, and 7). The two groups are then merged such that 0 is adjacent to 4, 1–5, 2–6, and 3–7.

The interconnection algorithm can also be derived from a cyclic shift of the addresses of PEs. That is, number PEs starting from 0, with each PE number consisting of n bits where the total number of PEs is 2^n. Then, to determine the destination of PE after a shuffle, shift the n-bit address of the PE left once cyclically. The shuffle shown in Figure 12.9 is thus derived as following:

PE	Source	Destination	PE
0	000	000	0
1	001	010	2
2	010	100	4
3	011	110	6
4	100	001	1
5	101	011	3
6	110	101	5
7	111	111	7

This transformation can be described by

$$\text{Shuffle}(i) = 2i \quad \text{if } i < N/2$$

$$= 2i - N + 1 \quad \text{if } i \geq N/2, \qquad (12.4)$$

where N is the number of PEs and is a power of 2.

The shuffle network of Figure 12.9 is a single-stage network. If an operation requires multiple shuffles to complete, the network is used multiple times. That is, the data recirculate through the network. Figure 12.10 shows a multistage shuffle network for eight PEs in which each stage corresponds to one shuffle. That is, the data inserted into the first stage ripples through the

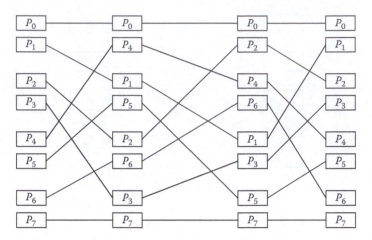

Figure 12.10   Movement of data in a perfect shuffle network.

multiple stages rather than recirculating through a single stage. In general, multistage network implementations provide faster computations at the expense of increased hardware compared with single-stage network implementations. The following example illustrates this further.

In the network of Figure 12.10, the first shuffle makes the vector elements that were originally $2^{(n-1)}$ distance apart adjacent to each other; the next shuffle brings the elements originally $2^{(n-2)}$ distance apart adjacent to each other, and so on. In general, the ith shuffle brings the elements that were $2^{(n-i)}$ distance apart adjacent to each other. In addition to the shuffle network, if the PEs are connected such that adjacent PEs can exchange data, the combined network can be used efficiently in several computations.

Figure 12.11 shows a network of eight PEs. In addition to the perfect shuffle, adjacent PEs are connected through a function block capable of performing several operations on the data from the PEs connected to it. Several possible functions are shown in Figure 12.11b. The add function returns the sum of the two data values to the top PE. The order function rearranges data in increasing or decreasing order. The swap function exchanges the data values and the pass through function retains them as they are.

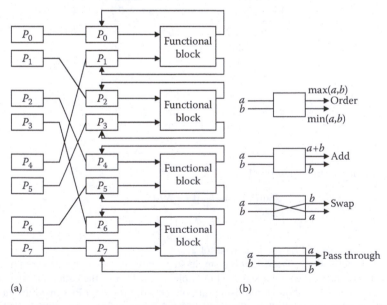

Figure 12.11   Processors connected by perfect shuffle. (a) Network. (b) Representative functions.

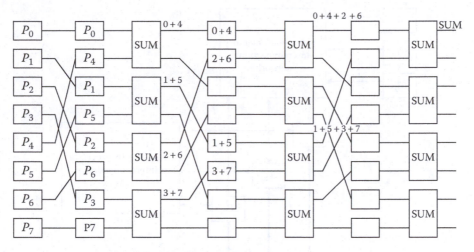

Figure 12.12 Accumulation of *N* items using perfect shuffle.

Example 12.3

Figure 12.12 shows the utility of the network of Figure 12.11 in the accumulation of *N* data values in the PEs. Each stage in the network shuffles the data and adds the neighboring elements with the sum appearing on the upper output of the SUM block. All the four SUM blocks are required in the first stage. Only the first and the third SUM blocks are needed in the second stage and only the first one in the last stage, for this computation. If each shuffle-and-add cycle consumes *T* time units, this addition requires only $\log_2 N \times T$ time units compared to $N \times T$ time units required by a sequential process. Also note that the CP will have to issue only one instruction (ADD) to perform this addition.

The addition of *N* elements can also be accomplished by the single-stage network of Figure 12.11, where each functional block will be a SUM block. Now, the data has to circulate through the network $\log_2 N$ times. Thus, the CP will have to execute the following program rather than a single instruction (as in the implementation with multistage network):

```
for i = 1 to log₂N
   shuffle
   add
endfor
```

Assuming *T* time units for shuffle and add as before, the single-stage network implementation requires an execution time of $T \times \log_2 N +$ fetch/decode time for instructions in the loop body each time through the loop + the loop control overhead.

12.3 MIMD

Figure 12.13 shows an MIMD structure consisting of *p* memory bocks, *n* PEs, and *m* input/ output channels. The processor-to-memory IN enables the connection of a processor to any of the memory blocks. In addition to establishing the processor–memory connections, the network should also have a memory-mapping mechanism that performs a logical-to-physical address mapping. The processor-to-I/O IN enables the connection of an I/O channel to any of the processors. The processor-to-processor IN is more of an interrupt network than a data exchange network, since the majority of data exchanges can be performed through the memory-to-processor interconnection.

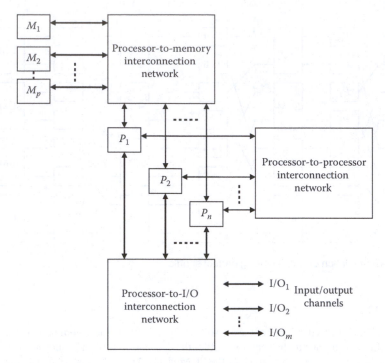

Figure 12.13   MIMD structure.

The most important characteristic of the multiprocessor systems discussed in this chapter is that all the processors function independently. That is, unlike the SIMD systems in which all the processors execute the same instruction at any given instant of time, each processor in a multiprocessor system can be executing a different instruction at any instant of time. For this reason, Flynn classified them as MIMD computers.

As mentioned earlier, the need for parallel execution arises since the device technology limits the speed of execution of any single processor. SIMD systems increased the performance and the speed manifold simply due to data parallelism. But, such parallel processing improves performance only in a limited case of applications that can be organized into a series of repetitive operations on uniformly structured data. Since a number of applications cannot be represented in this manner, SIMD systems are not a panacea. This led to the evolution of a more general form, the MIMD architectures where each PE has its own arithmetic/logic unit (ALU) and control unit and, if necessary, its own memory and I/O devices. Thus, each PE is a computer system in itself, capable of performing a processing task totally independent of other PEs. The PEs are interconnected in some manner to allow exchange of data and programs and to synchronize their activities.

The major advantages of MIMD systems are as follows:

1. Reliability: If any processor fails, its workload can be taken over by another processor thus incorporating graceful degradation and better fault tolerance in the system.
2. High performance: Consider an ideal scenario where all the N processors are working on some useful computation. At times of such peak performance, the processing speed of an MIMD is N times that of a single-processor system. However, such peak performance is difficult to achieve due to the overhead involved with MIMD operation. The overhead is due to
   a. Communication between processors
   b. Synchronization of the work of one processor with that of another processor
   c. Wastage of processor time if any processor runs out of tasks to do
   d. Processor scheduling (i.e., allocation of tasks to the processors)

A *task* is an entity to which a processor is assigned. That is, a task is a program, a function, or a procedure in execution on a given processor. *Process* is simply another word for a task. A processor or a PE is a hardware resource on which tasks are executed. A processor executes several tasks one after another. The sequence of tasks performed by a given processor in succession forms a thread. Thus, the path of execution of a processor through a number of tasks is called a thread. Multiprocessors provide for simultaneous presence of a number of threads of execution in an application.

Example 12.4

Consider Figure 12.14 in which each block represents a task and each task has a unique number. As required by the application, task 2 can be executed only after task 1 is completed and task 3 can be executed only after task 1 and task 2 are both completed. Thus, the line through tasks 1, 2, and 3 forms a thread (A) of execution. Two other threads (B and C) are also shown. These threads limit the execution of tasks to a specific serial manner. Although task 2 has to follow task 1 and task 3 has to follow task 2, note that task 4 can be executed in parallel with task 2 or task 3. Similarly, task 6 can be executed simultaneously with task 1, and so on. Suppose the MIMD has three processors and if each task takes the same amount of time to execute, the seven tasks shown in the figure can be executed in the following manner.

Time Slot	Processor 1	Processor 2	Processor 3
1	Task 1	Task 6	
2	Task 2	Task 4	Task 7
3	Task 3		Task 5

Initially, since there are only two tasks that can be executed in parallel, they are allocated to processors 1 and 2 (arbitrarily) and processor 3 sits idle. At the end of the first time slot, all three processors get a task each from the three different threads. Thus, tasks 2, 4, and 7 are executed in parallel. Finally, tasks 3 and 5 are again executed in parallel with one of the processors sitting idle due to the lack of tasks at that instant.

Figure 12.14 implies that the application at hand is partitioned into seven tasks and exhibits a degree of parallelism of 3 since at the most three tasks can be executed simultaneously. It is assumed that there is no interaction between the tasks in this figure. In practice, tasks communicate with each other since each task depends on the results produced by other tasks. Obviously, the communication between tasks reduces to zero if all the tasks are combined into a single task and run on a single processor (i.e., SISD mode).

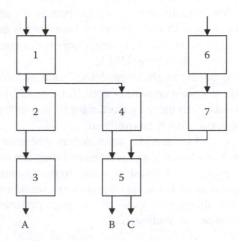

Figure 12.14 Threads of execution.

The R/C ratio, where R is the length of the run time of the task and C is the communication overhead produced by that task, signifies task granularity. This ratio is a measure of how much overhead is produced per unit of computation. A high R/C ratio implies that the communication overhead is insignificant compared with computation time, while a low R/C ratio implies that the communication overhead dominates computation time and hence a poorer performance.

High R/C ratios imply coarse-grain parallelism, while low R/C ratios result in fine-grain parallelism. The general tendency to obtain maximum performance is to resort to the finest possible granularity, thus providing for the highest degree of parallelism. However, care should be taken to see that this maximum parallelism does not lead to maximum overhead. Thus, a trade-off is required to reach an optimum level of granularity.

12.3.1 MIMD Organization

As mentioned earlier, in an MIMD system each PE works independently of the others. Processor 1 is said to be working independently of processors 2, 3, …, N at any instant, if and only if the task being executed by processor 1 has no interactions with the tasks executed by processors 2, 3, …, N and vice versa at that instant. However, the results from the tasks executed on a processor X now may be needed by processor Y sometime in the future. To make this possible, each processor must have the capability to communicate the results of the task it performs to other tasks requiring them. This is done by sending the results directly to a requesting process or storing them in a shared-memory (that is a memory to which each processor has equal and easy access) area. These communication models have resulted in two popular MIMD organizations. These are

1. Shared memory architecture
2. Message passing architecture

12.3.1.1 Shared-Memory Architecture

Figure 12.15a shows the structure of a shared-memory MIMD. Here, any processor i can access any memory module j through the IN. The results of the computations are stored in the memory by the processor that executed that task. If these results are required by any other task, they can be easily accessed from the memory. Note that each processor is a full-fledged SISD capable of fetching instructions from the memory and executing them on the data retrieved from the memory. No processor has a local memory of its own.

This is called *tightly coupled* architecture, since the processors are interconnected such that the interchange of data between them through the shared memory is quite rapid. This is the main advantage of this architecture. Also, the memory access time is the same for all the processors and hence the name *uniform memory architecture* (UMA).

If the processors in the system are nonhomogeneous, data transformations will be needed during exchange. For instance, if the system consists of both 16-bit and 32-bit processors and the shared memory consists of 32-bit words, each memory word must be converted into two words for the use of 16-bit processors and vice versa. This is an overhead.

Another problem is the memory contention, which occurs whenever two or more processors try to access the same shared-memory block. Since a memory block can be accessed only by one processor at a time, all the other processors requesting access to that memory block must wait until the first processor is through using it. In addition, if two processors simultaneously request the access to the same memory block, one of the processors should be given preference over the other. Memory organization concepts are discussed in Section 12.2.

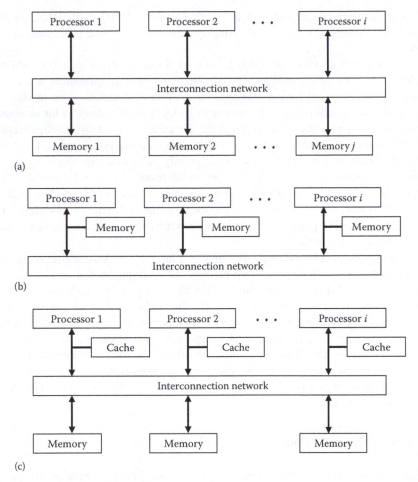

Figure 12.15 MIMD structures. (a) Shared-memory system. (b) Message-passing system. (c) MIMD with processors having personal caches.

12.3.1.2 Message-Passing Architecture

This is the other extreme, where there is no shared memory at all in the system. Each processor has a (local) memory block attached to it. The conglomeration of all local memories is the total memory that the system possesses. Figure 12.15b shows a block diagram of this configuration, also known as loosely coupled or distributed-memory MIMD system. If data exchange is required between two processors in this configuration, the requesting processor i sends a message to processor j, in whose local memory the required data are stored. In reply to this request, the processor j (as soon it can) reads the requested data from its local memory and passes it on to processor i through the IN. Thus, the communication between processors occurs through message passing.

The requested processor usually finishes its task at hand and then accesses its memory for the requested data and passes it on to the IN. The IN routes it toward the requesting processor. All this time, the requesting processor sits idle waiting for the data, thus incurring a large overhead. The memory access time varies between the processors and hence, these architectures are known as *Nonuniform Memory Architectures* (NUMA). Thus, a tightly coupled MIMD offers more rapid data interchange between processors than a loosely coupled MIMD, while the memory contention problem is not present in a message-passing system since only one processor has access to a memory block.

Shared memory architectures are also known as *multiprocessor* systems, while message-passing architectures are called *multicomputer* systems. These architectures are two extremes. MIMD systems in practice may have a reasonable mix of the two architectures as shown in Figure 12.15c. In this structure, each processor operates in its local environment as far as possible. Interprocessor communication can be either through the shared memory or by message passing.

Several variations of this memory architecture have been used. For instance, the data diffusion machine (DDM) (Hagersten et al. 1992) uses a cache-only memory architecture (COMA) in which all system memory resides in large caches attached to the processors in order to reduce latency and network load. The IBM research parallel processor (RP3) consists of 512 nodes, each containing 4 MB of memory. The interconnection of nodes is such that the 512 memory modules can be used as one global shared memory or purely as local memories with message-passing mode of communication or a combination of the both.

MIMD systems can also be conceptually modeled as either private-address-space or shared-address-space machines. Both address-space models can be implemented on shared-memory and message-passing architectures. Private memory, shared-address-space machines are NUMA architectures that offer scalability benefits of message-passing architectures, with programming advantages of shared-memory architectures. An example of this type is the J-machine from MIT, which has small private memory attached to each of a large number of nodes but has a common address space across the whole system. The DASH machine from Stanford considers local memory as a cache for the large global address space, but the global memory is actually distributed. In general, the actual configuration of an MIMD system depends on the application characteristics for which the system has been designed.

12.3.1.3 *Memory Organization*

Two parameters of interest in MIMD memory system design are the bandwidth and the latency. For an MIMD system to be efficient, memory bandwidth must be high enough to provide for simultaneous operation of all the processors. When memory modules are shared, the memory contention must be minimized. In addition, the latency (which is the time elapsed between a processor's request for data from the memory and its receipt) must be minimized. This section examines memory organization techniques that reduce these problems to a minimum and tolerable level.

Memory latency is reduced by increasing the memory bandwidth, which in turn is accomplished by one or both of the following mechanisms:

1. By building the memory system with multiple independent memory modules, thus providing for concurrent accesses of the modules. Banked, interleaved, and a combination of the two addressing architectures have been used in such systems. A recent trend is to use multiport memory modules in the design to achieve concurrent access.
2. By reducing the memory access and cycle times utilizing memory devices from the highest speed technology available. This is usually accompanied by high price. An alternative is to use cache memories in the design.

To understand the first method, consider an MIMD system with N processors and a shared-memory unit. In the worst case, all but one processor may be waiting to get access to the memory and not doing any useful computation, since only one processor can access the memory at a given instant of time. This bottlenecks the overall performance of the system. A solution to this problem is to organize memory such that more than one simultaneous access to the memory is possible.

Example 12.5

Figure 12.16a shows an MIMD structure in which N memory modules are connected to N processors through a crossbar IN. All the N memory modules can be accessed simultaneously by N different processors through the crossbar. To make the best possible use of such a design, all the instructions to be executed by one processor are kept in one memory module. Thus, a given processor accesses a given memory block for as long a time as possible, and the concurrency of memory accesses can be maintained over a longer duration of time. This mode of operation requires the banked memory architecture. If an interleaved memory architecture is used, consecutive addresses lie in different memory modules. Thus, the instructions corresponding to a task would be spread over several memory modules. If two tasks require the same code segment,

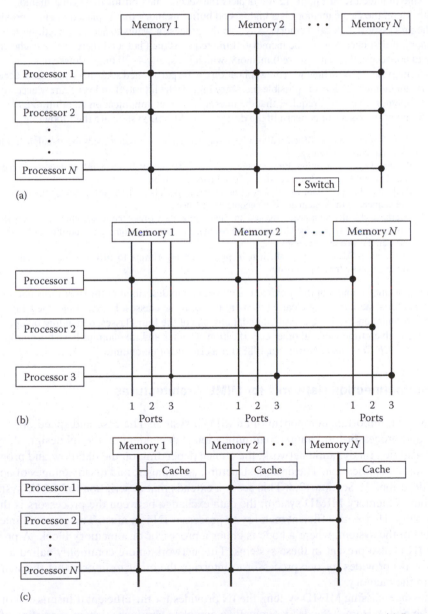

Figure 12.16 Processor–memory interconnection. (a) Crossbar. (b) Multiport memories. (c) Processors with personal caches.

it is possible to allow simultaneous access to the code segment, as long as one task starts slightly (at least one instruction cycle time) earlier than the other. Thus, processors accessing the code march one behind the other spreading the memory access to different modules and minimizing contention.

Figure 12.16b shows the use of multiport memories. Here, each memory module is a three-port memory device. All the three ports can be active simultaneously, reading and writing data to and from the memory block. The only restriction is that only one port can write data into a memory location. If two or more ports try to access the same location for writing, only the highest priority port will succeed. Thus, multiport memories have the contention resolution logic built into them and provide for concurrent access, at the expense of complex hardware. Large multiport memories are still expensive, because of their hardware complexity.

The architecture of Figure 12.16c depicts the second method for increasing memory bandwidth. Here, the cache memory is a high-speed buffer memory that is placed in close proximity to the processor (i.e., local memory). Anytime the processor wants to access something from the memory, it first checks its cache memory. If the required data is found there (i.e., a cache "hit"), it need not access the main (shared) memory, which is usually 4–20 times slower. The success of this strategy depends on how well the application can be partitioned such that a processor accesses its private memory as long as possible (i.e., very high cache hit ratio) and very rarely accesses the shared memory. This also requires that the interprocessor communication be minimized.

Some of the issues of concern in the design of an MIMD system are the following:

1. Processor scheduling: Efficient allocation of processors to processing needs in a dynamic fashion as the computation progresses.
2. Processor synchronization: Prevention of processors trying to change a unit of data simultaneously and obeying the precedence constraints in data manipulation.
3. IN design: The processor-to-memory or processor-to-peripheral IN is still probably the most expensive element of the system and can become a bottleneck.
4. Overhead: Ideally an n processor system should provide n times the throughput of a uniprocessor. This is not true in practice because of the overhead processing required to coordinate the activities between the various processors.
5. Partitioning: Identifying parallelism in processing algorithms to invoke concurrent processing streams is not a trivial problem.

It is important to note that the architecture classification described in this section is not unique. A computer system may not clearly belong to one of these classes. For example, the Cray series of supercomputers could be classified under any one of the four classes, depending on operating mode at a given time. Several other classification schemes and taxonomies have been proposed. Refer to *IEEE Computer* (November 1988) for a critique of taxonomies.

12.3.2 Interconnection Networks for MIMD Architectures

The IN is an important component of an MIMD system. The ease and speed of processor-to-processor and processor-to-memory communication is dependent on the IN design. A system can use either a static or a dynamic network, the choice depending on the dataflow and program characteristics of the application. The design, structure, advantages, and disadvantages of a number of INs were described in Section 12.3. This section extends that description to MIMD systems.

In a shared-memory MIMD system, the data exchange between the processors is through the shared memory. Hence, an efficient memory to processor IN is a must. This network interconnects the "nodes" in the system, where a node is either a processor or a memory block. A processor-to-processor IN is also present in these systems. This network (more commonly called a synchronization network) provides for one processor to interrupt the other to inform that the shared data is available in the memory.

In a message-passing MIMD system, the IN provides for the efficient transmission of messages between the nodes. Here, a "node" is typically a complete computer system consisting of a processor, memory, and I/O devices.

The most common interconnection structures used in MIMD systems are the following:

1. Bus
2. Loop or ring
3. Mesh
4. Hypercube
5. Crossbar
6. Multistage switching networks

Details of loop, mesh, hypercube, and crossbar networks were provided in the previous section as applied to SIMD systems. These networks are used in MIMD system design also, except that the communication occurs in an asynchronous manner, rather than the synchronous communication mode of SIMD systems. The rest of this section highlights the characteristics of these networks as applied to MIMD systems and covers bus and multistage switching networks in detail.

12.3.2.1 Bus Network

Bus networks are simple to build and provide the least cost among the three types of dynamic networks discussed earlier. They also offer the lowest performance. The bandwidth of the bus is defined as the product of its clock frequency and the width of the data path. The bus bandwidth must be large enough to accommodate the communication needs of all the nodes connected to it. Since the bandwidth available on the network for each node decreases as the number of nodes in the network increases, bus networks are suitable for interconnecting a small number of nodes.

The bus bandwidth can be increased by increasing the clock frequency. But technological advances that make higher bus clock rates possible also provide faster processors. Hence, the ratio of processor speed to bus bandwidth is likely to remain the same, thus limiting the number of processors that can be connected to a single-bus structure.

The length of the bus also affects the bus bandwidth since the physical parameters such as capacitance, inductance, and signal degradation are proportional to the length of wires. In addition, the capacitive and inductive effects grow with the bus frequency, thus limiting the bandwidth.

Figure 12.17 shows a shared-memory MIMD system. The global memory is connected to a bus to which several nodes are connected. Each node consists of a processor, its local memory, cache, and I/O devices. In the absence of cache and local memories, all nodes try to access the shared memory through the single bus. For such a structure to provide maximum performance, both the

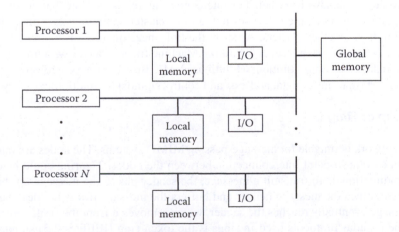

Figure 12.17 Shared-memories, shared-bus MIMD architecture.

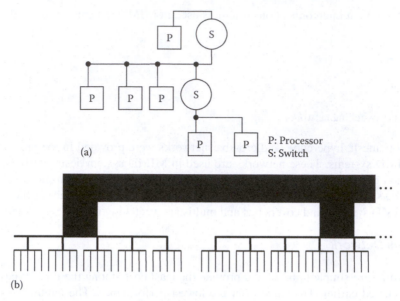

Figure 12.18 Interconnection structures. (a) Bus window. (b) Fat tree.

shared bus and the shared memory should have high enough bandwidths. These bottlenecks can be reduced if the application is partitioned such that a majority of memory references by a processor are to its local memory and cache blocks, thus reducing the traffic on the common (shared) bus and the shared memory. Of course, the presence of multiple caches in the system brings in the problem of cache coherency.

If we use a multiport memory system for the shared memory, then a multiple-bus interconnection structure can be used, with each port of the memory connected to a bus. This structure reduces the number of processors on each bus.

Another alternative is a bus window scheme shown in Figure 12.18a. Here, a set of processors is connected to a bus with a switch (i.e., a bus window) and all such buses are connected to form the overall system. The message transmission characteristics are identical to those of global bus, except that multiple-bus segments are available. Messages can be retransmitted over the paths on which they were received or on other paths.

Figure 12.18b shows a fat tree network that is gaining popularity. Here, communication links are fatter (i.e., have higher bandwidth) when they interconnect more nodes. Note that in practice, applications are partitioned such that the processes in the same cluster communicate with each other more often than with those in other clusters. As such, the links near the root of the tree must be thinner compared with the ones near the leaves. Thinking Machine Incorporated CM-5 uses the fat tree IN.

Several standard bus configurations (Multibus, VME Bus, etc.) have evolved over the years. They offer support (in terms of data, address, and control signals) for multiprocessor system design.

12.3.2.2 Loop or Ring

The ring network is suitable for message-passing MIMD systems. The nodes are interconnected by a ring with a point-to-point interconnection between the nodes. The ring could be either unidirectional or bidirectional. To transmit a message, the sender places the message on the ring. Each node in turn examines the message header and buffers the message if it is the designated destination. The message eventually reaches the sender, which removes it from the ring.

One of the popular protocols used in rings is the token ring (IEEE 802.5) standard. A token (which is a unique bit pattern) circulates over the ring. When a node wants to transmit a message, it accepts the token (i.e., prevents it from moving to the next node) and places its message on the ring. Once the message is accepted by the receiver and reaches the sender, the sender removes the message and places the token on the ring. Thus, a node can be a transmitter only when it has the token.

Since the interconnections in the ring are point-to-point, the physical parameters can be controlled more readily, unlike bus interconnections, especially when very high bandwidths are needed.

One disadvantage of token ring is that each node adds a 1-bit delay to the message transmission. Thus, the delay increases as the number of nodes in the system increases. If the network is viewed as a pipeline with a long delay, the bandwidth of the network can be effectively utilized. To accommodate this mode of operation, the nodes usually overlap their computations with the message transmission.

One other way to increase the transmission rate is to provide for the transmission of a new message as soon as the current message is received by the destination node, rather than waiting until the message reaches the sender, where it is removed.

12.3.2.3 Mesh Network

The mesh networks are ideal for applications with very high near-neighbor interactions. If the application requires a large number of global interactions, the efficiency of the computation goes down, since the global communications require multiple hops through the network. One way to improve the performance is to augment the mesh network with another global network. MasPar architectures and Intel iPSC architectures utilize such global interconnects.

12.3.2.4 Hypercube Network

One advantage of the hypercube networks is that routing is straightforward and the network provides multiple paths for message transmission from each node. Also, the network can be partitioned into hypercubes of lower dimensions, and hence, multiple applications utilizing smaller networks can be simultaneously implemented. For instance, a 4D hypercube with 16 processing nodes can be used as two 3D hypercubes, four 2D hypercubes, and so on.

One disadvantage of the hypercube is its scalability, since the number of nodes has to be increased in powers of 2. That is, to increase the number of nodes from 32 to 33, the network needs to be expanded from a 5D to a 6D network consisting of 64 nodes. In fact, the Intel Touchstone project has switched over to mesh networks from hypercubes because of this scalability issue.

12.3.2.5 Crossbar Network

The crossbar network offers multiple simultaneous communications with the least amount of contention but at a very high hardware complexity. The number of memory blocks in the system is at least equal to the number of processors. Each processor to memory path has just one crosspoint delay.

The hardware complexity and the cost of the crossbar are proportional to the number of crosspoints. Since there are N^2 crosspoints in an $(N \times N)$ crossbar, the crossbar becomes expensive for large values of N.

12.3.2.6 Multistage Networks

Multistage switching networks offer a cost/performance compromise between the two extremes of bus and crossbar networks. A large number of multistage networks have been proposed over the past few years. Some examples are omega, baseline, banyan, and benes networks. These networks differ in their topology, operating mode, control strategy, and the type of switches used. They are

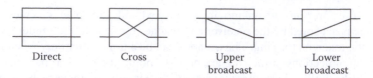

Direct	Cross	Upper broadcast	Lower broadcast

Figure 12.19 2 × 2 crossbar.

capable of connecting any source (input) node to any destination (output) node. But, they differ in the number of different N-to-N interconnection patterns they can achieve. Here, N is the number of nodes in the system.

These networks are typically composed of 2-input, 2-output switches (See Figure 12.19) arranged in $\log_2 N$ stages. Thus, the cost of the network is of the order of $N \log_2 N$ compared with N^2 of the crossbar. Communication paths between nodes in these networks are of equal length (i.e., $\log_2 N$). The latency is thus $\log_2 N$ times that of a crossbar, in general. In practice, large crossbars have longer cycle times compared with those of the small switches used in multistage networks.

The majority of multistage networks are based on the perfect shuffle interconnection scheme.

12.3.2.7 Omega Network

The Omega network (Almasi and Gottlieb 1989) is the simplest of the multistage networks. An N-input, N-output Omega interconnection topology is shown in Figure 12.20. It consists of $\log_2$ N stages. The perfect shuffle interconnection is used between the stages. Each stage contains N/2 2-input, 2-output switches. These switches can perform the four functions shown in Figure 12.19.

The network employs the packet switching mode of communication, where each packet is composed of data and the destination address. The address is read at each switch and the packet is forwarded to the next switch until it arrives at the destination node. The routing algorithm follows a simple scheme. Starting from the source, each switch examines the leading bit of the destination address and removes that bit. If the bit is 1, then the message exits the switch from the lower port; otherwise, from the upper port.

Figure 12.20 shows an example in which the message is sent from node (001) to node (110). Since the first bit of the destination address is 1, the switch in the first stage routes the packet to its lower port. Similarly, the switch in the second stage also routes it to its lower port since the second bit of the address is also 1. The switch in the third stage routes it to its upper port since the third bit of the address is 0. The path is shown in as solid line in the figure.

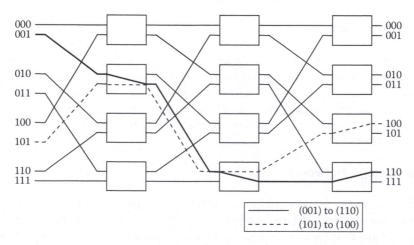

Figure 12.20 Omega network.

The aforementioned routing algorithm implies a distributed control strategy, since the control function is distributed among the switches. An alternative control strategy would be centralized control, in which the controller sends control signals to each switch based on the routing required. It may operate either on a stage by stage basis, starting from the first to the last, or it may reset the whole network each time, depending on the communication requirements.

Figure 12.20 also shows (in dotted lines) the path from node (101) to node (100). Note that the link between stages 1 and 2 is common to both the paths shown in the figure. Thus, if nodes (001) and (101) send their packets simultaneously to the corresponding destination nodes, one of the packets is "blocked" until the common link is free. Hence, this network is called a "blocking" network.

Note that there is only one unique path in this network from each input node to an output node. This is a disadvantage since the message transmission gets blocked, even if one link in the path is part of the path for another transmission in progress. One way to reduce the delays due to blocking is to provide buffers in switching elements so that the packets can be retained locally until the blocked links are free. The switches can also be designed to "combine" messages bound for the same destination. Recall that a crossbar is a nonblocking network, but it is much more expensive than an omega network.

The progress in hardware technology has resulted in the availability of fast processors and memory devices, two components of an MIMD system. Standard bus systems such as Multibus and VME bus allow the implementation of bus-based MIMD systems. Although hardware modules to implement other interconnection (loop, crossbar, etc.) structures are now appearing off-the-shelf, the design and implementation of an appropriate interconnection structure is still the crucial and expensive part of the MIMD design.

12.4 CACHE COHERENCE

Consider the multiprocessor system of Figure 12.15c in which each processor has a local (private) memory. The local memory can be viewed to be a cache. As the computation proceeds, each processor updates its cache. However, updates to a private cache are not visible to other processors. Thus, if one processor updates its cache entry corresponding to a shared data item, the other caches containing that data item will not be updated and hence, the corresponding processors operate with stale data. This problem wherein the value of a data item is not consistent throughout the memory system is known as cache incoherency. Hardware and software schemes should then be applied to insure that all processors see the most recent value of the data item all the time. This is the process of making the caches coherent.

Two popular mechanisms for updating cache entries are *write-through* and *write-back*. In write-through, a processor updating the cache also simultaneously updates the corresponding entry in the main memory. In write-back, an updated cache block is written back to the main memory just before that block is replaced in the cache.

The write-back mechanism clearly does not solve the problem of cache incoherency in a multiprocessor system. Write-through keeps the data coherent in a single-processor environment. But, consider a multiprocessor system in which processors 1 and 2 both load block A from the main memory into their caches. Suppose processor 1 makes some changes to this block in its cache and writes-through to the main memory. Processor 2 will still see stale data in its cache since it was not updated. Two possible solutions are as follows: (1) Update all caches that contain the shared data when the write-through takes place. Such a write-through will create an enormous overhead on the memory system. (2) Invalidate the corresponding entry in other processor caches when a write-through occurs. This forces the other processors to copy the updated data into their caches when needed later.

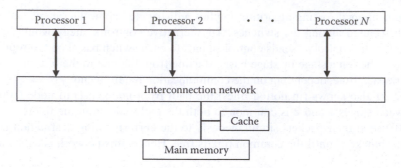

Figure 12.21 System with only one cache associated with the main (shared) memory.

Cache coherence is an active area of research. Several cache coherency schemes have evolved over the years. Some popular schemes are outlined below.

1. The least complex scheme for achieving cache coherency is not to use private caches. In Figure 12.21, the cache is associated with the shared-memory system rather than with each processor. Any memory write by a processor will update the common cache (if it is present in the cache) and will be seen by other processors. This is a simple solution but has the major disadvantage of high cache contention since all the processors require access to the common cache.

2. Another simple solution is to stay with the private cache architecture of Figure 12.21, but to only cache nonshared data items. Shared data items are tagged as noncached and stored only in the common memory. The advantage of this method is that each processor now has its own private cache for nonshared data, thus providing a higher bandwidth. One disadvantage of this scheme is that the programmer and/ or compiler has to tag data items as cached or noncached. It would be preferred that cache coherency schemes were transparent to the user. Further, access to shared items could result in high contention.

3. Cache flushing is a modification of the previous scheme in which the shared data is allowed to be cached only when it is known that only one processor will be accessing the data. After the shared data in the cache are accessed and the processor is through using it, it issues a flush-cache instruction that causes all the modified data in the cache to be written back to the main memory and the corresponding cache locations to be invalidated. This scheme has the advantage of allowing shared areas to be cached; but it has the disadvantage of the extra time consumption for the flush-cache instruction. It also requires program code modification to flush the cache.

4. The aforementioned coherency schemes eliminate private caches, limit what may be cached, or require programmer's intervention. A caching scheme that avoids these problems is bus watching or bus snooping. Bus snooping schemes incorporate hardware that monitors the shared bus for data LOAD and STORE into each processor's cache controller as shown in Figure 12.22. The snoopy cache controller controls the status of data contained within its cache based on the LOAD and STORE seen on the bus.

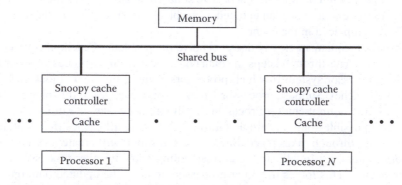

Figure 12.22 Snoopy bus controller on a shared bus.

If the caches in this architecture are write-through, then every STORE to cache is written-through simultaneously to the main memory. In this case, the snoopy controller sees all STOREs and take actions based on that. Typically, if a STORE is made to a locally cached block and the block is also cached in one or more remote caches, the snoopy controllers in remote caches will either update or invalidate the blocks in their caches. The choice of updating or invalidating remote caches will have its effect on the performance. The primary difference is the time involved to update cache entries versus merely changing the status of a remotely cached block. Secondly, as the number of processors increases the shared bus may become saturated. Note that for every STORE, main memory must be accessed, and for every STORE hit, additional bus overhead is generated. LOADs are performed with no additional overhead.

For instance, the following explanation of the Illinois cache coherence protocol (Papamarcos and Patel 1984) is based on the explanation in Archibald and Baer (1986). Each cache holds a cache state per block. The cache state is one of the following:

1. Invalid: The data for this block is not cached.
2. Valid exclusive: The data for this block is valid, clean (identical to the data held in main memory), and is the only cached copy of the block in the system.
3. Shared: The data for this block is valid, clean, and there are possibly other cached copies of this block in the system.
4. Dirty: The data for this block is valid, modified relative to the main memory, and is the only cached copy of the block in the system. Initially, all blocks are invalid in all caches. Cache states change according to bus transactions as shown in Figure 12.23. Cache state transitions by the requesting processor are shown as solid lines. Cache state transitions of snooping processors are shown as dotted lines. For instance, on a read, if no snooping processor has a copy of the block, the requesting processor will transition to state valid exclusive. If some snooping processor does have a copy of the block, then the requesting processor will transition to state shared. In addition, all snooping processors with a copy of the block will observe the read and transition to state shared. If some snooping processor has the block in state dirty, it will write the data to main memory at the same time.
5. A solution more appropriate for bus organized multiprocessor systems has been proposed by Goodman (1983). In this scheme, an invalidate request is broadcast only when a block is written in cache for the first time. The updated block is simultaneously written-through to the main memory. Only if a block in cache is written to more than once it is necessary to write it back before replacing it. Thus, the first STORE causes a write-through to the main memory and it also

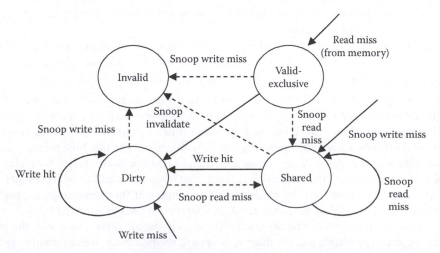

Figure 12.23 Illinois cache state transition diagram. (From Archibald, J. and Baer, J.L., *ACM Trans. Comput. Syst.*, 4(4), 273, 1986.)

invalidates the remotely cached copies of that data. Subsequent STOREs do not get written to the main memory, and since all other cache copies are marked invalid, copies to other caches are not required. When any processor executes a LOAD for this data, the cache controller locates the unit (main memory or cache) that has a valid copy of the data. If the data is in a block marked dirty, then the cache supplies the data and writes the block to the memory. This technique is called write-once.

In a write-back cache scheme, since a modified cache block is loaded to the main memory only when that block is to be replaced, it conserves the bandwidth on the shared bus and thus is generally faster than write-through. But this added throughput is at the cost of a more complex bus-watching mechanism. In addition to the cache controller watching a bus, it must also maintain ownership information for each cached block, allowing only one copy of the cached block at a time to be writable. This type of protocol is called an ownership protocol.

An ownership protocol works in general, as follows. Each block of data has one owner. If main memory or a cache owns a block, all other copies of that block are read-only (RO). When a processor needs to write to a RO block, a broadcast to the main memory and all caches is made in an attempt to find any modified copies. If a modified copy exists in another cache, it is written to main memory, copied to the cache requesting read–write privileges, and then the privileges are granted to the requesting cache.

This section has addressed the most primitive cache coherency schemes. Cache coherence has been an active area of research and has resulted in several other schemes.

12.5 DATAFLOW ARCHITECTURES

Dataflow architectures tend to maximize the concurrency of operations (parallelism) by breaking the processing activity into sets of the most primitive operations possible. Further, the computations in a dataflow machine are *data driven*. That is, an operation is performed as and when its operands are available. This is unlike the *control-driven* machines we have described so far, where the required data are gathered when an instruction needs them. The sequence of operation in a dataflow machine obeys the precedence constraint imposed by the algorithm used rather than by the control statements in the program. A dataflow architecture assumes that a number of functional units are available, that as many of these functional units as possible are invoked at any given time, and these functional units are purely functional in the sense that they induce no side effects on either the data or the computation sequence.

The dataflow diagram of Figure 12.24 shows the computation of the roots of a quadratic equation. Assuming that a, b, and c values are available, $(-b)$, (b^2) (ac), and $(2a)$ can be computed immediately, followed by the computation of $(4ac)$, $(b^2 - 4ac)$, and $\sqrt{(b^2 - 4ac)}$ in that order. After this, $-b + \sqrt{(b^2 - 4ac)}$ and $-b - \sqrt{(b^2 - 4ac)}$ can be simultaneously computed followed by the simultaneous computation of the two roots. Note that the only requirement is that the operands be ready before an operation can be invoked. No other time or sequence constraints are imposed.

Figure 12.25 shows a schematic view of a dataflow machine. The machine memory consists of a series of cells where each cell contains an opcode and two operands. When both operands are ready, the cell is presented to the arbitration network that assigns the cell to either a functional unit (for operations) or a decision unit (for predicates). The outputs of functional units are presented to the distribution network, which stores the result in appropriate cells as directed by the control network. A very high throughput can be achieved if the algorithms are represented with the maximum degree of concurrency possible and the three networks of the processor are designed to bring about fast communication between the memory, functional, and decision units.

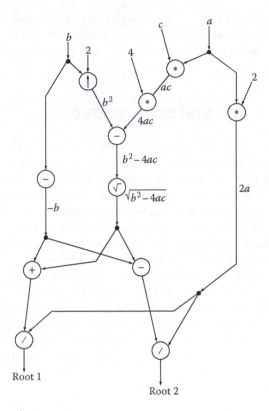

Figure 12.24 A dataflow graph.

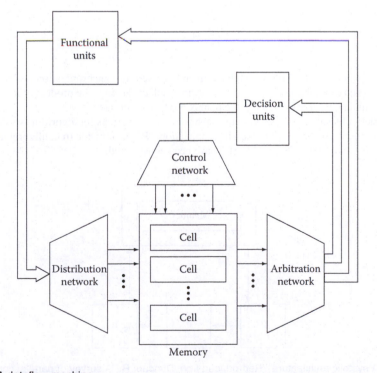

Figure 12.25 A dataflow machine.

Two experimental dataflow machines (at the University of Utah and in Toulouse, France) have been built. The dataflow project at the Massachusetts Institute of Technology has concentrated on the design of languages and representation techniques and feasibility evaluation of dataflow concepts through simulation.

12.6 SYSTOLIC ARCHITECTURES

Kung (1982) proposed systolic architectures as a means of solving problems of special-purpose systems that must often balance intensive computations with demanding I/O bandwidths. Systolic arrays (architectures) are pipelined multiprocessors in which data is pulsed in a rhythmic fashion from memory through a network of processors and the results returned to the memory (see Figure 12.26). A global clock and explicit timing delay synchronize this pipeline dataflow, which consists of operands obtained from memory and partial results to be used by each processor. The processors are interconnected by regular, local interconnections. During each time interval, the processors execute a short, time-invariant sequence of instructions.

Systolic architectures address the performance requirements of special-purpose systems by achieving significant parallel computation and by avoiding I/O and memory bandwidth bottlenecks. A high degree of parallelism is achieved by pipelining data through multiple processors, typically arranged in a 2D fashion. Systolic architectures maximize the computations performed on a datum once it has been fetched from the memory or an external device. Once a datum enters the systolic array, it is passed to any processor that needs it without an intervening store to memory.

Example 12.6

Figure 12.27 shows how a simple systolic array would calculate the product of two matrices

$$A = \begin{bmatrix} a & b \\ c & d \end{bmatrix} \quad \text{and} \quad B = \begin{bmatrix} e & f \\ g & h \end{bmatrix}.$$

The zero inputs shown moving through the array are used for synchronization. Each processor begins with an accumulator set to zero and, during each cycle, adds the product of its two inputs to the accumulator. After five cycles, the matrix product is complete.

A variety of special-purpose systems have used systolic arrays for algorithm-specific architectures, particularly for signal processing applications. Programmable (reconfigurable) systolic architectures (Intel iWarp, Saxpy Matrix) have been constructed.

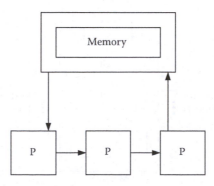

Figure 12.26 A systolic architecture. (Reproduced from Duncan, R., A survey of parallel computer architectures, *IEEE Comput.*, 23(2), 5–16, Copyright 1990. With permission from IEEE.)

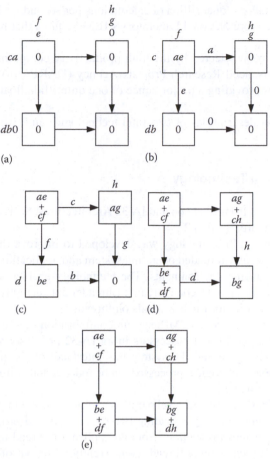

Figure 12.27 Systolic matrix multiplication. (a) Initialization. (b) Cycle 1. (c) Cycle 2. (d) Cycle 3. (e) Cycle 4. (Reproduced from Duncan, R., A survey of parallel computer architectures, *IEEE Comput.*, 23(2), 5–16, Copyright 1990. With permission from IEEE.)

12.7 EXAMPLE SYSTEMS

Several of the example systems described earlier in this book operate in SIMD and MIMD modes to certain extent. The pipelining (MISD) structure is used extensively in all modern-day machines. The Cray series of architectures described earlier in the book operate in all the four modes of Flynn's classification, depending on the context of execution. In this section, we concentrate on examples of SIMD and MIMD architectures, known as *supercomputers*.

The traditional definition of supercomputers is that they are the most powerful computers available at any given time. This is a dynamic definition, in that today's supercomputers will not be considered supercomputers in a few years. Almost all supercomputers today are designed for high-speed floating-point operations. On the basis of their size, they can be further classified as high-end, midrange, and single-user systems. These systems employ SIMD, MISD, and MIMD techniques extensively.

Several supercomputer architectures have been attempted since the 1960s to overcome the SISD throughput bottleneck. The majority of these architectures have been implemented in quantities of one or two. Nevertheless, these research and development efforts have contributed immensely to the area of computer architecture.

The top 500 list of supercomputers was published in November 2012 at http://www.top500.org/lists/2012/11/. Oak Ridge National Laboratory's Titan–Cray XK7 is at the top with a Linpack performance of 17.59 petaflops (quadrillion of calculations per second, or Pflops). The number two system is Lawrence Livermore National Laboratory's IBM Sequoia that has been at the top of the list since June 2012.

Cray and IBM have been selected by the high productivity computing systems (HPCS) program of the Defense Advanced Research Projects Agency (DARPA) to develop more powerful and easier to use systems providing a performance of one quintillion floating-point operations per second (exaflops).

We will provide brief descriptions of two Intel technologies and two supercomputer systems from Cray in this section.

12.7.1 Hyper-Threading Technology

This section is extracted from Intel® 64 and IA-32 Architectures, Software Developer's Manual, Volume 1: Basic Architecture, Section 2.2.8.

Intel Hyper-Threading (HT) technology was developed to improve the performance of IA-32 processors when executing multithreaded operating system and application code or single-threaded applications under multitasking environments. The technology enables a single physical processor to execute two or more separate code streams (threads) concurrently using shared execution resources. Refer to Section 15.2 for further details on threads.

Unlike a traditional multiprocessor (MP) system configuration that uses two or more separate physical IA-32 processors, the logical processors in an IA-32 processor supporting HT share the core resources (execution engine and the system bus interface) of the physical processor. After power-up and initialization, each logical processor can be independently directed to execute a specified thread, interrupted, or halted.

HT leverages the process and thread-level parallelism found in contemporary operating systems and high-performance applications by providing two or more logical processors on a single chip. Each logical processor executes instructions from an application thread using the resources in the processor core. The core executes these threads concurrently, using out-of-order instruction scheduling to maximize the use of execution units during each clock cycle.

HT is supported by specific members of the Intel Pentium 4 and Xeon processor families. In its first implementation in Intel Xeon processor, HT makes a single physical processor appear as two logical processors. The two logical processors each have a complete set of architectural registers while sharing one single physical processor's resources. By maintaining the architecture state of two processors, an HT-capable processor looks like two processors to software, including operating system and application code.

Figure 12.28 shows a typical bus-based symmetric multiprocessor (SMP) based on processors supporting HT. Each logical processor can execute a software thread, allowing a maximum of two software threads to execute simultaneously on one physical processor. The two software threads execute simultaneously, meaning that in the same clock cycle, an "add" operation from logical processor 0 and another "add" operation and load from logical processor 1 can be executed simultaneously by the execution engine. The physical execution resources are shared and the architecture state is duplicated for each logical processor. This minimizes the die area cost of implementing HT while still achieving performance gains for multithreaded applications or multitasking workloads.

The performance potential due to HT is due to

- The fact that operating systems and user programs can schedule processes or threads to execute simultaneously on the logical processors in each physical processor.
- The ability to use on-chip execution resources at a higher level than when only a single thread is consuming the execution resources; higher level of resource utilization can lead to higher system throughput.

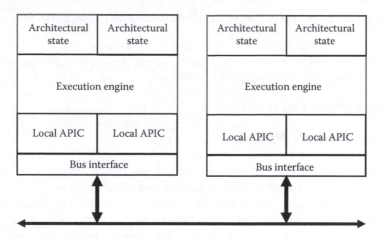

Figure 12.28 Hyper-threading technology on SMP.

The majority of microarchitecture resources in a physical processor are shared between the logical processors. Only a few small data structures were replicated for each logical processor.

The architectural state is replicated for each logical processor. The architecture state consists of registers that are used by the operating system and application code to control program behavior and store data for computations. This state includes the eight general-purpose registers, the control registers, machine state registers, debug registers, and others. There are a few exceptions, most notably the memory type range registers (MTRRs) and the performance monitoring resources.

Other resources such as instruction pointers and register renaming tables were replicated to simultaneously track execution and state changes of the two logical processors. The return stack predictor is replicated to improve branch prediction of return instructions.

In addition, a few buffers (e.g., the 2-entry instruction streaming buffers) were replicated to reduce complexity.

Several buffers are shared by limiting the use of each logical processor to half the entries. These are referred to as partitioned resources. Reasons for this partitioning include:

- Operational fairness
- Permitting the ability to allow operations from one logical processor to bypass operations of the other logical processor that may have stalled

For example, a cache miss, a branch misprediction, or instruction dependencies may prevent a logical processor from making forward progress for some number of cycles. The partitioning prevents the stalled logical processor from blocking forward progress.

In general, the buffers for staging instructions between major pipe stages are partitioned. These buffers include microoperation (μop) queues after the execution trace cache, the queues after the register rename stage, the reorder buffer that stages instructions for retirement, and the load and store buffers.

In the case of load and store buffers, partitioning also provided an easier implementation to maintain memory ordering for each logical processor and detect memory ordering violations.

Most resources in a physical processor are fully shared to improve the dynamic utilization of the resource, including caches and all the execution units. Some shared resources that are linearly addressed, like the DTLB, include a logical processor ID bit to distinguish whether the entry belongs to one logical processor or the other.

The first-level cache can operate in two modes depending on a context-ID bit:

- Shared mode: The L1 data cache is fully shared by two logical processors.
- Adaptive mode: In adaptive mode, memory accesses using the page directory are mapped identically across logical processors sharing the L1 data cache.

The other resources are fully shared.

12.7.1.1 Microarchitecture Pipeline

This section describes the HT microarchitecture and how instructions from the two logical processors are handled between the front end and the back end of the pipeline.

Although instructions originating from two programs or two threads execute simultaneously and not necessarily in program order in the execution core and memory hierarchy, the front end and back end contain several selection points to select between instructions from the two logical processors. All selection points alternate between the two logical processors unless one logical processor cannot make use of a pipeline stage. In this case, the other logical processor has full use of every cycle of the pipeline stage. Reasons why a logical processor may not use a pipeline stage include cache misses, branch mispredictions, and instruction dependencies.

12.7.1.2 Front-End Pipeline

The execution trace cache is shared between two logical processors. Execution trace cache access is arbitrated by the two logical processors every clock. If a cache line is fetched for one logical processor in one clock cycle, the next clock cycle a line would be fetched for the other logical processor provided that both logical processors are requesting access to the trace cache.

If one logical processor is stalled or is unable to use the execution trace cache, the other logical processor can use the full bandwidth of the trace cache until the initial logical processor's instruction fetches return from the L2 cache.

After fetching the instructions and building traces of μops, the μops are placed in a queue. This queue decouples the execution trace cache from the register rename pipeline stage. As described earlier, if both logical processors are active, the queue is partitioned so that both logical processors can make independent forward progress.

12.7.1.3 Execution Core

The core can dispatch up to six μops per cycle, provided the μops are ready to execute. Once the μops are placed in the queues waiting for execution, there is no distinction between instructions from the two logical processors. The execution core and memory hierarchy is also oblivious to which instructions belong to which logical processor.

After execution, instructions are placed in the reorder buffer. The reorder buffer decouples the execution stage from the retirement stage. The reorder buffer is partitioned such that each uses half the entries.

12.7.1.4 Retirement

The retirement logic tracks when instructions from the two logical processors are ready to be retired. It retires the instruction in program order for each logical processor by alternating between the two logical processors. If one logical processor is not ready to retire any instructions, then all retirement bandwidth is dedicated to the other logical processor.

Once stores have retired, the processor needs to write the store data into the level-one data cache. Selection logic alternates between the two logical processors to commit store data to the cache.

12.7.2 Intel SIMD Technology

This section is extracted from Intel 64 and IA-32 Architectures, Optimization Reference Manual, section 2.10. Intel introduced SIMD computation capabilities (see Figure 12.29) to their architectures with the MMX technology. MMX technology allows SIMD computations to be performed on packed byte, word, and doubleword integers. The integers are contained in a set of eight 64-bit registers called MMX registers (see Figure 12.30).

The Pentium III processor extended the SIMD computation model with the introduction of the Streaming SIMD Extensions (SSE). SSE allows SIMD computations to be performed on operands that contain four packed single-precision floating-point data elements. The operands can be in memory or in a set of eight 128-bit XMM registers (see Figure 12.30). SSE also extended the instruction set to include additional 64-bit MMX instructions.

Figure 12.29 shows a typical SIMD computation. Two sets of four packed data elements (X1, X2, X3, and X4 and Y1, Y2, Y3, and Y4) are operated on in parallel, with the same operation being performed on each corresponding pair of data elements (X1 and Y1, X2 and Y2, X3 and Y3, and X4 and Y4). The results of the four parallel computations are sorted as a set of four packed data elements.

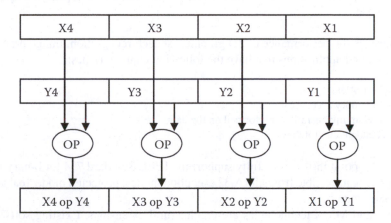

Figure 12.29 Typical SIMD operations. (From *Intel® 64 and IA-32 Architectures Optimization Reference Manual*, Volume A.)

128-bit XMM registers	64-bit MMX registers
XMM7	MM7
XMM6	MM6
XMM5	MM5
XMM4	MM4
XMM3	MM3
XMM2	MM2
XMM1	MM1
XMM0	MM0

Figure 12.30 SIMD instruction register usage. (From *Intel® 64 and IA-32 Architectures Optimization Reference Manual*, Volume A.)

The Pentium 4 processor further extended the SIMD computation model with the introduction of Streaming SIMD Extensions 2 (SSE2), Streaming SIMD Extensions 3 (SSE3), and Intel Xeon processor 5100 series introduced Supplemental Streaming SIMD Extensions 3 (SSSE3).

SSE2 works with operands in either memory or in the XMM registers. The technology extends SIMD computations to process packed double-precision floating-point data elements and 128-bit packed integers. There are 144 instructions in SSE2 that operate on two packed double-precision floating-point data elements or on 16 packed byte, 8 packed word, 4 doubleword, and 2 quadword integers.

SSE3 enhances x87, SSE, and SSE2 by providing 13 instructions that can accelerate application performance in specific areas. These include video processing, complex arithmetic, and thread synchronization. SSE3 complements SSE and SSE2 with instructions that process SIMD data asymmetrically, facilitate horizontal computation, and help avoid loading cache line splits. See Figure 12.30.

SSSE3 provides additional enhancement for SIMD computation with 32 instructions for digital video and signal processing.

SSE4.1, SSE4.2, and AESNI are additional SIMD extensions that provide acceleration for applications in media processing, text/lexical processing, and block encryption/decryption.

The SIMD extensions operate the same way in Intel 64 architecture as in IA-32 architecture, with the following enhancements:

- 128-bit SIMD instructions referencing XMM register can access 16 XMM registers in 64-bit mode.
- Instructions that reference 32-bit general-purpose registers can access 16 general-purpose registers in 64-bit mode.

SIMD improves the performance of 3D graphics, speech recognition, image processing, scientific applications, and applications that have the following characteristics:

- Inherently parallel
- Recurring memory access patterns
- Localized recurring operations performed on the data
- Data-independent control flow

SIMD floating-point instructions fully support the IEEE Standard 754 for binary floating-point arithmetic. They are accessible from all IA-32 execution modes: protected mode, real address mode, and virtual 8086 mode.

SSE, SSE2, and MMX technologies are architectural extensions. Existing software will continue to run correctly, without modification on Intel microprocessors that incorporate these technologies. Existing software will also run correctly in the presence of applications that incorporate SIMD technologies.

SSE and SSE2 instructions also introduced cacheability and memory ordering instructions that can improve cache usage and application performance.

In summary, MMX technology introduced the following:

- 64-bit MMX registers.
- Support for SIMD operations on packed byte, word, and doubleword integers MMX instructions are useful for multimedia and communications software.

SSE introduced the following:

- 128-bit XMM registers
- 128-bit data type with four packed single-precision floating-point operands
- Data prefetch instructions

- Non-temporal store instructions and other cacheability and memory ordering instructions
- Extra 64-bit SIMD integer support
- SSE instructions useful for 3D geometry, 3D rendering, speech recognition, and video encoding and decoding

SSE2 added the following:

- 128-bit data type with two packed double-precision floating-point operands
- 128-bit data types for SIMD integer operation on 16-byte, 8-word, 4-doubleword, or 2-quadword integers
- Support for SIMD arithmetic on 64-bit integer operands
- Instructions for converting between new and existing data types
- Extended support for data shuffling
- Extended support for cacheability and memory ordering operations

SSE2 instructions are useful for 3D graphics, video decoding/encoding, and encryption.
    SSE3 added the following:

- SIMD floating-point instructions for asymmetric and horizontal computation
- A special-purpose 128-bit load instruction to avoid cache line splits
- An x87 FPU instruction to convert to integer independent of the floating-point control word (FCW)
- Instructions to support thread synchronization

SSE3 instructions are useful for scientific, video and multithreaded applications.
    The SSSE3 introduced 32 new instructions to accelerate 8 types of computations on packed integers. These include the following:

- Twelve instructions that perform horizontal addition or subtraction operations
- Six instructions that evaluate the absolute values
- Two instructions that perform multiply and add operations and speed up the evaluation of dot products
- Two instructions that accelerate packed integer multiply operations and produce integer values with scaling
- Two instructions that perform a byte-wise, in-place shuffle according to the second shuffle control operand
- Six instructions that negate packed integers in the destination operand if the signs of the corresponding element in the source operand is less than zero
- Two instructions that align data from the composite of two operands

SSE4.1 introduces 47 new instructions to accelerate video, imaging, and 3D applications. SSE4.1 also improves compiler vectorization and significantly increases support for packed dword computation. These include the following:

- Two instructions perform packed dword multiplies.
- Two instructions perform floating-point dot products with input/output selects.
- One instruction provides a streaming hint for WC loads.
- Six instructions simplify packed blending.
- Eight instructions expand support for packed integer MIN/MAX.
- Four instructions support floating-point round with selectable rounding mode and precision exception override.
- Seven instructions improve data insertion and extractions from XMM registers.
- Twelve instructions improve packed integer format conversions (sign and zero extensions).
- One instruction improves SAD (sum absolute difference) generation for small block sizes.
- One instruction aids horizontal searching operations of word integers.

- One instruction improves masked comparisons.
- One instruction adds qword packed equality comparisons.
- One instruction adds dword packing with unsigned saturation.

SSE4.2 introduces seven new instructions. These include the following:

- A 128-bit SIMD integer instruction for comparing 64-bit integer data elements.
- Four string/text processing instructions providing a rich set of primitives; these primitives can accelerate the following:
    - Basic and advanced string library functions from strlen, strcmp, to strcspn
    - Delimiter processing, token extraction for lexing of text streams
    - Parser, schema validation including XML processing
- A general-purpose instruction for accelerating cyclic redundancy checksum signature calculations.
- A general-purpose instruction for calculating bit count population of integer numbers.

AESNI introduces seven new instructions, six of them are primitives for accelerating algorithms based on AES encryption/decryption standard, referred to as AESNI.

The PCLMULQDQ instruction accelerates general-purpose block encryption, which can perform carry-less multiplication for two binary numbers up to 64-bit wide.

Typically, algorithm based on AES standard involves transformation of block data over multiple iterations via several primitives. The AES standard supports cipher key of sizes 128, 192, and 256 bits. The respective cipher key sizes correspond to 10, 12, and 14 rounds of iteration.

AES encryption involves processing 128-bit input data (plaintext) through a finite number of iterative operations, referred to as "AES round," into a 128-bit encrypted block (ciphertext). Decryption follows the reverse direction of iterative operation using the "equivalent inverse cipher" instead of the "inverse cipher."

The cryptographic processing at each round involves two input data, one is the "state," the other is the "round key." Each round uses a different "round key." The round keys are derived from the cipher key using a "key schedule" algorithm. The "key schedule" algorithm is independent of the data processing of encryption/decryption and can be carried out independently from the encryption/decryption phase.

The AES extensions provide two primitives to accelerate AES rounds on encryption, two primitives for AES rounds on decryption using the equivalent inverse cipher, and two instructions to support the AES key expansion procedure.

Intel Advanced Vector Extensions (AVX) offers comprehensive architectural enhancements over previous generations of SSE. Intel AVX introduces the following architectural enhancements:

- Support for 256-bit wide vectors and SIMD register set
- 256-bit floating-point instruction set enhancement with up to 2X performance gain relative to 128-bit SSE
- Instruction syntax support for generalized three-operand syntax to improve instruction programming flexibility and efficient encoding of new instruction extensions
- Enhancement of legacy 128-bit SIMD instruction extensions to support three-operand syntax and to simplify compiler vectorization of high-level language expressions
- Support flexible deployment of 256-bit AVX code, 128-bit AVX code, legacy 128-bit code, and scalar code

12.7.3 Cray XT4*

The Cray XT4 system offers a new level of scalable computing where a single powerful computing system handles the most complex problems. Every component is engineered to run mas-

---

* This section is extracted from http://www.craysupercomputers.com/downloads/CrayXT4/CrayXT4_Datasheet.pdf

sively parallel processing (MPP) applications to completion reliably and fast. The operating system and management system are tightly integrated and designed for ease of operation at massive scale. Scalable performance analysis and debugging tools allow for rapid testing and fine tuning of applications. Highly scalable global I/O performance ensures high efficiency for applications that require rapid I/O access for large datasets.

The XT4 system brings new levels of scalability and sustained performance to high-performance computing (HPC). Engineered to meet the demanding needs of capability class HPC applications, each feature and function is selected so as to enable larger problems, faster solutions, and a greater return on investment. Designed to support the most challenging HPC workloads, the XT4 supercomputer delivers scalable power for the toughest computing challenges. Every aspect of the XT4 is engineered to deliver superior performance for massively parallel applications, including the following:

1. Scalable PEs each with their own high-performance AMD processors and memory
2. High bandwidth, low-latency interconnect
3. MPP-optimized operating system
4. Standards-based programming environment
5. Sophisticated reliability, availability, serviceability (RAS) and system management features
6. High-speed, highly reliable I/O system

The basic building block of the XT4 is a PE. Each PE is composed of one AMD Opteron processor (single, dual, or quad core) coupled with its own memory and dedicated communication resource. This design eliminates the scheduling complexities and asymmetric performance problems associated with clusters of shared-memory processors (SMP). It ensures that performance is uniform across distributed-memory processes, an absolute requirement for scalable algorithms.

Each XT4 compute blade includes four compute PEs for high scalability in a small footprint. Service blades include two service PEs and provide direct I/O connectivity. The AMD Opteron microprocessor offers a number of advantages for superior performance and scalability. It is on-chip, highly associative data cache supports aggressive out-of-order execution and can issue up to nine instructions simultaneously. The integrated memory controller eliminates the need for a separate memory controller chip, providing an extremely low-latency path to local memory (less than 60 ns). This is a significant performance advantage, particularly for algorithms that require irregular memory access. The 128-bit wide memory controller provides 10.6–12.8 GB/s local memory bandwidth per AMD Opteron, or more than one byte per FLOP. This balance brings a performance advantage to algorithms that stress local memory bandwidth. The HyperTransport technology enables a 6.4 GB/s direct connection between the processor and the XT4 interconnect, removing the PCI bottleneck inherent in most interconnects.

Each XT4 PE can be configured with from 1 to 8 GB DDR2 memory. Memory on compute PEs is unbuffered, which provides applications with the lowest possible memory latency.

The XT4 system (Figure 12.31) incorporates a high-bandwidth, low-latency interconnect composed of Cray SeaStar2 chips and high-speed links based on HyperTransport and proprietary protocols. The interconnect directly connects all PEs in a XT4 system in a 3D torus topology, eliminating the cost and complexity of external switches. This improves reliability and allows systems to economically scale to tens of thousands of nodes and well beyond the capacity of fat tree switches. As the backbone of the XT4 system, the interconnect carries all message-passing traffic as well as all I/O traffic to the global file system.

The Cray SeaStar2 (Figure 12.32) chip combines communication processing and high-speed routing on a single device. Each communication chip is composed of a HyperTransport link, a direct memory access (DMA) engine, a communication and management processor (PowerPC 440), a high-speed interconnect router, and a service port. The router in the Cray SeaStar2 chip provides six high-speed network links that connect to six neighbors in the 3D torus. The peak bidirectional

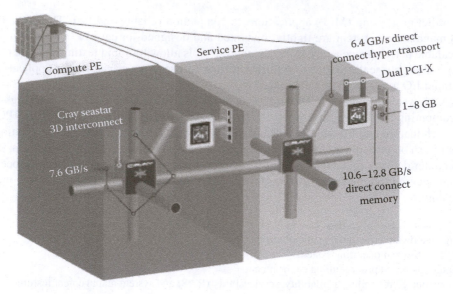

Figure 12.31 Cray XT4 architecture.

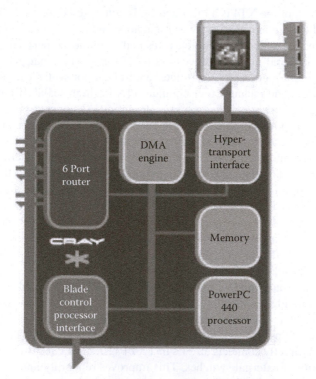

Figure 12.32 Cray SeaStar2.

bandwidth of each link is 7.6 GB/s with a sustained bandwidth in excess of 6 GB/s. The router also includes reliable link protocol with error correction and retransmission. The DMA engine and the PowerPC 440 processor work together to off-load message preparation and demultiplexing tasks from the AMD processor, leaving it free to focus exclusively on computing tasks. Logic within the SeaStar2 efficiently matches the message-passing interface (MPI) send and receive operations, eliminating the need for the large, applications-robbing memory buffers required on typical cluster-based systems. The DMA engine and the XT4 operating system work together to minimize latency by providing a path directly from the application to the communication hardware without the traps and interrupts associated with traversing a protected kernel.

Each link on the chip runs a reliability protocol that supports cyclic redundancy check (CRC) and automatic retransmission in hardware. In the presence of a bad connection, a link can be configured to run in a degraded mode while still providing connectivity.

The Cray SeaStar2 chip provides a service port that bridges between the separate management network and the Cray SeaStar2 local bus. This service port allows the management system to access all registers and memory in the system and facilitates booting, maintenance, and system monitoring.

The XT4 operating system UNICOS/lc is designed to run large complex applications and scale efficiently to 120,000 processor cores. As in previous generation MPP systems from Cray, UNICOS/lc consists of two primary components—a microkernel for compute PEs and a full-featured operating system for the service PEs.

The XT4 microkernel runs on the compute PEs and provides a computational environment that minimizes system overhead, critical to allowing the systems to scale to thousands of processors. The microkernel interacts with an application process in a very limited way, including managing virtual memory addressing, providing memory protection, and performing basic scheduling. The special lightweight design means that there is virtually nothing that stands between a user's scalable application and the bare hardware. This microkernel architecture ensures reproducible run times for MPP jobs, supports fine-grain synchronization at scale, and ensures high-performance, low-latency MPI and shared-memory (SHMEM) communication.

Service PEs run a full Linux distribution. They can be configured to provide login, I/O, system, or network services.

Login PEs offer the programmer the look and feel of a Linux-based environment with full access to the programming environment and all of the standard Linux utilities, commands, and shells to make program development both easy and portable.

Network PEs provide high-speed connectivity with other systems. I/O PEs provide scalable connectivity to the global, parallel file system. System PEs are used to run global system services such as the system database. System services can be scaled to fit the size of the system or the specific needs of the users.

Jobs are submitted interactively from login PEs using the XT4 job launch command, or through the PBS Pro batch program, which is tightly integrated with the system PE scheduler. Jobs are scheduled on dedicated sets of compute PEs and the system administrator can define batch and interactive partitions. The system provides accounting for parallel jobs as single entities with aggregated resource usage.

The XT4 system maintains a single root file system across all nodes, ensuring that modifications are immediately visible throughout the system without transmitting changes to each individual PE. Fast boot times ensure that software upgrades can be completed quickly, with minimal downtime.

Designed around open system standards, the XT4 is easy to program. The system's single PE architecture and microkernel-based operating system ensure that system-induced performance issues are eliminated, allowing the user to focus exclusively on their application.

The XT4 programming environment includes tools designed to complement and enhance each other, resulting in a rich, easy-to-use programming environment that facilitates the development of scalable applications. The AMD processor's native support for 32-bit and 64-bit applications and full x86–64 compatibility makes the XT4system compatible with a vast quantity of existing compilers and libraries, including optimized C, C++, and Fortran90 compilers and high-performance math libraries such as optimized versions of BLAS, FFTs, LAPACK, ScaLAPACK, and SuperLU.

Communication libraries include MPI and SHMEM. The MPI implementation is compliant with the MPI 2.0 standard and is optimized to take advantage of the scalable interconnect, offering scalable message-passing performance to tens of thousands of PEs. The SHMEM library is

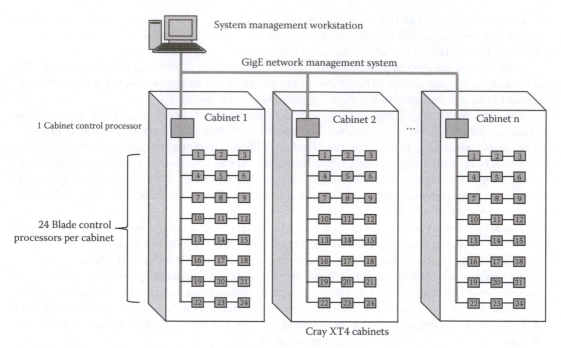

Figure 12.33   Cray XT4 system. (From http://cray.com/downloads/Cray_XT4_ Datasheet.pdf)

compatible with previous Cray systems and operates directly over the Cray SeaStar2 chip to ensure uncompromised communications performance.

Cray Apprentice performance analysis tools are also included with the XT4. They allow users to analyze resource utilization throughout their code and can help uncover load-balance issues when executing in parallel.

The Cray RAS and Management System (CRMS) (see Figure 12.33) integrates hardware and software components to provide system monitoring, fault identification, and recovery. An independent system with its own CPs and supervisory network, the CRMS monitors and manages all of the major hardware and software components in the XT4. In addition to providing recovery services in the event of a hardware or software failure, CRMS controls power-up, power-down, and boot sequences; manages the interconnect; and displays the machine state to the system administrator.

CRMS is an independent system with its own processors and supervisory network. The services CRMS provides do not take resources from running applications. When a component fails, CRMS can continue to provide fault identification and recovery services and allow the functional parts of the system to continue operating.

The XT4 is designed for high reliability. Redundancy is built in for critical components and single points of failure are minimized. For example, the system could lose an I/O PE, without losing the job that was using it. An AMD processor or local memory could fail and yet jobs routed through that node can continue uninterrupted. The system boards contain no moving parts, further enhancing overall reliability.

The XT4 processor and I/O boards use socketed components wherever possible. The SeaStar2 chip, the RAS processor module, the DIMMs, the voltage regulator modules (VRMs), and the AMD processors are all field replaceable and upgradeable. All components have redundant power, including redundant VRMs on all system blades.

The XT4 I/O subsystem scales to meet the bandwidth needs of even the most data-intensive applications. The I/O architecture consists of storage arrays connected directly to I/O PEs that reside on the high-speed interconnect. The Lustre file system manages the striping of file operations across

Table 12.1   Cray XT4 Configurations

	Cray XT4 System Sample Configurations			
	6 Cabinets	24 Cabinets	96 Cabinets	320 Cabinets
Compute PEs	548	2260	9108	30,508
Service PEs	14	22	54	106
Peak (TFLOPS)	5.6[a]	23.4[a]	94.6[a]	318[a]
Max memory (TB)	4.3	17.7	71.2	239
Aggregate memory bandwidth (TB/s)	7 TB/s[b]	29 TB/s[b]	116 TB/s[b]	390 TB/s[b]
Interconnect topology	$6 \times 12 \times 8$	$12 \times 12 \times 16$	$24 \times 16 \times 24$	$40 \times 32 \times 24$
Peak bisection bandwidth (TB/s)	1.4	2.9	8.7	19.4
Floor space (Tiles)	12	72	336	1,200

[a] Based on 26 GHz AMD dual core processor.
[b] Based on 600 MHz DDR2 memory system.

these arrays. This highly scalable I/O architecture enables customers to configure the XT4 with desired bandwidth by selecting the appropriate number of arrays and service PEs. It gives users and applications access to a high-performance file system with a global name space. To maximize I/O performance, Lustre is integrated directly into applications running on the system microkernel. Data move directly between applications space and the Lustre servers on the I/O PEs without the need for an intervening data copy through the lightweight kernel. The XT4 combines the scalability of a microkernel-based operating system with the I/O performance normally associated with large-scale SMP servers. Table 12.1 shows the performance characteristics of various configurations of XT4.

12.7.4 Cray XK7*

Cray XK7 supercomputer system, the latest Cray system in the XK family, offers a new level of scalable computing. It is scalable to 500,000 scalar processors and 50 petaflops of hybrid peak performance because of its combination of the high-performance Gemini interconnect system, AMD's multi-core scalar processors, and NVIDIA's powerful many-core Graphics Processing Unit (GPU) processors. Every component in XK7 is engineered to run MPP applications to completion, reliably and fast. The Oak Ridge National Laboratory's Titan system, a version of XK7, was the world's fastest computer in 2012. It achieved a sustained computing capability of 17.5 million billion mathematical calculations per second.

The XK7 system brings new levels of resiliency, scalability, and sustained performance to HPC. Each feature and function is selected in order to enable larger problems, faster solutions, and a greater return on investment. Designed to support the most challenging HPC workloads, the XK7 supercomputer delivers scalable power for the toughest computing challenges. Every aspect of the XK7 is engineered to deliver superior performance for massively parallel applications, including the following:

- Adaptive hybrid computing with the intra-node scalability that creates a system geared for any computing challenge.
- Cray XK7 supports a full range of powerful scalar tools, libraries, compilers, operating system, and third-party software so it enjoys scalable programming capabilities.
- Integrated Hardware Supervisory System (HSS) that is an independent system with its own CPs and supervisory network.
- The Gemini interconnect is designed for large systems in which failures are to be expected and applications must run to successful completion in the presence of errors.
- Extreme scale and cluster compatibility in one system.

* This section is extracted from http://www.cray.com/Assets/PDF/products/xk/CrayXK7Brochure.pdf.

- Support for other file system and data management services.
- Combined with standard air-or liquid-cooled high efficiency cabinet and optional phase-change liquid exchange (ECOphlex) technology, the XK7 system can reduce cooling costs and increase flexibility in datacenter design.
- Allows current Cray XE6 users to add Cray XK7 technology, so as new technologies become available, one can take advantage of these next-generation compute processors, I/O technologies, and interconnect without replacing the entire Cray XK7 system.

The XK7 operating system and management system are tightly integrated and designed for ease of operation at massive scale. Scalable performance analysis and debugging tools allow for rapid testing and fine tuning of applications. Highly scalable global I/O performance ensures high efficiency for the most data-intensive applications.

The basic building block of the XK7 is a compute node shown in Figure 12.34. It combines AMD's 16-core Opteron 6200 Series processor and NVIDIA Tesla K20 GPU Accelerator. It creates a hybrid unit with the intra-node scalability, power-efficiency of acceleration, and flexibility to run applications with either scalar or accelerator components.

Each XK7 node can be configured with 16 GB or 32 GB DDR3 memory. Memory on compute nodes is registered and memory controllers provide x4 device correction, ensuring reliable memory performance. Each compute node has an AMD Opteron 6200 Series processor with four channels of DDR3 memory and an NVIDIA Tesla K20 GPU Computing Accelerator with 6GB of

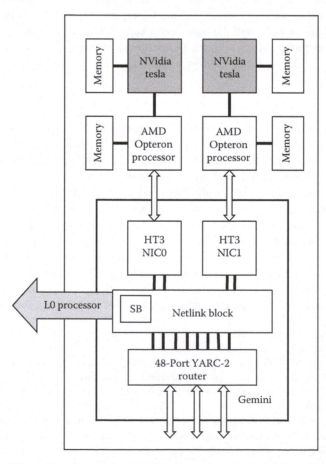

Figure 12.34 Cray XK7 node. (From CrayXK7Brochure, http://www.cray.com/Assets/PDF/products/xk/ CrayXK7Brochure.pdf.)

GDDR5 memory. Each XK7 blade is composed of four compute nodes for high scalability with up to 64 AMD processor cores per blade.

GUDA platform is the next-generation parallel computing platform and programming model that enables dramatic increases in computing performance by harnessing the power of the GPU. NVIDIA Tesla K20 GPU is designed for HPC. Based on CUDA GPU architecture codenamed "Kepler," it supports many must-have features for technical and enterprise computing. These features include ECC protection for uncompromised accuracy and data reliability, support for C++, and double-precision floating-point performance. The CUDA GPU architecture of the NVIDIA Tesla processor incorporates error correcting code for memories and double-precision floating-point units.

The Cray XK7 I/O subsystem scales to meet the bandwidth needs of even the most data-intensive applications. Each Cray XIO service blade provides four multipurpose I/O nodes, each with a six-core AMD Opteron Series 2000 processor coupled to 16 GB of DDR2 memory and a PCI-express GEN2 interface. Additionally, the Cray XIO service blade provides 32 GB/s of peak I/O bandwidth and supports connectivity to networks and storage devices using Ethernet, Fiber Channel (FC), or InfiniBand interfaces.

The Cray user data storage architecture consists of RAID6 arrays connected directly to Cray XIO nodes or via external SANs with complete multipath failover. A full line of FC-attached disk arrays with support for FC and SATA disk drives is arming the XK7. In addition, it is ordered with a parallel file system. Lustre is the file system that is designed for the world's largest and most complex computing environments. The Lustre file system redefines high performance, scaling to tens of thousands of nodes and petabytes of storage with groundbreaking I/O and metadata throughput. The Cray Data Virtualization Service allows support for Network File System (NFS), external Lustre, and/or any other file system. Lustre manages the striping of file operations across these arrays. This highly scalable I/O architecture allows for configuring bandwidth and data capacity by selecting the appropriate number of arrays and service nodes.

Each hybrid compute node in the Cray XK7 is interfaced to the Gemini interconnection through HyperTransport 3.0 technology that at the same time bypasses the PCI bottlenecks and provides a peak of over 20 GB/s of injection bandwidth per node. The Gemini interconnect is capable of tens of millions of MPI messages per second. The XK7 interconnectivity is maintained via the proven 3D torus topology that provides powerful bisection and global bandwidth characteristics and support for dynamic routing of messages. The 3D torus topology allows the addition of nodes to a system linearly without degrading the performance by having connection between nodes short and direct and reducing the latency of the links. A Gemini chip also has 48 switch ports with 160 GB/s internal switching capacity per chip.

The XK7 comes with two cooling techniques: the air-cooled technique with air flow of 3000 cfm (1.41 m^3/s) bottom intake and top exhaust and the optional ECOphlex liquid cooling that is an efficient and environmentally friendly computer component cooling technology developed by Cray and shipped with Cray system since 2008 (Cray XT5).

The system comes with the latest Cray Linux Environment (CLE) that has two different modes. The Extreme Scalability Mode (ESM) that supports uncompromised scalability and the Cluster Compatibility Mode (CCM) that comes with standardized communication layer to support compatibility with no compromise. Real-world applications have proven that the optimized mode scales to 250,000 cores. CLE also brings with it some additional reliability features, including Node Knowledge and Reconfiguration (NodeKARE), a diagnostic capability that makes sure jobs are running on healthy nodes. If a program terminates abnormally, NodeKARE automatically runs diagnostics on all involved compute nodes and removes any unhealthy ones from the compute pool.

The Cray XK7 has scalable programming capabilities. It is a fully integrated programming environment with variety of tools for programmers that allows maximum scalability and performance like FORTRAN, C, and C++ compilers and libraries with high-performance-optimized math libraries. It also supports parallel programming models like MPI, CUDA, shared-memory access library, Unified Parallel C (UPC), Coarray Fortran, OpenMP, and OpenACC.

Cray XK7 is an independent system with its own CPs and supervisory network called HSS that integrates hardware and software components to provide system monitoring, fault identification, and recovery. It monitors and manages all major hardware and software components; provides recovery services in the event of a hardware or software failure; controls power-up, power-down, and boot sequences; manages the interconnect; reroutes around failed interconnect links; and displays the machine state to the system administrator.

12.8 SUMMARY

This chapter provided the details on a popular architecture classification scheme. As shown, practical machines do not fall neatly into one of the classifications in this scheme, but span several classifications according to the mode of operation. Two supercomputer architectures (SIMD and MIMD) were detailed along with brief descriptions of two technologies facilitating implementation of these architectures and a commercially available supercomputer. The dataflow architectures described in this chapter provide structures accommodating a very fine grain parallelism while utilizing the data-driven paradigm. The systolic architectures are specialized MIMD architectures. The major aim of all the advanced architectures is to exploit the parallelism inherent in processing algorithms. The architecture parallelism spans from the fine-grain parallelism provided by dataflow architectures to coarse-grained parallelism provided by distributed systems.

PROBLEMS

12.1 The following are common operations on matrices: column sum, row sum, transpose, inverse, addition, and multiplication. Assume $N \times N$ matrix for each operation. Develop procedures suitable for an SIMD with N PEs. Specify how the matrices are represented in the N memory blocks of the SIMD.

12.2 Solve Problem 12.1 for an SIMD with M PEs where $M < N$ and N is not a multiple of M.

12.3 What is the maximum number of hops needed to route a single data item on
   a. A 64×64 processor array?
   b. A 4096-node hypercube?

12.4 Show that an n-cube has the same topology as an $n \times n$ array with toroidal edge connections.

12.5 List the desired characteristics of a compiler that performs parallelization of the sequential code for an SIMD.

12.6 Trace the evolution of the Connection Machine series (CM-1, CM-2, and CM-5), in terms of architectural characteristics and intended applications.

12.7 *Systolic array processors* and *associative processors* are versions of SIMD architecture. Investigate the architectural differences.

12.8 It is required to design a general-purpose multiprocessor system using processor and memory elements. Identify the minimum set of characteristics that each element needs to satisfy.

12.9 Four 16-bit processors are connected to four 64K × 16 memory banks through a crossbar network. The processors can access 64K memory directly. Derive the characteristics of the crossbar network to efficiently run this multiprocessor system. Show the hardware details. Describe the memory mapping needed. How is it done?

12.10 Study any multiprocessor system you have access to, to answer the following:
   a. What constitutes a "task?"
   b. What is the minimum task switching time?
   c. What synchronization primitives are implemented?

12.11 An algorithm requires access to each row of an $N \times N$ matrix. Show the storage of matrix to minimize the access time if the multiprocessor consists of N processors and N memory banks interconnected by
   a. Crossbar
   b. Bus
12.12 Repeat Problem 12.11, if the algorithm accesses both rows and columns of the matrix.
12.13 Study the PCI and USB bus architectures. What support do they provide for configuring multiprocessor systems?
12.14 How are SIMD and MIMD architectures different from computer networks?
12.15 Investigate the characteristics of grid computing.
12.16 How secure is hyper-threading given that two threads can reside in the same core and use the same cache?
12.17 What are the drawbacks of the vectorization feature in SIMD technology?
12.18 Compare the Cray XK7 super computer and the IBM Blue Gene/Q in terms of their architecture of the compute unit and performance.

BIBLIOGRAPHY

Almasi, G.S. and Gottlieb, A., *Highly Parallel Computing*, Redwood City, CA: Benjamin Cummings, 1989.

Anderson, G.A. and Jensen, E.D., Computer interconnection structures: Taxonomy, characteristics, and examples, *ACM Computing Surveys*, 7(4), 197–213, December 1975.

Archibald, J. and Baer, J.L., Cache coherence protocols: Evaluation using a multiprocessor simulation model, *ACM Transactions on Computer Systems*, 4(4), 273–298, November 1986.

Benes, V.E., On rearrangeable three-stage connecting networks, *The Bell System Technical Journal*, 41(5), 1481–1492, September 1962.

Benes, V.E., *Mathematical Theory of Communication Networks and Telephone Traffic*, New York: Academic Press, 1965.

Bhuyan, L.N. and Agrawal, D.P., Generalized hypercube and hyperbus structures for a computer network, *IEEE Transactions on Computers*, C-33, 1, April 1984.

Censier, L.M. and Feautrier, P., A new solution to coherence problems in multicache systems, *IEEE Transaction on Computers*, C-27, 1112–1118, December 1978.

Chen, P.-Y., Lowrie, D.H., and Yew, P.-C., Interconnection networks using shuffles, *IEEE Computer*, 33, 55–64, December 1981.

Clos, C., A study of nonblocking switching networks, *The Bell System Technical Journal*, 32, 406–424, March 1953.

Cray XK7, http://www.cray.com/Assets/PDF/products/xk/CrayXK7Brochure.pdf

Dennis, J.B., First version of a data flow procedure language, in *Lecture Notes in Computer Science*, Berlin, Germany: Springer Verlag, 1974, pp. 362–376.

Dennis, J.B. and Misunas, D.P., A preliminary data flow architecture for a basic data flow processor, *Proceedings of the Second Symposium on Computer Architecture*, New York, 1975, pp. 126–376.

Duncan, R., A survey of parallel computer architectures, *IEEE Computer*, 23(2), 5–16, 1990.

Eggers, J.S. and Katz, R.H., Evaluating the performance of four snooping cache coherency protocols, *Proceedings of Sixteenth Annual IEEE International Symposium on Architecture*, Honolulu, HI, Vol. 17, June 1988, pp. 3–14.

Eisner, C., Hoover, R., Nation, W., Nelson, K., Shitsevalov, I., and Valk, K., A methodology for formal design of hardware control with application to cache coherence protocols, *Proceedings of 37th Conference on Design Automation*, Los Angeles, CA, 2000, pp. 724–729.

El-Rewini, H. and Lewis, T.G., Scheduling parallel program tasks onto arbitrary target machines, *Journal of Parallel and Distributed Computing*, 9, 138–153, June 1990.

Gallant, J., Cache coherency protocols, *EDN*, 36(6), March 14, 1991, pp. 41–50.

Gara, A. et al., Overview of the Blue Gene/L system architecture, *IBM Journal of Research and Development*, 49(2/3), 195–212, 2005.

Goodman, J.R., Using cache memory to reduce processor memory traffic, *10th International Symposium on Computer Architecture*, Stockholm, Sweden, June 1983.

Hagersten, E., Landin, A., and Haridi, S., DDM—A cache-only memory architecture, *IEEE Computer*, 25(9), 44–54, September 1992.

Hillis, W.D., *The Connection Machine*, Cambridge, MA: MIT Press, 1985.

Hillis, W.D., The connection machine, *Scientific American*, 256(6), 108–115, June 1987.

Kim, D., Chaudhuri, M., and Heinrich, M., Leveraging cache coherence in active memory systems, *Proceedings of the 16th International Conference on Supercomputing*, New York, 2002, pp. 2–13.

Kung, H.T., Why systolic architectures? *IEEE Computer*, 15(1), 37–46, January 1982.

Papamarcos, M. and Patel, J., A low overhead coherence solution for multiprocessors with private cache memories, *Proceedings of the 11th International Symposium on Computer Architecture*, Ann Arbor, MI, 1984, pp. 348–354.

Parhami, B., *Introduction to Parallel Processing: Algorithms and Architectures*, Hingham, MA: Plenum, 1999.

Shiva, S.G., *Advanced Computer Architectures*, Boca Raton, FL: Taylor & Francis, 2006.

Siegel, H.J., *Interconnection Networks for Large Scale Parallel Processing*, Lexington, MA: Lexington Books, 1985.

Skillicorn, D.B., A taxonomy for computer architectures, *IEEE Computer*, 21(11), 46–57, 1985.

Skillicorn, D.B. and Talia, D., *Programming Languages for Parallel Processing*, Los Alamitos, CA: IEEE Computer Society Press, 1996.

Stone, H.S., Parallel processing with the perfect Shuffle, *IEEE Transactions on Computers*, C-20, 153–161, February 1971.

Stone, H.S., *High-Performance Computer Architecture*, New York: Addison-Wesley, 1990.

Veen, A.H., Data-flow machine architecture, *ACM Computing Surveys*, 18(4), 365–396, December 1986.

CHAPTER 13

Embedded Systems

An embedded system is a special-purpose system that embeds a computer and is designed to perform one or a few predefined tasks, usually with very specific requirements. That is, the computer system is completely encapsulated by the device it controls. For instance, a traffic light controller is a system that is designed to perform that single function. The heart of the system is the processor (computer) embedded in it. In contrast, our desktop systems are personal computers that perform general-purpose computation tasks. Since embedded processors are dedicated to specific tasks, their design can be optimized to reduce the size and cost of the system. They are often mass produced, thus multiplying the cost savings. Almost all systems we use in our daily lives today embed computers. Some of the examples are the following:

1. MP3 players
2. Handheld computers and personal digital assistants (PDAs)
3. Cellular telephones
4. Automatic teller machines (ATMs)
5. Automobile engine controllers and antilock brake controllers
6. Home thermostats, ovens, and security systems
7. Video game consoles
8. Computer peripherals such as routers and printers
9. Guidance and control systems for aircraft and missiles

Probably the first mass-produced embedded processor was the Autonetics D-17 guidance computer built in 1961 for the Minuteman missile. It was built from discrete transistor logic devices with a hard disk main memory. In 1966, the D-17 was replaced by a new design that was the first high-volume use of integrated circuits (ICs). In 1978, National Engineering Manufacturers Association released the standard for a programmable *microcontroller*. The definition included single-board computers, numerical controllers, and sequential controllers that perform event-based instructions.

The progress in hardware technology to very large scale integration (VLSI) provided the capability to build chips with enormous processing power and functionality at a very low cost. For instance, the first microprocessor (Intel 4004), which was used in calculators and other small systems, required external memory and support chips. By the 1980s, most of such external system components had been integrated into the same chip as the processor, resulting in *microcontrollers*, the first generation of processors for embedded systems. Embedded systems of today are classified as *microcontrollers*, embedded processors, and *systems on a chip* (SoC), depending on the context, their functionality, and complexity. The SoC is an application-specific integrated circuit (ASIC), which uses an intellectual property (IP) processor architecture. The distinction between these classes continues to blur due to the progress in hardware and software technologies. As such, the concepts discussed in this chapter apply equally well to all the three classes.

Example 13.1

Consider the design of a traffic light controller. The simplest of the design implements a fixed red–yellow–green light sequence with each light staying on for a predetermined time period. There are no inputs to this system. It has three outputs corresponding to each light. This system can be designed as a sequential circuit with three states and can be implemented using small- and medium-scale ICs. Alternatively, a programmable logic device (programmable logic array [PLA] or programmable array logic [PAL]) can be used along with flip-flops to reduce the part count. If the volume justifies, an FPGA can be used to implement the controller circuit.

Let us enhance the functionality of the traffic light controller to add the capability to sense the number of cars passing through the intersection in all directions and adjust the red–yellow–green periods based on the traffic count. We now need sensors on each street and corresponding inputs to the controller circuit and the outputs to handle all the lights. The control algorithm will be more complex. Although the hardware can be implemented using PLA, PAL, or field programmable gate array (FPGA), it is more advantageous to use a processor. The processors designed for such applications are *microcontrollers*, which are single-chip devices available from various vendors. They allow programming to implement and change the control algorithm, contain a small amount of RAM and ROM, provide for inputs and outputs through their interfaces, and provide timing circuits.

If the complexity of the system increases further, where a single-chip microcontroller solution is not adequate, a SoC solution is used if the quantities needed justify the cost of the system. Many central processing unit (CPU) architectures are used in embedded designs today (ARM, MIPS, Coldfire/68k, PowerPC, Intel x86 and 8051, PIC microcontroller, Atmel AVR, Renesas H8, SH, V850, etc.). Very-high-volume applications use either FPGAs or SoC. Operating systems (OSs) such as DOS, Linux, NetBSD, or embedded real-time operating systems (RTOS) such as QNX or Inferno are used in designing embedded systems.

We outline embedded system characteristics in the next section, followed by a brief introduction to software architectures in Section 13.2. RTOS are introduced in Section 13.3. Brief descriptions of two commercially available architectures are provided in Section 13.4. Chapter 14 provides further details on SoC and describes another class of embedded architectures, the mobile processors.

13.1 CHARACTERISTICS

As mentioned earlier, embedded systems are special-purpose machines designed to do one or more specific tasks. The application may also have real-time performance constraints that must be met. The major emphasis in embedded system design is to reduce hardware complexity and cost. The performance requirements dictate these aspects.

The input and output devices (switches, motors, lights, etc.) attached to the embedded processor through its input/output (I/O) interfaces are called *field devices*. The processor is usually integrated into the housing of the field devices or can even be a part of a circuit board of the field device. The processor examines the status of the input field devices, carries out the control plan (i.e., executes the program), and produces the responses to control the output field devices. Each such cycle of activity is called a *scan*. During each scan all inputs are tested, the control plan is evaluated, and the outputs are updated. The control program is stored in ROM or flash memory chips. A small amount of RAM supports the program operation.

Embedded systems may reside in hostile environments and are expected to run continuously for years. In addition, they need to recover by themselves if an error occurs. To facilitate this, unreliable mechanical moving parts such as disk drives, switches, and buttons are avoided as far as possible. The software is thoroughly tested before deployment. Special timing circuits such as watchdog timers are used.

A *watchdog timer* is an error recovery mechanism and is initiated with certain value at the beginning of a scan or at defined time intervals. It counts down as the scan progresses. The processor periodically notifies the watchdog to reset. If such a notification is not received by the watchdog before its value reaches zero or a predefined value, it assumes that a fault condition has occurred and resets the processor, thus bringing the system back into normal operation. Watchdog timers may also take the systems into a safety state, by turning off potentially dangerous subsystems until the fault is cleared.

Early systems used assembly language for developing applications. Recent systems use high-level languages such as C and Java or their versions for embedded systems, along with embedded assembly-level code for critical sections of the code. In addition to the compilers, assemblers, and debuggers, software designers use in-circuit emulators (ICE), cyclic redundancy check (CRC) checkers, and other tools to develop embedded system software. An ICE is a special hardware device that replaces or plugs into the embedded processor and facilitates loading and debugging of experimental code in the system.

All embedded systems have start-up firmware that runs a self-test before starting the application code. The self-test covers CPU, RAM, ROM, peripherals, and power supplies. Passing of self-test is usually indicated by LEDs or other visual means, providing simple diagnostics to technicians and users. In addition, safety tests are run within a "safety interval," to assure that the system is still reliable.

13.2 SOFTWARE ARCHITECTURES

This section provides brief descriptions of the most common software architectures for embedded systems.

13.2.1 Simple Control Loop (Round Robin)

In this architecture, the control software consists of a simple loop. Within the loop, calls are made to subroutines. Each subroutine handles a part of the control function and/or manages a part of hardware. A state machine model for software is utilized to represent the set of states that the system can be in and how it changes between them. This is the simplest of the software architectures and is employed in small devices with a stand-alone microcontroller dedicated to a simple task. If the system imposes timing constraints, the loops and modules need to be designed to meet those constraints. Since the software is controlled by the main loop, interrupt handling and adding of new features becomes difficult.

Example 13.2

The code for a controller servicing several devices is shown in the following:

```
Controller
{While (TRUE)
 Service A;
 Service B;
 Service C;
 .
 .
 Service X;
}
```

The devices are serviced in the order A, B, C,...., X in each scan. If additional devices are to be included, it is simple to include additional calls into this main loop. If the controller is servicing C, and if B requires service, B has to wait until the next scan. As long as such delays can be tolerated, this architecture is good enough to handle most simple control tasks.

Example 13.3

Consider a multimeter that can measure current (amps), resistance (ohms), and voltage (volts). The type of measurement needed is selected by a switch on the multimeter. The control software for this device is shown in the following:

```
Multimeter Controller
{While (TRUE)
 {Position = Read switch position;
 Switch (Position)
 {Case of current:
       Read Amps;
       Output;
       Break;
 Case of Voltage:
       Read Volts;
       Output;
       Break;
 Case of Resistance:
       Read Ohms;
       Output;
       Break;
 }
 }
}
```

This program services only the selected device in each scan. Thus, the scan period is dictated by the slowest of the devices.

13.2.2 Interrupt-Controlled Loop (Round Robin with Interrupts)

In interrupt-controlled architecture, the tasks performed by the system are triggered by events such as a timer trigger and an input port receiving a data byte. These systems run a simple task in a main loop also. Each interrupt is handled by an interrupt or event handler (i.e., interrupt service routine).

Example 13.4

The following code shows the software for a controller with three devices to handle. There are three interrupt flags one for each device, set by the hardware when the corresponding device interrupts. The main loop checks for the interrupt flag and services the device whose flag is set.

```
Multimeter Controller
{FlagA = FALSE;
FlagB = FALSE;
FlagC = FALSE;
InterruptHandler_A
  {Service Device A;
  FlagA = TRUE.
  }
InterruptHandler_B
  {Service Device B;
  FlagB = TRUE.
  }
```

```
InterruptHandler_C
  {Service Device C;
  FlagC = TRUE.
  }
Main ()
{
While (TRUE)
 {
  if (FlagA)
     {FlagA = FALSE;
     Perform I/O from A;
     }
  if (FlagB)
     {FlagB = FALSE;
     Perform I/O from B;
     }
 if (FlagC)
    {FlagC = FALSE;
    Perform I/O from C;
    }
  }
}
```

When an interrupt occurs, the controller gets out of the main loop and handles the interrupt. The interrupts can be assigned appropriate priorities. The execution time of the interrupt handlers needs to be short to keep the interrupt latency to a minimum. Longer tasks are usually added to a queue structure in the interrupt handler to be processed in the main loop later. This architecture corresponds to a multitasking kernel with discrete processes.

13.2.3 Real-Time Operating System-Based Architectures

There are two RTOS-based software architectures. In the *non-preemptive* multitasking (function queue scheduling) architecture, the programmer implements the control algorithm as a series of tasks, with each task running in its own environment. The tasks are arranged as an event queue, and a loop processes the events one at a time. Adding new functionality is easier (include a new task or adding to the queue-interpreter). In *preemptive* multitasking architecture, the system uses an RTOS allowing the application programmers to concentrate on device functionality rather than OS services. Refer to books by Simon (1999), Vahid and Givargis (2002), and Lewis (2002) listed in the Bibliography section for further details on this architecture. A brief description of OSs follows.

13.3 OPERATING SYSTEM

Each executing program is a *task* under the control of the OS. If an OS can execute multiple tasks simultaneously, it is said to be *multitasking*. The use of a multitasking OS simplifies the design of complex software applications by enabling the application be partitioned into a set of smaller and more manageable tasks. The multitasking OS allows execution of these simpler tasks and facilitates inter-task communication as needed. This mode of operation also removes complex timing and sequencing details from the application code and transfers that responsibility to the OS.

The major functions of the OS are the following:

1. Keeping track of the status of all the resources (processors, memories, switches, and I/O devices) at any instant of time
2. Assigning tasks to processors in a justifiable manner (i.e., to maximize processor utilization)

3. Spawning or creating new processes such that they can be executed in parallel or independent of each other
4. When all the spawned processes are completed, collecting their individual results and passing them to other processes as required

13.3.1 Multitasking versus Concurrency

Consider a single-processor system that can execute only a single task at a time. If the application is partitioned into multiple tasks, and if there is a strict serial dependency among tasks, then only one task can be run and in the order in which their dependency dictates. In this context, there is no advantage for a multitasking OS. But, if the application allows partitioning such that multiple tasks can be run simultaneously, then even in a single-processor system, multitasking OS provides an advantage.

Example 13.5

Figure 13.1 shows a timing diagram depicting the execution of three tasks on a single-processor system. Here, we assume that each task is given a time slot and the processor executes tasks in the order shown in each time slot. In practice, such round robin uniform scheduling of tasks may not be possible. A task might require a resource (peripheral, register, etc.) that may not be available. Then the task has to be suspended either by the OS or by the task itself until the resource is available.

The *scheduler* (part of the OS kernel) decides which task should be executing at any particular time. It can suspend and resume a task several times during the task lifetime. In addition to being suspended involuntarily by the RTOS kernel, a task may choose to suspend itself. This is done when the task wants to either delay (i.e., sleep) for a certain period or wait (i.e., block) for some resource to become available or an event to occur.

A task execution environment is shown in Figure 13.2. At T1 task 1 is executing. At T2 the kernel suspends Task 1 and starts (or resumes) Task 2. Let us assume that Task 2 locks the

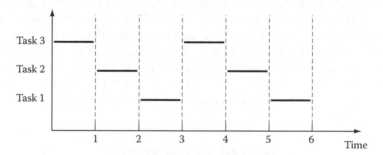

Figure 13.1 Multitasking.

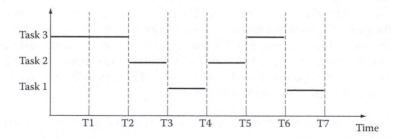

Figure 13.2 Task scheduling.

input device 1 for its exclusive access. At T3 the kernel suspends Task 2 and resumes Task 3. Suppose Task 3 tries to access input device 1, it finds that the device has been locked and hence cannot continue execution. Hence, Task 3 suspends itself at T4. Task 2 may resume at T4, complete its operation with device 1, and release it. The kernel resumes Task 1 at T5, followed by Task 3 at T6. Task 3 can now access device 1 and continues to execute until suspended by the kernel.

Switching tasks is equivalent to interrupt servicing. Recall that the processor status was saved while entering the interrupt service routine and resaved during the return from the interrupt service. Along the same lines, when a task is blocked or suspended, the resources it is holding (contents of various registers, peripherals, etc.—the context) must be maintained to allow the task to resume properly. The OS kernel is responsible for ensuring this. The process of saving the context of a task being suspended and restoring the context of a task being resumed is called context switching.

Since real-time systems are designed to provide a timely response to real-world events, the RTOS scheduling policy must ensure that the deadlines imposed by the system requirements are met. To achieve this objective, we typically assign a priority to each task. The scheduling policy of the RTOS ensures that the highest priority task that can execute at a particular time is the task scheduled to execute.

Example 13.6

Consider a real-time system equipped with a keypad and LCD. It is required that the user must get visual feedback of each key press within 50 ms. Thus, if the user cannot see that the key press has been accepted within 50 ms, the system will be awkward to use. A response between 0 and 50 ms would be acceptable.

The system performs a control function that samples a set of sensors (inputs), executes the control algorithm, and produces the outputs to operate a set of valves. It is required that the control cycle is executed every 4 ms. A timer provides the timing trigger every 4 ms.

Two tasks are required for this application, one to handle the key presses and the other to handle the control cycle. Note that the control task should have the higher priority.

```
KeyHandlerTask
{
 //Key handling is implemented using an infinite loop
 for(;;)
 {
  Suspend waiting for a key press;
  Process the key press;
 }
}
ControlTask
{
 for(;;)
 {
  Suspend waiting for 4 ms since the beginning of the previous Cycle;
  Sample the sensors;
  Perform control algorithm;
  Output;
 }
}
```

In addition, we will assume that the kernel has an idle task to indicate that the system is idle waiting:

```
Idle Task
{
For(;;)
 No operation;
}
```

The idle task is always in a state where it is able to execute. Hence, to begin with idle task is executing. Now, if a key press is detected, the key handler task is executed. If the timer issues the timing trigger while idling, the control task is executed. If a key press occurs while the control task is running, processing of the key press is deferred until the control cycle is completed, because of the priorities assigned to these tasks.

13.3.2 Process Handling

This section illustrates several aspects of process creation and handling. We will assume a multiprocessor system context in this description. The concepts discussed here are equally applicable to a single-processor context, except that all the work is done by the single processor rather than shared between the multiple processors.

Example 13.7

Consider, for example, the element by element addition of two vectors A and B to create the vector C. That is,

$$c_i = a_i + b_i \quad \text{for} \quad i = 1 \text{ to } n. \tag{13.1}$$

It is clear that the computation consists of n additions that can be executed independent of each other. Thus, an MIMD with n processors can do this computation in one addition time. If the number of processors m is less than n, the first m tasks are allocated to the available processors and the remaining tasks are held in a queue. When a processor completes the execution of the task allocated to it, a new task (from the queue) is allocated to it. This mode of operation continues until all the tasks are completed, as represented by the following algorithm:

1. /* Spawn $n - 1$ processes each with a distinct process identification number k. Each spawned process starts at "label" */
   for $k = 1$ to $n - 1$
   FORK label(k);
2. /*The process that executed FORK is assigned $k = n$. This is the only process that reaches here, the other spawned processes jump directly to "label" */
   $k = n$;
3. /* add kth element of each vector; n different processes perform this operation, not necessarily in parallel */
   label: $c[k] = a[k] + b[k]$;
4. /* terminate the n processes created by FORK; only 1 process continues after this point */
   JOIN n;

The new aspects of this algorithm are the FORK and JOIN constructs. They are two of the typical commands to any multiprocessing OS used to create and synchronize tasks (processes). The FORK command requests the OS to create a new process with a distinct process identification number (k in this example). The program segment corresponding to the new process starts at the statement marked "label."

Note that initially the entire algorithm earlier constitutes one process. It is first allocated to one of the free processors in the system. This processor through the execution of the FORK-loop (step 1) requests the OS to create ($n - 1$) tasks, after which it continues with step 2. Thus, after the execution of step 2, there are n processes in all waiting to be executed. The process with $k = n$ continues on the processor that is already active (i.e., the processor that spawned the other processes). The remaining ($n - 1$) processes created by the OS enter a process queue. The processes waiting in the queue are allocated processors as processors become available.

In this example, the kth process adds kth elements of A and B creating the kth element of C. The program segment corresponding to each process ends with the JOIN statement. The JOIN command can be viewed as the inverse of the FORK command. It has a counter associated with it that starts off at 0. A processor executing the JOIN increments the counter by 1 and compares it to n. If the value of the counter is not equal to n, the processor cannot execute any further, and hence it terminates the process and returns to the available pool of processors for subsequent allocation of tasks. If the value of the counter is n, the process, unlike the others that were terminated, continues execution beyond the JOIN command. Thus, JOIN ensures that the n processes spawned earlier have been completed before proceeding further in the program.

Several aspects of the previous algorithm are worth noting:

1. This algorithm does not use the number of processors m as a parameter. As such it works on systems with any number of processors. The overall execution time depends on the number of processors available.
2. OS functions of creating and queuing tasks, allocating them to processors, etc., are done by another processor (or a process), which is not visible in the previous discussion.
3. The procedure for creating processes requires O(N) time. This time can be reduced to $O(\log_2 N)$ by a more complex algorithm that makes each new process perform fork, thus executing more than one fork simultaneously.
4. Step 1 of the algorithm was executed only $(n-1)$ times, thus the nth process was not specifically created by a FORK but was given to the processor that was active already. Alternatively, the FORK-loop could have been executed n times. Then the processor executing the loop must have been deallocated and brought to the pool of available processors. A new process could have then been allocated to it. Thus, the previous procedure eliminates the overhead of deallocation and allocation of one process to a processor. Typically, creation and allocation of tasks result in a considerable overhead requiring execution of about 50–500 instructions.
5. Each process in this example performs an addition and hence is equivalent to three instructions (Load A, Add B, Store C). Thus, the process creation and allocation overhead mentioned earlier is of the order of 10–100 times the useful work performed by the process.

13.3.3 Synchronization Mechanisms

There was no explicit interprocess communication in the previous example. In practice, the various processes in the system need to communicate with each other. In general, an application will have a set of *shared data* items between all the tasks it comprises, and each task will have its own private data items and stacks and queues. It is important that the shared data items are accessed properly among the competing tasks.

Since the processes executing on various processors are independent of each other and the relative speeds of execution of these processes cannot be easily estimated, a well-defined synchronization between processes is needed for the communication (and hence the results of computation) to be correct. That is, processes operate in a cooperative manner and a sequence control mechanism is needed to ensure the ordering of operations. Also, processes compete with each other to gain access to shared data items (and other resources). An access control mechanism is needed to maintain orderly access. This section describes the most primitive synchronization techniques used for access and sequence control.

Example 13.8

Consider two processes $P1$ and $P2$ to be executed on two different processors. Let S be a shared variable in the memory. The sequence of instructions in the two processes is as follows:

*P*1:
1. MOV Reg1,[S]/* The first operand is the destination */
2. INC Reg1/* Increment Reg1 */
3. MOV [S],Reg1

*P*2:
1'. MOV Reg2,[S]
2'. ADD Reg2,#2/* Add 2 to Reg2 */
3'. MOV [S],Reg2

The only thing we can be sure of as far as the order of execution of these instructions is concerned is that instruction 2(2′) is executed after instruction 1(1′), and instruction 3(3′) is executed after instructions 1(1′) and 2(2′). Thus, the following three cases for the actual order of execution are possible:

1. 1 2 3 1′2′3′ or 1′2′3′1 2 3: Location S finally has the value 3 in it (assuming $S = 0$, initially).
2. 1 1′2 2′3 3′: Location S finally has value 2 in it.
3. 1 1′2′2 3′3: Location S finally has value 1 in it.

The desired answer is attained only in the first case and that is possible only when process $P1$ is executed in its entirety before process $P2$ or vice versa. That is, $P1$ and $P2$ need to execute in a mutually exclusive manner. The segments of code that are to be executed in a mutually exclusive manner are called critical sections. Thus, in $P1$ and $P2$ of this example, the critical section is the complete code corresponding to each process.

Example 13.9

As another example consider the code shown in Figure 13.3 to compute the sum of all the elements of an n-element array A. Here, each of the n processes spawned by the fork-loop adds an element of A to the shared variable SUM. Obviously, only one process should be updating SUM at a given time. Imagine process 1 reads the value of SUM and has not yet stored the updated value. Meanwhile, if process 2 reads SUM and updates are done by process 1 followed by process 2, the resulting SUM would be erroneous. To obtain the correct value of SUM, we should make sure that once process 1 reads SUM, no other process can access SUM until process 1 writes the updated value. This is the mutual exclusion of processes.

The mutual exclusion is accomplished by **LOCK** and **UNLOCK** constructs. This pair of constructs has a flag associated with it. When a process executes **LOCK**, it checks the value of the flag. If the flag is ON, it implies that some other process has accessed SUM, and hence the process waits until the flag is OFF. If the flag is not ON the process sets it ON and gains access to SUM, updates it, and then executes the **UNLOCK**, which clears the flag. Thus, the **LOCK/UNLOCK** brings about the synchronization of the processes.

Note that during the **LOCK** operation, the functions of fetching the flag, checking its value, and updating it must all be done in an indivisible manner. That is, no other process should have access to the flag until these operations are complete. Such an indivisible operation is brought about by a hardware primitive known as **TEST _ AND _ SET**.

TEST _ AND _ SET: The use of **TEST _ AND _ SET** primitive is shown in Figure 13.4. Here, K is a shared memory variable that can have a value of either 0 or 1. If it is 0, the **TEST _ AND _ SET** returns a 0 and sets K to 1. The process enters its critical section. If K is 1, **TEST _ AND _ SET** returns 1 thus locking the process from entering the critical section. When the process is through executing its critical section, it resets K to 0 thus allowing a waiting process access into its critical section.

```
              SUM = 0;
              for k = 1 to n-1 /*spawn n-1 processes */
              FORK label(k);
              endfor;
              k = n; /*the nth process */
              /*n different processes with distinct process id numbers reach
              this point */
Label:        LOCK (flag)
              SUM = SUM + A(k);
              UNLOCK (flag);
              JOIN n; /*terminate the N processes and gather results*
```

Figure 13.3 Synchronization using LOCK/UNLOCK.

```
P1: while not(TEST_AND_SET(K));
        ┌─────────────────────────────┐
        │     critical section of P1; │
        └─────────────────────────────┘
    K = 0;
P2: while not(TEST_AND_SET(K));
        ┌─────────────────────────────┐
        │     critical section pf P2; │
        └─────────────────────────────┘
    K = 0;
/* The body of TEST_AND_SET procedure */
    TEST_AND_SET(K)
    {
        temp = k;
        K = 1;
        return(temp);
    }
```

Figure 13.4 Use of TEST _ AND _ SET.

In Figure 13.3, the critical section is not part of the while loop. The range of the while loop is a single statement (terminated by ";"). The while loop makes the process wait until K goes to 0, to enter the critical section. There are two modes of implementing the wait: busy waiting (or spin lock) and task switching. In the first mode, the process stays active and repeatedly checks the value of K until it is 0. Thus, if several processes are busy waiting, they keep the corresponding processors busy but no useful work gets done. In the second mode, the blocked process is enqueued and the processor is switched to a new task. Although this mode allows better utilization of processors, the task-switching overhead is usually very high, unless special hardware support is provided.

Several processes could be performing the test-and-set operation simultaneously, thus competing to access the shared resource K. Hence, the process of examining K and setting it must be indivisible in the sense that K cannot be accessed by any other process until the test-and-set is completed once. TEST _ AND _ SET is usually an instruction in the instruction set of the processor and is the minimal hardware support needed to build other high-level synchronization primitives.

The TEST _ AND _ SET effectively serializes the two processes so that they execute in a mutually exclusive manner. Once a process is in the critical section, other processes are blocked from entering it by the TEST _ AND _ SET. Thus, the mutual exclusion is brought about by serializing the execution of processes and hence, affecting the parallelism. In addition, the blocked processes incur overhead due to busy waiting or task switching.

Semaphores: Dijkstra (1965) introduced the concept of semaphores and defined two high-level synchronization primitives P and V based on the semaphore variable S. They are defined in the following:

$P(S)$ or WAIT(S)
    If $S = 0$, the process invoking P is delayed until $S > 0$.
    If $S > 0$, $S = S - 1$ and the process invoking P enters the critical section.
$V(S)$ or SIGNAL(S)
    $S = S + 1$

S is initialized to 1. The testing and decrementing of S in $P(S)$ is indivisible. So is the incrementing of S in $V(S)$. Figure 13.5 shows the use of P and V to synchronize processes $P1$ and $P2$.

S is a binary variable in the implementation of P and V earlier. If S is an integer variable, it is called a counting semaphore. If it is initialized to M, then M processes can be in the critical section at any time.

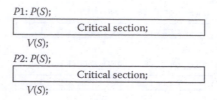

Figure 13.5 Synchronization using *P* and *V*.

FETCH _ AND _ ADD: FETCH _ AND _ ADD is similar to TEST _ AND _ SET in its implementation but is nonblocking synchronization primitive. That is, it allows the operation of several processes in the critical section in parallel yet nonconflicting manner. FETCH _ AND _ ADD is shown in the following:

```
FETCH_AND_ADD(S, T)
        {Temp = S;
         S = S + T;}
    Return Temp;
```

Two parameters are passed to FETCH _ AND _ ADD: *S* the shared variable and *T* an integer. If *S* is initialized to 0 and two processes *P*1 and *P*2 make a call to FETCH _ AND _ ADD at roughly the same time, the one reaching the FETCH _ AND _ ADD first receives the original value of *S* and the second one receives (*S* + *T*). The two processes can execute further independently, although the updating of *S* is effectively serialized. In general, the FETCH _ AND _ ADD gives each contending process a unique number and allows them to execute simultaneously, unlike the TEST _ AND _ SET that serializes the contending processes.

As another example, the instruction FETCH _ AND _ ADD(SUM, INCREMENT) provides for the addition of INCREMENT to SUM by several processes simultaneously and results in the correct value of the SUM without the use of LOCK/UNLOCK.

FETCH _ AND _ ADD is useful for cases in which the same variable is accessed by several contending processes. FETCH _ AND _ ADD on different variables are done sequentially if those variables reside in the same memory. The implementation cost of FETCH _ AND _ ADD is high. As such it is limited to environments in which updates become a bottleneck because 10s of processes are contending for access to the shared variable. If only one process requests access to the shared variable at a time, TEST _ AND _ SET is more economical.

The processes in a message-passing Multiple instruction stream, multiple data stream (MIMD) are automatically synchronized since a message cannot be received before it is sent. Message-processing protocols deal with the problems of missing or overwritten messages and sharing of resources. This section has described the most primitive synchronizing mechanisms. Refer to Silberschatz et al. (2003), Stallings (2005), and Tanenbaum and Woodhull (2006) for further details.

13.3.4 Scheduling

Recall that a parallel program is a collection of tasks. These tasks may run serially or in parallel. An optimal schedule determines the allocation of tasks to processors of the MIMD system and the execution order of the tasks, so as to achieve the shortest execution time. The scheduling problem in general is NP complete. But, several constrained models have evolved over the years and currently this problem is an active area of research in parallel computing.

Scheduling techniques can be classified into two groups: *static* and *dynamic*. In static scheduling, each task is allocated to a particular processor based on the analysis of the precedence constraints imposed by the tasks at hand. Each time the task is executed, it is allocated to that predetermined processor. Obviously, this method does not take into consideration the nondeterministic nature of tasks brought about by conditional branches and loops in the program. The target of the conditional branch and the upper bounds of the loops are not known until the program execution begins. Thus, static scheduling will not be optimal.

In dynamic scheduling, tasks are allocated to processors based on the execution characteristics. Usually some load-balancing heuristic is employed in determining optimal allocation. Since the scheduler has only the knowledge of local information about the program at any instant of time, finding the global optimum is difficult. Another disadvantage is the increased overhead since the schedule has to be determined while the tasks are running. Refer to Adam et al. (1974), Bashir et al. (1983), and El-Rewini and Lewis (1990) for further details.

13.3.5 Real-Time Operating Systems

RTOS offers the services described earlier in this section, except that its operating environment tends to be much simpler than that of a general-purpose system. The kernel is the core component within an OS. It is typical to consider services such as memory management, network software support, and debugging as not part of the kernel, although these services are provided by the OS. Yet, RTOS and real-time kernel designations are interchangeably used in practice. There are several differences in the operation of an RTOS compared to a non-RTOS.

In a desktop environment, for example, the OS is invoked and takes control of the system as soon as the power is turned on. The OS allows invocation of other applications and facilitates the compilation, linking, and loading of new applications. In a microcontroller, the start-up software starts the application, which in turn calls upon the RTOS as needed. The RTOS and the applications are more closely intertwined.

When a fault condition results in crashing the microcontroller, the RTOS also goes down with it and the system has to be restarted. In a general OS environment, application fault conditions do not bring the OS down.

In general, the OS cannot be configured for the application. RTOS allows itself be configured to include only the needed services for the particular application. This allows more optimized usage of memory and other resources, a main consideration in building embedded systems. Chapter 14 provides additional details on OSs.

13.4 EXAMPLE SYSTEMS

This section provides a brief description of two systems used in embedded applications. The first is a popular microcontroller family and the second is a processor core used as an IP in several embedded applications.

13.4.1 8051 Family of Microcontrollers

As mentioned earlier, a microcontroller is basically an entire computer on a single chip. Usually this includes a CPU, ROM, RAM, Parallel I/O, and serial I/O counters. The prime use of microcontrollers is to control the operation of a machine using a fixed program that is stored in ROM and does not change over the lifetime of the system.

A family of microcontrollers is a group of devices that share the same basic elements and have the same basic group of instructions. Several microcontroller families are available in the market today. The leaders are probably Motorola 6811, Microchip PIC, and the Intel 8051 families. The following sections give a brief overview of the 8051 microcontroller family and one of its most recent derivatives is DS89C450 from Dallas Semiconductor.

Table 13.1 MCS-51 ICs

Part Number	On-Chip Code Memory	On-Chip Data Memory (bytes)	Timers
8031	0K	128	3
8032	0K	256	2
8051	4K ROM	128	2
8751	4K EPROM	128	2
8052	8K ROM	256	3
8752	8K EPROM	256	3

The term "8051" loosely refers to the MCS-51 family of microcontrollers started by Intel in 1980. The 8051 family is composed of more than 300 different ICs. Each microcontroller in the family boasts a complement of features suited to a particular design setting. Table 13.1 summarizes the differences among popular 8051 family chips. The 8052 is an enhanced 8051, with an extra timer and more RAM and ROM. The 8031 and 8032 are identical to the 8051 and 8052, except that the ROM area is unused and program code must be stored in an external EPROM or other memory chips. The 87C series has the advantage of providing EPROM instead of ROM.

Intel 8051 was extremely popular in the 1980s and early 1990s, but currently it has largely been superseded by a wide range of enhanced derivatives with 8051-compatible processor cores produced by several independent manufacturers including Atmel, Dallas Semiconductor, Cypress Semiconductor, Silicon Labs, NXP (formerly Philips Semiconductor), Texas Instruments, and Winbond.

Dallas Semiconductor offers several families of 8051-compatible microcontrollers including secure microcontroller, high-speed microcontroller, and ultra-high-speed flash microcontroller families. While preserving instruction set and object code compatibility, these families provide additional architectural features as well as enhanced performance and power consumption when compared to older 8051 members. One of the latest 8051-derivitive microcontrollers from Dallas Semiconductor is DS89C450, which uses the ultra-high-speed core. Following is an overview of the detailed architecture and characteristics of the DS89C450 as presented by Dallas Semiconductor manual.

13.4.1.1 DS89C450 Ultra-High-Speed Flash Microcontroller

The DS89C450 is a fully static CMOS microcontroller that maintains pin and software compatibility with standard 8051. In general, software developed for existing 8051-based systems works on the DS89C450 without modification, with the exception of critical timing routines, as the DS89C450 performs its instructions much faster for any given crystal selection. In addition, the DS89C450 can be used as a drop-in replacement for an older 8051 microcontroller without any circuit modification in most cases.

DS89C450's newly designed processor core executes instructions up to 12 times faster than the original 8051 at the similar crystal speed. It can also operate at a maximum clock rate of 33 MHz that, combined with the 12 times speed, allows for a maximum performance of 33 million instructions per second (MIPS). Besides greater speed, the DS89C450 offers many added hardware features to 8051 standard resources. It includes 1 KB of data RAM, a second full hardware serial port, seven additional interrupts, two extra levels of interrupt priority, programmable watchdog timer, brownout monitor, and power-fail reset. Furthermore, the DS89C450 provides several peripherals and hardware features including three 16-bit timer/counters, two full-duplex serial ports, five levels of interrupt priority, dual data pointers, and 256 bytes of direct RAM and 1 KB of extra MOVX RAM. These and other architectural features of the DS89C450 are detailed in this section, which was extracted from the DS89C450 user manual.

13.4.1.2 *Internal Hardware Architecture*

As mentioned earlier, a microcontroller is a highly integrated chip that contains a CPU, some form of memory, I/O ports, and timers. A more detailed view, specific to the DS89C450, is shown in Figure 13.6. The CPU controls the activity. The instructions and data travel back and forth from the CPU to memory over the data bus. Communication with the outside world takes place through the I/O ports. Timers provide real-time information interrupts to the processor. These terms as well as other microcontroller features are discussed next.

CPU: As discussed previously, the CPU administers all activity in the microcontroller system and performs all operations of data. It executes program instructions including arithmetic (addition, subtraction), logic (AND, OR, NOT), data transfer, and program branching operations. An external crystal provides a timing reference for clocking the CPU.

Address/data bus: The device addresses a 64 KB program and 64 KB data memory area that resides in a combination of internal and external memory. When external memory is accessed, ports 0 and 2 are used as a multiplexed address and data bus.

Memory: Ultra-high-speed flash microcontrollers use several distinct memory areas including internal registers (scratchpad RAM), program memory, and data memory. The registers are located on-chip but the program and data memory spaces can be internal, external, or both. The DS89C450 uses a memory-addressing scheme that separates program memory from data memory such that the 16-bit address bus can address each memory area up to maximum of 64 KB. The program and data segments can be overlapped since they are accessed in different manners. The DS89C450 has 64 KB of on-chip program memory (internal ROM), 1 KB of on-chip data memory space (SRAM), and 256 byte of internal registers (internal RAM). It also has the capability to address 64 KB of external RAM and 64 KB of external ROM. If the maximum address of on-chip program or data

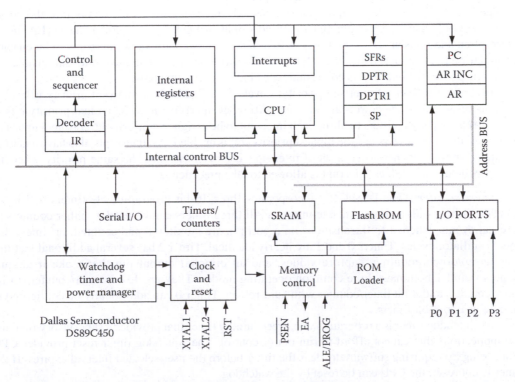

Figure 13.6 Block diagram of the DS89C450 core. (Courtesy of maxim-ic.com)

memory is exceeded, the DS89C450 performs an external memory access using the expanded memory bus. More details about memory map and organization will be presented later when we discuss the DS89C450 programming model.

I/O parallel ports: Without some way to exchange information with the "outside world," a computer is useless. While PCs have fairly standardized I/O connections (COM1, LTP1, etc.), microcontrollers excel in providing much more adaptable I/Os.

The DS89C450 offers four 8-bit parallel I/O ports labeled *P*0, *P*1, *P*2, and *P*3. Each port appears as a special function register (SFR) that can be addressed as a byte or eight individual bit locations. Each I/O port can be used as a general-purpose, bidirectional parallel I/O port. Data written to the port latch serve to set both the level and the direction of the data on the pin.

Serial ports: Serial ports transfer single bits of data one after another, taking at least eight transfers to exchange a byte. The DS89C450 provides two universal asynchronous receiver/transmitters (UARTs) that are controlled and accessed by SFRs. Each UART has an address that is used to read and write the value contained in the UART. The same address is used for both read and write operations, while the read and write operations are distinguished by the instruction.

Interrupts: An interrupt is defined as a signal informing a program that an event has occurred. When a program receives an interrupt signal, it takes a specified action (which can be to ignore the signal). Interrupt signals can cause a program to suspend itself temporarily to service the interrupt by following a special set of events or routines called interrupt handlers. Interrupts are a mechanism of the microcontroller, which enables it to respond to some events at the moment when they occur, regardless of what the microcontroller is doing at the time. This is very important because it provides connection between a microcontroller and the environment that surrounds it. Generally, each interrupt changes the program flow and interrupts it, and after executing an interrupt subprogram (interrupt routine), it continues from that same point on.

The DS89C450 provides 13 interrupt sources. All interrupts, with the exception of the power fail, are controlled by a series combination of individual enable bits and a global enable (EA) in the interrupt-enable register (IE.7). Setting EA to a logic 1 allows individual interrupts to be enabled while setting EA to a logic 0 disables all interrupts regardless of the individual interrupt-enable settings. The power-fail interrupt is controlled by its individual enable only.

There are five levels of interrupt priority: Levels 4–0. The highest interrupt priority is level 4, which is reserved for the power-fail interrupt. All other interrupts have individual priority bits in the interrupt priority registers to allow each interrupt to be assigned a priority level from 3 to 0. The power-fail interrupt always has the highest priority if it is enabled. All interrupts also have a natural hierarchy. In this manner, when a set of interrupts has been assigned the same priority, a second hierarchy determines which interrupt is allowed to take precedence.

Timers/counters: The DS89C450 incorporates three 16-bit programmable timers and has a watchdog timer with a programmable interval. All three timers can be used as either counters of external events, where 1-to-0 transitions on a port pin are monitored and counted, or timers that count oscillator cycles. Timers 0 and 1 are nearly identical. Timer 2 has several additional features such as up/down counting, capture values, and an optional output pin that make it unique. Timers 0 and 1 both have three common operating modes. They are 13-bit timer/counter, 16-bit timer/counter, and 8-bit timer/counter with auto-reload. Timer 0 can additionally be configured to operate as two 8-bit timers.

The watchdog timer is a programmable, free-running timer that provides a supervisory function for applications that cannot afford to run out of control. The watchdog timer reset provides CPU monitoring by requiring software to clear the timer before the user-selected interval expires. If the timer is not reset, the CPU can be reset by the watchdog.

Timing control: The DS89C450 microcontroller provides an on-chip oscillator for use with an external crystal. This can be bypassed by injecting a clock source into the XTAL1 pin. The clock source is used to create machine cycle timing (four clocks), ALE, PSEN, watchdog, timer, and serial baud rate timing. In addition, an on-chip ring oscillator can be used to provide an approximately 10 MHz clock source. A frequency multiplier feature is included, which can be selected by SFR control to multiply the input clock source by either two or four. This allows lower-frequency (and cost) crystals to be used while still allowing internal operation up to the full 33 MHz limit.

Power monitor: A bandgap reference and analog circuitry are incorporated to monitor the power-supply conditions. When VCC begins to drop out of tolerance, the power monitor issues an optional early warning power-fail interrupt. If power continues to fall, the power monitor invokes a reset condition. This remains until power returns to normal operating voltage.

Programming model: This section provides a programmer's overview of the DS89C450 microcontroller core. It includes information on the memory map, SFRs, addressing modes, and instruction set.

Memory map: It is critical to understand the memory layout of the DS89C450 architecture to program the device. The complete memory map is shown in Figure 13.7.

Registers are located in 256 bytes of on-chip scratchpad RAM labeled "INTERNAL REGISTERS" (Figure 13.7), which can be divided into two subareas of 128 bytes each. Separate classes of instructions are used to access the registers and the program/data memory.

The upper 128 bytes are overlapped with the 128 bytes of SFRs in the memory map. Indirect addressing is used to access the upper 128 bytes of scratchpad RAM, while the SFR area is accessed using direct addressing. SFRs will be discussed in more detail later in this section.

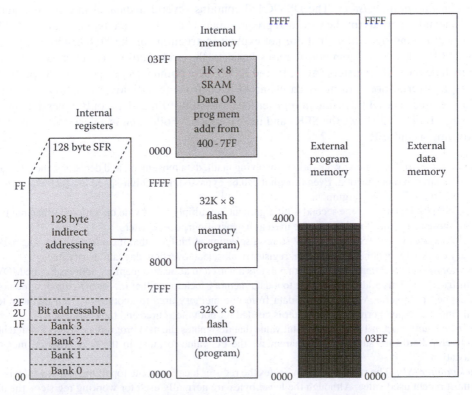

Figure 13.7 Internal memory map. (Courtesy of maxim-ic.com)

The lower 128 bytes can be accessed using direct or indirect addressing. It contains 16 bytes (128 bits) of bit-addressable data memory allowing bit access of character and integer variables stored in that area. It also contains four banks of eight working registers, which are general-purpose RAM locations that can be addressed within the selected bank by any instructions that use R0–R7. The register bank selection is controlled through the program status register in the SFR area. The contents of the working registers can be used for indirect addressing of the upper 128 bytes of scratchpad RAM.

The internal 1 KB SRAM is usable as data, program, or merged program/data memory. Upon a power-on reset, the internal 1 KB memory is disabled and transparent to both program and data memory maps. When the SRAM is enabled as internal data memory, the memory is addressed through MOVX accesses to the first 1 KB (0000h–03FFh) of internal SRAM. When the SRAM is configured as program memory, the memory is addressed through MOVC accesses to the second 1 KB (4000h–07FFh) of internal SRAM.

Program memory is the area from which all instructions are fetched. It is inherently read only. On-chip program memory begins at address 0000h and ends at FFFFh (64 KB) on the DS89C450. Exceeding the maximum address of on-chip program memory causes the device to access off-chip memory. The maximum on-chip decoded address is selectable by software using the ROMSIZE feature. Software can cause the DS89C430 to behave like a device with less on-chip memory. This is beneficial when overlapping external memory is used. The maximum memory size is dynamically variable. Thus, a portion of memory can be removed from the memory map to access off-chip memory and then be restored to access on-chip memory. In fact, all the on-chip memory can be removed from the memory map allowing the full 64 KB memory space to be addressed from off-chip memory.

Special-function registers: The DS89C450 contains several dedicated internal registers that provide special functions for the CPU and programmer. These dedicated registers are called SFRs. All peripherals and operations that are not explicit instructions in the DS89C450 are controlled through SFRs. The most common features basic to the architecture are mapped to the SFRs. These include the CPU registers (ACC, B, and PSW), data pointers, stack pointer, I/O ports, timer/counters, and serial ports. In many cases, an SFR controls an individual function or reports the function's status. The SFRs reside in register locations 80h–FFh and are only accessible by direct addressing. Table 13.2 shows the SFRs and their locations. Following is a description for some of the most important SFRs:

Accumulator (ACC): For many operations involving math, data movement, and decisions, the ACC acts as a source and destination. Even though it can be bypassed, most high-speed instructions need the use of the ACC as one argument.

B register (B): It is used as the second 8-bit argument in multiply and divide operations. When not used for these tasks, the B register can be used as a general-purpose register.

Program status word (PSW): The PSW stores a selection of bit flags that include the carry flag, auxiliary carry flag, general-purpose flag, register bank select, overflow flag, and parity flag.

Data pointers (DPTR and DPTR1): Data pointers are used to allocate a memory address for the MOVX instructions. This address can point to a data memory location, either on- or off-chip or a memory-mapped peripheral. When moving data from one memory area to another or from memory to a memory-mapped peripheral, a pointer is needed for both the source and the destination.

Program counter (PC): The PC is a 16-bit value that designates the next program address to be fetched. On-chip hardware automatically increments the PC value to move to the next program memory location.

Stack pointer (SP): The stack pointer indicates the register location at the top of the stack, which is the most recent used value. Although the lower bytes are normally used for working registers, the user can place the stack anywhere in the scratchpad RAM by setting the stack pointer to the desired location.

Table 13.2   SFR Register Map

Register	Address	Register	Address	Register	Address
P0	80h	CKMOD	96h	STATUS	C5h
SP	81h	SCON0	98h	TA	C7h
DPL	82h	SBUF0	99h	T2CON	C8h
DPH	83h	ACON	9Dh	T2MOD	C9h
DPL1	84h	P2	A0h	RCAP2L	CAh
DPH1	85h	IE	A8h	RCAP2H	CBh
DPS	86h	SADDR0	A9h	TL2	CCh
PCON	87h	SADDR1	AAh	TH2	CDh
TCON	88h	P3	B0h	PSW	D0h
TMOD	89h	IP1	B1h	FCNTL	D5h
TL0	8Ah	IP0	B8h	FDATA	D6h
TL1	8Bh	SADEN0	B9h	WDCON	D8h
TH0	8Ch	SADEN1	Bah	ACC	E0h
TH1	8Dh	SCON1	C0h	EIE	E8h
CKCON	8Eh	SBUF1	C1h	B	F0h
P1	90h	ROMSIZE	C2h	EIP1	F1h
EXIF	91h	PMR	C4h	EIP0	F8h

Source: Courtesy of maxim-ic.com

Instruction set: All instructions are 100% binary compatible with the industry standard 8051 and are only different in the number of machine cycles used for the instructions. Refer to DS89C450 manuals for the complete instruction set. Instructions occupy 1, 2, or 3 bytes. Most instructions have the following format:

```
opcode <destination>, <source>
```

However, based on the addressing mode, the destination and/or the source fields are omitted from some instructions:

Addressing modes: The DS89C450 microcontroller supports eight addressing modes:

1. Register addressing
2. Direct addressing
3. Register indirect addressing
4. Immediate addressing
5. Register indirect addressing with displacement
6. Relative addressing
7. Page addressing
8. Extended addressing

Five of the eight addressing modes are used to address operands. The remainder three are used for program control and branching. Each mode of addressing is summarized next. Note that many instructions (such as ADD) have multiple-addressing modes available.

Register addressing: Register addressing is used for operands that are located in one of the eight working registers (R7–R0), as determined by the current register bank select bits. A register bank is selected using 2 bits in the PSW. Two examples of register addressing are provided in the following:

ADD A, R3	Add register R3 to ACC.
INC R5	Increment the value in register R5.

In the first example, the value in R3 is the source of the operation. In the latter, R5 is the destination.

Direct addressing: Direct addressing is the mode used to access the entire lower 128 byte of scratchpad RAM and the SFR area. It is commonly used to move the value from one register to another. Two examples are shown in the following:

MOV 72h, 74h	Move the value in register 74 to register 72.
MOV 90h, 20h	Move the value in register 20 to the SFR at 90h (port 1).

Note that there is no instruction difference between a RAM access and an SFR access. Direct addressing also extends to bit addressing. There is a group of instructions that explicitly use bits. The address information provided to such an instruction is the bit location, rather than the register address. An example of direct bit addressing is as follows:

MOV C, 0B7h	Move the contents of bit B7 to the carry flag.

Register indirect addressing: This mode is used to access the scratchpad RAM locations above 7Fh. It can also be used to reach the lower RAM (0h–7Fh), if needed. The address is supplied by the contents of the working register specified in the instruction. Thus, one instruction can be used to reach many values by altering the contents of the designated working register. Note that, in general, only R0 and R1 can be used as pointers. An example of register indirect addressing is as follows:

MOV A, @R0	Move contents of RAM location whose address is held by R0 into ACC.
MOV @R1, B	Move contents of B into RAM location whose address is held by R1.

Immediate addressing: Immediate addressing is used when one of the operands is predetermined and coded into the software. This mode is commonly used to initialize SFRs and to mask particular bits without affecting others. An example is as follows:

ORL A, #30h	Logical OR the ACC with 30h

Register indirect with displacement: Register indirect addressing with displacement is used to access data in lookup tables in program memory space. The location is created using a base address with an index. The base address can be either the PC or the DPTR. The index is the ACC. The result is stored in the ACC. An example is as follows:

MOVC A, @A +DPTR	Load the ACC with the contents of program memory pointed to by the contents of the DPTR plus the value in the ACC.

Relative addressing: Relative addressing is used to determine a destination address for the conditional branch. Each of these instructions includes an 8-bit value that contains a 2s complement address offset (−127 to +128), which is added to the PC to determine the destination address. This destination is branched to when the tested condition is true. The PC points to the program memory location immediately following the branch instruction when the offset is added. If the tested condition is not true, the next instruction is performed. An example is as follows:

JZ $−15	Branch to the location (PC + 2)−15 if the contents of the ACC = 0.

Page addressing: Page addressing is used by the branching instructions to specify a destination address within the same 2 KB block as the next contiguous instruction. The full 16-bit address is calculated by taking the five highest-order bits for the next instruction (PC + 2) and concatenating them with the lowest order 11-bit field contained in the current instruction. An example is as follows:

0800h ACALL 100h	Call to the subroutine at address 100h plus the current page address (800h + 100h).

Extended addressing: Extended addressing is used by the branching instructions to specify a 16-bit destination address within the 64 KB address space. The destination address is fixed in the software as an absolute value. An example is as follows:

LJMP 0F712h	Jump to address 0F712h.

13.4.2 ARM (Advanced RISC Machine) Microprocessor

This section gives an overview of advanced RISC machine (ARM), one of the popular architectures utilized in building embedded systems today. Acorn Computers in England introduced ARM (Acorn RISC Machine, then renamed to Advanced RISC Machine) between 1983 and 1985. This was the first reduced instruction set computer (RISC) microprocessor developed for commercial purposes. Today, ARM has become the leader in the microprocessor market accounting for a large percentage of all 32-bit embedded CPUs. The ARM architecture has evolved to a point where it supports implementations across a wide spectrum of performance points and establishing it as the dominant architecture across many market segments. The architectural simplicity of ARM processors has traditionally led to very small implementations, and small implementations allow devices with very low power consumption. Implementation size, performance, and very low power consumption remain key attributes in the development of the ARM architecture.

Seven major versions of ARM architecture (ARMv1–ARMv7) exist today, each with its own instruction set. Of these, the first three versions are now obsolete. Versions can be qualified with variant letters to specify additional instructions and other functionality that are included as an architecture extension.

The valid architecture variants of ARMv4, ARMv5, and ARMv6 are as follows: ARMv4, ARMv4T, ARMv5T, ARMv5TE, ARMv5TEJ, ARMv6, ARMv6K, and ARM v6T2.

The recently released ARMv7 architecture design is divided into three profiles*:

ARMv7-A	Application profile. Implements a traditional ARM architecture with multiple modes and supporting a virtual memory system architecture (VSMA) based on MMU. Supports the ARM and Thumb instruction sets.
ARMv7-R	Real-time profile. Implements a traditional ARM architecture with multiple modes and supporting a protected memory system architecture (PMSA) based on an MPU. Supports the ARM and Thumb instruction sets.
ARMv7-M	Microcontroller profile. Implements a programmers' model designed for fast interrupt processing, with hardware stacking of registers and support for writing interrupt handlers in high-level languages.

The ARM Cortex processor families are the first products developed on ARMv7 architecture. The ARM Cortex-A8, A9, and A15 processors are based on one of the profile of ARMv7, called ARMv7-A as discussed previously. Chapter 14 discusses in detail about ARM Cortex-A8, A9, and A15 processors.

---

* For more information on these ARM versions and the indication of each variant letter, refer to the *ARM Architecture Reference Manual*, from which this section is extracted. Reproduced with permission from ARM Limited. Copyright © 2013 ARM Limited.

13.4.2.1 Internal Hardware Architecture

The ARM architecture is probably the most widely used 16- or 32-bit embedded RISC solution in the world. It maintains a good balance of high performance, low code size, low power consumption, and low silicon area. ARM architecture incorporates the following RISC architecture features:

1. Large uniform register file
2. Load/store architecture that allows data-processing operations to operate on register contents instead of directly operating on memory contents.
3. Simple addressing modes
4. Fixed-length instructions to simplify instruction decode

Moreover, ARM provides additional features such as autoincrement and autodecrement addressing modes to optimize loops, and loading and storing multiple instructions to maximize data throughput. ARM also gives the developer full control over both ALU and shifter in every data-processing instruction and allows conditional execution of all instructions to maximize execution throughput.

This section describes the internal structures of two basic organizations of ARM processor core: the three-stage pipeline used by earlier ARM versions developed before 1995 and the higher-performance five-stage pipeline.

13.4.2.2 ARM Three-Stage Pipeline Organization

The primary elements of ARM three-stage pipeline are shown in Figure 13.8 and explained in the following:

Register bank: It is used to hold the state of the processor. It has two read ports and one write port, which can each be used to access any register. It has an additional read port and an additional write port that give special access to the PC.

Barrel shifter: It can shift or rotate one operand by any number of bits. In other words, it performs logical shift left, logical shift right, arithmetic shift right, and rotate right operations.

Arithmetic and logic unit: This is the part of the processor that performs the arithmetic and logic functions required by the instruction set.

Address register and incrementer: It selects and holds all memory addresses and generates sequential addresses when required.

Data registers: They are used to store data to and from memory.

Instruction decoder and its control logic: This block contains the mechanisms to decode the instruction and the control logic.

13.4.2.3 Single-Cycle Instruction Execution

In single-cycle data processing, two register operands are accessed; the value on the B bus is shifted and combined with the value on the A bus in the ALU. Then the result is written into the register bank. The PC value is stored in the address register and incremented by the incrementer, and the incremented value is copied into the PC in the register bank and also into the address register to be used as the address for the next instruction fetch.

A simple three-stage pipeline ARM processor has the following stages:

1. *Fetch*: The instruction is fetched from the memory and placed in the instruction pipeline.
2. *Decode*: The instruction is decoded and the data path control signals are produced for the next cycle.
3. *Execute*: The register bank is read, an operand shifted, and the ALU result is generated and written back into a destination register.

The three-stage pipeline operation for single-cycle instructions is shown in Figure 13.9.

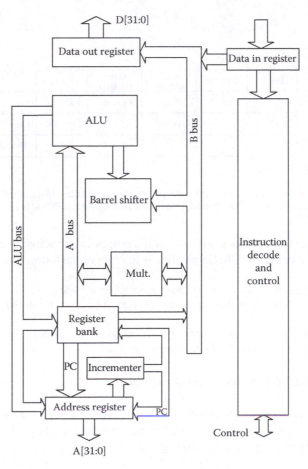

Figure 13.8 Three-stage pipeline ARM organization. (Adapted from *ARM Architecture Reference Manual*, ARM Limited, http://www.arm.com/miscPDFs/14128.pdf)

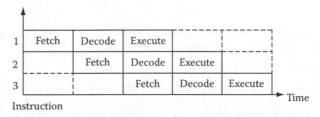

Figure 13.9 ARM single-cycle instruction three-stage pipeline operation. (Adapted from *ARM Architecture Reference Manual*, ARM Limited, http://www.arm.com/miscPDFs/14128.pdf)

When a multicycle instruction is executed the flow is less regular, as shown in Figure 13.10. This shows a sequence of single-cycle ADD instructions with a data store (multicycle) instruction, STR, following the first ADD. The light shading represents the cycles that access main memory. The data path is involved in all execute cycles, the address calculation, and the data transfer. The decode logic always generates the control signals for the data path to use in the next cycle. In addition to the explicit decode cycles, it also generates the control for the data transfer during the address calculation cycle of the STR.

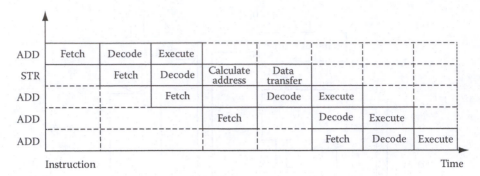

ADD	Fetch	Decode	Execute						
STR		Fetch	Decode	Calculate address	Data transfer				
ADD			Fetch			Decode	Execute		
ADD				Fetch			Decode	Execute	
ADD							Fetch	Decode	Execute

Instruction   Time

Figure 13.10 ARM multicycle instruction three-stage pipeline operation. (Adapted from *ARM Architecture Reference Manual*, ARM Limited, http://www.arm.com/miscPDFs/14128.pdf)

Thus, in this instruction sequence, all parts of the processor are active in every cycle. The simplest way to examine breaks in the ARM pipeline is to observe the following:

1. All instructions occupy the data path for one or more adjacent cycles.
2. For each cycle that an instruction occupies the data path, it occupies the decode logic in the immediately preceding cycle.
3. During the first data path cycle, each instruction issues a fetch for the next instruction but one.
4. Branch instructions flush and refill the instruction pipeline.

13.4.2.4 ARM Five-Stage Pipeline Organization

Higher performance is the major criteria for the development of new processors. Although the three-stage pipeline is very cost-effective, higher performance needs a processor redesign. The time, T, required to execute a given program is given by

$$T = \frac{N_{\text{inst}} \times \text{CPI}}{f_{\text{clk}}}, \tag{13.2}$$

where
N_{inst} is the number of ARM instructions executed in the course of the program
CPI is the average number of clock cycles per instruction
f_{clk} is the clock frequency of the processor

Since N_{inst} is constant for a given program, there are only two ways to increase the performance:

1. *Increase the clock rate,* f_{clk}: This requires the logic in each pipeline stage be simplified and, so, the number of pipeline stages be increased.
2. *Reduce the average number of clock cycles per instruction, CPI*: This requires either that instructions that occupy more than one pipeline slot in a three-stage pipeline are re-implemented to occupy fewer slots or else that pipeline stalls caused by dependencies between instructions are decreased or a combination of both.

The basic problem with decreasing the CPI relative to a three-stage core is associated with the von Neumann bottleneck, that is, any stored-program computer with a single instruction, and data memory will have its performance restricted by the available memory bandwidth. A three-stage ARM core accesses memory on almost every clock cycle either to fetch an instruction or to transfer

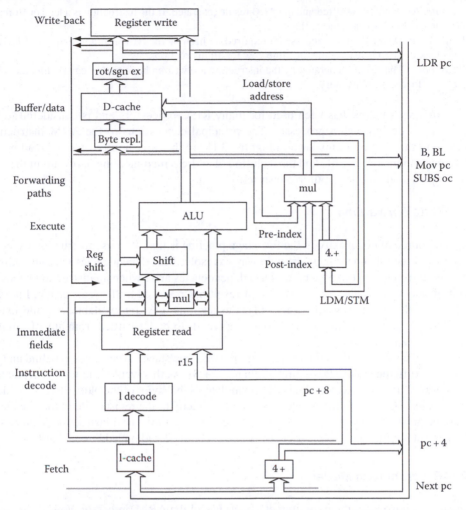

Write-back — Register write

LDR pc

rot/sgn ex

Load/store address

Buffer/data — D-cache

Byte repl.

B, BL
Mov pc
SUBS oc

Forwarding paths

mul

Execute

ALU

Pre-index

Reg shift — Shift

4.+

Post-index

mul

LDM/STM

Immediate fields — Register read

Instruction decode

r15

pc + 8

l decode

Fetch — l-cache

4 +

pc + 4

Next pc

Figure 13.11 ARM9TDMI five-stage pipeline organization. (Adapted from *ARM Architecture Reference Manual*, ARM Limited, http://www.arm.com/miscPDFs/14128.pdf)

data. Simply tightening up on the few cycles where the memory is not used will yield only a small performance gain. To get a significantly better CPI, the memory system must deliver more than one value in each clock cycle either by delivering more than 32 bits per cycle from a single memory or by having separate memories for instruction and data accesses.

The higher-performance ARM cores employ a five-stage pipeline (Figure 13.11) and have separate instruction and data memories. Breaking instruction execution down into five components rather than three decreases the maximum work, which must be completed in a clock cycle, and hence allows a higher clock frequency to be used. The separate instruction and data memories that may be separate caches connected to a unified instruction and data main memory allow a significant reduction in the core's CPI. The five-stage pipeline has the following stages:

1. *Fetch*: The instruction is fetched from memory and placed in the instruction pipeline.
2. *Decode*: The instruction is decoded and register operands read from the register file. There are three operand read ports in the register file, so most ARM instructions can source all their operands in one cycle.

3. *Execute*: An operand is shifted and the ALU result generated. If the instruction is a load or store, the memory address is generated in the ALU.
4. *Buffer/data*: Data memory is accessed if required. Otherwise the ALU result is simply buffered for one clock cycle to give the same pipeline flow for all instructions.
5. *Write-back*: The results generated by the instruction are written back to the register file, including any data loaded from memory.

This five-stage pipeline has been used for many RISC processors and is considered to be the "classic" way to design such a processor. The principal concessions to the ARM instruction set architecture in the organization shown in Figure 13.11 are the three source operand read ports and two write ports in the register file and the inclusion of address incrementing hardware in the execute stage to support load and store multiple instructions.

13.4.2.5 Data Forwarding

A major source of complexity in the five-stage pipeline is that, because instruction execution is spread across three pipeline stages, the only way to resolve data dependencies without stalling the pipeline is to introduce *forwarding* paths. Data dependencies arise when an instruction needs to use the result of one of its predecessors before that result has returned to the register file. Forwarding paths allow results to be passed between stages as soon as they are available, and the five-stage ARM pipeline requires each of the three source operands to be forwarded from any of three intermediate result registers.

The five-stage pipeline reads the instruction operands one stage earlier in the pipeline and would naturally get a different value (PC + 4 rather than PC + 8). As this would lead to unacceptable code incompatibilities, the five-stage pipeline ARMs emulate the behavior of the older three-stage designs. The incremented PC value from the fetch stage is fed directly to the register file in the decode stage, bypassing the pipeline register between the two stages. PC + 4 for the next instruction is equal to PC + 8 for the current instruction, so the correct r15 value is obtained without additional hardware.

13.4.2.6 Programming Model

This section introduces the programmer's overview of the ARM processor. It provides information about data types, memory, register set, and exceptions.

Data types: ARM processors support the following three data types:

1. *Bytes*: 8-bit signed and unsigned bytes.
2. *Half-words*: 16-bit signed and unsigned half-words aligned on 2-byte boundaries.
3. *Words*: 32-bit signed and unsigned words aligned on 4-byte boundaries.

Memory: The ARM architecture has a single flat address space of 2^{32} byte with address range of 0 to $2^{32} - 1$. Based on different ARM versions, this address space can consist of 2^{30} 32-bit word (word aligned), 2^{31} 16-bit half-word (half-word aligned), or 2^{32} 8-bit byte. Words aligned address space means that each address consists of 4 bytes while half-word-aligned address space means that each address consists of only 2 bytes.

Arm supports the little-Endian memory system. However, it can also be configured to work with the big-Endian memory system. In a little-Endian memory system, a byte or half-word at a word-aligned address is the least significant byte or half-word within the word at that address. Similarly, a byte at a half-word-aligned address is the least significant byte within the half-word at that address. In a big-Endian memory system, a byte or half-word at a word-aligned address is the most significant byte or half-word within the word at that address. Similarly, the byte at a

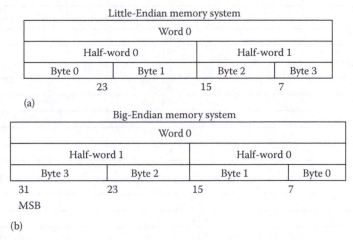

Figure 13.12 Endian memory systems. (a) Little-Endian memory system. (b) Big-Endian memory system.

half-word-aligned address is the most significant byte within the half-word at that address. Both systems are illustrated in Figure 13.12.

13.4.2.7 Processor Modes

The ARM architecture supports the following seven processor modes:

1. USER (usr): Normal program execution mode
2. FAST INTERRUPT (fiq): Supports high-speed data transfer or channel process
3. INTERRUPT (irq): Used for general-purpose interrupt handling
4. SUPERVISOR (svc): A protected mode for the OS
5. ABORT (abt): Implements virtual memory and/or memory protection
6. UNDEFINED (und): Supports software emulation of hardware coprocessors
7. SYSTEM (sys): Runs privileged OSs tasks

Most program application runs under USER mode, which does not allow the running program to access protected system resources or change processor modes. The other six modes are called *privileged modes* because they have full access to system resources and can change mode freely. The modes FIQ, IRQ, SUPERVISOR, ABORT, and UNDEFINED are called exception modes because they are entered when specific exception occurs. The SYSTEM mode is not entered by any exceptions and intended for use by OS tasks.

ARM register set: The ARM processor has 37 registers: 31 general-purpose registers and 6 status registers. All registers are 32 bits wide, and they are arranged in partially overlapping banks with a different register bank for each processor mode as shown in Figure 13.13.

The general-purpose registers are registers R0–R15. Registers R0–R7 refer to the same 32-bit physical registers in all processor modes, which means they are visible to all programs. Registers R8–R14, however, refer to different physical registers based on the current processor mode. This is why the total number of general-purpose registers is 31 not 15. Register R15 holds the PC, which is accessible in all processor modes.

Status registers include current program status register (CPSR) and five saved program status registers (SPSR). CPSR contains condition code flags (Negative, Zero, Carry, Overflow), interrupt-enable bits (I and F), the current processor mode, and other status and control information. The current process mode and interrupt-enable bits are not accessible in user process mode. Each exception mode has an SPSR to preserve the value of CPSR when the associated exception occurs.

User mode	System mode	Supervisor mode	Abort mode	Undefined mode	Interrupt mode	Fast interrupt mode
R0						
R1						
R2						
R3						
R4						
R5						
R6						
R7						
R8						R8_fiq
R9						R9_fiq
R10						R10_fiq
R11						R11_fiq
R12						R12_fiq
R13		R13_svc	R13_abt	R13_und	R13_irq	R13_fiq
R14		R14_svc	R14_abt	R14_und	R14_irq	R14_fiq
R15		R15_svc	R15_abt	R15_und	R15_irq	R15_fiq

CPSR					
	SPSR_svc	SPSR_abt	SPSR_und	SPSR_irq	SPSR_fiq

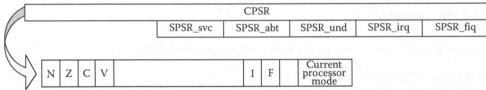

N	Z	C	V			I	F	Current processor mode

Figure 13.13   ARM register set. (From *ARM Architecture Reference Manual*, ARM Limited, http://www.arm.com/miscPDFs/14128.pdf)

Instruction set: Except for the compressed 16-bit thumb instructions, ARM instructions are exactly one word (32 bits) aligned on a 4-byte boundary. The ARM instruction set can be divided into six major categories:

1. *Branch instructions*: These instructions change the control flow of the program execution.
2. *Data-processing instructions*: These instructions perform calculations on the general-purpose registers. They include arithmetic/logic instructions and comparison instructions.
3. *Status register transfer instructions*: These instructions transfer the contents of the CPSR and SPSR to or from the general-purpose registers.
4. *Load and store instructions*: These copy memory values into registers (load instructions) or copy register values into memory (store instructions).
5. *Coprocessor instructions*: These start a coprocessor-specific internal operation, transfer coprocessor data to and from memory, and allow a coprocessor value to be transferred to or from an ARM register.
6. *Exception generating instructions*: These instructions are designed to cause specific exceptions to occur.

I/O system. The ARM handles peripherals such as disk controllers, network interfaces as memory-mapped devices with interrupt support. The internal registers in these devices appear as addressable locations within the ARM's memory map and may be read and written using the same load/store instructions as any other memory location. Peripherals may attract the processor's attention by making an interrupt request using either the normal interrupt (IRQ) or the fast interrupt (FIQ) input. Both interrupt inputs are level sensitive and maskable. Normally most interrupt sources share the IRQ input, with just one or two time-critical sources connected to the higher priority FIQ input. Some systems may include direct memory access (DMA) hardware external to the processor to handle high-bandwidth I/O traffic.

13.5 SUMMARY

As the hardware and software technologies have progressed to the level of providing cost-effective processors, embedded processors have become very common. This chapter introduced the concept of embedded processing and provided examples of two commercially available architectures. Software structures and OS characteristics suitable for this architecture type were also discussed.

PROBLEMS

13.1 A traffic light controller at an intersection has sensors to detect the presence of cars and turns the lights red, yellow, and green appropriately. What hardware facilities are needed in a microcontroller to be used in this application? What software architecture would you use?

13.2 Study the functionality of a microwave oven and determine the requirements for a microcontroller and the corresponding software architecture.

13.3 Study any microcontroller system you have access to, to answer the following:
   a. What constitutes a "task"?
   b. What is the minimum task-switching time?
   c. What synchronization primitives are implemented?

13.4 From the data book of a microcontroller you have access to, list the key characteristics (registers, memory architecture, instruction set, I/O interfaces, interrupt handling, etc.) of the basic version of the microcontroller. Trace the enhancement to the previous characteristics for subsequent versions in the family.

13.5 Find a SoC-based application using ARM architecture. List its salient characteristics. Which of the application requirements necessitated a SoC, rather than a microcontroller-based design?

13.6 Select real-time versions of (a) Linux and (b) Microsoft Windows OSs. Compare each to its non-real-time version, in terms of the operating characteristics and services provided.

13.7 Look up the definitions of the following:
   a. Periodic versus aperiodic computation
   b. Preemptive versus cooperative context switching
   c. Reentrancy
   d. Coroutines
   e. Blocking versus nonblocking interprocess communication

13.8 The fetch-and-add can be generalized to fetch-and-op(a,b), where a is a variable and b is an expression. This operation returns the value of a and replaces it with $(a$ op $b)$. What would be the advantages of this operation over the synchronization primitives described in this chapter?

13.9 Consider the following computation:

$$c_i = a_i * b_i + c_i * d_i,$$

where $i = 1$ to N. Write a high-level program for this computation using a fork/join.

13.10 Estimate the computation time for the program in the previous problem assuming the following time characteristics:

Operation	Execution Time
Addition	1
Multiplication	3
Fork	10
Join	4

Assume that the tasks start execution as soon as they are spawned and there is no waiting time.

13.11 What are the two instruction sets that most ARM processors implement? How are they different?

13.12 What is Unified Assembler Language (UAL)?

13.13 What development tools are available for ARM architectures?

BIBLIOGRAPHY

Adam, T., Chandy, K., and Dickson, J.A., Comparison of list schedulers for parallel processing systems, *Communication of ACM*, 17, 685–690, December 1974.

ARM Architecture Reference Manual, Copyright 1996–1998, 2000, 2004, 2005, ARM Limited, ARM DDI 01001.

ARM Architecture Reference Manual, ARM v7-A and ARM v7-R edition Errata markup, Copyright 1996–1998, 2000, 2004–2011, ARM Limited, ARM DDI 0406B_errata_2011_Q3 (ID 120611).

Bashir, A., Susarla, G., and Karavan, K., A statistical study of a task scheduling algorithm, *IEEE Transactions on Computers*, C-32(8), 774–777, August 1983.

Dijkstra, E.W., Co-operating sequential processes, *Programming Languages*, Genuys, F. (Ed.), London, U.K.: Academic Press, 1965.

DS89C430, DS89C440, DS89C450, Ultra-high-speed flash microcontrollers, http://www.maximintegrated.com/datasheet/index.mvp/id/4078.

EE Times (Weekly), Manhasset, NY: CMP Media, www.eetimes.com.

El-Rewini, H. and Lewis, T.G., Scheduling parallel program tasks onto arbitrary target machines, *Journal of Parallel and Distributed Computing*, 9(2), 138–153, June 1990.

Embedded Systems Design (Monthly), San Francisco, CA: CMP Media, www.embedded.com.

Furber, S., *ARM System-On-Chip Architecture*, 2nd edn., New York: Addison-Wesley, 2000.

Kallstrom, M. and Thakkar, S.S., Programming three parallel computers, *IEEE Software*, 5(1), 11–22, January 1988.

Lewis, D.W., *Fundamentals of Embedded Software*, Upper Saddle River, NJ: Prentice Hall, 2002.

Pont, M.J., *Embedded C*, London, U.K.: Addison-Wesley, 2002.

Shiva, S.G., *Advanced Computer Architectures*, Boca Raton, FL: Taylor & Francis, 2006.

Silberschatz, A., Galvin, P., and Gragne, G., *Operating System Concepts*, Addison-Wesley, 2003.

Simon, D.E., *An Embedded Software Primer*, Boston, MA: Addison-Wesley, 1999.

Stallings, W., *Operating Systems*, New York: Macmillan, 2005.

Tanenbaum, A. and Woodhull, A., *Operating Systems, Design and Implementation,* Upper Saddle River, NJ: Prentice Hall, 2006.

Vahid, S. and Givargis, T., *Embedded System Design: A Unified Hardware/Software Introduction*, New York: Wiley, 2002.

Williams, S.A., *Programming Models for Parallel Systems*, New York: Wiley, 1990.

Wolf, W., *Computers as Components—Principles of Embedded Computing System Design*, San Francisco, CA: Morgan Kaufmann, 2001.

Yeralan, S. and Emery, H., *Programming and Interfacing the 8051 Microcontroller in C and Assembly*, Gainesville, FL: Rigel Press, 2000.

CHAPTER 14

Mobile Processors and System on Chip

In the previous chapter, we discussed embedded systems in detail. These are computer systems that are tailored to perform specific functions within a larger system. Examples include MP3 players and traffic light controllers, systems simple in their functionality to complex systems such as smart cellular phones. In this chapter, we discuss two prominent paradigms in today's technology that extend the applicability of embedded systems: mobile processors and systems on a chip (SoC). Since the early 1970s, the microprocessors have continued to evolve at an impressive and steady rate. Their remarkable growth can aptly be warranted by the success of Moore's law, which states that the number of transistors on integrated circuits (ICs) doubles approximately every 2 years. Earlier IC technology allowed the fabrication of a small number of transistors on an IC chip, thus limiting the functionality provided by the chip. Microprocessors were then built using multiple interconnected chips.

With the advent of very large scale integrated (VLSI) technology, we were able to integrate most functionalities of the processor and other support components into a single chip. Intel 4004, released in 1971, marked a significant feat in microprocessor technology as being the first central processing unit (CPU) built on a single chip. It was a 4-bit CPU manufactured using the 10 μm fabrication process and consisted of approximately 2300 transistors. The 4004 was followed by Intel 8008, which was the first 8-bit CPU and consisted of 3500 transistors. As of 2012, Intel's Ivy Bridge series of microprocessors utilize the 22 nm silicon fabrication process and consist of over a billion transistors.

The magnitude of increase in performance achieved by the microprocessor industry over the last three decades makes it one of the fastest evolving technologies. This can also be observed by their ubiquitous use in our day-to-day lives in the form of personal computers (PCs) to smartphones.

Initially, advancements in processor evolution were aimed toward increasing their speed and functionality by increasing the number of transistors that could be housed in a single die. However, as we began to approach the limitations of silicon at the atomic level, the advancements in evolution shifted from merely increasing the clock cycles and number of transistors to building multicore processors. As more applications embedded processors into systems, optimizing the processors for power usage and heat emission became important. As these systems became more mobile (cell phones, tablets, etc.), a new form of high-performance computing known as *mobile* computing came into being.

Currently, microprocessors can be broadly categorized into two categories based on their size, power consumption, and performance. These categories are *desktop* and *mobile* processors. Desktop processors are aimed toward delivering higher performance at the expense of high-power usage and heat output. Mobile processors on the other hand are aimed toward providing slightly less processing power while using significantly low power and generating low heat. These processors are one of the main reasons for the success for tablets, netbooks, ultrabooks, smartphones, etc., which we all use today.

The concept of system in a package (SiP) came into being when multiple chips were put together on a board forming the system. As we moved into VLSI era, we could integrate all functionalities into a chip resulting in the current concept of system on chip (SoC). Electronic devices such as cell phones, audio players, and digital cameras are nowadays delivered to consumers with a whole computer with a processing unit, sufficient memory, input/output, and possibly some other components, all packaged in a single chip that fits in a small and increasingly complex device. An SoC thus is a packaging of all of system components in a single IC chip that functions as a complete system that has its own input/output in addition to processing and storage. The SoC designs usually consume less power and have a lower cost and higher reliability than the multichip systems that they replace. SoC's only real disadvantage is their lack of flexibility in the sense that, in a desktop PC, components (CPU, GPU, RAM, etc.) can be replaced, whereas this is not the situation with SoCs. It is important to note that the modern microprocessors are endowed with such high functionality and capabilities; most manufacturers call their processors SoCs.

Figure 14.1 shows the Toshiba TC35661SBG Bluetooth SoC, a single-chip controller that supports Bluetooth classic and offers the new low-energy standards. It provides a minimum external component architecture and an easy Bluetooth integration for various industrial, medical, and automotive applications with its multiple input/output options like UART, USB2.0, I2C, I2S, SPI, and GPIO.

An abstract SoC architecture consists of but is not limited to the following components:

- A microcontroller, microprocessor, or DSP core(s)
- Memory blocks like ROM, RAM, EEPROM, and flash memory
- Timing sources including oscillators and phase-locked loops
- Peripherals including counter-timers, real-time timers, and power-on reset generators
- External interfaces to increase connectivity and portability including industry standards such as USB, FireWire, Ethernet, USART, and SPI
- Analog interfaces including analog-to-digital converters and digital-to-analog converters
- Power management circuits and voltage regulators
- Other application-specific components like the RF block component in Figure 14.1

In 2011, Microsoft also has announced support of SoC architectures from Intel-, AMD-, and ARM-based systems indicating the acceptance of such model by the audience and the experts from the field. There is also a huge move by IC manufacturers toward producing SoCs. For example, the Apple™ "A" series of SoCs that are designed based on ARM SoCs are used in Apple's consumer electronic devices, such as the iPod™, iPad™, iPhone™, and Apple TV™.

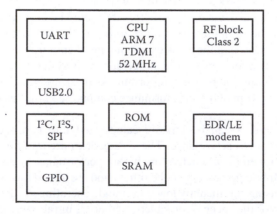

Figure 14.1 Toshiba TC35661SBG Bluetooth SoC. (Adapted from http://www.toshiba-components.com/bluetooth/index.html)

Section 14.1 provides details of Apple iPhone systems as an example of mobile processor and SoC architectures. Section 14.2 describes another SoC-based system (Raspberry Pi™), and Section 14.3 extends the description of ARM architectures provided in Chapter 13.

14.1 APPLE iPHONE 4S

The iPhone is a series of smartphones that are designed by Apple Inc. These devices serve as good examples of the applicability of SoC and mobile processors to create a commercial computing and communication device. These devices use a single multitouch display as the primary user interaction device. For input of text, they use a virtual keyboard instead of a physical one. These devices run a mobile operating system known as iOS™, which is also developed by Apple Inc. The first smartphone in this series was called the iPhone and was released in June 2007. In this section, we will provide the details of the fifth generation of iPhone known as the iPhone 4S, which was released in October 2011. This generation of iPhone provides a series of hardware and software improvements and additions over the previous generation of iPhone (iPhone 4). In particular, the iPhone 4S introduced a voice recognition system known as Siri™ and a cloud-based storage service called the iCloud™. Figure 14.2 (by iFixit, http://www.ifixit.com) shows the printed circuit board (PCB) layout of the iPhone 4S. Some of the prominent chips on this PCB are as follows:

- Apple A5 dual-core processor
- Qualcomm RTR8605 multiband/mode RF transceiver
- Skyworks 77464-20 load-insensitive power amplifier
- Avago ACPM-7181 power amplifier
- TriQuint TQM9M9030 surface acoustic wave (SAW) filter
- TriQuint TQM666052 PA-duplexer module
- TI 343S0538 touch screen controller
- STMicro AGD8 2135 LUSDI gyroscope
- STMicro 8134 33DH 00D35 three-axis accelerometer

The underside of this PCB contains the flash memory among other chips. We provide a brief description of the function of these chips and other hardware and software characteristics of the iPhone 4S next.

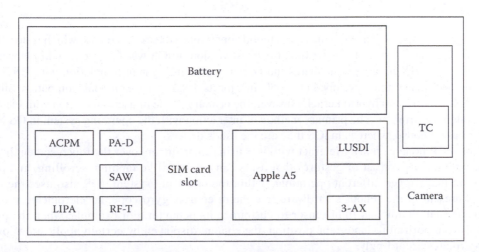

Figure 14.2 iPhone 4S PCB layout. (Adapted from http://www.ifixit.com/)

14.1.1 Hardware

The iPhone 4S is powered by the Apple A5 SoC. This SoC incorporates a dual-core ARM Cortex A9 mobile processor, which is based on the ARMv7 instruction set and an Imagination Technologies' PowerVR SGX543 graphics processing unit (GPU) as two of its major components. This GPU features pixel, vertex, and geometry shader hardware, which supports OpenGL ES 2.0. It also includes an image signal processor. The iPhone has 512 MB of DDR2 RAM and comes in three models based on its flash storage capacity of 16, 32, or 64 GB.

The display of the iPhone includes a 3.5 in. diagonal widescreen LED backlit IPS TFT LCD. It has a resolution of 640×960 at 326 pixels per inch (PPI) and offers a contrast ratio of 800:1 with a maximum brightness of 500 cd/m^2. The front (display) and back of the iPhone are covered with a fingerprint-resistant oleophobic coating.

The iPhone 4S features two cameras (one on the front and the other on the back) for taking pictures, recording videos, and making video calls (a feature known as FaceTime in iOS). The front-facing camera uses a 0.3 megapixel VGA sensor, which is capable of recording video at 480p resolution. The rear-facing camera includes an 8-megapixel backside illuminated sensor and is capable of recording video at 1080p resolution (HD) at up to 30 fps. It also includes an IR filter and a wider f/2.4 aperture and provides features like autofocus, facial recognition (stills only), image stabilization, and photo and video geotagging.

In terms of connectively, this generation of iPhone provides Wi-Fi access based on 802.11 b/g/n and supports Bluetooth 4.0. The iPhone 4S supports both CDMA and GSM cellular networks and is a world phone so both CDMA and GSM users can roam internationally on GSM networks. The iPhone also includes a global position system (GPS) to locate its position on Earth.

In addition to the conventional mode of user inputs like touch screen and microphones, this iPhone also includes a series of sensors like three-axis gyroscope, accelerometer, proximity sensor, and an ambient light sensor. The gyroscope along with three-axis accelerometer is used to determine the angular position of the iPhone with respect to the ground. The proximity sensor is used to determine when the phone is held close to the ear during calls so that the display can be turned off to preserve power and to prevent accidental user input through the touch screen display. The ambient light sensor is used to adjust the display brightness relative to the ambient light. The iPhone uses a built-in rechargeable lithium-ion battery that provides talk time of up to 8 h and standby time up to 200 h.

14.1.2 Software

The iPhone series of smartphones use a mobile operating system called iOS, which is developed by Apple Inc. At the time of this writing, the latest version of iOS was 6.0.1 released in November 2012. iOS is UNIX based, and it shares the Darwin operating system foundation with OS X, the desktop operation system developed by Apple Inc. for its desktop and notebook computers. Similar to OS X, the iOS uses a hybrid kernel; however, the primary mode of user interaction with the operating system is different. The primary method of interaction with the operating system in OS X is a pointer-based mouse, whereas in iOS they are the user's fingers.

Users of iPhone primarily interact with iOS using their fingers over the touch screen display. These interactions consist of gestures that users can perform like tapping, scrolling, and pinch and zoom that trigger different operations within the operating system. iOS also uses the other sensors present in the iPhone's hardware like the three-axis gyroscope, accelerometer, proximity sensor, and an ambient light sensor to customize its behavior. For example, when the phone is tilted from portrait to landscape position, the content displayed by certain applications on the display rotates accordingly.

The iOS is structured to include four layers of abstraction. These are the Core OS layer, the Core Services layer, the Media layer, and the Cocoa Touch layer. The default user interface that includes the multitouch gestures is performed by the outermost Cocoa Touch layer. This layer also includes support for accelerometer, camera, and localization. The Media layer enables certain multimedia operations like video playback, audio mixing and recording, image rendering, and core animation. The Core Services layer provides support for networking, threads, location, and embedded SQLite database. The Core OS layer contains the low-level features upon which other technologies are built on. This includes core Bluetooth framework, security framework, power management, and TCP/IP.

The home screen rendered on the iPhone by the iOS consists of a series of application icons that provide certain features to its users. For example, these applications include phone, maps, messages, and calendar. The user can open applications by tapping on them with their fingers and uses gestures mentioned earlier to navigate through them and operate accordingly.

Third-party developers can build native applications for iOS using the software development kit (SDK) provided by Apple called iOS SDK. Developed applications can be submitted to an online application store called App Store through which potential users can purchase and install them.

14.2 ARM v7-A APPLICATION PROFILE

ARM processors are used widely in different handheld devices, particularly PDAs and smartphones. Some examples include Microsoft Surface, Apple iPad, Apple iPhone, and iPod. Different versions of ARM processors are discussed in Chapter 13 under the section ARM, and among them, ARMv7 is the most recent. ARMv7 provides three profiles as discussed, which include ARMv7-A application profile, ARMv7-R real-time profile, and ARMv7-M microcontroller profile. In this section, we discuss in detail about ARMv7-A profile, its unique features, and the different applications.

ARMv7-A application profile consists of the following features:

1. ARM Thumb-2 instruction set for overall code density comparable with Thumb and performance comparable with ARM instructions.
2. Thumb Execution Environment (ThumbEE) to provide execution environment acceleration.
3. Advanced SIMD architecture extension to accelerate the performance of multimedia applications such as 3D graphics and image processing. The Advanced SIMD architecture extension, its associated implementations, and supporting software are commonly referred to as NEON technology.
4. Vector Floating-Point v3 (VFPv3) architecture for floating-point computation that is fully compliant with the IEEE 754 Standard.
5. Security Extensions architecture for enhanced security features that facilitates the development of secure applications.

14.2.1 ARM Thumb-2 Instruction Set

The Thumb instruction set is a re-encoded subset of the ARM instruction set. Thumb is designed to increase the performance of ARM implementations that use a 16-bit or narrower memory data bus and to allow better code density than provided by the ARM instruction set. T variants of the ARM architecture incorporate both a full 32-bit ARM instruction set and the 16-bit Thumb instruction set. Every Thumb instruction is encoded in 16 bits.

Thumb does not alter the underlying programmers' model of the ARM architecture. It merely presents restricted access to it. All Thumb data-processing instructions operate on full 32-bit values, and full 32-bit addresses are produced by both data-access instructions and instruction fetches. When the processor is executing Thumb instructions, eight general-purpose integer registers are available, R0–R7, which are the same physical registers as R0–R7 when executing ARM instructions.

Some Thumb instructions also access the program counter (ARM register 15), the link register (ARM register 14), and the stack pointer (ARM register 13). Further instructions allow limited access to ARM registers 8–15, which are known as the high registers.

Thumb instruction set consists of the following.

14.2.1.1 Exceptions

Exceptions generated during Thumb execution switch to ARM execution before executing the exception handler.

14.2.1.2 Branch Instructions

Thumb supports six types of branch instructions:

1. A conditional branch to allow forward and backward branches of up to 256 bytes (−256 to + 254).
2. An unconditional branch that allows a forward or backward branch of up to 2 kB (−2048 to +2046).
3. A branch with link (subroutine call) is supported with a pair of instructions that allow forward and backward branches of up to 4 MB ($-2^{22} <= offset <= +2^{22} - 2$).
4. A branch with link and exchange uses a pair of instructions, similar to branch with link, but additionally switches to ARM code execution.
5. A branch and exchange instruction branches to an address in a register and optionally switches to ARM code execution.
6. A second form of branch with link and exchange instruction performs a subroutine call to an address in a register and optionally switches to ARM code execution.

14.2.1.3 Data-Processing Instructions

Thumb data-processing instructions are a subset of the ARM data-processing instructions. They are divided into two sets. The first set can only operate on the low registers, r0–r7. The second set can operate on the high registers, r8–r15, or on a mixture of low and high registers.

14.2.1.4 Load and Store Register Instructions

Thumb supports eight types of load and store register instructions. Two basic addressing modes register plus register and register plus 5-bit immediate are available, and these allow the load and store of words, half-words, and bytes and also the load of signed half-words and bytes. If an immediate offset is used, it is scaled by 4 for word access and 2 for half-word accesses. In addition, three special instructions allow words to be loaded using the PC as a base with a 1 KB (word-aligned) immediate offset and words to be loaded and stored with the stack pointer (R13) as the base and a 1 KB (word-aligned) immediate offset.

14.2.1.5 Load and Store Multiple Instructions

Thumb supports four types of load and store multiple instructions. Two instructions, LDMIA and STMIA, are designed to support block copy. They have a fixed increment after addressing mode from a base register. The other two instructions, PUSH and POP, also have a fixed addressing mode. They implement a full descending stack, and the stack pointer (R13) is used as the base register. All four instructions update the base register after transfer, and all can transfer any or all of the lower eight registers. PUSH can also stack the return address, and POP can load the PC.

Thumb-2 is an enhancement to the 16-bit Thumb instruction set. It adds 32-bit instructions that can be freely intermixed with 16-bit instructions in a program. The additional 32-bit instructions enable Thumb-2 to cover the functionality of the ARM instruction set. The 32-bit instructions enable Thumb-2 to combine the code density of earlier versions of Thumb, with performance of the ARM instruction. The most important difference between the Thumb-2 instruction set and the ARM instruction set is that most 32-bit Thumb instructions are unconditional, whereas most ARM instructions can be conditional. Thumb-2 introduces a conditional execution instruction, IT, that is a logical if-then-else function that you can apply to the following instructions to make them conditional.

14.2.2 Thumb Execution Environment

The ThumbEE is a variant of the Thumb instruction set that is designed as a target for dynamically generated code. However, it cannot interwork freely with the ARM and Thumb instruction sets.

In general, instructions in ThumbEE are identical to Thumb instructions, with the following exceptions:

1. A small number of instructions are affected by modifications to transitions from ThumbEE state.
2. A substantial number of instructions have a null check on the base register before any other operation takes place but are identical (or almost identical) in all other respects.
3. Three Thumb instructions, BLX (immediate), 16-bit LDM, and 16-bit STM, are removed in ThumbEE state. The encoding corresponding to BLX (immediate) in Thumb is UNDEFINED in ThumbEE state. 16-bit LDM and STM are replaced by new instructions.
4. Two new 32-bit instructions, ENTERX and LEAVEX, are introduced in both the Thumb instruction set and the ThumbEE instruction set.

14.2.3 Advanced SIMD and VFP Extensions

Advanced SIMD performs packed single instruction, multiple data (SIMD) operations, either integer or single-precision floating point. VFP performs single-precision or double-precision floating-point operations. VFP is a floating-point coprocessor extension to the instruction set architectures. There have been three main versions of VFP to date:

1. VFPv1 is obsolete.
2. VFPv2 is an optional extension to the ARM instruction set in the ARMv5TE, ARMv5TEJ, ARMv6, and ARMv6K architectures and the ARM and Thumb instruction sets in the ARMv6T2 architecture.
3. VFPv3 is an optional extension to the ARM, Thumb, and ThumbEE instruction sets in the ARMv7-A profile.

Advanced SIMD and VFPv3 use the same register set. This is distinct from the ARM core register set. These registers are generally referred to as the extension registers. The extension register set consists of either 32 or 16 double-word registers, as follows:

a. If VFPv2 is implemented, it consists of 16 double-word registers.
b. If VFPv3 is implemented, it consists of either 32 or 16 double-word registers. Where necessary, the terms VFPv3-D32 and VFPv3-D16 are used to distinguish between these two implementation options, and the term VFPv3 is used to cover both options.
c. If Advanced SIMD is implemented, it consists of 32 double-word registers. If both Advanced SIMD and VFPv3 are implemented, VFPv3 must be implemented in its VFPv3-D32 form. The Advanced SIMD and VFP views of the extension register set are not identical. Advanced SIMD can view this register set as sixteen 128-bit quad-word registers (Q0–Q15) and thirty-two 64-bit double-word registers (D0–D31). This view is also available in VFPv3 and can be used simultaneously.

14.2.4 Security Extension Architecture*

The Security Extensions integrate hardware security features into the architecture, to facilitate the development of secure applications. The Security Extensions define two security states, secure state and nonsecure state. All code execution takes place either in secure state or in nonsecure state. Each security state operates in its own virtual memory address space, and many system controls can be set independently in each of the security states. All of the processor modes that are available in a system that does not implement the Security Extensions are available in each of the security states. The Security Extensions also define an additional processor mode, monitor mode that provides a bridge between code running in nonsecure state and code running in secure state.

The fundamental mechanism that determines the security state is the secure configuration register (SCR) SCR.NS bit. For all modes other than monitor mode, the SCR.NS bit determines the security state for code execution. Code executing in monitor mode is executed in the secure state regardless of the value of the SCR.NS bit. Code can change the SCR only if it is executing in the secure state. The general-purpose registers and the processor status registers are not banked between the secure and the nonsecure states. When execution switches between the nonsecure and secure security states, ARM expects that the values of these registers are switched by a kernel running mostly in monitor mode.

The above section was extracted from the *ARM Architecture Reference Manual*, ARM v7-A and ARM v7-R edition Errata markup and the *Cortex-A8 Technical Reference Manual*.

14.2.5 Cortex-A8 Processor*

This section gives a brief overview of the Cortex-A8 microprocessor. The ARM Cortex-A8 processor developed by ARM is the highest performance and most power-efficient ARM processor to meet the power and performance needs of mobile Internet devices. The processor is used in a wide variety of end devices used by users ranging from high-end smartphones to netbooks, DTVs, printers, storage networking, and set-top box. For detailed technical information about the processor, refer to the Cortex-A8 Technical Reference Manual, from which this section is extracted.

Cortex-A8 processor is a 32-bit RISC architecture with 16 registers and Harvard memory architecture, and it can run at speeds between 600 MHz and over 1 GHz. Cortex-A8 uses the ARMv7-A architecture, which includes a 13-stage integer pipeline and a 10-stage NEON pipeline, which is useful for accelerating multimedia and signal processing applications. The structure of the Cortex-A8 processor is shown in Figure 14.3, and the main components of the processor consist of instruction fetch, instruction decode, instruction execute, load/store, L2 cache, NEON, and ETM.

Instruction fetch: The instruction fetch unit predicts the instruction stream, fetches instruction from the L1 instruction cache, and places the fetched instruction into a buffer for consumption by the decode pipeline.

Instruction decode: The instruction decode unit decodes all ARM instructions discussed in Section 13.4.2.7. The decode unit also handles the sequencing of exceptions, debug events, reset initialization, memory built-in self-test (MBIST), and wait-for-interrupt and other unusual events.

Instruction execute: The instruction execute unit executes all integer arithmetic and logic unit (ALU) and multiply operation including flag generation, generates the virtual addresses for loads and stores, supplies formatted data for stores, also forwards data and flags, processes branches and other changes of instruction steam, and evaluates instruction condition codes. It consists of two symmetric ALU pipelines, an address generator for load and store instructions and multiply pipeline.

---

* This section reproduced with permission from ARM Limited. Copyright © 2013 ARM Limited.

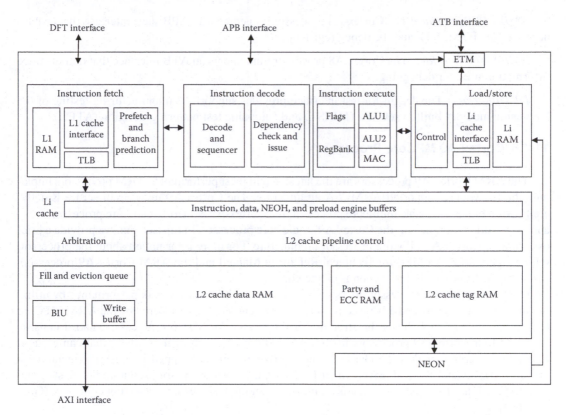

Figure 14.3 Cortex A8 block diagram. (From *Cortex-A8 Revision: r3p2, Technical Reference Manual*, Copyright 2006–2010, ARM Limited, ARM DDI 0344K (ID060510). Reproduced with permission from ARM Limited. Copyright © 2013 ARM Limited.)

Load/store: The load/store unit encompasses the entire L1 data side memory system and the integer load/store pipeline. The pipeline accepts one load or store per cycle that can be present in either pipeline 0 or pipeline 1. This gives the processor flexibility when scheduling load and store instructions.

L2 cache: The L2 cache unit includes the L2 cache and the *buffer interface unit* (BIU). It services L1 cache misses from both the instruction fetch unit and the load/store unit.

NEON: The NEON unit includes the full 10-stage NEON pipeline that decodes and executes the Advanced SIMD media instruction set. The NEON unit includes the NEON instruction queue, the NEON load data queue, two pipelines of NEON decode logic, three execution pipelines for Advanced SIMD integer instructions, two execution pipelines for Advanced SIMD floating-point instructions, one execution pipeline for Advanced SIMD and VFP load/store instructions, and the VFP engine for full execution of the VFPv3 data-processing instruction set.

ETM: The ETM unit is a nonintrusive trace macrocell that filters and compresses an instruction and data trace for use in system debugging and system profiling. The ETM unit has an external interface outside of the processor called the *advanced trace bus* (ATB) interface. AMBA AXI interface, AMBA APB interface, AMBA ATB interface, and DFT interface are the external interfaces the processor consists.

AMBA AXI interface: The AXI bus interface is the main interface to the system bus. It supports 64-bit or 128-bit wide input and output data buses. It also supports multiple outstanding requests on the AXI bus. It performs L2 cache fills and noncacheable accesses for both instructions and data.

AMBA APB interface: The Cortex-A8 processor implements an APB slave interface that enables access to the ETM, CTI, and the debug registers.

AMBA ATB interface: The Cortex-A8 processor implements an ATB interface that outputs trace information used for debugging.

DFT interface: The design-for-test interface provides support for manufacturing testing of the core using memory built-in self-test (MBIST) and automatic test pattern generation (ATPG).

14.2.6 Cortex-A9 MPCore Processor*

The ARM Cortex-A9 processors are the latest and highest performance ARM processors implementing the widely supported ARMv7-A architecture. The Cortex-A9 microarchitecture is delivered within either a scalable multicore processor, the Cortex-A9 MPCore multicore processor, or, as a more traditional processor, the Cortex-A9 single core processor. In this section, we provide a brief overview of Cortex-A9 MPCore processor architecture. For more information about the processor, refer to the Cortex-A9 MPCore Technical Reference Manual and the ARM Cortex-A9 processors white paper, from which this section is extracted.

Cortex-A9 MPCore processor offers mobile devices with increased peak performance by utilizing the design flexibility and advanced power management techniques offered by the ARM MPCore technology. Figure 14.4 shows the structure of the Cortex-A9 MPCore processor, and it consists of one to four Cortex-A9 processors in a cluster and a *snoop control unit* (SCU) that can be used to ensure coherency within the cluster; a set of private memory-mapped peripherals, including a global timer; a watchdog and private timer for each Cortex-A9 processor present in the cluster; and an integrated interrupt controller that is an implementation of the generic interrupt controller (GIC) architecture.

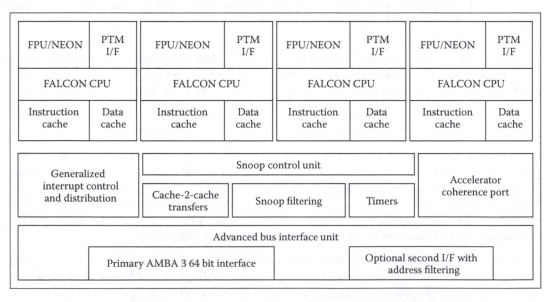

Figure 14.4 Cortex A9 MPCore processor. (From *The ARM Cortex-A9 Processors*, White Paper, ARM, Document Revision 2.0 September 2009. Reproduced with permission from ARM Limited. Copyright © 2013 ARM Limited.)

Individual Cortex-A9 processors in the Cortex-A9 MPCore cluster can be implemented with their own hardware configurations. The major configuration options that impact Cortex-A9 MPCore System Integration are as follows:

1. One or two AXI master port interfaces, with address filtering capabilities
2. An optional accelerator coherency port (ACP) suitable for coherent memory transfers
3. A configurable number of interrupt lines

SCU: The SCU is the central intelligence in the ARM's multicore technology and is responsible for managing the interconnect, arbitration, communication, cache-2-cache and system memory transfers, cache coherence, and other multicore capabilities for all MPCore technology-enabled processors.

ACP: The ACP is an optional AXI 64-bit slave port that can be connected to noncached AXI master peripherals, such as DMA engine or cryptographic engine.

GIC: It is responsible for centralizing all interrupt sources before dispatching them to each individual Cortex-A9 processor. There is one interrupt interface per Cortex-A9 processor.

Advanced bus interface unit: Supporting the design configuration of either a single or dual 64-bit AMBA 3 AXI master interface, the processor can provide, at CPU speed, full load balancing of transactions capable of exceeding 12 GB/s into the system interconnect. Alternatively, the second interface may define a transaction filter to a subset of the global address space so presenting the system design with the flexibility to partition the address space immediately within the processor fabric.

Floating-point unit (FPU): The FPU provides high-performance single- and double-precision floating-point instructions compatible with ARM VFPv3 architecture that is software compatible with previous generations of ARM floating-point coprocessor. Cortex-A9 FPU is capable of significantly enhancing solutions with rich graphics, 3D, imaging, and scientific computation.

NEON media processing engine (MPE): MPE can be used with either of the Cortex-A9 processors and provides an engine that offers both the performance and functionality of the Cortex-A9 FPU plus an implementation of the ARM NEON Advanced SIMD instruction set. The MPE extends the Cortex-A9 processor's FPU to provide a quad-MAX and additional 64-bit and 128-bit integer and 32-bit floating-point data quantities every cycle. The MPE also supports structured load/store capabilities to eliminate shuffling data between algorithm format to machine formats.

Program trace macrocell (PTM): PTM provides program-flow trace capabilities for either of the Cortex-A9 processors and provides full visibility into the processor's actual instruction flow. The PTM includes visibility over all code branches and program-flow changes with cycle counting enabling profiling analysis.

14.2.7 Cortex-A15 Processor*

In this section, we provide a brief overview of the Cortex-A15 MPCore processor, which is today's high-performance engine for highly connected device. This processor delivers unprecedented flexibility and processing capability and is designed with advanced power reduction techniques and enables compelling products in a wide range of new and existing ARM markets ranging from mobile computing, high-end digital home, servers, and wireless infrastructure. The Cortex-A15 MPCore processor is built around the ARMv7-A architecture and ensures full software compatibility with the rest of the highly acclaimed Cortex-A processors. This enables immediate access to an established developer and software ecosystem including Android, Adobe Flash Player, Java Platform Standard Edition (Java SE), JavaFX, Linux, Microsoft Windows Embedded, Symbian, and Ubuntu, along with more than 1000 ARM Connected Community members providing application software, hardware and software development tools, middleware, and SoC design services. The Cortex-A15 MPCore processor will deliver twice the performance of today's smartphones based on the Cortex-A9 processor.

The applications of Cortex-A15 processor consist of advanced smartphones, mobile computing, high-end digital home entertainment, wireless infrastructure, and low-power servers. The growing complexity of the Web2.0 centric devices is creating the requirement for devices to support multiple software personalities and combine disparate functionality. For this reason, the Cortex-A15 MPCore processor introduces new technology from ARM that enables efficient handling of the complex software environments including full hardware virtualization, large physical address extensions (LPAE) addressing up to 1 TB of memory, as well as error correction capability for fault-tolerance and soft-fault recovery. For more information about the processor, refer to the *Cortex-A15 MPCore Technical Reference Manual* and the official ARM website (http://www.arm.com), from which this section is extracted.

Figure 14.5 shows the block diagram of Cortex-A15 MPCore processor, and the main components of the processor consist of instruction fetch, instruction decode, instruction dispatch, integer execute, load/store unit, L2 memory system, NEON and VFP unit, GIC, generic timer, debug, and trace.

Instruction fetch: The instruction fetch unit fetches instructions from the L1 instruction cache and delivers up to three instructions per cycle to the instruction decode unit. It supports static and dynamic branch prediction.

Instruction decode: The instruction decode unit decodes the instruction that includes ARM, Thumb, ThumbEE, Advanced SIMD, CP14, and CP15. The instruction decode unit also performs register renaming to facilitate out-of-order execution by removing *write-after-write (WAW) and write-after-read (WAR)* hazards. A loop buffer provides additional power savings while executing small instruction loop.

Instruction dispatch: The instruction dispatch unit controls when the decoded instructions can be dispatched to the execution pipelines and when the returned results can be retired.

Integer execute: The integer execute unit includes two symmetric ALU pipelines, integer multiply-accumulate pipeline, iterative integer divide hardware, branch and instruction condition code resolution logic, and result forwarding and comparator logic.

Load/store unit: The load/store unit executes load and store instructions and encompasses the L1 data side memory system. It also services memory coherency requests from the L2 memory system.

L2 memory system: The L2 memory system services L1 instruction and data cache misses from each processor. It handles requests on the AMBA 4 ACE master interface and AXI3 ACP slave interface.

NEON and VFP unit: NEON technology is the implementation of the Advanced SIMD extension to the ARMv7-A architecture. It provides support for integer and floating-point vector operations. This technology extends the processor functionality to provide support for the ARMv7 Advanced SIMDv2 instruction set. VFP is the vector floating-point coprocessor extension to the ARMv7-A architecture. It provides a low-cost high-performance floating-point computation. VFP extends the processor functionality to provide support for the ARMv7 VFPv4 instruction set.

GIC: The GIC provides support for handling multiple interrupt sources. It provides masking of interrupts, prioritization of interrupts, distribution of the interrupts to the target processors, tracking the status of interrupts, generation of interrupts by software, support for Security Extensions, and support for virtualization extensions.

Generic timer: The generic timer provides the ability to schedule events and trigger interrupts. It can trigger events after a period of time has elapsed. It provides a physical counter that contains the

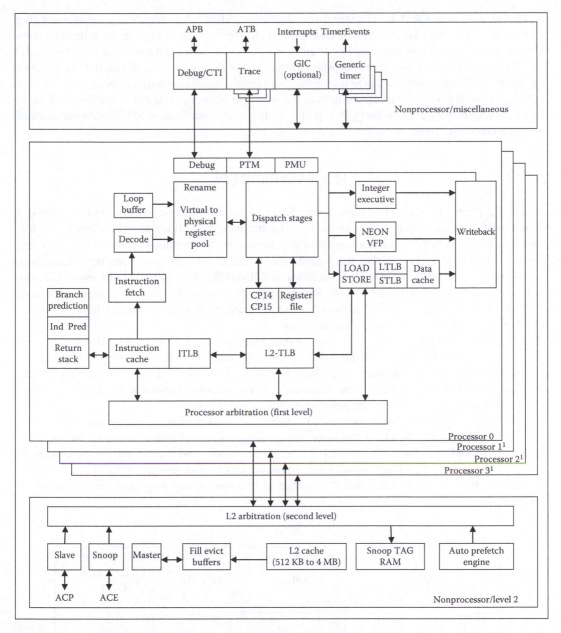

Figure 14.5 Top level functional diagram of the Cortex A15 MPCore processor. (From *Cortex-A15 MPCore Revision: r3p2, Technical Reference Manual*, Copyright 2011–2012, ARM Limited, ARM DDI 0438G (ID080412). Reproduced with permission from ARM Limited. Copyright © 2013 ARM Limited.)

count value of the system counter, a virtual counter that indicates virtual line, generation of timer events as interrupt outputs, generation of event streams, and support for virtualization extensions.

Debug and trace: The debug unit assists in debugging software running on the processor. This unit can be used in combination with a software debugger program to debug application software, operating systems, and hardware systems based on an ARM processor. It enables to stop program execution, examine and alter process and coprocessor state, examine and alter memory and input/output peripheral state, and restart the processor.

In addition to the ARM architecture that is licensable, ARM offers several microprocessor core designs, including the ARM7, ARM9, and ARM11. Companies often license these designs from ARM to manufacture and integrate into their own SoC. Other manufacturers like Intel and AMD have their own SoC designs that can as well be employed in designing general and specific purpose applications. Manufacturers are now delivering more highly integrated purpose-built SoCs ranging from industrial robotics and in-car infotainment systems to be transplanted in electronic devices. In the following section, the Raspberry Pi is proposed as a simple application to illustrate the usage of SoC as a general-purpose computer.

14.3 RASPBERRY Pi

The Raspberry Pi is an ARM GNU/Linux SoC in a box that is currently offered in market in a twentieth of an average laptop price. In the size of a credit card, the Raspberry Pi offers all the features that a PC has but offered in a way cheaper price due to the limitations it has in the speed and memory size. The Raspberry Pi shown in Figure 14.6 (by Paul Beech, http://www.raspberrypi.org/faqs) has a Broadcom BCM2835 SoC, which includes an ARM1176JZF-S 700 MHz processor. It comes with 256 MB of RAM and 512 MB. The Raspberry Pi does not include a hard drive or solid-state drive however; it uses an SD card for booting and a secondary storage. It comes with a video core 4 GPU that is capable of Blu-ray quality playback.

The Raspberry Pi uses Linux kernel-based operating systems. There is currently an optimized for the Raspberry Pi hardware called Raspbian that is a Debian-based free operating system that can be downloaded from www.raspbian.org. There is also a list of operating systems running, ported, or in the process of being ported to the Raspberry Pi. Third-party manufacturers started to offer a number of the Raspberry Pi specific peripheral devices and cases.

The Raspberry Pi users will need an SD card with an operating system preloaded before they can boot the Raspberry Pi. This palm-sized device supports digital audio via the HDMI port and analog stereo audio via their 3.5 mm jack. A monitor, a mouse, and a keyboard can be connected as input/output devices. Ethernet port is also provided to provide connectivity. The Raspberry Pi supports Python as the educational language in addition to any language that compiles for ARMv6.

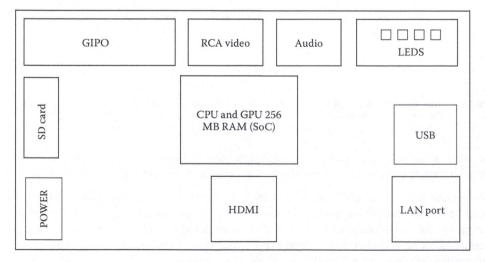

Figure 14.6 Raspberry Pi System components. (Adapted from Paul Beech, http://www.raspberrypi.org/faqs)

14.4 SUMMARY

This chapter discusses some of the extensions of embedded systems through the introduction of mobile processors and SoC. The majority of the electronic devices we work with today are powered by some kind of a mobile processor or SoC. This chapter covered the hardware and software components of representative devices such as Apple iPhone and Raspberry Pi. ARM Cortex A8, A9, and A15, architectures suitable for mobile processors, are also explained in brief.

PROBLEMS

14.1 List some of the unique features of ARM processor that makes it the most popular embedded architecture today.

14.2 List some of the real-world examples/applications of the ARM processors.

14.3 Briefly explain the different profiles of ARM v7-A architecture.

14.4 List the main components of Cortex-A8 processor.

14.5 Name the processors developed by Intel to compete with ARM's Cortex-A8, A9, and A15 processors.

14.6 Compare ARM's Cortex A9 with Intel's Atom N270 for the following features: number of cores, frequency, and power consumption.

14.7 In ARMv7, a processor in ARM state can enter Thumb state by executing which instructions?

14.8 Investigate about different processor modes available in Cortex-A8.

14.9 Look for information about three major clock domains that Cortex-A8 process has.

14.10 Investigate real-world examples of where SoCs can be found in different fields. Mention their area of application, the SoC components, and any other components that are connected to the SoC but not actually built in the SoC itself.

14.11 How is SoC different from SiP?

14.12 SoC examples that are mentioned in this chapter are all based on ARM architecture mobile processors. Is there any non-ARM-based SoC? If yes, how are they different from ARM architecture?

14.13 Look up the details of (a) a multiprocessor system on chip (MPSoC), (b) a chip-scale package (CSP), and (c) a wafer-level chip-scale package (WL-CSP).

BIBLIOGRAPHY

1974—Digital watch is first system-on-chip integrated circuit, http://www.computerhistory.org/semiconductor/timeline/1974-digital-watch-is-first-system-on-chip-integrated-circuit-52.html, retrieved on February 17, 2013.

ARM Architecture Reference Manual, Copyright 1996–1998, 2000, 2004, 2005, ARM Limited, ARM DDI 01001.

ARM Architecture Reference Manual, ARM v7-A and ARM v7-R edition Errata markup, Copyright 1996–1998, 2000, 2004–2011, ARM Limited, ARM DDI 0406B_errata_2011_Q3 (ID 120611).

Badawy, W. and Jullien, G., *System-on-Chip for Real-Time Applications*, Kluwer, October 31, 2002.

Cortex-A8 Processor, http://www.arm.com/, retrieved on February 17, 2013.

Cortex-A8 Technical Reference Manual, ARM Limited, 2010.

Cortex-A9 MPCore Technical Reference Manual, Copyright 2008–2012, ARM Limited, Revision: r4p1, 2008–2012.

Cortex-A9 Technical Reference Manual, Copyright 2008–2012, ARM Limited, Revision: r4p1, 2008–2012.

http://en.wikipedia.org/wiki/ARM_Cortex-A#Architecture, retrieved on February 17, 2013.

http://www.arm.com, retrieved on February 17, 2013.

Raspberry Pi, An ARM GNU/Linux box for $25. Take a byte! http://www.raspberrypi.org/, retrieved on February 17, 2013.

SoC vs. CPU, The battle for the future of computing, http://www.extremetech.com/computing/126235-soc-vs-cpu-the-battle-for-the-future-of-computing, retrieved on February 17, 2013.

Technical Reference Manual, ARM Limited, Revision: r3p2, 2011–2012.

White Paper, *The ARM Cortex-A9 Processors*, ARM Limited, Document Revision 2.0, September 2009.

CHAPTER 15

Computer Networks and Distributed Processing

In single instruction stream, multiple data (SIMD) and multiple instruction stream, multiple data (MIMD) architectures, several processors are connected with each other and memory blocks through the interconnection network, in a *tightly coupled* manner. These systems typically reside in one or more cabinets in a room rather than widely dispersed geographically. They cannot be easily expanded in small increments and typically employ only one type of processor and hence are not suitable for environments with an array of specialized applications. They usually employ a fixed interconnection topology, thereby restricting the users when the applications dictate a different more efficient topology. Some modern SIMD and MIMD systems have addressed these shortfalls by using heterogeneous processing nodes and being scalable to a fairly large number of nodes. They have also merged the SIMD and MIMD concepts, as evidenced by the evolution of the Thinking Machines' Connection Machine series. The earlier machines in the series were SIMDs while the CM-5 operates in both the modes.

Computer networks are the most common multicomputer architectures today. Figure 15.1 shows he structure of a computer network. It is essentially a message-passing MIMD system, except that the nodes are loosely coupled, by the communication network. Each node (host) is an independent computer system. The user can access the resources at the other nodes through the network. The important concept is that the users execute their applications at the nodes they are connected to, as far as possible. Users submit their jobs to other nodes when resources to execute their jobs are not available at their nodes. The Internet is the best example of the worldwide network.

With the introduction of microprocessors, local networks have become very popular because this system architecture provides a dedicated processor for local processing while providing the possibilities of sharing the resources with other nodes. Various networks have been built using large-scale machines, minicomputers, and microcomputers. A number of topologies for communication networks exist.

In a computer network, if a node fails, the resources at that node are no longer available. If, on the other hand, each node in a network is a general-purpose resource, the processing can continue even if a node fails, although at a reduced rate. A distributed processing system is one in which the *processing*, *data*, and *control* are distributed. That is, there are several general-purpose processors distributed geographically; no one processor will be a master controller at any time and there is no central database; rather it is a combination of all the sub-databases on all the machines. Although no such ideal distributed processing system exists to the author's knowledge, systems that adopt the distributed processing concepts to various degrees exist. The advantages of distributed processing systems are

1. Efficient processing, since each node performs the processing for which it is most suited
2. Dynamic reconfiguration of the system architecture to suit the processing loads
3. Graceful degradation of the system in case of failure of a node and redundancy if needed

The next step in this progression of building powerful computing environments is grid computing. A *grid* is a network of computers that act as a single "virtual" computer system. Utilizing specialized scheduling software, grids identify resources and allocate them to tasks for processing on the fly.

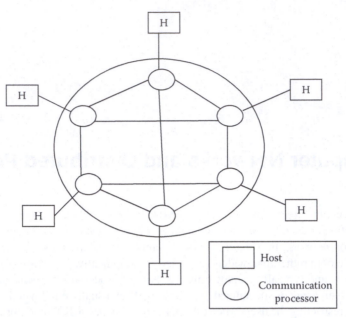

Figure 15.1   A computer network.

Resource requests are processed wherever it is most convenient, or wherever a particular function resides, and there is no centralized control. Grid computing exploits the underlying technologies of distributed computing, job scheduling, and workload management, which have been around more than a decade. The recent trends in hardware (commodity servers, blade servers, storage networks, high-speed networks, etc.) and software (Linux, web services, open-source technologies, etc.) have contributed to make grid computing practical. The Hewlett–Packard Adaptive Enterprise Initiative, IBM's on-demand computing effort, and Sun Microsystems' Network One framework are examples of commercial grid computing products.

Section 15.1 provides the details of networks. Section 15.2 addresses distributed processing. Section 15.3 covers grid architectures and Section 15.4 covers cloud computing.

15.1 COMPUTER NETWORKS

As mentioned earlier, a computer network is simply a system of interconnected computing devices that share information and resources among each other. The term "computing devices" includes traditional personal computers (PCs) and laptops as well as personal digital assistants (PDAs), web TVs, and smartphones. In this section, we use the word "computers" to include any of these devices.

The connection among network computers is not necessarily via a copper wire. Fiber optics, microwaves, infrared, and communication satellites can also be used. Computer networks involve several primary components:

1. Hosts: The computing devices connected to the networks are called hosts or end systems.
2. Links: Communication links are the paths that the transmitted information takes from the sending to the receiving hosts.
3. Routers: A router takes information arriving on one of its incoming communication links and forwards it through one of its outgoing communication links.
4. Bridges: A bridge reduces the amount of traffic on a network by dividing data into segments.
5. Protocols: Protocols such as Internet protocol (IP) and transmission control protocol (TCP) control the sending and receiving information within the network.

15.1.1 Network Architecture

Computer networks can be divided based on their underlying architecture (design) into two main categories: *client/server* architecture and *peer-to-peer* architecture. In the client/server architecture, each computer on the network is either a client or a server. Servers are powerful computers that control data and manage network traffic. Clients, on the other hand, rely on servers for resources to run applications. A typical client/server network arrangement is shown in Figure 15.2.

In peer-to-peer architecture, each host on the network has equivalent capabilities and responsibilities. There is no fixed division into clients and servers. A typical peer-to-peer network arrangement is shown in Figure 15.3. Peer-to-peer networks are generally simpler than client/server networks, but they usually do not offer the same performance under heavy loads.

15.1.2 Network Reference Models

A network reference model is a layered, abstract description for communications and computer network protocol design used to visualize and clearly describe the structure of a network. This section gives an overview of the two major network reference models, the OSI model and the TCP/IP

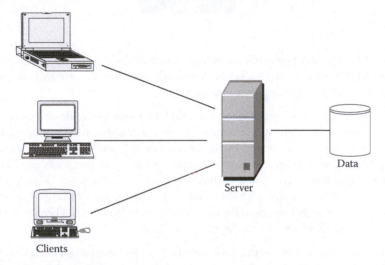

Figure 15.2 Client/server network architecture.

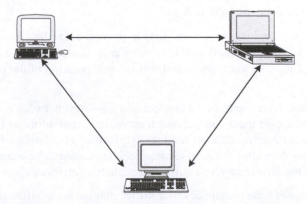

Figure 15.3 Peer-to-peer network architecture.

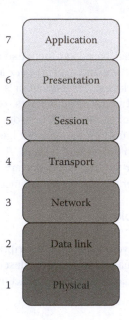

Figure 15.4 OSI model layers.

model. While the OSI model protocols are seldom used any more, the model itself is still valid. The TCP/IP model, on the other hand, is not commonly used but its protocols are widely spread (Tanenbaum, 1995). Both models are described next.

OSI model: The open system interconnection (OSI) reference model describes how information transfers from one end-user application into another end-user application through a network. This model is based on a proposal developed by the International Standards Organization (ISO) in the late 1970s, as a first attempt to international standardization of the protocols used in the different layers. This model has seven layers, and each has a specific functionality. Figure 15.4 shows the order of these layers, while Figure 15.5 illustrates transferring data through the different layers from the sending host through a router to the receiving host. Following is a brief description of each layer's functionality starting from the top layer.

Layer 7: The application layer is the main interface for the user(s) to interact with the application and therefore the network. It provides network services, such as simple mail transfer protocol (SMTP), file transfer protocol (FTP), and telnet, that allow the end users to access the information on the network. One widely used application protocol is hypertext transfer protocol (HTTP), which is the basis for the World Wide Web (WWW).

Layer 6: The presentation layer transforms data to provide a standard interface for the application layer. It converts local representation of data to its canonical form and vice versa. Data compression, MIME encoding, data encryption, and similar manipulation of the presentation are done at this layer.

Layer 5: The session layer controls the connections (sessions) between computers and defines the format of the data sent over these connections. It establishes, maintains, and ends sessions across the network and provides synchronization services by planning checkpoints in the data stream so if one session fails, only data after the most recent checkpoint need to be transmitted. This layer is also responsible for name identification so only the designated parties can join the session.

Layer 4: The transport layer subdivides user buffer into network-buffer-sized datagrams (segments) and enforces desired transmission control. It controls the reliability of a given link through

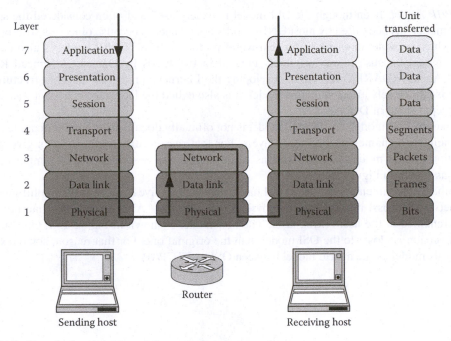

Figure 15.5 Transferring data through the seven layers.

flow control, segmentation/desegmentation, and error control. It also provides error-checking mechanism to guarantee error-free data delivery with no losses or duplications and provides acknowledgment of successful transmissions and retransmission requests if some packets do not arrive error-free. The two best-known transport protocols are TCP and user datagram protocol (UDP).

Layer 3: The network layer is responsible for routing (directing) datagrams from one host to another and for managing network problems such as packet switching and data congestion. It also translates logical network address and names to their physical address. If the router cannot send datagrams as large as the source computer sends, the network layer may have to break large datagrams into smaller packets and the network layer at the host receiving the packet will have to reassemble the fragmented datagram. The best-known example of network-layer protocol is the IP. The IP identifies each host with a 32-bit (4-byte) IP address written as four dot-separated decimal numbers between 0 and 255, for example, 123.145.67.240. The first 3 bytes of the IP identify the network and the remaining bytes identify the host on that network.

Layer 2: The data-link layer defines the format of data on the network and provides the functional and procedural ways to transfer data from one network element to an adjacent network element and detect and probably correct errors that may happen in the physical layer. The input data are broken down by the sender node into a few hundred or a few thousand byte data frames (packets) that are transmitted sequentially to the receiver. A network data frame includes checksum, source and destination address, and data. At the receiving and sending hosts, the data-link layer handles data between the physical and the network layers. At the sending host, it turns packets from the network layer into raw bits to send to the physical layer. At the sending end, it turns the received raw data from the physical layer into data frames for delivery to the network layer.

Layer 1: The physical layer defines all the physical transmission medium specifications that include voltages, cables, and pin layout to successfully transmit raw bit stream over that medium. All media are functionally equivalent; the key difference is in the convenience and cost of installation and maintenance.

TCP/IP model: Even though the OSI model is widely used and often considered the standard, TCP/IP model has been used by most UNIX workstation vendors because of its less strictly defined layers, which provide an easier fit for real-world protocols. The TCP/IP model is also called the Internet model because it was originally created in the 1970s by Defense Advanced Research Projects Agency (DARPA) for use in developing the Internet's protocols, and the structure of the Internet is still closely affected by this model. It is also called the DoD model since it was designed for the Department of Defense.

Unlike the OSI model, the TCP/IP model is not officially documented; not all documents agree on the number or names of the model layers. For that reason, different textbooks give different description of this model. Earlier versions of the model are described with a simple four-layer scheme as shown in Figure 15.6.

In modern documentations, the model has evolved into a five-layer model by splitting the first layer (network access) into two layers: physical and data link. Also the Internet/Internetworking layer is renamed to the networking layer. The new model is shown in Figure 15.7. Obviously, this evolved version is closer to the OSI model than the original one. For that reason, some text books refer to this model as the hybrid model between OSI and TCP/IP.

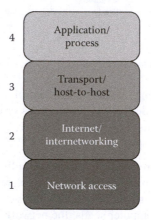

Figure 15.6 The original four-layer TCP/IP model.

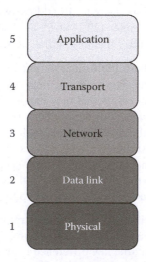

Figure 15.7 Five-layer TCP/IP model.

15.1.3 Network Standardization

Standards are considered one of the fundamentals of any technology and computer networking is no exception. Communication among different computers would be rather impossible if all network vendors and suppliers have their own ways and do not agree on important aspects.

In the area of computer networks, standards are set by several organizations including ISO and their American representative American National Standards Institute (ANSI), National Institute of Standards and Technology (NIST), and Institute of Electrical and Electronics Engineers (IEEE), the largest professional organization in the world. This section discusses IEEE 802, one of IEEE's popular standards that deal with computer networks.

IEEE 802 Standards: IEEE 802 refers to a family of IEEE standards dealing with local area networks (LANs) and metropolitan area networks (MANs), discussed later in the chapter. The number 802 was simply the next free number IEEE could assign, though "802" is sometimes associated with the date the first meeting was held—February 1980.

The services and protocols specified in IEEE 802 are designated to the lower two layers (physical and data link) of the OSI reference model mentioned earlier. IEEE 802 standards divide the data-link layer into two subgroups:

- Logical link control (LLC): It manages data-link communication and defines the use of logical interface points called service access points (SAP). It is responsible for sequencing frames and controlling frame traffic.
- Media access control (MAC): It provides shared access to the physical layer for the computer's network interface card (NIC), recognizes frame addresses and checks for frame errors, and ensures delivering error-free data between two computers on the network.

The IEEE 802 standard family is maintained by the IEEE 802 LAN/MAN Standards Committee (LMSC). A collection of working groups, listed in Table 15.1, produce standards in different areas in networking. Some of these standards have become more successful than others. Ethernet family, wireless LAN, token ring, bridging, and virtual bridged LANs are the most commonly used standards.

15.1.4 Computer Network Types

The most common types of computer networks classified by their scope or scale are

1. LAN—local area network
2. WLAN—wireless local area network
3. WAN—wide area network
4. MAN—metropolitan area network
5. CAN—campus area network
6. DAN—desk area network
7. PAN—personal area network

This section examines three of these common network types: LAN, WAN, and WLAN.

Local area network: LANs are privately owned networks that are limited to a relatively small spatial area such as a room, a single building, or an aircraft. They are commonly used to connect PCs and workstations in schools, houses, and company offices to share resources such as printers and exchange information.

Current LANs are most likely to be based on switched IEEE 802.3 Ethernet technology, running at 10–1000 Mbps. LAN topology is one of its distinguishing characteristics in addition to limited size and high transfer rate. Various possible topologies for LANs are shown in Figure 15.8.

Table 15.1   802 Working Groups

Number	Area
802.1	LAN protocols
802.2	LLC
802.3	Ethernet
802.4	Token bus (disbanded)
802.5	Token ring
802.6	MANs (disbanded)
802.7	Broadband LAN using coaxial cable (disbanded)
802.8	Fiber-optic TAG (disbanded)
802.9	Integrated services LAN (disbanded)
802.10	Interoperable LAN security (disbanded)
802.11	WLAN
802.12	Demand priority
802.13	Not used (unlucky 13!)
802.14	Cable modems (disbanded)
802.15	Wireless PAN
802.16	Broadband wireless access
802.17	Resilient packet ring
802.18	Radio regulatory TAG
802.19	Coexistence TAG
802.20	Mobile broadband wireless access
802.21	Media independent handoff
802.22	Wireless regional area network

Source: IEEE 802 LAN/MAN Standards Committee, http://www.
ieee802.org/dots.shtml

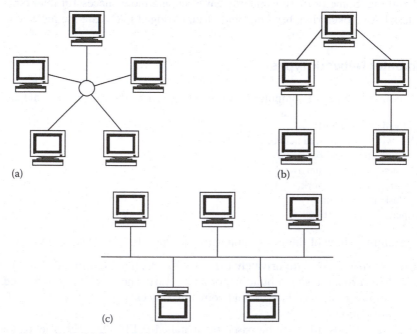

Figure 15.8   LAN topology. (a) Star. (b) Ring. (c) Bus.

Wide area network: In contrast to the LAN, WAN covers a broad area, often a country or continent and uses communications circuits to connect the intermediate nodes. WANs are used to connect LANs and other types of networks together, so that users and computers in one location can communicate with users and computers in other locations. A WAN can be viewed as a geographically dispersed collection of LANs connected with routers or switching circuits. Transmission rates are typically slower than LANs in the range of 2–625 Mbps or sometimes considerably more.

There are two basic types of WANs, central and distributed. Central WANs consist of a server or group of servers in a central location and client computers connected to the central server(s) that provides most or all of the functionality of the network. Distributed WANs, on the other hand, consist of client and server computers that are distributed throughout the WAN with the functionality of the network being distributed throughout the WAN.

Several WANs have been built, including public packet networks, large corporate networks, airline reservation networks, banking networks, and military networks. The largest and most well-known example of a WAN is the Internet, which spans the Earth.

Wireless local area network: WLAN is a wireless LAN connecting two or more computers without using wires, which gives users the mobility to move around within a broad coverage area and still be connected to the network. WLAN is usually based on IEEE 802.11 high-speed Wi-Fi technology. This network has recently become very common in offices and homes owing to the increasing popularity of laptops.

15.1.5 Internet and WWW

When we see the term "computer network," the first thing that comes to our mind is the Internet and the WWW. Neither the Internet nor the WWW is a computer network. The Internet is not a single network but a network of networks and the WWW is a distributed system that runs on top of the Internet.

The terms Internet and WWW are used interchangeably. However, they are not synonyms. Internet is a worldwide publicly accessible network of interconnected computer networks, linked by copper wires, fiber-optic cables, wireless connections, and so on, while the web is a collection of interconnected documents and other resources, linked by hyperlinks and URLs.

The Internet was the outcome of some visionary thinking by people in the early 1960s who foresaw great potential in connecting computers to share information on research and development in scientific and military fields. J.C.R. Licklider of MIT proposed the idea of a global network of computers in his 1960 paper, Man–Computer Symbiosis. In 1962, he was appointed head of the US Department of Defense's DARPA information processing office and formed a research group to develop his idea. The Internet, then known as ARPANET, was first brought online in 1969 when the first ARPANET link was established connecting four major computers at universities in the southwestern United States (UCLA, Stanford Research Institute, UCSB, and the University of Utah).

The development of the TCP/IP architecture in the 1970s was a big jump in the Internet advancement. TCP/IP was adopted by the Defense Department in 1980 replacing the earlier network control protocol (NCP) and universally adopted by 1983. In the period between 1970 and 1980, the Internet predecessors had only four main applications: e-mail, news groups, file transfer, and remote login.

Because of its limited application and difficulty to use, the Internet was mostly limited to academic and government use until late 1980 when a new application, the WWW, changed all that and attracted millions of new, nonacademic users to the net. Tim Berners-Lee and others at the European Laboratory for Particle Physics, more popularly known as CERN, proposed a new protocol for information distribution based on hypertext—a system of embedding links in text to link to other text. Along with the Mosaic browser, written by Marc Andreessen at the National Center for Supercomputer Applications in 1993, the WWW made it possible for a site to contain a number of pages of different information types including text, pictures, sound, and video, with embedded links that transport users to different pages with a simple click.

15.2 DISTRIBUTED PROCESSING

In distributed computing systems, a computational task is split into smaller chunks (or subtasks), which can be performed at the same time independently of each other by the various processors in the system. That is, the subtasks should be able to execute independently of each other. If a task cannot be split up in this way, it is not suitable for running on a distributed system.

Consider the computation of the column sum of an $N \times N$ matrix. This computation can be split into N tasks, where each task computes the sum of the elements of a column. The computational task is started on one of the processors in the distributed system, which creates and distributes additional tasks. Once the computation is complete, the results are gathered by one processor to generate the column sum. A *task* corresponds to the program that computes the sum, the data elements of the column assigned to it, and the memory space to run the program and return the result. There are several requirements for this scheme to work. The first part of the job is to split the computation into N tasks and package the appropriate column with each task. These tasks are then needed to be distributed to various nodes. The computing nodes should send the result back to one node that puts the results together. Also note that, if the computation performed by a task is minimal compared to the overhead to put the task together and coordinate its execution, we may not gain much by this mode of operation. In this example, we might have to package the computation of several columns into each task to balance the overhead.

The simplest of the distributed computing models (*client/server*) consists of two main entities: a server and many clients. The server generates the *work packages*, which are passed onto worker *clients*. The clients perform the task detailed in the work package and pass the *completed work package* to the server.

Table 15.2 shows the protocol between the clients and the server to execute the column sum computation task. We assume that N is a multiple of M. Each subtask corresponds to computing the column sum of M columns. The server has three states: initialize, create, and provide subtasks to clients and receive the results and assemble them. The client has two states: request a task and wait and complete the task and go back to request mode.

The server and the clients are the nodes in the interconnection network forming the distributed system. The network transports the messages between the clients and the server. The ubiquitous

Table 15.2   Client/Server Protocol

Server	Client
• Initialize	• Initialize
1. $K = 0$	1. Request the work package from the server
2. Wait for a message	2. Wait
• Receive request for work package	
1. Generate a work package, with columns K to $K + M$	
2. Send the work package to client	• Work package received
3. $K = K + M$	1. Compute the M column sums
4. Wait for a message	2. Return completed work package to server
• Receive complete work package	3. Go to Initialize
1. Assemble into appropriate columns of the results matrix	
2. If all the N columns' results are in, output the result and stop, else, continue waiting for a message	

Internet has become such a transport medium. If a web server is permanently online, it can be set up to distribute work packages. In this environment, clients make an HTTP request to a URL, and the server responds with a work package.

In general, the client and the server are not homogeneous. As such, we need to make sure that the messages are translated appropriately at both ends, to bring about the communication. In a general distributed system, all the nodes should be capable of being a server or a client and this mode of operation can be changed dynamically.

15.2.1 Processes and Threads

The two main characteristics of a *process* (or a *task*) are that it is a unit of resource ownership and a unit of dispatching. A process is allocated a virtual address space to hold the process image and it has control of some resources such as files and I/O devices. As a unit of dispatching, a process is an execution path through one or more programs. It has an execution state and a dispatching priority and the execution may be interleaved with other processes. If the resource ownership and dispatching characteristics are treated as being independent, the resource ownership is usually referred to as a process or task, and the unit of dispatching is usually referred to as a thread or a lightweight process. Thus, a thread also has an execution state (blocked, running, ready). The thread context is saved when it is not running. It has an execution stack and some static storage for local variables. It has access to the memory address space and resources of its process. All threads of a process share the resources of the process. When one thread alters a (nonprivate) memory item, all other threads of the process see that a file opened with one thread is available to others.

If implemented correctly, threads have several advantages over multiple process implementations. It takes less time to create a new thread than a process, since the newly created thread uses the process address space. It takes less time to terminate a thread compared with terminating a process and less time to switch between two threads within the same process. The communication overhead is considerably reduced since the threads share all the resources of the process.

Depending on the operating system capabilities, four thread/process combinations are possible: single process with a single thread (MS-DOS), single process and multiple threads, multiple processes with a single thread in each (UNIX), and multiple processes with multiple threads in each process (Solaris). Figure 15.9 shows these combinations.

A program with several activities that are not dependent upon each other can be implemented with multiple threads, one for each activity. Since one activity does not have to wait for the other to complete, the application's responsiveness is improved. When an application is designed to run on a

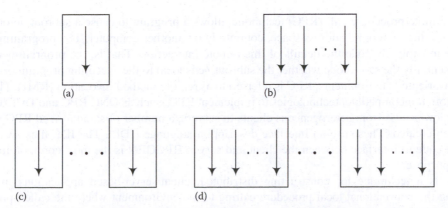

Figure 15.9 Threads and processes. (a) One process/one thread. (b) One process/multiple threads. (c) Multiple process/one thread per process. (d) Multiple process/multiple threads per process.

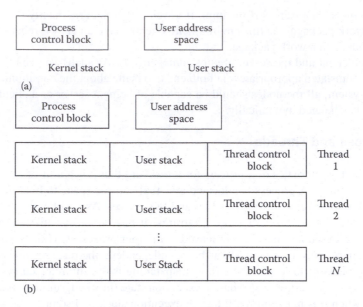

Figure 15.10 Single and multithread implementation. (a) Single thread. (b) Multiple threads.

multiprocessor system, the concurrency requirements of the application are translated into threads (and processes), and the program is not dependent on the number of available processors. The performance of the application improves transparently as processors are added to the system. Thus, applications with a high degree of parallelism can run much faster when implemented with threads on a multiprocessor. Multithreaded programs are more adaptive to variations in user demands than single-threaded programs. Multithreaded implementations also offer improved program structure, because of the multiple independent units of execution compared with a single, monolithic thread. They also use fewer system resources.

Figure 15.10 shows the process models. Typical resources of a process implementation are the process control block, user address space, and user and kernel stacks. In a multithreaded process, all threads utilize the same process control block and the user address space. Each thread will have its own control block and user/kernel stacks.

15.2.2 Remote Procedure Call

The remote procedure call (RPC) technique allows a program to cause a subroutine or procedure to execute in another address space, commonly on another computer. The programmer does not have to explicitly code the details of this remote interaction. That is, the programmer would write essentially the same code whether the subroutine is local to the executing program or remote. In object-oriented environments, RPC is referred to as remote method invocation (RMI). There are several (often incompatible) technologies to implement RPC, such as ONC RPC and DCE/RPC.

To allow access from heterogeneous clients to servers, a number of standardized RPC systems have been created. These use an interface description language (IDL). The IDL files are used to generate code to interface between the client and server. RPCGEN is the most common tool used for this purpose.

RPC is a technique for constructing distributed, client/server-based applications. It allows extending the conventional local procedure calling to the environment where the called procedure need not exist in the same address space as the calling procedure. The two processes may be on the same system, or they may be on different systems with a network connecting them. Through RPC, a programmer of distributed applications can avoid the details of the interface with the network. Since the RPC is transport independent, it isolates the application from the physical and logical elements of the data communication mechanisms and also enables the application using a variety of transports.

As in a local function call, when an RPC is made, the calling arguments are passed to the remote procedure and the caller waits for a response to be returned from the remote procedure. The client makes an RPC, sending a request to the server and waits. The thread is blocked from processing until either a reply is received or it times out. When the request arrives, the server calls a dispatch routine that performs the requested service and sends the reply to the client. After the RPC is completed, the client continues. Three steps are needed to develop an RPC application:

1. Specifying the client/server communication protocol
2. Client program development
3. Server program development

The communication protocol is implemented by generated stubs. These stubs, RPC, and other libraries are then linked. A protocol compiler such as RPCGEN is used to define and generate the protocol. The protocol identifies the name of the service procedures and data types of parameters and return arguments. The protocol compiler reads the definition and automatically generates client and server stubs. The RPCGEN uses its own language (RPC language—RPCL) and typically generates four files: the client stub, the server stub, external data representation (XDR) filters, and the header file needed for XDR filters. The XDR is a data abstraction needed for machine-independent communication.

The client and application code must communicate via procedures and data types specified in the protocol. The server side has to register the procedures that may be called by the client and receive and return any data required for processing. The client applications call the remote procedure, pass any required data, and receive the returned data.

The concept of RPC was first described in 1976 as RFC-707 and Xerox implemented it as "Courier" in 1981. Sun's UNIX-based RPC (ONC RPC), first used as the basis for Sun's Network File System (NFS), is still widely used on several platforms. Apollo Computer's Network Computing System (NCS) RPC was used as the foundation of DCE/RPC in the open software foundation's (OSF) Distributed Computing Environment (DCE). Microsoft adopted DCE/RPC as the basis of the Microsoft RPC (MSRPC) mechanism and implemented DCOM on top of it. In 1990s, Xerox PARC's ILU, and the Object Management Group's Common Object Request Broker Architecture (CORBA), offered an RPC paradigm based on distributed objects with an inheritance mechanism. Microsoft.NET Remoting offers RPC facilities for distributed systems implemented on the Windows platform. Java's Java RMI API provides similar functionality to standard UNIX RPC methods.

There are two common implementations of the Java RMI API. The original implementation depends on Java virtual machine (JVM) class representation mechanisms and supports making calls from one JVM to another. The protocol underlying this Java-only implementation is known as Java Remote Method Protocol (JRMP). For non-JVM context, a CORBA version was later developed. Usage of the term RMI may denote solely the programming interface or may signify both the API and JRMP, whereas the term RMI-IIOP denotes the RMI interface, delegating most of the functionality to the supporting CORBA implementation. The original RMI API was generalized somewhat to support different implementations, such as an HTTP transport. Additionally, work was done to CORBA, adding a pass by value capability, to support the RMI interface. Still, the RMI-IIOP and JRMP implementations are not fully identical in their interfaces.

15.2.3 Process Synchronization and Mutual Exclusion

When multiple processors are executing in a distributed system, it is essential that their activities are coordinated to make sure they start and stop appropriately and access the data items in an

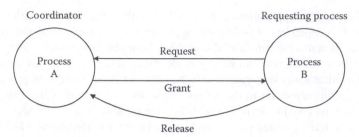

Figure 15.11   Centralized mutual exclusion.

exclusive manner. For instance, a process may run only to a certain point, at which it will stop and wait for another process to complete certain computations. A device or a location in memory may be shared by several processes and hence requires exclusive access. Processes have to coordinate among themselves to ensure that the access is fair and exclusive. A set of process synchronization techniques are available for such coordinated execution among processes. As we have seen earlier in this book, in centralized systems it was common to enforce exclusive access to shared code. Test-and-set locks, semaphores, and condition variables were used to bring about the mutual exclusion.

We will now discuss some popular algorithms to achieve mutual exclusion in distributed systems. A unique identifier is used to represent each critical resource. This identifier (typically, name and a number) is recognizable by all the processes and is passed on as a parameter with all requests.

Central server algorithm: The central server algorithm simulates a single processor system. One of the processes in the distributed system is first elected as the coordinator (see Figure 15.11). When a process needs to enter a critical section, it sends a request (with the identification of the critical section) to the coordinator. If no other process is currently in the critical section, the coordinator sends back a grant and marks that process as using the critical section. If another process has already claimed the critical section, the server does not reply, and hence the requesting process gets blocked and enters the queue of processes, requesting that critical section. When a process exits the critical section, it sends a release to the coordinator. The coordinator then sends a grant to the next process in the queue of processes, waiting for the critical section. This algorithm is fair, in that it processes the requests in the order they are received, and easy to implement and verify. The major drawback is that the coordinator becomes the single point of failure. The centralized server can also become a bottleneck in the system.

Distributed mutual exclusion: The distributed mutual exclusion algorithm was proposed by Ricart and Agrawala (1981). In this algorithm, when a process wants to enter a critical section, it composes a message consisting of its identifier, identifier of the critical section, and current time (i.e., time stamp) and sends the request to all other processes in the group. The requesting process waits until all the processes in the group give permission to enter the critical section. A process receiving the request takes one of the three possible actions. If it is not a contender for the critical section, it sends its permission to the requesting process. If the receiver is already in the critical section, it does not reply but adds the requester to the local queue of requests. If the receiver is a contender for the critical section and has sent its request, it compares the time stamp in the received message with the one that it has sent out and the earliest time stamp wins. If the receiver is the loser, it sends the permission to the requester. If the receiver is the winner, it does not reply but adds the requester to the queue. When the process exits the critical section, it sends permission to all the processes in the queue and deletes the processes from the queue.

In Figure 15.12a, two processes, A and C, request access to the same critical section. Process A sends its request with a time stamp of 18 and process C sends its request with a time stamp of 22. Since process B is not interested in the critical section, it immediately sends back permission to both A and C. Process A is interested in the critical section. It sees that the time stamp of process C was later than its own. Thus, process A wins. It queues a request from C (Figure 15.12b).

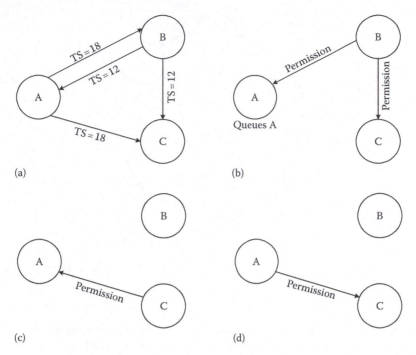

Figure 15.12 Distributed mutual exclusion. (a) Request sent. (b) A wins. (c) A enters critical section. (d) C enters critical section. *Note:* TS, time stamp.

Process C is also interested in the critical section. When it compares its time stamp with that in the message it received from process A, it sees that it did not win and hence it sends permission to process A and continues to wait for all the processes to give it permission to enter the critical section (Figure 15.12c). As soon as process C sends permission to process A, process A will have received permissions from the entire group and can enter the critical section. When process A exits the critical section, it examines its queue of pending permissions, finds process C in that queue, and sends permission to enter the critical section (Figure 15.12d). Now process C has received permission from all the processes and enters the critical section.

This algorithm requires a total ordering of all events in the system and that the messages are reliable. One drawback of this algorithm is that the single point of failure of the previous algorithm is now replaced with *n* points of failure. This can be remedied by having the sender always send a reply to a message (a YES or a NO). If the request or the reply is lost, the sender will time out and retry. The other drawback of this algorithm is its heavy message traffic.

Token ring algorithm: This algorithm assumes that there is a group of processes with no inherent ordering of processes but that some ordering can be imposed on the group. For example, we can identify each process by its machine address and process ID to obtain an ordering. Using this imposed ordering, a logical ring can be constructed in software. Each process is assigned a position in the ring and each process must know its neighboring process in the ring. Figure 15.13 shows a ring with *n* nodes. The ring is initialized by giving a token to process 0. The token circulates around the ring (process *k* passes it to process (*k* + 1) mod *n*). When a process acquires the token, it checks to see if it is attempting to enter the critical section. If so, it enters and does its work. On exit, it passes the token to its neighbor. If a process is not interested in entering a critical section, it simply passes the token along. Only one process has the token at a time and it must have the token to work on a critical section, so mutual exclusion is guaranteed. Order is also well defined, so starvation cannot occur. The biggest drawback of this algorithm is that if a token is lost, it will have to be generated. Determining that a token is lost can be difficult.

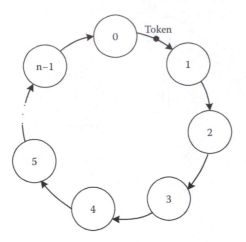

Figure 15.13   Token ring.

15.2.4 Election Algorithms

From the earlier discussion, we see that one process acts as a coordinator. It may not matter which process does this, but there should be a group agreement on only one. We assume that all processes are exactly the same with no distinguishing characteristics. Each process can obtain a unique identifier (typically, a machine address and process ID) and each process knows of every other process but does not know which is up and which is down.

Bully algorithm: The bully algorithm selects the process with the largest identifier as the coordinator as follows:

1. When a process k detects that the coordinator is not responding to requests, it initiates an election, which has three steps:
   k sends an *election* message to all the processes with higher numbers. If none of the processes respond, then k will take over as the coordinator. If one of the processes responds, then the job of process k is done.
2. When a process receives an election message from a lower-numbered process at any time, it sends a replay (OK) back and holds an election (unless it is already holding one).
3. A process announces its election by sending all processes a message telling them that it is the new coordinator.
4. When a process that was down recovers, it holds an election.

Ring algorithm: The ring algorithm uses the same ring arrangement as in the token ring mutual exclusion algorithm, but does not employ a token. Processes are physically or logically ordered so that each knows its successor. If any process detects failure, it constructs an election message with its process ID and sends it to its successor. If the successor is down, it skips over it and sends the message to the next party. This process is repeated until a running process is located. At each step, the process adds its own process ID to the list in the message. Eventually, the message comes back to the process that started it. The process sees its ID in the list and changes the message type to *coordinator*. The list is circulated again, with each process selecting the highest numbered ID in the list to act as the coordinator. When the coordinator message has circulated fully, it is deleted. Multiple messages may circulate if multiple processes detected failure. Although this creates additional overhead, it produces the same result.

15.3 GRID COMPUTING

There are several definitions of "grid computing architecture." Some define clustered servers that share a common data pool as grids. Others define large distributed networked environments that make use of thousands of heterogeneous information systems (IS) and storage subsystems as grids. There are definitions that fall somewhere between these two ends of the spectrum. A grid in general is a network architecture for connecting computing and storage resources. It is based on standards that allow heterogeneous systems and applications to share computing and storage resources transparently. A computational grid is a hardware and software infrastructure that provides dependable, consistent, pervasive, and inexpensive access to high-end computational capabilities (Foster and Kesselman, 2004). The resource sharing is not primarily file exchange but rather direct access to computers, software, data, and other resources, as required by a range of collaborative problem-solving and resource-brokering strategies emerging in industry, science, and engineering. This sharing is necessarily, highly controlled, with resource providers and consumers defining clearly and carefully just what is shared, who is allowed to share, and the conditions under which the sharing occurs (Foster et al., 2001). Thus, a grid is a system that coordinates resources that are not subject to centralized control; uses standard, open, general-purpose protocols and interfaces; and delivers nontrivial qualities of service.

Figure 15.14 shows a possible example of a grid computing model that illustrates the grid concept with a control node in the middle to manage the interconnectivity between the grid components. There are multiple design variations when designing a grid; however, the main idea is to have machines that are more loosely coupled and geographically separated and acting together to achieve a single task.

A grid consists of networks of computers, storage, and other devices that can pool and share resources and the grid architecture provisions those resources to users and/or applications. All the elements in the grid resource pool are considered to be virtual (i.e., they can be activated when needed).

Provisioning involves locating computing and storage resources and making them available to requestors (users/applications). Some ways to provision resources are reconfiguring to meet the new workload, repartitioning an existing server environment to provide additional storage, activating installed but not active components (CPU, memory, or storage) on a local server to provide local access to additional resources, accessing additional computing and storage resources that can be found and exploited over a LAN or across a WAN, and finding and exploiting additional computing and storage resources using a grid.

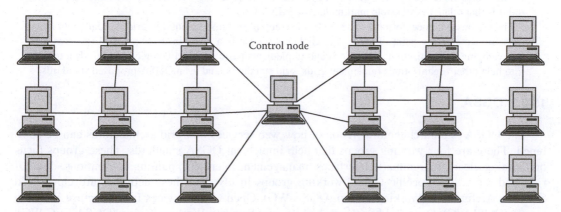

Figure 15.14 A grid computing model.

All computing and storage resources are considered virtual and become real when they are activated. Two approaches to exploiting virtualized resources on a local server environment are manual or programmatic reconfiguration and activation of computing/storage/memory modules that have been shipped with systems (i.e., activated and paid for when used). Virtualized resources can also be found on a network. For instance, clustered systems provide access to additional CPU power when needed (utilizing a tightly coupled network), and grid software can find and exploit loosely coupled virtual resources on LAN or WAN. In addition, virtualized services can be acquired externally using a utility computing model.

Grids are generally deployed at the department level (departmental grid), across an enterprise (intergrids), and across multiple enterprises or between multiple organizations (extragrids). Sun Microsystems' scalable virtual computing concept identifies three grid levels:

1. Cluster grid (departmental computing): Simplest grid deployment, maximum utilization of departmental resources, resources allocated based on priorities
2. Enterprise grid (enterprise computing): Resources shared within the enterprise, policies ensure computing on demand, gives multiple groups seamless access to enterprise resources
3. Global grid (Internet computing): Resources shared over the Internet, global view of distributed datasets, growth path for enterprise grids

In 2001, the global grid forum (the GGF—the primary grid standards organization) put forward an architectural view of how grids and web services could be joined. This architecture is called the Open Grid Services Architecture (OGSA). Since the GGF announced its OGSA view, there has been strong progress in the articulation of web services and grid standards. The following organizations are the leading standards organizations involved in articulating and implementing web services and service-oriented architectures (SOAs) over grid architecture:

- The GGF is the primary standards setting organization for grid computing. The GGF works closely with OASIS (described later) as well as with the Distributed Management Task Force (DMTF) to help build interoperable web services and management infrastructure for grid environments.
- The Organization for the Advancement of Structured Information Standards (OASIS) is very active in setting standards for web services and works closely with the GGF to integrate web services standards with grid standards.
- The DMTF works with the GGF to help implement DMTF management standards such as the DMTF's Common Information Model (CIM) and Web-Based Enterprise Management (WBEM) standards on grid architecture.
- The World Wide Web Consortium (W3C) is also active in setting web services standards (and standards that relate to extensible markup language [XML]).
- The Globus Alliance (formerly The Globus Project) is also instrumental in grid standards but from an implementation point of view. The Globus Alliance is a multi-institutional grid research and development organization. It develops and implements basic grid technologies and builds a toolkit to help other organizations implement grids, grid standards, and even OGSA-proposed standards.

15.3.1 OGSA

The OGSA is an architectural vision of how web services and grid architectures can be combined. There are four working groups that help implement OGSA standards. These groups focus on defining clear programmatic interfaces, management interfaces, naming conventions, directories, and more. The specific OGSA working groups involved in these activities are Open Grid Services Architecture Working Group (OGSA-WG), Open Grid Services Infrastructure Working Group (OGSI-WG), Open Grid Services Architecture Security Working Group (OGSA-SEC-WG), and Database Access and Integration Services Working Group (DAIS-WG). The OGSI is an implementation/test bed of OGSA.

There are several standards involved in building an SOA and underlying grid architecture that can support business process management. These standards form the basic architectural building blocks that allow applications and databases to execute service requests. Moreover, these standards also make it possible to deploy business process management software that enables IS executives to manage business process flow. The most important grid and grid-related standards include

1. Program-to-program communications (SOAP, WSDL, and UDDI)
2. Data sharing (XML)
3. Messaging (SOAP, WS-Addressing, MTOM (for attachments))
4. Reliable Messaging (WS-Reliable Messaging)
5. Managing workload (WS-Management)
6. Transaction-handling (WS-Coordination, WS-Atomic Transaction, WS-Business Activity)
7. Managing resources (WS-RF or web services resource framework)
8. Establishing Security (WS-Security, WS-Secure Conversation, WS-Trust, WS-Federation, Web Services Security Kerberos Binding)
9. Handling metadata (WSDL, UDDI, WS-Policy)
10. Building and integrating web services architecture over a grid (see OGSA)
11. Orchestration (standards used to abstract business processes from application logic and data sources and set up the rules that allow business processes to interact)
12. Overlaying business process flow (business process engineering language for web services—BPEL4WS)
13. Triggering process flow events (WS-Notification)

Grids are being used in a variety of scientific and commercial applications such as aerospace and automotive (for collaborative design and modeling), architecture (engineering and construction), electronics (design and testing), finance (stock/portfolio analysis, risk management), life sciences (data mining, pharmaceuticals), manufacturing (inter-/intra-team collaborative design, process management), and media/entertainment (digital animation). Some of the most famous scientific and research grids include the following:

1. The SETI@home Project—Thousands of Internet PCs used for the search for extraterrestrial life.
2. The Mersenne Project—The Great Internet Mersenne Prime Search (GIMPS) is a worldwide mathematics research project.
3. The NASA Information Power Grid (IPG)—The IPG joins supercomputers and storage devices owned by participating organizations into a single, seamless computing environment. This project will allow the government, researchers, and industry to amass computing power and facilitate information exchange among NASA scientists).
4. The Oxford e-Science Grid—Oxford University's "e-Science" project addresses scientific distributed global collaborations that require access to very large data collections, very large-scale computing resources, and high-performance visualization back to the individual user scientists.
5. The Intel-United Devices Cancer Research Project—This project is a grid-based research project designed to uncover new cancer drugs through the use of organizations and individuals willing to donate excess PC processing power. This excess power is applied to the grid infrastructure and used to operate specialized software. The research focuses on proteins that have been determined to be a possible target for cancer therapy.

The largest grid effort currently underway is the "TeraGrid" scientific research project. The TeraGrid was launched by the United States' National Science Foundation in August 2001 as a multiyear effort to build the world's largest grid infrastructure for scientific computing. In 2004, the TeraGrid will include 20 teraflops of computing power, almost one petabyte of data, and high-resolution visualization environments for modeling and simulation. The supporting grid network is expected to operate at 40 Gbits/s.

Although the preponderance of compute grids has been in the scientific, research, and educational communities, there is a strong growth of compute grids in commercial environments.

At the end of 2003, the Office of Science of the U.S. Department of Energy published a report called "Facilities for the Future of Science: A 20 Year Outlook" located at http://www.er.doe.gov/Sub/Facilities_for_future/20 Year-Outlook-screen.pdf. This report details how the US government will use ultrascale computing (a very high-speed grid approach with very powerful servers/supercomputers) to encourage discovery in the public sector.

India has undertaken the building of a national "I-Grid" (information grid). India's Centre for Development of Advanced Computing—makers of India's PARAM Padma supercomputers—sees its role as helping India to carve out a niche in the global arena of advanced information technology, to further expand the frontiers of high-performance computing, and to utilize resulting intellectual property for the benefit of society "by converting it into an exciting business opportunity and establishing a self-sustaining and wealth creating operation."

The United Kingdom has created e-Science centers that utilize data grids for scientific research projects. The National e-Science Centre coordinates projects with regional centers located in Belfast, Cambridge, Cardiff, London, Manchester, Newcastle, Oxford, and Southampton, as well as with sites in Daresbury and Rutherford Appleton. These centers provide facilities to scientists and researchers who wish to collaborate on very large, data-intensive projects. This project is one of dozens of governmental grid projects within the United Kingdom.

Numerous other examples of government grids can be found at http://www.grids-center.org/news/news_deployment.asp.

15.4 CLOUD COMPUTING

The delivery of computing as a utility has evolved into multiple paradigms. Grid computing described earlier is one of the paradigms that promised to deliver computing as a service and played a role in the advent of a more recent and a broader paradigm known as cloud computing. The cloud computing paradigm emerged as a result of convergence of various service models like distributed computing, grid computing and utility computing, and a group of hardware and software technologies like hardware virtualization, SOA, and Web 2.0. The term "cloud computing" has become more popular and widely used to describe any service that is provided via the Internet. Nevertheless, terms like cloud computing, grid computing, and utility computing are alternately employed when referring to the same paradigm. This section provides a brief description of cloud computing.

Amazon is one of the most popular cloud service providers (CSPs) today. Amazon's Elastic Compute Cloud (EC2) platform provides computing, storage facility (through its servers), applications, and services. The users are charged only for the resources that they actually use and EC2 allows consumers to quickly scale-up or scale-down capacity according to their computing requirement. EC2 consumers pay for *instance* of use, that is, the compute capacity by the hour with no long-term commitment. Figure 15.15 shows a high-level illustration of the Amazon EC2. The figure shows that different consumers can access and share the same cloud resources using HTTP. The major cloud component in addition to the cloud database servers and the application server is the server that does load balancing among different servers. Different techniques might be used to secure the cloud including firewalls.

According to the National Institute of Standards and Technology (NIST), cloud computing is a model for enabling ubiquitous, convenient, on-demand network access to a shared pool of configurable computing resources that can be rapidly provisioned and released with minimal management effort or service provider interaction. The NIST's model is composed of five characteristics, three service models, and four deployment models as shown in Figure 15.16.

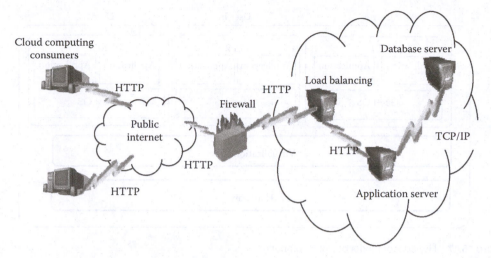

Figure 15.15 Amazon EC2 cloud computing.

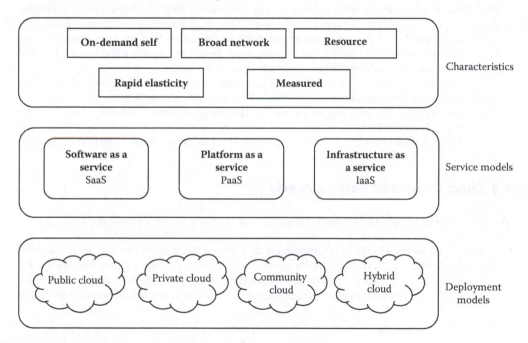

Figure 15.16 NIST cloud computing broad definition.

15.4.1 Cloud Computing Characteristics

The cloud computing definition involves five characteristics that every cloud computing model enjoys:

- *On-demand self-service*: Every action from the end-user side to the cloud and back to the end user happens automatically and without human interaction.
- *Broad network access*: Cloud computing services are accessed by clients through a network connection using standard mechanisms.
- *Resource pooling*: Cloud computing services including storage (*data centers*), processing, memory, and network bandwidth are pooled to serve multiple consumers at the same time in a *multitenancy*

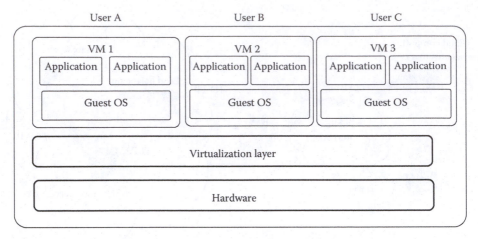

Figure 15.17   Hypervisor model of cloud computing.

model that is achieved by using a *hypervisor model* shown in Figure 15.17 to support multiple *virtual machines* for multiple consumers at the same time. In cloud computing there is a comparatively *location independence* sense that appears because consumers have no control over the cloud resources exact location although providers sometimes specify the numbers and locations of their data centers and the country and/or the states where they are located.

*   *Rapid elasticity*: In cloud computing different resources are allocated to consumers elastically by service providers. These resources are scalable to meet consumers' requirements at any time.
*   *Measured service*: CSPs apply *pay per use* metrics that are appropriate to the type of service or the resource that is allocated. These metrics aim to measure resource usage, similar to utility companies who charge their consumers for power consumption of electricity monthly on kilowatt per hour basis.

15.4.2 Cloud Computing Service Models

As illustrated in Figure 15.18, cloud computing services are divided into three major areas:

*   *Software as a service (SaaS)*: In this service model, CSPs provide applications that are running on a cloud infrastructure to the consumers to be accessed from anywhere using various devices such as PCs and smartphones. The client interface is provided by the CSP and is usually built using a web technology like ASP and PHP that can be accessed using a web browser or a program interface that can be

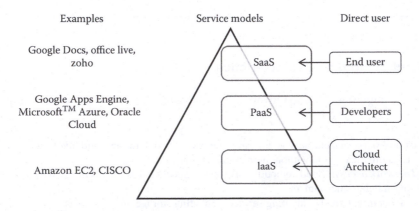

Figure 15.18   SaaS, PaaS, and IaaS examples and users.

accessed by running the program. Consumers in this service model have no control over the platforms where these applications are built or the infrastructure where these platforms and applications reside. However, consumers with different roles can have different configurations when they access and use that application. A good example of a SaaS model is the web-based Google docs-based application.

- *Platform as a service (PaaS)*: In this model, CSPs provide the platform where consumers can deploy their software systems (programming languages, service, libraries, etc.) onto the cloud using tools provided by the CSP. Consumers here have no control over the infrastructure of the cloud. They can build their own software using the infrastructure and the platforms provided. Examples of PaaS are Google Apps Engine, Microsoft Azure, and Oracle Cloud.
- *Infrastructure as a service (IaaS)*: The infrastructure provided to consumers here includes processing, storage, networks, and other resources. Consumers then can mount platform, software, or even an operating system of their choice. In IaaS, consumers do not have any control over the cloud infrastructure; it is only over their mounted platform and operating system configuration and some network settings. Examples of IaaS are Amazon EC2, CISCO Cloud, and IBM Cloud.

There are other service models composed of the earlier three standard models. Computing as a service (CaaS) implies that cloud computing service itself is composed of one or more of the standard service models. Storage as a service (STaaS) is mainly an IaaS. Similarly, desktop as a service (DaaS) and API as a service (APIaas) are both (PaaS).

15.4.3 Cloud Computing Deployment Models

There are four different deployment models for cloud computing each with its own targeted consumers, owners, operators, and service managers:

- A *private cloud* is owned and operated by a single organization. It can be managed internally within the organization or by a third party and hosted internally by the organization that is known as on-premises private cloud or externally using some CSP but strictly used by a single organization and its consumers that is known as off-premises private cloud.
- A *public cloud* that is opened for public use but owned, managed, and operated by an organization. Here, the cloud serves other various businesses, specific domains such as academia and health care, or government uses.
- A *community cloud* is owned, managed, and operated by multiple organizations and is used by a specific community. This community can be formed of consumers of different organizations.
- A *hybrid cloud* is composed of two or more of the previously mentioned cloud computing deployment models. Each cloud is operated and managed separately, but all the component clouds work as an entity.

15.4.4 Amazon EC2

Cloud computing service providers strive to provide consumers with solutions that are elastic, scalable, and secure. They offer services in the form of SaaS, PaaS, and IaaS or any combination of them. Consumers have freedom to get one service from one provider and contract with other providers to get the other services. Amazon EC2 provides consumers with an IaaS that is resizable, elastic, reliable, easily configured with minimal effort, secure, and controllable.

Amazon EC2 IaaS locations are distributed over eight regions: three in the United States (Northern Virginia, Oregon, and Northern California), one in Europe (Ireland), two in Asia (Singapore, Tokyo), one in South America (Sao Paulo), and Amazon Web Services (AWS) GovCloud that is designed specifically for US government agencies and customers that are located in the northwestern region of the United States. Each region consists of one or more availability zones (AZs). By being able to run instances on multiple AZs in one region or different regions and preventing application failure when an AZ fails, EC2 Service Level Agreement is able to offer 99.95% availability.

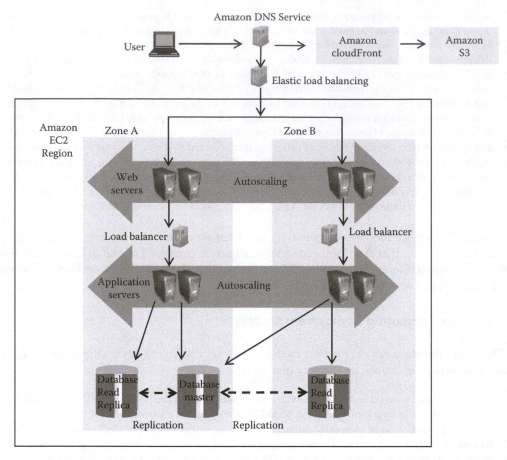

Figure 15.19 AWS for web application hosting. (From http://media.amazonwebservices.com/architecturecenter/ AWS_ac_ra_web_01.pdf.)

Figure 15.19 illustrates how application hosting operates on Amazon EC2. Users' requested pages are served through a domain name system (DNS) called Amazon Route 53. It connects user requests to infrastructure running in AWSs—such as an Amazon EC2 instance, an Amazon Elastic Load Balancer, or an Amazon Simple Storage Service (Amazon S3). Amazon Route 53 can also be used to route users to infrastructure outside of AWS.

S3 is a standalone AWS that can be used to store and retrieve any amount of data at any time, from anywhere. It enjoys the high-scalability characteristic that any of the AWS has. This storage as a service facility that AWS provided can be used as a content storage, backups, archiving, and/ or static website hosting.

For static or streaming content, requests are handled by Amazon CloudFront that is a web service used for content delivery by integrating with the other AWSs like EC2 and S3. CloudFront brings the static content to the nearest edge location so that your popular static content (e.g., your site's logo, navigational images, cascading style sheets, JavaScript code) will be available at a nearby edge location for a better performance. Dynamic or interactive content requests to an AWS region go through an elastic load balancing that handles HTTP requests and automatically distributes incoming application traffic across multiple Amazon EC2 instances across AZs.

Every AZ consists of an auto scaling group of web servers followed by a load balancer again to balance web servers' requests within the AZ. This is followed by an auto scaling group of application servers where all requests are executed to fetch data from database. For higher availability, Amazon EC2 provides multiple regions and AZs, two layers of load balancing, auto scaling groups of web and application servers, and the master–slave replication to avoid failure in the database layer.

Amazon EC2 PaaS provides consumers with a variety of tools to get their applications implemented in a secure environment. Consumers are empowered with bundling tools to upload their own operating systems or use an operating system from a list of most commonly used operating systems, such as Red Hat Enterprise Linux, Windows Server, Oracle Enterprise Linux, SUSE Linux Enterprise, Amazon Linux AMI, Ubuntu, Fedora, Gentoo Linux, or Debian. EC2's PaaS is also equipped with a pool of software such as database systems, application server tools, business intelligence tools, and content management systems. All these operating systems and software tools are available for EC2 consumers in a single place that is known as AWS Marketplace.

Moving to the cloud is a widespread concern among organizations from different domains. Issues like feasibility study, performance, privacy, and security are considered by cloud adopters when making such a decision. There is a big pool of CSPs to choose from (Amazon EC2, Microsoft Azure, Google Apps Engine, IBM Cloud, Oracle Cloud, etc.) and the list is growing. It is the cloud computing service consumer's responsibility to choose the suitable service, based on their requirements.

15.5 SUMMARY

This chapter provided a brief introduction to four architecture types: computer networks, distributed systems, grids, and clouds. The Internet has made these architectures cost-effective. In fact, we now say that the network is the computer. These architectures have allowed us to build complex systems that are open and scalable. These features have also contributed to the increased requirements for security and reliability.

PROBLEMS

15.1 Describe the characteristics of any commercial distributed system you have access to. What services are provided? How do the servers and clients interact?

15.2 We would like to utilize the processing capability of idle workstations in a company. Outline a design. There are commercial applications to enable this function. Investigate their operating characteristics.

15.3 What are the functions of an Internet router and an Ethernet bridge?

15.4 Satellite communication channels are used to form computer networks. What are the main drawbacks of this interconnection scheme?

15.5 How are domain names related to IP addresses on the Internet? How do we make sure that no two computers in the Internet have the same address?

15.6 How are the functions of hubs, repeaters, switches, routers, and gateways different?

15.7 Investigate the protocol layer definition of TCP/IP applications such as telnet and FTP, as compared with the OSI model.

15.8 Why is a distributed system less secure than a central system?

15.9 How does a firewall protect the machine connected to it?

15.10 Look up the details of

CSMA/CD

Bluetooth

Wi-Fi

MAC

ISDN

15.11 Investigate three commercial cloud computing services: Google Apps Engine, IBM Cloud, and Oracle Cloud in terms of differences in their service model(s) and deployment model(s)?

15.12 Discuss cloud computing scalability, multi-tenancy, and transparency characteristics.

15.13 How does load balancing help mitigate the attacks on a cloud computing service?

BIBLIOGRAPHY

Coulouris, G., Dollimore, J., and Kindberg, T., *Distributed Systems*: *Concepts and Design*, 5th edn., Reading, MA: Addison Wesley, 2011.

Forrest, S., Hofmeyr, S.A., Somayaji, A., and Longstaff, T.A., A sense of self for UNIX processes, *Proceedings of IEEE Symposium on Research in Security and Privacy*, Oakland, CA, 1996.

Foster, I. and Kesselman, C., *The GRID2*: *Blueprint for a New Computing Infrastructure*, San Francisco, CA: Morgan Kaufmann, 2004.

Foster, I., Kesselman, C., and Tuecke, S., The anatomy of the grid: Enabling scalable virtual organizations, *International Journal of Supercomputer Applications*, 15(3), 200–222, 2001.

Grid Computing, *The DoD SoftwareTech News*, April 2004.

IEEE 802, Wikipedia, the free encyclopedia, http://en.wikipedia.org/wiki/IEEE_802.

Intel Corporation, Toward TeraFLOP performance: An update on the Intel/DARPA Touchstone Program, Supercomputer Systems Division Report, November, 1990.

Joseph, J. and Fellenstein, C., *Grid Computing*, 1st edn., Upper Saddle River, NJ: IBM Press, 2004.

Kurose, J.F. and Ross, K.W., *Computer Networking*: *A Top-Down Approach Featuring the Internet*, 6th edn., Boston, MA: Addison Wesley, 2012.

Magoules, F., Pan, J., Tan, K., and Kumar, A., *Introduction to Grid Computing* (Chapman & Hall/CRC Numerical Analysis and Scientific Computing Series), 1st edn., Boca Raton, FL: CRC Press, 2009.

Mell, P. and Grance, T., NIST, *The NIST Definition of Cloud Computing*, http://csrc.nist.gov/publications/nistpubs/800-145/SP800-145.pdf.

Ricart, G. and Agrawala, A.K., An optimal algorithm for mutual exclusion in computer networks, *Communications of ACM*, 24(1), 9–17, 1981.

Shiva, S.G., *Advanced Computer Architectures*, Boca Raton, FL: Taylor & Francis, 2006.

Silberschatz, A., Galvin, P.B., and Gagne, G., *Operating System Concepts*, Reading, MA: Wiley, 2012.

Stallings, W., *Operating Systems*: *Internals and Design Principles,* Upper Saddle River, NJ: Prentice Hall, 2011.

Tanenbaum, A.S., *Distributed Operating Systems*, Englewood Cliffs, NJ: Prentice Hall, 1995.

Tel, G., *Introduction to Distributed Algorithms*, Cambridge, U.K.: Cambridge University Press, 2000.

Varia, J. and Mathew, S., *Amazon Web Services Overview*, https://d36cz9buwru1tt.cloudfront.net/AWS_Overview.pdf.

Wilkinson, B., *Grid Computing: Techniques and Applications (Chapman & Hall/CRC Computational Science)*, 1st edn., Boca Raton, FL: Chapman & Hall/CRC, 2009.

Zomaya, A. (Ed.), *Parallel and Distributed Computing Handbook*, New York: McGraw-Hill, 1996.

CHAPTER 16

Performance Evaluation

In previous chapters, we have highlighted the parameters for the performance evaluation of various components and architectural features of computer systems. Performance evaluation and estimate are necessary either for acquiring a new system or for evaluating the enhancements made (or to be made) to an existing system. Ideally, it is best to develop the target application on the system to be evaluated to determine its performance. The next best thing to do is to *simulate* the target application on an existing system. In practice, these modes of evaluation are not always possible and may prove to be not cost-effective. As such, analytical methods of evaluating performance and determining costs are necessary. As the systems get complex, the analytical methods become unwieldy. Benchmarking is used in practice to evaluate complex systems. This chapter introduces the most common analytical techniques and benchmarking. It also provides a brief introduction to program optimization techniques.

The major aim of any system design is to maximize the *performance-to-cost* ratio of the target system. That is, maximizing the performance while minimizing the cost. The three major aspects of a computer system from the performance maximization point of view are the following:

1. The processor bandwidth (fastest execution of instructions)
2. The memory bandwidth (fastest instruction/data retrieval and storage)
3. The input/output (I/O) bandwidth (maximum throughput)

An application is said to be *processor bound, memory bound, or I/O bound*, depending on which of these aspects limit its performance.

The total time to execute a program is the product of the number of instructions in the program, the number of cycles (major or minor, in the context of ASC) per instruction, and the time per cycle. The processor design contributes to the last two items while the program design contributes to the first. In addition, we have seen the features, such as pipelining, superscalar execution, branch prediction, and so on, contributing to the enhancement of program performance.

The memory bandwidth is dependent on the cache and virtual structures, shared versus message passing structures, speeds of memory components, and the architecture of the program to allow best utilization of cache and virtual memory schemes.

The I/O bandwidth is a function of device speeds, bus and interconnects speeds, the control structures such as DMA, peripheral processors, and so on, and the associated protocols.

If we have the luxury of building a computer system to optimize performance of a particular application, a holistic approach can be used, as mentioned in the previous chapter. Suppose that the application allows the development of processing algorithms with a degree of parallelism A. The degree of parallelism is simply the number of computations that can be executed concurrently. Further, if the language used to code the algorithm allows the representation of algorithms with a degree of parallelism L, the compilers produce an object code that retains a degree of parallelism C and the hardware structure of the machine has a degree of parallelism H, then, for the processing to be most efficient, the following relation must be satisfied:

$$H \geq C \geq L \geq A. \tag{16.1}$$

Here, the objective is to minimize the computation time of the application at hand. Hence, the processing structure that offers the least computation time is the most efficient one.

For the architecture to be most efficient, the development of the application algorithms, programming languages, the compiler, the operating system, and the hardware structures must proceed together. This mode of development is only possible for very few special purpose applications. In the development of general-purpose architectures, however, the application characteristics cannot be easily taken into account. But, the development of other components should proceed concurrently, as far as possible.

Development of algorithms with a high degree of parallelism is application dependent and basically a human endeavor. A great deal of research has been devoted to developing languages that contain parallel processing constructs, thereby enabling the coding of parallel algorithms. Compilers for these parallel processing languages retain the parallelism expressed in the source code during the compilation process, thus producing parallel object code. In addition, compilers that extract parallelism from a serial program (thus producing a parallel object code) have been developed. Progress in hardware technology has yielded a large number of hardware structures that can be used in executing parallel code.

According to Amdahl's law, the system speedup is maximized when the performance of the most frequently used component of the system is maximized. It is stated as

$$S = \frac{1}{(1-f) + \dfrac{f'}{k}}, \tag{16.2}$$

where
   S is the overall system speedup
   f represents the fraction of the work performed by the enhanced component
   k is the speedup of the enhanced component

16.1 PERFORMANCE MEASURES

Several measures of performance have been used in the evaluation of computer systems. The most common ones are million instructions per second (MIPS), million operations per second (MOPS), million floating-point operations per second (MFLOPS or megaflops), billion floating-point operations per second (GFLOPS or gigaflops), and million logical inferences per second (MLIPS). Machines capable of trillion floating-point operations per second (teraflops) are now available. The measure used depends on the type of operations one is interested in, for the particular application for which the machine is being evaluated. We will use the MIPS measure to illustrate various aspects of performance measurement in this section. These concepts are equally valid for other performance measures also.

Example 16.1

Consider the SHL instruction of ASC from Chapter 6. Its instruction cycle requires four minor cycles. Assuming that each minor cycle corresponds to a nanosecond, ASC can complete

$$\frac{1}{(4 \times 10^{-9})} \text{ SHL instructions per second (IPS)}$$

$$= 0.25 \times 10^9 \text{ IPS}$$

$$= 0.25 \times 10^3 \text{ or 250 MIPS.}$$

On the other hand, the LDA* (Load Indirect) instruction of ASC requires 12 cycles. If only this instruction is executed, the ASC MIPS rating will be 250/3 or 83.3 MIPS. A salesman trying to sell ASC will have the tendency to quote the MIPS rating to be 250 while a critical evaluator will use the 83.3 MIPS rating. Neither of these ratings is useful in practice to evaluate the machine, since an application will use several other instructions of ASC.

Example 16.2

Let us suppose that an application uses just the following four instructions:

Instruction	Speed (Minor Cycles)
ADD	8
SHIFT	4
LDA*	12
STA*	12

Then the average (or the arithmetic mean) speed is (8 + 4 + 12 + 12)/4 = 9 cycles, resulting in a MIPS rating of (1/9) ×10³ or 111.11 MIPS. We can extend this analysis to include all the instructions of ASC rather than just the four instructions used earlier. But, this measure is also not realistic since the frequency of instruction usage (i.e., the instruction mix) depends on the application characteristics. Thus, the rating has to be based on the mix of operations representative of their occurrence in the application.

Example 16.3

Consider the following instruction mix:

Instruction	Speed (Cycles)	Occurrence (%)
ADD	8	30
SHIFT	4	20
LOAD	12	30
STORE	12	20

The weighted average (weighted *arithmetic mean*) instruction speed is (8 × 0.3 + 4 × 0.2 + 12 × 0.3 + 12 × 0.2) = 9.2 cycles. Assuming one nanosecond per cycle, the machine performs (1/9.2) × 10³ or 108.69 MIPS. This rating is more representative of the machine performance than the maximum rating (250 MIPS) computed by using the speed of execution (4 cycles) of the fastest instruction (SHIFT).

Thus, the performance rating could be either the *peak* rate (i.e., the MIPS rating the central processing unit (CPU) cannot exceed) or the more realistic average or *sustained* rate. In addition, a comparative rating that compares the average rate of the machine to that of other well-known machines (e.g., IBM MIPS and VAX MIPS) is also used. A better measure than the arithmetic mean to compare relative performances is the *geometric mean*.

The geometric mean of a set of positive data is defined as the nth root of the product of all the members of the set, where n is the number of members. That is, the geometric mean of a set $\{a_1, a_2, \ldots, a_n\}$ is

$$\left(\prod_{i=1}^{n} a_i \right)^{\frac{1}{n}} = \sqrt[n]{a_1 \cdot a_2 \ldots a_n}. \tag{16.3}$$

The arithmetic mean is relevant any time several quantities add together to produce a total. The arithmetic mean answers the question, "if all the quantities had the same value, what would that value have to be in order to achieve the same total?" The geometric mean on the other hand, is relevant any time several quantities multiply together to produce a product. The geometric mean answers the question, "if all the quantities had the same value, what would that value have to be in order to achieve the same product?"

The geometric mean of a dataset is always less than or equal to the set's arithmetic mean (the two means are equal if and only if all members of the dataset are equal).

The geometric mean is useful to determine "average factors." For example, if a stock rose 10% in the first year, 20% in the second year, and fell 15% in the third year, then we compute the geometric mean of the factors 1.10, 1.20, and 0.85 as $(1.10 \times 1.20 \times 0.85)^{1/3} = 1.0391$, and we conclude that the stock rose 3.91% per year, on average. Using arithmetic mean in this calculation is incorrect since the data are multiplicative.

Example 16.4

Assume that ASC instruction speeds have been improved resulting in two versions ASC2 and ASC3. The following table shows the cycles needed for the four instructions on all the three versions (the original machine is ASC1):

Instruction	ASC1	ASC1 Normalized	ASC2	ASC2 Normalized	ASC3	ASC3 Normalized
ADD	8	1	6	6/8	4	4/8
SHIFT	4	1	4	4/4	2	2/4
LDA*	12	1	8	8/12	8	8/12
STA*	12	1	8	8/12	12	12/12
Geometric mean		**1**		**0.806**		**0.66**

ASC2 and ASC3 speeds are normalized with respect to those of ASC1. The last row shows the geometric means for all the three machines. The ratio of geometric means is an indication of the relative performances. The relative performances are as follows:

ASC2/ASC1 = 0.806
ASC3/ASC1 = 0.66
ASC3/ASC2 = 0.66/0.806 = 0.819

These ratios remain consistent, no matter which of the machines is used as a reference.

Another measure commonly used is the *harmonic mean*. The harmonic mean of a group of terms is the number of terms divided by the sum of the terms' reciprocals. The harmonic mean H of the positive real numbers $a_1, \ldots, a_n$ is defined as

$$H = \frac{n}{\dfrac{1}{a_1} + \dfrac{1}{a_2} + \cdots + \dfrac{1}{a_n}}. \tag{16.4}$$

This measure is useful when the measures are "rates" such as instructions per second. It is useful for the environments with known workloads.

In certain situations, the harmonic mean provides the correct notion of "average." For instance, if one travels at 40 km/h for half the *distance* of a trip and at 60 km/h for the other half, then the average speed for the trip is given by the harmonic mean of 40 and 60, which is 48; that is, the total amount of time for the trip is the same as if you traveled the entire trip at 48 km/h. Note, however, that if one traveled for half the *time* at one speed and the other half at another, the arithmetic mean (50 km/h) would provide the correct notion of "average."

The harmonic mean is never larger than the arithmetic and geometric means. It is equivalent to a weighted arithmetic mean with each value's weight being the reciprocal of the value.

Since the harmonic mean of a list of numbers tends strongly toward the least elements of the list, it tends (compared to the arithmetic mean) to mitigate the impact of large outliers and aggravate the impact of small ones.

Example 16.5

Consider an application with four modules. The performance of each module is monitored and the following table lists the average number of instructions executed per second for each module. The modules are enhanced to improve the performance and the new instructions per second ratings are as shown. Note that the performance of module 4 has actually worsened. The harmonic means of both the versions is shown in the table.

Module	Original	Improved
1	400	300
2	220	210
3	370	300
4	110	120
Harmonic mean	212.3	202.4

According to the harmonic mean analysis, the overall performance has improved by about 5%.

16.2 COST FACTOR

The unit cost of the machine is usually expressed as dollars per MIPS (or MFLOPS). It is important to note that the cost comparison should be performed on architectures of approximately the same performance level. For example, if the application at hand requires a performance level of N MIPS, it is usually an overkill to select an architecture that delivers M MIPS, where M is far greater than N, even though the unit cost of the latter system is lower. On the other hand, an architecture that offers N/X MIPS at a lower unit cost would be better for the application at hand if it is possible to attain N MIPS by using Y such systems (where $Y \geq X$) with a lower total cost compared with the architecture delivering N MIPS. Of course, if multiple units of an N/X–MIPS machine cannot be configured to deliver N MIPS, then it is not a candidate for comparison. This is obviously an oversimplification, since configuring multiple machines to form a system typically requires other considerations such as partitioning of application into subtasks, reprogramming the sequential application into parallel form, overhead introduced by the communication between multiple processors, and so on. These considerations are discussed later in this book.

The cost of a computer system is a composite of its software and hardware costs. The cost of hardware has fallen rapidly as the hardware technology progressed, while the software costs are steadily rising as the software complexity grew, despite the availability of sophisticated software engineering tools. If this trend continues, the cost of software would dictate the cost of the system while the hardware would come free once the software is purchased.

The cost of either hardware or software is dependent on two factors: an upfront development cost and per unit manufacturing cost. The development cost is amortized over the life of the system and distributed to each unit produced. Thus, as the number of systems produced increases, the development component of the cost decreases.

The production cost characteristics of the hardware and software differ. Production of each unit of hardware requires assembly and testing, and hence, the cost of these operations will never be zero even if the cost of hardware components tends to be negligible. In the case of software, if we assume that there are no changes to the software once it is developed, resulting in zero maintenance costs, the production cost becomes almost zero as the number of units produced is large. This is because producing a copy of the software system and testing it to make sure it is an accurate copy of the original (by bit-by-bit comparison) is not an expensive operation. However, the assumption of zero maintenance costs is not realistic, since the software system always undergoes changes and enhancements are requested by the users on a continual basis.

There are other effects of progress in hardware and software technologies on the cost of the system. Each technology provides a certain level of performance, and as the performance requirements increase, we exhaust the capability of a technology and hence will have to move to a new technology. Here, we are assuming that the progress in technology is user driven. In practice, the technology is also driving the user's requirements in the sense that the progress in technology provides systems with higher performance at lower cost levels thereby making older systems obsolete faster than before. That means that the life-spans of systems are getting shorter, bringing an additional burden of recuperating development costs over a shorter period of time.

The cost considerations thus lead to the following guideline for a system architect: make the architecture as general purpose as possible in order to make it suitable for a large number of applications, thus increasing the number of units sold and reducing the cost per unit.

16.3 BENCHMARKS

All the analytical techniques used in estimating the performance are approximations. As the complexity of the system increases, most of these techniques become unwieldy. A practical method for estimating the performance in such cases is by using *benchmarks*.

Benchmarks are standardized batteries of programs run on a machine to estimate its performance. The results of running a benchmark on a given machine can then be compared with those on a known or standard machine, using criteria such as CPU and memory utilization, throughput and device utilization, and so on.

Benchmarks are useful in evaluating hardware as well as software and single processor as well as multiprocessor systems. They are also useful in comparing the performance of a system before and after certain changes are made. As a high-level language host, the computer should execute efficiently those features of a programming language that are most frequently used in actual programs. This ability is often measured by benchmarks. Benchmarks are considered to be representative of classes of applications envisioned for the architecture. We provide a brief description of some common benchmarks as follows:

Real world/application benchmarks: They use system- or user-level software code drawn from real algorithms or full applications, commonly used in system-level benchmarking. They usually have large code and data storage requirements.

Derived benchmarks: They are also called "algorithm-based benchmarks." They extract the key algorithms and generate realistic datasets from real-world applications. They are used for debugging, internal engineering, and competitive analysis.

Single processor benchmarks: They are low-level benchmarks used to measure performance parameters that characterize the basic architecture of the computer. These hardware/compiler parameters predict the timing and performance of the more complex kernels and applications. They are used to measure the theoretical parameters that describe the overhead or potential bottleneck, or the properties of some part of hardware.

Kernel benchmarks: They are code fragments extracted from real programs in which the code fragment is responsible for most of the execution time. They have the advantage of small code size and long execution time. Examples are *Linpack* and *Lawrence Livermore loops*.

The Linpack (LINear algebra PACKage) measures the MFLOPS rating of the machine in solving a system of linear equations, in double precision arithmetic, in a FORTRAN environment through Basic Linear Algebra Subroutines (BLAS). This benchmark was developed at the Argonne National Laboratory in 1984 to evaluate the performance of supercomputers. C and Java versions of the benchmark suite are now available.

The Lawrence Livermore loops measure the MFLOPS rating in executing 24 common FORTRAN loops operating on datasets with 1001 or fewer elements.

Local benchmarks: These are programs that are site specific. That is, they include in-house applications that are not widely available. Since the user is most interested in the performance of the machine for his/her applications, local benchmarks are the best means of evaluation.

Partial benchmarks: These are partial traces of programs. It is in general difficult to reproduce these benchmarks when the portion of benchmarks that was traced is unknown.

UNIX utility and application benchmarks: These are programs that are widely employed by the UNIX user community. The *SPEC* (system/standard performance evaluation cooperative effort) Benchmark suite belongs to this category and consists of 10 scenarios taken from a variety of science and engineering applications. This suite developed by a consortium of about 60 computer vendors is for the evaluation of workstation performance. The performance rating is provided in *SPEC marks*. There are three main groups working on distinct aspects of performance evaluation. The open systems group (OSG) concentrates on desktop, work station, and file server environments. The graphics performance characterization group (GPCG) concentrates on graphic-intensive and multimedia systems. The high-performance computing group (HPG) concentrates on multiprocessor systems and supercomputers. These groups select applications that represent typical workloads for corresponding environments. The I/O and non-CPU intensive parts of the applications are removed from each application to obtain its "kernel." The composite of these kernels forms the benchmark suite.

Synthetic benchmarks: These are small programs constructed specially for benchmarking. They do not perform any useful computation, but statistically approximate the average characteristics of real programs. Examples are *Whetstone* and *Dhrystone* benchmarks.

Whetstone benchmark: In its original form this benchmark set published in 1976 was developed in ALGOL 60. Whetstone reflects mostly numerical computing, using a substantial amount of floating-point arithmetic. It is now chiefly used in a FORTRAN version. Its main characteristics are as follows:

- A high degree of floating-point data and operations, since the benchmark is meant to represent numeric programs.
- A high percentage of execution time is spent in mathematical library functions.
- Use of very few local variables, since the issue of local versus global variables was hardly being discussed when these benchmarks were developed.
- Instead of local variables, a large number of global variables are used. Therefore, a compiler in which the most heavily used global variables are used as register variables (as in C) will boost the Whetstone performance.

Since the benchmark consists of nine small loops, Whetstone has an extremely high code locality. Thus, a near 100% hit rate can be expected even for fairly small instruction caches.

The distribution of the different statement types in this benchmark was determined in 1970. As such, the benchmark cannot be expected to reflect the features of more modern programming languages (e.g., record and pointer data types). Also, recent publications on the interaction between programming languages and architecture have examined more subtle aspects of program behavior (e.g., the locality of data references—local vs. global) that were not explicitly considered in earlier studies.

Dhrystone benchmark: In early efforts dealing with the performance of different computer architectures, performance was usually measured using some collection of programs that happened to be available to the user. However, following the pioneering work of Knuth in the early 1970s, an increasing number of publications have been providing statistical data about the actual usage of programming language features. The Dhrystone benchmark program set is based on these statistics, particularly in systems programming. This benchmark suite contains a measurable quantity of floating-point operations. A considerable percentage of execution time is spent in string functions. In C compilers, this number goes up to 40%. Unlike Whetstone, Dhrystone contains hardly any loops within the main measurement loop. Therefore, for processors with small instruction caches, almost all the memory accesses are cache misses. But as the cache becomes larger, all the accesses become cache hits. Only a small amount of global data is manipulated and the data size cannot be scaled.

Parallel benchmarks: These are for evaluating parallel computer architectures. The 1985 workshop at the National Institute of Standards (NIST) recommended the following suite for parallel computers: Linpack, Whetstone, Dhrystone, Livermore loops, Fermi National accelerator Laboratory codes used in equipment procurement, NASA/Ames benchmark of 12 Fortran subroutines, John Rice's numerical problem set, and Raul Mendez's benchmarks for Japanese machines.

Stanford small programs: Concurrent with the development of the first RISC systems, John Hennessy and Peter Nye at Stanford's Computer systems laboratory collected a set of small C programs. These programs became popular because they were the basis for the first comparisons of RISC and CISC processors. They have now been collected into one C program containing eight integer programs (Permutations, Towers of Hanoi, Eight queens, Integer matrix multiplication, Puzzle, Quicksort, Bubble sort, and Tree sort) and two floating-point programs (matrix multiplication and fast Fourier transform).

PERFECT: The PERFormance Evaluation for Cost-effective Transformations benchmark suite consists of 13 Fortran subroutines spanning four application areas (signal processing, engineering design, physical and chemical modeling, and fluid dynamics). This suite consists of complete applications (with the input/output portions removed) and hence constitutes significant measures of performance.

SLALOM: The Scalable, Language-independent, Ames Laboratory, One-minute Measurement is designed to measure the parallel computer performance as a function of problem size. The benchmark always runs in 1 min. The speed of the system under test is determined by the amount of computation performed in 1 min.

Transaction processing benchmarks: Transaction processing servers perform a large number of concurrent short-duration activities. These transactions typically involve disk I/O and communications. IBM introduced a benchmark (TP1) in the 1980s for evaluating transaction processing mainframes. Several other benchmarks to augment TP1 were proposed later. The Transaction Processing Performance Council (TPC), a consortium of 30+ enterprise system vendors, has released several versions of their TPC benchmark suites.

GeekBench: It is a cross-platform benchmark tool that measures processor and memory performance. GeekBench uses four types of workloads to measure the performance: floating-point performance, integer performance, memory performance, and streaming performance.

CineBench: A real-world cross-platform benchmark application that evaluates computer system performance using two main components, its graphics card and processor. CineBench conducts two main tests on a system. The first test involves CPU performance using the complete processing power to render a photorealistic 3D scene. The second test involves the graphic hardware performance using a complex 3D animation playback scene provided by CineBench to test the time capabilities of the graphic card being able to render the scene, measured in frames per second (fps).

Mobile benchmarks: Mobile benchmarks come in a variety of performance tests. Mobile benchmarks involve both the hardware and software of the mobile device, as well as the performance of external software representing classes of applications envisioned for the mobile architecture. Mobile benchmarks test CPU usage, disk storage, memory usage, and graphics. An example of a mobile hardware benchmark is PassMark Android and iOS, whereas an example of a mobile application benchmark is Gomez.

PassMark Android and iOS benchmark: PassMark provides two separate performance evaluations for the Android and iPhone mobile devices. Android and iOS main benchmark tests are conducted in the following suites:

- CPU suite measures mainly the mathematical operations, encryption, and compression.
- Disk suite measures the reading and writing of internal and external file storage.
- Memory suite measures the reading and writing to memory locations.
- Graphic suite measure simple, complex vector and images, as well as complex scenes.

Gomez benchmark: Dealing with the performance of different mobile architectures, performance relative to the speed and accuracy of web applications for mobile devices is critical also. The Gomez benchmark set is based on statistics, particularly in web and mobile site performance. This benchmark suite measures the quality, performance, and availability of business processes relative to a mobile application.

Security benchmarks: Hardware and software has become increasingly interconnected via the Internet. Security benchmarks are programs designed to measure application information security status. These programs typically evaluate the probability of inappropriate actions processed within a subject application. Security benchmarks differ from traditional benchmarks, due to their nature of containing sensitive information and potential damage to reputation, including ethical and legal issues. Examples are *CIS* and *WINE* benchmarks.

CIS benchmark: CIS benchmark was established to promote the best practices and increase the security and privacy, along with the integrity of transactions using interconnected systems. Its main characteristics are to resolve the following:

- Secure configuration of the target system to recommend technical control for hardening software application, operating systems, and network devices.
- Security metrics and definitions used to collect and analyze security process performance and outcomes.

The security benchmark is relatively a new developing area of concern. Ongoing research is currently being conducted to make way for improvements in the area of security benchmarking. Recent publications on the development of public data storage to assist organizations with benchmarking security-related occurrences have examined a concern in security (e.g., data sharing) that impacts the ethical and legal ramifications of security benchmarking.

WINE benchmark: Computer security efforts are generally large and complex, considering the nature of the impact a security breach can have on a system. The WINE benchmark is in the early stages of developing a way to share security-related breaches. WINE is proposed to reduce the propagation of attack incidents, when defenses are available for implementation. The WINE benchmark attempts to cover the complete lifecycle of security threats.

SELinux Android benchmark: Both the Apple iPhone and Android contain an operating system and software stack for mobile devices, which are susceptible to an attack. Security-Enhanced Linux (SELinux) is a Linux-based feature using Linux Security Modules (LSM) in the Linux Kernel. Research has provided an implementation of SELinux as a security feature in the Android mobile device to harden the Android system and enforce low-level controls. This security feature can be used as a benchmark to reduce the damage from a successful attack within mobile Android device.

There are many other benchmark suites in use and more are being developed. It is important to note that the benchmarks provide only a broad performance guideline. It is the responsibility of the user to select the benchmark that comes close to his application and further evaluate the machine based on scenarios expected in the application for which the machine is being evaluated.

16.4 CODE OPTIMIZATION

As mentioned earlier, the first step in optimizing the performance of a program is to select the appropriate algorithm. The algorithm is then coded using an appropriate language for which optimizing compilers exist. Application developers (especially for supercomputer systems) spend enormous amount of time tweaking the code produced by compilers to optimize its performance. In fact, the code tweaking also extends back to the source code (after observing the code produced by the compilers). This section provides a brief description of some common code tweaking techniques.

Scalar renaming: It is typical for programmers to use a scalar variable repeatedly as shown by the following loop:

Example 16.6

$$\text{for } i = 1, n$$

$$x = A[i] + B[i]$$

$$Y[i] = 2 * x$$

$$x = \frac{C[i]}{D[i]}$$

$$P = x + 2$$

$$\text{endfor.}$$

If the second instance of x is renamed as shown in the following, the two code segments become data independent. The data-independent code segments can then be handled as two loops with each loop running on a separate processor concurrently.

$$\text{for } i = 1, n$$

$$x = A[i] + B[i]$$

$$Y[i] = 2 * x$$

$$xx = \frac{C[i]}{D[i]}$$

$$P = xx + 2$$

$$\text{endfor.}$$

Scalar expansion: In the following code segment, x is assigned a value and then used in a subsequent statement.

Example 16.7

$$\text{for } i = 1, n$$

$$x = A[i] + B[i]$$

$$Y[i] = 2*x$$

$$\text{endfor.}$$

If the scalar x is expanded into a vector as shown in the following, the two statements can be made independent, thus allowing a better vectorization.

$$\text{for } i = 1, n$$

$$x[i] = A[i] + B[i]$$

$$Y[i] = 2*x[i]$$

$$\text{endfor.}$$

Loop unrolling: For a loop of small vector length it is more efficient to eliminate the loop construct and expand out all iterations of the loop.

Example 16.8

The loop

$$\text{for } I = 1 \text{ to } 3$$

$$x[I] = a[I] + b[I]$$

$$\text{endfor.}$$

is unrolled into the following:

$$x[1] = a[1] + b[1]$$

$$x[2] = a[2] + b[2]$$

$$x[3] = a[3] + b[3].$$

This eliminates the looping overhead and allows the three computations to be performed independently. (In this case, the computations at each iteration are not dependent on each other. If this is not the case, the computations must be partitioned into nondependent sets.)

Loop fusion or jamming: Two or more loops that are executed the same number of times using the same indices can be combined into one loop.

Example 16.9

Consider the following code segment:

$$\text{for } i = 1, n$$

$$X[i] = Y[i]*Z[i]$$

$$\text{endfor}$$

$$\text{for } i = 1, n$$

$$M[i] = P[i] + X[i]$$

$$\text{endfor.}$$

Note that each loop would be equivalent to a vector instruction. X is stored back into the memory by the first instruction and then retrieved by the second. If these loops are fused as shown in the following, the memory traffic can be reduced to

for $i = 1, n$

$X[i] = Y[i]*Z[i]$

$M[i] = P[i] + X[i]$

endfor.

This assumes that there are enough vector registers available in the processor to retain X. If the processor allows chaining, the earlier loop can be reduced to

for $i = 1, n$

$M[i] = P[i] + Y[i]*Z[i]$

endfor.

Loop distribution: If the loop body contains dependent (i.e., statements that are data dependent) and nondependent code, a way to minimize the effect of the dependency is to break the loop into two, one containing the dependent code and the other nondependent code.

Force maximum work into inner loop: Since maximizing the vector length increases the speed of execution, the inner loop should always be made the longest. Further, dependency conflicts are avoided by shifting dependencies in an inner loop to an outer loop, if possible.

Subprogram in-lining: For small subprograms, the overhead of control transfer takes longer than the actual subprogram execution. Calling a subprogram might consume about 10–15 clock cycles when no arguments passed and one argument might nearly double that overhead. In such cases, it is better to move the subprogram code into the calling program.

16.5 SUMMARY

The performance parameters for various components and architectural features were discussed in previous chapters. Various performance enhancement techniques were also described in those chapters. This chapter provided a brief introduction to some common analytical techniques for performance evaluation, cost factor, and the most common benchmarks. In addition to the performance and cost, other factors considered in evaluating architectures are generality (how wide is the range of applications suited for this architecture), ease of use, and expandability or scalability. One feature that is receiving considerable attention now is the openness of the architecture. The architecture is said to be open if the designers publish the architecture details such that others can easily integrate standard hardware and software systems to it.

PROBLEMS

16.1 Look up the specifications of three personal computer systems in the catalog of a computer store. What are the important performance characteristics highlighted by the catalog description?

16.2 You are planning to buy a personal computer system for your home use. List the performance characteristics you would look for. Formulate the criteria to compare and select a system.

16.3 You are tasked to buy a computer system to handle a large database application for your company. List the performance characteristics you would look for. Formulate the criteria to compare and select a system.

16.4 Select a processor family you have access to and estimate the speedup obtained by using a floating-point coprocessor in the system, using an appropriate benchmark program.

16.5 Create a list of all the performance enhancements discussed in this book using ASC as the reference.

16.6 An application uses the following ASC instruction mix: ADD (20%), SHR (15%), LDA* (20%), STA (15%), TDX (15%), and LDX (15%).

    a. Compute the average speed and MIPS rating for this application assuming a 2 MHz clock.

    b. What should be the clock frequency to run this application at 2 MIPS?

16.7 Compute the geometric mean for the Example 16.4 with ASC3 as the reference. Are the relative performances same as in Example 16.4?

16.8 The following are common operations on matrices: column sum, row sum, transpose, inverse, addition, and multiplication. Examine the algorithm for each operation and develop vectorized procedures for each assuming a vector processor capable of performing vector add, subtract, reciprocal, and multiply operations.

16.9 Discuss the special array operations needed on vector processors to enable parallel execution of conditional operations, that is, vectorizing loops containing branches.

16.10 Assume that a vector processor operates 20 times faster on vector code than scalar code. If only x% of the program is vectorizable, what is the value of x, for the machine to execute the program twice as fast as the scalar processor?

16.11 Discuss in brief one benchmark tool focused toward evaluating the performance of mobile hardware.

16.12 Investigate benchmarks used to evaluate modern composite computing models like cloud computing, grid computing, and green computing. Is there any single benchmark to evaluate performance of the complete model?

16.13 Compare the capabilities of security benchmarks like CIS, WINE, and/or SELinux Android Benchmark.

BIBLIOGRAPHY

Dumitras, T. and Shou, D., Toward a standard benchmark for computer security research: The worldwide intelligence network environment *(WINE), Proceedings of the First Workshop/Building Analysis Datasets and Gathering Experience Returns for Security (BADGERS)*, Salzburg, Austria, 2011, pp. 89–96.

Grace, R., *The Benchmark Book*, Upper Saddle River, NJ: Prentice-Hall, 1996.

Harmonic mean, Wikipedia, the free encyclopedia, http://en.wikipedia.org/wiki/Harmonic_mean.

Harmonic mean, Wolfram Math World, http://mathworld.wolfram.com/HarmonicMean.html.

Hennessy, J.L. and Patterson, D.A., *Computer Architecture: A quantitative Approach*, San Francisco, CA: Morgan Kaufmann, 1996.

Kogge, P.M., *The Architecture of Pipelined Computers*, New York: McGraw-Hill, 1981.

Levesque, J.M. and Williamson, J.L., *A Guidebook to Fortran on Supercomputers*, San Diego, CA: Academic Press, Inc., 1989.

Lilja, D.J., *Measuring Computer Performance: A Practitioner's Guide*, New York: Cambridge University Press, 2000.

Moore, D.S., *The Basic Practice of Statistics*, New York: W.H. Freeman, 2006.

Ortega, J.M., *Introduction to Parallel and Vector Solution of Linear Systems*, New York: Plenum Press, 1988.

Polychronopoulos, C.D., *Parallel Programming and Compilers*, Boston, MA: Kluwer Academic Publishers, 1988.

Price, W.J., A benchmarking tutorial, *IEEE Microcomputer*, October 1989, 28–43.

Shabtai, A., Fledel, Y., and Elovici, Y. Securing android-powered mobile devices using SELinux, *IEEE Security and Privacy Magazine*, 8(3), 36–44, 2010.

Shiva, S.G., *Advanced Computer Architectures*, Boca Raton, FL: Taylor & Francis, 2006.

SPEC Benchmarks, Standard Performance Evaluation Corporation, http://www.spec.org.

Stone, H.S., *High-performance Computer Architecture*, New York: Addison-Wesley, 1990.

TPC Benchmarks, Transaction Processing Performance Council, http://www.tpc.org.

Weicker, R.P., An overview of common benchmarks, *IEEE Computer*, December 1990, 65–75.

Appendix A: Details of Representative Integrated Circuits

Pertinent details of representative integrated circuits (ICs) taken from vendor's manuals are reprinted in this appendix. Refer to vendor's manuals for further details on these and other ICs. The ICs detailed here are grouped under

1. Gates, decoders, and other ICs useful in combinational circuit design
2. Flip-flops, registers, and others ICs useful in sequential circuit design
3. Memory ICs

Most of the details provided here are from the TTL technology. No attempt is made to provide the details of the most up-to-date ICs. Because of the rapid changes in IC technology, it is hard to provide the latest details in a book. Refer to the current manuals from IC vendors for the latest information. Some of the ICs detailed here may no longer be available. Usually, alternative form of these ICs or a later version in another technology can be found from the same vendor or from another source. Nevertheless, the details given here are representative of the characteristics a designer would seek.

A.1 GATES, DECODERS, AND OTHER ICs USEFUL IN COMBINATIONAL CIRCUIT DESIGN

Figure A.1 shows the details of several gate ICs and an IC with dual two-wide, two-input AND-OR-Invert circuits. Figure A.2 shows a BCD to decimal decoder. Details of a 4-bit adder are shown in Figure A.3. Figure A.4 shows a data selector/multiplexer, and Figure A.5 shows a 4-line to 16-line decoder/demultiplexer.

The 74155 (Figure A.6) is a dual 1-of-4 decoder/demultiplexer with common address inputs and separate Enable inputs. Each decoder section, when enabled, will accept the binary weighted address input (A_0, A_1) and provide four mutually exclusive active-LOW outputs ($\overline{0-3}$). When the enable requirements of each decoder are not met, all outputs of that decoder are HIGH.

Both decoder sections have a two-input enable gate. For decoder "a," the enable gate requires one active-HIGH input and one active-LOW input ($E_a \cdot \overline{E}_a$). Decoder "a" can accept either true or complemented data in demultiplexing applications by using the $\overline{E}_a$ or E_a inputs, respectively. The enable gate of decoder "b" requires two active-LOW inputs ($\overline{E}_a \cdot \overline{E}_b$). The device can be used as a 1-of-8 decoder/demultiplexer by tying E_a to $\overline{E}_b$ and relabeling the common connection address as (A_2), forming the common enable by connecting the remaining $\overline{E}_b$ and $\overline{E}_a$.

A.2 FLIP-FLOPS, REGISTERS, AND OTHER ICs USEFUL IN SEQUENTIAL CIRCUIT DESIGN

Figure A.7 shows the details of a D flip-flop IC. A 4-bit latch IC is shown in Figure A.8. Figure A.9 shows an IC with dual JK flip-flops.

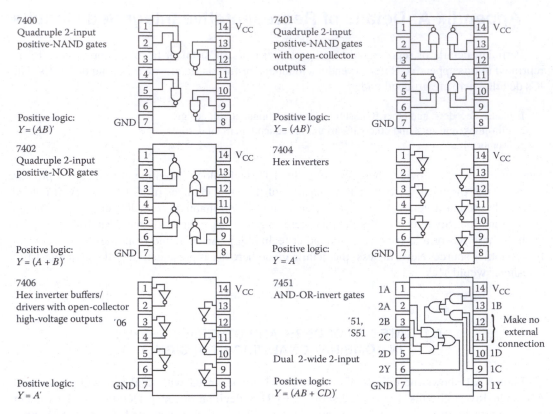

Figure A.1 Representative gate ICs.

The 7490 (Figure A.10) is a 4-bit, ripple-type decade counter. The device consists of four master–slave flip-flops internally connected to provide a divide-by-two section and a divide-by-five section. Each section has a separate clock input to initiate state changes of the counter on the high-to-low clock transition. State changes of the Q outputs do not occur simultaneously because of internal ripple delays. Therefore, decoded output signals are subject to decoding spikes and should not be used for clocks or strobes.

A gated AND asynchronous master reset ($MR_1 \cdot MR_2$) is provided, which overrides both clocks and resets (clears) all the flip-flops. Also provided is a gated AND asynchronous master set ($MS_1 \cdot MS_2$), which overrides the clocks and the MR inputs, setting the outputs to nine (HLLH).

Since the output from the divide-by-two section is not internally connected to the succeeding stages, the device may be operated in various counting modes. In a BCD (8421) counter, the $\overline{CP}_1$ input must be externally connected to the Q_0 output. The CP_0 input receives the incoming count producing a BCD count sequence. In a symmetrical bi-quinary divide-by-ten counter, the Q_3 output must be connected externally to the $\overline{CP}_0$ input. The input count is then applied to the CP_1 input and a divide-by-ten square wave is obtained at output Q_0. To operate as a divide-by-two and a divide-by-five counter, no external interconnections are required. The first flip-flop is used as a binary element for the divide-by-two function ($\overline{CP}_0$ as the input Q_0 as the output). The $\overline{CP}_1$ input is used to obtain a divide-by-five operation at the Q_3 output.

Two Shift Register ICs with various capabilities are shown in Figures A.11 and A.12. The 7495 (Figure A.12) is a 4-bit shift register with serial and parallel synchronous operating modes. It has serial data (D_S) and four parallel data (D_0–D_3) inputs and four parallel outputs (Q_0–Q_3). The serial

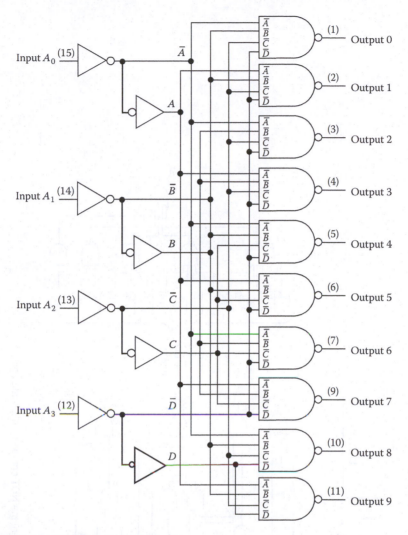

Figure A.2 7442 4-to-10 BCD to decimal decoder.

or parallel mode of operation is controlled by a mode select input (S) and two clock inputs ($\overline{CP}_1$ and $\overline{CP}_2$). The serial (shift right) or parallel data transfers occur synchronously with the high-to-low transition of the selected clock input.

When the mode select input (S) is high, $\overline{CP}_2$ is enabled. A high-to-low transition on enabled $\overline{CP}_2$ loads parallel data from the D_0–D_1 inputs into the register. When S is low, CP_1 is enabled. A high-to-low transition on enabled $\overline{CP}_1$ shifts the data from serial input D_S to Q_0 and transfers the data in Q_0 to Q_1, Q_1 to Q_2, and Q_2 to Q_3, respectively (shift right). Shift left is accomplished by externally connecting Q_3 to D_2, Q_2 to D_1, Q_1 to D_0, and operating the 7495 in the parallel mode (S = high).

In normal operations, the mode select (S) should change states only when both clock inputs are low. However, changing S from high-to-low while $\overline{CP}_2$ is low, or changing S from low-to-high while $\overline{CP}_1$ is low will not cause any changes on the register outputs. Figure A.13 shows the details of a synchronous decode counter.

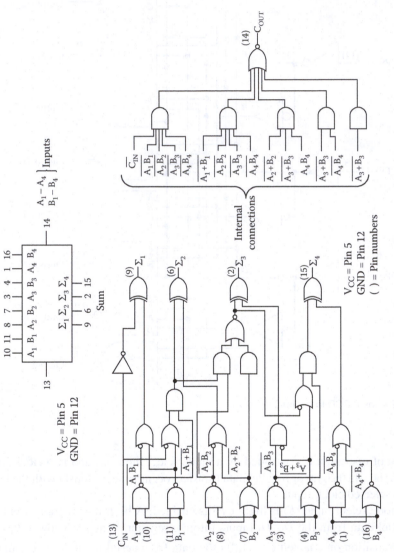

Figure A.3 7483 4-bit adder. *Note:* "‾" is used to represent the NOT operation in IC catalogs.

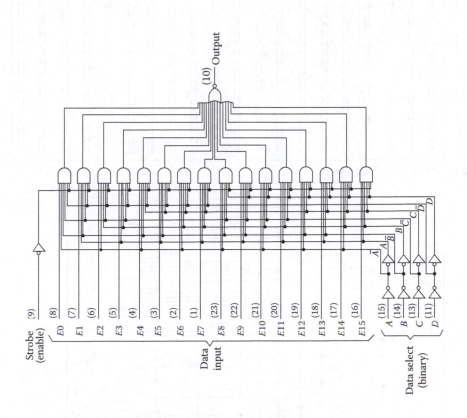

'150
Function table

Inputs					Output
Select				Strobe S	W
D	C	B	A		
X	X	X	X	H	H
L	L	L	L	L	$\overline{E0}$
L	L	L	H	L	$\overline{E1}$
L	L	H	L	L	$\overline{E2}$
L	L	H	H	L	$\overline{E3}$
L	H	L	L	L	$\overline{E4}$
L	H	L	H	L	$\overline{E5}$
L	H	H	L	L	$\overline{E6}$
L	H	H	H	L	$\overline{E7}$
H	L	L	L	L	$\overline{E8}$
H	L	L	H	L	$\overline{E9}$
H	L	H	L	L	$\overline{E10}$
H	L	H	H	L	$\overline{E11}$
H	H	L	L	L	$\overline{E12}$
H	H	L	H	L	$\overline{E13}$
H	H	H	L	L	$\overline{E14}$
H	H	H	H	L	$\overline{E15}$

Figure A.4 74150 Data selector/multiplexer.

Function table

Inputs						Outputs															
G1	G2	D	C	B	A	0	1	2	3	4	5	6	7	8	9	10	11	12	13	14	15
L	L	L	L	L	L	L	H	H	H	H	H	H	H	H	H	H	H	H	H	H	H
L	L	L	L	L	H	H	L	H	H	H	H	H	H	H	H	H	H	H	H	H	H
L	L	L	L	H	L	H	H	L	H	H	H	H	H	H	H	H	H	H	H	H	H
L	L	L	L	H	H	H	H	H	L	H	H	H	H	H	H	H	H	H	H	H	H
L	L	L	H	L	L	H	H	H	H	L	H	H	H	H	H	H	H	H	H	H	H
L	L	L	H	L	H	H	H	H	H	H	L	H	H	H	H	H	H	H	H	H	H
L	L	L	H	H	L	H	H	H	H	H	H	L	H	H	H	H	H	H	H	H	H
L	L	L	H	H	H	H	H	H	H	H	H	H	L	H	H	H	H	H	H	H	H
L	L	H	L	L	L	H	H	H	H	H	H	H	H	L	H	H	H	H	H	H	H
L	L	H	L	L	H	H	H	H	H	H	H	H	H	H	L	H	H	H	H	H	H
L	L	H	L	H	L	H	H	H	H	H	H	H	H	H	H	L	H	H	H	H	H
L	L	H	L	H	H	H	H	H	H	H	H	H	H	H	H	H	L	H	H	H	H
L	L	H	H	L	L	H	H	H	H	H	H	H	H	H	H	H	H	L	H	H	H
L	L	H	H	L	H	H	H	H	H	H	H	H	H	H	H	H	H	H	L	H	H
L	L	H	H	H	L	H	H	H	H	H	H	H	H	H	H	H	H	H	H	L	H
L	L	H	H	H	H	H	H	H	H	H	H	H	H	H	H	H	H	H	H	H	L
L	H	X	X	X	X	H	H	H	H	H	H	H	H	H	H	H	H	H	H	H	H
H	L	X	X	X	X	H	H	H	H	H	H	H	H	H	H	H	H	H	H	H	H
H	H	X	X	X	X	H	H	H	H	H	H	H	H	H	H	H	H	H	H	H	H

H, high level; L, low level; X, irrelevant.

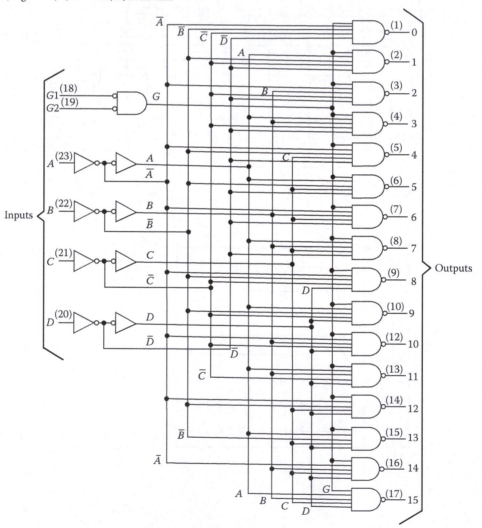

Figure A.5   74154 4-line to 16-line decoder/demultiplexer.

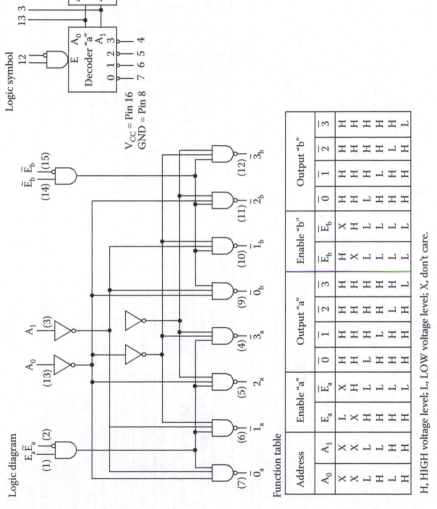

Figure A.6 74155 dual 2-line to 4-line decoder/demultiplexer.

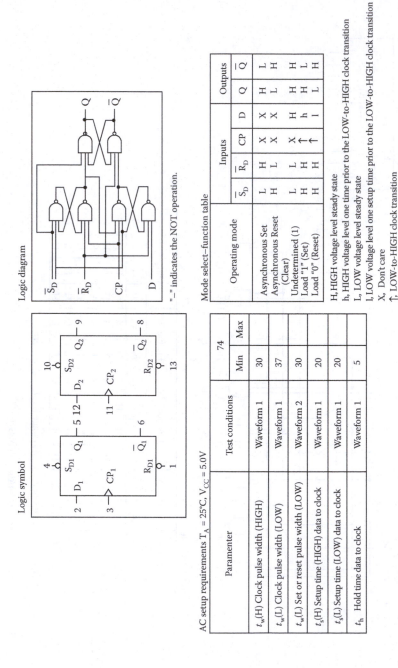

Logic symbol

Logic diagram

"‾" indicates the NOT operation.

AC setup requirements $T_A = 25°C$, $V_{CC} = 5.0V$

Parameter	Test conditions	74 Min	74 Max
$t_w(H)$ Clock pulse width (HIGH)	Waveform 1	30	
$t_w(L)$ Clock pulse width (LOW)	Waveform 1	37	
$t_w(L)$ Set or reset pulse width (LOW)	Waveform 2	30	
$t_s(H)$ Setup time (HIGH) data to clock	Waveform 1	20	
$t_s(L)$ Setup time (LOW) data to clock	Waveform 1	20	
t_h Hold time data to clock	Waveform 1	5	

Mode select–function table

Operating mode	Inputs $\bar{S}_D$	$\bar{R}_D$	CP	D	Outputs Q	$\bar{Q}$
Asynchronous Set	L	H	X	X	H	L
Asynchronous Reset (Clear)	H	L	X	X	L	H
Undetermined (1)	L	L	X	X	H	H
Load "1" (Set)	H	H	↑	h	H	L
Load "0" (Reset)	H	H	↑	l	L	H

H, HIGH voltage level steady state
h, HIGH voltage level one time prior to the LOW-to-HIGH clock transition
L, LOW voltage level steady state
l, LOW voltage level one setup time prior to the LOW-to-HIGH clock transition
X, Don't care
↑, LOW-to-HIGH clock transition

Figure A.7 7474 D flip-flop IC. *Note:* Both outputs will be HIGH while both $\bar{S}_D$ and $\bar{R}_D$ are LOW, but the output states are unpredictable it $\bar{S}_D$ and $\bar{R}_D$ go to HIGH simultaneously.

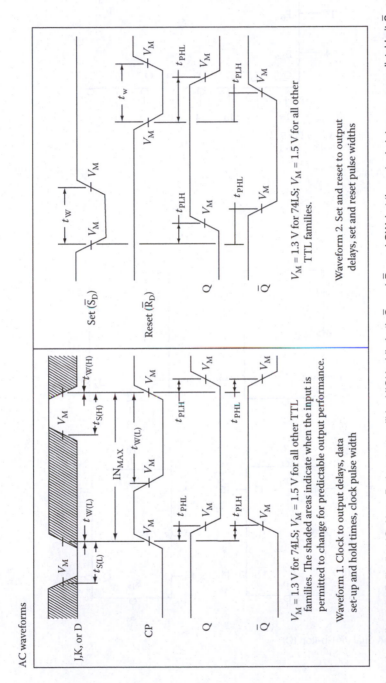

Figure A.7 (continued) 7474 D flip-flop IC. *Note:* Both outputs will be HIGH while both $\bar{S}_D$ and $\bar{R}_D$ are LOW, but the output states are unpredictable it $\bar{S}_D$ and $\bar{R}_D$ go to HIGH simultaneously.

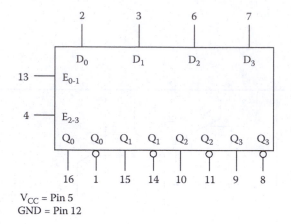

Figure A.8 7475 A latch IC.

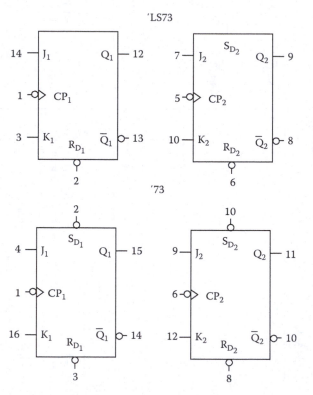

Figure A.9 7473 A dual *JK* flip-flop IC.

Logic symbol

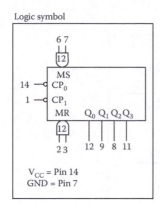

Mode selection–function table

RESET/SET INPUTS				OUTPUTS			
MR_1	MR_2	MS_1	MS_2	Q_0	Q_1	Q_2	Q_3
H	H	L	X	L	L	L	L
H	H	X	L	L	L	L	L
X	X	H	H	L	L	L	H
L	X	L	X	COUNT			
X	L	X	L	COUNT			
L	X	X	L	COUNT			
H	L	L	X	COUNT			

V_{CC} = Pin 14
GND = Pin 7

H, HIGH voltage level; L, LOW voltage level;
X, don't care.

Logic symbol

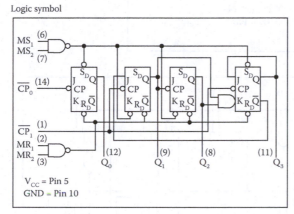

BCD count sequence-
function table

COUNT	OUTPUTS			
	Q_0	Q_1	Q_2	Q_3
0	L	L	L	L
1	H	L	L	L
2	L	H	L	L
3	H	H	L	L
4	L	L	H	L
5	H	L	H	L
6	L	H	H	L
7	H	H	H	L
8	L	L	L	H
9	H	L	L	H

Output Q_0 connected to input $\overline{CP}_1$

V_{CC} = Pin 5
GND = Pin 10

Figure A.10 7490 8-bit shift register.

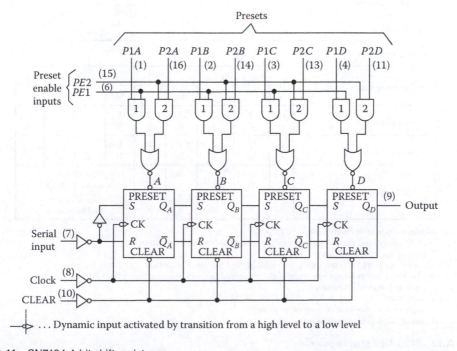

Figure A.11 SN7494 4-bit shift registers.

Logic symbol

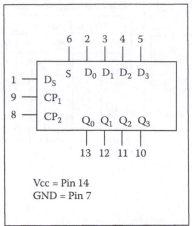

Logic diagram

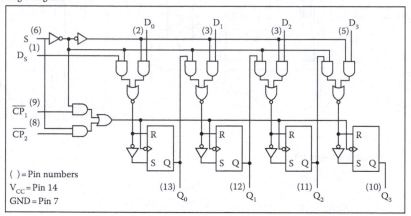

Function table

Operating mode	Inputs					Outputs			
	S	CP_1	CP_2	D_S	D_N	Q_0	Q_1	Q_2	Q_3
Parallel load	H	X	↓	X	l	L	L	L	L
	H	X	↓	X	h	H	H	H	H
Shift right	L	↓	X	l	X	L	q_0	q_1	q_2
	L	↓	X	h	X	H	q_0	q_1	q_2
Mode change	↑	L	X	X	X	No change			
	↑	H	X	X	X	undetermined			
	↓	X	L	X	X	No change			
	↓	X	H	X	X	undetermined			

H, high voltage level steady state
h, high voltage level one setup time prior to the high-to-low clock transition
L, low voltage level steady state
l, low voltage level one setup time prior to the high-to-low clock transition
q, lower case letters indicate the state of the referenced output one set-up time prior to the
  high-to-low clock transition
X, don't care
↓ , high-to-low transition of clock or mode select
↑ , low-to-high transition of mode select

Figure A.12   7495 4-bit shift register.

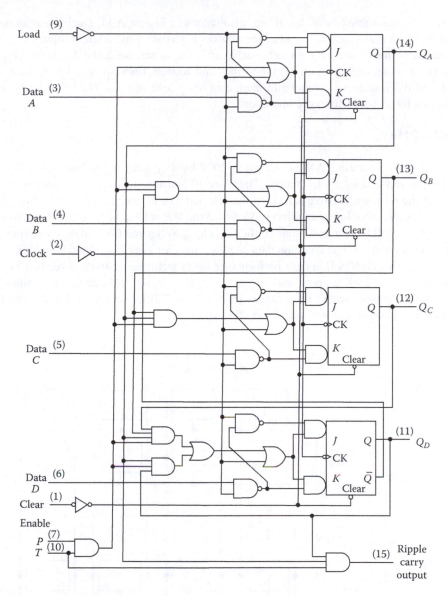

Figure A.13   74160 synchronous decode counter.

A.3 MEMORY ICS

A.3.1 Signetics 74S189

This 64-bit RAM (Figure A.14) is organized as 16 words of 4 bits each. There are four address lines (A_0–A_3), four data input lines (I_1–I_4), and four data output lines (D_1–D_4). Note that the data output lines are active-low. Therefore, the output will be the complement of the data in the selected word. If the low-active chip enable (CE) is high, the data outputs assume the high impedance state. When the write enable (WE) is low, the data from the input lines are written into the addressed location. When the WE is high, the data are read from the addressed location. The operation of the ICs is summarized in the truth table.

The timing characteristics of this IC are also shown in Figure A.14. During a read operation, the data appear on the output T_{AA} ns after the address is stable on the address inputs. T_{CE} indicates the time required for the output data to be stable after $\overline{CE}$ is activated, the T_{CD} is the chip disable time. During a write, once data on the input lines and address lines are stabilized, the $\overline{CE}$ is first activated, and $\overline{WE}$ is activated after a minimum data setup time of T_{WSC}. The $\overline{WE}$ must be active for at least T_{WP} ns for a successful write operation.

A.3.2 Intel 2114

This is a 4096-bit static RAM organized as 1024 by 4 (Figure A.15). Internally, the memory cells are organized in a 64-by-64 matrix. There are 10 address lines (A_0–A_9). Address bits A_3–A_8 select one of the 64 rows. A 4-bit portion of the selected row is selected by address bits A_0, A_1, A_2, and A_9. There is an active-low chip select ($\overline{CS}$). When the $\overline{WE}$ is low, the IC is put into a write mode; otherwise, the IC will be in a read mode (if the $\overline{CS}$ is low). When the $\overline{CS}$ is high, the outputs assume the high impedance state. The common data I/O lines thus are controlled by $\overline{CS}$ and $\overline{WE}$.

This device uses HMOS II, a high-performance MOS technology, and is directly TTL compatible in all respects: inputs, outputs, and single 5 V power supply. This device is available in five versions. The maximum access time ranges from 100 to 250 ns depending on the version. The maximum current consumption ranges from 40 to 70 mA.

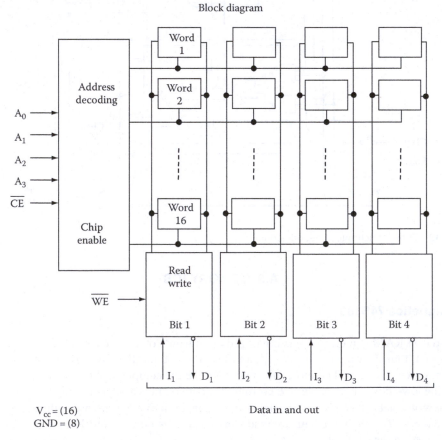

Block diagram

Figure A.14 74S189 64-bit *RAM*.

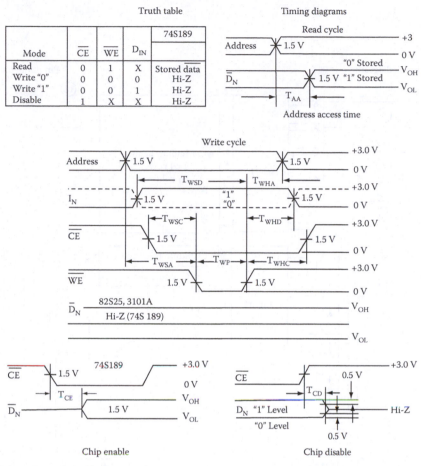

Figure A.14 (continued) 74S189 64-bit *RAM*.

A.3.3 Texas Instruments TMS4116

This is a 16K × 1 dynamic NMOS RAM (Figure A.16). It has a data input (*D*), a data output (*Q*), and a read/write control ($\overline{W}$) input. To decode 16K, 14 address lines are required. The IC provides only seven address lines (A_0–A_6). The 14 address bits are multiplexed onto these seven address lines using row address select ($\overline{RAS}$) and column address select ($\overline{CAS}$) inputs. Although this multiplexing decreases the speed of the RAM operation, it minimizes the number of pins on the IC. Three power supplies (12 V, +5 V, and −5 V) are required.

The operation of the IC is illustrated in Figure A.16a. The memory cells are organized in a 128 × 128 array. The 7 low-order address bits select a row. During a read, the data from the selected row are transferred to 128 sense/refresh amplifiers. The 7 high-order address bits then select one of the 128 sense amplifiers and connect it to the data output line. At the same time, the data on the sense amplifiers are refreshed (i.e., the capacitors are charged) and rewritten to the proper row in the memory array. During a write, the data in the sense amplifiers are changed to new data values just before the rewrite operation. Thus, at each read or write cycle, a row of the memory is refreshed.

The timing diagrams for read, refresh, and write cycles are shown in Figure A.16b through d. When $\overline{RAS}$ and $\overline{CAS}$ are both high, *Q* will be at a high impedance state. To begin a cycle, the 7 low-order address bits are first placed on the address lines and the $\overline{RAS}$ is driven low. The 4116 then latches the row address. The high-order 7 bits of the address are then placed on address lines

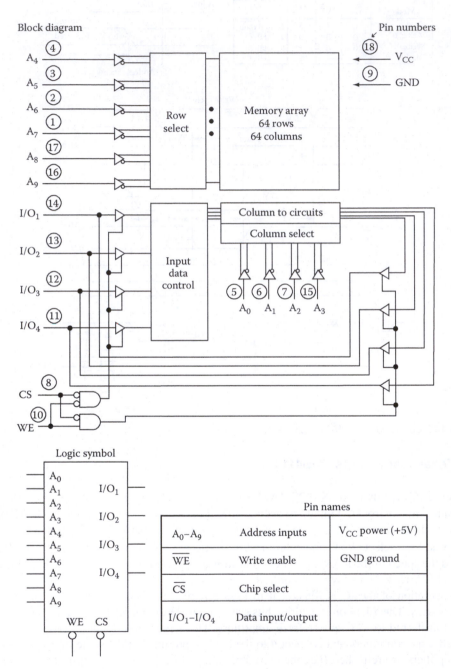

Figure A.15   Intel 2114.

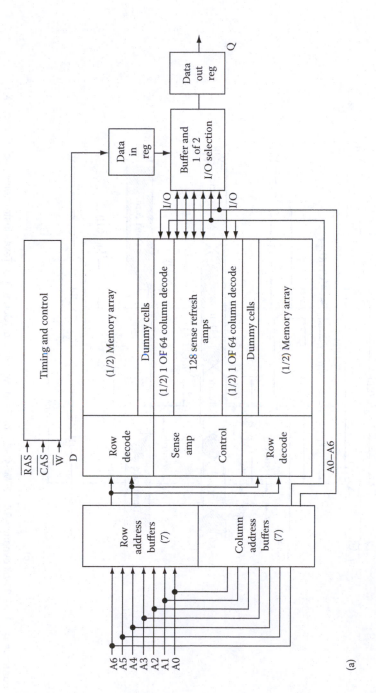

Figure A.16 A 16384-bit dynamic *RAM* (TMS4116). (a) Functional block diagram. (Courtesy of Texas Instruments, Inc., Dallas, TX.)

(continued)

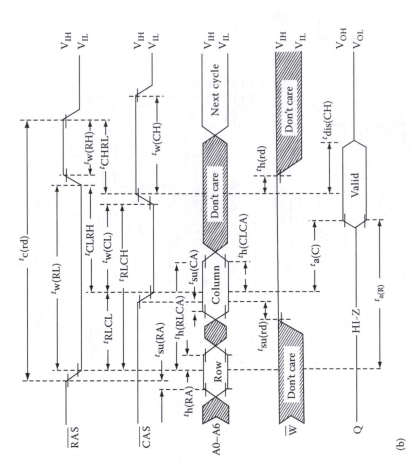

Figure A.16 (continued) A 16384-bit dynamic *RAM* (TMS4116). (b) Read cycle timing. (Courtesy of Texas Instruments, Inc., Dallas, TX.)

(b)

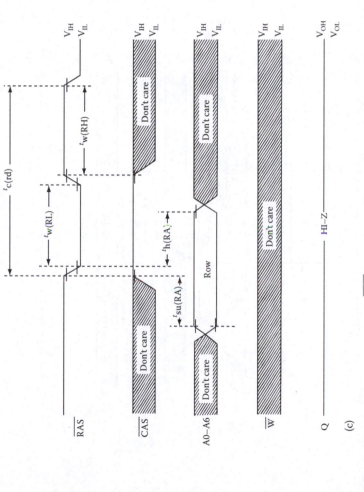

Figure A.16 (continued)   A 16384-bit dynamic *RAM* (TMS4116). (c) $\overline{RAS}$-only refresh timing. (Courtesy of Texas Instruments, Inc., Dallas, TX.)

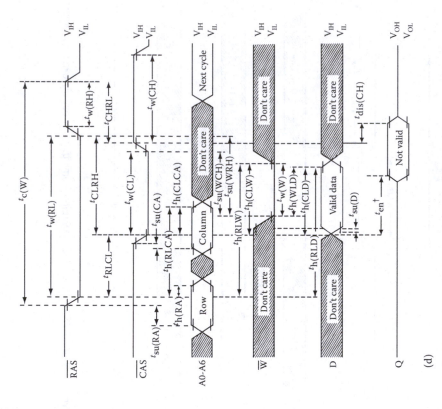

Figure A.16 (continued) A 16384-bit dynamic *RAM* (TMS4116). (d) Write cycle timing. (Courtesy of Texas Instruments, Inc., Dallas, TX.)

and the $\overline{CAS}$ is driven low, thus selecting the required bit. For a write cycle, D must have valid data when the $\overline{CAS}$ goes low and the $\overline{W}$ should go low just after $\overline{CAS}$ does. For a read cycle, Q will be valid after $t_{a(c)}$ from the start of $\overline{CAS}$ and will remain valid until $\overline{RAS}$ and $\overline{CAS}$ go high.

The data in the memory must be refreshed every 2 ms. Refreshing is done by reading a row of data into the sense amplifiers and rewriting it. Only $\overline{RAS}$ is required to perform the refresh cycle. A 7-bit counter can be used to refresh all the rows in the memory. The counter counts up every 2 ms. The 7-bit output of the counter becomes the row address at each cycle.

Two modes of refresh are possible: *burst* and *periodic*. In a burst mode, all rows are refreshed every 2 ms. Thus, if each refresh cycle takes about 450 ns, in a burst refresh mode, the first ($128 \times 450 = 57{,}600$ ns) 57.6 μs will be consumed by the refresh and 1942.4 μs will be available for the read and write. In a periodic mode, there will be one refresh cycle every ($2/128 =$) 15.626 μs. The first 450 ns at each 15.625 μs interval will be taken for the refresh.

Several dynamic memory controllers are available off-the-shelf. These controllers generate appropriate signals to refresh the dynamic memory module in the system. One such controller is described next.

A.3.4 INTEL 8202A Dynamic RAM Controller

This device (Figure A.17) provides all the signals needed to control a 64K dynamic *RAM* of the TMS4116 type. It performs address multiplexing and generates *RAS* and *CAS* signals. It contains a refresh counter and timer. The device has several other features that make it useful in building microcomputer systems. In what follows, we will concentrate on the basic features needed to control a dynamic *RAM*. The reader should refer to the manufacturers' manuals for further details.

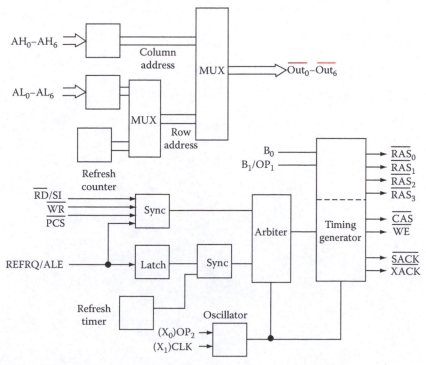

Figure A.17 Dynamic RAM controller (Intel 8202A). (Courtesy of Intel Corp.)

The outputs $(\overline{OUT}_0 - \overline{OUT}_6)$ are functions of either the 14-bit address inputs (AL_0–AL_6 and AH_0–AH_6) or the refresh counter outputs. The outputs of the 8202A are directly connected to the address inputs of the 4116. The $\overline{WE}$ is connected to the $\overline{W}$ of the memory. There is a chip select input ($\overline{PCS}$) and a clock ($\overline{CLK}$) input. In addition to the $\overline{CAS}$, four $\overline{RAS}$ signals are generated by the device. These multiple $\overline{RAS}$ signals are useful in selecting a bank of the memory when larger memory systems are built using dynamic memory ICs. Signals such as $\overline{RD}$, $\overline{WR}$, $\overline{XACK}$, and $\overline{SACK}$ are compatible with the control signals produced by microprocessors (such as Intel 8088).

BIBLIOGRAPHY

Bipolar Memory Data Manual, Sunnyvale, CA: Signetics, 1987.
Bipolar Microcomputer Components Data Book, Dallas, TX: Texas Instruments, 1987.
FAST TTL Logic Series Data Handbook, Sunnyvale, CA: Phillips Semiconductors, 1992.
Intel 8202 Datasheet, http://archive.org/stream/intel-8202-datasheet/8202DynamicRamController#page/n0/mode/2up
TTL Data Manual, Sunnyvale, CA: Signetics, 1987.

Appendix B: Stack Implementation

A last-in/first-out (LIFO) stack is a versatile structure useful in a variety of operations in a computer system. It is used for address and data manipulation, return address storage and parameter passing during subroutine call and return, and arithmetic operations in an ALU. It is a set of storage locations or registers organized in a LIFO manner. A coin box (Figure B.1) is the most popular example of a LIFO stack. Coins are inserted and retrieved from the same end (top) of the coin box. PUSHing a coin moves the stack of coins down one level, the new coin occupying the top level (TL). POPing the coin box retrieves the coin on the top level. The second level (SL) coin becomes the new top level after the POP.

In a LIFO stack (or simply "stack"):

PUSH implies all the levels move down by one; TL ← data.
POP implies pops out ← TL; TL ← SL; all other levels move up.

Two popular implementations of the stack are

1. RAM-based implementation
2. Shift-register-based implementation

In a RAM-based implementation, a special register called a *stack pointer* (SP) is used to hold the address of the top level of the stack. The stack is built in a reserved area in the memory. The PUSH operation then corresponds to the following:

SP ← SP + 1.
MBR ← data.
MAR ← SP.
WRITE MEMORY.

The POP operation corresponds to the following:

MAR ← SP.
READ MEMORY.

Output ← MBR. "Output" is the destination for data from the Top level.

SP ← SP − 1.

In this implementation, the stack grows toward higher address memory locations as items are PUSHed into it. The data do not actually move between the levels during PUSH and POP operations.

Figure B.2 portrays the shift-register-based implementation of an n-level stack. Each stack level can hold an m-bit datum. Data are PUSHed into stack by using a SHIFT RIGHT signal and POPed out of stack by using a SHIFT LEFT signal. There is movement of data between levels in this implementation.

Shift-register-based implementations are faster than RAM-based stacks because no memory access is needed. RAM-based implementations are popular because no additional hardware is needed (other than SP) to implement a stack. Instructions to PUSH and POP registers and memory locations must be added to the instruction set once a stack is included in the design.

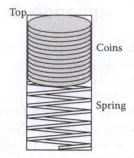

Figure B.1   A last-in/first-out (LIFO) stack.

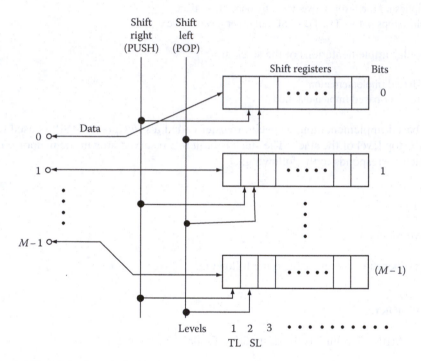

Figure B.2   A shift-register-based stack.

Index